Innovations in Dryland Agriculture

Muhammad Farooq • Kadambot H.M. Siddique
Editors

Innovations in Dryland Agriculture

Springer

Editors
Muhammad Farooq
Department of Agronomy
University of Agriculture
Faisalabad, Pakistan

Kadambot H.M. Siddique
The University of Western Australia
Institute of Agriculture
The University of Western Australia
Perth, WA, Australia

ISBN 978-3-319-47927-9 ISBN 978-3-319-47928-6 (eBook)
DOI 10.1007/978-3-319-47928-6

Library of Congress Control Number: 2016959536

Printed on acid-free paper

This Springer imprint is published by Springer Nature
The registered company is Springer International Publishing AG
The registered company address is: Gewerbestrasse 11, 6330 Cham, Switzerland

Foreword

Given the pressure on the world's ecosystems, dryland agriculture—a sector that has been neglected in the past—is becoming increasingly important in meeting global food requirements in the future. Drylands cover more than 40 % of the world's land surface and are home to 2.5 billion people, or one-third of the global population, about 50 % of this population living in poverty and 90 % living in developing nations. Half of the inhabitants in the dryland depend directly on rainfed or irrigated farming and pastoralism. Poverty, food insecurity, biodiversity loss, frequent droughts and environmental degradation are widespread in these areas. The effects of climate change will lead to even more water scarcity and declining crop yields, leaving the people of these regions excessively vulnerable in the absence of appropriate technologies and risk management strategies.

In recent decades, food production has fallen significantly in most dryland areas, while food demand has increased due to high levels of population growth. The challenge for global agriculture to produce 70 % more food and fibre by 2050 cannot be achieved without implementing more sustainable farming methods and responding to climate change in dry areas. With half of the population of dryland regions depending on agriculture, the economic development of these regions is inextricably tied to the performance of the agriculture sector.

Drylands differ from humid lands in several ways. Yet, development pathways for drylands are often driven by a distorted idea of how drylands should or could exist, often modelled on more humid areas. Notions of "greening the desert" are developed from a misunderstanding of dryland ecology and have led to many harmful policies and investments. Furthermore, misrepresentation of drought and water scarcity in the drylands diverts attention from sustainable and adaptive management, capable of being supported by limited resources, towards unsustainable practices that are ecologically harmful. Rather than adapting development strategies to fit the drylands, considerable effort has been expended trying to adapt drylands to fit the development strategies.

The sustainability of dryland agriculture must consider (1) practices that maintain soil organic matter and restore soils degraded by past practices, (2) crop cultivars and species that can withstand climatic abnormalities (3) and growers who can tailor crop management practices according to the local climatic conditions. Practices such as residue burning and fallow period tillage have reduced soil organic carbon levels by as much as 60 %. Current no-till dryland systems with intensified crop rotations have stabilized soil carbon and increased soil organic matter accretion.

Livestock production in drylands is expected to play a major role in response to the demands for animal protein production over the coming decades. However, livestock production in these areas is prone to several risks and threats. Nonetheless, sustainability in livestock production in dryland areas can be achieved by raising shrubs and native plant combinations in silvopastoral systems with strategies to promote self-herding and by the careful selection of animal species.

Despite the availability of several recent publications on various aspects of dryland agriculture, no single book covers the basic concepts, elements, potential benefits, experiences, challenges and innovations in dryland agriculture. This book is a timely effort to fill the gap. The book describes various elements of dryland agriculture, highlights associated breeding and modelling efforts, analyses the experiences and challenges in dryland agriculture in different regions and proposes some practical innovations and new areas of research in this critical area of agriculture.

This volume edited by Associate Professor Muhammed Farooq and Professor Kadambot Siddique will be a ready reference on dryland agriculture to develop environmentally sustainable and profitable food production systems in this region.

Chennai, India M.S. Swaminathan

Preface

Agriculture provides the basic life requirements for the society, drives economic development and contributes to poverty reduction particularly in the developing world. To achieve the Sustainable Development Goals (SDG) of eradicating hunger and poverty, growth in the agricultural sector is essential with the projected doubling of food production needed over the next two to three decades.

The drylands, covering more than 40 % of the world's landmass, may be ideal sites to achieve this targeted growth in food production. However climate change is threatening global food production systems, and the situation is predicted to worsen. Changes in climate patterns have affected the people living in the drylands the most as they largely depend on agriculture for their livelihoods. It follows that research and development innovations in dryland agriculture will provide resilience to climate-related shifts in these areas.

Despite the topographical and climatic constraints in dry areas, subsistence food production remains a predominant activity for the majority of the people living in these areas. This strong evidence together with the holistic and integrative approaches to ecosystem sustainability suggests that the concerns about land degradation, biodiversity loss and fading cultural colours of 2.5 billion people can be confronted with renewed optimism.

Although many publications on issues, challenges and pragmatic options to improve the productivity and sustainability of dryland agriculture are available, no single book provides a holistic view and recent innovations in this critical area of agriculture. With increasing global concern and interest in dryland agriculture, we felt it timely to assemble and synthesize the latest developments and innovations in dryland agricultural research and development. This book covers the basic concepts and history, components and elements, breeding and modelling efforts and potential benefits, experiences, challenges and innovations relevant to agriculture in dryland areas around world. This book is divided into five sections and 20 chapters as detailed below.

1. Introduction
   - Chapter 1 describes the basic concepts, origin and a brief history of dryland agriculture.
   - Chapter 2 states the challenges and researchable issues in dryland agriculture.
2. Elements of Dryland Agriculture
   - Chapter 3 collates the conventional and innovative water harvesting techniques in dry environments.
   - Chapter 4 describes the weed problem in dryland agriculture systems and proposes strategies for integrated weed management.
   - Chapter 5 discusses the nutrient management perspectives in dryland agriculture and suggests strategies for improving nutrient use efficiency in dryland agriculture systems.
   - Chapter 6 describes the insect pests in drylands and proposes innovative options for integrated management of insect pests in dry environments.
   - Chapter 7 covers the epidemiology and management of fungal diseases in dry environments.
   - Chapter 8 describes integrated crop–livestock production in drylands.
3. Modeling and Crop Improvement for Dryland Agriculture
   - Chapter 9 introduces the application of modelling in dryland agricultural systems.
   - Chapter 10 covers breeding and genetic enhancement strategies for dryland crops.
4. Dryland Agriculture: Some Case Studies
   - Chapter 11 discusses Australian experiences of dryland agriculture.
   - Chapter 12 analyses the experiences, issues and options in pasturelands of Australia's drylands.
   - Chapter 13 covers the experiences, challenges and options regarding dryland agriculture in South Asia.
   - Chapter 14 describes the experiences from Northwest China regarding the sustainable management of drylands.
   - Chapter 15 illustrates the experiences in dryland agriculture from the Great Plains region of the United States and Canada.
   - Chapter 16 describes the experiences in dryland agriculture from Eastern and Southern Africa.
5. Innovations in Dryland Agriculture
   - Chapter 17 covers the sustainable use of soil and other natural resources in relation to agronomic productivity and environment quality. It also addresses

soil C sequestration potential in drylands and its management in diverse soils and agroecosystems.

- Chapter 18 discusses the potential applications of microbiology in dryland agriculture.
- Chapter 19 provides an overview of dryland salinity—its causes, forms and management options. It also examines how climate change may affect both its future extent and the viability of management and recovery options.
- Chapter 20 discusses some case studies from different dryland countries where supplemental irrigation has successfully increased land and water productivity.

In developing this book, we have collaborated with authors from many countries and with vast experience to cover the different aspects of dryland agriculture. We thank all the authors for their contributions, help and cooperation during the manuscript writing and revision process. We are grateful to Professor M.S. Swaminathan for his encouragement and writing the foreword. We also thank Ms Melanie van Overbeek, Assistant Editor, Agronomy, Springer Dordrecht, the Netherlands, for her patience and trust in us during this book project. We thank the University of Agriculture Faisalabad and the University of Western Australia for continued support.

Faisalabad, Pakistan — Muhammad Farooq
Perth, Australia — Kadambot H.M. Siddique

Contents

Contributors

Akhtar Ali Environment, Agriculture and Natural Resources Division, Asian Development Bank, Mandaluyong City, Metro Manila, Philippines

Brett L. Allen United States Department of Agriculture-Agricultural Research Service, Sidney, MT, USA

Abdullah M. Al-Sadi Department of Crop Sciences, College of Agricultural and Marine Sciences, Sultan Qaboos University, Muscat, Oman

Tilahun Amede International Crops Research Institute for the Semiarid Tropics (ICRISAT), Addis Ababa, Ethiopia

Saseendran Anapalli United States Department of Agriculture-Agricultural Research Service, Stoneville, MS, USA

Walter K. Anderson The University of Western Australia Institute of Agriculture, The University of Western Australia, Perth, WA, Australia

Muhammad Shakeel Arshad Department of Agronomy, University of Agriculture, Faisalabad, Pakistan

E.G. Barrett-Lennard School of Plant Biology, The University of Western Australia, Crawley, WA, Australia

Department of Agriculture and Food, South Perth, WA, Australia

Dominique Blache School of Agriculture and Environment M085 and The UWA Institute of Agriculture, Faculty of Science, The University of Western Australia, Perth, Australia

Robert E. Blackshaw Agriculture and Agri-Food Canada, Lethbridge, AB, Canada

S. Desai Central Research Institute for Dryland Agriculture, Hyderabad, India

Peter de Voil Queensland Alliance for Agriculture and Food Innovation (QAAFI), The University of Queensland, Toowoomba, QLD, Australia

Muhammad Farooq Department of Agronomy, University of Agriculture, Faisalabad, Pakistan

The UWA Institute of Agriculture, The University of Western Australia, Perth, WA, Australia

R.J. George Department of Agriculture and Food, Bunbury, WA, Australia

M. Gilfedder CSIRO Dutton Park Laboratories, Brisbane, QLD, Australia

Muhammad Dildar Gogi Integrated Pest Management Laboratory, Department of Entomology, University of Agriculture, Faisalabad, Pakistan

K.A. Gopinath Indian Council of Agricultural Research-Central Research Institute for Dryland Agriculture, Hyderabad, India

Minakshi Grover Central Research Institute for Dryland Agriculture, Hyderabad, India

Yu Guan State Key Laboratory of Grassland Agroecosystem, Institute of Arid Agroecology, School of Life Sciences, Lanzhou University, Lanzhou, China

Ann Hamblin The UWA Institute of Agriculture, The University of Western Australia, Perth, WA, Australia

Neil C. Hansen Department of Plant and Wildlife Sciences, Brigham Young University, Provo, UT, USA

Feng-Min Li State Key Laboratory of Grassland Agroecosystem, Institute of Arid Agroecology, School of Life Sciences, Lanzhou University, Lanzhou, China

Xiao-Gang Li State Key Laboratory of Grassland Agroecosystem, Institute of Arid Agroecology, School of Life Sciences, Lanzhou University, Lanzhou, China

Drew J. Lyon Department of Crop and Soil Sciences, Washington State University, Pullman, WA, USA

Stephen Machado Columbia Basin Agricultural Research Center, Oregon State University, Pendleton, OR, USA

Graeme B. Martin School of Agriculture and Environment M085 and The UWA Institute of Agriculture, Faculty of Science, The University of Western Australia, Perth, Australia

D.J. McFarlane The University of Western Australia, Crawley, WA, Australia

Reza Mohammadi Dryland Agricultural Research Institute (DARI), Kermanshah, Iran

Hafsa Naheed The University of Agriculture, Peshawar, Pakistan

Vinay Nangia International Center for Agricultural Research in the Dry Areas (ICARDA), Amman, Jordan

Ahmad Nawaz Department of Agronomy, University of Agriculture, Faisalabad, Pakistan

Ahmad Nawaz Integrated Pest Management Laboratory, Department of Entomology, University of Agriculture, Faisalabad, Pakistan

Theib Oweis International Center for Agricultural Research in the Dry Areas (ICARDA), Amman, Jordan

B. Power Commonwealth Scientific and Industrial Research Organisation, Toowoomba, QLD, Australia

B.M.K. Raju Indian Council of Agricultural Research-Central Research Institute for Dryland Agriculture, Hyderabad, India

C.A. Rama Rao Indian Council of Agricultural Research-Central Research Institute for Dryland Agriculture, Hyderabad, India

R. Rejani Indian Council of Agricultural Research-Central Research Institute for Dryland Agriculture, Hyderabad, India

Dean K. Revell School of Agriculture and Environment M085 and The UWA Institute of Agriculture, Faculty of Science, The University of Western Australia, Perth, Australia

Revell Science, Duncraig, WA, Australia

Daniel Rodriguez Queensland Alliance for Agriculture and Food Innovation (QAAFI), The University of Queensland, Toowoomba, QLD, Australia

Muhammad Sanaullah Institute of Soil and Environmental Sciences, University of Agriculture, Faisalabad, Pakistan

Kadambot H.M. Siddique The University of Western Australia Institute of Agriculture, The University of Western Australia, Perth, WA, Australia

Quahir Sohail International Wheat and Maize Improvement Center (CIMMYT), International Winter Wheat Improvement Program, Ankara, Turkey

Cherukumalli Srinivasa Rao Indian Council of Agricultural Research-Central Research Institute for Dryland Agriculture, Hyderabad, India

David Stephens Australian Export Grains Innovation Centre, South Perth, WA, Australia

Bob A. Stewart Dryland Agriculture Institute, West Texas A & M University, Canyon, TX, USA

Muhammad Sufyan Biological Control Laboratory, Department of Entomology, University of Agriculture, Faisalabad, Pakistan

Sushil Thapa Dryland Agriculture Institute, West Texas A & M University, Canyon, TX, USA

Admassu Tsegaye College of Development Studies, Addis Ababa University, Addis Ababa, Ethiopia

G. Venkatesh Indian Council of Agricultural Research-Central Research Institute for Dryland Agriculture, Hyderabad, India

B. Venkateswarlu Vasantrao Naik Marathwada Krishi Vidyapeeth, Parbhani, India

Philip E. Vercoe School of Agriculture and Environment M085 and The UWA Institute of Agriculture, Faculty of Science, The University of Western Australia, Perth, Australia

V. Visha Kumari Indian Council of Agricultural Research-Central Research Institute for Dryland Agriculture, Hyderabad, India

Michael Walsh University of Sydney, Plant Breeding Institute, Narrabri, NSW, Australia

You-Cai Xiong State Key Laboratory of Grassland Agroecosystem, Institute of Arid Agroecology, School of Life Sciences, Lanzhou University, Lanzhou, China

Attila Yazar Irrigation and Agricultural Structures Department, Çukurova University, Adana, Turkey

Feng Zhang State Key Laboratory of Grassland Agroecosystem, Institute of Arid Agroecology, School of Life Sciences, Lanzhou University, Lanzhou, China

Part I
Introduction

Dryland Farming: Concept, Origin and Brief History

Bob A. Stewart and Sushil Thapa

1 Introduction

Dryland farming and dryland agriculture are often used synonymously. However, the raising of livestock on native vegetation can be considered dryland agriculture. In contrast, dryland farming requires conversion of a natural ecosystem to an agro-ecosystem, usually by tillage, and then growing introduced crops. Dryland farming is often used synonymously with rainfed farming although they can be vastly different. While both exclude irrigation, dryland farming emphasizes water conservation, sustainable crop yields, limited fertilizer and other inputs, and wind and water erosion constraints. Rainfed farming often deals with disposal of excess water and water erosion constraints, and strives for maximum crop yields using high levels of inputs (Stewart and Burnett 1987). Dryland farming occurs primarily in semiarid areas where annual precipitation is <25–50 % of the potential evapotranspiration (ET) demands (Stewart and Peterson 2015).

Dryland farming has probably been practiced since the beginning of farming. China has a recorded history of farming for more than 8000 years and people in dryland areas cultivated their land for cropping (Li 2007). Other regions in the world have long histories of dryland farming. During the seventh century, Tunis had more than a million ha of olive trees in full fruitage in the absence of irrigation with annual precipitation of about 225 mm (Shaw 1909). Koohafkan and Stewart (2008) briefly reviewed the development of dryland farming in Australia, China, Ethiopia, India, Mediterranean Regions, North America and West Africa. Drylands account for about 40 percent of the world's total land area and are home to about one-third of the population. Dryland farming is of growing importance worldwide due to the increasing demand for food and fiber. This chapter, however, is mostly limited to the

B.A. Stewart (✉) • S. Thapa
Dryland Agriculture Institute, West Texas A & M University, Canyon, TX, USA
e-mail: bstewart@mail.wtamu.edu

M. Farooq, K.H.M. Siddique (eds.), *Innovations in Dryland Agriculture*,
DOI 10.1007/978-3-319-47928-6_1

development and practice of dryland farming in North America since its beginning in the latter half of the nineteenth century.

2 Dryland Farming in North America

Even though the American Indians began farming on the North American continent approximately 7000 years ago and in the southwest United States more than 4000 years ago (Hurt 2003), modern dryland farming originated in the middle of nineteenth century. In 1847, Brigham Young, an American leader in the Latter Day Saint movement, led a group of settlers from Illinois to the Great Salt Lake Valley that later became part of Utah. The early endeavors of these pioneers were devoted to the construction of irrigation systems. Widtsoe (1920) tells how some of these settlers transitioned to growing crops without irrigation and summarizes the results of early day experiments of dryland farming. Gradually, the pioneers were convinced that farming without irrigation was possible, but the small population was so busy with their small irrigated farms that no serious efforts at dryland farming were attempted during the first several years. Early day publications indicate that dry farming occurred occasionally as early as 1854 or 1855, but it was not until about 1863 that the first dryland farming of any consequence occurred in Utah. A number of Scandinavian immigrants settled in what is now Bear River City, Utah and irrigated their farms with the alkali water of Malad Creek. When the crops failed, the starving settlers plowed up sagebrush land in desperation, planted grain and awaited the results. Modest yields were obtained and dryland farming became an established practice in that part of the Great Salt Lake Valley. Major John Wesley Powell saw the ripened fields of grain in the hot, dry sand and made special mention of them in his report about arid lands in the west (Powell 1879).

While it appears certain that John A. Widtsoe was the first to write definitively about dryland farming in North America, his book "Dry Farming" (Widtsoe 1920) was not published for a decade after the books "Arid Agriculture" (Buffum 1909) and "Dry Land Farming" (Shaw 1909). Widtsoe's book was clearly written many years earlier because the preface was dated June 1, 1910 and Shaw (1909) acknowledged that "he was indebted to the respective authors of the excellent books, "Arid Agriculture" and "Dry Farming" which were freely consulted in its preparation". Though Buffam did not acknowledge Widtsoe in his book, he had a chapter entitled Dry Farming, which contained much of the information that Widtsoe used in coining the name Dry Farming to describe farming in areas previously considered not suitable. Widtsoe (1920) stated "dry farming, as at present understood, is the profitable production of useful crops, without irrigation, on lands that receive annually a rainfall of 20 in. (500 mm) or less. In districts of torrential rains, high winds, unfavorable distribution of the rainfall, or other water-dissipating factors, the term "dry farming" is also properly applied to farming without irrigation under an annual precipitation of 25 (625 mm) or even 30 inches (750 mm). There is no sharp demarcation between dry and humid farming." He stressed, however, that dry farming always implies farming under a comparatively small annual rainfall. Even though he coined the term dry farming, Widtsoe (1920) said it was a misnomer. In reality, it is farming under drier conditions than those prevailing in the countries in which scientific agriculture originated. Widtsoe (1920) proposed that districts receiving less than 250 mm of annual atmospheric precipitation be designated arid, between 250 and 500 mm semiarid, between 500 and 750 mm subhumid and more than 750 mm humid. These values are commonly used by many todays. Based on those values, Widtsoe (1920) said that instead of dry farming, it would perhaps have been better to use arid farming, semiarid farming, humid farming and irrigation farming but 'dry farming' already in general use that it seemed unwise to suggest any change. Therefore, dry farming had for its purpose the reclamation, for the use of man, of vast areas of the world that had previously been considered unsuitable for growing crops.

Shaw (1909) used the term Dry Land Farming and defined it by both what it means and what it does not mean. He stated that it does not mean growing crops without water which would be absurd, but growing them with less water than would be successful without resorting to special methods of cultivation. It does not mean the growing of crops in all areas where precipitation falls. The degree of precipitation essential to growing crops is highly variable, since it is influenced by the soil and the nature of the evaporation. It does not mean growing tilled crops every year on the same land. Under some conditions, it is possible to grow only one crop in two years. It does not mean growing crops to the exclusion of livestock. Dry farming means (1) growing crops under semiarid conditions, (2) growing crops where moisture is normally deficient, (3) growing crops where moisture is temporarily deficient and (4) growing special crops by special methods. Shortage of moisture supply is the dominant thought that underlies any definition that may be framed regarding dryland farming.

Buffum (1909) wrote an early handbook for farmers and stockmen dealing with dryland farming. He titled his book "Arid Agriculture" because he felt Dry Farming was an untruth because no crop can be grown without moisture. Even though he objected to the name Dry Farming, he defined it as farming where annual precipitation is not considered sufficient for the production of profitable crops. Buffum (1909) further stated that "dry farming is usually carried on where the rainfall of one season is not sufficient and the moisture must be saved up for a longer period."

Even though Widtsoe (1920), Shaw (1909) and Buffum (1909) differed slightly in their preferred term for this new way of farming, they essentially used the same definition. More importantly, they all agreed that tillage was the key to success. Plowing in the fall as deep as possible was the general recommendation, usually 18–25 cm deep. This was to make a large enough reservoir to absorb and hold the moisture. Plowing was to be followed by disking in the fall or spring, followed by harrowing. Widtsoe (1920) stated "the all-important practice for the dry-farmer who is entering upon the growing season is cultivation. The soil must be covered continually with a deep layer of dry loose soil, which because of its looseness and dryness makes evaporation difficult." It is clear that they understood capillary water movement and that having loose soil on the surface with large pores would stop upward water movement. Widtsoe (1920) also understood the benefit of spreading straw or other litter over the soil. He reported the results of two studies showing reductions in evaporation of 22 and 38 percent with straw mulch on the surface. At the time, however, Widtsoe (1920) stated: "on the modern dry-farms, which are large in area, the artificial mulching of soils cannot become a very extensive practice, but it is well to bear the principle in mind."

The belief that deep plowing followed by shallow disking and harrowing to form a dust mulch on the soil surface was also strongly supported by H.W. Campbell who came from Vermont to South Dakota in 1879. Campbell had been inspired by the writings of Jethro Tull who had worked in England in the mid-1750s and was a strong advocate of intensive tillage. Tull concluded that thorough soil tillage produced crops that in some cases could not be produced by the addition of manure, and came to the erroneous conclusion that "tillage is manure" (Widtsoe 1920). While Tull's conclusion was incorrect, tillage does accelerate the decomposition of soil organic matter and increase soil fertility. This was certainly a benefit of tillage particularly before the widespread use of chemical fertilizer.

Campbell became a strong advocate of the dry-farm movement, particularly in the Great Plains that was previously called the Great American Desert and considered unsuitable for farming. Campbell grew wheat and other crops with considerable success in the late 1880s and in about 1895 began to publish Campbell's "Soil Culture and Farm Journal" (Widtsoe 1920). His writings were widely followed and led to the famous book "Soil Culture Manual" (Campbell 1907). Widtsoe (1920) summarized the Campbell system essential features as "the storage of water in the soil is imperative for the production of crops in dry years. This may be accomplished by proper tillage. Disk the land immediately after harvest; follow as soon as possible with the plow; follow the plow with the subsurface packer, and follow the packer with the smoothing harrow. Disk the land again as soon as possible in the spring and stir the soil deeply and carefully after every rain. Sow thinly in the fall with a drill. If the grain is too thick in the spring, harrow it out. To make sure of a crop, the land should be "summer tilled", which means that clean summer fallow should be practiced every other year, or as often as may be necessary."

With more and more successes of dryland farming, a wave of support developed. In 1907, the first Dry Farming Congress was held in Denver, Colorado. The USDA and state agricultural experiment stations also established dryland experiment stations throughout the Great Plains, some as early as 1903, that probably spurred dryland farming. In 1914, the USDA Division of Dryland Agriculture established field stations at 22 locations in the Great Plains (Fig. 1). World War I, coupled with high wheat prices and the development of power machinery, led to rapid expansion in cultivated land and large-scale production of wheat and other crops. Even marginal soils that should have been left in grass were plowed and seeded to wheat. This expansion in cultivated land took place primarily from 1915–1925 (Burnett et al. 1985). While various parts of the Great Plains experienced drought during the expansion of cultivated land, the Plains as a whole received average or better rainfall before 1930. The widespread drought of the 1930s, coupled with severe economic conditions and widespread wind erosion that was exacerbated by the dust-mulch

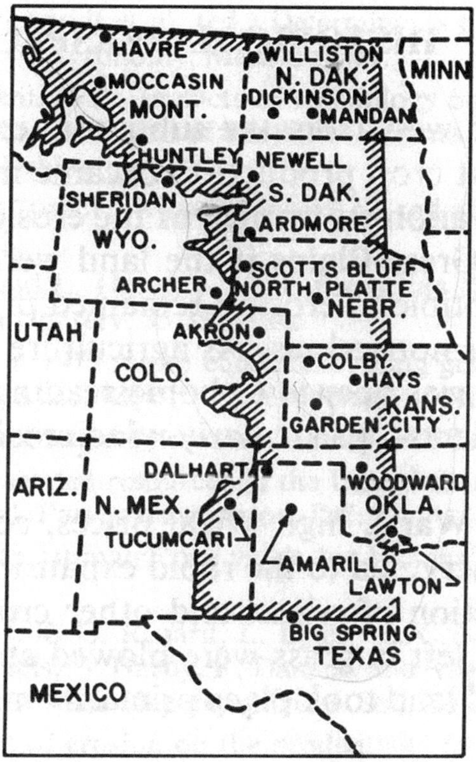

Fig. 1 Locations where dryland cropping experiments were conducted from 1903 to 1938 by USDA and State Agricultural Experiment Stations (Adapted from Cole and Mathews 1940)

tillage methods advocated by Campbell, resulted in the infamous Dust Bowl considered by many as the worst ecological manmade disaster in U.S. history. Surveys made during and shortly after the drought years of the 1930s in the southern Great Plains suggested that 43 percent of the area had serious wind-erosion damage. It was estimated that 2.6 million ha in the Southern Plains were removed from cultivation because of erosion but that about 10.3 million ha of less erodible soils made it through the drought cycle in sufficient condition to return to cultivation in the 1940s. With increased rainfall and higher crop prices in the early 1940s, the tendency to convert more grassland to cropland returned (Burnett et al. 1985).

3 The Dust Bowl Mandated a Change in Tillage

Extensive wind erosion during the Dust Bowl of the 1930s made it clear that changes had to be made. The intensive tillage that had worked so well in the beginning had greatly diminished the soil organic matter content leading to poor soil structure, lower infiltration, reduced water-holding capacity and uncontrollable wind erosion. Charles Noble of Alberta, Canada was one of the first to search for a cure. Noble concluded as early as 1916 that the once-favored "dust mulch" popularized by H.W. Campbell had to be abandoned (MacEwan 1983). This was not an easy decision for Noble to make because he had been a staunch disciple of the Campbell method embracing fine cultivation that inevitably pulverized the soil. However, Noble had observed neighbors that were using alternate strips of fallow and wheat as a means of controlling wind erosion. The standing stubble on the fallow prevented wind erosion on the fallowed land. Prairie farmers were challenged as never before to find some way of cultivating their fields without burying or sacrificing the ugly but precious stubble. Implements like the plow buried the stubble and all other vegetable residues completely, and the disc reduced the usefulness of the residue and buried much of it. Noble, working with agronomy teachers and others developed an important directive of southern Alberta, Canada; an important directive that was in stark contrast to the Campbell Soil Culture method. Their directives were (i) do not pulverize the soil with discs or harrows, (ii) use such implements as rod weeders, duck foot cultivators and spring tooth harrows that do not pulverize but leave a rough surface, (iii) if the shifting is small, a covering of manure or straw will sometimes save a whole field, (iv) reduce the extent of bare land by growing winter rye on part of the summer fallow, (v) consider working the farm in alternate strips of fallow and crop and (vi) co-operate with your neighbors–soil drifting can only be controlled by community action.

The strip cropping with fallow and wheat that started in 1918 was attracting attention and was sometimes considered plowless summer fallow or stubble mulching. Noble sensed the benefits from what was to become known as stubble mulch or trash cover and was searching for a way to cultivate for the destruction of weeds without destroying or burying the stubble and other vegetable residue from the previous crop. It would be unsightly but saving the soil was vastly more important than the house-cleaned appearance of a farm field (MacEwan 1983). This would be a vast departure from the plowing of the past. For generations, the plow was the symbol of farming. Good plowing was good farming; good farming was good plowing. Even the seal of the USDA has the moldboard plow as its centerpiece.

It was not until 1935 that the Noble Blade Cultivator was developed and played a crucial role in arresting soil erosion in the Canadian Plains and the western parts of the U.S. (MacEwan 1983). Noble was on a trip to California when he happened to see a California farmer using a straight blade tool to cut into the subsoil to loosen his sugar beets as an aid to lift them. The blade heaved the soil and disturbed weeds with little effect on the general appearance of the field. Noble immediately envisioned a machine that could cultivate the soil after harvesting a wheat crop to control weeds without depriving the land of its surface trash and standing stubble. He went directly to the home of a friend that had a shop attached and, with an old 3 m road grader, a borrowed forge and anvil, and being somewhat of a blacksmith, Noble constructed the first prototype. He tested it in a California orange grove and, while it did not work perfectly, Noble was pleased enough to take it to an advanced stage. He cut his planned stay in California short, loaded the crude cultivator onto a trailer, and headed home. Before it was time to start summer-fallowing, he had constructed four blade implements by pounding out new frames and blades making each one better than the last. Summer-fallowing that season was done with the four blades, and Noble was happy with the result. Neighbors and friends came to inspect the fields to satisfy themselves that the weeds had been controlled and that the stubble was still standing. Although Noble had not thought about manufacturing the plow, the plea from neighbors for one of his blades was so great that about 50 blade cultivators were constructed for local sale in 1937. The timing could not have been better because the worst dust storm in the history of the Plains area around Regina and Calgary, Canada hit on June 2, 1937 (MacEwan 1983). Nineteen of the 50 cultivators were purchased by the newly-formed USDA Soil Conservation Service and several others went to Canadian experiment stations for testing. In time, the straight bar was replaced by a V-shaped blade, but the effect was about the same, undercutting weeds and leaving stubble on the surface to blunt the force of wind and to catch and hold snow. Figure 2 shows a typical Noble Blade Cultivator. The V-shaped blades, often called sweeps, varied in size but the goal was always the same—to leave most of the crop residues on the surface. As a general guideline, about 75 % of surface residue cover remained on the surface following one operation. Unger et al. (2012) summarized the percentages of surface residue cover remaining after one pass with various implements.

Prior to the Noble Blade Cultivator, Fred Hoeme, a Hooker, Oklahoma farmer located in the center of the Dust Bowl area, developed a heavy-duty chisel plow in 1933 in an attempt to control wind erosion. He and his sons built and sold about 2000 of them from their homestead. In 1938, W.T. Graham purchased the manufacturing and distribution rights and established manufacturing in Amarillo, TX. The Graham-Hoeme plow was marketed as the "The Plow to Save the Plains" and was sold worldwide (Fig. 3). The chisel plow helped control wind erosion during the seven-year drought of the 1950s when precipitation was even lower than that during the Dust Bowl years of the 1930s. It was estimated that about half of the farmers in the Great Plains owned chisel plows. The chisels were often replaced with sweeps to achieve results similar to the Noble Blade Cultivator.

Fig. 2 Noble blade plow that ran about 10 cm below the soil surface to till the soil to cut roots of weeds without depriving the land of surface trash and standing stubble

Fig. 3 A Graham-Hoeme plow developed in 1933 and marketed as the "Plow to Save the Plains"

The Graham-Hoeme plow and the Noble Blade Cultivator brought an end to the emphasis on deep tillage and dust mulch that was earlier believed essential for dryland farming. These two implements were the forerunners of tillage implements used worldwide in dryland farming areas. In 2000, the Graham-Hoeme plow was dedicated as a National Historic Agricultural Engineering Landmark by the American Society of Agricultural and Biological Engineers and chisel plows and historic plaques are exhibited at Hooker, Oklahoma and at the USDA Conservation and Production Research Laboratory, Bushland, Texas.

4 Changing Tillage to Control Wind Erosion and Water Saving

Although Mr. Noble recognized that keeping residue on the surface would have a positive effect on soil water, particularly by trapping more snow, his primary focus was on controlling wind erosion. There is every indication that Mr. Hoeme focused entirely on controlling wind erosion. However, it became evident that the change from deep plowing and dust mulching to stubble mulching by using sweep and chisel plows resulted in more stored soil water in the soil profile during the fallow periods and increased yields of subsequent crops. Countless studies worldwide have shown that soil becomes air-dry to whatever depth the soil is plowed resulting in large losses of water. Therefore, farmers moved away from using dust mulch that had been widely promoted. Theoretically, dust mulch reduces water loss because plowing the soil to create the mulch forms large pores. Water cannot move from small pores into large pores by capillary action so the mulch prevents upward movement of water from the wetter soil below the mulch. However, by the time a soil can be tilled to create dust mulch, it has already dried to the point that capillary water movement has essentially stopped. Crop residues on the soil surface have other

Table 1 Progress in wheat–fallow cropping at the USDA Central Great Plains Research Station, Akron, Colorado

Years	Tillage	Number tillage operations	Fallow water storage[a]		Wheat yield (Mg ha^{-1})
			(mm)	(% precipitation)	
1916–1930	Maximum tillage: plow, harrow (dust mulch)	7–10	102	19	1.07
1931–1945	Conventional tillage: shallow disk, rodweeder	5–7	118	24	1.16
1946–1960	Improved conventional tillage: begin stubble mulch in 1957	4–6	137	27	1.73
1961–1975	Stubble mulch: begin minimum tillage with herbicides in 1969	2–3	157	33	2.16
1976–1990	Minimum tillage (projected estimate): begin no-tillage in 1983	0–1	183	40	2.69

Adapted from Greb et al. (1979)
[a]Fallow period of 14 months, mid-July to second mid-September

benefits such as preventing capillary action and keeping the soil cooler. Therefore, less intensive tillage and maintaining crop residues clearly increase water in the soil profile during fallow periods. Widtsoe (1920) recognized the benefit of having straw or manure on the soil surface but did not think it could be done in a practical way.

Peterson et al. (2012) summarized some studies that focused on increasing precipitation use efficiency in dryland agroecosystems. A long-term study that began in 1916 at the U.S. Central Great Plains Research Station, Akron, Colorado is a classic example of how a greater percentage of the precipitation occurring during a fallow period can be stored in the soil profile by decreasing tillage intensity and maintaining more crop residues on the soil surface. The results from 1916 to 1975 are presented in Table 1 along with projections for future years. Between 1916 and 1975, the number of tillage operations during the 14-month fallow period decreased dramatically resulting in more water storage in the soil profile for use by the subsequent wheat crop. Increasing soil water storage during the fallow period from 102 mm (19 % fallow efficiency) under maximum tillage to 157 mm (33 % fallow efficiency) using stubble mulch tillage more than doubled the yield of the following wheat crop (Table 1). Soil water storage increased because fewer tillage operations dried the soil less often and the newly-adopted tillage implements left more crop residue on the soil surface. Greb et al. (1979) predicted that adoption of no-till practices would increase fallow efficiency to 40 %. However, during the 1980s and 1990s, fallow efficiencies were generally less than 40 %, regardless of the climatic zone where the data were collected (Peterson et al. 2012). This is most likely due to insufficient crop residues being produced in dryland farming areas for the required surface mulch. Greb et al. (1967) and Unger (1978) showed that surface residue greatly increased fallow water storage, but residue amounts in excess of 6 Mg ha^{-1} were needed to achieve fallow efficiencies greater than 35–40 %. In the west-central Great Plains at wheat harvest, the maximum residue accumulation point in the system cycle, residue amounts commonly range from 2.2 to 5.6 Mg ha^{-1} (Peterson et al. 2012). The authors concluded that crop residues may not be feasible in dryland areas to increase fallow efficiency percentages above about 35 %. Nevertheless, the move away from intensive tillage in dryland areas to subsurface tillage and no-tillage systems has greatly reduced wind erosion and simultaneously increased precipitation use efficiency. This has significantly reduced wind erosion, slowed soil organic matter decline, increased crop yields and improved sustainability.

4.1 Conservation Agriculture

In recent years, there has been a worldwide focus on reducing tillage not only in dryland farming areas but in all climatic regions where crops are grown. The Food and Agriculture Organization (FAO) of the United Nations is the leading organization pushing this system, known as conservation agriculture (CA), but many other international, national and local entities are solidly behind the movement. While current crop production systems have often resulted in soil degradation and in extreme cases desertification, the adoption of CA technology has led to a reversion of these processes (Friedrich et al. 2012). CA is an approach to managing agroecosystems for improved and sustained productivity, increased profits and food security while preserving and enhancing the resource base and the environment (Friedrich et al. 2012). The authors characterized CA by three linked principles, namely:

1. Continuous no- or minimal mechanical soil disturbance (i.e., no-tillage and direct sowing or broadcasting of crop seeds, and direct placing of planting material in the soil; minimum soil disturbance from cultivation, harvest operation or farm traffic, in special cases limited strip tillage).
2. Permanent organic soil cover, especially by crop residues, crops and cover crops.
3. Diversification of crop species grown in sequence or associations through rotations or, in the case of perennial crops, associations of plants including a balanced mix of legume and non-legume crops.

CA principles are universally applicable to all agricultural landscapes and land uses with locally-adapted practices. However, the benefits and success of implementing CA successfully are perhaps the least in dryland farming areas because these areas are generally hotter and drier than more favored areas. Stewart et al. (1991) developed an index of temperature and precipitation that compares different climatic regions to the difficulty of developing sustainable agricultural systems, which are significantly more difficult in dryland farming regions (Fig. 4). However, CA grew out of dryland farming. Friedrich et al. (2012) stated that tillage, particularly in fragile ecosystems, was questioned for the first time in the 1930s when the Dust Bowl devastated wide areas of the mid-west United States. Concepts for reducing tillage

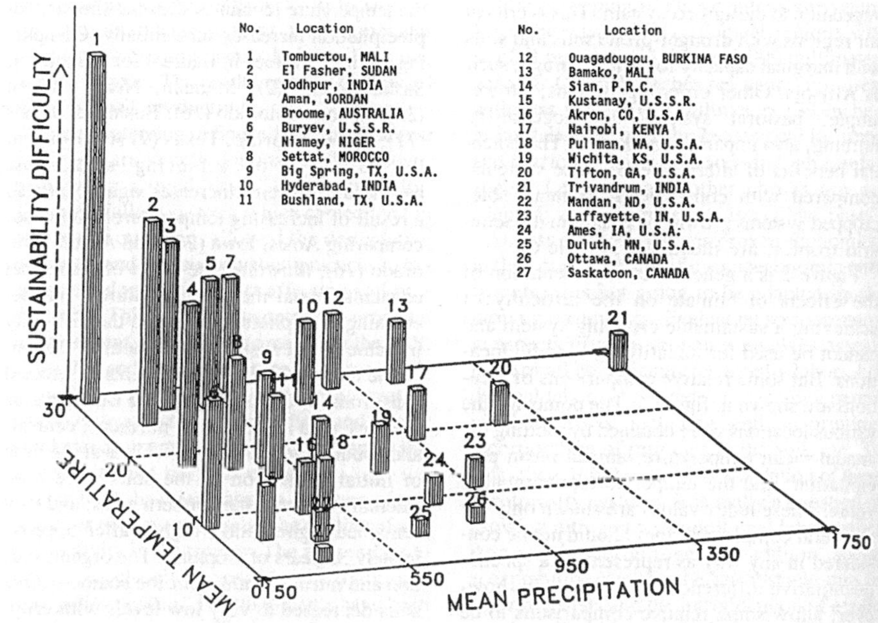

Fig. 4 Mean annual temperature and precipitation for various locations and temperature/precipitation as and "index" indicating the difficulty of developing a sustainable agriculture system (Stewart et al. 1991)

and keeping the soil covered with crop residues were developed and the term conservation tillage was introduced to reflect such practices aimed at soil protection. As discussed above, concepts for reducing tillage and keeping soil covered were developed and the term conservation tillage was introduced to reflect such practices aimed at soil protection. Unger (2006) stated that conservation tillage is any combination of tillage and planting practices that generally reduces the loss of soil and water relative to losses with conventional tillage. McCarthy et al. (1993) reported that conservation tillage is any tillage/planting system which leaves at least 30 percent of the field surface covered with crop residue after planting has been completed, and that this amount of cover will reduce erosion by at least 50 percent compared to bare, fallow soil.

In dryland farming areas, even when no-tillage is used between crops, there are often sufficient crop residues remaining after planting the next crop to cover 30 % of the soil surface. This is particularly true when only one crop is produced every two years, or two crops every three years, that result in long fallow periods between crops. As a result, current definitions of conservation tillage often do not include the 30 % minimum cover requirement. For example, CTIC (2015) states "conservation tillage is a system of crop production with little, if any, tillage. It increases the residue from the crop that remains in the field after harvest through planting. This results in increased natural recycling of crop residues." The emphasis is on leaving last year's crop residue on the surface before and during planting operations to provide cover for the soil at a critical time of the year. The residue is left on the surface by reducing tillage operations and turning the soil less, even when the amount of residue is less than desired, particularly during drought years. Although this meets the first principle of CA in terms of minimal soil disturbance, it does not meet the second principle regarding soil cover based on the criteria that FAO uses to collect global data for adoption of CA. Global data of CA adoption are not officially reported, but collected from local farmer and interest groups. The data are assembled and published by FAO. For data collection, FAO has quantified the CA definition as follows (Friedrich et al. 2012):

1. Minimum Soil Disturbance: Minimum soil disturbance refers to low disturbance no-tillage and direct seeding. The disturbed area must be less than 15 cm wide or less than 25 % of the cropped area (whichever is lower). There should be no periodic tillage that disturbs a greater area than the aforementioned limits. Strip tillage is allowed if the disturbed area is less than the set limits.
2. Organic Soil Cover: Three categories are distinguished: 30–60 %, >60–90 % and >90 % ground cover, measured immediately after the direct seeding operation. Areas with less than 30 % cover are not considered CA.
3. Crop Rotations/Associations: Rotation/association should involve at least three different crops. However, repetitive wheat or maize cropping is not an exclusion factor for the purpose of this data collection, but rotation/association is recorded where practiced.

Therefore, even though CA largely stemmed from dryland farming, many dryland farming cropping systems do not qualify as CA systems due to the lack of sufficient crop residues. Limited precipitation and shorter growing periods allow only a small amount of crop residue to be returned to the soil. To make things worse, many dryland farming areas, especially in developing countries, have high rates of poverty and many crop residues are removed from the land for animal feed or for cooking and heating fuel. This can cause already low soil organic matter concentrations to become even lower resulting in lower soil quality. As already illustrated in Fig. 4, developing a sustainable cropping system becomes increasingly difficult in marginal climatic regions. However, adoption of reduced tillage and continuous cropping can increase dryland nitrogen (N) storage to a depth of 20 cm compared to a conventional farming practice, and hence, a no-till system provides better opportunities to sustain crop yields by growing more crops and to conserve soil carbon (C) and N more than the traditional tillage practice in the drylands of the U.S. Great Plains (Sainju et al., 2006). In the U.S. Northern Great Plains, continuous cropping with no-tillage improved soil and crop management practices and increased dryland soil C sequestration from 0 to 15 cm depth by 233 kg C ha^{-1} $year^{-1}$ compared to a loss of 141 kg C ha^{-1} $year^{-1}$ with conventional tillage (Halvorson et al. 2002). CA systems allow producers to increase cropping intensity because no-till conserves surface residues and retains water in the soil profile more than conventional tillage practices (Aase and Pikul 2000; Farhani et al. 1998). Similarly, data collected from Bushland, TX in the southern Great Plains showed that precipitation storage efficiency increased from 15 % with disk tillage to 35 % with no-till (Unger and Wiese 1979).

Table 2 Extent of adoption of conservation agriculture (CA) worldwide (countries with more than 100,000 ha in 2011)

Country	CA area (ha)	Country	CA area (ha)
United States	26,500,000	South Africa	368,000
Argentina	25,553,000	Venezuela	300,000
Brazil	25,502,000	France	200,000
Australia	17,000,000	Zambia	200,000
Canada	13,481,000	Chile	180,000
Russia	4,500,000	New Zealand	162,000
China	3,100,000	Finland	160,000
Paraguay	2,400,000	Mozambique	152,000
Kazakhstan	1,600,000	United Kingdom	150,000
Bolivia	706,000	Zimbabwe	139,300
Uruguay	655,000	Columbia	127,000
Spain	650,100	Others	409,400
Ukraine	600,000	Total	124,794,840

Adapted from Friedrich et al. (2012)

CA adoption worldwide has increased rapidly in recent years from 2.8 M ha worldwide in 1973/1974 to 6.2 M ha in 1983/1984, 38 M ha in 1996/1997, 45 M ha in 1999 and 72 M ha in 2003 (Friedrich et al. 2012). They stated that for the last 11 years reported, CA systems have expanded at an average rate of 7 M ha per year to increase from 45 M ha in 1999 to about 125 M ha in 2011 showing the increased interest of farmers in this production system. Countries with more than 100,000 ha of CA in 2011 are listed in Table 2.

The worldwide adoption of CA shows the dramatic shift away from tillage that was considered necessary and beneficial at the beginning of dryland farming in the mid-1800s. However, even with the increased adoption of CA, 125 M ha is less than 10 percent of the World's arable land (1396 M ha) (FAOSTAT 2015). Friedrich et al. (2012) consider that the main barriers to the adoption of CA practices continue to be: knowledge on how to do it (know how); mindset (tradition, prejudice); inadequate policies, for example, commodity-based subsidies (EU, US) and direct farm payments (EU); lack of availability of appropriate equipment and machines (many countries of the world); and lack of suitable herbicides to facilitate weed and vegetation management (especially for large-scale farms in developing countries). They also concluded that the roots of CA origins lie more in the farming communities than in the scientific community, and its spread has been largely farmer-driven. Experience and empirical evidence across many countries has shown that the rapid adoption and spread of CA requires a change in commitment and behavior of all concerned stakeholders. For the farmers, a mechanism to experiment, learn and adapt is a prerequisite. For policy makers and institutional leaders, the transformation of tillage systems to CA systems requires a thorough understanding of the large and longer-term economic, social and environmental benefits that the CA paradigm offers to producers and society at large.

4.2 Major Crops for Dryland Farming

Wheat (*Triticum aestivum* L.) was the leading crop in the beginning of dryland farming in the US (Widtsoe 1920) and remains so today in most regions. The crop is not highly drought-resistant and does not tolerate prolonged drought. Wheat is considered a drought avoidance crop which is well suited to dryland regions because it can take advantage of soil moisture that accumulates during fallow periods, winter and early spring, and matures before the hot and dry, late summer when planted at the appropriate time (Hansen et al. 2012). Similarly, under water-stressed growing conditions, wheat plants produce smaller cells reducing the height of the culm, the size of the leaves and the stomatal openings, and hence are less adversely affected by drought (Arnon 1972). Maize (*Zea mays* L.) was also a major crop in the beginning along with sorghum (*Sorghum bicolor* L.). Maize lost favor in semiarid regions because it was not considered as drought tolerant as wheat and sorghum. This is interesting because it was considered the most successful dryland crop in the early days of dryland farming (Buffum 1909; Shaw 1909; Widtsoe 1920). Widtsoe (1920) stated that of all the crops tried, maize was the most uniformly successful under extremely dry conditions and he felt that the dryland community had yet to realize the value of maize as a dryland crop. He stressed, however, that the value in dry years was as a fodder crop because grain may not be produced. These early writers stated that maize grown in dry areas should be planted in hills of multiple plants per hill. Sorghum is grown largely in regions of Africa, Asia, North America and Australia that are too dry for successful maize production. Sorghum produces higher grain yield than maize under dry and hot climatic conditions because it remains dormant during the period of severe water stress and resumes growth when favorable conditions reappear (Leonard and Martin 1963).

Cotton (*Gossypium hirsutum* L.) is also a major crop in dryland areas where there is a sufficient frost-free growing season of at least 180 to 200 days, ample sunlight and enough growing heat units accumulated during the growing season. Sunflower (*Helianthus annuus* L.) is cultivated in some dryland areas; it has a deep rooting system capable of extracting water and nutrients to a depth of 3 m, which contributes to its adaptation to dryland environments (Jones and Johnson 1983). Since there is a close relationship between plant available water in the soil profile during the reproductive period and total biomass production, plants with high water use efficiency (WUE) are important for determining crop yield under dryland environments (Blum 2009). Crop yields have increased significantly in most dryland farming regions over the years. For example, in Whitman County, Washington winter and spring wheat yields have increased from less than 2000 kg ha^{-1} to about 5000 kg ha^{-1} over 73 years (Fig. 5), an average of 48 kg ha^{-1} year^{-1}. However, the data points indicate that yields may have plateaued during the 1980s. There are many reasons for the increased yields, but there is little doubt that the main driver has been reducing water loss by evaporation and runoff so more water is available for the plants. This is because the units of water to produce a unit of biomass for crops have changed little, if any. Widtsoe (1920) reported that German scientists, in

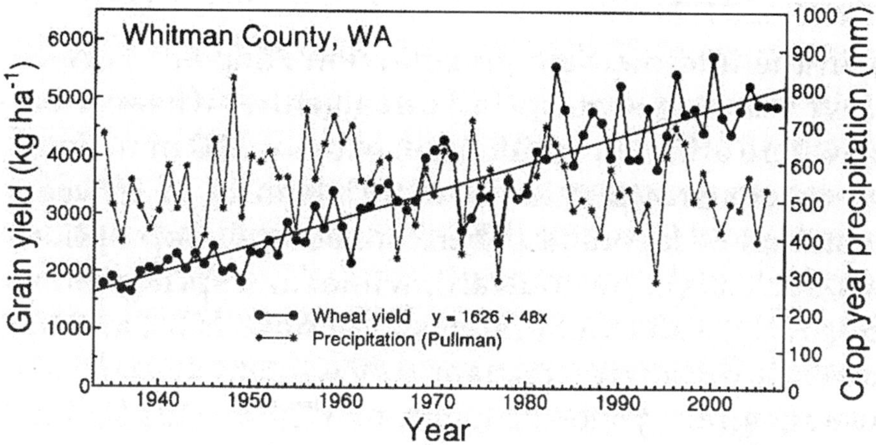

Fig. 5 Long-term county-wide dryland wheat grain yield in Whitman County, Washington, superimposed with crop-year precipitation from Pullman, WA (Source: Schillinger et al. 2010)

the late 1800s, reported that 233 units of water were required for one unit of corn dry matter and 338 units for a unit of wheat dry matter. The studies were done in a humid environment so it can be assumed that most of the water use was for transpiration. Sinclair and Weiss (2010) reported that a C_4 grass growing in a somewhat average transpiration environment (examples include maize and sorghum) have a transpiration rate of approximately 220 units water for each unit of plant growth. A C_4 species (includes wheat) growing in the same environment will transpire about 330 units of water for each unit of dry matter produced. Stewart and Peterson (2015), based on field data, recently estimated that the transpiration rates of maize and grain sorghum are generally between 225 and 275 g of water per g of biomass production in the U.S. Great Plains region. Therefore, increased yields over time like those shown in Fig. 5 cannot be attributed to more water efficient crop cultivars since no significant change in the units of transpiration required to produce a unit of biomass for a given species appears to have occurred. Accumulation of plant biomass is directly related to water availability (Sinclair 2009). Sinclair states that the difference in vapor pressure inside and outside the leaf (VPD) controls water loss through the stomata. VPD in arid regions is large because the vapor pressure of the atmosphere is very low relative to humid areas. For a given environment, VPD cannot be controlled—it is what it is. Sinclair further stated "despite claims that crop yields will be substantially increased by the application of biotechnology, the physical linkage between growth and transpiration imposes a barrier that is not amenable to genetic alteration." While many plant scientists disagree, there is little evidence to repute it (Gurian-Sherman 2012). Therefore, the increased yields shown in Fig. 5 are most likely due to improved water management that increased ET and to improved cultivars with higher harvest index (HI) values. HI is the weight of grain divided by the weight of aboveground biomass which increased significantly in wheat with the development of short-straw cultivars. Fischer et al. (2014) also concluded that there is little hard evidence that plant breeding has increased transpiration efficiencies and that higher HI values account for much of the increased yield in recent years. However, Berry et al. (2007) concluded that there appears to be little scope for further increases in HI beyond 0.55 to 0.60 in crops that bear aboveground grain because such crops depend on a stable structure to distribute leaf area, support grain and prevent lodging.

4.3 Potential to Enhance Water Productivity in Dryland Farming

Wani et al. (2012) stated that there is vast untapped potential to increase yields in rainfed areas with appropriate soil and water interventions. They indicated that a linear relationship is generally assumed between biomass growth and ET, which describes water productivity between 1000 and 3000 $m^3 t^{-1}$ (1000 to 3000 kg water per 1 kg grain) for grain production (Fig. 6). Below yield levels of 3 t ha^{-1}, however, they state that this relationship does not hold true and this coincides with the yield levels of small and marginal farmers in dryland areas. Water productivities range from 5000 to 8000 $m^3 t^{-1}$ when grain yields are as low as 1 t ha^{-1}. Wani et al. (2012) state that evidence from water balance analyses in farmers' fields around the world shows that only a small fraction, less than 30 % of precipitation, is used as plant transpiration to support plant growth.

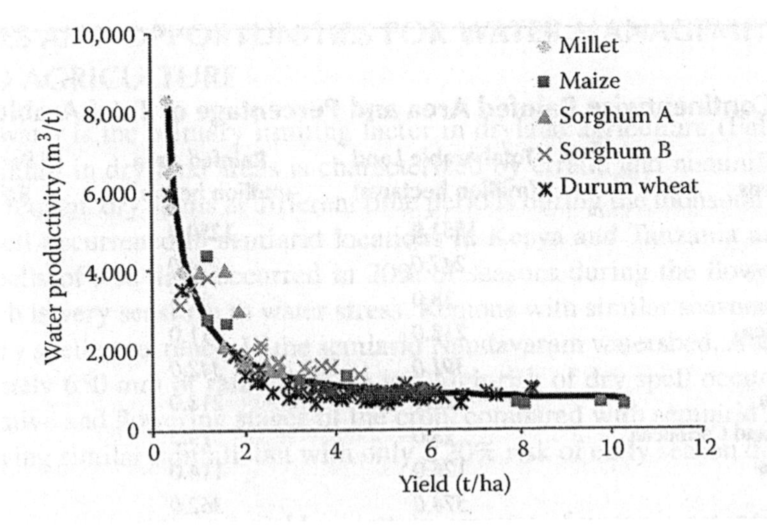

Fig. 6 Yield relationship between water productivity and yield of cereal crops in different climatic conditions and management (Source: Wani et al. 2012)

A graphical model of the water balance for a wide range of climatic conditions (Ponce 1995) is presented in Fig. 7. This model was based on the range of climates in the Sertao region of Brazil, but it exemplifies important differences in how water is used over a range of climatic zones. Although Fig. 7 is only conceptual, and ignores drainage regardless of the amount of precipitation, it does illustrate important points. Most importantly, it shows that only a small proportion of annual precipitation in arid regions is used for ET and that a large proportion is lost as evaporation during the time that crops are not being produced. Furthermore, a large part of ET is lost as evaporation from the soil surface during the growing season such that the amount of water used in arid areas for transpiration is small, and it is this water that increases biomass. Studies have shown that for the semiarid southern U.S. Great Plains, about 50 % of annual precipitation is used for ET and about 50 % for transpiration, so 25 % or even less is transpired. Wani et al. (2012) stated that in arid areas as much as 90 % of precipitation can be lost as evaporation and 10 % for transpiration. These estimates support the concept illustrated in Fig. 7. While these low percentages of precipitation used for transpiration support the view of Wani et al. (2012), that there is untapped potential for using water more efficiently, they also point out the challenges of successfully tapping into this potential. There are three basic strategies for increasing crop yields in dryland cropping systems: (1) increase the capture of precipitation by reducing runoff and storing it in the soil profile for later use by the crop for ET, (2) increase to the fullest extent feasible the portion of ET that is used for transpiration relative to that lost by evaporation from the soil surface and (3) ration water use so that early vegetative growth does not use all of the plant available water in the soil profile so that some is available during reproductive and grain-filling periods, particularly for grain crops. These strategies led to the more than doubling of wheat yields in Whitman County, Washington discussed earlier with Fig. 5. Although the average annual precipitation has remained fairly constant, a greater percentage is likely to have been stored in the soil profile for use by subsequent crops, and the mulch left on the soil surface reduced evaporation during the growing season, so that a higher percentage of annual precipitation was used for transpiration. As illustrated in Fig. 7, increasing yields in semiarid regions depends on increasing the portion of precipitation that is used for ET, and increasing the portion of ET that is used for transpiration.

Stewart and Peterson (2015) stated that the grain yield (GY) of a crop can be expressed by:

$$GY = ET \times T/ET \times 1/TR \times HI \tag{1}$$

where GY is kg ha^{-1} of dry grain yield, ET is kg ha^{-1} of ET (water use by evaporation from the soil surface and transpiration by the crop between seeding and harvest), T/ET is the fraction of ET transpired by the crop, TR is the transpiration ratio (kg water transpired per kg of aboveground biomass) and HI is harvest index (kg dry grain/kg aboveground biomass). All weights are dry weights, so GY values should be adjusted before comparing with field grain weights that usually have 12–15 % water content. While this equation applies to all situations where grain

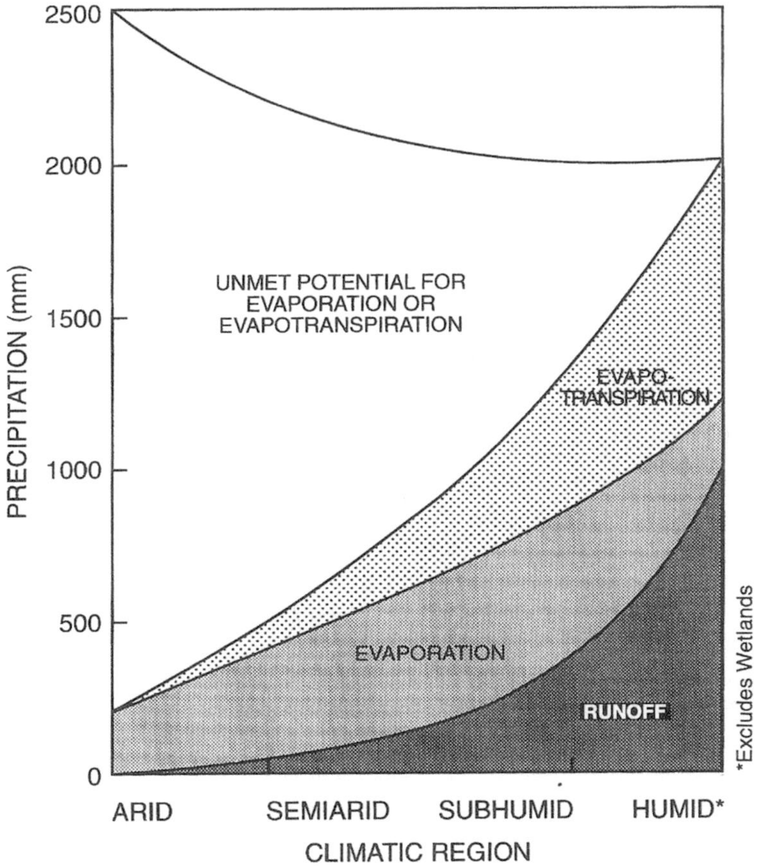

Fig. 7 A graphical model of the water balance for a wide range of climatic conditions (Source: Ponce 1995)

crops are produced, the range of values for each of the components becomes considerably greater and more variable in dryland farming regions. Average annual precipitation in dryland farming regions is generally less than 500 mm, so ET amounts are commonly below 250 mm even when annual cropping is practiced, and perhaps about 350 mm when a fallow period is used prior to seeding a crop. The T/ET values usually range between 0.50 and 0.75 with the lower values associated with limited water. Sinclair and Weiss (2010) stated that the TR for crops like maize and sorghum is about 220 when grown in a somewhat average transpiration environment, but could be about 280 for an arid environment and as low as 160 for a humid transpiration environment. HI values can also vary considerably. Prihar and Stewart (1990) reported upper values of HI for maize, grain sorghum, and wheat of about 0.60, 0.56, and 0.47, respectively. However, the lower values depended mainly on the amount of stress. Under dryland farming conditions, HI values are often below 0.30 and under extreme water stress, values can equal or approach zero.

Others, including Passioura (1977), French and Schultz (1984), and Connor et al. (2011), have developed equations that link grain yield with T, ET and HI. Eq. (1), however, includes all four factors that affect grain yield; since changing any one factor almost always changes the other factors, it can be easily used to evaluate how changing management practices will likely affect grain yields. Stewart and Peterson (2015) discussed each of the components for dryland regions and strategies for their improvement. Since Eq. (1) is linear, increasing any one component by 5 % will increase grain yield by 5 %. Therefore, large increases in yield only occur when all of the components increase and this generally happens when ET increases. Stewart and Peterson (2015) presented hypothetical component values for maize growing in areas of annual precipitation ranging from 500 to 1000 mm that showed doubling ET increased grain yield 4-fold. Increasing ET usually favors the other components. However, some common strategies used for growing grain crops in dryland farming areas where ET amounts are low often result in some components becoming less favorable. For example, when growing season ET is expected to be low, producers commonly reduce plant populations significantly. If this is not done, plant available water in the soil profile is used early for vegetative growth leaving little or no water for the reproductive and grain-filling stages. Looking at Eq. (1), however, reducing plant populations sometimes results in failure to exploit all of the plant available water in the soil profile. Lower plant populations also reduce canopy cover such that the increased bare ground increases evaporation from the soil surface resulting in a lower T/ET component. In addition, with the plants further apart, the microclimate is less favorable so TR increases because more units of water must be transpired to produce a unit of biomass. Thus, reducing plant populations in dryland farming areas is considered an essential practice, but it tends to make every component in Eq. (1) less favorable apart from HI which in reality means reducing the plant population is to avert risk of failure rather than to increase yield. Another common strategy in dryland farming is to use skip rows or leave wider spaces between rows. Again, this practice makes all of the components in Eq. (1) less favorable apart from HI. As a result, it is exceedingly difficult to increase yields in dryland farming regions where ET amounts are low. Eq. (1) also illustrates why mulch is important for dryland farming because it increases soil water storage during a fallow period thereby increasing ET, and it significantly increases the T/ET component leading to a better canopy that improves the microclimate that tends to reduce TR. Without question, reducing tillage and leaving more crop residues on the soil surface has been the main reason for increased yields in dryland farming regions like those shown in Fig. 5 for Whitman County, Washington. In many dryland farming regions, however, there are not sufficient crop residues produced for adequate mulch, or the residues are used for animal feed or fuel.

A relatively new strategy for dryland farming is growing grain sorghum or maize plants in clumps rather than equally spaced in rows (Bandaru et al. 2006; Kapanigowda et al. 2010; Krishnareddy et al. 2010). The hypothesis is that clumped plants, particularly grain sorghum, will produce fewer tillers that use water and nutrients, produce little or no grain when water is limited, improve the microclimate to lower vapor pressure deficit, and fully use plant available soil water during the growing season. A possible disadvantage is that less shading of the soil surface might lower the T/ET component so it is recommended that clump planting is combined with mulching. Although growing maize or grain sorghum in clumps is not expected to increase yield substantially, the practice has potential because it is easily done with existing equipment with no additional input cost. The use of clumps as well as skip rows is based on the principle of non-uniformity (Loomis 1983; Connor et al. 2011). These authors stated that when a soil resource such as water is limiting, non-uniform treatment of the land or crop can be an advantage. Where soil resources are non-limiting, however, they stated that uniform cropping will provide the greatest efficiency in light interception and photosynthesis. It is of interest that both Shaw (1909) and Buffum (1909) emphasized in their early writings that maize grown for grain in dryland areas with less than about 400 mm average annual precipitation should be planted in hills with at least four stalks. They stressed that pollination significantly improved in maize plants growing in hills.

Stewart and Peterson (2015) suggested that the best strategy for crop production in dryland farming areas may be to grow forage rather than grain crops. A careful analysis of Eq. (1) supports this suggestion because the HI component is eliminated and concern about crop failure is diminished. Therefore, a larger plant population can be used that will more fully utilize growing season precipitation and stored soil water, shade more of the soil surface to improve the T/ET component, and improve the microclimate that tends to decrease the TR value. The forage can be removed as hay or silage, or grazed by animals which is perhaps the best strategy because most of the nutrients are recycled to the soil with the manure. Therefore, in cases where it is feasible, the best way to capture the potential of dryland farming areas is to develop crop–livestock systems.

4.4 Conservation Bench Terraces

The primary goal in dryland farming regions is to capture precipitation where it falls, use it immediately for a growing crop or store it in the soil profile for later use by plants. There are situations, however, where it can be advantageous to harvest water from a larger area and concentrate it on a smaller area. Oweis et al. (2012) and Koohafkan and Stewart (2008) summarized some of the benefits and constraints of water harvesting. Water harvesting is perhaps best used in areas where average annual precipitation is too low for commonly-accepted dryland farming practices and where crops can only be grown after concentrating runoff water. Therefore, water harvesting is generally not considered a common dryland farming practice. An exception is the conservation bench terraces or Zingg terraces developed in the U.S. as a type of rainfall multiplier for use on large-scale mechanized farms (Zingg and Hauser 1959). As depicted in Fig. 8, bench terraces use part of the land surface as a catchment to provide additional runoff onto level terraces where crops are grown. However, the catchment areas can also be used for cropping but generally

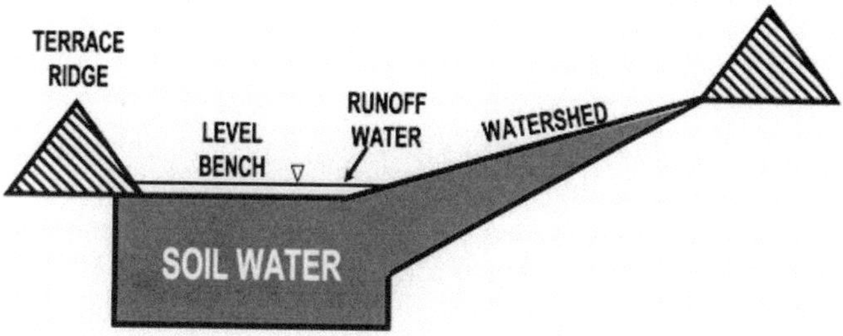

Fig. 8 Cross-sectional diagram of a conservation bench terrace that concentrates runoff water from an adjacent catchment area to a level bench (Modified from Baumhardt et al. 2008)

only one crop every two years or two crops in three years to allow for long fallow periods between crops. Conservation bench terraces have also been successfully developed in semiarid and subhumid regions of India. Sharda et al. (2002) conducted a study in India and, compared to the conventional system of sloping borders, the conservation bench terrace reduced surface runoff from 36.3 to 7.4 % of rainfall and soil erosion from 10.1 to 1.19 t ha^{-1} $year^{-1}$.

Water running off the catchment area is retained on a level bench for annual cropping. The terraces are designed for use on sloping lands of about 1–2 % with the catchment areas about twice as wide as the level terrace. The width of the terraces is usually at least 25 m to be compatible with tillage and harvesting equipment. However, the width can be constrained by the slope because wide level terraces require large amounts of soil movement to make the level terrace which make it cost prohibitive and create fertility problems on some soils because soil has to be removed from one side of the level terrace and to the other. The hypothesis is that the catchment area can still be farmed without serious yield reductions and the runoff water can be captured on the level terrace to make it more productive. For example, assuming average annual runoff from a field is 50 mm and the catchment area is twice as wide as the level terrace, the level terrace would not lose any water from runoff and would receive 100 mm from the catchment area so the level terrace would have 150 mm more water for crop production than the catchment area. Although the concept has worked well in some areas, the practice has not been widely accepted because of the constraints already listed. Unger and Stewart (1983) briefly reviewed the design of experimental conservation bench terraces at several locations in the U.S. Great Plains. The constraints are greater for large-scale producers as the equipment is on a larger scale and the construction of terraces and land leveling has become more costly. In recent years, controlling water erosion has shifted away from using terraces to some form of conservation tillage that leaves sufficient mulch on the surface to increase infiltration and reduce runoff. Conservation bench terraces, however, are effective and should be considered when feasible, particularly for small holder farmers.

5 Gravel and Plastic Mulches

Although the use of crop residues for mulch is the focus of dryland farming, particularly on large-scale operations, other forms of mulch play an important role in producing crops in low precipitation areas in some countries such as China. In some marginal areas of China, where annual precipitation is 200 to 300 mm, farmers have survived for many generations (Liang et al., 2012) by applying a 10 cm thick layer of porous gravel on the soil surface. This lessens the risk of crop failure, which commonly occurs as a result of low precipitation and high evaporation. Such gravel-based systems have existed for at least 200 to 300 years with some evidence that they date back 400 to 500 years. Even today, new areas are being developed in several counties of Gansu Province and other areas of northwest China. Liang et al. (2012) estimated that 170,000 ha of cropland in these areas are being farmed using gravel mulches. Although such methods have severe labor constraints, they clearly illustrate the effectiveness of mulch for conserving water in dryland areas. The gravel mulch is effective in reducing soil surface evaporation and runoff, improving water infiltration and increasing soil temperature (Xie et al. 2010).

In contrast to the historical use of gravel for mulch, plastic film is increasingly being used in crop production systems. Although frequently used in developed countries with high-income crops, it has been widely used in China in dryland farming areas. With limited available land for agricultural expansion, China has used plastic film mulch extensively to reduce soil evaporation and weed populations, increase soil temperature and enhance crop transpiration. It has been widely used for the production of cereals, vegetables, oilseeds, fruits and other crops. China uses 40 % of the world's plastic mulch (Ingman et al. 2015). In 2011, 1.2 M t of plastic film was applied to 19.8 M ha of cropland in China, equivalent to 12.2 % of the arable land (Ma 2013; Zhang 2013). While it is unlikely that plastic film will be used on large-scale commercial farms due to water runoff problems and other constraints, its use has greatly increased crop production on smallholder farms in China. This practice can capture the potential of low rainfall regions because it drastically reduces the evaporation of water from the soil surface during the growing season so that a high percentage of ET is used by plants as transpiration that is directly related to biomass production. In 2012 in Gansu Province, 92 % of the 896,000 ha of maize was produced using plastic mulch (personal communication, Dr. Fan Tinglu, Gansu Academy Agricultural Science, Langzhou). Several field experiments have shown that plastic mulch can modify soil moisture and temperature, both of which control near-surface biological processes including seed germination, plant growth and insect population dynamics (Anikwe et al. 2007; Wang et al. 2011). Plastic film usage does present some environmental problems as it is difficult to dispose of after it is no longer usable. This is commonly called baise wuran, the 'white pollution' in China. Biodegradable films are being developed that may alleviate this problem in the future. This practice has not been widely used for dryland farming outside of China and is generally not considered feasible for large commercial farms. However, it demonstrates the effectiveness of mulch for capturing potential for increasing crop production in dryland farming regions.

5.1 Future and Challenges of Dryland Farming

The challenge for global agriculture in the twenty-first century is to produce 70 % more food and fiber by 2050 for a growing and more prosperous population while implementing more sustainable methods of farming and responding to climate change (FAO 2009). As incomes rise, people tend to eat fewer grains and increase their consumption of meat and other animal-based products. Dryland farming areas must play an important role in meeting this challenge.

Dryland farming areas are largely semiarid and have unique characteristics. Bowden (1979) lists four management keys that must be understood and applied as new strategies and practices are developed:

1. No growing season is or will be nearly the same in precipitation amount, kind, or range, or in temperature average, range, or extremes, as the previous growing season. Although this is true for all rainfed cropping systems, it requires absolute attention in dryland farming. Crop cultivation requires adjustment every year, which leads to the second key.
2. Crops cannot be planned or managed in the same manner from season to season. Most of the practices in either humid or arid areas have some predictability on an annual basis. In semiarid climates, however, even highly mechanized, technically advanced, commercial farms such as those in the High Plains of North America or the outback of Western Australia do not have sufficiently stable production for the individual or government to count on for a given production figure for the following season.
3. Soil and water resources do not remain the same for any extensive period of time once agriculture is introduced. Most semiarid cropland suitable for cultivation is developed under grass or scattered shrubs on relatively flat topography. The competition for water and nutrients to produce crops requires removal of the protective cover. Because appropriate crops are annual and dependent on precipitation, severe drought often leaves the soil highly vulnerable to wind erosion.
4. There is abundant sunshine due to many cloud-free days. Abundant sunshine means higher temperatures that induce rapid growth, but it also demands careful management of soil water. It is possible for a grain crop to mature rapidly due to several weeks of sun-drenched, rainless conditions and desiccate just days before ripening. It is also possible for a few millimeters of precipitation to occur at almost the last moment to produce a good grain crop.

The four key points outlined above have always been challenging, but they are even more so today in view of the increased environmental concerns and the threat of climate change. The challenges are to increase resilience and to achieve and maintain the sustainability of dryland agriculture. The public at large is increasingly demanding sustainable systems, although their demands are not always clear. There are many definitions of sustainable systems, sustainable agriculture, sustainability and other terms that are difficult to interpret and often not feasible to implement.

The American Society of Agronomy (1989) stated "A sustainable agriculture is one that, over the long term, enhances environmental quality and the resource base on which agriculture depends; provides for basic human food and fiber needs; is economically viable; and enhances the quality of life for farmers and society as a whole."

There are clearly times that dryland farming regions fall short of meeting all the requirements for a sustainable system. The infamous Dust Bowl that occurred in the U.S. and Canada during the 1930s is considered by many as the worst ecological disaster in North America and it was directly associated with dryland farming. Some of the removed cropland was returned to grassland, and drastic changes were made in tillage and other management practices to improve sustainability. There is little doubt that considerable areas of dryland farming are economically viable only because of government policies that provide subsidies or insurance programs to provide some income during drought years. There are no standard guidelines that can be applied to determine the sustainability of dryland farming regions because it will depend on local, regional and country conditions. However, the adoption of conservation agriculture that is based on continuous minimum mechanical soil disturbance, permanent organic soil cover and diversification of crop species grown in sequences and/or associations should be applied to the fullest extent feasible in all dryland farming systems. These principles will enhance or maintain soil physical, chemical and biological properties to assure continued crop production and to prevent wind and water erosion from degradation of the environment. Therefore, successful dryland farming requires favorable soils that can hold moisture and supply nutrients to growing plants, crop cultivars that can withstand climatic abnormalities, and growers who can tailor crop management practices according to the local climatic conditions.

References

Aase JK, Pikul JL (2000) Water use in a modified summer fallow system on semiarid northern Great Plains. Agric Water Manage 43:345–357

American Society of Agronomy (1989) Decision reached on sustainable agriculture. Agronomy News, January, p 15, Madison, WI

Anikwe MAN, Mbah CN, Ezeaku PI, Onyia VN (2007) Tillage and plastic mulch effects on soil properties and growth and yield of cocoyam (*Colocasia esculanta*) on an ultisol in southeastern Nigeria. Soil Tillage Res. 93(2):264–272

Arnon I (1972) Crop Production in Dry Regions, Systematic treatment of principal crops, vol II. Leonard Hill, London

Bandaru V, Stewart BA, Baumhardt RL, Ambati S, Robinson CA, Schlegel A (2006) Growing dryland grain sorghum in clumps to reduce vegetative growth and increase yield. Agron J 98:1109–1120

Baumhardt RL, Jones OR, Schwartz RC (2008) Long-term effects of profile-modifying deep plowing on soil properties and crop yield. Soil Sci Soc Am J 72:677–682. doi:10.2136/sssaj2007.0122

Berry PM, Sylvester-Bradley R, Berry S (2007) Ideotype design for lodging-proof wheat. Euphytica 154:165–179

Blum A (2009) Effective use of water (EUW) and not water-use efficiency (WUE) is the target of crop yield improvement under drought stress. Field Crops Res. 112:119–123
Bowden L (1979) Development of present dryland farming systems. In: Hall AE, Cannell GH, Lawton HW (eds) Agriculture in semi-arid environments. Springer, Berlin, pp. 45–72
Buffum BC (1909) Arid agriculture; A handbook for the western farmer and stockman. 443 pp. Published by the author. Available at: https://archive.org/details/aridagricultureh00buffrich. Accessed June 18, 2015
Burnett E, Stewart BA, Black AL (1985) Regional effects of soil erosion on crop productivity–Great Plains. In: Follett RF, Stewart BA (eds) Soil erosion and crop productivity. ASA-CSSA-SSSA, Madison, pp. 285–304
Campbell HW (1907) Soil culture manual. The Campbell Soil Culture Co. Inc. Lincoln, 320 pp. Available at: https://books.google.com/books?hl=en&lr=&id=pWNBAQAAMAAJ&oi=fnd&pg=PA5&dq=H.W.+Campbell+1907+Soil+Culture+Handbook&ots=y24pu1W-On&sig=n9Ng3uc329EVDvw1NYVIL6RjkNQ#v=onepage&q=H.W.%20Campbell%201907%20Soil%20Culture%20Handbook&f=false
Cole JS, Mathews OR (1940) Relation of the depth to which the soil is wet at seeding time to the yield of spring wheat on the Great Plains. USDA Circ. 563. Washington, DC
Connor DJ, Loomis RS, Cassman KG (2011) Crop ecology: productivity and management in agricultural systems, 2 edn. Cambridge University Press, Cambridge
Conservation Technology Information Center (CTIC) (2015) Conservation tillage facts. Conservation Agriculture Systems Alliance, West Lafayette
FAO (2009) Global agriculture towards 2050. Food and Agriculture Organization of the United Nations, Rome
FAOSTAT (2015) Statistics division. Food and Agriculture Organization of the United Nations, Rome Available at http://faostat3.fao.org/home/E
Farhani HJ, Peterson GA, Westfall DG (1998) Dryland cropping intensification: a fundamental solution to efficient use of precipitation. Adv. Agron. 64:197–223
Fisher T, Byerlee D, Edmeades G (2014) Crop yields and global food security: will yield increase continue to feed the world? Australian Centre for International Agricultural Research, Canberra
Friedrich T, Derpsch R, Kassam A (2012) Overview of the global spread of conservation agriculture. Field actions science reports. Reconciling poverty eradication and protection of the environment. Special issue, 6. Available at: http://factsreports.revues.org/1941
French RJ, Schultz JE (1984) Water use efficiency of wheat in a Mediterranean type environment. 1. The relation between yield, water use and climate. Aust J Agric Res 35:743–764
Greb BW, Smika DE, Black AL (1967) Effects of straw-mulch rates on soil water storage during summer fallow in the Great Plains. Soil Sci Soc Am Proc 31:556–559
Greb BW, Smika DE, Welsh JR (1979) Technology and wheat yields in the central Great Plains: Experiment station advances. J Soil Water Conserv 34:264–268
Gurian-Sherman D (2012) High and dry: why genetic engineering is not solving agriculture's drought problem in a thirsty world. UCS Publications., Cambridge. Available at: www.ucsusa.org/assets/documents/food_and_agriculture/high-and-dry-report.pdf. Accessed 2 July 2015
Halvorsen AD, Peterson GA, Ruele CA (2002) Tillage system and crop rotation effects on dryland crop yields and soil carbon in the central Great Plains. Agron. J. 94:1429–1436. doi:10.2134/agronj2002.1429
Hansen NC, Allen BL, Baumhardt RL, Lyon DJ (2012) Research achievements and adoption of no-till, dryland cropping in the semi-arid U.S. Great Plains. Field Crops Res. 132:196–203
Hurt RD (2003) Agriculture, American Indian. Dictionary of American History. Available at: http://www.encyclopedia.com/doc/1G2-3401800063.html
Ingman M, Santelmann MV, Tilt B (2015) Agricultural water conservation in China: plastic mulch and traditional irrigation. Ecosystem Health and Sustainability 1(4):12. doi:org/10.1890/EHS14-0018.1

Jones OR, Johnson WC (1983) Cropping practices: Southern Great Plains. In: Dregne HE, Willis WO (eds) Dryland agriculture. Agronomy monograph no. 23. ASA, CSSA, SSSA, Madison, WI, pp 365–385

Kapanigowda M, Stewart BA, Howell TA, Kadasrivenkata H, Baumhardt RL (2010) Growing maize in clumps as a strategy for marginal climatic conditions. Field Crops Res 118:115–125

Koohafkan P, Stewart BA (2008) Water and Cereals in Drylands. The Food and Agriculture Organization of the United Nations and Earthscan. London; Sterling, 110 pp

Krishnareddy S, Stewart BA, Payne WA, Robinson CA (2010) Grain sorghum tiller production in clump and uniform planting geometries. J Crop Improv 24:1–11

Leonard WH, Martin JH (1963) Cereal crops. McMillan, Inc., New York

Li S (2007) Dryland agriculture in China. In: 1017 pp. Science Press, Beijing

Liang W, Gao W, Xu Q, Huang G (2012) Historical and present usage of *Shaitan* gravel mulch for crop production in arid and semiarid regions of northwestern China. In: Lal R, Stewart BA (eds) Soil water and agronomic productivity, Advances in soil science. CRC Press, Boca Raton, pp. 477–496

Loomis RS (1983) Crop manipulations for efficient water use. In: Taylor HM, Jordan WR, Sinclair TR (eds) Limitations in efficient water use in crop production. American Society of Agronomy, Crop Science Society of America, Soil Science Society of America, Madison, pp. 345–360

Ma JT (2013) China statistical yearbook. China Statistical Press. Beijing

MacEwan, G. (1983) Charles Noble: Guardian of the Soil. Western Producer Prairie Books. Saskatoon, Saskatchewan, Canada, 208 pp

McCarthy, J.R., D.L. Pfost, and H.D. Currence (1993) Conservation tillage and residue management to reduce soil erosion. University of Missouri Extension. G1650. Columbia, MO. Available at: http://extension.missouri.edu/p/G1650

Oweis TY, Prinz D, Hachum AY (2012) Rainwater harvesting for agriculture in the dry areas. CRC Press. Boca Raton, 237 pp

Passioura JB (1977) Grain yield, harvest index and yield of wheat. J Aust Inst Agric Sci 43:117–120

Peterson GA, Westfall DG, Hansen NC (2012) Enhancing precipitation-use efficiency in the World's dryland agroecosystems. In: Lal R, Stewart BA (eds) Soil water and agronomic productivity, Advances in soil science. CRC Press, Boca Raton, pp. 455–476

Ponce VM (1995) Management of droughts and floods in the semiarid Brazilian Northeast – the case for conservation. J Soil Water Conserv 50:422–431

Powell JW (1879) Report on the lands of the arid region of the United States with a more detailed account of the land of Utah with maps. U.S. Geographical and Geological Survey of the Rocky Mountain Region, Government Printing Office, Washington, DC

Prihar SS, Stewart BA (1990) Using upper-bound slope through origin to estimate genetic harvest index. Agron J 82:1160–1165

Sainju UM, Lenssen AW, Caesar-Tonthat T, Waddell J (2006) Tillage and crop rotation effects on dryland soil and residue carbon and nitrogen. Soil Sci Soc Am J 70:668–678

Schillinger WF, Papendick RI, McCool DK (2010) Soil and water challenges for Pacific Northwest agriculture. In: Zobeck TM, Schillinger WF (eds) Soil and water conservation advances in the United States. SSSA Special Publication 60. SSSA, M-adison, pp 47–79

Sharda VN, Juyal GP, Singh PN (2002) Hydrologic and sedimentologic behaviour of a conservation bench terrace system in a sub-humid climate. Trans ASAE 45(5):1433–1441

Shaw T (1909) Dry land farming. The Pioneer Company, St. Paul, 502 pp. Available at: http://archive.org/stream/drylandfarming00shawrich/drylandfarmin00shawrich_djvu.txt. Accessed 17 June 2015

Sinclair TR (2009) Taking measure of biofuel limits. Am Sci 97(5):400–407. doi:10.1511/2009.80.1

Sinclair TR, Weiss A (2010) Principles of ecology in plant production, 2nd edn. CAB International, Cambridge, MA

Stewart BA, Burnett E (1987) Water conservation technology in rainfed and dryland agriculture. In: Jordan WR (ed) Water and water policy in world food supplies. Texas A&M University Press, College Station, pp. 355–359
Stewart BA, Lal R, El-Swaify SA (1991) Sustaining the resource base of an expanding world agriculture. In: Lal R, Pierce FJ (eds) Soil management for sustainability. Soil and Water Conservation Society, Ankeny, pp. 125–144
Stewart BA, Peterson GA (2015) Managing green water in dryland agriculture. Agron. J. 107:1544–1553. doi:10.2134/agronj14.0038
Unger PW (1978) Straw-mulch rate effect on soil water storage and sorghum yield. Soil Sci Soc Am. J 42:486–491
Unger PW (2006) Soil and water conservation handbook: policies, practices, conditions, and terms. Haworth Food & Agricultural Products Press, New York
Unger PW, Stewart BA (1983) Soil management for efficient water use. In: Taylor HM, Jordan WR, Sinclair TR (eds) Limitations to efficient water use in crop production. ASA, CSSA, and SSSA, Madison, WI, pp 419–460
Unger PW, Wiese AF (1979) Managing irrigated winter residues for water storage and subsequent dryland grain sorghum production. Soil Sci Soc Am J 42:582–588
Unger PW, Baumhardt RL, Arriaga FJ (2012) Mulch tillage for conserving soil water. In: Lal R, Stewart BA (eds) Soil water and agronomic productivity, Advances in soil science. CRC Press, Boca Raton, pp. 427–453
Wang Y, Xie Z, Malhi SS, Vera CL, Zhang Y, Guo Z (2011) Effects of gravel-sand mulch, plastic mulch and ridge and furrow rainfall harvesting system combinations on water use efficiency, soil temperature and watermelon yield in a semiarid Loess Plateau of northwestern China. Agri Water Manage 101:88–92
Wani AP, Garg KK, Singh AK, Rockström J (2012) Sustainable management of scarce water resources in tropical rainfed agriculture. In: Lal R, Stewart BA (eds) Soil water and agronomic productivity, Advances in soil science. CRC Press, Boca Raton, pp. 347–408
Widtsoe JA (1920) Dry-Farming: A System of Agriculture for Countries under Low Rainfall. Macmillan Company, New York
Xie Z, Wang Y, Guodong C, Malhi S, Vera CL, Guo Z, Zhang Y (2010) Particle-size effects on soil temperature, evaporation, water use efficiency and watermelon yield in fields mulched with gravel and sand in semi-arid Loess Plateau of northwest China. Agric Water Manage 97(6):917–923. doi:10.1016/j.agwat.2010.01.023
Zhang SY (2013) China rural statistical yearbook. China Statistical Press, Beijing
Zingg AW, Hauser VL (1959) Terrace benching to save potential runoff for semiarid land. Agron J 51:289–292

Research and Developmental Issues in Dryland Agriculture

Muhammad Farooq and Kadambot H.M. Siddique

1 Introduction

Drylands cover more than 40 % of the terrestrial land surface and are not on the margins of the 'economically productive' world; rather, they are vast areas often lying right in the center, and contribute about 40 % to global net primary productivity (Grace et al. 2006; GLP 2005; MEA 2005). Dryland ecosystems are extremely diverse and include the Mediterranean systems, the cold deserts of Chile and Mongolia, the Sahel and Sahara of Africa, the Arctic Circle, and the high altitude drylands of Afghanistan and Iran (Fig. 1). More than two and a half billion people (40 % of Africans, 39 % of Asians and 30 % of South Americans) inhabit of dryland areas, which is more than 38 % of the world population (GLP 2005; MEA 2005).

The term 'dryland agriculture' is often used interchangeably with 'rainfed agriculture'. However, rainfed agriculture is synonymous with non-irrigated agriculture which includes rainfed drylands and rainfed wetlands. Therefore, dryland agriculture is component of rainfed agriculture (Stewart et al. 2006).

Drylands are defined in terms of water deficit, as areas where mean annual precipitation is less than half of the potential evapotranspiration (FAO 1993, 2004). These areas include hyper-arid, arid, semi-arid and dry sub-humid areas (FAO 2004) and receive less than 200 mm, less than 250 mm, 200–500 mm of total annual rainfall and 500–700 mm total annual rainfall, respectively (Table 1, FAO 2004). In

M. Farooq (✉)
Department of Agronomy, University of Agriculture, Faisalabad 38040, Pakistan

The UWA Institute of Agriculture, The University of Western Australia,
Perth, WA 6001, Australia
e-mail: farooqcp@gmail.com

K.H.M. Siddique
The University of Western Australia Institute of Agriculture,
The University of Western Australia, MO82, LB 5005, Perth, WA, 6001 Australia

M. Farooq, K.H.M. Siddique (eds.), *Innovations in Dryland Agriculture*,
DOI 10.1007/978-3-319-47928-6_2

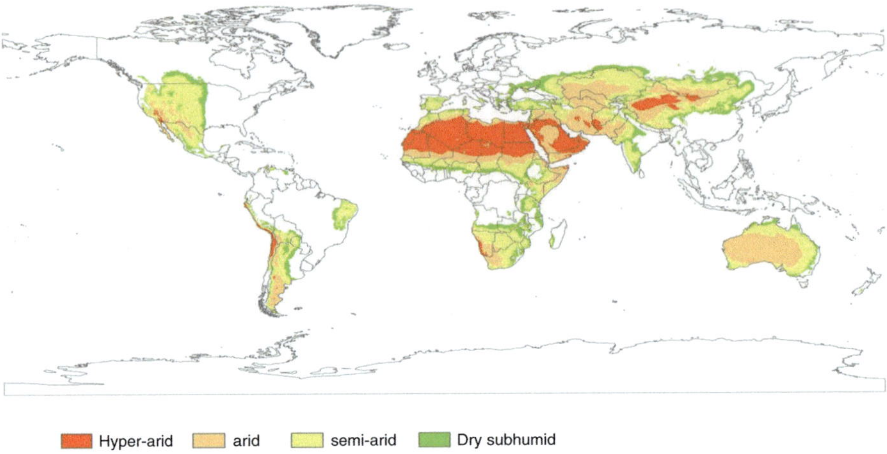

Fig. 1 Global distribution of drylands (Source: USAID (2014))

Table 1 Major indicators and characteristics of drylands

Climate type	[a]Aridity index	Average annual rainfall (mm) and variability	Growing season (days) and typical crops	Pastoral systems	Examples of biomes
Hyper-arid	< 0.05	150 mm	0 days (unless irrigated)	Very limited, fodder available only for short periods (<4 months)	Desert
		Inter-annual variability 100 %	No rainfed crops		
Arid	0.05–0.20	150–250 mm	< 60 days	Marginal pasture, available for short periods. Mainly small stock and cattle in transhumance systems	Desert, xeric shrub-land, desert scrub
		Inter-annual variability 50–100 %	No rainfed crops		
Semi-arid	0.20–0.50	250–500 mm	60–119 days	Large and small stock	Savanna, steppe
		Inter-annual variability 25–50 %	Bulrush millet, sorghum, sesame		
Dry sub-humid	0.50–0.65	500–700 mm	120–179 days	Large and small stock	Open woodland, savanna, steppe
		Inter-annual variability < 25 %	Maize, bean, groundnut, pea, barley, wheat, tef		

Source: FAO (2004)
[a]Precipitation (P)/Potential evapotranspiration (PET)

addition to water deficit, drylands are characterized by erratic distribution with common periodic droughts (Zurayk and Haidar 2002). Dryland soils, characterized by moisture deficit, and low levels of soil organic matter and biological activity, often havepoor fertility (USAID 2014). When inappropriately used for agriculture, these soils are prone to rapid fertility loss, erosion, desertification, and salinization.

Drylands span all continents (Fig. 1) in areas where rainfall is highly variable, droughts are common and water is the principal limiting factor for agriculture. The global population is expected to reach ten billion by 2050. This increase, together with diet changes to include more animal products, is projected to i increase food demand by up to 70 % (UNESCO 2012). Dryland agriculture can play a role in meeting this enormous challenge (Stewart et al. 2006). In this chapter, the challenges for dryand agriculture are discussed, and researchable issues to improve the productivity of drylands on a sustainable basis are suggested.

2 Challenges for Dryland Agriculture

Dryland ecosystems, are characterized by recurrent but unpredictable droughts, high temperatures, variable rainfall and low soil fertility. Therefore, agriculture in dryland areas is fragile. The stress on land resources to meet the basic demands of humans is expected to further increase as population increases in the coming decades. In this context, the challenges for dryland agriculture are discussed in this section.

2.1 Water Deficit

Water deficit is the most critical determinant of the success or failure of crop production in drylands (Falkenmark et al. 1990). Reduced plant available soil moisture occurs as a result of a low quantity of water supplied (rainfall) or changes in rainwater partitioning (reduced in infiltration and water retention in the soil). The variability of rainfall in time and space is a common feature of drylands (Falkenmark and Rockström 2004; Adnan et al. 2009).

Generally, dryland production is possible where the growing season does not fall under the temporal distribution of water deficit i.e. precipitation during the growing season is sufficient to meet the crop's water requirement. However, water deficit becomes severe in cropping systems and growing season where precipitation is less than evaporation (Baumhhardt and Salinas-Garcia 2006; Adnan et al. 2009). The difference between evaporation demand and precipitation received determines the crop water deficit. It is argued that in dry sub-humid and semi-arid areas, the amount of rainfall is not the limiting factor rather its extreme variability, such as high rainfall intensity, few rainy days, and uneven spatial ortemporal distribution (Klaij and

Table 2 Types of droughts and underlying causes in semi-arid and dry sub-humid tropical environments

	Dry spell	Drought
i. Meteorological drought		
Frequency	Two out of three years	Once every 10 years
Impact	Yield reduction	Complete crop failure
Cause	Rainfall deficit of 2- to 5-week periods during crop growth	Seasonal rainfall below the minimum seasonal plant water requirement
ii. Agricultural drought		
Frequency	More than two out of three years	Once every 10 years
Impact	Yield reduction or complete crop failure	Complete crop failure
Cause	Low plant water availability and poor plant water uptake capacity	Poor rainfall partitioning, leading to seasonal soil moisture deficit for producing harvest (where poor partitioning refers to a high proportion of runoff and non-productive evaporation relative to oil water infiltration at the surface)

Source: Falkenmark and Rockström (2004)

Vachaud 1992; Hatibu et al. 2003), In arid and hyper-arid regions, absolute water scarcity is the prevalent limiting factor due to the crop water demand being higher than the rainfall (Hatibu et al. 2003).

Drought is a recurring phenomenon in drylands (Table 2). Agricultural drought is linked with meteorological characteristics including erratic rainfall, soil moisture deficits, differences between actual and potential evapotranspiration, and lower groundwater tables and/or reservoir levels (Stewart and Peterson 2015). Such events are likely to increase with the current climate shift (IPCC 2007; Adnan et al. 2009). Although agricultural droughts are observed in all regions, the impact of drought on drylands is usually more severe as the amount of rainfall received is far below that of potential evapotranspiration (Stewart and Peterson 2015). However, there are management-induced dry spells and droughts where the rainfall received is not stored and utilized in a productive way. This cases can be prevented by improved management practices than blaming on droughts (Rockström 2003).

2.2 Weather Variability

Weather variability is a major issue in the drylands and includes the unpredictable nature of various weather elements such as precipitation, temperature and winds. This variability is directly related the availability of water for crop use, its temporal and spatial distribution, and the success of crop production.

The changing climate may increase the intensity and the frequency of current risks and the probability of extreme events and new hazards (Nicholls and Lowe 2006). For instance, climatic studies have predicted a decline in overall rainfall with increased episodes of drought (Rockström 2003; Peterson et al. 2006) and increased high-temperature events (Peterson et al. 2006) in the drylands in the future.

Climate change is dictating the changes in climate variability and the frequency, intensity, spatial extent, duration, and timing of extreme weather and climate events (IPCC 2012). Increasing episodes of drought and heat stress are projected for the rest of this century, which are expected to have adverse effects over and above the impacts due to changes in mean variables alone (IPCC 2012). Key ecosystem processes are seasonally sensitive to climate variability and the timing of climate variability may be just as important as its magnitude for plant productivity (Craine et al. 2012). For instance, in drylands, crop production occurs in areas receiving sufficient precipitation to support crop production. However, precipitation is usually unpredictable between years and within growing seasons, such that crops can fail due to inadequate water, even with good management (Peterson et al. 2006). Droughts, desertification, and water shortages are permanent features of life in drylands and under such conditions, any increase in evapotranspiration has a severe impact on agriculture, horticulture, forestry and human activities (Sen Roy and Singh 2002). The vulnerability of arid regions is further accentuated by low levels of socio economic development, as is the case in India (Singh and Gurjar 2011).

2.3 Erosion

Soil erosion is a leading cause of land degradation in drylands. For instance, in Sub-Saharan Africa soil erosion and degradation are considered a bigger problem than climatic variability and drought (Peterson et al. 2006). Possible factors responsible for severe soil erosion includes, injudicious land usage for crop production, monoculture and lack of crop rotation, excessive tillage and clean fallowing, disconnecting arable farming from livestock, intensive livestock grazing without knowledge of the grazing capability of rangelands and inappropriate removal of forest vegetative cover (Grimm et al. 2002; Irshad et al. 2007).

Soil erosion is a form of land degradation leading to the removal of topsoil (typically the layer with the most effect on plant production and thus food production) by water and/or wind. Topsoil contains organic matter, provides micro- and macronutrients to plants, and is responsible for soil structural stability—the determining factor for the provision of water to plants (Rojas et al. 2016). The soil is eroded in two steps i.e. soil detachment and soil transport. Raindrops are the main reason for soil detachment—also known as sheet erosion (Lal 2003). However, it is hard to identify soil erosion at the starting phase. By the time farmers identify soil erosion, the land has most likely lost its productivity (Lal 2003). Soil transport mainly happens through the wind or air (McCarthy et al. 1993).

Wind erosion reduces crop yields in drylands and large and unprotected fields (e.g., through soil being mobilized and relocated from farmland and 'sandblasting' of standing crops) (Rojas et al. 2016). Sometimes soil reaches as much as 100 t ha^{-1} due to extreme events, i.e. storms (Grimm et al. 2002; Verheijen et al. 2009). In the

Mediterranean region, soil erosion in some areas is so severe that some soils are close to being rendered unproductive.

Erosion reduces soil fertility by physically removing organic matter and nutrients from the soil. Soil degradation has a negative impact on water infiltration thereby reducing crop production and soil sustainability. Globally, management factors for soil degradation in dryland regions are often similar (Peterson et al. 2006). The impact of soil erosion on crop production varies across soils and ecoregions and depend on soil management, cropping systems, soil conservation measures and technological inputs. The resistance of soils to degradation (soil resilience) and the degree to which soils degrade (soil erodibility) also differ by soil type (Scherr 1999).

The sustainable development of agroecosystems is at risk due to accelerated soil erosion resulting from the pressure of environmental degradation and human intrusions of land exploitation. Soil erosion disturbs the natural balance and reduces production potential (Pimentel et al. 1995). Accelerated soil erosion causes loss of biodiversity which further fuels erosion and the soil becomes devoid of beneficial production factors such as carbon recycling (Chapin et al. 1997). Loss of biodiversity is a reason for the reduced replenishment of resources and such soils are more prone to irreversible changes (Chapin et al. 1997).

2.4 *Nutrient Mining*

The deficiency of nutrients in drylands is the second main reason for low yield output as, fertilizers are not applied after sowing in dryland agriculture and the rate of fertilizer application is quite low due to weather vulnerability. Crop residues are removed from the fields to feed livestock and not incorporated, resulting in widespread nutrient deficiencies in the soils of dryland agriculture (Rojas et al. 2016).

Nutrient mining is the removal of soil nutrients by continuous cropping without adequate supplementation of inorganic and/or organic fertilizers and manures. Nutrient mining reduces crop yields in drylands due to soil fertility losses and is a key link to environmental damage and land degradation (Henao and Baanante 2006). Over time, developing countries face the net depletion of macro nutrients stock while more developed areas that replenish soils with macronutrients unintentionally cause excessive micronutrient mining (Emmett et al. 1997). The situation in drylands with marginal soils is even alarming, which are already under pressure due to various environmental impacts and increased human intrusions (Rojas et al. 2016).

In the case of Africa, dryland soils are depleted of nutrients and soil organic matter due to inappropriate practices of cultivation, deforestation and overgrazing, continuous cropping and non-judicious and inadequate fertilizer use (Hartemink 1997; De Jager et al. 2001). Nutrient balances which consider system inputs and outputs have been used to estimate the magnitude and extent of nutrient mining. From 2002 to 2004, 85 % of agricultural land in Africa had annual nutrient mining rates greater than 35 kg (N, P and K) ha^{-1} and 40 % had annual rates greater than 60 kg ha^{-1} (Blum 2013).

2.5 Institutional Role

Institutional challenges are a hurdle in the sustainable development of dryland agriculture; there are three types, *viz.* institutional, policy and legal related issues (Liniger et al. 2011). Institutional issues include inapplicable local and national political agendas, absence of operational capacity, unclear and overlapping delineation of responsibilities, bad governance (Liniger et al. 2011), the lack of a credit system for growers and the agro-based industry (Pinto 1987; Bevan et al. 1999) and the lack of weather forecasting systems (Odjugo 2010). Legal framework/policy constraints may arise due to the non-implementation of sustainable land management laws because law implementation is costly, difficult, and may cause hostile interactions between land users and law enforcement agencies (Liniger et al. 2011).

To improve and manage the dryland regions, an institutional setup is needed to deliver laws and policies regarding the sustainability and management of these areas. Financial provision to the institutes for research and development of natural resources or ways to protect these dryland regions from further degradation is also needed (Rosegrant et al. 2002).

3 Researchable Issues in Dryland Agriculture

Drylands are fragile ecosystems. Major researchable issues in dryland agriculture include building resilience and reducing vulnerability of the people living on marginal lands, improving crop varieties and livestock breeds, integrating crop–livestock systems, conservation agriculture, the diversification of food production systems, management of natural resources especially the water, increasing investment in institutional support and agricultural research, and taking an integrated agroecosystem approach to these actions. These researchable are discussed below.

3.1 Rainwater Harvesting and the Efficient Use of Water

As rainfall is very low and erratic in dry areas is a recurrent problem, increasing the availability of water in the root zone of the crop is paramount for sustainable agriculture production in the drylands (Lal 2001). This approach has a positive effect and helps to improve the yield per unit of rainfall in dryland areas (Kurukulasuriya et al. 2006).

Rain water capture, infiltration into the root zone, and the efficient use of available water are necessary pragmatic strategies for improving crop yields in dryland areas. However, there is a broad spectrum of integrated land and water management options to achieve these aims (Rockström et al. 2010; Table 3). For instance, some techniques focus on capturing more water, e.g. external water harvesting systems,

Table 3 Rainwater management strategies and corresponding management options to improve yields and water productivity

Rainwater management strategy		Purpose	Management options
Increase plant water availability	External water harvesting systems	Mitigate dry spells, protect springs, recharge groundwater, enable off-season irrigation, permit multiple uses of water	Surface microdams, subsurface tanks, farm ponds, percolation dams and tanks, diversion and recharging structures
	In situ water harvesting systems, soil and water conservation	Concentrate rainfall through runoff to cropped area or other use	Bunds, ridges, broad-beds and furrows, microbasins, runoff strips
		Maximize rainfall infiltration	Terracing, contour cultivation, conservation agriculture, dead furrows, staggered trenches
	Evaporation management	Reduce non-productive evaporation	Dry planting, mulching, conservation agriculture, intercropping, windbreaks, agroforestry, early plant vigor, vegetative bunds
Increase plant water uptake capacity	Integrated soil, crop and water management	Increase proportion of water balance flowing as productive transpiration	Conservation agriculture, dry planting (early), improved crop varieties, optimum crop geometry, soil fertility management, optimum crop rotation, intercropping, pest control, organic matter management

Source: Rockström et al. (2010)

while others, such as mulching and drip irrigation, aim to increase water productivity directly as a good crop canopy can help reduce water loss through soil evaporation (Rockström 2003).

Water harvesting is a hydro-agronomic term which comprises techniques and methods to collect and conserve water from surface run off and rainfall (Siegert 1994). Water harvesting may involve the capture of local farmland rainfall (*in situ* water harvesting) or the capture of rainfall received away from the farmland (*ex situ* water harvesting) (Oweis and Hachum 2001). Water harvesting differs from conventional water conservation practices as it does not deprive the farmland of its share (Reij et al. 1988). Rain-water harvesting allows water to be conserved for later use during dry spells in the cropping season (Rockström 2003; Awulachew 2010). Water harvesting has been practiced successfully for millennia in parts of the world, yet the potential of water harvesting remains largely unknown, unacknowledged and unappreciated. Water harvesting offers opportunities for the drylands in the developing world (Rockström 2003).

Rainfall is the most important natural resource in drier environments. Low rainfall, water scarcity and land degradation severely inhibit the production capacities of agricultural lands in arid and semi-arid environments. Improving the efficiency of rainwater use is, therefore, critical in these water-scarce areas with rapidly expanding, poor populations living in a fragile environment and facing food insecurity and depleted natural resources bases. Water harvesting has become increasingly important for improving the management of water resources in such dry environments. The ultimate goal of *in situ* water harvesting is a sustainable and environmentally-friendly system of agricultural production is to complement rather than replace the existing water use system. Improved systems must be socially acceptable as well as more productive. Water harvesting interventions form part of a plan for integrated land and water resources development which takes into consideration the necessary technical, agronomic, socioeconomic and institutional aspects and inputs (Oweis et al. 2001).

Supplemental irrigation systems should be promoted in drylands, especially for small-scale farmers (Fan et al. 2000; Fox et al. 2005). Policy frameworks and institutional structures similar to those for irrigated systems will be needed to implement supplemental irrigation systems successfully in the drylands.

3.2 *Crop Diversification*

Traditional monocropping may be a risky option in light of the predicted climatic changes, eventually leading to a severe decline in agricultural productivity (Huggins et al. 2015). In this regard, crop diversification has the potential to increase the sustainable intensification of agriculture in dryland areas. Crop diversification widens the variety of crops in a system or extends its niche. It helps to save the system from various risks such as crop failure and pest attack (Lin 2011). An alternative crop or species must be adapted to the chosen environment or agroecosystem i.e. the crop or species must be able to tolerate the harsh environment and climate, variation specifically the water deficit.

Crop diversification is an excellent opportunity to enhance income and improve the social and overall livelihood of people living in dryland areas (Lin 2011). In dryland areas, the inclusion of legumes in a cereal system may be a good option for improving system sustainability (Peterson et al. 2006). While diversification is an effective way to mitigate risks and increase incomes in dryland ecosystems, the prevailing global eating habits, market scenarios, dwindling economies, unpredictable climatic changes, and traditional cultivation has shifted the scope of agriculture from diversification to intensified monocropping, particularly in the drylands. This requires extensive research on biological and socio-economic perspectives.

3.3 Conservation Agriculture

Conservation agriculture (CA) —a suite of three key technologies *viz.* minimum soil disturbance, stubble retention and diversified crop rotation—offers a system for sustainable agriculture production (Farooq et al. 2011). CA is a good option for successful crop production in drylands because it improves soil organic matter and conserves water making it available for the plant when it is needed (Lyon et al. 2004; Thomas et al. 2007; Bayala et al. 2012).

CA is practiced in different dryland agro ecosystem - and has the potential to reduce the threat of food insecurity in the Middle East, North and Sub-Saharan Africa, and West and Central Asia (CGIAR 2013). Losses such as soil degradation through erosion and water through runoff can be reduced by adopting CA practices such as reduced tillage and maintaining soil cover (Serraj and Siddique 2012). CA benefits farmers in economic terms by reducing costs (plowing, labor), improving water use efficiency and water availability in the root zone and reducing water loss due to soil evaporation because the soil is covered with residues from the previous crop. In CA, the inclusion of legumes in the rotation and mulch (previous crop residues) helps to restore the fertility status of soil (Marongwe et al. 2011).

Constraints to the successful adoption of CA include mindset of the famers, unavailability of appropriate seeding machinery, poor farmer knowledge on CA benefits, weeds and disease issues during initial stages, and the lack of collaboration between farmers, extension workers and institutions (Farooq and Siddique 2015). Generally, CA systems are better adapted to the drylands because CA triggers increase in infiltration resulting in more effective rainfall use, less surface runoff, less soil erosion and improve in soil water-holding capacity. In CA systems, crops are more likely to produce better yields than those under conventional tillage (Stewart 2007; Friedrich et al. 2012).

3.4 Mixed Crop–Livestock Systems

Livestock production plays and will continue to play an important role in dryland agriculture. In this regard, ruminant production in dryland areas is expected to expand with the increased in the demand for animal protein production in the coming decades. The importance of mixed crop–livestock systems and grazing of annual forages varies between regions and climatic conditions (Thornton 2014). However, with an increase in the total annual rainfall from 200 to more than 500 mm, grazing and production of drought tolerant crops (such as sorghum and millets) is replaced with other crops like maize and wheat (Schiere et al. 2006).

Most dryland systems in Sub-Saharan Africa integrate crop and livestock production with the productivities of livestock, croplands and rangelands intricately linked in these systems (Powell et al. 2004). Maintaining a balance between the food and feed supply, nutrient inputs and outputs, and human and livestock populations is critical for sustaining the productivity of livestock, croplands and rangelands. In addition to the biophysical response of crops and livestock to additional nutrient inputs, the innovative approaches must be evaluated to ensure that they are accessible and affordable to growers, and to determine how these and other inputs can help to reduce the risks associated with erratic weather variability (Powell et al. 2004).

Ruminant production is the predominant form of livestock production in Sub-Saharan Africa. Despite it being a risky business. Although, options to manage the risk are available, they tend to be specific solutions for specific factors causing risk rather than integrated and innovative approaches that simultaneously manages and reduces all of the risk factors associated with the uncertainties of animal production in dryland (Martínez et al. 2014). One innovative solution to sustainable livestock production in rangelands involves an integrative approach, combining shrubs and native plants in silvopastoral systems with strategies to promote self-herding (or traditional herding) and careful selection of animal genotypes. This approach needs to be refined for each system and location, underpinned by sound principles, to ensure that the interaction between genotype, environment and management is optimized to maximize productivity and minimize the impact on the environment.

3.5 *Policy Options for Improvement*

Agriculture in dryland regions is facing several policy management issues, agriculture research problems, and institutional challenges. The collaboration of researchers with practitioners is desired as both have to adopt an integrated approach because social and ecological issues are interlinked (Reynolds et al. 2007). Short-term policies to solve ecological issues are not beneficial because they do not have potential to resolve problems faced by the inhabitants of dryland regions (CGIAR 2013).

Policies are needed which address market availability problems faced by farmers. Markets need to be located in the vicinity of farmers. There is poor market policy support for dryland crops (CGIAR 2013). Policies should be formulated to address serious problems such as land degradation, water scarcity, and food insecurity. Failing to resolve these issues will lead to more poverty and poor nutrition, loss of biodiversity, and more land degradation. A long-term and unified action plan by all of the stakeholders is desired (Bantilan et al. 2006).

The United Nations Convention established a policy to alleviate desertification and suggested diverting funds at the national and global level for the betterment of livelihoods in dryland areas (UNCCD 2007). However, efforts should be initiated to improve the social condition of the inhabitants of drylands.

3.6 The Ecosystem Approach: Collaboration for Integration

Dryland agriculture is considered as agro ecosystem, comprised of grassland, forest and arable land. Ecosystem management in dryland regions will only be beneficial and successful if it protects and improves the sustainability and profitability of the ecosystem. An ecosystem approach involves decision making with more efficient tools and policies for dryland crop production, and the provision of stakeholders with knowledge on the benefits of policies and risks of investment, development and management for drylands (White et al. 2002).

An ecosystem approach to dryland agriculture may help to monitor, assess and address the actual needs and requirements of the dryland population. This should provide indicators to stakeholders regarding the integrated socioeconomic and environmental impacts for the assessment of development, investment domains and management options for dryland regions.

Each dryland ecosystems ha several native microbial, plant and animal species equipped with special strategies to cope with the extreme weather conditions in these ecosystems. Such adaptive traits may have some global implications in the context of predicted climate change (Bonkoungou 2007). However, research efforts should be initiated to develop balanced ecological strategies for sustaining the productivity of dryland ecosystems.

4 Conclusion

Dryland areas occupy almost half of Earth's land surface and are central to the provision of most pulses and some cereals. Drylands also meet our protein needs through legumes and livestock, as these regions provide pasture and natural vegetation without additional input costs. However, dryland agriculture is deteriorating day by day due to increased human activities, climate extreme, growing needs for food and through natural such as wind and water. The main challenges which threaten reduce yields or cause total crop failure include moisture deficits, unpredictable weather, soil fertility losses, policy negligence and nutrient deficiencies. To resolve these challenges, a collaborative interdisciplinary ecosystem approach of the researchers, extension agents, farmers, and research and policy institutions is desired. The formulation and implementation of appropriate policies for drylands may help to improve farm income on a sustainable basis. In this regard, campaigns for CA, rain water harvesting and mixed crop–livestock systems may be helpful. Installation and provision of weather prediction systems may also help to avoid climatic calamities.

References

Adnan S, Rashed M, Azmat HAK (2009) Water balance conditions of Potohar and Baluchistan Plateau from 1931 to 2008. World Appl Sci J 2:169–169

Awulachew SB (2010) Irrigation potential in Ethopia: constraints and opportunities for enhancing this system. International Water Management Institute. https://ethiopianagriculture.files.wordpress.com/2010/11/ethiopia-irrigation-diagnostic-july-20101.pdf

Bantilan MCS, Anand Babu P, Anupama GV, Deepthi H, Padmaja R (2006) Dryland agriculture: dynamics, challenges and priorities, Research Bulletin No. 20. Patancheru 502 324. International Crops Research Institute for the Semi-Arid Tropics, Patancheru

Baumhardt RL, Salinas-Garcia J (2006) Dryland agriculture in Mexico and the U.S. Southern Great Plains. In: GA P, PW U, WA P (eds) Dryland agriculture, Agronomy Monograph 23, 2nd edn. ASA, CSSA, and SSSA, Madison

Bayala J, Sileshi GW, Coe R, Kalinganire A, Tchoundjeu Z, Sinclair F, Garrity D (2012) Cereal yield response to conservation agriculture practices in drylands of West Africa: a quantitative synthesis. J Arid Environ 78:13–25

Bevan D, Collier P, Gunning JW (1999) The political economy of poverty, equity, and growth: *Nigeria and Indonesia*. Oxford University Press, New York

Blum WEH (2013) Soil and land resources for agricultural production: general trends and future scenarios-a worldwide perspective. Int Soil Water Cons Res 1:1–14

Bonkoungou EG (2007) Biodiversity in drylands: challenges and opportunities for conservation and sustainable use. IUCN World Conservation Union, Gland

CGIAR (2013) New research approaches to improve dryland agriculture to deliver more prosperous future. CGIAR Research Program on Dryland Agricultural Production Systems. http://www.icarda.org/publications-resources/drylans-systems

Chapin FS III, Walker BH, Hobbs PJ, Hooper DU, Lawton JH, Sala OE, Tilman D (1997) Biotic control over the functioning of ecosystems. Science 277:500–504

Craine JM, Nippert JB, Elmore AJ, Skibbe AM, Hutchinson SL, Brunsell NA (2012) Timing of climate variability and grassland productivity. Proc Natl Acad Sci U S A 109:3401–3405

De Jager A, Onduru D, van Wijk MS, Vlaming J, Gachini GN (2001) Assessing sustainability of low external input farm management systems with the nutrient monitoring approach: a case study in Kenya. Agric Syst 69:99–118

Emmett BA, Cosby BJ, Ferrier RC, Jenkins A, Tietema A, Wright RF (1997) Modelling the ecosystem effects of nitrogen deposition: simulation of nitrogen saturation in a Sitka spruce forest, Aber, Wales, UK. Biogeochemistry 38:129–148

Falkenmark M, Rockström J (2004) Balancing water for humans and nature. The new approach in ecohydrology. Earthscan, London 247 pp

Falkenmark M, Lundqvist J, Widstrand C (1990) Coping with water scarcity–implications of biomass strategy for communities and policies. Int J Water Resour Dev 6:29–43

Fan S, Hazell P, Haque P (2000) Targeting public investments by agro-ecological zone to achieve growth and poverty alleviation goals in rural India. Food Policy 25:411–428

FAO (1993) Key aspects of strategies for the sustainable development of drylands. Food and Agriculture Organization, Rome http://www.fao.org/docrep/t0752e/t0752e00.htm

FAO (2004) Carbon sequestration in dryland soils, World soil resources reports No. 102. Food and Agriculture Organization, Rome http://www.fao.org/3/a-Y5738e.pdf

Farooq M, Siddique KHM (2015) Conservation agriculture: concepts, brief history and impacts on agricultural systems. In: Farooq M, KHM S (eds) Conservation agriculture. Springer International Publishing, Cham

Farooq M, Flower K, Jabran K, Wahid A, Siddique KHM (2011) Crop yield and weed management in rainfed conservation agriculture. Soil Tillage Res 117:172–183

Fox P, Rockström J, Barron J (2005) Risk analysis and economic viability of water harvesting for supplemental irrigation in semi-arid Burkina Faso and Kenya. Agric Syst 83:231–250

Friedrich T, Derpsch R, Kassam A (2012) Overview of the global spread of conservation agriculture. Field Actions Science Reports. Reconciling Poverty Eradication and Protection of the Environment. Special Issue 6. Available at: http://factsreports.revues.org/1941

GLP (2005) Science plan and implementation strategy. Global Land Project, International Geosphere Biosphere Program, Report No. 53/International Human Dimensions Programme Report No. 19, IGBP Secretariat, Stockholm

Grace J, Jose JS, Meir P, Miranda HS, Montes RA (2006) Productivity and carbon fluxes of tropical savannas. J Biogeogr 33:387–400

Grimm M, Robert J, Montanarella L (2002) Soil erosion risk in Europe, European Commission Directorate General Joint Research Centre, Institute for Environment and Sustainability, European Soil Bureau

Hartemink AE (1997) Soil fertility decline in some major soil groupings under permanent cropping in Tanga Region, Tanzania. Geoderma 75:215–229

Hatibu N, Young MDB, Gowing JW, Mahoo HF, Mzirai OB (2003) Developing improved dryland cropping systems for maize in semi-arid Tanzania. Part 1: experimental evidence of the benefits of rainwater harvesting. J Exp Agri 39:279–292

Henao J, Baanante C (2006) Agricultural production and soil nutrient mining in Africa: implications for resource conservation and policy development. International Fertilizer Development Center (IFDC), Muscle Shoals

Huggins D, Pan B, Schillinger W, Young F, Machado S, Painter K (2015) Crop diversity and intensity in Pacific Northwest dryland cropping systems. In: Borrelli K, Laursen DD, Eigenbrode S, Mahler B, Stokes B (eds) Regional approaches to climate change for northwest agriculture, climate science northwest farmers can use. REACCH, Idaho, pp 38–41. Available at: https://www.reacchpna.org/sites/default/files/REACCHreportYr3.pdf

IPCC (2007) Climate change 2007: impacts, adaptation, and vulnerability. Contribution of working group II to the third assessment report of the Intergovernmental Panel on Climate Change. Cambridge University Press, Cambridge, UK

IPCC (2012) Managing the risks of extreme events and disasters to advance climate change adaptation. In: CB F, Barros V, Stocker TF et al (eds) A special report of working groups I and II of the Intergovernmental Panel on Climate Change. Cambridge University Press, Cambridge, UK

Irshad M, Inoue M, Ashraf M, Faridullah HK, Delower M, Tsunekawa A (2007) Land desertification an emerging threat to environment and food security of Pakistan. J Appl Sci 8:1199–1205

Klaij MC, Vachaud G (1992) Seasonal water balance of a sandy soil in Niger cropped with pearl millet, based on profile moisture measurements. Agric Water Manag 21:313–330

Kurukulasuriya P, Mendelsohn R, Hassan R, Benhin J, Deressa T, Diop M, Eid HM, Fosu KY, Gbetibouo G, Jain S, Ali M, Renneth M, Jane KM, El-Marsafaway S, Ernest M, Samiha O, Mathie O, Isidor S, David M, Niggol SS, Ariel D (2006) Will African agriculture survive climate change? World Bank Econ Rev 20:367–388

Lal R (2001) World cropland soils as source or sink for atmospheric carbon. Adv Agron 71:145–191

Lal R (2003) Soil erosion and the global carbon budget. Environ Int 29:437–450

Lin BB (2011) Resilience in agriculture through crop diversification: adaptive management for environmental change. Bioscience 61:183–193

Liniger HP, Studer MR, Hauert C, Gurtner M (2011) Sustainable land management in practice-guidelines and best practice for Sub-Saharan Africa. WOCAT and FAO. http://www.fao.org/docrep/014/i1861e/i1861e00.pdf

Lyon D, Bruce S, Vyn T, Peterson G (2004) Achievements and future challenges in conservation tillage. In: 4th international crop science congress, 26 Sep–1 Oct 2004, www.cropscience.org.au, Brisbane

Marongwe LS, Kwazira K, Jenrich M, Thierfelder C, Kassam A, Friedrich T (2011) An African success: the case of conservation agriculture in Zimbabwe. Int J Agric Sustain 9:153–161

Martínez J, Cajas YS, León JD, Osorio NW (2014) Silvopastoral systems enhance soil quality in grasslands of Colombia. Appl Environ Soil Sci. doi:10.1155/2014/359736

McCarthy JR, Pfost DL, Currence HD (1993) Conservation tillage and residue management to reduce soil erosion, Agric Publ G-1650. University of Missouri, Missouri
MEA (2005) Ecosystems and human well-being: desertification synthesis. Millennium ecosystem assessment. World Resources Institute, Washington, DC
Nicholls RJ, Lowe JA (2006) Climate stabilization and impacts of sea-level rise. In: HJ S, Cramer W, Nakicenovic N, Wigley T, Yohe G (eds) Avoiding dangerous climate change. Cambridge University Press, Cambridge, UK
Odjugo PAO (2010) General overview of climate change impacts in Nigeria. J Hum Ecol 29:47–55
Oweis T, Hachum A (2001) Reducing peak supplemental irrigation demand by extending sowing dates. Agric Water Manag 50:109–123
Oweis T, Prinz D, Hachum A (2001) Water harvesting: indigenous knowledge for the future of the drier environments. The International Center for Agriculture Research in the dry areas, Aleppo, Syria.
Peterson GA, Unger PW, Payne WA, Anderson RL, Baumhardt RL (2006) Dryland agriculture research issues. In: Peterson GA, Unger PW, Payne WA (eds) Dryland agriculture, Agronomy Monograph No. 23. ASA-CSSA SSSA, Madison, WI, 2nd edn
Pimentel D, Harvey C, Resosudarmo P, Sinclair K, Kurz D, McNair M, Crist S, Sphpritz L, Fitton L, Saffouri R, Blair R (1995) Environmental and economic costs of soil erosion and conservation benefits. Science 267:1117–1123
Pinto B (1987) Nigeria during and after the oil boom: a policy comparison with Indonesia. World Bank Econ Rev 1:419–445
Powell JM, Pearson RA, Hiernaux PH (2004) Crop–livestock interactions in the West African drylands. Agron J 96:469–483
Reij C, Mulder P, Begemann L (1988) Water harvesting for plant production, Technical Paper No. 91. World Bank, Washington, DC
Reynolds JF, Smith DMS, Lambin EF, Turner BL, Mortimore M, Batterbury SPJ, Downing TE, Dowlatabadi H, Fernández RJ, Herrick JE, Huber-Sannwald E, Jiang H, Leemans R, Lynam T, Maestre FT, Ayarza M, Walker B (2007) Global desertification: building a science for dryland development. Science 316:847–851
Rockström J (2003) Water for food and nature in drought-prone tropics: vapour shift in rain-fed agriculture. Royal Soc Trans B Biol Sci 358:1997–2009
Rockström J, Karlberg L, Wani SP, Barron J, Hatibu N, Oweis T, Bruggeman A, Farahani J, Qiang Z (2010) Managing water in rainfed agriculture—the need for a paradigm shift. Agric Water Manag 97:543–550
Rojas RV, Achouri M, Maroulis J, Lucrezia C (2016) Healthy soils: a prerequisite for sustainable food security. Environ Earth Sci 75:180
Rosegrant MW, Cai X, Cline SA (2002) World water and food to 2025: dealing with scarcity. International Food Policy Research Institute, Washington, DC
Scherr SJ (1999) Soil degradation. A threat to developing-country food security by 2020? Food, agriculture, and the environment discussion paper 27. International Food Policy Research Institute, Washington, DC
Schiere JB, Baumhardt RL, Van Keulen H, Whitbread AM, Bruinsma AS, Goodchild AV, Gregorini P, Slingerland MA, Wiedemann-Hartwell B (2006) Mixed crop–livestock systems in semi-arid regions. In: GA P, Unger PW, Payne WA (eds) Dryland agriculture, Agronomy Monograph No. 23, 2nd edn. ASA, CSSA, and SSSA, Madison
Sen Roy S, Singh RB (2002) Climate variability, extreme events and agriculture productivity in mountain regions. Oxford/IBH Publication, New Delhi
Serraj R, Siddique KHM (2012) Conservation agriculture in dry areas. Field Crop Res 132:1–6
Siegert K (1994) Introduction to water harvesting: some basic principles for planning, design and monitoring. In: Water harvesting for improved agricultural production. In: Proceedings of the FAO Expert Consultation, 21–25 November 1993, Cairo. Water Report 3. Food and Agriculture Organization, Rome

Singh RB, Gurjar AK (2011) Climate change vulnerability of agriculture in arid region of Rajasthan, India. Ann National Assoc Geographers India 31:31–38

Stewart BA (2007) Water conservation and water use efficiency in drylands. In: Stewart B, Asfary A, Belloum A, Steiner K, Friedrich T (eds) Proceedings of the International Workshop on Conservation Agriculture for Sustainable Land Management to Improve the Livelihood of People in Dry Areas, 7–9 May 2007. ACSAD and GTZ, Damascus, Syria

Stewart BA, Peterson GA (2015) Managing green water in dryland agriculture. Agron J 107:1544–1553

Stewart BA, Koohafkan P, Ramamoorthy K (2006) Dryland agriculture defined and its importance to the world. In: GA P, Unger PW, Payne WA (eds) Dryland agriculture, Agron. Monogr. 23, 2nd edn. ASA, CSSA, and SSSA, Madison

Thomas GA, Titmarsh GW, Freebairn DM, Radford BJ (2007) No-tillage and conservation farming practices in grain growing areas of Queensland–a review of 40 years of development. Aust J Exp Agri 47:887–898

Thornton PK (2014) Livestock and mixed crop-livestock systems. In: Sebastian K (ed) Atlas of African agriculture research & development. International Food Policy Research Institute (IFPRI), Washington, DC, USA

UNCCD (2007) Convention to combat desertification – the 10-year strategy. The United Nations Convention to Combat Desertification. http://www.unccd.int/Lists/OfficialDocuments/cop8/16add1eng.pdf

UNESCO (2012) Managing water under uncertainty and risk, United Nations World Water Development Rep. 4. UNESCO World Water Assessment Programme. UNESCO, Paris

USAID (2014) Small-scale dryland agriculture. The United States Agency for International Development, Washington, DC http://pdf.usaid.gov/pdf_docs/PA00JR32.pdf

Verheijen FGA, Jones RJA, Rickson RJ, Smith CJ (2009) Tolerable versus actual soil erosion rates in Europe. Earth-Sci Rev 94(1–4):23–38

White RP, Tunstall D, Henninger N (2002) An ecosystem approach to drylands: building support for new development policies. World Research Institute, Washington, DC, USA

Zurayk R, Haidar M (2002) The dryland predicament: natural capital, global forces. Dev Outreach 3:35–39

Part II
Elements of Dryland Agriculture

Water Harvesting in Dry Environments

Attila Yazar and Akhtar Ali

1 Introduction

Drylands cover approximately 40 % of the world's land area and support more than two billion people, 90 % of whom live in developing countries, with relatively low amounts of precipitation in the form of rainfall or snow (UN 2011). In these regions, precipitation is insufficient to support low-risk crop production. The non-uniform distribution of precipitation, in arid and semiarid regions, usually results in frequent drought periods during crop growth which severely stress growing crops thereby reducing yields, sometimes to crop failure. Today, rain is the cheapest and often the only source of water for agricultural purposes, albeit not always reliable. In many dry regions of the world, there is no alternative but a better and more effective use of rain to increase and secure food production. This is the essence and potential significance of runoff farming in a hungry world (Koohafkan and Stewart 2008).

Global water demand has been rising over the past century and is projected to increase further due to population growth and the need for increased food production (Kummu et al. 2010; Lasage and Verburg 2015) and change in lifestyle. The global food challenge is tremendous to feed an additional three billion people in 2050 and to address the hidden food gap i.e. eradicating malnutrition among 800 million of the world's poorest people (Conway 1997; Molden 2007; De Fries and Rosenzweig 2010). This food challenge is, to a large extent, a development challenge as 95 % of

A. Yazar (✉)
Irrigation and Agricultural Structures Department, Çukurova University, Adana, Turkey
e-mail: yazarat@cu.edu.tr

A. Ali
Environment, Agriculture and Natural Resources Division, Asian Development Bank, Mandaluyong City, Metro Manila, Philippines
e-mail: akhtar_rn@yahoo.com

M. Farooq, K.H.M. Siddique (eds.), *Innovations in Dryland Agriculture*,
DOI 10.1007/978-3-319-47928-6_3

population growth occurs in developing countries. A large proportion of these countries is hosted in tropical climates characterized by highly unreliable rainfall and frequent meteorological droughts and dry spells. Most of the poorest people in the world (65–70 %) are among the 1.1 billion farmers who make their living from agriculture (FAOSTAT 2012).

Rainfall is the most important natural resource in dry environments. In arid and semiarid regions, precipitation is generally lower than potential evaporation and non-uniform in distribution, resulting in frequent drought periods during the crop growing season, and usually comes in intense showers, resulting in surface runoff and uncontrolled rill and gully erosion (Oweis and Hachum 2009). Despite its scarcity, rainfall is generally poorly managed with much of it lost through runoff and evaporation. Capturing rainwater and making effective use of it is crucial for any integrated research and development project. Water harvesting (WH) can play an important role in fulfilling the objectives of such projects. As the water shortage in dry areas is a recurring crisis, people need information on how to capture and use every available drop of water efficiently. WH is an effective and economical means of achieving this objective and information on its various systems and techniques is in great demand (Oweis et al. 2012).

Water harvesting is a method of water collection that has historically been applied in arid and semiarid regions where rainfall is either not sufficient to sustain good crop growth or where, due to the erratic nature of precipitation, the risk of crop failure is very high (Prinz 1996) is now being employed all over the world. As new developments are made, more and more regions employ WH to help offset pressures on existing water resources. This resurgence in popularity comes on the crest of a new wave of environmentalism and drive toward sustainable development where the focus is on renewable sources of water collection. WH is aimed at reducing the pressure that development, and the consequences thereof, has placed on what is now considered a limited resource, our water (Prinz 2002).

This chapter deals with the methods and techniques of water harvesting to make more water available to humans, animals and for irrigation purposes, in places where rainwater is the only source of water. Details are provided on the main factors for selection of reliable water harvesting technique that is sustainable under local circumstances, including physical (hydrologic, terrain, and technical), cultural acceptability and socioeconomic (institutional and economic) factors. The aim is to compile a synthesis of experiences that can provide insight into rainwater harvesting opportunities which address human wellbeing while continuing to sustain a range of ecosystem services.

2 History of Water Harvesting

The act of harvesting rainwater, floodwater and groundwater has been practiced for thousands of years, from the most rudimentary techniques to large, complex methods such as the Roman aqueducts. For many cultures, water harvesting was an effective way to meet their water needs at a time when alternative sources of water for drinking and agricultural purposes were not readily available (Oweis et al. 2001).

Historically, many settlements have been situated in arid and semiarid climates, such as the Middle East, Northern Africa and Western Asia. These cultures largely depended on subsistence farming with few other opportunities to generate income. Water harvesting became widespread in many of these regions and, although various methods were devised almost universally, each emerging culture established their own unique way of collecting or diverting runoff for productive purposes (Prinz 1996; Oweis et al. 2001; Oweis et al. 2012).

Many communities in arid and semiarid regions have been harvesting water for almost as long as humans have engaged in agriculture (Bruins et al. 1986). Examples of water harvesting structures are known from Babylonians, Israel, Tunisia, China and the Americas (Frasier 1980; Boers and Ben-Asher 1982; Li 2005; Ouessar et al. 2004). Such structures have received renewed attention with the implementation of policies to increase food production since the droughts and food crisis in sub-Saharan Africa in the 1970s and 1980s (Critchley et al. 1992; Prinz and Singh 2000; Denison and Wotshela 2012).

The Middle East was one of the first regions in the world to practice harvesting water for consumption in both domestic and agricultural realms. WH structures found in this part of the world date back more than 9000 years to Southern Mesopotamia where simple WH structures were used as early as 4500 BC (Prinz 1996; Oweis et al. 2012). In the Negev Desert region, now modern day Israel, runoff irrigation farming has been practiced since the 10th century BC. This form of WH was used throughout Roman rule and well into the Byzantine era. In North Yemen, a system dating back to at least 1000 BC diverted enough floodwater to irrigate 20 000 ha of agricultural crops that may have fed as many as 300 000 people. This method of floodwater management is still in use today, making this region one of the few places where runoff agriculture has been continuously practiced since the earliest settlement (Prinz 1996; Oweis et al. 2001). Similarly, floodwater systems have been used in regions of modern-day Pakistan and Saudi Arabia, both varying the design and process to meet the needs of their climate and terrain.

Africa, Northern Africa in particular, has a long history of WH, where the technique was often devised to match the terrain of each region. Historically known as the granary of the Roman Empire, in Libya, runoff irrigation was often used to grow barley, wheat, olive, grapes, figs and dates in this arid region of the continent. This form of WH also allowed for sheep, pig and cattle farming. The farming system lasted for over 400 years and sustained a large stationary population, often wealthy, which created enough crops to generate a surplus (Prinz 1996).

Many other WH methods employed in Northern Africa are still used today including rainwater storage ponds called 'lacs collinaires' in Algeria, Meskat and Mgoud harvesting systems in Tunisia, Caag and Gawan systems in Somalia, and the Zay system in Burkina Faso. To the east, in Tanzania, water harvesting has been a mainstay with rural farmers using rainwater harvesting to irrigate their crops for centuries. People who rely completely on rainwater for their survival have, over the centuries, developed indigenous techniques to harvest rainwater. These methods include Majauba (excavated bund basins for rice production in the lake zone), Vinyungu (raised broad basins in the Iringa region) and Ndira (water storage structures in the Kilimanjaro region) (Mbilinyi et al. 2005).

In Asia, many communities have emerged and thrived in harsh arid regions where their social life has evolved around water scarcity and indigenous water harvesting techniques. India is one such nation where the ordering of certain social groups has been arranged around water scarcity. The national annual average rainfall in India is 1200 mm, yet regional variations can be as high as 10,000 mm per year in the northeast and as low as 150 mm per year in the desert regions (Krishiworld 2006). In the cool arid region of the Spiti valley, situated in the northern province of Himachal Pradesh, an intricate system of channels called Kuls has been devised to harvest meltwater from glaciers. This water is then delivered to local villages in the valley where the harvested water is used for irrigation purposes, turning this desert-like valley into one where agriculture is the mainstay (Rainwater Harvesting 2006).

Ancient WH techniques are not restricted solely to the old world. In the Americas, structures left behind by the Mayan civilization indicate a long history of WH. The Mayans used structures known as Chultuns, an early type of cistern with the capacity to hold 20 to 45 m^3, to harvest clean drinking water. Furthermore, Aguadas, an artificially dug rainwater reservoir designed to hold 10 to 150 million liters of water, and Aquaditas, small artificial reservoirs that could store 100 to 50,000 liters of water, were commonplace (Gnadlinger 2000).

Despite these historical accounts, water harvesting has, within the last few centuries, experienced a decline in implementation and practice due to: the decline of central powers (e.g. the Byzantine empire in the Middle East) due to political shifts; warfare including civil war; economic changes, e.g. in competitiveness on local or export markets; social changes, including availability of cheap labor; aspirations or attitudes of the people involved in water harvesting; climatic change (increasing aridity, change in rainfall regime); increasing salinity; decreasing soil fertility; and soil erosion (Prinz 1996; Denison and Wotshela 2012; Oweis et al. 2012; Lasage and Verburg 2015).

3 The Concept of Water Harvesting

Water is important for life, raising food, social and economic development, and sustained environmental services. WH can alleviate drought stress in arid and semi-arid environments and significantly contribute to water livelihood and environmental management by augmenting domestic water supplies, stabilizing crop yield and supporting fragile ecosystems. WH has been practiced for centuries in many dry regions of the world. In arid and semiarid regions, where low and poorly distributed rainfall makes crop production impossible, WH can make crop production possible. As such, WH has been employed for thousands of years to irrigate and restore productivity to land, provide drinking water to both humans and animals, minimize risk in drought-prone areas, increase groundwater recharge and reduce storm water discharge (Rainwater Harvesting 2006). Today, WH is used for crop irrigation, groundwater recharge and water storage for future use in drought-prone areas.

WH works by concentrating rainwater from a large catchment area to a small target area. It can significantly increase plant production in drought-prone areas by concentrating rainfall/runoff in parts of the total area (Prinz 1996; Oweis et al. 1999; Rockström 2002). WH collects water from (1) sources where water is widely dispersed and quickly changes location or form and becomes unavailable or (2) where it occurs in quantities and at locations where it is unusable unless intervention can gather the water to locations where it provides benefits (Pereira et al. 2002).

WH has been defined in several ways, with more general definitions being (1) the collection of runoff for its productive use (Critchley et al. 1991) and (2) the process of collecting and concentrating water from runoff into a run-on area where it is either directly applied to the cropping area and stored in the soil profile for immediate use by the crop (Prinz and Singh 2000). Here, we consider it in the broadest sense, as an umbrella term covering a wide range of techniques and methodologies to collect and conserve various forms of runoff water, originating from ephemeral water flows generated during rainstorms. In this sense, we adopt a similar approach to the definition by Critchley et al. (1991) of water harvesting as "the collection of water for its productive use". WH focuses on improving the productive use of rainwater on the local scale (field to subcatchment scale) before the runoff water leaves the geographical unit in question. The aim is to mitigate the effects of temporal water shortages to cover both domestic and agricultural needs. In terms of upgrading rainfed agriculture, WH can be categorized into three broad objectives (Rockström 2002): the systems that (i) improve infiltration of rainwater into the soil, (ii) prolong the duration of soil water availability in the soil and (iii) store surface and subsurface runoff for later use.

WH incorporates a broad set of techniques and methodologies that can be grouped into three main domains (Rockström 2002; Prinz 2002):

1. *In situ* water conservation (soil and water conservation)
2. Concentration of runoff to crops in the field, at a field (runoff farming) or catchment (floodwater harvesting) scale
3. Collection and storage of runoff water into different structures (soil, ponds, dams, tanks etc.) for supplemental irrigation.

This simple categorization is useful in terms of water management as it distinguishes between three distinctly-different hydrological situations in a farmer's field. *In situ* soil water conservation aims to make the best use of the rain that falls on the field, i.e., maximize rainfall infiltration into the soil and its storage in the root zone of the crop. The crop still lives at the mercy of the rainfall. Runoff concentration systems add surface water from outside the cultivated land at two scales, either from local sheet, rill and small gully runoff (i.e., runoff generated immediately adjacent up to several hundred meters from the land) or from gully flow at catchment scale (i.e., storm flow in larger gullies). Storage systems collect runoff water at different scales from field to catchment scale in various types of structures or in the soil for supplemental irrigation (Rockström 2002; Oweis and Hachum 2006).

The goals of water harvesting are summarized by Prinz (1996) as:

1. Restoring the productivity of land which suffers from insufficient rainfall
2. Increasing yields of rainfed farming (dry farming)
3. Minimizing the risk of crop failure in drought-prone areas
4. Combating desertification by tree plantation
5. Supplying drinking water to humans and animals.

In regions with annual precipitation between 100 and 700 mm, low-cost water harvesting may provide an interesting alternative if irrigation water from other sources is not readily available or is too costly. In summer rainfall areas, the minimum precipitation for WH is around 200 mm year^{-1}. In areas with more than 600–700 mm annual rainfall, WH can prolong the cropping season. In comparison to pumping water, water harvesting saves energy and maintenance costs. These advantages are countered by the problem of unreliable rainfall, which can partly be overcome by interim storage. Modern hydrological tools (e.g. calculation of rainfall probability and water yield) allow more precise determination of the catchment area (Prinz 2002).

WH supports a flourishing agriculture in many dry regions where rainfall is low and erratic in distribution (Oweis et al. 1999; Oweis et al. 2001; Ali et al. 2010). Water harvesting is particularly advantageous in the following circumstances:

- In a dry environment where low and poorly-distributed rainfall normally makes agricultural production impossible. Provided other production factors such as soils and crops are favorable, WH can make farming possible even in the absence of other water resources.
- In rainfed areas, where crops can be produced but with low yields and a high risk of failure. Here WH systems can provide enough water to supplement rainfall and thereby increase and stabilize production.
- In areas where water supply for domestic and animal production is not sufficient. These needs can be satisfied with WH.
- In arid land suffering from desertification, where potential production is diminishing due to lack of proper management. Providing water to these lands through WH can improve the vegetative cover and help to halt environmental degradation (Prinz 1996; Rockström 2002).

The specific benefits listed above lead to many other non-tangible and indirect socioeconomic gains. These include stabilization of rural communities, reduced migration of rural people to cities, use and improvement of local skills, and improvement in the standard of living of millions of poor people living in the drought-stricken areas (Prinz 1996; Rockström 2002; Mekdaschi and Liniger 2013).

The main components of a WH system include catchment area, storage facility and the target area (Fig. 1; Oweis et al. 2001) as described below:

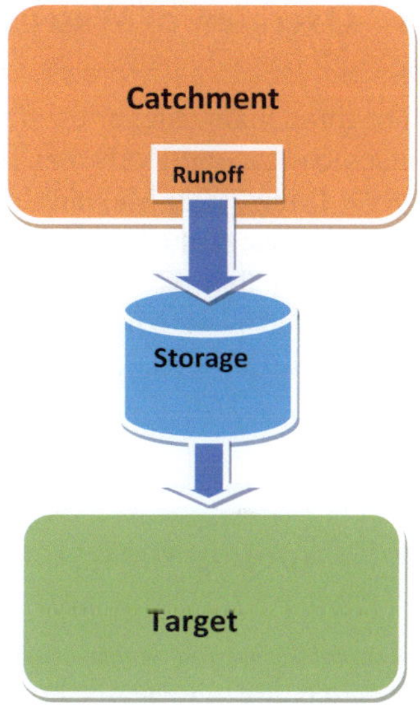

Fig. 1 The main components of a water harvesting system

- **Catchment area** (runoff area) is the part of the land that contributes some or its entire share of rainwater to a target area outside its boundaries. The catchment area can be as small as a few square meters or as large as several square kilometers. It can be agricultural, rocky or marginal land, or even a rooftop or paved road. A runoff area (catchment) with a sufficiently high runoff coefficient (impermeability) would be optimal.
- **Storage facility** is the place where runoff water is held from the time it is collected until it is used. Storage can be in surface reservoirs, subsurface reservoirs such as cisterns, the soil profile as soil moisture and in groundwater aquifers.
- **Target area (run-on)** is the area where the harvested water is used. In agricultural production, the target is the plan for the animal, while in domestic use, it is the human being or the enterprise and it needs. A run-on area, where accumulated water is stored and/or used. In most cases, the runoff is used to irrigate agricultural crops with the water then stored in the soil profile. A high storage capacity of the soil (i.e. medium-textured soils) and sufficient soil depth (>1 m) are prerequisites here. The water retention capacity has to be high enough to supply the crops with water until the next rainfall event.

The most important parameters to take into consideration in practicing WH include (i) rainfall distribution, (ii) rainfall intensity, (iii) runoff characteristics of the catchment, (iv) water storage capacity of soils, (v) cisterns or reservoirs, (vi) agricultural crops, (vii) available technologies and (viii) socioeconomic conditions (Tauer and Humborg 1992; Mekdaschi and Liniger 2013).

4 Overview of Water Harvesting Techniques

The three main categories of water harvesting include microcatchment harvesting, macrocatchment/floodwater harvesting and groundwater harvesting (Oweis et al. 1999; Prinz 2002; Mekdaschi and Liniger 2013) and are practiced in many water-scarce areas of the world. Within these three main categories there are numerous techniques that have been developed to match local conditions and need which could be adapted to benefit people in other water-scarce areas. Each category has its own subset of techniques and methods that can be used to maximize the benefits (Fig. 2).

4.1 Rooftop Water Harvesting

Harvesting of rainwater can be from roofs of private, public or commercial buildings (e.g. greenhouses, schools). Rooftop WH are getting more and more popular in both developed and emerging economy countries to secure /improve water supply for domestic use such as sanitation or garden irrigation (Mekdaschi and Liniger 2013). Rooftop due to small surface area (catchment area) and high runoff

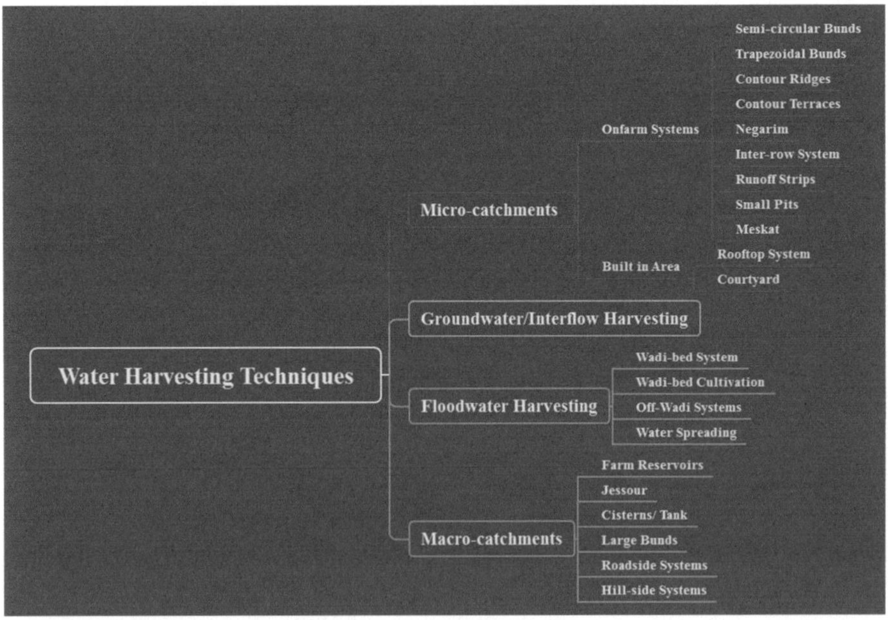

Fig. 2 Water Harvesting Techniques

Fig. 3 Rooftop water harvesting in Beypazari in Turkey (top left), Brazil (top right) and Ethiopia (bottom) (Source: Akhtar Ali personal communication)

efficiency can translate 80–90 % of incident rainfall into runoff. Most of the rainwater is only available for a short period of time, either during the rainfall or immediately afterward. This method involves a catchment area from one or more roofs, a storage tank, and gutters and pipes to guide the water from the roof into the storage tank. A tap is often attached to the tank to access the stored water (Fig. 3). Rope and pully arrangements or simple rope and bucket arrangements also work. This method is suitable for humid, semiarid and arid regions. Rooftop WH is practiced in many countries while in others it is not common; the main concerns are with traditional roofing materials, water quality, and the cost of the storage tank. Social traditions also discourage rooftop WH in some areas. There is concern over whether the water is clean enough for drinking as pollutants in the atmosphere can be present in rainfall. Today, water harvesters must be wary of pesticide contamination, high mineral levels, bacteria and other impurities in their runoff water (Palmback 2004).

Increasing pressures from water scarcity and rising awareness, development of roofing materials, and innovative ideas for water storage have made roof water harvesting a serious, feasible and attractive option for water resource availability. An increase in the adoption of rooftop WH requires low-tech and low-cost means of collecting water from roofs and construction of suitable storage facilities (Pereira et al. 2002). To maintain water quality, rooftop WH systems should have appropriate screens and purification systems built into the infrastructure to remove leaves and twigs from the water and to purify the water prior to use (O'Hogain et al. 2011). A rooftop WH system installed in 2007 in Tekke village of the Beypazari Municipalities included some filtering equipment so that the stored water could be used for drinking and kitchen use.

4.2 Microcatchment Water Harvesting

Microcatchment water harvesting (MCWH) is a method of collecting surface runoff from a small catchment area and storing it in the root zone of an adjacent infiltration basin with the plant (Boers and Ben-Asher 1982). Water productivity in water harvesting catchments is usually related to that measured in small plots (Frasier 1984). The size of the catchment affects runoff such that, under the same hydrological conditions, a small area may generate up to 50 % of rainfall as runoff compared with only 5 % of rainfall runoff from river basins (Stern 1979). The higher the runoff generated per unit area from a small catchment forms the basis of microcatchment water harvesting as an alternative option (Ali et al. 2010).

Common MCWH structures include continuous and intermittent contour ridges, semi-circular bunds, pitting, natural depressions, inter-row water harvesting, Meskat types, vellarany types, contour bench terraces, eyebrow terraces or hill slope microcatchments, contour strips and negarim basins (Prinz 1996; Hatibu and Mahoo 1999; Yazar et al. 2014). Small earthen or stone-made structures are constructed across the land slope along the contour. Construction along the contour ensures smooth water spreading. Land slopes between 2 and 8 % are considered suitable for MCWH, but these structures have been constructed on flat land and slopes up to 20 %. Rectangular or hexagonal shapes are suitable for this purpose. Frequent damage and high maintenance costs are the main limitations to the development of MCWH on steep slopes (Prinz 2002).

Microcatchments involve a distinct division of a runoff-generating catchment area (generally less than 1000 m^2) and a cultivation basin where runoff is concentrated and stored in the root zone and productively used by plants (Hatibu and Mahoo 1999; Oweis and Hachum 2009). The distance between the catchment area and runoff receiving area is usually less than 100 m. There are multiple advantages to this WH system in that the design is simple and cheap, there is higher runoff efficiency than the larger-scale WH systems, they often prevent or reduce erosion, and they can be implemented on almost any slope or level plane (Prinz 1996). The use of MCWH will depend on local conditions and the type of crop that receives the runoff water.

Fig. 4 Negarim microcatchment WH for trees; Negarim plot covered with plastic sheet and an infiltration pit was constructed around the pistachio tree (Sources: Rocheteau et al. 1988 *left*; Yazar et al. 2014, *right*)

4.2.1 Negarims

Negarims are regular squares made of soil bunds turned from the contour to concentrate water runoff at the lower corner of the square. In this corner, an infiltration basin is made with a planting pit in the center of this basin (Fig. 4). The Negarim consists of a catchment area and a cropped area. Runoff collects from the catchment area and flows into the cropped area where it ponds, infiltrates and is stored in the soil (Critchley et al. 1991; Yazar et al. 2014). The technique requires deep soils up to 2 m to store the harvested runoff and is mainly used for tree cultivation in dry areas with seasonal rainfall as low as 150 mm. When used for fruit trees, it is designed to provide sufficient moisture to a producing tree. When used in uneven topography, it is recommended that blocks of negarims are subdivided into smaller units (Hatibu and Mahoo 1999).

4.2.2 Semi-circular Bunds

This is a network of earth bunds shaped as half-circles with the tips upside and on the contour (Fig. 5). They can be used for trees, fodder and to improve range productivity. They vary depending on the crop type, soil and rainfall amounts. Semi-circular bunds are used in areas of 200–750 mm rainfall, deep soils and low slopes. They require even topography. The space between the tips on consecutive bunds is used for discharge of excess runoff. The top width of the bunds is usually 10 cm and the height may be uniform where the topography is flat. The side slopes are 1:1 although flatter sides have also been used. As the slope increases, the height is increased accordingly from the tip to the lowest point. The minimum height at the tip is 10 cm (Oweis et al. 2001; Anschuetz et al. 2003).

Fig. 5 Semi-circular bunds for bushes and trees (Sources: Oweis and Hachum 2009, ICARDA; http://www.actionagainsthunger.org)

4.2.3 Contour Ridges

Contour ridges consist of bunds or ridges, constructed along the contour line at an interval of 5–20 m. A 1–2 m strip upstream of the ridge is allocated for cultivation, and the rest constitutes the catchment. The height of the ridges varies according to the slope and the expected depth of the runoff water retained behind it (Fig. 6). The bunds may be strengthened by stones. This is a simple technique which can be implemented by farmers. Bunds can be formed manually, with animal-driven equipment, or with tractors fitted with suitable implements. Ridges may be constructed on a wide range of slopes from 1 to 50 %. Contour ridges are important for supporting the regeneration of forage crops, grasses and trees on mild to steep slopes in the step. In the semiarid tropics, they are used for growing sorghum, millet, cowpea and beans. The system is sometimes combined with other techniques (such as the zai system) or with *in situ* water conservation techniques (such as tied-ridge system) in the semiarid tropics (Oweis et al. 2001; Mugabe 2004; Adgo et al. 2013).

Contour ridges are used in row crops in areas with 350–700 mm of annual rainfall and require even topography. The use of contour ridges in Jordanian and Syrian deserts (annual rainfall ~150 mm) successfully raised fodder shrubs and cereal crops. Furrows are dug on a contour and ridges formed immediately on the lower side. The ridges are spaced 1–2 m apart and are usually 15–20 cm high. This forms

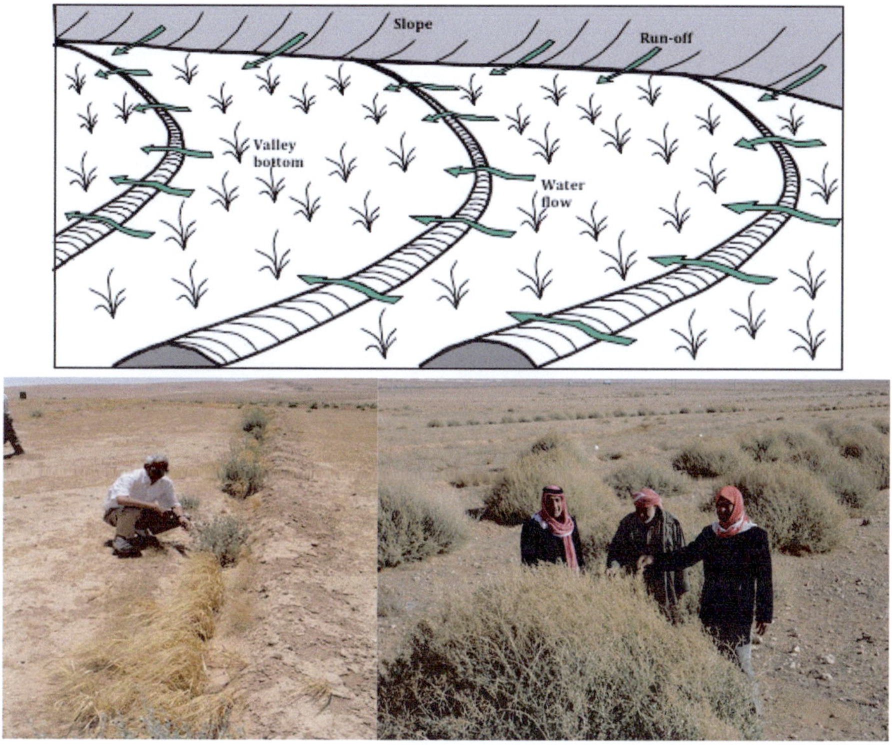

Fig. 6 Contour ridges: a schematic sketch; combination of shrub and barley in Jordan (*bottom left*) and shrub in Syria (*bottom right*) (Source http://www.ricehub.org/RT/land-development/inland-valleys/water-control-structure/; Aktar Ali, personal communication)

a catchment which produces runoff that collects at the furrow. A cereal crop is planted on the lower side of the furrow and a pulse on the upper side. The furrows are tied every 5 m to ensure that, in the case of a defective spirit level, runoff does not flow laterally and concentrate on one side causing erosion. The technique is, therefore, suitable in areas with a well-developed use of animal power (Hai 1998; Oweis et al. 2001).

4.2.4 Trapezoidal Bunds

This technique is suitable for areas with 250–500 mm of annual rainfall. It consists of large structures enclosing up to 1 ha and impounding large amounts of runoff from an external area. Crops are planted in the cropping area enclosed by soil bunds. The impounding bunds are laid on the contour but staggered down the slope to allow for the release of excess runoff. Excess runoff is discharged from the tips of the bunds. The most suitable slopes are 0.25–1.5 % on even topography and non-cracking soils such as black cotton soil. The maximum bund height is 0.6 m

Fig. 7 Contour stone bunds in Kenya (*left*) and Syria (*right*) (Source: http://www.infonet-biovision.org/EnvironmentalHealth/introduction-soil-conservation-measures; Akhtar Ali)

decreasing to 0.2 m at the tips. The technique can be used for trees and grass but it is best suited for row crops where manual work is the mode of cultivation. The standard design method is used to size the required catchment area (Oweis et al. 2001).

4.2.5 Contour Stone Bunds

These are made of stones laid on the contour on up to 2 % slope in areas with 200–750 mm rainfall. They are suitable on stoney land, and are used to slow down runoff and filter out the soil, and to increase infiltration of runoff water. Contour stone bunds can be used without spillways and have low construction and maintenance requirements. This technique is well suited to small-scale application on farmer's fields and, given an adequate supply of stones, can be implemented quickly and cheaply over large areas.

The spacing between stone bunds is normally 15–30 m but should be decreased as the slope increases. The minimum height is 25 cm with a base width of 35–40 cm set into a 5–10 cm deep trench which acts as a key. On slopes less than 1 %, bunds are spaced at 20 m and on 1–2 %, 15 m. Bunds are made with a good mix of large and small stones to ensure that the runoff is allowed to pass through slowly (Fig. 7). The small stones are normally placed upstream and the large stones downstream (Hai 1998).

4.2.6 Bench Terraces

Terraces of different types have been extensively and often successfully used in watershed management for soil and water conservation on hilly land, up to 40 % slope. Bench terraces are perhaps the oldest type of terrace. They are used primarily in areas where the supply of agricultural land is limited and where population pressure has forced cultivation up steep slopes. Early bench terraces were constructed by carrying soil from the uphill side of a strip to the lower side so that a level step or bench was formed. The steep slopes below the terraces were stabilized by vegetation or neatly-fitted stonework (Fig. 8). The construction of terraces has continued in recent years, particularly in countries with limited land and high population pressure.

Terraces reduce both the amount and velocity of water moving across the soil surface, which greatly reduces soil erosion. Terracing thus permits more intensive cropping than would otherwise be possible. A horizontal surface in terrace reduces runoff and maximizes water infiltration into the soil. If the soil surface is kept tilled and free of vegetation except for the desirable crop, almost all of the rain falling on the terrace will be used for crop growth. In regions with low rainfall, soil water can be stored for long periods provided the soil surface is kept tilled or mulched and vegetation free. Thus, it may be possible to store up to two years of rainfall to obtain one cereal crop. If there are several terraces down a hillslope, it may be possible to grow a crop on alternate terraces each year (Prinz 2002). Infiltration terraces can support soil water storage to make a crop viable each year where total rainfall is sufficient. It can also provide for perennial tree crops which develop extensive root systems, as typically occurs for olive trees in semiarid Mediterranean areas. In some cases, the cropped terraces are downslope of a runoff area and may infiltrate more water than that provided directly by rainfall, as occurs in the Meskats which are traditional in Tunisia (Missaoui 1996).

Fig. 8 Bench terraces: Rice paddies in China. Source: WILDCHINA (2012)

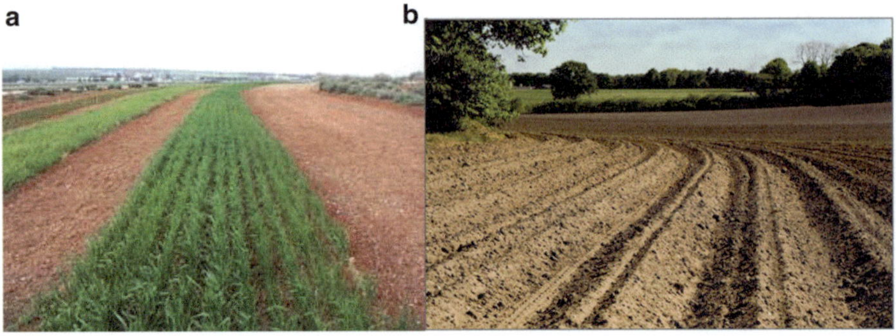

Fig. 9 Contour runoff strips (**a**) and contour furrows (**b**) (Source: www.icarda.org)

Despite several benefits, the use of terraces has decreased in recent years for several reasons: costly to construct and maintain, and more difficult to farm, particularly with large equipment. The construction of terraces may result in soil fertility problems because topsoil is buried or moved downslope. Terraces are also subject to failure during large, intensive rainfall events, resulting in considerable damage that is costly to repair.

Conservation bench terraces (CBTs) or Zingg terraces are a type of rainfall multiplier. They use part of the land surface as a catchment to provide runoff onto level terraces on which crops are grown. The method is particularly appropriate for large-scale mechanized farming such as the wheat/sorghum farmlands in the southwest of the United States of America, where this method was pioneered by Zingg in 1955 (Rockström 2000; WOCAT 2007; Al Ali et al. 2008).

4.2.7 Runoff Strips and Contour Furrows

This technique is applied on gentle slopes and is used to support field crops in drier environments (such as barley in the *badia*) where production is usually risky or has low yields. The farm is divided into strips following contour lines where one strip is used as a catchment and the strip downstream is cropped. The cropped strip should not be too wide (1–3 m), and the catchment width should be determined with a view to providing the required runoff water to the cropped area. The same cropped strips are cultivated every year (Fig. 9a, b). Clearing and compaction may be implemented to improve runoff (Oweis et al. 2001). Contour furrows are variations on the theme of surface manipulation that require less soil movement than conservation bench terraces and are more likely to be used by small farmers, or in lower rainfall areas. Cropping is usually intermittent on strips or in rows with the catchment area left fallow. The principle is the same as conservation bench terraces, that is, to collect runoff from the catchment to improve soil moisture in the cropped area.

Fig. 10 The Meskat and Menka type water harvesting system in Tunisia (Source: http://www.rainwaterharvesting.org)

4.2.8 Meskat

The Meskat system is the main traditional water harvesting practice widely used in the region of Tunisian Sahel. In Tunisia, the Meskat and the Jessour systems have a long tradition and are still practiced. The Meskat microcatchment system consists of an impluvium called Meskat, about 500 m^2 in size, and a Menka or cropping area of about 250 m^2. Thus, the catchment to cropping area ratio (CCR) is 2:1. Both are surrounded by 20 cm high bunds equipped with spillways to let runoff flow into the Menka plots (Fig. 10) (Prinz 2001; Majdoub et al. 2014).

The Tunisian Sahel provides a typical territory where intensive WH practices were made by peasants to manage aridity and to improve crop yields. The rainfall is erratic and insufficient for rainfed crops, thus, runoff harvesting practices are designed to collect supplementary water. In this region, inhabitants constructed sequences of small earthen dikes in the foothills, in the gently sloped areas, and across watershed in order to intercept the surface runoff from the surrounding hilly slopes on the upstream sides, generally used as rangeland. This traditional system is commonly known as olive grove Majrouf or Meskat system. The earthen dike, traditionally called Isser or Tabia, is managed by the spillway routing the remaining runoff to the downstream structures and is extended by embankments on the lateral sides. The plots equipped with these embankments, known as Mankaa, are used for cropping fruit species, especially olive trees. The hilly bare upstream area, with moderate to steep slope commonly called Meskat, produces surface runoff needed to ensure olive oil production in this dry region. The Meskats are arranged with channels along the slope that diverts flow to cropped trees in the Menka. This system contributes to olive trees by improving water availability and soil fertility (Majdoub et al. 2014).

Fig. 11 Floodwater harvesting in Balochistan, Pakistan: (**a**) water diversion structures on ephemeral stream (*top*) and (**b**) spreading of floodwater on the agricultural land (*bottom*) (Source: Akhtar Ali)

4.3 Floodwater Harvesting

Flood flows are a feature of all landscapes, including regions of water scarcity. A large part of the annual flow may occur in one or two floods, but the flow is often so large that the water passes through the region and cannot be used where the rain fell. Some advantage can be gained from these large flows by encouraging them to spread across flat areas (Fig. 11). If water can be retained on flat surfaces for a day or two, the upper soil layers may be saturated or water may percolate downwards to replenish the local aquifer. In both these circumstances, the water thus 'harvested' is available for later use, in the first case for crop growth or to support grazing and, in the second case, for whatever purpose groundwater is used (Prinz 1996; Missaoui 1996). Where the terrain is suitable, it may be possible to restrict flow in the river channel and force water to flow over the floodplain or into an old floodway. Flood spreading by these means is most easily achieved when the upstream–downstream gradient of the river valley is quite small. With small longitudinal gradients, water forced onto the floodplain tends to flow in the downstream direction very slowly, allowing maximum time for infiltration to occur (Pereira et al. 2002).

Often referred to as water spreading or spate irrigation, flood water harvesting (FWH) is involved in the collection and storage of creek flow for irrigation use (Prinz and Singh 2000). The main characteristics of FWH are a turbulent channel of

Fig. 12 Macrocatchment water harvesting (Source: Akhtar Ali)

water flow harvested either by diversion or spreading within a channel bed/valley floor where runoff is stored in the soil profile (Critchley et al. 1991). Two categories of FWH are macrocatchments and large catchments.

4.3.1 Macrocatchment Water Harvesting

Macrocatchments, sometimes called medium-sized catchments, are characterized by large flood zones that are situated outside the cropping area. Often farmers must use structures such as dams or bunds to divert, transfer, collect and store the runoff. Such systems are often difficult to differentiate from conventional irrigation systems and are considered FWH as long as the harvested water is available year around (Mbilinyi et al. 2005). Examples of macrocatchments include stone dams, large semi-circular hoops, trapezoidal bunds, hillside conduit systems and cultivated reservoirs, which from 1000 m^2 to 200 ha (Prinz 1996).

4.3.2 Large Catchment Water Harvesting

Large catchment water harvesting (LCWH) comprises catchments of many square kilometers, from which runoff water flows through a large stream bed (also known as Wadi) necessitating a more complex dam structure and distribution network (Fig. 12). There are two main forms of LCWH: Floodwater Harvesting (FWH) within a stream bed and Floodwater Diversion. FWH involves blocking the water flow in order to flood the valley of an entire floodplain and force the water to infiltrate the ground for crop production or pasture improvement. Floodwater diversion is a method where the river, stream (wadi) or creek bed water is diverted from its natural course and used to flood nearby areas as an irrigation method (Majdoub et al. 2014).

The 'Jessour' (or jesr) technique has long been known as a way of exploiting surface runoff water for agriculture in arid regions. It is a typical system in the highlands of southeastern regions of Tunisia and is the foundation of agricultural activities in this zone. The system retains water in dams made from earth or stone, perpendicular to the runoff, behind which crops, mainly fruit trees, are cultivated. The dam stops and stores the runoff to supply the crops. Jessour is generally used in mountainous areas where they are often built into wadis, but they are also constructed on plains. The dams encourage infiltration of rainwater which not only intensifies agricultural production but recharges the groundwater. During extreme rainfall, part of the disastrous runoff can be retained behind the dams to reduce the potential damage caused by floods (Majdoub et al. 2014).

4.4 Groundwater Harvesting

Groundwater harvesting (GWH) encompasses all methods, traditional or contemporary, of harvesting water from the ground for productive use. It has also been used as a storage method for the other forms of water harvesting outlined above, with many of these techniques requiring a particular terrain to divert the water from its original source so it can seep into the ground for crop use. Traditional methods of GWH have used underground dams, sand dams, wells, cisterns and aquifers (Rockström 2002).

Extraction of subsurface water flow, either from soil water trapped in shallow sand layers or from the water table, constitutes an interesting form of water harvesting. Storing water underground is attractive as it reduces evaporation losses and often contributes to high-quality water thanks to filtration, e.g., through sand (even though soil and geological characteristics will determine the level of filtration capacity). Sand dams and subsurface dams, where water is trapped behind small dam walls in sandy riverbeds, is an efficient and cheap form of water harvesting (SIWI 2001).

Wells are probably the most common GWH technique; they tap into the water table from a hole excavated on the surface (Figs. 13, 14). Wells have been employed as a source of water for thousands of years, dating back to 8100–7500 BC. Like other forms of water harvesting, wells have been adapted to meet the needs of individuals living in specific regions. Technology has also increased the returns from wells, making it easier to obtain water (Prinz 2002).

Cisterns are manmade caves or underground constructions to store water. Often the walls of these cisterns are plastered to prevent water loss, deep percolation and/or evaporation (Prinz and Singh 2000). The underground cistern supplies water for domestic and irrigation purposes in drought-prone areas. There are two variants to this cistern, one being shaped like a bottle, the other in a circular form. Both are constructed in a similar fashion with the ground excavated to form the shape of the cistern. The surface is covered with polyethylene or concrete plastering to avoid seepage. Both cisterns are expensive and difficult to build, and often too complex for farmers to construct themselves (Alem 2003).

Fig. 13 An ancient groundwater harvesting system in Kathmandu, Nepal (Source: Akhtar Ali)

Fig. 14 Groundwater harvesting well in Pakistan (Source: Akhtar Ali)

Aquifers occur when underground layers of water seep into permeable rock or other materials such as sand, gravel, silt or clay. They generally occupy large areas under the earth's surface and will often supply water sources such as streams, rivers and springs. Aquifers are often on the receiving end of water harvesting, in that they can store harvested rainwater. Recently, awareness of depleting aquifers has spurred an increase in WH techniques that aim to directly recharge these depleting resources. Many forms of rainwater harvesting collect water and store it underground for

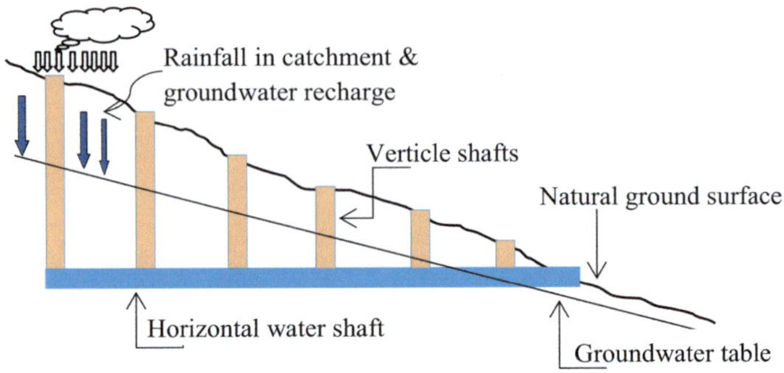

Fig. 15 Groundwater harvesting Qanat by sketch

future use. Not only does this recharge depleting groundwater sources, it raises the declining water table and helps to augment the water supply (Prinz and Singh 2000).

Qanats are an example of how an aquifer can be accessed to provide fresh water. Widely used in Iran, Pakistan, North Africa and Spain, the Qanat consists of a horizontal tunnel that taps underground water in an alluvial fan, bringing it to the surface due to gravitational effects (Fig. 15). Qanat tunnels are on a 1–2 % incline up to 30 km long. Many are still maintained and steadily deliver water for agricultural production and villages for drinking water supply (Prinz and Singh 2000).

5 Main Consideration for Selection of Water Harvesting Techniques

To reliably select a water harvesting technique suitable for local circumstances, the characteristics of the different techniques need to be reviewed and include those that are physical (hydrologic, terrain and technical), culturally acceptable and socioeconomic (institutional and economic) in nature (Critchley et al. 1991; Ngigi et al. 2005; Bewket 2007; Lasage et al. 2008; Tumbo et al. 2011).

Water harvesting measures should be technically applicable under the physical circumstances in the field. However, it is important to account for cultural acceptance of the technique and the need for simple governance after implementation. Governance is necessary if the available water needs to be shared among many people in one village or between several villages. There are many examples of water harvesting projects that have failed to meet targets due to the complexity of governance, or because they were not acceptable to the population for cultural, environmental or economic reasons. The resources needed for construction (physical, labor, knowledge, capital) and their effects on the surrounding environment and hydrological conditions (quality and quantity) also need to be taken into account. Water quality is especially important when a structure provides water for domestic use. Water quality is less relevant when used only for irrigation (Lasage and Verburg 2015).

There is sufficient freshwater available each year to fulfill the needs of the current population on Earth. However, in certain regions and countries, the annual renewable supply of water per person is less than 500 m^3 (Qadir et al. 2007). The need for WH, as mentioned above, arises from many factors such as low rainfall and uneven distribution, high losses due to evaporation and runoff, and increased demand for water due to population growth (Abu-Awwad and Shatanawi 1997; Prinz 2002). With a large proportion of the human race living in arid and semiarid regions, the applicability of WH to increase water access in these areas should be investigated.

As WH becomes an important strategy to deal with water scarcity or water stress, it is important to consider the factors involved in selecting the appropriate WH method to maximize hydrological returns. It is tempting to assume that a system which works in one area will also work in another, seemingly similar zone. However, there may be technical dissimilarities such as the availability of stone or intensity of rainfall, and distinct socioeconomic differences (Critchley et al. 1991).

Several critical factors that need to be considered when selecting the appropriate WH method are outlined below.

5.1 Rainfall

WH depends on limited and uncertain rainfall, thus understanding the dynamics of precipitation within the environment will influence which WH method fits best in each context (Qadir et al. 2007). Various factors should be taken into account regarding precipitation including:

1. Number of days where the rain exceeds the threshold rainfall of the catchment, on a weekly or monthly basis.
2. Probability of occurrence (in years) of the mean monthly rainfall.
3. Probability and reoccurrence of the minimum and maximum monthly rainfall.
4. Frequency distribution of storms of different specific intensities (Prinz and Singh 2000).

Rainfall in arid and semiarid regions is highly erratic, and most rain falls as intensive, generally convective storms, with very high rainfall intensity and extreme spatial and temporal variability. The result is very high risk for annual droughts and intra-seasonal dry spells. The annual variation of rainfall can typically range from a low of one-third of the long-term average to a high of approximately double, meaning that a high rainfall year can have some six times higher rainfall than a dry year. Statistically, in a semiarid region, severe crop reductions caused by a dry spell occur once or twice in five years, and total crop failure by annual droughts occurs once every ten years (Rockström 2002). This means that poor distribution of rainfall over time often constitutes a more common cause of crop failure than absolute water scarcity due to low cumulative annual rainfall. This is why it is important to distinguish between droughts and dry spells. An agricultural drought occurs when the cumulative plant available soil water is significantly lower than the cumulative crop water requirement, i.e.; there is absolute water scarcity. A dry spell is a short period of water stress, often only for a couple of weeks during crop growth. A short period of water stress can have a serious effect on crop yield if it occurs during water-sensitive development stages such as flowering (Rockström and de Rouw 1997).

5.2 Land Use or Vegetation Cover

Working to reduce erosion and redirect runoff into appropriate catchments can lead to high labor inputs due to the need to keep the catchment area free of vegetation so that it is as efficient as possible. The vegetation of the selected area will heavily influence runoff, infiltration and retention levels, and must be taken into account prior to implementation to reduce high labor costs in the future. Maintenance of the catchment system must also be understood when selecting the size of the catchment. The system may be damaged during heavy rainstorms or require regular maintenance which could prove problematic in the future (Qadir et al. 2007).

5.3 Topography and Terrain Profile

Topography is an important aspect of WH as the slope of the terrain and gradient will greatly impact the size and type of catchment area of the WH system (Prinz and Singh 2000).

5.4 Soil Type and Soil Depth

The soils of a WH site must function as a water storage facility, a medium for plant growth and a water collection surface. These soil conditions must be met in a single area. As such, soil depth and texture are important physical elements for understanding how to design a successful WH system. Soil type and depth help judge percolation and infiltration rates, potential for runoff and the storage potential of water within the soil itself (Prinz and Singh 2000).

5.5 Hydrology and Water Resources

Hydrology monitors the available water sources involved in storage, production and runoff of the WH system, which will aid in the informed selection of an appropriate WH technique for the proposed site (Prinz and Singh 2000). The hydrological characteristics of a region are determined largely by its climate, topography, soil and geology. Key climatic factors are the amount, intensity and frequency of rainfall, and the effects of temperature and humidity on evapotranspiration. Determining probable maximum precipitation, forecasting precipitation, estimating evapotranspiration and determining rainfall–runoff relationships can be problematic in arid and semiarid areas (Oweis et al. 2012).

5.6 Socioeconomic and Infrastructure Conditions

There are several social, cultural and economic factors that are important to consider when selecting the appropriate WH technique.

- People's Priorities—when opting to introduce WH methods to a specific area. WH aims to increase the availability of water resources for productive use and it is, therefore, important that the WH infrastructure meets the needs of the individuals using it.
- Participation—when implementing projects surrounding WH. For example, when development schemes are implemented by governments or non-government organizations (NGOs), it is imperative that the community, farmer or individual be involved in the process from beginning to end. This helps create a sense of ownership of the project within the community. Knowledge plays an important role here for individuals involved in the WH scheme as they need to fully understand how it operates. One potentially negative effect of implementing complex WH technologies is that those left to use it are unfamiliar with the technology and thus unable to properly maintain it (Oweis and Hachum 2006).
- Adoption of Systems—indicates the importance of selecting the appropriate WH method for each site. Widespread adoption of WH techniques by the local population is the only way that significant areas can be treated at a reasonable cost on a sustainable basis (Critchley et al. 1991).
- Area Differences—it is not always possible to implement the same WH system in different areas. This is due to subtle but important differences that exist between sites that can cause a WH system to be successful in one region and fail in another (Critchley et al. 1991).
- Land Tenure—not having full ownership of the land which one lives can cause an individual to be reluctant to invest in a WH scheme that would only benefit the user in the short term (Critchley et al. 1991).

- Land Use Management—how land, both communal and private, is managed and used can determine the effectiveness of the WH strategy being proposed or implemented. Effective land management is important as conflicts and disputes over water rights, and land ownership and use can arise (Oweis and Hachum 2006).
- Environmental and Ecological Impacts—ecosystems are often fragile and can be adversely affected if the water table is tampered with. It is important to pay attention to these factors, understanding where the water flows and how it affects the surrounding ecology before implementing any kind of WH system. Some negative impacts that WH can potentially have on the existing environment are the reduction of valuable cropland that would be occupied by the catchment area. The catchment often requires a large area and thus occupies valuable crop land (Qadir et al. 2007). However, today the technology exists to allow for WH to occur on a larger scale, allowing for various commercial uses such as plant nurseries, garden centers, vehicle washing plants, agricultural uses and for use in washrooms and urinal flushing in public buildings (Rainwater Harvesting 2006).

6 Design Characteristics of Microcatchment Water Harvesting

Microcatchment water harvesting is a viable means of providing food, fiber and drinking water if the systems are designed to fit the local physical, economic and social environment. To design a successful microcatchment system, certain physical, technical and socioeconomic elements must be investigated.

6.1 Physical and Technical Design Characteristics

For a microcatchment water harvesting system (MCWH) to be successful, certain physical and technical design characteristics must be considered. In any WH system, the runoff collected during rain events must fulfill the needs of the crop during the growing period in dry, hot weather. The catchment area must have a smooth soil surface with sufficient slope to generate runoff during the precipitation events. The soils of the infiltration basin must have sufficient depth with a texture suitable for infiltrating, retaining and storing runoff water. If the physical system is poorly designed and managed, soil erosion, flooding and insufficient water to meet the needs of the crops will occur. Frasier (1984) commented that no universal water harvesting technique exists because each location has unique conditions that influence the design of the optimum system. Some important physical and technical design characteristics for consideration are precipitation, soil type, slope, runoff and catchment area ratios, runoff efficiency, agronomic features and plant species (Renner and Frasier 1995b; Oweis et al. 2001; Mekdaschi and Liniger 2013).

6.1.1 Physical Design Considerations

Precipitation

WH is of significant interest in arid and semiarid regions where crop growth is restricted by infrequent or limited precipitation. These areas have unpredictable precipitation patterns and quantities. The quantity and frequency of rainfall needed for WH to be feasible are debatable. A few authors suggest 80 mm, others 250 mm while some contend that WH is possible between 200 and 300 mm under unusual conditions. Several researchers claim that WH may be physically possible but not economically feasible in areas where precipitation is less than 50–80 mm (National Academy of Sciences 1974).

The frequency of rain and probability of certain intensities and amounts is more important than the annual quantity. It is usually desirable to look at monthly or growing season precipitation quantities as opposed to annual precipitation amounts. It is easier to design the necessary size of the WH structures with information on rainfall patterns during the growing season rather than annual quantities. When possible, analyzing a ten-year minimum climate record is recommended. If the rainfall quantity fluctuates greatly, discarding the year with the greatest and year with the least seasonal rainfall may help in determining a realistic estimate of potential rainfall amounts (Frasier 1984).

Soils

The soils of an MCWH site must function as a water storage facility, a medium for plant growth and a water collection surface. These soil conditions must be met in a single area. Because of this, soil depth and texture are important physical elements necessary to understand how to design a successful MCWH system. Each microcatchment stores runoff water in the soil profile of the collection area making soil depth very important. The soil of the infiltration basin must be deep enough to hold water with a minimum depth of 1.5 to 2.0 m. The soil profile should be approximately the same depth as the plant roots as a shallow soil depth may not store sufficient water for extended plant growth while a deeper profile allows water to travel below the reach of the roots (Renner and Frasier 1995b; Prinz and Singh 2000).

For MCWH systems, the soil texture must have good water holding and infiltration capacities. The soil of a microcatchment must allow rain to quickly infiltrate but with sufficient size soil pores for proper aeration. With sandy soils, there may be insufficient water holding capacity in the infiltration basin to sustain crop growth.

With very fine-textured soils, water may not infiltrate and is lost to evaporation. High clay content soils with low infiltration may be suitable if infiltration is increased in the infiltration area by the addition of organic material. For microcatchments, in addition to having good water holding and infiltration properties, the soil texture must be able to generate runoff in the catchment area. An ideal soil for the catchment area will seal or form a crust that becomes impermeable during rainfall. With sandy soils, the runoff from the catchment area may be low, making the area unsuitable for WH (Renner and Frasier 1995b; Prinz and Singh 2000).

For microcatchment structures, another important quality of soil texture to consider is erosion potential. Highly-erodible soils do not make durable crusts and should be avoided. In some situations, if the clay content of the runoff area soil is between 5 and 35 %, compaction of the area may increase the runoff efficiency (Pratt 1980). Microcatchments on clay soils with medium to fine texture are the best at generating sufficient runoff and are not susceptible to erosion (Shanan and Schick 1980).

Slope

In microcatchment systems, the slope is an important design element of the catchment area which affects the quantity and quality of water generated. Slopes that are too steep may erode and produce high amounts of sediment in the runoff water. On flat land, water is lost by retention in small depressions. The retained water either infiltrates into the soil or is evaporated into the atmosphere and lost. The most efficient WH systems are usually on slopes of 3–5 % (Prinz and Singh 2000).

6.1.2 Technical Design Considerations

Runoff/Runon Area Ratios

For microcatchments, the ratio of the runoff area to the infiltration basin area (C:CA) is an important technical design consideration. The catchment to application area ratio (C:CA) represents the degree of concentration of rainfall/runoff in water harvesting systems and compares the size of the catchment with that of the application area (Mekdaschi and Liniger 2013). It is generally used where runoff is stored in the soil for plant production. This ratio must be balanced to facilitate collecting the correct amounts of water for maximum crop production without producing inundation or flooding. Major factors for consideration in determining the proper catchment/cultivated area ratio are the climate (precipitation quantity and timing), geomorphology (slope and soil texture, soil depth), crop type (water requirements) and runoff efficiency of the collecting area. The ratio should consider both abnormally high and low rainfall quantities and erratic timing of precipitation events (Renner and Frasier 1995b; Prinz 2002; Oweis et al. 2012).

There is no universally-standard catchment area to infiltration basin ratio. Generally, the ratio varies between 1:1 and 20:1 depending on site conditions, precipitation characteristics and crop water requirements (Frasier 1984). However, Carter and Miller (1991) suggested a ratio between 17:1 and 50:1 for their microcatchments in Bostwana. They adjusted their ratio from 50:1 to 17:1 due to the high runoff volume and small water-holding capacity of the soils. The C:CA ratio should compensate for the possibility that insufficient water will be collected during periods of below average rainfall. This ratio should depend on the average rainfall of the lowest precipitation year. However, researchers have different ideas of what the proper ratio should be in relation to precipitation. With 500–600 mm of rain in tropical areas, some have suggested that the C:CA ratio should between 5:1 and 20:1. For arid regions with 200 to 300 mm of rainfall, the C:CA ratio should be between 10:1 to 30:1 (Renner and Frasier 1995b). With smaller catchments on a relatively steep slope, a higher percentage of runoff to rainfall is collected because less water is lost by soil depressions. A larger C:CA ratio such as 17:1 may require an elevated border around the infiltration basin to retain the collected runoff until it can infiltrate the soil. Under some situations, this may create water-logging and erosion problems.

Runoff Efficiency

For designing microcatchment structures, an important component is the catchment runoff efficiency which is defined as the percentage of total rainfall which is harvested as runoff. The efficiency of the system—the amount of runoff collected in relation to precipitation—depends on storm duration and intensity, and antecedent soil water (Frasier 1975). Typical runoff efficiencies range from 30 to 50 % of average monthly precipitation (Ali et al. 2010).

Agronomic Features

The application of mulch and organic matter is a notable feature of the technical design considerations for microcatchment water harvesting systems. The addition of mulch and organic matter in the infiltration basin area can significantly improve soil structure, and increase fertility and infiltration while decreasing soil water evaporation (Renner and Frasier 1995b). The use of vertical mulching has increased infiltration rates in microcatchment structures (Fairbourn 1975).

Plant Species

Selecting the right plant species is one of the most important technical design considerations for microcatchments. According to several studies, certain types of trees and crops are best for microcatchment and water harvesting systems in general.

Because water harvesting systems increase the amount of water availability to crops, certain types of plants can survive drought periods and produce greater yields. With these systems, plant species that can endure periods of intermittently wet or dry soil will benefit. Plants used in water harvesting systems should have water requirements which match local rainfall patterns (Renner and Frasier 1995b). Sorghum and pearl millet are excellent choices for planting in water harvesting systems; sorghum endures drought and water-logging while millet tolerates drought but not waterlogging. Maize is inappropriate for water harvesting systems because it does not tolerate drought or water-logging. Several nut trees, such as pistachio and almonds, and olives have been used successfully in microcatchment systems (Tubeileh et al. 2009; Yazar et al. 2014).

6.2 Economic Design Characteristics

In addition to being technically sound, water harvesting systems must be economically feasible for the local population. That is, crop or tree production must have greater benefits than costs (capital and labor). In other situations, WH is economically beneficial for local farmers because it is the only feasible method of farming for degraded land which lacks irrigation water. Frequently, the local population will view WH differently if water is a means of survival as opposed to a method of profit (Renner and Frasier 1995a; Mekdaschi and Liniger 2013).

6.2.1 Labor Requirement

Water harvesting systems are usually labor intensive to construct and maintain. Depending on the type of WH systems, the capital costs and labor requirements vary greatly. Some WH systems have high material construction costs and low labor needs or vice versa (Frasier 1988). In some instances, the amount of labor required to maintain a WH system can be higher than that for construction. Maintenance labor can be high since the system may need to be inspected and repaired as necessary, especially after major rainfall events (Renner and Frasier 1995a).

6.2.2 Costs Versus Benefits

Proposed WH systems should be closely analyzed to evaluate if the potential costs versus benefits are economically feasible for the local population. To make a WH project successful, the system must provide both economic and non-economic benefits. The economic benefits may be monetary profits from the sale of production items such as crops or trees. Local people may not immediately see the non-economic benefits of reduced silting in local rivers, soil conservation and aquifer recharge which may come from WH. Some other less noticeable advantages of rainwater collection in communities may be sustainable agriculture, heightened self-reliance and future famine relief. For new WH systems, designers need to acknowledge that increases in gross income may be negated by additional construction and maintenance costs (Oron et al. 1983; Oweis et al. 2012). For agricultural WH, the costs must include construction, maintenance, seed, tillage and fertilizer for the system (Pacey and Cullis 1986).

Microcatchment systems are considered by some to be the most economical runoff farming system because they are less expensive to construct and maintain. The costs are lower because microcatchments do not require terraces, conduits, water conveyance channels or water storage structures. They can also be constructed without high technological inputs so only a small capital investment is required. Microcatchments are less likely to be destroyed during heavy storms so reconstruction costs are generally lower (Renner and Frasier 1995a; Prinz 2002).

6.3 Social Design Characteristics

Microcatchment water harvesting is a viable means of providing food, fiber and drinking water to communities if the systems are designed to fit the local physical, economic and social environments. MCWH systems, when properly designed, have several socioeconomic advantages for the small farmer: (1) microcatchments are a simple technology to understand so they are easily transferable to the local population, (2) since microcatchments are small-scale structures, they have lower economic and labor costs for construction and maintenance than large-scale irrigation systems, and (3) at their small scale, microcatchments are not apportioned among many users so conflicts over water rights are minimized. These systems are usually independently used and small scale, eliminating the need for communal-regulating groups (Renner and Frasier 1995a; Oweis et al. 2012).

If water harvesting is to be successful then the socioeconomic direct and peripheral effects on the local population should be monitored. In planning and designing a WH project, impoverished groups and individuals should be identified to reduce the possibility of increased economic inequality within the society (Critchley et al. 1992). Improperly designed WH systems can lead to heightened inequalities in a society. For example, impoverished village herders may be excluded from certain farming land which was previously lost to pasture, resulting in animal deaths or loss of weight. This may cause economic hardship to the herders during important animal market periods. WH systems should be compatible with and supported by the existing social structure (Bruins et al. 1986). To do this, WH structures must be designed to incorporate the socioeconomic characteristics and to monitor the resulting impacts on the local population. To do so, the needs of the local population must be identified so that they feel that the proposed rainwater collection program is the best alternative for their intended uses and situation. Local participation, involvement of women, incentives versus need and appropriate technology are important attributes in the selection of a WH system (Renner and Frasier 1995a; Mekdaschi and Liniger 2013).

6.3.1 Local Participation

A project's success depends on the attention paid to social issues from the outset. The social aspects of a community must be understood in order to enlist local participation. In some villages, it may be difficult to get community members to give their input during planning and development phases. With good local participation, the design of a WH system can be changed and improved to better meet the needs of the local population. Participation throughout the phases of a WH project depends on many socioeconomic factors. It is also affected by the scale of the WH project; a very large project may need to form a communal organization to organize and run the project. As a result, individual participation may be minimal. Greater farmer participation may occur after the communal organization is decentralized. Microcatchments can have higher individual participation because of the small scale and need for minimal communal organization. The most successful WH systems are those where communities work in small, collective groups. One benefit of these collective systems is that farmers can be trained during work sessions by a community extension service. By encouraging communal action, groups of poor farmers also have the same benefits as rich farmers in relation to accessing credit and marketing (Pacey and Cullis 1986; Renner and Frasier 1995a; Oweis et al. 2012).

6.3.2 Involvement of Women

In designing WH projects, the role of women is often overlooked. In semiarid regions of Africa, women are the majority of the workforce and, because of the seasonal labor migration of men, are often the head of the household. This is also the case for women in developing nations in South America and Asia as men journey to urban areas to work on a seasonal basis. Agricultural extension services are oriented toward working with men rather than women. Very few women work as extension agents for agricultural services outside of home economics. This is unfortunate since so many women in developing nations own farms or gardens and could greatly benefit from WH. In many areas of the developing world, the women farm a small, degraded portion of their family's holding (Renner and Frasier 1995a; Oweis et al. 2012).

6.3.3 Incentives Versus Need

If a WH project is installed in an area which lacks water for agricultural production or reforestation, the system may be accepted by the local population. Without a locally-perceived need for increased water, the project will end in disaster. Unless the local community believes that the project is best for their needs and demand, the project will fail. In general, reliance on food aid or other incentives to motivate the local participation is not productive (Reig et al. 1988). Many incentives for participation offered to a local community are meaningless either for social or economic reasons (Renner and Frasier 1995a).

6.3.4 Appropriate Technology

Water harvesting techniques may be unsuccessful because the systems are deemed inappropriate for the social environment in the area due to (1) the technical design being too difficult for farmers to comprehend making it difficult for them to maintain the structures, or (2) the system is incompatible with patterns of local food production, has high labor requirements or relies on machinery. Machinery usually requires some degree of technical knowledge to maintain and without regular maintenance will fail. Where a WH project relies on machinery, the system will fail because the local population or government usually cannot afford the necessary maintenance.

Although WH techniques may be appropriate for certain communities, the system may be inappropriate for certain indigenous cultures. In marginal areas, governments should be aware of the social and environmental implication of introducing WH projects. Indigenous WH techniques may be more appropriate than introduced systems. For instance, in Burkina Faso, development project directors decided to abandon an introduced WH technique and adopt an indigenous technology of constructing stone bunds in fields. They improved the microcatchments by building the bunds on the contour which allowed water to be collected, spread evenly over the fields and percolate slowly through the bunds (Renner and Frasier 1995a). By modifying the indigenous microcatchment system, they satisfied local environmental characteristics while increasing crop production. Due to the small scale of microcatchments, modification of these traditional systems often augments crop production making the technique even more adaptable by the local population (Oweis et al. 2012; Mekdaschi and Liniger 2013).

Microcatchment water harvesting is an important technique for sustainable agriculture in developing countries as long as the important socioeconomic design elements are incorporated into these systems. Professionals implementing WH systems should analyze the economic benefits and social acceptability of their projects. Most scientific research on WH has been primarily technical and has ignored ways of extending these systems to farmers and communities. Increased research on the socioeconomic aspects of WH should be performed and widely published. All WH projects should evaluate and monitor their activities and learn how new technologies influence socioeconomic factors. If professionals in this field increase their awareness of the socioeconomic elements of implementing a WH project, more success will occur in the field and this viable technology will help increase agricultural production while reducing consumption of water resources (Renner and Frasier 1995a; UNEP-SEI 2009).

7 Supplemental Irrigation and Water Harvesting

In dry areas, it is water, not land, which is the most limiting resource for improved agricultural production. Maximizing water productivity, not yield per unit of land, is a better strategy for dry farming systems. Under such conditions, more efficient water management techniques must be adopted. Supplemental irrigation (SI) is a highly-efficient practice with great potential for increasing agricultural production and improving livelihoods in dry rainfed areas. In drier environments, most rainwater is lost by evaporation such that rainwater productivity is extremely low. Water harvesting can improve agriculture by directing and concentrating rainwater through runoff to the plants and other beneficial uses. Over 50 % of lost water can be recovered at very little cost. However, socioeconomic and environmental benefits of this practice are far more important than increasing agricultural water productivity (Oweis and Hachum 2006).

Supplemental irrigation is defined as the application of a limited amount of water to the crop when rainfall fails to provide sufficient water for plant growth to increase and stabilize yields. According to Oweis et al. (1999), the characteristics of SI in rainfed areas include: (1) water is applied to rainfed crops which are normally produced without irrigation, (2) water is applied only when rainfall is inadequate because rainfall is a prime source of water for rainfed crops, (3) the amount and timing of SI are not meant to provide water stress conditions over the growing season, but to provide enough water during the critical stages of crop growth to ensure optimal yield in terms of yield per unit of water.

SI during dry spells with microcatchment rainwater harvesting could improve the soil water content in the rooting zone by up to 30 % (Biazin et al. 2012). Harvested water from a small pond increased sorghum harvests by 41 % and, when combined with added fertilizer, by 180 % (Fox and Rockström 2000).

SI systems are affordable for small-scale farmers (Fox et al. 2005). However, policy frameworks, institutional structures and human capacities similar to those for full irrigation infrastructure are required to successfully apply SI in rainfed agriculture. Rainfed agriculture has traditionally been managed at the field scale. SI systems, with storage capacities generally in ranging from 100–10,000 m^3, are small in comparison to irrigation storage but require planning and management at the catchment scale as capturing local runoff may impact other water users and ecosystems. Legal frameworks and water rights pertaining to the collection of local surface runoff are required, as are human capacities for planning, constructing and maintaining the storage systems for SI. Moreover, farmers must take responsibility for the operation and management of the system. SI systems also can be used in small vegetable gardens during dry seasons to produce fully-irrigated cash crops. SI is a key strategy, which is underused, for unlocking rainfed yield potential and water productivity (Oweis et al. 2001; Rockström et al. 2010).

8 Benefits of Water Harvesting

Rainwater harvesting (RWH) is as old as human settlements. It is practiced in many different forms both in rural and urban landscapes, often in small-scale, decentralized ways. A growing number of cases describing the multiple benefits of rainwater harvesting have emerged. There is evidence of increased human wellbeing, and sustained or enhanced ecosystem services by water harvesting intervention from developing countries such as India, various sub-Sahara African, transition countries such as Brazil and China through to developed countries such as USA, Australia (UNEP-SEI 2009; Oweis et al. 2012).

There are numerous positive benefits for harvesting rainwater. The technology is low cost and highly decentralized which empowers individuals and communities to manage their water. It has been used to improve access to water and sanitation at the local level. In agriculture, rainwater harvesting has demonstrated the potential to double food production by 100 % compared to the 10 % increase from irrigation. Rainfed agriculture is practiced on 80 % of the world's agricultural land area and generates 65–70 % of the world's staple food. For instance, more than 95 % of the farmland is rainfed in Africa and almost 90 % in Latin America (UNEP-SEI 2009; Mekdaschi and Liniger 2013).

Macro- and micro-catchment rainwater harvesting systems have a variable but positive impact on soil moisture regimes and crop yields (Walker et al. 2005; Wei et al. 2005; Mupangwa et al. 2006). Li and Gong (2002) and Tian et al. (2003) found that MCWH of ridges and furrows with plastic mulch increased tuber yields of potatoes by 158–175 % for two years and corn yield by 1.9 times due to higher water use efficiency (WUE). The plastic used to mulch the ridges poses environmental problems so biodegradable plastic film should be used (Wang et al. 2008). Aftab et al. (2012) concluded that WH systems were a relatively low-cost option for temporary access to a water source. RWH minimizes some of the problems associated with irrigation, such as competition for water among various uses and users, low water use efficiency, and environmental degradation. RWH is a simple, cheap and environmentally-friendly technology that can easily be managed with limited technical skills (Ngigi 2003).

Besides the increase in agricultural productivity, RWH technologies also enhance household food security and rising incomes (Mutekwa and Kusangaya 2006). Specifically, Vohland and Barry (2009) concluded that *in situ* RWH practices improved hydrological indicators such as infiltration and groundwater recharge, improved soil nutrients, increased biomass production and improved soil temperatures (Li et al. 2000). RWH practices enhance floral diversity, modify the spatial structure of the ecosystem and increase animal biodiversity as more biomass is available for food and shelter (Rockström et al. 2004).

Experimental plots by Abu-Zreig and Tamimi (2011) of *in situ* WH with a sand ditch significantly reduced runoff and sediment loss by 46 and 60 %, respectively, while infiltration and soil moisture increased. RWH techniques such as Jessour in Tunisia and the Middle East decreased the amount and velocity of runoff consequently reducing soil erosion, and ameliorating the soil water storage capacity and soil fertility (Schiettecatte et al. 2005; Al-Seekh and Mohammad 2009). Glendenning and Vervoort (2010) discovered that approximately 7 % of rainfall recharged groundwater in various rainwater harvesting structures in the Arvari River catchment in India. In addition to fostering the value of groundwater recharge, the technique of a small water impounding system (SWIP) has an equivalent value for flood prevention as well as trapping sediment to prevent a negative impact on downstream regions (Concepcion et al. 2006). The implementation of RWH increases the irrigation area which changes more blue water into green water. This has a positive impact on groundwater recharge but decreases streamflow downstream thereby increasing the resilience and sustainability of the groundwater system (Glendenning and Vervoort 2011).

Microcatchment water harvesting made more water available to trees and significantly improved the growth of *Tamarix ramosissima* in the semiarid loess region of China (Li 2005). Water harvesting can be attractive to farmers because it reduces the risk of crop loss from spatial or temporal drought, provides more options for extending the growing season, supplies more rainfall to offer a wider selection of crops to grow, and allows abandoned land to be cultivated (Tabor 1995). Hafif and Murni (2012) reported that the presence of a small farm reservoir (SFR) as a water harvesting technique in a tropical region, Indonesia, with a 7 m long × 3 m wide × 2.5 m deep in a 1.5 ha catchment area, increased the planting area of vegetables in the dry season by up to 650 %. The SFR also increased the intensity of vegetables and tobacco cultivation thus increasing the farmers' income from marginal land by as much as 37.5 %. Yields remain dependent on water supply in spring although the construction of a WH system has greatly enhanced the possibilities for growing olives in Tunisia (Fleskens et al. 2005). In Spain, almond yields doubled due to irrigation with spare water (Schwilch et al. 2012; WOCAT 2012).

The *in situ* technique of ridge and furrow rainwater-harvesting uses rainfall better by increasing soil moisture storage, but it cannot resolve the temporal problem of moisture deficits because this system cannot harvest rainfall in the dry season when crops need water most (Li and Gong 2002). Therefore, combining an *in situ* WH technique with supplemental irrigation in semiarid and arid regions can help improve agricultural production. Oweis and Taimeh (1996) found that runoff efficiency of the natural soil surface can be as high as 60 % in small catchment areas, but will drop for large areas and small storms. Compaction of the soil surface or covering it with plastic mulch improved the catchment's runoff efficiency, but water should be applied for high-value crops. Yazar et al. (2014) evaluated the performance of a small runoff-basin water-harvesting system (Negarim) under an arid environment in the southeastern Anatolia region of Turkey. One microcatchment area (36 m^2) and four surface treatment methods (hay covered, stone covered, plastic covered and compaction) were used on young pistachio trees. Surface treatments had a significant effect on plant height and stem diameter. Among the surface treatments, the stone cover was the least effective while the plastic cover was superior to other treatments on plant height. Plastic cover had a significantly different effect on stem diameter than the other treatments.

Ali et al. (2010) assessed the MCWH potential of a Mediterranean arid environment in Syria using both a runoff microcatchment and a soil water balance approach. Average annual rainfall and evapotranspiration of the studied environment were estimated as 111 and 1671 mm, respectively. This environment hardly supports vegetation without supplementary water. About 5000 MCWH basins were developed for shrub raising on a land slope between 2 and 5 % by using three different techniques. Runoff yield varied between 5 and 187 m^3 ha^{-3} for various rainfall events. Overall, the soil water balance approach predicted 38–57 % less water than the microcatchment runoff approach. Rainfall does not infiltrate properly due to a combination of human-induced land degradation and high-intensity rainfall events (Falkenmark et al. 2001). Hatibu et al. (2006) revealed that contrary to expectations, improving the RWH system by adding a storage pond may not increase productivity and the benefits do not occur when rainfall is very small (below normal). Therefore, it is critical to increasing the linkages of crops to livestock producers and profitable markets.

Numerous successful reports of water harvesting projects are mirrored by equally as many failures, which are mostly due to poor design, bad engineering, over-ambitious goals and poor communication between villagers and designers (Tabor 1995). Although ecologically appropriate, the traditionally-practiced WH system and runoff farming in Madhya Pradesh, India has been stopped due to certain governmental policies, e.g., the changeover to a cash crop and canal irrigation with the completion of the big Bargi dam (Pangare 1992). Monitoring and evaluation of currently promoted WH techniques are critical in order to determine the impact of technologies on the whole farm and the community (Mupangwa et al. 2006). Mishra et al. (2009) stated that unscientific use of irrigation water by farmers, along with poor management of the water tariff system in irrigation WH projects, has led to low returns. The water fees are insufficient to meet routine operation and maintenance costs. Besides poor management of the water fees system, poor planning in proper crop selection, especially during the dry season, led to a severe scarcity of irrigation water during advanced crop growth stages. Therefore, the planning of WH systems should carefully consider the relationship between the resources and the users (Oweis et al. 1999).

Increasing runoff yield using the compaction method of the catchment area has the side impact of serious soil erosion under high-intensity rainfall, thus, soil stabilization would be needed for future use (Li and Gong 2002). A study by Barron and Okwach (2005) implied that *in situ* WH through terracing did not meet the crop water demand even during the rainy season (total rainfall during the maize growing season >300 mm), but it should be supported by harvesting rainwater in an earth dam with a viable technical solution for supplemental irrigation and fertigation. An analysis by Fox et al. (2005) suggested a strong mutual dependence between the investment in WH for supplemental irrigation and the input of fertilizers. Collecting rainfall in storage tanks or earth ponds can be expensive and is subjected to water losses due to deep percolation and evaporation, particularly in arid and semiarid regions where the amount of evaporation largely exceeds rainfall. Therefore, an *in situ* WH technique where rainfall is collected and stored directly into the soil profile will be better (Abu-Zreig and Tamimi 2011). The MCWH system requires a large area to collect water and its construction requires more labor. Increased withdrawals of water in rainfed and irrigated agriculture may have negative implications on downstream water availability within a river basin scale, and this needs further investigation (Rockström et al. 2010; Glendenning and Vervoort 2010). Ngigi (2003) stated that the impacts of an RWH system in Ethiopia, Kenya, Tanzania and Uganda were still marginal because the adoption rate is low despite the success of a number of RWH systems.

9 Implementation and Operational Management of Rainwater Harvesting Systems

There is historical evidence that rainwater harvesting has been an important element of community development since the beginning of human settlements. Many cultures have developed their societies with the primary management of water resources as a cornerstone, developing more sophisticated ways of supplying water both for consumption and agriculture (UNEP-SEI 2009).

Rainfed agriculture is practiced on 80 % of the world's agricultural land area and generates 65–70 % of the world's staple food. For instance, more than 95 % of the farmland is rainfed in Africa and almost 90 % in Latin America (UNEP-SEI 2009). There are numerous positive benefits for harvesting rainwater. The technology is low cost and highly decentralized which empowers individuals and communities to manage their water. WH has been used to improve access to water and sanitation at the local level. In agriculture, rainwater harvesting has demonstrated the potential to double food production by 100 % compared to the 10 % increase from irrigation.

At the global level, there is no comprehensive assessment of the extent of implementation of rainwater harvesting technologies for specific uses. Nor is there any summarized information on how much land is currently under any type of *in situ* rainwater harvesting. For the specific application of conservation tillage, as no-tillage agriculture, national statistics have been aggregated by Hobbs et al. (2008). Their information suggests that, globally, only a small fraction of the land surface, about 95 million hectares, is currently under conservation or no-till agriculture. For irrigation and conservation tillage, the AQUASTAT database (AQUASTAT 2009) holds data for a selected number of countries. Unfortunately, the information on irrigation cannot directly be associated with rainwater harvesting systems for irrigation purposes as it differentiates between surface water and groundwater, which does not allow the separation of shallow groundwater from deep groundwater or surface water withdrawn from 'blue' water sources (lakes, waterways, large dams) from small-scale systems. The recent assessment of irrigated and rainfed land, completed in the Comprehensive Assessment of Water Management in Agriculture (CA

2007), did not differentiate between areas under rainwater harvested water supply and areas under other types of water supply for irrigation. This lack of global information on where and how much rainwater harvesting is currently in use makes it impossible to say how many people actually benefit from rainwater harvesting today. It also becomes challenging to summarize the global and/or regional benefits and costs in specific locations, countries or regions using rainwater harvesting for human wellbeing or ecosystem impacts arising from rainwater harvesting (UNEP-SEI 2009).

Introducing rainwater harvesting to improve soil ecosystem productivity in rainfed agriculture promises large social, economic and environmental paybacks, particularly in poverty reduction and economic development. Rainwater harvesting presents a low-cost approach for mediating dry spell impacts in rainfed agriculture. Remarkable success has been witnessed in poverty-stricken and drought-prone areas in India and Africa. In sub-Saharan Africa, the future of more than 90 % of rainfed farmers depends heavily on improved water security. In South Asia, about 70 % of agriculture is rainfed and some good work has been done in the design and successful demonstration of a range of water harvesting structures, for both drinking water supply and irrigation. In several other countries in the Middle East, Latin America and South East Asia, rainwater harvesting is a traditional practice in certain regions, but the transferability of these models and practices has been limited. One of the main problems is that the local institutions needed often are ineffective and inconsistent with the predominant governmental structures and institutional arrangements and policies prevailing in these countries (Samra 2005).

10 Uncertainties in Water Harvesting and Coping Mechanisms

Water harvesting largely depends on rainfall which is variable in space and time, amount and intensity, and its potential for runoff generation. These variabilities along with the impacts of climate change induce uncertainty and constrain farmers' crop production planning and investment. Frequent drought in some dry areas can also prohibit effective water harvesting operations. Variability of annual and mean monthly rainfall at Matrouh in northwestern Egypt is shown in Fig. 16a, b. Soil moisture also varies with location and time and can constrain crop production. Among other factors, soil moisture depends on the rainwater harvesting potential of the catchment and the properties of soil as a storage medium *vis-à-vis* the evapotranspiration needs of the crop.

Weather forecasting and dissemination of weather parameters (rainfall, temperature, wind, humidity) can help with the planning and timely actions of farmers and users of water harvesting technology. Technologies such as weather radars, weather forecasting models, and mobile communication can be used to disseminate information to the users of water harvesting.

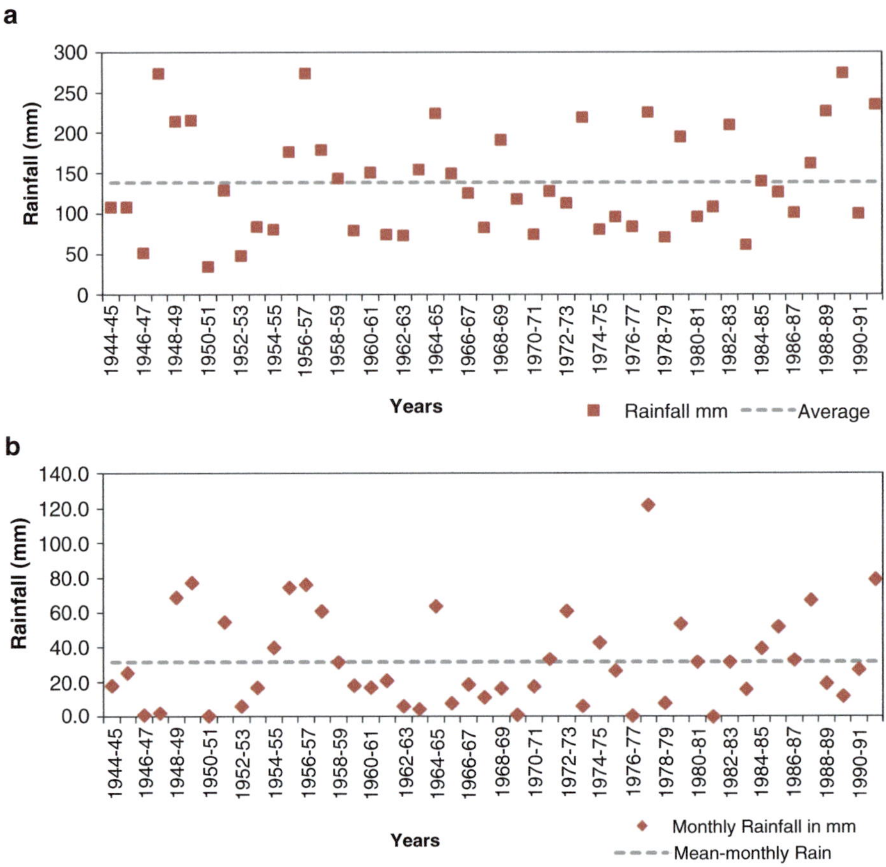

Fig. 16 (**a**) Variability of annual rainfall at Matrouh in Northwestern Egypt and (**b**) Variability of monthly, for December, rainfall at Matrouh in Northwestern Egypt

Water storage spurs economic growth and helps alleviate poverty by making water available when and where it is needed. In northwestern Egypt, farmers use stored water in the cistern to irrigate olive and fig trees at critical stages to match the rainfall variability. One rainwater cistern, one family and one unit of olive and fig orchard are an essential part of most of the people in northwestern Egypt. The rainwater harvesting and utilization project in Gansu province of the People's Republic of China harvested and stored rainwater for about 1.2 million people and their livestock from 1988 to 1996 (Gansu Bureau of Water Resources 1995). Several techniques are available to improve the runoff efficiency of a catchment (cleaning, compaction, covering patches with impervious materials) thereby improving the soil in the target area and reducing evaporation losses.

11 Investment in Water Harvesting

For developing countries alone, an estimated \$103 billion per year is needed to finance water, sanitation and wastewater treatment (Yepes 2008). Africa Infrastructure Country Diagnostic estimated that to close the infrastructure gap in water supply and sanitation and meet the corresponding millennium development goal targets in Africa within 10 years, an annual investment of approximately \$22 billion, equal to 2.58 percent of the gross domestic product, is required. Nearly \$15 billion of this is needed for capital expenditure with the remaining \$7 billion for operational expenditure. The study estimates that it will take \$3.4 billion per year to attain the region's goal of doubling the amount of land under irrigation, with 85 percent of the total going to capital works (Foster and Briceño-Garmendia 2010).

Globally, a modest investment is made in water. Countries are expected to share about 75 % of the total investment in the water sector. International financial institutions and donors roughly share 25 % of the investment. Asian Development Bank's (ADB's) water investment from 2006–2010 was \$10 billion and planned to invest \$2–2.5 billion annually from 2011–2020 in the Asia and Pacific region (ADB 2011). The World Bank Group committed more than US\$100 billion in 2009 to help countries who had cut spending in services during previous crises to maintain and expand infrastructure (Rodriguez et al. 2012). They noted that official development assistance for water and sanitation sharply increased from an average annual commitment of \$3.3 billion in 2002–2003 to \$8.3 billion in 2009–2010. The OECD (2012) estimates that by 2025, water will make up the lion's share of global infrastructure investment. For the OECD countries and Russia, China, India and Brazil, water spending will top \$1 trillion in 2025, nearly triple the amount needed.

A major chunk of the investment in the water sector has been allocated to water infrastructure and urban water supply and sanitation. Despite the small investment requirement, WH has not been a priority investment area. In rural areas, the investment varies from \$500–\$800 for rooftop water harvesting for domestic use, \$200–\$500 for smallholders farmers and around \$20,000 for a community tank (GTZ 2010). Being (i) a small investment, (ii) well-known technology, with (iii) simple governance and a high probability of sustainability, WH has potential for public–private partnerships. Microfinance Institutes in Nepal provided small credits (\$100–\$200) to diverse groups and individuals that can finance 15–20 % of the total costs for installation of rainwater harvesting (RAIN Foundation 2014). Nepali experiences show that farmers could repay their loan within three years due to the increased income and savings. Other financing modalities for WH can be explored to meet area- and community-specific requirements.

In an area with 155 mm average annual rainfall in northwest Egypt, World-Bank-financed water harvesting developed 8300 rainwater cisterns with a total storage capacity of 1.25 million cubic meters of freshwater to improve water and sanitation of 56,000 poor people and their livestock. The Matrouh Resource Management Project (1997–2003), at a total cost of \$27 million (\$10 million contribution by beneficiaries and \$17 million loan), supported water harvesting for olive and fig trees over 5000 Feddans (2.5 Feddans equals 1 ha) and conserving 10,000 Feddans of soil. The overall beneficiaries of the project were 139,000 people.

12 Challenges and Opportunities in Operationalizing Water Harvesting

The biggest challenge with rainwater harvesting is its exclusion in water policies in many countries. In many cases, water management is based on renewable water from surface and groundwater with little consideration of rainwater. Rainwater is taken as a 'free for all' resource and there had been an increase in its use over the last few years. This has resulted in over use, drastically reducing water available for downstream users including ecosystems. This has introduced water conflicts in some regions of the world. For sustainable use of water resources, it is critical that rainwater harvesting is included as a water source along with ground and surface water (UNEP-SEI 2009).

There is a lack of awareness and information sharing with regard to water harvesting. In fact, in many areas, water harvesting does not penetrate the grass-root level. Media can be used to disseminate information to everyone. In some cases, insufficient maintenance has constrained the adoption of water harvesting. Poor households face obstacles due to the lack of capital investment. Borrowing is difficult as scheduled bank procedures do not make it easy for the poor. Community self-financing schemes, if introduced, may facilitate the funding of rainwater harvesting technologies. Lehmann et al. (2010) reported that many countries such as Germany support the construction of rainwater harvesting systems for drinking and the promotion of decentralized seepage. The State Government of Delhi in India has introduced a financial incentive to promote rainwater harvesting. In Japan, a subsidy is provided to encourage people to install rainwater harvesting systems. For example; the Government of Sumida City offers subsidies up to 1 million Yen per rainwater utilization project. Other cities including Okinawa Prefecture, Takamatsu City, Toyota City, Kamakura City and Kawaguchi City have begun subsidizing or loaning funds to install rainwater systems. The success of the Gansu rainwater harvesting project has seen the construction of 12 million water tanks and small ponds in China storing 16 billion m^3 of water benefitting 36 million people and providing supplemental irrigation to 2.6 million ha to dryland supporting another 30 million people (UN-HABITAT 2015).

Water harvesting could be divided into systems that are managed by (i) individual households (rooftop, microcatchment, rainwater cistern etc.), (ii) communities, where communal management is important for operational management and sustainability, and (iii) the public sector including large-scale water harvesting dams. The first step to operationalize water harvesting could be the inclusion of water harvesting in the country's annual and mid-term development plans. The first two types of systems may need technical assistance and financial support to meet the capital investment cost. This could be supported through soft loans.

13 Conclusions and Future Research Thrusts

Water harvesting has been practiced successfully for millennia in parts of the world and some recent interventions have also had significant local impact. Yet the potential of water harvesting remains largely unknown, unacknowledged or unappreciated.

Rainwater harvesting can serve as an opportunity to enhance ecosystem productivity thereby improving livelihoods, human wellbeing and economies. Therefore, ecosystem services are fundamental to human wellbeing and are the basis of rural livelihoods in particular for the poor. Regarding the benefits of water harvesting, the Stockholm Environment Institute (UNEP-SEI 2009) summarized the key findings as:

- Water harvesting has increased synergies in landscape management and human wellbeing. These synergies are particularly obvious when rainwater harvesting improves rainfed agriculture, is applied in watershed management, and when rainwater harvesting interventions address household water supply in urban and rural areas.
- Rainwater harvesting has often been a neglected opportunity in water resources management, with only surface water and groundwater sources conventionally considered. Managing rainfall also presents new management opportunities, including rainwater harvesting.
- Improved water supply enhanced agricultural production through the adoption of rainwater harvesting can be attained with relatively low investments in fairly short time spans (5–10 years).
- Rainwater harvesting is a coping strategy for variable rainfall. In the future, climate change will increase rainfall variability and evaporation, and population growth will increase the demand for ecosystem services, particularly water. Rainwater harvesting will become a key intervention to cope with water scarcity.
- Awareness of ecosystem services must be increased by practitioners and policy makers alike, to realize the potential of rainwater harvesting and ecosystem benefits for human wellbeing.

Enabling policies for rainwater harvesting uptake and implementation are the first step to increased adoption. To move from a centralized to a decentralized water system, for example, is not an impossible task but one that needs sustained efforts in the rationalization, planning, implementation and adjustment to the system. It is recommended that responsible global bodies take on the task of assisting countries to mainstream rainwater harvesting in their policy agendas. This effort should be supported by education, technical exchanges and capacity building efforts which are institutionalized to assists countries who are ready to venture onward with a change in the historical paradigm and culture for water availability and climate change protection. However, such changes should be undertaken from a position of understanding and knowledge of the potential benefits and risks of using rainwater harvesting, including the human benefits and environmental costs of diverting flows from surface and ground waters (UNEP-SEI 2009).

A key strategy is to minimize the risk for dry-spell-induced crop failures, which requires an emphasis on water harvesting systems for supplemental irrigation. Large-scale adoption of water harvesting systems will require a paradigm shift in integrated water resources management in which rainfall is regarded as the entry point for the governance of fresh water, thus incorporating green water resources (sustaining rainfed agriculture and terrestrial ecosystem) and blue water resources (local runoff) (Rockström et al. 2010).

It should be emphasized that the ultimate goal of *in situ* water harvesting is a sustainable and environmentally-friendly system of agricultural production. The aim is to complement rather than replace the existing water use system. Improved systems must be socially acceptable as well as more productive. It is highly recommended that water harvesting interventions form part of a plan for integrated land and water resources development and that such a plan takes into consideration all the necessary technical, agronomic, socioeconomic and institutional aspects and inputs (Oweis et al. 2001).

Finally, it is time to scale-up the 'good practices' of water harvesting that have survived or emerged from new experiences, after decades of an almost exclusive focus on mastering fresh water flows in rivers and lakes through investments in irrigation infrastructure. Water harvesting offers under-exploited opportunities for the predominantly rainfed farming systems in the drylands of the developing world. It works best in precisely those areas where rural poverty is worst. When practiced well, its impact is to simultaneously reduce hunger and alleviate poverty, as well as improve the resilience of the environment. Investment in rainwater harvesting is important for meeting not only the Millennium Development Goals on reducing hunger, but also on reducing poverty and ensuring environmental sustainability.

References

Abu-Awwad AM, Shatanawi MR (1997) Water harvesting and infiltration in arid areas affected by surface crust: examples from Jordan. J Arid Environ 37:443–452

Abu-Zreig M, Tamimi A (2011) Field evaluation of sand ditch water harvesting technique in Jordan. Agric Water Manag 98:1291–1296

ADB (2011) Water Operational Plan 2011–2020. Asian Development Bank, Mundaluyong City

Adgo E, Teshome A, Mati B (2013) Impacts of long-term soil and water conservation on agricultural productivity: the case of Anjenie watershed, Ethiopia. Agric Water Manag 117:55–61

Aftab TB, Hasnain SA, Iqbal SR (2012) Save water and safe water: evaluation of design and storage period on water quality of rainwater harvesting system. J Environ Earth Sci 2:106–111

Al Ali Y, Touma J, Zante P, Nasri S, Albergel J (2008) Water and sediment balances of a contour bench terracing system in a semi-arid cultivated zone (El Gouazine, central Tunisia). Hydrol Sci J 53(4):883–892

Alem G (2003) Water harvesting: a water security strategy for mitigating the impact of drought in Ethiopia. Water Security in the 21st century, pp 1–5

Ali A, Yazar A, Abdul Aal A, Oweis T, Hayek P (2010) Micro-catchment water harvesting potential of an arid environment. Agric Water Manag 98:96–104

Al-Seekh AH, Mohammad AG (2009) The effect of water harvesting techniques on runoff, sedimentation, and soil properties. Environ Manag 44:37–45

Anschuetz J, Kome A, Nederlof M, de Neef R, van de Ven T (2003) Water harvesting and soil moisture retention. Agrodok 13, Wageningen: Agromisa Foundation. Available: agromisa.zenopx. Nl

AQUASTAT (2009) AQUASTAT online. FAO, Rome http://www.fao.org/nr/water/aquastat/dbase/index.stm

Barron J, Okwach G (2005) Run-off water harvesting for dry spell mitigation in maize (*Zea mays* L.): Results from on-farm research in semi-arid Kenya. Agric Water Manag 74:1–21

Bewket W (2007) Soil and water conservation intervention with conventional technologies in northwestern highlands of Ethiopia: acceptance and adoption by farmers. Land Use Policy 24:404–416

Biazin B, Sterk G, Temesgen M, Abdulkedira A, Stroosnijder L (2012) Rainwater harvesting and management in rainfed agricultural systems in sub-Saharan Africa, a review. Phys Chem Earth 47–48:139–151

Boers Th M, Ben-Asher J (1982) A review of rainwater harvesting. Agric Water Manag 5:145–158

Bruins HJ, Evenarl M, Nessler U (1986) Rainwater-harvesting agriculture for food production in arid zones: the challenge of the African famine. Appl Geogr 6:13–32

CA (Comprehensive Assessment of Water Management in Agriculture) (2007) In: Molden D (ed) Water for food, water for life. A Comprehensive assessment of water management in agriculture. International Water Management Institute (IWMI). Earthscan, London

Carter DC, Miller S (1991) Three years' experience with an on-farm macrocatchment water harvesting system in Botswana. Agric Water Manag 19:191–203

Concepcion RN, Contreras SM, Sanidad WB, Gesite AB, Nilo GP, Salandanan KA, Manalang KM, de Vera SV (2006) Enhancing multi-functionality of agriculture through rainwater harvesting system. Paddy Water Environ 4:235–243

Conway G (1997) The doubly green revolution. Food for all in the twenty-first century. Penguin Books, London, p 334

Critchley W, Siegert K, Chapman C, Finkel M (1991) Water harvesting. A manual for the design and construction of water harvesting schemes for plant production. FAO report 9210403. FAO, Rome, 133 p

Critchley W, Reij C, Seznec A (1992) Water harvesting for plant production. Volume II: case studies and conclusions for Sub-Saharan Africa'. World Bank technical paper number 157, Africa Technical Department series, Washington, DC

De Fries R, Rosenzweig C (2010) Toward a whole-landscape approach for sustainable land use in the tropics. PNAS 107(46):19627–19632

Denison JA, Wotshela L (2012) An overview of ındigenous, indigenised and contemporary water harvesting and conservation practices in South Africa. Irrig Drain 61(Suppl 2):7–23

Fairbourn ML (1975) Field evaluation of microwatershed and vertical mulch systems. In: Proceedings of the water harvesting symposium. U.S. Department of Agriculture ARS W-22

Falkenmark M, Fox P, Persson G, Rockström J (2001) Water harvesting for upgrading of rainfed agriculture: problem analysis and research needs. Stockholm International Water Institute (SIWI) Report 11, Stockholm

FAOSTAT (2012) Database. Food and Agriculture Organization (FAO), Rome. http://faostat.fao.org/. Accessed June 2012

Fleskens L, Stroosnijder L, Ouessar M, Graaff JD (2005) Evaluation of the on-site impact of water harvesting in southern Tunisia. J Arid Environ 62:613–630

Foster V, Briceño-Garmendia C (eds) (2010) Africa's infrastructure: a time for transformation. World Bank, Washington, DC

Fox P, Rockström J (2000) Water-harvesting for supplementary irrigation of cereal crops to overcome intra-seasonal dry-spells in the Sahel. Phys Chem Earth (B) 25(3):289–296

Fox P, Rockström J, Barron J (2005) Risk analysis and economic viability of water harvesting for supplemental irrigation in semi-arid Burkina Faso and Kenya. Agric Syst 83(3):231–250

Frasier GW (1975) Water harvesting for livestock, wildlife and domestic use. In: Proceedings of the water harvesting symposium. Phoenix, Arizona. U.S. Department of Agriculture ARS W-22

Frasier GW (1980) Harvesting water for agricultural, wildlife and domestic uses. J Soil Water Conserv 5(3):125–128

Frasier GW (1984) Water harvesting, including new techniques of maximizing rainfall use in semi arid areas. In: Davis TJ (ed) Proceedings of the fourth agricultural Sector Symposium. World Bank, Washington, DC

Frasier GW (1988) Technical, economic and social consideration of water harvesting and runoff farming. In: Arid Lands, Today and Tomorrow. Proceedings of an international research and development conference. Westview Press, Boulder, Colorado

Gansu Bureau of Water Resources (1995) Introduction to the experiment and extension on rainwater utilization in Gansu. In: Proceedings of the 1st China National Conference on Rainwater Utilization, Lanzhou, 1996. (in Chinese) reported in Zhu Qiang, Li Yuanhong, Rainwater Harvesting in the Loess Plateau of Gansu, China and Its Significance, Gansu Research Institute for Water Conservancy. Lanzhou 730000, China www.cpatsa.embrapa.br/catalogo/doc/wordwide/2_3_Zhu_Qiang.doc. Accessed 22 Aug 2015

Glendenning CJ, Vervoort RW (2010) Hydrological impacts of rainwater harvesting (RWH) in a case study catchment: the Arvari River, Rajasthan, India. Part 1: field-scale impacts. Agric Water Manag 98:331–342

Glendenning CJ, Vervoort RW (2011) Hydrological impacts of rainwater harvesting (RWH) in a case study catchment: the Arvari River, Rajasthan, India. Part 2: catchment-scale impacts. Agric Water Manag 98:715–730

Gnadlinger J (2000) Rainwater harvesting in rural areas. 2nd World Water Forum J March 16–22. http://www.irpaa.org.br/colheita/index.htm

GTZ (2010) Small-scale water Entrepreneurs in Nepal http://www.gtz.de/

Hafif B, Murni AM (2012) Small farm reservoir as a supplemental irrigation source for crops planted on marginal land. AGRIVITA J Agricult Sci 34(1):44–49

Hai MT (1998) Water harvesting – an illustrative manual for development of micro catchment techniques for crop production in dry areas. Technical handbook no. 16. Regional Land Management Unit, Swedish International Development Authority: Nairobi, Kenya

Hatibu N, Mahoo H (1999) Rainwater harvesting technologies for agricultural production: a case for Dodoma, Tanzania. In: Kaumbutho PG, Simalenga TE (eds) Conservation tillage with animal traction. A resource book of the Animal Traction Network for Eastern and Southern Africa (ATNESA). Harare. Zimbabwe, 173 p

Hatibu N, Mutabazi K, Senkondob EM, Msangi ASK (2006) Economics of rainwater harvesting for crop enterprises in semi-arid areas of East Africa. Agric Water Manag 80:74–86

Hobbs P, Sayre K, Gupta R (2008) The role of conservation agriculture in sustainable agriculture. Philos Trans R Soc B 363:543–555

Koohafkan P, Stewart BA (2008) Water and Cereals in Drylands. Earhtscan, London

Krishiworld (2006) Water resources. The Pulse of Indian Agriculture, Krishiworld. http://www.krishiworld.com/html/water_resources1.html

Kummu M, Ward PJ, de Moel H, Varis O (2010) Is physical water scarcity a new phenomenon? Global assessment of water shortage over the last two millennia. Environ Res Lett 5(034006):10 doi:1748-9326/10/034006

Lasage R, Verburg PH (2015) Evaluation of small scale water harvesting techniques for semi-arid environments. J Arid Environ 118:48–57

Lasage R, Aerts JCJH, Mutiso G-CM, de Vries AC (2008) Potential for community-based adaptations to droughts: Sand dams in Kitui Kenya. Phys Chem Earth 33:67–73

Lehmann C, Raquel Tsukada R, Lourete A (2010) Low-cost technologies towards achieving the millennium development goals: the case of rainwater harvesting. International Policy Centre for Inclusive Growth, Research Brief No. 12, Esplanada dos Ministérios, Bloco O, 7° andar 70052-900 Brasilia, DF – Brazil

Li JL (2005) High effective rainwater harvesting technique and benefit evaluation for semiarid area. Agric Res Arid Areas 23:230–232

Li XY, Gong JD (2002) Compacted microcatchments with local earth materials for rainwater harvesting in the semiarid region of China. J Hydrol 257:134–144

Li FR, Cook S, Geballe GT, Burch WR Jr (2000) Rainwater harvesting agriculture: an integrated system for water management on rainfed land in China's semiarid areas. Ambio 29:447–483

Majdoub R, Khlifi S, Salem AB, M'Sadak Y (2014) Impacts of the Meskat water-harvesting system on soil horizon thickness, organic matter, and canopy volume of olive tree in Tunisia. Desalin Water Treat 52(10–12):2157–2164

Mbilinyi BP, Tumbo SD, Mahoo HF, Senkondo EM, Hatibu N (2005) Indigenous knowledge as decision support tool in rainwater harvesting. Phys Chem Earth 30:792–798

Mekdaschi SR, Liniger H (2013) Water harvesting: guidelines to good practice. Centre for Development and Environment (CDE), Bern; Rainwater Harvesting Implementation Network (RAIN), Amsterdam; MetaMeta, Wageningen; The International Fund for Agricultural Development (IFAD), Rome

Mishra A, Adhikary AK, Panda SN (2009) Optimal size of auxiliary storage reservoir for rainwater harvesting and better crop planning in a minor irrigation project. Water Resour Manag 23:265–288

Missaoui H (1996) Soil and water conservation in Tunisia. In: Pereira LS et al (eds) Sustainability of irrigated agriculture. NATO ASI series, vol 312, pp 121–135

Molden D (2007) Water for food, water for life: comprehensive assessment of water management in agriculture. IWMI, Colombo

Mugabe FT (2004) Evaluation of the benefits of infiltration puts on soil moisture in semi arid Zimbabwe. J Agron 3:188–190

MupangwaW, Love D, Twomlow S (2006) Soil-water conservation and rainwater harvesting strategies in the semi-arid Mzingwane Catchment, Limpopo Basin, Zimbabwe. Phys Chem Earth 31:893–900

Mutekwa V, Kusangaya S (2006) Contribution of rainwater harvesting technologies to rural livelihoods in Zimbabwe: the case of Ngundu ward in Chivi District. Water SA 32(3) July 2006 African Journals Online (AJOL)

National Academy of Sciences (1974) More water for arid lands. Promising Technologies and Research Opportunities, Washington, DC

Ngigi SN (2003) What is the limit of up-scaling rainwater harvesting in a river basin? Phys Chem Earth 28:943–956

Ngigi SN, Savanije HHG, Rockström J, de Vries FWTP (2005) Agrohydrological evaluation of on farm rainwater storage systems for supplemental irrigation in Laikipia district, Kenya. Agric Water Manag 73:21–41

OECD (Organisation for Economic Co-operation and Development) (2012) Infrastructure to 2030: telecom, land, transport, water and electricity. 2006. OECD Publishing, Paris

O'Hogain S, McCarton L, McIntyre N, Pender J, Reid J (2011) Physicochemical and microbiological quality of harvested rainwater from an agricultural installation in Ireland. Water Environ J 26:1–6

Oron G, Ben-Asher J, Issar A, Boers T (1983) Economic evaluation of water harvesting in microcatchments. Water Resour Res 19:1099–1105

Ouessar M, Sghaier M, Mahdhi N, Abdelli F, De Graaff J, Chaieb H, Yahyaoui H, Gabriels D (2004) An integrated approach for impact assessment of water harvesting techniques in dry areas: the case of Oued Oum Zessar watershed (Tunisia). Environ Monit Assess 99:127–140

Oweis T, Hachum A (2006) Water harvesting and supplemental irrigation for improved water productivity of dry farming systems in West Asia and North Africa. Agric Water Manag 80:57–73

Oweis T, Hachum A (2009) Water harvesting for improved rainfed agriculture in the dry environments. In: Wani SP et al (eds) Rainfed agriculture: unlocking the potential. CAB International, Wallingford, p 164

Oweis T, Taimeh AY (1996) Evaluation of a small basin water-harvesting system in the arid region of Jordan. Water Resour Manag 10:21–34

Oweis T, Hachum A, Kijne J (1999) Water harvesting and supplemental irrigation for improved water use efficiency in the dry areas. SWIM Paper 7, International Water Management Institute, Colombo, Sri Lanka

Oweis T, Prinz D, Hachum A (2001) Water harvesting: indigenous knowledge for the future of the drier environments. ICARDA, Aleppo 40 p

Oweis T, Prinz D, Hachum A (2012). Rainwater harvesting for agriculture in the dry areas. CRC Press, Taylor and Francis Group, Leiden, 366 p

Pacey A, Cullis A (1986) Rainwater harvesting: the collection of rainfall and runoff in rural areas. Intermediate Technology Publications, London

Palmback J (2004) Traditional water harvesting. University of Wisconsin – Eau Claire, USA. http://academic.evergreen.edu/g/grossmaz/PALMBAJP/

Pangare G (1992) Traditional water harvesting on way out. Econ Polit Wkly 27:505–507

Pereira LS, Cordery I, Iacovides I (2002) Coping with water scarcity. International Hydrological Programme (IHP)-VI. Technical documents in hydrology no. 58 UNESCO, Paris, 272 p

Pratt RC (1980) Water harvesting: an alternative irrigation method for desert gardeners. Desert Plants 2:131–134

Prinz D (1996). Water Harvesting: Past and Future. In: Pereira LS (ed) Sustainability of irrigated agriculture. In: Proceedings of the NATO advanced research workshop, Vimeiro, 21–26.03.1994, Balkema, Rotterdam, pp 135–144

Prinz D (2001) Water harvesting for afforestation in dry areas. In: Proceedings of the 10th international conference on rainwater catchment systems, Mannheim, 10–14 September 2001, pp 195–198

Prinz D (2002) The role of water harvesting in alleviating water scarcity in arid areas. Keynote lecture, Proceedings, International conference on water resources management in arid regions, 23–27 March, 2002, Kuwait Institute for Scientific Research, Kuwait, vol III, pp 107–122

Prinz D, Singh AK (2000) Technological potential for improvements of Water Harvesting Study for the World Commission on Dams, Cape Town, South Africa. Report: Dams and Development

Qadir M, Sharma BR, Bruggeman A, Choukr-Allah R, Karajeh F (2007) Non-conventional water resources and opportunities for water augmentation to achieve food security in water scarce countries. Agric Water Manag 87:2–22

RAIN Foundation (2014) Increasing access to water: scaling up rain water harvesting practices through microfinance. Barentszplein 7 1013 NJ Amsterdam, the Netherlands. http://wwwrainfoundationorg/wp-content/uploads/2015/05/140307-2-pager-RHW-and-MF_DEFpdf. Accessed 30 July 2015

Rainwater Harvesting (2006) Centre for Science and Environment http://www.rainwaterharvesting.org/methods/methods.htm

Reig C, Mulder P, Begemann L (1988) Water harvesting for plant production. The World Bank, Washington, DC

Renner HF, Frasier G (1995a) Microcatchment water harvesting for agricultural production: Part I: physical and technical considerations. Rangelands 17:72–78

Renner HF, Frasier G (1995b) Microcatchment water harvesting for agricultural production: Part II: Socio-Economic Considerations. Rangelands 17(3):79–82

Rocheteau D, Weber F, and Field-Juma A (1988) Agroforestry in Dryland Africa. International Council for Research in Agroforestry (ICRAF), Nairobi, Kenya

Rockström J (2000) Water resources management in smallholder farms in Eastern and Southern Africa: an overview. Phys Chem Earth Part B Hydrol Oceans Atmos 25(3):279–288
Rockström J (2002) Potential of rainwater harvesting to reduce pressure on freshwater resources international water conference, dialogue on water, food and environment Hanoi, Vietnam, October 14–16, 2002
Rockström J, de Rouw A (1997) Water, nutrients, and slope position in On-farm pearl millet cultivation in the Sahel. Plant Soil 195:311–327
Rockström J, Folke C, Gordon L, Hatibu N, Jewitt G, deVries FP, Rwehumbiza F, Sally H, Savenije H, Schulze R (2004) A watershed approach to upgrade rainfed agriculture in waters scarce regions through Water System Innovations: an integrated research initiative on water for food and rural livelihoods in balance with ecosystem functions. Phys Chem Earth 29:1109–1118
Rockström J, Karlberg L, Wani SP, Barron J, Hatibu N, Oweis T, Bruggeman A, Farahani J, Qiang Z (2010) Managing water in rainfed agricultured the need for a paradigm shift. Agric Water Manag 97:543–550
Rodriguez DJ, van den Berg C, McMahon A (2012) Water papers, investing in water infrastructure: capital, operations and maintenance. The World Bank 1818 H Street NW Washington DC 20433. Paper is also available at http://www.worldbank.org/water
Samra JS (2005) Policy and institutional processes of participatory watershed management in India: Past lessons learnt and future strategies. In: Sharma BR, Samra JS, Scott CA, Wani SP (eds) Watershed Management Challenges: Improving productivity, resources and livelihoods. International Water Management Institute, Colombo, pp. 116–128
Schiettecatte W, Ouessar M, Gabriels D, Tanghe S, Heirman S, Abdelli F (2005) Impact of water harvesting techniques on soil and water conservation: a case study on a micro catchment in Southeastern Tunisia. J Arid Environ 61:297–313
Schwilch G, Hessel R, Verzandvoort S (eds) (2012) Desire for Greener Land. Options for sustainable land management in drylands. University of Bern – CDE, Alterra – Wageningen UR, ISRIC – World Soil Information and CTA – Technical Centre for Agricultural and Rural Cooperation, Bern/Wageningen
Shanan L, Schick AP (1980) A hydrological model for the Negev Desert Highlands: effects of infiltration, runoff, and ancient agriculture. Hydrol Sci Bull 25:269–282
SIWI (2001) Water harvesting for upgrading of rainfed agriculture. Problem analysis and research needs, SIWI report no 11. SIWI, Stockholm, p. 97
Stern PH (1979) Small–scale irrigation. Intermediate Technology Publication, London
Tabor JA (1995) Improving crop yields in the Sahel by means of water harvesting. J Arid Environ 30:83–106
Tauer W, Humborg G (1992) Runoff irrigation in the Sahel zone: remote sensing and geographic information systems for determining potential sites. Verlag Josef Margraf, Weikersheim
Tian Y, Su D, Li F, Li X (2003) Effect of rainwater harvesting with ridge and furrow on yield of potato in semiarid areas. Field Crop Res 84:385–391
Tubeileh A, Bruggeman A, Turkelboom F (2009) Effect of water harvesting on growth of young olive trees in degraded Syrian dryland. Environ Dev Sustain 11:1073–1090
Tumbo SD, Mutabazi KD, Byakugila MM, Mahoo HF (2011) An empirical framework for scaling-out of water system innovations: lessons from diffusion of water system innovations in the Makanya catchment in Northern Tanzania. Agric Water Manag 98:1761–1773
UN (2011) Global drylands: a UN system-wide response. Environment Management Group
UN HABITAT (2015) UN-Habitat global country activities report: 2015 – increasing synergy for greater national ownership. UN
UNEP-SEI (2009) Rainwater harvesting: a lifeline for human wellbeing, Stockholm Environment Institute and United Nations Environment Programme, 67 p. ISBN 978-92-807-3019-7
Vohland K, Barry B (2009) A review of in situ rainwater harvesting (RWH) practices modifying landscape functions in African drylands. Agric Ecosyst Environ 131:119–127
Walker S, Tsubo M, Hensley M (2005) Quantifying risk for water harvesting under semi-arid conditions. Part II. Crop yield simulation. Agric Water Manag 76:94–107

Wang Q, Zhang E, Li F, Li F (2008) Runoff efficiency and the technique of microwater harvesting with ridges and furrows, for potato production in semi-arid areas. Water Resour Manag 22:1431–1443

Wei H, Li JL, Liang TG (2005) Study on the estimation of precipitation resources for rainwater harvesting agriculture in semiarid land of China. Agric Water Manag 71:33–45

WILDCHINA (ed) (2012) Cross border journey. Trip notes from Vietnam and China. http://www.sswm.info/content/bunds

WOCAT (2007) In: Liniger H, Critchley W (eds) Where the land is greener– case studies and analysis of soil and water conservation initiatives worldwide

WOCAT (World Overview of Conservation Approaches and Technologies) (2012) WOCAT Database: Technologies. http://cdewocat.unibe.ch/wocatQT/index.php

Yazar A, Kuzucu M, Celik I, Sezen SM, Jacobsen S-E (2014) Water harvesting for improved water productivity in dry environments of the Mediterranean region. Case study: Pistachio in Turkey. J Agron Crop Sci 200(2014):361–370

Yepes T (2008) Investment needs for infrastructure in developing countries 2008–2015. Latin America and Caribbean Region, World Bank. Unpublished

Weed Management in Dryland Cropping Systems

Michael Walsh

1 Introduction

Approximately 40 % of the world's total land area is generally described as regions in which crop production is constrained by low annual rainfall (<500 mm) (ACIAR 2002; FAO 2000; Oram 1980). FAO defines drylands as areas with a growing period of 1–179 days (FAO 2000). The United Nations Convention to Combat Desertification (UNCCD 2015) defines drylands as areas where the ratio of precipitation to potential evapotranspiration ranges from 0.05 to 0.65. Regardless of the definition, crop production within these regions is primarily restricted by rainfall. Although rainfall is the primary constraint to cropping in dryland areas soil fertility, land degradation, poor resources and subsistence farming limit crop production to 4 % of the world's dryland regions (Koohafkan and Stewart 2012).

Dryland production regions in Mediterranean type climates of hot dry summers and coil moist winter-spring growing seasons are characterised by cereal dominated cropping systems (Fig. 1). Across these regions consistently the dominating influence on crop production is climate, specifically rainfall amount and frequency. Dryland crop production is practised in a moisture limiting environment and weed control practice must fit within the context of soil water conservation. The increasing need to conserve soil moisture along with improving soil structure and nutrient retention continues to drive the widespread adoption of conservation cropping practices based on residue retention and restricted cultivation (FAO 2015; Kassam et al. 2012; Llewellyn et al. 2012). The adoption of these systems has been facilitated by the availability of highly effective in-crop selective herbicides.

Frequently soils of dryland cropping regions are unproductive and do not support the full crop production potential thereby creating the opportunity for weed invasion

---

M. Walsh (✉)
Plant Breeding Institute, University of Sydney, Narrabri, NSW 2390, Australia
e-mail: m.j.walsh@sydney.edu.au

M. Farooq, K.H.M. Siddique (eds.), *Innovations in Dryland Agriculture*,
DOI 10.1007/978-3-319-47928-6_4

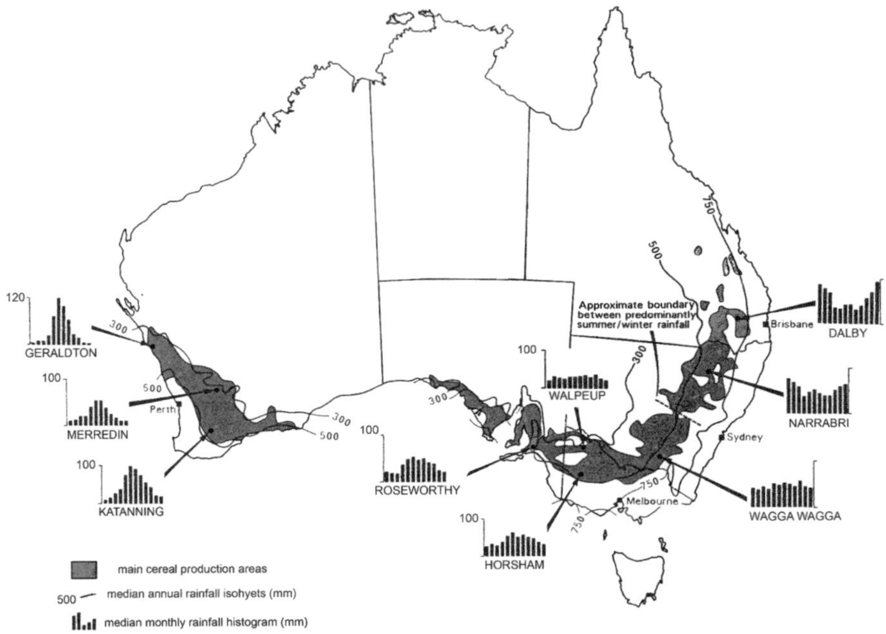

Fig. 1 Variable long-term average monthly rainfall distributions for nine centres across the Australian dryland cropping region (Adapted from Cramb 2000)

and competition. While there is a diversity of soil types across the dryland crop production areas, characteristically soils are low in natural fertility, organic matter and water-holding capacity. Across large areas, these issues are combined with high salinity and soil pH extremes (alkalinity or acidity) creating hostile soil environments for crop production. Consequently, the average wheat yields across the grain belt are only 2 t ha^{-1}. Erosion-prone soils have been a further barrier to cropping, however, the adoption of reduced tillage crop establishment systems and stubble retention have combined to minimise the soil erosion problem. Fertilizer inputs are an annual necessity and soil physical and chemical amelioration are frequently required to allow the growth of productive and weed competitive crops.

Increasing global population growth is driving demand for greater food production. However, for crop producers in dryland cropping regions particularly in developed and developing countries, low yield potentials, decreasing produce prices and increasing input costs are severely impacting the viability of crop production. Consequently, these enterprises are increasing in size as financial pressures force growers to rely on the economies of scale to remain viable. These production systems are now highly mechanised to address the need for timeliness of seeding, weed control and harvest operations in typically short growing seasons.

Current and future cropping practices will more frequently lie within the conservation production approach, based on reduced tillage and crop residue retention.

The increasing need to improve soil structure, retain nutrients, and conserve soil moisture will continue to drive the widespread adoption of conservation cropping practices (FAO 2015; Kassam et al. 2012; Llewellyn et al. 2012). These systems are now proven to be more robust and sustainable, providing some stability in spite of highly variable climatic conditions. Thus, even though conservation cropping has led to herbicide reliance and subsequently herbicide resistance, this approach has provided substantial gains in productivity and reverting to less conservative systems for the sake of weed control is no longer an option.

Australia boasts arguably the most effective dryland crop production systems in the world. Despite very low and infrequent rainfall events and low natural soil fertility levels, Australian cropping systems have evolved, adapted and even flourished. Although cropping systems are generally successful, there have at times been major production constraints that needed addressing. Recently the most significant of these has been evolution of herbicide resistance and the need to weed control. Faced with the most dramatic and extensive herbicide resistance problems in the world, the Australian weed control industry has been forced to develop strategies that mitigate and mange this problem. In this chapter, recent developments in weed management in Australia are highlighted, which may serve as a warning and an opportunity for weed managers in dryland production regions of the world.

2 Weed Control in Australian Dryland Cropping Systems

There are at present six recognised opportunities annually for targeting weeds in dryland crop production systems (Fig. 1). The attributes of these opportunities as they have been implemented in Australian dryland production systems are discussed below in detail.

2.1 *Out of Season Weed Control*

Successful crop production in water limited environments is fundamentally dependant on the effective supply of soil water for crop plant growth. Wherever possible rainfall including out of season (summer) rainfall should be conserved, particularly in dryland cropping regions where summer rainfall dominates (Fig. 1). Summer rain stimulates the emergence of annual weeds (e.g. *Conyza* spp., *Echinochloa* spp., *Chloris virgata* etc.) and although these weeds do not interfere directly with crop growth they must be controlled to prevent the use of soil moisture and nutrient reserves for subsequent crop growth (Hunt et al. 2013) (Fig. 2). Prior to the introduction of non-selective herbicides (e.g. glyphosate, paraquat/diquat) in the 1970s, cultivation was used to control any summer rainfall stimulated weed emergence, as well as for seedbed preparation. Summer weed control is now almost exclusively

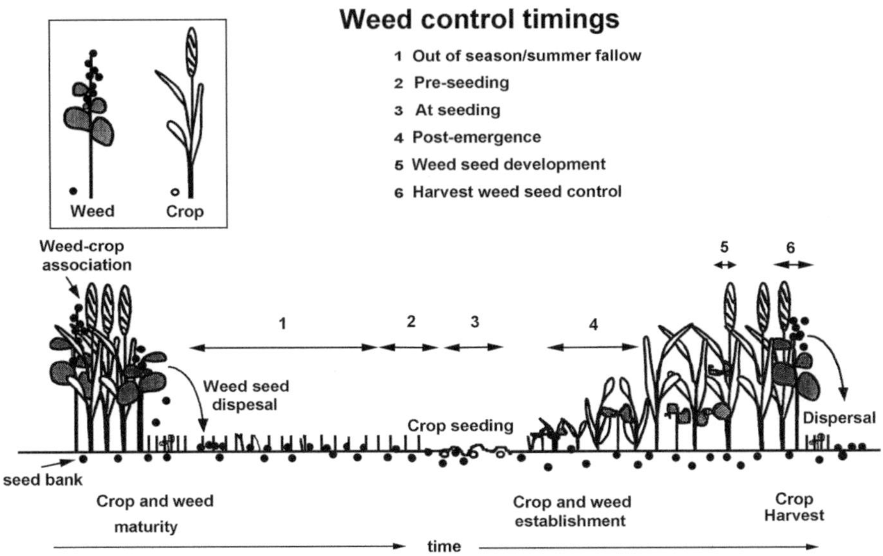

Fig. 2 Timing of routine weed control practices in annual dryland crop production systems (Adapted from Cousens and Mortimer 1995)

achieved through the use of non-selective herbicides applied as whole paddock treatments. Frequently though summer rainfall events are light and sporadic resulting reduced emergence of weed populations. In these instances, targeted spray applications using boom mounted weed detection systems (e.g. Weed Seeker and WeedIT) are often used in place of whole field spray treatments (Blackshaw et al. 1998; López-Granados 2011). This technology comprises optical sensors that turn on nozzles only when green weeds are detected, greatly reducing total herbicide use. The choice between whole paddock treatments and spot spray technology typically depends on the amount of out of season rainfall and the size of subsequent weed infestations (Keller et al. 2014).

Elevated soil moisture and soil temperature levels support the rapid establishment and growth of summer weeds. However, summer herbicidal weed control is hampered by the prevalence of hot, dry conditions that reduce treatment efficacy by interfering with the application, uptake and translocation of herbicides to plant target sites (Jursík et al. 2011; Lubbers et al. 2007; Mahan et al. 2004). Therefore, in large cropping programs there are significant constraints on the effectiveness of blanket herbicide treatments in summer fallow situations. For efficacy as well as economic reasons, there is strong demand for the introduction and use of weed detection and identification systems and it is likely that spot spray treatments will in time become commonplace.

2.2 Pre-seeding Weed Control

Non-selective herbicides. The introduction of non-selective herbicides, paraquat/diquat and glyphosate, in the 1970's created the opportunity for effective pre-seeding weed control eliminating the need for pre-seeding cultivations. These herbicides in particular greatly facilitated the rapid adoption of conservation cropping systems based on reduced tillage and residue retention (Derpsch et al. 2010; Llewellyn et al. 2012). In reduced tillage cropping systems, the only soil disturbance occurs at seeding where knife point or disc seeding systems provide some cultivation for seed placement and fertilizer incorporation. Presently about 80 % of the Australian crop is seeded in a one pass reduced tillage seeding system (Llewellyn et al. 2012). The rapid and widespread adoption of reduced tillage seeding systems has revolutionised global dryland cropping systems with large production benefits realised from the conservation of soil moisture and nutrients as well as from the timeliness of crop seeding.

2.3 At Seeding Weed Control

Pre-emergence herbicides. In Australia and elsewhere the widely adopted 'no-till' seeding systems utilise knife point seeding tynes with trailing press wheels that accurately place crop seed in a protected environment at the bottom of a 15–20 cm furrow. During seeding the knife point provides soil disturbance to depths of up to 30 cm allowing the deep placement of fertilizer below crop seed (Derpsch et al. 2014). The trailing press wheels compact the bottom of the furrow around the seed to improve soil-seed contact. Post-seeding rainfall is concentrated around the seed in the bottom furrow creating an ideal micro-climate for the germination and seedling establishment. There is considerable "soil-throw" created during furrow formation with knife-point seeding tynes. This dynamic of some soil disturbance in combination with furrow formation creates an ideal environment for crop seedling establishment as well as the effective incorporation of pre-emergence herbicides (Chauhan et al. 2006a, b). The removal of tillage for weed control at seeding in combination with the widespread occurrence of resistance evolution to post emergence herbicides has placed considerable emphasis on achieving effective weed control with pre-emergence applied soil residual herbicides. This focus on pre-emergence herbicides for weed control has however, come at a cost and is now inevitably leading to the evolution of resistance in weed populations (Boutsalis et al. 2012; Owen et al. 2014). Although pre-emergence herbicide resistance levels are not yet at those for post-emergence herbicides they are impacting the ability of producers to reliably control in-crop weeds. Residual herbicides are most effective when they are soil incorporated and although this can be achieved by rainfall in dryland environments incorporation is most reliably achieved through cultivation (Chauhan et al. 2006b). This need for effectively incorporated pre-emergence herbicide treatments in Australia has prevented the adoption of more conservative disc seed systems that is potentially limiting the potential benefits of further reductions in soil disturbance (D'Emden et al. 2008).

Crop competition. Although crop competition occurs throughout the growing season, predominantly, the opportunities for enhancing the competitive effects of crops on weed populations are implemented at crop seeding. Enhancing crop competition is aimed at maximising utilization of crop resource to the detriment of weed populations. In the absence of control, weeds compete for the essential primary resources of nutrients, water, and sunlight, reducing their availability for wheat crops (Roush and Radosevich 1985). Crop cultivar, seed size, seeding rate, row spacing, row orientation, and fertiliser placement can all be adjusted to ensure establishing crop seedlings have a competitive advantage over weeds (Andrew et al. 2015; Blackshaw 2004; Borger et al. 2009; Lemerle et al. 2004; Lemerle et al. 2001; Lutman et al. 2013; Yenish and Young 2004; Zerner et al. 2008). Enhanced crop competition offers the potential for substantial weed control advantages and importantly yield increases. In Australia, increased crop competition through higher wheat plant densities (150–200 plants/m^2) has consistently resulted in substantial (> 50 %) reductions in plant growth and seed production of the dominant weed species, annual ryegrass (Lemerle et al. 2004), wild radish (Walsh and Minkey 2006), wild oats (Radford et al. 1980) and brome grass (Gill et al. 1987). Typically in Australia, enhanced crop competition thorough increase in plant densities has a positive impact on grain yield without compromising grain quality (Anderson et al. 2004).

Enhanced crop competition cannot be considered a standalone weed control treatment but when combined with other weed control practices, the additional impact on weed populations can be critical for weed control. For example, enhanced wheat crop competition will routinely increase the efficacy of selective herbicides in controlling crop-weed populations (Kim et al. 2002). Importantly, this competition can lead to the control of weed populations that are resistant to the applied herbicide. For example, a 2,4-D resistant wild radish population was controlled when 2,4-D was applied at the recommended rate to resistant plants present within a competitive wheat crop (Walsh et al. 2009). As well as complementing herbicide activity, enhanced wheat competition will likely improve the efficacy of harvest weed seed control (HWSC) strategies. Annual weed species infesting global wheat production systems are typically not shade tolerant (Gommers et al. 2013) and as indicated from competition studies, grow poorly when shaded (Zerner et al. 2008). When competing with wheat for light, the likely response for shade intolerant weed species is a more upright growth habit (Morgan et al. 2002; Vandenbussche et al. 2005). This erect growth habit will undoubtedly lead to higher proportions of total seed production being located above harvester cutting height and increasing subsequent exposure to HWSC methods. Clearly then, the combined benefits of higher yield potential and enhanced weed control ensure that agronomic weed management should be standard practice throughout global wheat production systems.

2.4 Post Emergence Weed Control

The early crop establishment phase is the most critical for weed control in annual crops as it is during this period that weed competition causes maximum yield loss. It is also the period during which weed seedlings are generally most susceptible to weed control strategies. Consequently, for many decades early post-emergence weed control has been the focus of global herbicide companies for the development for in-crop selective weed control treatments. However, the resultant highly effective herbicidal weed control and then subsequent resistance evolution has resulted now in the severe circumstance of multiple herbicide resistance and limited in-crop weed control options. The Australian experience should serve as an example of how herbicides alone as not the solution for long term effective weed control in dryland cropping systems.

In Australian dryland cropping systems, there are remarkably few problematic weed species that dominate nationally. Annual ryegrass is by far the most important (Alemseged et al. 2001; Llewellyn et al. 2015) with this competitive annual grass abundant throughout the western and eastern Australian winter cropping regions. Initially this species was introduced and nurtured across these regions as a highly productive pasture plant for sheep production (Kloot 1983). However, the change in emphasis from livestock to crop production systems in the 1990's meant that endemic and well adapted annual ryegrass populations were now a weed in cereal dominated crop production systems. Although not previously established in the sub-tropical northern wheat belt zone, annual ryegrass is now encroaching on this region as well. Other major grass weeds, important across much of the Australian grain belt, are wild oats (principally *Avena fatua* and *A. ludoviciana*) (Martin et al. 1988; Paterson 1976) and brome grass (*B. diandrus* and *B. rigidus*). The principal broadleaf weed of Australian cropping is the highly competitive wild radish (*Raphanus raphanistrum*) (Alemseged et al. 2001; Cheam and Code 1995), a serious weed in Western Australia and a localised problem throughout other regions of the Australian wheat belt (Walsh and Minkey 2006). These weed species infest cropping systems spread across a diverse range of environments and production practices, and provide clear evidence of their genetic diversity resulting in adaptation and importantly their evolutionary potential (Kercher and Conner 1996; Kloot 1983; Kon and Blacklow 1989; Menchari et al. 2007).

The combination of huge numbers of genetically diverse annual weed species present in crop-production systems with minimal crop diversity, and strong herbicide reliance has provided a remarkable example of evolutionary selection for herbicide resistance. Although herbicide-resistant weed populations have evolved in many parts of the world (Heap 2015), nowhere else has this evolution been more dramatic than across the Australian dryland crop production region. In Australia, there has been widespread and frequent evolution of herbicide-resistant populations in several weed species including annual ryegrass (Boutsalis et al. 2012; Broster and Pratley 2006; Owen et al. 2014), wild radish (Owen et al. 2015; Walsh et al. 2007) wild oats (Mansooji et al. 1992; Owen et al. 2012b), barley grass (Owen et al.

2012a; Powles 1986; Tucker and Powles 1991), and brome grass (Owen et al. 2012b). Resistance in wild radish and annual ryegrass populations continues to increase in frequency and distribution where in the majority of populations, resistance extends simultaneously across many herbicide chemistries (multiple herbicide resistance) (Owen et al. 2014, 2015). The management of multi-resistant populations of these species is now a major focus across most cropping regions of Australia.

The consequence of herbicide resistance is that once highly effective herbicides are no longer of any use. The widespread occurrence of multiple resistant weed populations and the subsequent loss of essential herbicide resources, threatens the sustainability of conservation agriculture. Globally, the herbicide industry has shifted focus from herbicide discovery programs to gene trait development. Combined with the increasing difficulty in identification and development of new herbicide molecules, the introduction of new herbicide modes of action has ceased since the 1990s. Since that time, introduced herbicides have all focussed on existing modes of action. Thus in the absence of new herbicides, producers for the foreseeable future will be relying on existing herbicide technologies. With herbicide resistance continuing to erode these herbicide resources, Australian crop producers are losing the ability to effectively control weeds in conservation cropping systems. Thus the sustainability of the current and any future herbicide resources is dependent on the development of alternate, highly effective weed-control technologies.

3 Weed Seed Development

Bitter experience with herbicide resistant annual weed populations has resulted in the widespread acceptance that it is necessary to target weed seed production wherever feasible. The recognition that, the major proportion of an in-crop annual ryegrass population results from the previous season's seed production, (Gill and Holmes 1997; McGowan 1970; Monaghan 1980; Pearce and Holmes 1976; Reeves and Smith 1975) has led to a focus on minimising weed seed production. Crop-topping, where non-selective herbicides (glyphosate or paraquat) applied over a crop to target weed seed production, was developed following the successful use of this practice in pasture phases (spray-topping) to control weed seed production (Gill and Holmes 1997; Mayfield and Presser 1998). In situations where the crop has matured while weed seeds remain immature there is a period of time, 'window of opportunity' during which weed seeds can effectively be targeted without incurring crop yield loss. Weed seed production is frequently reduced by up to 90 % for a range of weed species including annual ryegrass (Gill and Holmes 1997; Steadman et al. 2006), wild radish (Walsh and Powles 2009), *Vulpia bromoides* (Leys et al. 1991), barley grass (Powles 1986) and brome grass (*Bromus* spp.) (Dowling and Nicol 1993). However, the reproductive phase of annual ryegrass is strongly influenced by seasonal conditions and often the period of seed maturation overlaps that of the crop (Aitken 1966). Frequently then the timing of crop-topping treatments is a trade-off between acceptable levels of yield loss and effective weed seed control. A consequence is that there frequently is little or no 'window of opportunity' for the application of effective crop-topping treatments. Despite this crop-topping is routinely used in non-cereal crops (e.g. lupins, peas, chickpeas) where some yield loss is often tolerated in preference to maximising weed seed control.

4 Harvest Weed Seed Control

The biological attribute (weakness) of seed retention at maturity in annual ryegrass, wild radish and other annual weed species, means that at crop maturity seed heads remain intact and at a height that enables weed seeds to be collected (harvested) during grain crop harvest (Fig. 3). For example, in field crops a large proportion (~80 %) of total annual ryegrass seed production can be collected during a typical commercial grain harvest (Blanco-Moreno et al. 2004; Walsh and Powles 2014). Grain harvest presents an opportunity to target weed seed production, thereby minimizing replenishment/increase of the weed seed bank. HWSC systems have been developed, principally in Australia, that target and destroy weed seeds during commercial grain crop harvest preventing inputs into the seed bank (Walsh et al. 2013) (Fig. 2).

A number of HWSC systems have been developed for the specific purpose of targeting weed seed during crop harvest, restricting contributions to the seed bank thereby reducing subsequent in-crop interference (Walsh and Newman 2007; Walsh

Fig. 3 Upright and intact annual ryegrass seed heads in mature cereal crop

et al., 2013). With the vast majority of weed seeds exiting in the chaff fraction during harvest a number of commercially available systems are now available that target this weed seed bearing fraction. These include, chaff collection and subsequent burning, grazing or mulching (chaff cart), concentration in a narrow windrow with straw residues for subsequent burning (narrow windrow burning), concentration of chaff into narrow rows (chaff lining), chaff collected and baled along with straw residues (Bale Direct System) and mechanical destruction during harvest (Harrington Seed Destructor)

As HWSC systems target the weed seed bearing chaff fraction, these will all be similarly effective in in targeting weed seed during crop harvest. With the assumption of efficient harvester setup and operation, weed seeds will predominantly exit the harvester in the chaff fraction which logically then is the focus for HWSC systems. An extensive evaluation of three HWSC systems (chaff cart, narrow windrow burn and HSD) for their efficacy on annual ryegrass was conducted across the Australian dryland cropping region in 2010 and 2011. These treatments, imposed during harvest, were assessed for their efficacy on subsequent autumn emergence of annual ryegrass. Across 25 sites chaff cart, narrow windrow burning and HSD treatments similarly reduced ($p < 0.05$) annual ryegrass emergence in comparison to the control treatment (Table 1). In this study the overall impact of HWSC treatments was varied and likely influenced by pre-harvest seed shed as well as pre-existing seed banks and early season rainfall. When averaged across all 25 sites, HWSC treatments resulted in a 56 % reduction in subsequent autumn annual of ryegrass emergence. However, there was considerable variation in HWSC efficacy between sites with large reductions in annual ryegrass emergence of up to 90 % (Arthurton) contrasting with reductions as low as 29 % (Rand) (Table 1). Thus although HWSC systems will be similarly effective on annual ryegrass populations, large variations in population responses can be expected due to seasonal conditions and background seedbank levels

HWSC is now established as an effective weed control practice with Australian crop producers. It is estimated that currently almost one-third of Australian growers are routinely using some form of HWSC to target their crop weed problems. However, although these systems have proven their efficacy on annual ryegrass and wild radish (Walsh et al. 2013) their efficacy on the other dominant weed species of Australian cropping, wild oats and brome grass, may be limited by poor seed retention at crop harvest (Walsh and Powles 2014). Therefore, given that HWSC is now a routine form of weed control the challenge for researchers and the industry is to increase the efficacy of these systems on other weed species.

5 Weed Management in Future Crop Production Systems

The opportunity for substantial cost savings combined with the potential for introducing novel weed control technologies is driving the demand for site-specific weed management control. However, this approach to weed control requires suitable weed detection and identification technology that currently is not commercially

Table 1 Autumn annual ryegrass plant densities for untreated control and HWSC treatments at 25 sites in Western (2010) and Eastern (2011) Australia (Walsh et al. 2014)

	Control	Chaff cart	Windrow burn	Harrington Seed Destructor	LSD
Location	Annual ryegrass (plants m^{-2})				(P 0.05)
Binnu	21	12	7	11	5
Tenindewa	52	–[a]	17	21	31
Mingenew	33	17	23	18	5
Buntine	222	82	58	73	54
Wongan Hills	24	9	9	12	10
Wyalkatchem	106	54	45	52	30
Corrigin	11	4	3	3	5
Kondinin	208	98	133	139	66
Holt Rock	277	111	145	102	38
Kojonup	254	81	–	64	113
Broomehill	20	9	7	10	7
Minnipa	329	259	174	245	74
Minnipa	209	–	84	62	128
Cummins	425	145	144	162	236
Bute	89	43	46	44	25
Arthurton	12	3	1	2	9
Pinnaroo	174	93	55	55	80
Underbool	0.2	–	0.1	0.1	0.1
Dimboola	14	5	7	8	5
Dookie	4619	–	1663	2079	2027
Rand	238	161	170	148	61
Old Junee	2	–	1	1	1
Harden	1117	–	726	726	378
Peak Hill	358	–	179	158	65
Coonamble	208	–	98	109	77

[a]treatment not included at this site

available for in-crop use. The options currently available are based on spectral reflectance that can reasonably accurately detect green leaf material (Scotford and Miller 2005). Obviously, these systems are not suitable for in-crop use but have been successful used for many years to control weeds in fallows phases. Another limitation to the adoption of site-specific weed management is that this approach only becomes economically viable once low weed densities (< 1.0 plants m-2) have been achieved. However, a strong focus on weed control efficacy driven by diminishing herbicide resources is helping to deliver lower than ever weed population densities Australian in dryland cropping systems. For example, in-crop wild radish populations across many areas of the Western Australian wheat belt are well below one plant 10 m^{-2} with many farmers opting to hand weed areas in preference to the cost of applying herbicides. Thus for these growers the demand is now for effective site-specific weed management systems.

Given the potential cost savings of site-specific versus whole field weed control there is now considerable effort being focussed on the development of weed detection and mapping systems. The capacity to accurately detect and map low-density weed populations within a crop creates the opportunity for the use of range of alternate control tactics. In low weed density situations, because of the small areas involved, and therefore, the reduced impact on crop yields, detected weeds can be aggressively targeted. For example, non-selective herbicides, tillage treatments, even hand weeding all become viable options. Additionally, the ability to strategically target low weed densities creates the potential for the introduction of more novel and unique weed control technologies such as electrocution (Vigneault et al. 1990), flaming (Hoyle et al. 2012), microwaves (Brodie et al. 2012), infrared (Ascard 1998) and lasers (Marx et al. 2012). There is now considerable investment in the area of weed identification and mapping occurring on many fronts ranging from vehicle mounted to UAV and even satellite systems.

6 Conclusion

Many tools are currently available for weed management in dryland cropping systems. Additionally, there opportunities for the development of new options and techniques particular in the advent of automatipsite-specific However, all tools have both costs and benefits, and it is the balance that will determine their use within particular farming systems. Herbicides are viewed as effective, relatively cheap tools compared with some other practices, and while their benefits are significant, the failure to integrate herbicides with other methods has led to their downfall in many areas due to herbicide resistance.

Considerable opportunities exist for designing cropping systems that exhibit IWM and are economically and biologically sustainable. These will become increasingly important as herbicide resistance becomes more widespread. The length of the system and the types of crops and control methods are generally more extensive than currently used by many farmers. For example, the ability to grow niche crops such as durum wheat requires high fertility systems and very low levels of weeds if grain protein and yield are to be maximised. Practices such as green manuring of legume crops, grown specifically to harvest nitrogen in the year preceding the durum crop, provide an excellent way to achieve high levels of weed control and the quality standards required for the crop.

Do farmers practice IWM? In the main, herbicide technology dominates weed management strategies, although many farmers implement, deliberately or otherwise, some IWM principles. However, only when confronted with problems, such as herbicide resistance or rapidly increasing populations of recalcitrant weeds, is serious attention given to alternative weed management tools.

Even the best made systems apply selection pressure to weeds to evolve in response to the control strategies imposed. For this reason, and because of the natural human desire for simplicity, weeds will continue to persist and evolve to prevail in cropping systems throughout Australia. Dynamic and integrated strategies are needed to achieve a balance between crop and economic productivity and weed infestation.

References

ACIAR (2002) Improving water-use efficiency in dryland cropping. Australian Centre for International Agricultural Research, Canberra

Aitken Y (1966) The flowering responses of crop and pasture species in Australia I. Factors affecting development in the field of *Lolium* species (*L. rigidum* Gaud., *L. perenne* L., *L. multiflorum* Lam.). Aust J Agric Res 17:821–839

Alemseged Y, Jones RE, Medd RW (2001) A farmer survey of weed management and herbicide resistance problems of winter crops in Australia. Plant Prot Q 16:21–25

Anderson WK, Sharma DL, Shackley BJ, D'Antuono MF (2004) Rainfall, sowing time, soil type and cultivar influence optimum plant population for wheat in Western Australia. Aust J Agric Res 55:921–930

Andrew IKS, Storkey J, Sparkes DL (2015) A review of the potential for competitive cereal cultivars as a tool in integrated weed management. Weed Res n/a-n/a

Ascard J (1998) Comparison of flaming and infrared radiation techniques for thermal weed control. Weed Res 38:69–76

Blackshaw RE (2004) Application method of nitrogen fertilizer affects weed growth and competition with winter wheat. Weed Biol Manag 4:103–113

Blackshaw RE, Molnar LJ, Lindwall CW (1998) Merits of a weed-sensing sprayer to control weeds in conservation fallow and cropping systems. Weed Sci 46:120–126

Blanco-Moreno JM, Chamorro L, Masalles RM, Recasens J, Sans FX (2004) Spatial distribution of *Lolium rigidum* seedlings following seed dispersal by combine harvesters. Weed Res 44:375–387

Borger CPD, Hashem A, Pathan S (2009) Manipulating crop row orientation to suppress weeds and increase crop yield. Weed Sci 58:174–178

Boutsalis P, Gill GS, Preston C (2012) Incidence of herbicide resistance in rigid ryegrass (*Lolium rigidum*) across Southeastern Australia. Weed Technol 26:391–398

Brodie G, Ryan C, Lancaster C (2012) Microwave technologies as part of an integrated weed management strategy: a review. Int J Agron. doi:10.1155/2012/636905

Broster JC, Pratley J (2006) A decade of monitoring herbicide resistance in *Lolium rigidum* in Australia. Aust J Exp Agric 46:1151–1160

Chauhan BS, Gill G, Preston C (2006a) Tillage systems affect trifluralin bioavailability in soil. Weed Sci 54:941–947

Chauhan BS, Gill GS, Preston C (2006b) Tillage system effects on weed ecology, herbicide activity and persistence: a review. Aust J Exp Agric 46:1557–1570

Cheam AH, Code GR (1995) The biology of Australian weeds. 24. *Raphanus raphanistrum* L. Plant Prot Q 10:2–13

Cousens R, Mortimer M (1995) Dynamics of weed populations. Cambridge University Press, Cambridge, UK, p 332

Cramb J (2000) Global circulations affecting weather in south-western Australia. In: Anderson WK, Garlinge JK (eds) The wheat book: principles and practices. Department of Agriculture, Western Australia, pp 3–21

D'Emden FH, Llewellyn RS, Burton MP (2008) Factors influencing adoption of conservation tillage in Australian cropping regions. Aust J Agric Resour Econ 52:169–182

Derpsch R, Friedrich T, Kassam A, Hongwen L (2010) Current status of adotopion of no-tillage farming in the world and some of its main benefits. Int J Agric Biol Eng 3:1–25

Derpsch R, Franzluebbers AJ, Duiker SW, Reicosky DC, Koeller K, Friedrich T, Sturny WG, Sá JCM, Weiss K (2014) Why do we need to standardize no-tillage research? Soil Tillage Res 137:16–22
Dowling PM, Nicol HI (1993) Control of annual grasses by spraytopping and the effect on triticale grain yield. Aust J Agric Res 44:1959–1969
FAO (2000) Land resource potential and constraints at regional and country levels. FAO, Rome
FAO (2015) Introduction to conservation agriculture (its principles & benefits). http://teca.fao.org/technology/introduction-conservation-agriculture-its-principles-benefits. Accessed Aug 2015
Gill GS, Holmes JE (1997) Efficacy of cultural control methods for combating herbicide-resistant *Lolium rigidum*. Pestic Sci 51:352–358
Gill G, Poole M, Holmes J (1987) Competition between wheat and brome grass in Western Australia. Aust J Exp Agric 27:291–294
Gommers CMM, Visser EJW, Onge KRS, Voesenek LACJ, Pierik R (2013) Shade tolerance: when growing tall is not an option. Trends Plant Sci 18:65–71
Heap IM (2015). The international survey of herbicide resistant weeds. http://www.weedscience.com. Accessed 31 Aug 2015
Hunt JR, Browne C, McBeath TM, Verburg K, Craig S, Whitbread AM (2013) Summer fallow weed control and residue management impacts on winter crop yield though soil water and N accumulation in a winter-dominant, low rainfall region of southern Australia. Crop Pasture Sci 64:922–934
Hoyle JA, McElroy JS, Rose JJ (2012) Weed control using an enclosed thermal heating apparatus. Weed Technol 26:699–707
Jursík M, Soukup J, Holec J, Andr J (2011) Important aspects of chemical weed control: environmental factors affecting herbicide efficacy. Listy Cukrovarnické a Reparské 127:348
Kassam A, Friedrich T, Derpsch R, Lahmar R, Mrabet R, Basch G, González-Sánchez EJ, Serraj R (2012) Conservation agriculture in the dry Mediterranean climate. Field Crops Res 132:7–17
Keller M, Gutjahr C, Möhring J, Weis M, Sökefeld M, Gerhards R (2014) Estimating economic thresholds for site-specific weed control using manual weed counts and sensor technology: an example based on three winter wheat trials. Pest Manage Sci 70:200–211
Kercher S, Conner JK (1996) Patterns of genetic variability within and among populations of wild radish, *Raphanus raphanistrum* (Brassicaceae). Am J Bot 83:1416–1421
Kim DS, Brain P, Marshall EJP, Caseley JC (2002) Modelling herbicide dose and weed density effects on crop: weed competition. Weed Res 42:1–13
Kloot P (1983) The Genus *Lolium* in Australia. Aust J Bot 31:421–435
Kon KF, Blacklow WM (1989) Identification, distribution and population variability of great brome (*Bromus diandrus* Roth.) and rigid brome (*Bromus rigidus* Roth.). Aust J Agric Res 39:1039–1050
Koohafkan P, Stewart BA (2012) Water and cereals in Drylands. Taylor and Francis, London
Lemerle D, Gill GS, Murphy CE, Walker SR, Cousens RD, Mokhtari S, Peltzer SJ, Coleman R, Luckett DJ (2001) Genetic improvement and agronomy for enhanced wheat competitiveness with weeds. Aust J Agric Res 52:527–548
Lemerle D, Cousens RD, Gill GS, Peltzer SJ, Moerkerk M, Murphy CE, Collins D, Cullis BR (2004) Reliability of higher seeding rates of wheat for increased competitiveness with weeds in low rainfall environments. J Agric Sci 142:395–409
Leys AR, Cullis BR, Plater B (1991) Effect of spraytopping applications of paraquat and glyphosate on the nutritive value and regeneration of vulpia (*Vulpia bromoides* (L.) S.F. Gray). Australia J Agric Res 42:1405–1415
Llewellyn RS, D'Emden FH, Kuehne G (2012) Extensive use of no-tillage in grain growing regions of Australia. Field Crops Res 132:204–212
Llewellyn RS, Ronning D, Ouzman J, Walker SR, Mayfield A, Clarke M (2015) Impact of weeds on Australian grain production and adoption of tillage practices. CSIRO, Australia 160 p
López-Granados F (2011) Weed detection for site-specific weed management: mapping and real-time approaches. Weed Res. 51:1–11
Lubbers MD, Stahlman PW, Al-Khatib K (2007) Fluroxypyr efficacy is affected by relative humidity and soil moisture. Weed Sci 55:260–263

Lutman PJW, Moss SR, Cook S, Welham SJ (2013) A review of the effects of crop agronomy on the management of *Alopecurus myosuroides*. Weed Res 53:299–313
Mahan JR, Dotray PA, Light GG (2004) Thermal dependence of enzyme function and inhibition; implications for, herbicide efficacy and tolerance. Physiol Plant 120:187–195
Mansooji AM, Holtum JA, Boutsalis P, Matthews JM, Powles SB (1992) Resistance to aryloxyphenoxypropionate herbicides in two wild oat species (*Avena fatua* and *Avena sterilis* ssp. ludoviciana). Weed Sci 40:599–605
Martin R, McMillan M, Cook J (1988) Survey of farm management practices of the northern wheat belt of New South Wales. Aust J Exp Agric 28:499–509
Marx C, Barcikowski S, Hustedt M, Haferkamp H, Rath T (2012) Design and application of a weed damage model for laser-based weed control. Biosyst Eng 113:148–157
Mayfield A, Presser R (1998) Crop-topping in pulses with paraquat and glyphosate for control of annual ryegrass. In: Proceedings of the agribusiness crop updates. Department of Agriculture Western Australia, Perth, pp 20–21
McGowan A (1970) Comparative germination patterns of annual grasses in north-eastern Victoria. Aust J Exp Agric 10:401–404
Menchari Y, Délye C, Le Corre V (2007) Genetic variation and population structure in black-grass (*Alopecurus myosuroides* Huds.), a successful, herbicide-resistant, annual grass weed of winter cereal fields. Mol Ecol 16:3161–3172
Monaghan NM (1980) The biology and control of *Lolium rigidum* as a weed of wheat. Weed Res 20:117–121
Morgan PW, Finlayson SA, Childs KL, Mullet JE, Rooney WL (2002) Opportunities to improve adaptability and yield in grasses: lessons from sorghum. Crop Sci 42:1791–1799
Oram P (1980) What are the worlds resources and constraints for dryland agriculture. In: Proceedings of the international congress dryland farming. Australia, Adelaide, pp 17–78
Owen MJ, Goggin DE, Powles SB (2012a) Identification of resistance to either paraquat or ALS-inhibiting herbicides in two Western Australian *Hordeum leporinum* biotypes. Pest Manage Sci 68:757–763
Owen MJ, Goggin DE, Powles SB (2012b) Non-target-site-based resistance to ALS-inhibiting herbicides in six *Bromus rigidus* populations from Western Australian cropping fields. Pest Manage Sci 68:1077–1082
Owen MJ, Martinez NJ, Powles SB (2014) Multiple herbicide-resistant Lolium rigidum (annual ryegrass) now dominates across the Western Australian grain belt. Weed Res 54:314–324
Owen MJ, Martinez NJ, Powles SB (2015) Multiple herbicide-resistant wild radish (Raphanus raphanistrum) populations dominate Western Australian cropping fields. Crop Pasture Sci 66:1079–1085
Paterson JG (1976) The distribution of *Avena* species naturalized in Western Australia. J Appl Ecol 13:257
Pearce GA, Holmes JE (1976) The control of annual ryegrass. J Agric Western Australia 17:77–82
Powles SB (1986) Appearance of a biotype of the weed, *Hordeum glaucum* Steud., resistant to the herbicide paraquat. Weed Res 26:167–172
Radford BJ, Wilson BJ, Cartledge O, Watkins FB (1980) Effect of wheat seeding rate on wild oat competition. Aust J Exp Agric Anim Husb 20:77–81
Reeves TG, Smith IS (1975) Pasture management and cultural methods for the control of annual ryegrass (*Lolium rigidum*) in wheat. Aust J Exp Agric Anim Husb 15:527–530
Roush ML, Radosevich SR (1985) Relationship between growth and competitiveness of four annual weeds. J Appl Ecol 22:895–905
Scotford IM, Miller PCH (2005) Applications of spectral reflectance techniques in Northern European cereal production: a review. Biosyst Eng 90:235–250
Steadman KJ, Eaton DM, Plummer JA, Ferris DG, Powles SB (2006) Late-season non-selective herbicide application reduces *Lolium rigidum* seed numbers, seed viability, and seedling fitness. Aust J Agric Res 57:133–141
Tucker ES, Powles SB (1991) A biotype of hare barley (*Hordeum leporinum*) resistant to paraquat and diquat. Weed Sci 39:159–162

UNCCD (2015) An introduction to the United Nations Convention to combat desertification http://www.unccd.int. Accessed Sept 2015

Vandenbussche F, Pierik R, Millenaar FF, Voesenek LACJ, Van Der Straeten D (2005) Reaching out of the shade. Curr Opin Plant Biol 8:462–468

Vigneault C, Benoit DL, McLaughlin NB (1990) Energy aspects of weed electrocution. Rev Weed Sci 5:15–26

Walsh MJ, Minkey DM (2006) Wild radish (*Raphanus raphanistrum* L.) development and seed production in response to time of emergence, crop topping and sowing rate of wheat. Plant Prot Q 21(1):25–29

Walsh MJ, Powles SB (2009) impact of crop-topping and swathing on the viable seed production of wild radish (*Raphanus raphanistrum*). Crop Pasture Sci 60:667–674

Walsh MJ, Powles SB (2014) High seed retention at maturity of annual weeds infesting crop fields highlights the potential for harvest weed seed control. Weed Technol 28:486–493

Walsh MJ, Newman P (2007) Burning narrow windrows for weed seed destruction. Field Crop Res 104:24–40

Walsh MJ, Owen MJ, Powles SB (2007) Frequency and distribution of herbicide resistance in *Raphanus raphanistrum* populations randomly collected across the Western Australia wheatbelt. Weed Res 47:542–550

Walsh MJ, Maguire N, Powles SB (2009) Combined effects of wheat competition and 2.4-D amine on phenoxy herbicide resistant *Raphanus raphanistrum* populations. Weed Res 49:316–325

Walsh MJ, Newman P, Powles SB (2013) Targeting weed seeds in-crop: a new weed control paradigm for global agriculture. Weed Technol 27:431–436

Walsh MJ, Aves C, Powles SB (2014) Evaluation of harvest weed seed control systems. In: Barker M (ed) 19th Australasian weeds conference. Tasmanian Weed Science Society, Hobart, pp 288–291

Yenish JP, Young FL (2004) Winter wheat competition against jointed goatgrass (*Aegilops cylindrica*) as influenced by wheat plant height, seeding rate, and seed size. Weed Sci 52:996–1001

Zerner MC, Gill GS, Vandeleur RK (2008) Effect of height on the competitive ability of wheat with oats. Agron J 100:1729–1734

Nutrient Management in Dryland Agriculture Systems

Ahmad Nawaz and Muhammad Farooq

1 Introduction

Dryland agriculture is being followed on various soil types and in different climatic regions across the globe. Extreme variability in rainfall events and low fertility of soils are the major biophysical factors limiting the productivity of dryland regions which accounts for 40 % of the total earth surface land area (Turner et al. 2011). Therefore, nutrient management in dryland regions is of prime concern. Fertilizer input in dryland agriculture may vary with soil and climate. However, the actual fertilizer use is less than the optimal in dryland areas in the world. Indeed, cultivated crops in dryland areas deplete the soil nutrients more than the amount of nutrients that is returned back to soil which causes negative balance of nutrients thus making the soil deficient in nitrogen (N), phosphorous (P), and potassium (K) (Irshad et al. 2007). For example, in Pakistan, about 100 % of the dryland soils are deficient in N, 90 % in P, and 20–40 % in K (Shah and Arshad 2006). The deficiencies of boron (B), zinc (Zn), and iron (Fe) are also prevalent on 60 %, 67–71 % and 21–25 % soils, respectively (Rashid 1994). The losses of crop nutrients are usually high in dryland regions due to leaching, soil erosion (IFPRI 1996), runoff, extreme weather events, and low soil moisture.

The productivity of crops in dryland areas in the world is very low owing to low use of fertilizers by the farmers. For instance, the average yield of cereals does not exceed 1.5 t ha^{-1} in the developing countries due to low use of fertilizer input.

A. Nawaz (✉)
Department of Agronomy, University of Agriculture, Faisalabad 38040, Pakistan
e-mail: ahmadnawaz2006@gmail.com

M. Farooq
Department of Agronomy, University of Agriculture, Faisalabad 38040, Pakistan

The UWA Institute of Agriculture, The University of Western Australia,
Perth, WA 6001, Australia

M. Farooq, K.H.M. Siddique (eds.), *Innovations in Dryland Agriculture*,
DOI 10.1007/978-3-319-47928-6_5

Table 1 Impact of various soil amendments on reduction of runoff, soil loss, NO_3-N and PO_4-P than control

Treatments	Runoff (%)	Eroded soil (%)	NO_3-N (%)	PO_4-P (%)
Control	0	0	0	0
Compost/manure + soil surfactant	22	72	55	72
Compost/animal or green manure + pigeon-pea hedges + soil surfactant	28	71	75	81
Compost/animal or green manure + mulch + pigeon-pea hedges	98	100	100	100

Source: Baptista et al. (2015)

Likewise, the productivity of various crops [sorghum (*Sorghum bicolor* (L.) Moench), pigeonpea (*Cajanus cajan* (L.) Millspaugh)] in the dryland soils of semi-arid tropics (India, East Africa, Australia, South America) is limited due to nutrient poor soils in these regions. The potential of dryland soil to produce a good yield of a specific crop is also limited due to widespread deficiency of micro and secondary nutrients such as Zn, B, and sulphur (S), in addition to deficiency of macronutrients and organic carbon (Rego et al. 2007; Sahrawat et al. 2007). The yield of various crops in the dryland regions can be profitability enhanced with the management of natural resources, the use of chemical fertilizers (IAEA 2005), and organic amendments. This also helps in substantial reduction in soil and nutrient losses (Table 1). In this chapter, the problems related to nutrient management in dryland agriculture systems has been discussed, and strategies to improve the nutrient management and crop productivity in the dryland soils are proposed.

2 Challenges to Nutrient Management in Dryland Agriculture Systems

Crop production systems in drylands are confronted by several challenges including water scarcity, weather variability, nutrient mining, low fertilizer use efficiency and soil erosion. In the following lines, challenges related to nutrient management in dryland systems has been described.

2.1 Monocropping and Nutrient Mining

In most of the dryland regions for the world, monocropping is quite common due to insufficient availability of soil moisture at seeding time. Thus, the soil remains fallow for half period of time during the year which left the soil vulnerable to nutrient losses which may be through soil erosion (wind and water), or through leaching or volatilization. Moreover, the crop residues of the monocrop are mostly used by the farmers as animal feed which also results in nutrient imbalances and nutrient mining. The other reasons for nutrient mining includes the low fertilizer use in dryland regions, and nutrient losses in the form of soil erosion and leaching (Henao and Baanante 2006). For example, the nutrient losses due to soil erosion ranges from 10 to 45 kg NPK ha^{-1} $year^{-1}$ in Africa. In Kenya, about 112, 2.5 and 70 kg ha^{-1} of N, P and K are lost every year (Smaling et al. 1993). In sub-Saharan Africa, the mining of crop nutrients has been identified as severe yield limiting factor in dryland agriculture. In 2002–2004, many African farmlands experienced a nutrient loss of 30 kg ha^{-1} per year (IFDC 2006).

In India, the crops such as pearl millet (*Pennisetum glaucum* (L.) R. Br.), sorghum, chickpea (*Cicer arietinum* L.), pigeonpea and groundnut (*Arachis hypogaea* L.) remove 72 kg ha^{-1} of N, P and K, while the actual annual addition to the soil is only 10–11 kg ha^{-1} (FAO 1988). Due to moisture deficiency in dryland soils, the application method and the placement of fertilizer in relation to rhizosphere are very critical. The top dressing or split application of fertilizer may not be a better choice under dryland conditions. Likewise, placement of fertilizer away from the root zone may also be less useful due to poor root development owing to deficiency of soil moisture.

2.2 *Low Fertilizer Use Efficiency and Low Fertilizer Input*

In dryland regions, the fertilizer use efficiency is very low than that that in irrigated agriculture (Saleem et al. 1996), owing to poor fertilizer and crop management practices, water stress at critical growth stages, and high weed infestation in high rainfall areas (FAO 1981). Imbalanced fertilizer application has been predicted as one of the most important factors responsible for low fertilizer use efficiency and it may cause a 20–50 % reduction in fertilizer use efficiency (Zia et al. 1997, 1998; Rashid et al. 2004), especially in dryland regions. In east Africa, depletion of soil fertility in dry areas has been reported due to imbalance between the nutrient inputs, their removal by crops and losses (Murage et al. 2000). Nambiar (1985) reported that addition of N with P, may decrease P level within dry soil by 21–26, 64 and 68 % in Alfisols, Vertisols and Oxisols, respectively. They also reported depletion of Zn, K and S in these soils when there was no external input from the artificial sources. A survey by Singh and Venkateswarlu (1985) in India showed widespread deficiency of Zn besides the deficiencies of N and P. In Bangladesh, N was found as most limiting plant nutrient in rainfed rice (*Oryza sativa* L.) followed by P, S, and K (BRRI 1984). In Nigeria, after growing maize (*Zea mays* L.) crop on an Alfisol, the yield without P input was decreased by 98 % (Kang and Osiname 1979). According to FAO, the fertilizer use in summer dryland crops was 201 % lower than irrigated

Table 2 Impact of variable water contents on the mineralization of N at different temperatures over time

Temperature (°C)	Time (week)	Amount of N mineralized (mg N kg^{-1}) at different water content (%)			
		8	15	22	29
15	2	2.5	5.7	9.2	10.6
	21	10.2	18.8	24.3	33.3
25	2	2.9	6.4	9.6	12.6
	21	18.9	29.6	37.6	50.6
35	2	3.9	10.7	14.4	18.3
	21	28.4	49.2	68.9	84.6

Source: Li et al. (1995)

summer crops due to unpredictable growing conditions and poor extension services provided by extension staff (FAO 1988), but fertilizer application at optimum rate enhanced the crop yield by 50–129 % (Tandon 1980). Beside the low fertilizer input in dryland regions than irrigated regions, the crop residues of dryland crops are used as fodder by farmers, which further worsen the situation.

2.3 *Water Scarcity*

Water scarcity in dryland regions, and extreme weather variability are major factors which affects the nutrients retention in soil, and their availability to the plants. Presence of water in root zone is essential for the uptake of nutrients by plants. If the soil water contents will be lower, the N mineralization (Table 2), and its uptake by the plant will be lower. In dryland regions, water shortage prevails regularly due to which the nutrients are not taken by the plants even they are present within the soil. In summer rainfall areas in India, efforts have been made to improve the water use efficiency. In a study, Kanwar and Mudhahar (1984) reported that the water use efficiency of prepetition received was enhanced when N was applied to pearl millet on an Alfisol in India. Fertilizer application to dryland soils enables the plants extract to soil moisture from deeper soil layers due to deeper root penetration. In dryland soils, soil moisture sometimes becomes a decisive factor for the crop responses to applied fertilizers. In two independent studies in India, the rate of N to which sorghum and wheat (*Triticum aestivum* L.) responded was completely dependent on the amount of moisture stored in soil profile (Meelu et al. 1976; Umrani and Patil 1983). In another study, application of N at the rate of 120 kg ha^{-1} to wheat on a clay loam soil with water storage capacity of 320 mm produced the same grain yield as was obtained with same fertilizer level under irrigated conditions (FAO 1988).

2.4 Soil Erosion

Soil erosion in dryland regions also causes substantial loss of nutrients. Tillage of dryland soils degrades the top soil layer (Izaurralde et al. 2006), which is followed by increase in soil bulk density, deterioration of soil structure and alteration in total soil porosity causing decrease in crop yield (Malhi et al. 1994). Soil erosion also depletes the soil nutrients thus causing nutrient deficiencies, and alters the soil chemical properties (Larney et al. 2000; Izaurralde et al. 2006) in dryland regions. In a study, Baptista et al. (2015) noted higher concentration of nitrate-N (2.20–4.83 mg L^{-1}) in runoff, and PO_4-P (5.27–18.8 mg g^{-1}) in eroded soil, which indicates how nutrients are lost through soil erosion and runoff in dryland regions. Wind and water erosion has also been major drivers to organic N losses from soil in dryland soils in China (Li et al. 2009b).

2.5 Soil Type and Nutrient Deficiencies

Successful crop production in drylands depends upon the soil management of nutrients (macronutrients and micronutrients) in the soil. The macronutrients include N, P, K, calcium (Ca), magnesium (Mg) and S. The response of crops to N grown in dryland area depends on the type of the soil, crop rotation and the crop season (Russell 1967; McDonald 1989; Mason et al. 1994). The main reason of N deficiency in dryland regions include the low soil organic matter as more than 95 % of the N in the soil exists in soil organic matter. In dryland regions, the losses of applied N on coarse texture soil are higher than clay soil owing to leaching losses. In many dryland soils, the deficiency of P is widespread due to P fixation (Russell 1967; Harmsen 1984; Impiglia and Ryan 1997), which necessities regular application of P in dryland soils. The K deficiencies in dryland crop production (especially) has rarely been reported, but may be present where the crop production is intensified (Ryan et al. 1997). The less deficiency of K in dryland soils is attributed to the presence of K-minerals in these soils which weather very slowly (Sekhon 1983).

In Australia, removal of crop nutrients in harvested products, nutrient loss through soil erosion, denitrification (from waterlogged Vertisols), and leaching (from light textured soils) has led towards severe deficiencies of N and P in dryland regions. The Zn and S deficiencies are reported on large area, but the deficiencies of K, manganese (Mn), copper (Cu) and molybdenum (Mo) (in ferrosols) has been reported on very small area in Australia (Birch and Bell 2011).

In drylands, the most frequently deficient micronutrients are Zn (on alkaline soil), Fe and Mn. Indeed, the most of the soils in the dryland regions have pH of 8.0 or above, and are highly alkaline and calcareous which reduced the availability of these trace nutrients. The deficiency of Zn and Mn has mostly been observed in dryland wheat. Deficiency of Zn in dryland soils (e.g., rice and maize) might be also attributed to high incident of solar radiations.

In several states of India, where dryland agriculture is being practiced, 73–95 % of fields were found deficient in S, 62–94 % in Zn and, 70–100 % in B (Sahrawat et al. 2010). The deficiency of Cu has also been reported in some dryland soil of southern Australia. The cropping intensity is also very decisive factor for determining the deficiency/toxicity of micro/trace nutrients in dryland soils. The soil analysis to find the trace nutrients requirement of crops is less satisfactory than the plant analysis (e.g., leaf) (El-Fouly 1983). The Fe deficiency in dryland soils is associated with the high calcareous nature of dryland soils and crop type. For example, sorghum is sensitive to Fe deficiency (Babaria and Patel 1981). In Australia, wheat has been found to suffer from Cu deficiency (Smith 1983).

2.6 Lack of Integrated Nutrient Management Approaches

Lack of integrated nutrient management in dryland soils is another issue which affects the nutrient use efficiency and the crop productivity. Farmers usually use chemical fertilizers as source of nutrients which is not as beneficial as an integrated nutrient management approach may be. Studies have reported decline in crop yield over time in dryland regions with the continuous use of the chemical fertilizers (Bekunda et al. 1997). The reasons for this yield decline includes (i) higher nutrient removal by crops than the added nutrients (Scaife 1971), (ii) enhanced nutrient losses through denitrification and volatilization, and (iii) decline in soil organic matter over time. For example, application of N fertilizer to monocropped sorghum (with crop residues removed from the field) enhanced the annual soil organic matter loss from 1.5 % without fertilizer to 1.9 and 2.6 % with moderate and high N rates, respectively (Pieri 1995). In a study, continuous application of chemical fertilizers decreased the grain yield of finger millet (*Eleusine coracana* L.) from an average of 2,880 kg ha^{-1} during the initial 5 years to 1490 kg ha^{-1} by the 19th year of the study (Gajanan et al. 1999).

In China, the excessive use of fertilizer N, inadequate use of K and P fertilizers, and low use of organic manures are the main reasons for low fertilizer use efficiency in dryland crops (Yu et al. 2007). Most of the farmers in China use ammonium bicarbonate as main N fertilizer source which has high losses due to ammonia volatilization, and thus has low N use efficiency than urea fertilizer (Wang et al. 2003).

3 Strategies for Improving Nutrient Management in Dryland Agriculture Systems

Nutrient management challenges in dryland crop production systems may be effectively tackled through integrated nutrient and crop management strategies. Conservation agriculture approach, water conservation, use of slow and controlled release fertilizers, and site-specific nutrient management options may also help reduce the nutrient losses in the dryland crop production systems. In the following lines, some pragmatic strategies for improving nutrient management in dryland agriculture systems are discussed.

3.1 Conservation Agriculture

The conservation agriculture (CA) has been identified as an alternative to conventional agriculture for sustaining the productivity of dryland regions through better management of soil, water and nutrient resources. Reduction in soil losses and production cost, and improvement in soil organic matter and plant available soil water has been reported in various dryland regions, due to implementation of CA (Lyon et al. 2004; Thomas et al. 2007). For example, in Ethiopia, long-term experiments have demonstrated that CA may enhance the crop yield (Mesfin et al. 2005; Burayu et al. 2006; Temesgen et al. 2009) by reducing the soil runoff and erosion (Oicha et al. 2010) with substantial improvement in soil organic matter and soil mineral nutrients (Burayu et al. 2006). In CA, surface mulch and legume cover improves the storage of nutrients and water, which enhances the soil organic matter and thus crops grown under CA requires less fertilizer (Lafond et al. 2008; Crabtree 2010; Lindwall and Sonntag 2010).

Numerous other studies have reported yield improvement of 20–120 % due to adoption of CA systems in dry Mediterranean climates than the traditional tillage systems (Mrabet 2000, 2002; Fernández-Ugalde et al. 2009; Fileccia 2009; Crabtree 2010; Piggin et al. 2011), due to improvement in soil moisture and nutrient availability. Indeed, after several years of adoption of CA, the soil has more amount of biological N and has greater ability to release N than the tilled soil which help enhancing the availability of nutrients to plants (Duiker and Beegle 2006; Lafond et al. 2008; Crabtree 2010), and reducing runoff.

3.2 Fertilizer Application at Right Time, Through Right Source, and at Right Place

For the best management of nutrition in dryland crops, the fertilizer should be applied at right time, at right place, through right application method at appropriate rate using the right source (Ryan and Sommer 2010). Janzen et al. (1999) opined that sensible use of fertilizers in dryland agriculture, based on soil tests, proper placement of fertilizer in the soil at or near the place of seeding can increase the yield and quality of dryland crops owing to substantial reduction in nutrient losses to the ground water and aerial environment.

The timing of N application in dryland soils is very important. It is the most beneficial when applied as basal dose; nonetheless split application from sowing to early leaf boot stage in wheat can be useful when season is longer than normal (Mason 1975), and more rainfall is expected. Based on several field experiments, Spratt and Chowdhury (1978) reported that split application of N was beneficial for sorghum, pearl millet, maize and upland rice under dryland conditions in India. Fillery and McInnes (1992) reported less than 50 % recovery of applied N in dryland wheat.

Optimum N application to wheat crop in dryland soils depends upon the input of N from soil organic matter, fixation by legumes, N losses through vitalization and leaching, and the crop demand. Westfall et al. (1996) reported that the N fertilization in dryland crop production must be managed carefully to gain high economic returns. In dryland crop production systems, the rate of N application is more important than its placement method. However, the placement becomes important with the increase in rainfall incidence. The subsurface N placement is better than the surface broadcast applications. In China, deep placement of N fertilizer is recommended for dryland soils because of the availability of surplus water in deeper soil layers. In deeper layers, more roots are present which may use the applied N very efficiently and thus help improving the fertilizer use efficiency (Li et al. 2009a). Other studies have also reported that deep placement of N fertilizer improves the N use efficiency and crop yield (Li et al. 1976b). Rees et al. (1997) reported 18 % N recovery from surface applied N, 33 % from the N mixed with 15 cm soil, and 36 % when N was applied in a pit (15 cm deep). In another study, Lu and Li (1987) found that deep placement of N and P fertilizer enhanced the fertilizer use efficiency and wheat yield by 20 %, and reduced the losses of ammonium-N through volatilization by 14–39 % for urea, and from 36 to 54 % for ammonium bicarbonate. Li (2007) reported a reduction in N loss through ammonia volatilization by 29 %, 15 % and 1 %, when the N fertilizer was surface applied, soil mixed, or was deeply placed, respectively.

In dryland regions, the early application of organic and P fertilizers is as useful as that of N (Li et al. 1976a). The usefulness of fertilizer application method may vary with crop type. For instance, Kanwar and Rego (1983) reported that split band application of urea in pearl millet and sorghum was better than broadcasted urea (incorporated or either left on surface). However, in sorghum, split application of N fertilizer (either broadcasted, band application) improved the grain yield than that of other N application methods (Kanwar and Rego 1983). Drilling the N and P fertilizers in the moist zone (5–7 cm below seed) improved the sorghum grain yield by 60 % than its application by the furrow side (Randhawa and Singh 1983).

Due to high pH buffering capacity of dryland soils, nitrate-N is the main form of N taken by plants in dryland soils. Major crops such as wheat and maize grown in dryland regions also respond better to nitrate-N than ammonium-N. Thus, it is not wise to use ammonium-based N fertilizer (e.g., urea) in dryland soils. However, if no other source of N fertilizer is available, pretreating of the dryland soil with a small quantity of NH_4^+ salt before application of a large amount of urea is beneficial to reduce the nitrification of ammonical N (Li et al. 2009a).

Foliar application of N is also an option in dryland soils because the farmers apply fertilizer when rain occurs which are often delayed. Foliar application of concentrated urea solution enhanced the dryland wheat yield by 31 % (FAO 1988). In dryland cotton (*Gossypium hirsutum* L.), application of urea at the rate of 25–30 kg ha^{-1} through soil and foliar application enhanced the yield by 13–51 % than sole soil application (Rao et al. 1973). Foliar application is usually pragmatic when the fertilizers are not applied as basal dose and the onward soil applications are also not possible due to some operational difficulties (Dargan et al. 1973).

Some other factors like fertilizer and cash availability, the relative return from other inputs used for crop production, and the environmental risk may affect the choice of fertilizer application to crops in drylands. Various site specific characteristic such as soil texture, soil organic matter, rainfall, soil tests for different nutrients, the crop type and tillage method may affect the efficiency of nutrient use in dryland areas (Ryan et al. 2010). In dryland soils, less rainfall and leaching losses necessities the same amount of N applications in spring and fall, top dressing in spring being better in relation to flexibility with rainfall. The responses to applied N increases with increase in rainfall and soil moisture availability in dryland regions. For example, response to N were highest at rainfall level of 350–550 mm in Syria, and they were lowest when rainfall was less than 250 mm. Likewise, the responses of N are more when applied after fallow than applied after cereal crop (Ryan and Sommer 2010). Deficiency of Zn and Fe, and the toxicity of B may affect the responses of crops towards applied N and P. Level of soil organic matter and the type of crop rotation in a specific dryland region also affects the responses of crops to N fertilization (Ryan et al., 2010), which should be taken care while devising the N application schedules for dryland crops. In the dryland areas, where urea is used as N fertilizer source, it should be mixed into soil, or should be applied under cooler conditions, or may be top-dressed before or during spring rains to avoid the losses through volatilization (Abdel Monem et al. 2010).

In calcareous soil, most of the soluble P sources become fixed. Under such conditions, long-term band placement of P is beneficial (Ryan and Sommer 2010). The N fertilizer use efficiency in dryland crops can be enhanced through innovative placement techniques and timing of application. For example, application of ammonia based N fertilizers in concentrated bands may restrict nitrification and leaching. Top dressing of N based fertilizers in favorable season or before a forecasted rainfall event can enhance crop yield and quality (Angus et al. 1994). Leaching losses of urea are usually less in dryland losses due to less soil moisture.

In dryland areas of Africa, the losses of N from applied N, regardless of N sources in gaseous form has been reported with increasing application rate (Bekunda et al. 1997). However, certain sources may yield better for some crops. For instance, the use of calcium ammonium nitrate was better than urea regarding the plant N uptake, and it eventually enhanced the yield of pearl millet (Mughogho et al. 1986). In West Africa, the N uptake by millet crop was about three times more from point-placed calcium ammonium nitrate against point-placed urea fertilizer, and the N uptake was 57 % less from the broadcasted calcium ammonium nitrate than the point-placed calcium ammonium nitrate. The split N application enhanced the N use efficiency (Uyovbisere and Lombin 1991).

The placement of the fertilizer in relation to seed position in soil can influence the uptake, crop response and crop germination and onward performance. Both the N and K should never be mixed with crop seeds because they can delay the germination and may cause seedling mortality, and the loss can be more on frequently drying soils (Carter 1967; Mason 1971). Band placement of N and K fertilizers below the seed at some distance is quite beneficial than other fertilizer placement methods (Nyborg and Hennig 1969). However, in the soils severely deficient in phosphate, mixing of phosphate fertilizer with the seeds of cereals at sowing is very useful (Loutit et al. 1968). The surface placement of phosphatic fertilizer is less effective in dryland soils due to its fixation. The availability may further decrease upon drying of the soil surface as happens in drylands. In India, substantial increase in dryland crop yields with the amendment of soil with micronutrients and the yield was further enhanced when both micronutrients and adequate N and P were applied in sorghum, mungbean (*Vigna radiata* (L.) Wilczek), maize, chickpea, pigeonpea, groundnut and castor bean (*Ricinus communis* L.). Supplementing crops with micronutrients also enhanced the rainwater productivity in these crops (Rego et al. 2005).

A crop usually obtains 45–70 % of its total N from soil which necessities that the application of N should be done on the basis of soil N supplying capacity in dryland soils. For dryland soils, the fertilizer application should be done in furrows below the crop seed, which will not only help to replenish crop nutrients but will also help to efficiently utilize the soil moisture. In P deficient soil, and the dry soils where the root penetration is not too much, the P fertilizers must be placed very near to young seedlings so that it can be readily absorbed by the plant roots. In a study at Agra, India, the mixing of P fertilizer with seed or band application 10 cm below the seed gave higher black gram yield than the broadcasting of P fertilizer. At Indore, the soybean (*Glycine max* (L.) Merr.) yields were much higher when P fertilizer was placed in furrow, or at/below the seed level (Sharma et al. 1979). In dryland soils, the foliar application of P is also beneficial. In cowpea (*Vigna unguiculata* L. Walp.), P use efficiency was enhanced when some of the P fertilizer was drilled into the soil at the time of sowing and the rest of P was foliage applied in the form of 5 % solution of triple superphosphate (Gill et al. 1971; Mohta and De 1971). In India, the recommended fertilizer application schedule for dryland crops consist of drilling the basal fertilizer application 5–10 cm below the upper soil curst. In the season with high rainfall, whole of K and P, while a portion of the N dose are used as basal dose in India. However, in dry spells of little or no rainfall, the full amount of fertilizers is applied as basal dose. Yield improvement from 340 to 1500 kg grain ha^{-1} has been reported in dryland crops in India due to adoption of this recommended fertilizer placement method (Venkateswarlu 1987).

3.3 Nutrient Application on the Basis of Soil Analysis

For estimating the deficiency and fulfill the crop nutrient needs in dryland regions, there is a need of careful soil analysis (Peverill et al. 1999). The soil nutrient status may change over time due to continuous nutrient uptake by the crops. For example, regular application of P is needed in the cereal–legume rotation in dryland soils (Ryan et al. 2008b) to harvest good crop yield. Likewise, K application (based on soil test and crop needs) to dryland soil may help to boost the crop productivity by improving root growth, enhancing the water use efficiency and regulating the stomatal closure. Soil and plant analysis may be helpful to correct the deficiency and toxicity of micronutrients (Reuter and Robinson 1997). For this, the most economical way to correct the micronutrient deficiencies in dryland soils is the small addition of these micronutrients to the base fertilizers.

3.4 Integrated Nutrient and Crop Management Approaches

Integrated use of inorganic and organic fertilizers in dryland regions may be the most beneficial than their separate use. For instance, in a 5-years study, Bouraima et al. (2015) reported that combined application of fertilizer with manure reduced runoff losses and prevented the soil erosion. Losses of total N and P were also reduced by 41.2 and 33.3 % with combine application of manure with chemical fertilizer. In another study, implementation of advanced moisture conservation practices and the optimum use of inorganic and organic fertilizers were beneficial for improving crop yields in dryland soils (Hadda and Arora 2006). Combine use of organic manures, crop residues and green manures in dryland soils together with mineral fertilizers may help build up macronutrient pool in these soils for sustaining the crop productivity on long term basis (Bationo et al. 1997). In dryland soils, the grain production of cereals is more when organic and the mineral fertilizers are used in combination (Palm et al. 1997). For instance, combine use of N with organic fertilizer enhanced the wheat grain yield by 10–41 %, and precipitation use efficiency by 10–42 % (Table 3). In Sudan, combine application of manure with chemical fertilizers sustained the productivity of sorghum (Sedogo 1993). Supplying 50 % of the recommended fertilizer through crop residues and 50 % through the loppings of a N fixing tree (*Leucaena leucocephala*) enhanced the yield of sorghum by 31–87 %, against the application of 25–50 kg N ha^{-1} through inorganic fertilizer source (AICRPDA 1999). The fertilizer is one of the most expensive input in dryland farming in China, and its increased use has already resulted in the 100 % increment in crop yield in rainfed regions. In China, the farm yard manure was regarded as main source of nutrients before 1970s in dryland agriculture. The fertilizer use is still increasing in dryland agriculture in china, but the ratio of N to P used by most of the famers is more against the recommended ratio (1:0.3) for dryland crops (Tong et al. 2003).

Table 3 Impact of nitrogen and various rates of organic fertilizers on percent increase in grain yield and precipitation use efficiency of wheat

Nitrogen rate (kg ha^{-1})	Organic fertilizer rate	Yield increase (%)	Increase in PUE (%)
00	No organic fertilizer (OF)	–	–
50	No organic fertilizer (OF)	10.1	9.7
50	OF (19 t ha^{-1})	34.2	34.1
50	OF (38 t ha^{-1})	37.5	37.1
50	OF (72 t ha^{-1})	39.3	39.1
50	OF (124 t ha^{-1})	39.7	41.0
50	OF (248 t ha^{-1})	41.1	41.7

Source: Cheng et al. 1987
PUE precipitation use efficiency

Fertilizer application together with other improved crop management practices may improve the yield of dryland crops. Implementation of water and soil conservation measures, before sowing and during the crop growth, optimum plant population and managing the insect-pests and weeds below economic threshold level might be useful to enhance the fertilizer use efficiency in dryland crops. The fertilizer use efficiency in dryland soils can also be enhanced through use of suitable crop cultivars, planting at appropriate time, fertilizer application through right method, addition of organic manures and through water and soil conservation techniques. Planting of improved crop cultivars utilize the N more efficiently than other traditional crop cultivars. For example, Tandon (1980) reported that improved crop management practices may enhance the sorghum yield by 10 % and fertilizer use efficiency by 3.5 %. Increase in fertilizer use efficiency has also been reported on the Vertisols of India in several crops by Randhawa and Singh (1983). They reported that maintenance of optimum plant population in pearl millet together with application at 25–30 kg ha^{-1} N provided the highest grain yield when the plots were kept weed free.

In India, several on-farm studies demonstrated that the yield of castor bean, mungbean, groundnut and maize was enhanced with the application of B, Zn and S, and the results were more apparent when they were combined with N and P applications (Sahrawat et al. 2010). In China, the application of fertilizer N with organic manures can significantly enhance the nitrogen use efficiency, crop yield and water use efficiency (Li et al. 2009a). They further reported that combine use of N and P fertilizer can enhance the nitrogen fertilizer recovery, nitrogen and water use efficiencies, and can reduce the losses of N through volatilization. In drylands of USA and Canada, where the crop production is limited due to low N use and less soil water availability, use of conservation tillage, harvesting the crop stubbles tall enough to trap snow, crop diversification, and with long term diversified crop rotations involving the legumes and cover crop helped increase the overwinter water storage, decrease the soil degradation and enhance the yield of dryland crops (Cutforth and McConkey 1997).

3.4.1 Diversified Crop Rotations

The losses of N in dryland soils through leaching, denitrification and volatilization can be reduced/replenished by incorporating the legume crops in rotation, as the legumes can contribute 40–80 kg N ha^{-1}. For example, under dryland conditions, the alfalfa (*Medicago sativa* L.) can fix about 200–250 kg N ha^{-1} per year, while pea can fix 71–91 kg N per growth cycle (Li et al. 1990, 1992). The applied N recovery from urea in combination with using the twigs and loppings of N fixing tress, and urea alone was in 1:1 ratio (Sharma et al. 2002). However, the success of biological nitrogen fixation by legumes depends on the correct match between the *Rhizobium* strain, the environment and the host legume variety (Beck 1992), and these factors should be considered while including the legume crops in crop rotations in dryland regions.

In some areas of the world, the root diseases of wheat are controlled through crop rotation which can also influence the responses of wheat to applied N (Rowland et al. 1988). Crop rotation with legumes on clay soil can reduce the input of applied N to greater extent in dryland regions (Anderson et al. 1995). Inclusion of forage legumes in cereal-based systems in dryland soils may lead towards the buildup of N enriched soil organic matter, which enhanced the potentially available N within soil pool. Interestingly, this increases in soil organic matter due to addition of legumes in crop rotation caused linear and parallel increase of total mineral, labile and biomass N forms (Ryan et al. 2008d).

During the fallow years, the legumes may help provide soil cover to reduce soil erosion and nutrient losses, and can fix the atmospheric nitrogen to benefit the following grain crop. In dryland areas, where summer rainfall exceeds 550 mm and the duration of rainfall is 90 days, mungbean or mashbean ((*Vigna mungo* (L.) Hepper) can be grown in the moonsoon season followed by barely or wheat in dry season. The yield advantage from the legume may be equivalent to 20–30 kg fertilizer N ha^{-1}, for the cool season crops with substantial improvement in fertilizer use efficiency (Singh and Venkateswarlu 1985). In another study, Giri and De (1981) reported that growing cowpea, (*Vigna aconitifolia* (Jacq) Marechal), guar (*Cyamopsis tetragonoloba* (L.) Taub.) and soybean for 55 days and then using them as green fodder, followed by growing of barely was equivalent to more than 40 kg ha^{-1} N applied to barely grown after a non-legume fodder crop. In some parts of Australia, the wheat is not provided by N fertilizer, and the N fixed by legumes is enough to achieve good wheat yields in some dryland regions (Smith 1983). Mixtures of legumes and non-legumes in dryland regions of the world may be useful to meet the feed requirements of the animals with substantial improvement in soil fertility. This improvement in soil fertility might be attributed to co-current transfer of N form the legume to non-legume crop in rhizosphere. Such type of intercropping also improves the N use efficiency (Eaglesham et al. 1981; Bandyopadhyay and De 1986).

Organic inputs from livestock manure, green manures and crop residues may enhance the fertilizer use efficiency as well as the yield of crops in dryland regions (Palm et al. 1997; Place et al. 2003). Few legume species not only fix the atmospheric

N but also enhance the availability of P thus enhancing the yield of dryland crops (Snapp and Silim 2002). The net profit from maize crop were enhanced by 50 % when synthetic fertilizer was used in a maize-legume intercrop or crop after a grain legume in rotation than the continuous crop of maize (Waddington and Karigwindi 2001). In a 4-year study in West Africa (Zimbabwe), the overall grain yield was the maximum for millet-groundnut and millet-cowpea rotations than the continuous pearl millet cropping supplied with 45 kg N ha^{-1} (Mukurumbira 1985). In Malawi, MacColl (1989) reported that the maize grain yield with no fertilizer application after pigeonpea was 2.8 t ha^{-1} more than a maize-maize rotation supplied with 35 kg N ha^{-1} on annual basis.

In a long term experiment on vertisol, Rego and Rao (2000) found that the concentration of N in the surface (0–15 cm) layer was enhanced by 125 mg N kg^{-1} of soil in a pigeonpea –based cropping system with not no external N input after 12 years. On the other hand, there was rapid decline in total N concentration in the traditional (rainy season fallow, post rainy season sorghum) and in non-legume based cropping systems. These improved management practices not only lessen soil loss through erosion, but also enhanced the accumulation of total N and carbon with substantial improvement in crop yield (Wani et al. 2003).

Pasture and forage legumes in a cereal-sheep ley farming system in Australia (Puckridge and French 1983) provide not only fodder for livestock, but also the mineralized N for the subsequent cereal crop (Hossain et al. 1996). Likewise, several other studies in USA has reported that the rotation of legumes with cereals improves the soil N and thus provides diversification of crops (Norwood 2000; Carpenter-Boggs et al. 2000). Few other crop rotations showed positive impacts of these rotations not only on the soil properties (especially soil organic matter), but also on water use efficiency and nutrient use (Pala et al. 2007; Ryan et al. 2008a, c).

3.4.2 Organic Manures and Crop Residues

Chemical fertilizers should be applied at optimal dose together with use of crop residues and organic manures. Indeed, most of the dryland soils are low in soil organic matter due to high temperature or low recovery of plant residues which are used by the animals, as fuel or as thatching material and are not returned to the soil. Any improvement in soil organic matter will improve soil physical properties as well as nutrient status of the soil (FAO 1988). Organic matter may be applied as bulky organic manures (crop residues, compost, farmyard manure) or municipal sludge or may be grown *in situ* as green manure crop and incorporated into the soil. Usually, the coarse textures dryland Alfisols has low organic matter and are subjected to crust formation which impacts the seedlings establishment and further plant growth. In this situation, the incorporation of crop residues after the harvest of grain crops may improve the fertilizer use efficiency and crop productivity (Hegde 1980). Soil organic matter contents were also enhanced from 0.55 to 0.90 %, and the yields of cowpea and pearl millet were enhanced by 14 and 19 %, respectively after 5 years of crop residue application. Restoration of crop residues in the soil enhanced the fertilizer use efficiency by 8–16 % relative to the application of fertilizer without crop residues (Hegde 1980).

Another pragmatic way to add the organic matter into the dryland soil is the growing of leguminous crops *in situ*, and then ploughing as green manure. Indeed, the green manures are primarily used as soil amendment to provide the soil with nutrients for upcoming crops, and are recognized as the most secure and clean source of organic fertilizer in conventional agriculture in China (Cao and Huang 2009), and other parts of world. Use of legume crops as green manures help enhancing the grain yield of the proceeding cereal crops by adding N, maintaining biological activity and soil fertility and decreasing the use of synthetic fertilizers (Evans et al. 2003; Walley et al. 2007; Espinoza et al. 2012; Seymour et al. 2012; Yu et al. 2013; Hayden et al. 2014). For instance, Hegde (1980) proposed the incorporation of cowpea as green manure to enhance the yield and fertilizer use efficiency of rainy season castor and sorghum on a coarse textured Alfisol of Hyderabad, India. In China, growing leguminous green manure crop prior to wheat improved the yield and water use efficiency of wheat by 13 and 28 %, respectively, and was very useful to maintain the soil water balance than fallow-wheat crop rotation (Zhang et al. 2016). In another study, green manuring with lentil (*Lens culinaris* L.) maintained the wheat productivity and offset the needs of N fertilizer after three cropping cycles than the conventional-wheat–fallow rotations (Allen et al. 2010).

The timing of manure application to the dryland soil is very important. In a 4-year study, application of farm yard manure prior to onset of moonsoon and application of mineral fertilizers about 2-month before sowing was the most beneficial (Singh et al. 1985). Studies in dryland regions of Kenya showed that the goat manure has superior soil fertility impact than chemical fertilizer (e.g., diammonium phosphate) (Onim et al. 1990), and it may be preferred in dryland crop production. In dryland regions, growing of multipurpose tress within crops may help to provide nutrient pool by capturing nutrients from the atmosphere and fixing them through biological nitrogen fixation, and extracting nutrients from deep soil horizons and their deposition in the soil through leaves and other plants parts after decomposition (Sanchez et al. 1997). However, they opined that in low rainfall regions, tree plantation within crops may negatively impact the crop growth by using the available soil moisture (Sanchez et al. 1997).

3.5 Water Conservation

Effective water management in drylands may enhance the availability and transformation of nutrients from fertilizers or from the soil. The organic N mineralization is proportional to the amount of water present in the soil, and the net mineralized nitrate-N is enhanced with increase in soil water contents under an optimal temperature range (Li et al. 2009b). Thus, the benefits of fertilizer application to the dryland crops can be enhanced through effective water conservation in dryland areas. In Burkina Faso, the grain yield of sorghum was highest at on-farm locations with the combination of tied ridges and fertilizer than with tied ridges and fertilizer alone (Nagy et al. 1990). Likewise, the sorghum yield enhanced from 118 to 388 kg ha^{-1} when the crop was planted on 1.5 m tied ridges, and to 1071 kg ha^{-1} when 50 kg N ha^{-1} was applied to the tied-ridges (Nyakatawa 1996).

Mulching in dryland soil may be beneficial to improve the water conservation. Mulching also modulates the soil temperature and soil moisture regimes, and thus help improving the soil nutrient utilization (Lal 1974). In two independent studies, higher utilization of N and P was reported with mulch in sorghum (Raghavulu and Singh 1982), and barely (*Hordeum vulgare* L.; Agarwal and De 1979). Application of N combined with mulching improved wheat yield after maize in dryland soil (Prihar et al. 1981). van Duivenbooden et al. (1999) also reported that improvement in soil water storage and its availability to crop plants at critical growth stages increases the fertilizer use efficiency and the efficiencies of other farm inputs. In India, Singah and Das (1995) reported higher yield of dryland sorghum in a deep soil having more stored water with N input of 50 kg ha^{-1} than a shallow soil where the response was only up to 25 kg ha^{-1}. Mid-season rainfall on a sandy soil determined the N use efficiency and yield of millet (Bationo et al. 1989). When mid-season rainfall was low, the millet yield was not affected with fertilizer N. However, when mid-season rainfall was more than the average, the application of N enhanced the grain yield of millet by 4–5 folds (Bationo et al. 1989). They also reported that response of millet to applied N was low in dry years which was enhanced in the years when rainfall was optimum. Application of N application at 30 kg N ha^{-1} provided the fertilizer N use efficiency as high as 25 kg grain kg^{-1} N (Bationo et al. 1989).

In China, two types of mulches haven been used widely in dryland agriculture, *viz.* straw mulch and plastic mulch. Use of plastic mulch conserve soil moisture, enhances the soil temperature, and availability of N by reducing the losses of N through volatilization, thus enhancing the N use efficiency and crop yield. In Gansu province in China, use of plastic mulch enhanced the yield of wheat by 1500 kg ha^{-1}. However, the environmental risks associated with the plastic mulch is the slow rate of decomposition, and eventually the mulch becomes a pollutant. The benefits of straw mulching are the same (e.g., water conservation and increase in nutrient use efficiency) as the plastic mulch, but straw mulch lowers the soil temperature and delays the germination of spring sown crops. When combined with conservation tillage, the benefits of straw mulching to conserve soil water in dryland area may increase several folds. Plastic mulch causes rapid decomposition of soil organic matter, while straw mulch lowers the soil organic matter decomposition due to low temperature.

Beside the total seasonal rainfall, the distribution of rainfall during the crop ontogeny also triggers the fertilizer N use efficiency. For example, in Canada, the rainfall during grain filling and at seeding was the most beneficial than at other growth stages (Campbell et al. 1988). The amount/distribution of rainfall in northern India during the vegetative and reproductive stages has been identified as a decisive factor to determine the fertilizer N use efficiency (Sandhu et al. 1992). The pattern of rainfall in dryland areas can also affect the effectiveness of fertilization application. For example, Singh et al. (1977) reported that the benefits from fertilizer N placed below the seeds were higher than its broadcast application when the rainfall occurred immediately after planting as compared to when rain was delayed in wheat.

Application of organic manures in dryland soil may be useful to improve water holding capacity of soil. For instance, Zhang et al. (1982) reported that application of organic manure (7.5 t ha^{-1}) to a dryland soil improved the water storage ~44.7 mm more in top 2-m soil layer than no manure application. In a similar study, Cheng et al. (1987) found 33 mm more water in 2-m soil layer with manure application than no manure application. Application of synthetic fertilizers along with organic fertilizers not only enhance the soil nutrients and crop nutrition, but also enhance soil organic carbon, soil structure, water holding capacity, water use efficiency and finally the crop yield of crops (Ma et al. 1984).

3.6 *Balanced Fertilization*

Most of the farmers in dryland regions of the world are just focusing on a single nutrient. Balanced fertilization in dryland crops may be useful to enhance the nutrient use efficiency. In India, several on-farm trials demonstrated that the yield and N use efficiency of dryland crops were increased by balanced fertilization. The agronomic efficiency of the applied N was enhanced by applying K and P fertilizers, by 10.3 kg pearl millet grain kg^{-1} N, 6.7 kg sorghum grain kg^{-1} N, and 19.5 kg maize grain kg^{-1} N. The N use efficiency in pearl millet, maize and sorghum was also enhanced by 6–20 % (Prasad 2009). In Bangalore- India, growing the yield and fertilizer N use efficiency of pearl millet grown on K-deficient red soil was more with NK application than NP application over time (Vasuki et al. 2009). In another study in northwestern India on loamy sand soil, the growth of pearl millet was substantially enhanced with K application (Yadav et al. 2007). In a study on sorghum, N application at 40 kg ha^{-1} and P application at 13 kg ha^{-1} enhanced the grain yield by 2.5 times more than the farmers' fertilizer use practice (use of single nutrient source). In another study, combine application of N and P fertilizers enhanced the wheat grain yield by 27.4–65.2 % on a P deficient soil (Table 4). Likewise, application of 50 kg N ha^{-1} and 13 kg P ha^{-1} improved the castor bean yield by 38 % against the suboptimal application (10 kg N ha^{-1} and 13 kg P ha^{-1}) (Sharma et al. 2007).

On a P deficient soil, the yield of wheat was 1125 kg ha^{-1} without fertilizer application. When, the N was applied at 135 kg ha^{-1}, wheat yield was reduced to 975 kg ha^{-1}. However, the yield was increased to 2775 kg ha^{-1} with the application of 20 kg ha^{-1} P. When both N and P were applied at the same rate together, the yield was increased to 4495 kg ha^{-1} (Li et al. 1978, 1979), which indicated the importance of balanced fertilization in dryland soils. Several other studies reported increase in N and P use efficiency and regulation of imbalances between N and P use in dryland soils due to combine application of N and P (Jin 1989; Wu 1989). In a study,

Table 4 Improvement in wheat yield (%) due to combine use of nitrogen and phosphorus at variable rates on a dryland soil deficient[a] in phosphorus supply

	Phosphorus (P_2O_5, kg ha^{-1})				
Nitrogen (kg ha^{-1})	0	17	34	68	102
0	–	27.4	38.4	40.3	40.6
34	7.7	31.0	45.2	47.3	52.7
68	3.9	33.0	46.1	63.0	64.9
102	4.4	43.9	54.1	62.3	65.1
136	23.2	41.1	56.3	63.5	65.2

Source: Li (2002)
P phosphorous
[a]Olsen-P = 5.7 mg kg^{-1} soil

Table 5 Influence of integrated use of nitrogen (N), phosphorus (P) and potash (K) on percent increase in maize yield under various precipitation years on a dryland soil

Precipitation	Increase in yield due to NP than N (%)	Increase in yield due to NPK than N (%)	Increase in yield due to NPK than NP (%)
Drought (<400 mm)	14.1	15.7	1.9
Normal (400–550)	10.7	19.6	10.0
High (550–650 mm)	12.5	24.1	13.2
Waterlogging (>650 mm)	34.7	44.0	14.2

Source: Ma et al. (2010)
N, P and K were applied at 150 kg N ha^{-1}, 17.9–25 kg P ha^{-1}, and 60 kg K ha^{-1}

combine use of N and P improved the maize grain yield by 14.1 % than the use of N alone. Application of NPK was most beneficial which enhanced the maize grain yield by 15.7 % than the use of N alone, while use of NPK than NP provided yield improvement of 1.9 % (Table 5).

3.7 *Use of Slow Release Fertilizers and Nitrification Inhibitors*

Use of slow release fertilizers and nitrification inhibitors in drylands regions might be useful to enhance the fertilizer use efficiency and reduce the nutrient losses. Slow release fertilizers (isobutylidene diurea, N-lignin, sulphur coated urea, shellac coated urea, urea briquettes, and urea super granules) and nitrification inhibitors [N-serve (2-chloro-6 (trichloromethyl pyridine), AM (2-amino, 4-chloro, 6 methyl pyridine), ST (2-sulphanileamide-thiazole) and dicyandiamide has widely tested in agriculture and offer pragmatic option to improve fertilizer use efficiency and reduce the nutrient losses in dryland agriculture. In dryland agriculture, losses of N due to nitrification are high which can be minimized by using the nitrification inhibitors. Indeed, nitrification inhibitors delay the microbial oxidation [conversion of NH_4^+ to nitrite (NO_2^-)] for some time (several weeks or months), and thus are very useful to reduce the microbial nitrification and subsequent denitrification (Weiske et al. 2001; Zerulla et al. 2001), in dryland agriculture. However, the efficacy of nitrification inhibitors to block nitrification and leaching is governed by several factors such as rate, method and time of application of nitrification inhibitors (Zaman and Blennerhassett 2010; Zaman and Nguyen 2012), crop type, crop geometry, irrigation, method of application of NH_4^+ based fertilizers (Sanz-Cobena et al. 2012), temperature, rainfall (Shepherd et al. 2012), soil pH, soil texture, soil mineral N and soil organic carbon (Barth et al. 2001; Shepherd et al. 2012).

4 Conclusions

The productivity of dryland crops can be substantially enhanced with integrated nutrient management techniques. The CA-bases cropping systems may be beneficial to conserve soil moisture and nutrients within soil. Application of chemical fertilizers should be on the basis of soil analysis. Besides chemical fertilizers, use of green manures, organic manures and compost should be encouraged and subsidized. The intercropping of legume and non-legume crops in dryland soils may be beneficial to enhance/sustain the soil fertility. Cover cropping during the rainy season may be useful to reduce the soil losses due to soil erosion. To reduce the loss of nutrients through runoff and erosion, the land leveling, contouring and terracing may be beneficial. Mulching should be done between crop rows to conserve the soil moisture and modulate the soil temperature. Use of slow release fertilizers and nitrification inhibitors may be another pragmatic option to improve the crop nutrition in dryland crops. Use of biofertilizers such as azotobacter, rhizobium, azospirilium, blue green algae, phosphate solubilizing organisms and vesicular arbuscular mycorrhiza may be an integrated part of integrated nutrient management in dryland agriculture. In dryland agriculture, the integrated plant nutrient management system necessities a holistic approach to manage crop nutrition through judicious and combined use of chemical fertilizers, organic manures (farmyard manure, compost, biogas slurry, crop residues, green manures, vermi-compost), biofertilizers and growing of leguminous crops in crop rotation.

References

Abdel Monem M, Lindsay WL, Sommer R, Ryan J (2010) Loss of nitrogen from urea applied to rain-fed wheat in varying rainfall zones in northern Syria. Nut Cycl Agroecosyst 86:357–366

Agarwal SK, De R (1979) Effects of rates of nitrogen, mulching and anti-transpirants on nutrient uptake of barely varieties under rainfed conditions. Indian J Agron 24:66–71

AICRPDA (All India Coordinated Research Project for Dryland Agriculture) 1999. Annual report of the all India coordinated research project for dryland agriculture, Central Research Institute for Dryland Agriculture, Hyderabad, India.

Allen BL, Pikulb JL, Waddellc JT, Cochrand VL (2010) Long-term lentil green-manure replacement for fallow in the semiarid northern Great Plains. Agron J 103:1292–1298
Anderson WK, Crosbie GB, Lemsom K (1995) Production practices for high protein, hard wheat in Western Australia. Aust J Exp Agric 35:589–595
Angus JF, Bowden JW, Keating BA (1994) Modelling nutrient responses in the field. Plant Soil 155:57–66
Babaria CJ, Patel CL (1981) Response and uptake of iron by sorghum to application of iron, farmyard manure and sulphur on calcareous soil. Gujrat Agric Uni Res J 6:121–124
Bandyopadhyay SK, De R (1986) N relationship in a legume-non-legume association grown in an intercropping system. Fert Res. 10:73–82
Baptista I, Ritsema C, Geissen V (2015) Effect of integrated water-nutrient management strategies on soil erosion mediated nutrient loss and crop productivity in Cabo Verde drylands. PLoS One 10:e0134244
Barth G, von Tucher S, Schmidhalter U (2001) Influence of soil parameters on the effect of 3,4-dimethylpyrazole-phosphate as a nitrification inhibitor. Biol Fertil Soils 34:98–102
Bationo A, Christianson CB, Baethgen WJE (1989) Plant density and nitrogen fertilizer effects on pearl millet production in a sandy soil of Niger. Agron J 82:290–295
Bationo A, Beelders CL, van Duivenbooden N, Buerkert AC, Seyni F (1997) The management of nutrients and water in the west african semi-arid tropics. In: Management of nutrients and water in rainfed arid and semi-arid areas Proceedings of a consultants meeting organized by the Joint FAO/IAEA Division of Nuclear Techniques in Food and Agriculture and held in Vienna, 26–29 May, 1997
Beck U (1992) Risk society: towards a new modernity. Sage, London
Bekunda MA, Bationo A, Sali H (1997) Soil fertility management in Africa: a review of selected research trials. In: Replenishing soil fertility in Africa, SSSA special publication number 51. Soil Science Society of America and American Society of Agronomy, Madison
Birch CJ, Bell LW (2011) Rainfed farming systems of North-Eastern Australia. In: Tow P, Cooper L, Partridge L, Birch C (eds) Rainfed farming systems. Springer, Heidelberg/London/New York
Bouraima AK, He B, Tian T (2015) Runoff, nitrogen (N) and phosphorus (P) losses from purple slope cropland soil under rating fertilization in Three Gorges Region. Environ Sci Pollut Res Int 23:4541–4550
BRRI (1984) Annual report. Bangladesh Rice Research Institute, Gazipur
Burayu W, Chinawong S, Suwanketnikom R, Mala T, Juntakool S (2006) Conservation tillage and crop rotation: win-win option for sustainable maize production in the dryland central Rift Valley of Ethiopia. Kamphaengsaen Academic J 4:48–60
Campbell CA, Zentner RP, Johnson PJ (1988) Effect of crop rotation and fertilization on the quantitative relationship between spring wheat yield and moisture use in southwestern Saskatchewan. Can J Soil Sci 68:1–16
Cao WD, Huang HX (2009) Ideas on restoration and development of green manures in China. (In Chinese.). Soil Fert Sci China 4:1–3
Carpenter-Boggs L, Pikul JL, Vigil MF (2000) Riedell WE. Soil nitrogen mineralization influenced by crop rotation and nitrogen fertilization Soil Sci Soc Am J 64:2038–2045
Carter OG (1967) Effect of chemical fertilizers on seedling establishment. Aust J Exp Agric Animal Husb 7:174–180
Cheng SY, Qiao CF, Han GM, Guo XD (1987) The effect of fertilizer application upon the improvement of soil moisture productive efficiency in wheat on the drylands. Agric Res Arid Areas 5:58–65
Crabtree B (2010) Search for sustainability with no-till bill in dryland agriculture. Crabtree Agricultural Consulting, Australia
Cutforth HW, McConkey BG (1997) Stubble height effects on microclimate, yield and water use eciency of spring wheat grown in a semi-arid climate on the Canadian Prairies. Can J Plant Sci 77:359–366

Dargan KS, Chandra H, Bhumbla DR (1973) In saline-alkali soil urea spray for more paddy. Indian Fmg 22:9–10

Duiker SW, Beegle DB (2006) Soil fertility distributions in long-term no-till chisel/disk and moldboard plow/disk systems. Soil Tillage Res 88:30–41

Eaglesham ARJ, Ayanaba A, Rao VR, Eskew DL (1981) Improving the nitrogen nutrition of maize by intercropping with cowpea. Soil Biol Biochem 13:169–171

El-Fouly MM (1983) Micronutrients in arid and semi-arid areas: levels in soils and plants and need for fertilizers with reference to Egypt. In: Nutrient balances and the need for fertilizers in semi-arid and arid regions. International Potash Institute, Berne

Espinoza S, Ovalle C, Zagal E, Matus I, Tay J, Peoples MB, del Pozo A (2012) Contribution of legumes to wheat productivity in Mediterranean environments of central Chile. Field Crops Res 133:150–159

Evans J, Scott G, Lemerle D, Kaiser A, Orchard B, Murray GM, Armstrong EL (2003) Impact of legume 'break' crops on the yield and grain quality of wheat and relationship with soil mineral N and crop N content. Crop Past Sci 54:777–788

FAO (1981) Crop production levels and fertilizer use. Fertilizer and PLANT NUTRITION. Bull. 2. FAO, Rome

FAO (1988) Efficient fertilizer use in summer rainfed areas. FAO Fertilizer and Plant Nutrition Bulletin 11, Fertilizer and Plant Nutrition Service, FAO Water and Land Development Divison, Rome, p 37.

Fernández-Ugalde O, Virto I, Bescansa P, Imaz MJ, Enrique A, Karlen DL (2009) No-tillage improvement of soil physical quality in calcareous, degradation prone, semiarid soils. Soil Tillage Res 106:29–35

Fileccia T (2009) Conservation Agriculture and Food Security in Kazakhstan. Working Paper, FAO Investment Centre Division, June 2009, Rome, FAO

Fillery IR, McInnes KJ (1992) Components of the fertilizer nitrogen balance for wheat production on duplex soils. Aust J Exp Agric 32:887–899

Gajanan GN, Hegde BR, Ganapathi P, Somashekhar K (1999) Organic manure for stabilizing productivity: experience with dryland finger millet, All India Coordinated Research Project on Dryland Agriculture. University of Agricultural Sciences, Bangalore

Gill AS, Pandey RK, Singh M, Abichandani CT (1971) Effect of soil and foliar application of phosphorous on seed yield of cowpea. Indian J Agron 16:303–304

Giri G, De R (1981) Short season fodder legume effects on the grain yield and nitrogen economy of barely grown under dryland conditions. J Agric Sci Camb 96:457–461

Hadda MS, Arora S (2006) Soil and nutrient management practices for sustaining crop yields under maize-wheat cropping sequence in sub-mountain Punjab, India. Soil Environ 25:1–5

Harmsen K (1984) Nitrogen fertilizer use in rainfed agriculture. Fertil Res 5:371–382

Hayden ZD, Ngouajio M, Brainard DC (2014) Rye–vetch mixture proportion tradeoffs: cover crop productivity, nitrogen accumulation, and weed suppression. Agron J 106:904–914

Hegde BR (1980) Agronomic practices suited to the dryland conditions in Karna-taka and Andhra Pradesh. In: Third FAI specialized training programme on management of rainfed areas. FAI, New Delhi

Henao J, Baanante C (2006) Agricultural production and soil nutrient mining in Africa: implications for resource conservation and policy development. IFDC-An International Center for Soil Fertility and Agricultural Development

Hossain SA, Strong WM, Waring SA, Dalal RC, Weston EJ (1996) Comparison of ley-based cropping systems at Warra, Queensland. II. Mineral nitrogen accumulation and availability to the subsequent wheat crop. Aust J Soil Res 34:289–297

IAEA (2005) Nutrient and water management practices for increasing crop production in rainfed arid/semi-arid areas. IAEA, Vienna. IAEA-TECDOC-1468; ISBN 92–0–109405–1

IFDC (2006) Agricultural production and soil nutrient mining in Africa: implications for resource conservation and policy development, IFDC Technical Bulletin. IFDC, Alabama

IFPRI (1996) Feeding the world, preventing poverty and protecting the earth: a 2020 vision. International Food Policy Research Institute, Washington, DC
Impiglia A, Ryan J (1997) Cereal quality and mineral nutrition in the West Asia and North Africa environments. In: Johnston AE (ed) Food security in the WANA region, the essential need for balanced fertilization. International Potash Institute, Basel
Irshad M, Inoue M, Ashraf M, Faridullah HK, Delower M, Tsunekawa A (2007) Land desertification – an emerging threat to environment and food security of Pakistan. J Applied Sci 7:1199–1205
Izaurralde RC, Malhi SS, Nyborg M, Solberg ED, Quiroga MC (2006) Crop performance and soil properties in two artificially eroded soil in North-Central Alberta. Agron J 98:1298–1311
Janzen HH, Desjardins RL, Asselin JMR, Grace B (1999) The health of our air: towards sustainable agriculture in Canada. Research Branch, Agriculture and Agric-Food Canada, Ottawa, Publication No. 1981/E.
Jin SL (1989) Balanced fertilization of spring wheat in the Hexi Corridor. In: Proceedings of the international symposium on balanced fertilization. Institute of Soil and Fertilizers, China's Academy of Agriculture Sciences, China Agriculture Press, Beijing, pp 179–184
Kang BT, Osiname OA (1979) Phosphorous response of maize grown on Alfisols of Southern Nigeria. Agron J 71:873–877
Kanwar JS, Mudhahar MS (1984) Fertilizer, sulphur and food production: research and policy implications for tropical countries. Fert News 30:37–54
Kanwar JS, Rego TJ (1983) Fertilizer use and watershed management in rainfed areas for increasing crop production. Fert News 28:33–42
Lafond GP, Walley H, Schoenau J, May WE, Holzafel CB, McKell J, Halford J (2008) Long-term vs. short-term conservation tillage. In: Proceedings of the 20th annual meeting and conference of the Saaskatchewan soil conservation association, 12–13 February, Regina, pp 28–43
Lal R (1974) Soil temperature, soil moisture and maize yield from mulched and unmulched tropical soils. Plant Soil 40:129–143
Larney FJ, Olson BM, Janzen HH, Lindwall CW (2000) Early impact of topsoil removal and soil amendments on crop productivity. Agron J 92:948–956
Li SX (2002) Ways and strategies for increasing fertilizer nitrogen efficiency in dryland soil. Acta Pedologica Sin 39:56–75
Li SX (2007) Dryland Agriculture in China. Science Press, Beijing
Li SX, Zhao BS, Li CW (1976a) The effect of fertilizer application time on wheat yields. In: Selected works of Shaanxi agricultural science and technology. Shaanxi Agricultural Ministry, Shaanxi People's Publishing House, Xi'an
Li SX, Zhao BS, Li CW (1976b) Increasing fertilizer efficiency by deep application. In: Selected works of Shaanxi agricultural science and technology. Shaanxi Agricultural Ministry, Shaanxi People's Publishing House, Xi'an
Li SX, Zhao BS, Li CW (1978) Rational application of phosphate fertilizer to wheat, part I. Acta Collegii Septentrionali Occidentali Agric 6:77–85
Li SX, Zhao BS, Li CW (1979) Rational application of phosphate fertilizer to wheat, part II. Acta Collegii Septentrionali Occidentali Agric 7:55–99
Li SX, Zhao BS, Li CW (1990) The effect of application of N, P and organic fertilizers on N-fixing performance of leguminous crops. Chinese J Soil Sci 21:13–15
Li SX, Yuan HL, He HX, Li SQ (1992) Assessing nitrogen fixation efficiency of two legume plants from nitrogen intake in three successively grown crops. Chinese J Soil Sci 23:130–133
Li SQ, Wang XQ, Gao YJ, Du JJ, Li SX (1995) Effects of fertilization on increase of transpiration and reduction of evaporation. In: Wang DS (ed) The principle of relationship between fertilizer and water of dry farming and its regulation technology. Agricultural Science and Technology Publishing House of China, Beijing

Li SX, Wang ZH, Hu TT, Gao YJ, Stewart BA (2009a) Nitrogen in dryland soils of China and its management. Adv Agron 101:123–181

Li SX, Wang ZH, Malhi SS, Li SQ, Gao YJ, Tian XH (2009b) Nutrient and water management effects on crop production, and nutrient and water use efficiency in dryland areas of China. Adv Agron 102:223–265

Lindwall CW, Sonntag B (2010) Landscape transformed: the history of conservation tillage and direct seeding. Knowledge Impact in Society. University of Saskatchewan, Sakskatoon

Loutit A, Stallwood P, Cox WJ (1968) Drilled versus topdressed superphosphate for cereal production. J West Aust Dep Agric 9:418–421

Lu, DQ, Li Y (1987) Enhancement of the rate of nitrogen utilization for increasing wheat production on Loess Plateau. In: Proceedings of the international symposium on dryland farming, Sponsored by Agronomy Society of China, Agricultural Academy of China and Norwest Agricultural University, September 18–22, 1987, Yangling, Northwest Agricultural University Press, Yangling, pp 236–248

Lyon D, Bruce S, Vyn T, Peterson G (2004) Achievements and future challenges in conservation tillage. In: 4th international crop science congress, 26 Sep-1 Oct, 2004, Brisbane

Ma YH, Zhang YP, Ca GL, Wan JL, Wan GX, Ren JX (1984) Studies on improvement of soil fertility—the effect of organic fertilizer on organic–mineral colloidal complex and water use efficiency of soil. Agric Res Arid Areas 2:56–71

Ma Q, Yu W, Shen S, Zhou H, Jiang Z, Xu Y (2010) Effects of fertilization on nutrient budget and nitrogen use efficiency of farmland soil under different precipitations in Northeastern China. Nutr Cycl Agroecosyst 88:315–327

MacColl D (1989) Studies on maize (*Zea mays* L.) at Bunda, Malawi: II. Yield in short rotation with legumes. Exp Agric 25:367–374

Malhi S, Izaurralde R, Nyborg M, Solberg E (1994) Influence of topsoil removal on soil fertility and barbey growth. J Soil Water Cons 49:96–101

Mason MG (1971) Effects of urea, ammonium nitrate and superphosphate on establishment of cereals, linseed and rape. Aust J Exp Agric Animal Husb 11:662–669

Mason MG (1975) Nitrogen fertilizers for cereal production. J Agric West Austr 16:3–10

Mason MG, Porter WM, Cox WJ (1994) Effect of an acidifying nitrogen fertilizer and lime on soil pH and wheat yields 2. Plant response Aust J Exp Agric 34:247–253

McDonald GK (1989) The contribution of nitrogen fertilizer to the nutrition of rainfed wheat crops in Australia: a review. Aust J Exp Agric 29:455–481

Meelu OP, Singh R, Sekhar GS (1976) Fertilizer use and crop production technology for rainfed areas of Punjab. Fert News 21:34–38

Mesfin T, Abebe G, Al-Tawaha ARM (2005) Effect of reduced tillage and crop residue ground cover on yield and water use efficiency of sorghum (*Sorghum bicolor* (L.) Moench) under semiarid conditions of Ethiopia. World J Agric Sci 1:152–160

Mohat NK, De R (1971) It pays to fertilize cowpeas with phosphorous through leaves. Indian Fmg 20:27–28

Mrabet R (2000) Differential response of wheat to tillage management systems under continuous cropping in a semiarid area of Morocco. Field Crops Res 66:165–174

Mrabet R (2002) Wheat yield and water use efficiency under contrasting residue and tillage management systems in a semiarid area of Morocco. Exper Agric 38:237–248

Mughogho SK, Bationo A, Christianson CB, Vlek PLG (1986) Management of nitrogen fertilizers for tropical African soils. In: Mokwunye U, PLG V (eds) Management of nitrogen and phosphorus fertilizers in Sub-Saharan Africa. Martinus Nijho Publishers, Dordrecht

Mukurumbira LM (1985) Effects of rate of fertilizer nitrogen and previous grain legume crop on maize yields. Zimbabwe Agric J 82:177–179

Murage EW, Karanja NK, Smithson PC, Woomer PL (2000) Diagnostic indicators of soil quality in productive and non-productive smallholders' fields of Kenya's Central Highlands. Agric Ecosyst Environ 79:1–8

Nagy JG, Ohm HW, Sawadogo S (1990) Farmer-researcher collaboration in developing cropping practices that involve the complementary use of mechanical tied ridge and fertilizer technology for sorghum in Burkina Faso. Exp Agric 26:161–169

Nambiar KKM (1985) All-India coordinated research project on long-term fertilizer experiments and its research achievements. Fert News 3:56–66

Norwood CA (2000) Dryland winter wheat as affected by previous crops. Agron J 92:121–127

Nyakatawa EZ (1996) Rain water and soil fertility management for sustainable cropping on sandy soils in semi-arid south-east Lowveld of Zimbabwe. J Sustain Agric 7:19–34

Nyborg M, Hennig AMF (1969) Field experiments with different placements of fertilizers for barley, flax and rapeseed. Can J Soil Sci 49:79–88

Oicha T, Cornelisc WM, Verplancke H, Nyssen J, Govaerts B, Behailu M, Haile M, Deckers J (2010) Short-term effects of conservation agriculture on Vertisols under tef (*Eragrostis tef* (Zucc.) Trotter) in the northern Ethiopian highlands. Soil Tillage Res 106:294–302

Onim JFM, Ochola P, Fitzhugh P, Oduor HA (1990) Potential of goat manure as a valuable fertilizer for small-scale farmers. In: Proceedings of the Eighth Small Ruminants Scientific Workshop. Small Ruminants Collaborative Research, ILRAD

Pala M, Ryan J, Zhang H, Singh M, Harris HC (2007) Water-use efficiency of wheat-based rotation systems in a Mediterranean environment. Agric Water Manage 93:136–144

Palm CA, Myers RJK, Nandwa SM (1997) Combined use of organic and inorganic nutrient sources for soil fertility maintenance and replenishment. In: Replenishing soil fertility in Africa, SSSA special publication number 51. Soil Science Society of America and American Society of Agronomy, Madison

Peverill KI, Sparrow LA, Reuter DJ (1999) Soil Analysis: an interpretation manual. CSIRO Publishing, Collingwood, p. 369

Pieri C (1995) Long-term soil management experiments in semiarid Francophone Africa. In: Lal R, Stewart BA (eds) Soil management: experimental basis for sustainability and environmental quality. CRC Press, Boca Raton

Piggin C, Haddad A, Khalil Y (2011) Development and promotion of zero-tillage in Iraq and Syria. In: Proceedings of the 5th world congress on conservation agriculture, Brisbane, 26–29 September, pp 304–305

Place F, Barrett CB, Freeman HA, Ramisch JJ, Vanlauwe B (2003) Prospects for integrated soil fertility management using organic and inorganic inputs: evidence from smallholder African agricultural systems. Food Policy 28:365–378

Prasad R (2009) Enhancing nutrient use efficiency-environmental benign strategies. Souvenir. Platinum Jubilee Celebrations of the Indian Society of Soil Science and 74th Annual Convention, New Delhi, pp 67–74

Prihar SS, Sandhu KS, Singh Y, Singh R (1981) Effect of N rates on dryland wheat in relation to mulching previous crop or fallow. Fert Res 2:211–219

Puckridge DW, French RJ (1983) The annual legume pasture in cereal ley farming systems of Southern Australia: a review. Agri Ecosyst Environ 9:229–267

Raghavulu P, Singh SP (1982) Effects of mulches and transpiration suppressants on yield, water use efficiency and uptake of nitrogen and phosphorous of sorghum under dryland conditions of north-western India. J Agri Sci Camb 98:103–109

Randhawa SS, Singh R (1983) Fertilizer management in rainfed areas- available technologies and future needs. Fert News 28:17–32

Rao SBP, Ramachandran K, Roybhowmik AB (1973) Foliar application of urea on cotton. Cotton Dev 3:1–12

Rashid A (1994) Annual report of micronutrient project. Land Resources Research Institute, NARC, Islamabad

Rashid A, Ryan J, Chaudhry MA (2004) Challenges and strategies for dryland agriculture in Pakistan. In: Rao SC, Ryan J (eds) Challenges and strategies of dryland agriculture, Crop Science Society of America and American Society of Agronomy. CSSA Special Publication No, vol 32

Rees RM, Roelcle M, Li SX, Wang XQ, Li SQ, Stockdale EA, McTaggart IP, Smith KA, Richter J (1997) The effect of fertilizer placement on nitrogen uptake and yield of wheat and maize in Chinese loess soil. Nutr Cycl Agroecosyst 47:81–91

Rego TJ, Rao VN (2000) Long-term effects of grain legumes on rainy-season sorghum productivity in a semi-arid tropical Vertisol. Exper. Agric. 36:205–221

Rego TJ, Wani SP, Sahrawat KL, Pardhasaradhi G (2005) Macro-benefits from boron, zinc and sulfur application in Indian SAT: a step for Grey to Green Revolution in agriculture, Global Theme on Agroecosystems Report no. 16. International Crops Research Institute for the Semi-Arid Tropics, Patancheru 24 pp

Rego TJ, Sahrawat KL, Wani SP, Pardhasaradhi G (2007) Widespread deficiencies of sulfur, boron, and zinc in Indian semi-arid tropical soils: on-farm crop responses. J Plant Nutr 30:1569–1583

Reuter DJ, Robinson JB (1997) Plant analysis: an interpretation manual. CSIRO Publishing, Collingwood, p. 572

Rowland I, Mason MG, Hamblin J (1988) Effect of lupins and wheat on the yield of subsequent wheat crops grown at several rates of applied nitrogen. Aust J Exp Agric 28:92–97

Russell JS (1967) Nitrogen fertilizer and wheat in a semi-arid environment. 1. Effect on yield. Aust J Exp Agric Animal Husb 7:453–462

Ryan J., Sommer R (2010) Fertilizer best management practices in the dryland Mediterranean area- concepts and perspectives. In: 19th world congress of soil science, soil solutions for a changing world 1–6 August 2010, Brisbane, pp 97–99.

Ryan J, Masri S, Masri Z (1997) Potassium status in soils and crops, recommendations and present use in Syria. In: Johnston AE (ed) Food security in the WANA region, the essential need for balanced fertilization. International Potash Institute, Basel

Ryan J, Singh M, Pala M (2008a) Long-term cereal-based rotation trials in the Mediterranean region: implications for cropping sustainability. Adv Agron 97:273–319

Ryan J, Ibrikci H, Rashid A, Masri S, Pala M, Matar A, Singh M (2008b) Residual and currently applied phosphorus in dryland cereal/legume rotations across a rainfall gradient in a Mediterranean ecosystem. Eur J Agron 28:126–137

Ryan J, Masri S, Ibrikci H, Singh M, Pala M, Harris HC (2008c) Implications of cereal-based crop rotations, nitrogen fertilization, and stubble grazing on so organic matter in a Mediterranean environment. Turk J Agric For 32:289–297

Ryan J, Masri S, Singh M, Pala M, Ibrikci H (2008d) Total and mineral nitrogen in a wheat-based rotation under dryland Mediterranean conditions. Basic Appl Dryland Res 2:34–46

Ryan J, Pala M, Harris H, Masri S, Singh M (2010) Rainfed wheat-based rotations under Mediterranean-type climatic conditions: crop sequences, N fertilization, and stubble grazing intensity in relation to cereal yields parameters. J Agric Sci Camb 148:205–216

Sahrawat KL, Wani SP, Rego TJ, Pardhasaradhi G, Murthy KVS (2007) Wide spread deficiencies of sulphur, boron and zinc in dryland soils of the Indian semi-arid tropics. Current Science 93(10):1–6

Sahrawat KL, Wani SP, Pardhasaradhi G, Murthy KVS (2010) Diagnosis of secondary and micronutrient deficiencies and their management in rainfed agroecosystems: case study from Indian semi-arid tropics. Commun Soil Sci Plant Anal 41:346–360

Saleem MT, Nahban H, DeBie C (1996) Fertilizer management and crop production in rainfed areas in Pakistan. In: Proceedings of IMPHOS seminar on crop production techniques and fertilizer management in rainfed agriculture in Southern Asia, New Delhi

Sanchez PA, Buresh RJ, Leakey RRB (1997) Trees, soils, and food security. Philos Trans R Soc London B 352:946–961

Sandhu KS, Benbi DK, Prihar SS, Saggar S (1992) Dryland wheat yield dependence on rainfall, applied N and mulching in preceding maize. Fert Res 32:229–237

Sanz-Cobena A, S´anchez-Mart´ın L, Garc´ıa-Torres L, Vallejo A (2012) Gaseous emissions of N_2O and NO and NO−3 leaching from urea applied with urease and nitrification inhibitors to a maize (*Zea mays*) crop. Agric Ecosyst Environ 149:64–73

Scaife MA (1971) The long-term effects of fertilizers, farmyard manure and leys at Mwanhala, western Tanzania. East Afr Agric Forest J 37:8–14
Sedogo PM (1993) Evolution des sols ferrugineux lessives sous culture: incidence des modes de gestion sur la fertilite. PhD thesis. University of Abidjan, Cote d'lvoire
Sekhon GS (1983) Potassium dynamics in the soils of semi-arid and arid areas In: Proceedings of 17th colloquium, International Potash Institute, pp 153–162
Seymour M, Kirkegaard JA, Peoples MB, White PF, French RJ (2012) Break-crop benefits to wheat in Western Australia–insights from over three decades of research. Crop Past Sci 63:1–16
Shah ZH, Arshad M (2006) Land degradation in Pakistan: a serious threat to environments and economic stability. Eco Services International. Available at: http://www.ecoweb/edi/060715.html. Accessed in 12 July 2016
Sharma HC, Singh UN, Singh RK (1979) Fertilizer use in drylands: problems, suggestions and field experiences to solve them in U.P. and M.P. In: Proceedings of FAI group discussion on fertilizer use in dry-lands, FAI, New Delhi, pp 92–113
Sharma KL, Srinivas K, Das SK, Vittal KPR, Kusuma GJ (2002) Conjunctive use of inorganic and organic sources of nitrogen for higher yield of sorghum in dryland Alsol. Indian J Dryland Agri Res Dev 17:79–88
Sharma KL, Vittal KPR, Ramakrishna YS, Srinivas K, Venkateswarlu B, Kusuma Grace J (2007) Fertilizer use constraints and management in rainfed areas with special emphasis on N use efficiency. In: Abrol YP, Raghuram N, Sachdev MS (eds) Agricultural nitrogen use & its environmental implications. IK International Publishing House Pvt. Ltd., New Delhi
Shepherd M, Wyatt J, Welten B (2012) Effect of soil type and rainfall on dicyandiamide concentrations in drainage from lysimeters. Soil Res 50:67–75
Singh RP, Das SK (1995) Agronomic aspects of plant nutrient management in rain dependent food crop production systems in India. In: Singh RP (ed) Sustainable development of dryland agriculture in India. Scientific Publishers, Jodhpur
Singh RP, Venkateswarlu J (1985) Role of All-India coordinated research project for dryland agricultural research development. Fert News 30:43–55
Singh Y, Singh R, Verma HN, Gill AS (1977) Growth and yield of dryland wheat as affected by methods of N application on clay loam and sandy loam soils. J Res (Punjab Agri Univ) 14:29–33
Singh RK, De R, Turkhede BB (1985) Time of application of farmyard manure and fertilizer nitrogen on the growth, yield and nutrient uptake of dryland wheat. J Agric Sci Camb 106:7–15
Smaling EMA, Stoorveogel JJ, Winmeijer PN (1993) Calculating soil nutrient balances in Africa at different scales II. District scale. Fert Res 35:237–250
Smith AN (1983) Fertilization. In: Soil factors in crop production in semi-arid environment. University of Queenland Press, St. Lucia
Snapp SS, Silim SN (2002) Farmer preferences and legume intensification for low nutrient environments. Plant Soil 245:181–192
Spratt ED, Chowdhury SL (1978) Improved cropping system for rainfed agriculture in India. Field Crop Res 1:103–106
Tandon HLS (1980) Fertilizer use in dryland agriculture, Fert Inform Bull No 13. European Nitrogen Service Programme, New Delhi
Temesgen M, Hoogmoed WB, Rockstrom J, Savenije HHG (2009) Conservation tillage implements and systems for smallholder farmers in semiarid Ethiopia. Soil Tillage Res 104:185–191
Thomas GA, Titmarsh GW, Freebairn DM, Radford BJ (2007) No-tillage and conservation farming practices in grain growing areas of Queensland – a review of 40 years of development. Aust J Exper Agric 47:887–898
Tong CL, Hall CAS, Wang HQ (2003) Land use change in rice, wheat and maize production in China (1961–1998). Agric Ecosyst Environ 95:523–536

Turner NC, Li FM, Xiong YC, Siddique KHM (2011) Agricultural ecosystem management in dry areas: challenges and solutions. Plant Soil 347:1–6
Umrani NK, Patil ND (1983) Advances in fertilizer management for rainfed sorghum. Fert News 28:57–61
Uyovbisere EO, Lombin G (1991) Efficient fertilizer use for increased crop production: -e sub-humid Nigeria experience. Fert Res 29:81–94
van Duivenbooden N, Pala M, Studer C, Bielders CL (1999) Efficient soil water use: the key to sustainable crop production in the dry areas of West Asia, and North and Sub-Saharan Africa. International Center for Agricultural Research in the Dry Areas, Aleppo, Syria, and International Crops Research Institute for the Semi-Arid Tropics, Patancheru
Vasuki N, Yogananda SB, Preethu DC, Sudhr K, Jayaprakash SM (2009) Impact of long term fertilizer application on soil quality, crop productivity and sustainability: two decades' experience. Department of Soil Science and Agricultural Chemistry, University of Agricultural Sciences, Bangalore, p 22
Venkateswarlu J (1987) Efficient resource management systems for drylands of India. Adv Soil Sci 7:165–221
Waddington SR, Karigwindi J (2001) Productivity and profitability of maize + groundnut rotations compared with continuous maize on smallholder farms in Zimbabwe. Exp Agric 37:83–98
Walley FL, Clayton GW, Miller PR, Carr PM, Lafond GP (2007) Nitrogen economy of pulse crop production in the Northern Great Plains. Agron J 99:1710–1718
Wang XB, Cai DX, Zhang JQ, Gao XK (2003) Nitrogen uptake by corn and N recovery in grain in dry farmland. Chinese Agric Sci 2:898–905
Wani SP, Pathak P, Jangawad LS, Eswaran H, Singh P (2003) Improved management of Vertisols in the semi-arid tropics for increased productivity and soil carbon sequestration. Soil Use Manage 19:217–222
Weiske A, Benckiser G, Herbert T, Ottow JCG (2001) Influence of the nitrification inhibitor 3,4-dimethylpyrazole phosphate (DMPP) in comparison to dicyandiamide (DCD) on nitrous oxide emissions, carbon dioxide fluxes and methane oxidation during 3 years of repeated application in field experiments. Biol Fertil Soils 34:109–117
Westfall DG, Havlin JL, Hergert GW, Raun WR (1996) Nitrogen management in dryland cropping systems. J Prod Agric 9:192–199
Wu JG (1989) The effects of combination of N, P and B application on increase of wheat production. In: Proceedings of the international symposium on balanced fertilization. Institute of Soil and Fertilizers, China's Academy of Agriculture Sciences, China Agriculture Press, Beijing, pp 357–360
Yadav SS, Singh S, Tikoo A, Yadava JS (2007) Studies on potash responses to field crops in light textured soils of Southern Haryana, India. e-ifc No. 13, p 4–7
Yu WT, Zhao XH, Zhang L, Ma Q (2007) Contribution of long-term fertilization to crop yield. Chinese J Ecol 26:2040–2044
Yu YL, Xue LH, Yang LZ (2013) Winter legumes in rice crop rotations reduces nitrogen loss, and improves rice yield and soil nitrogen supply. Agron Sustain Dev 34:633–640
Zaman M, Blennerhassett JD (2010) Effects of the different rates of urease and nitrification inhibitors on gaseous emissions of ammonia and nitrous oxide, nitrate leaching and pasture production from urine patches in an intensive grazed pasture system. Agric Ecosyst Environ 136:236–246
Zaman M, Nguyen ML (2012) How application timings of urease and nitrification inhibitors affect N losses from urine patches in pastoral system. Agric Ecosyst Environ 156:37–48
Zerulla W, Barth T, Dressel J, Erhardt K, Horchler von Locquenghien K, Pasda G, R¨adle M, Wissemeier AH (2001) 3,4-Dimethylpyrazole phosphate (DMPP) – a new nitrification inhibitor for agriculture and horticulture. Biol Fertil Soils 34:79–84
Zhang JY, Yang BQ, Wan TY (1982) Raising dryland agriculture into the modern level: investigation of Shanxi agricultural development. In: Proceedings of national symposium on dryland

farming of northern China. Shanxi Agronomy Science Society, Academy of Agricultural Sciences, Economic Institute Press, Beijing, pp 42–47

Zhang D, Yao P, Na Z, Cao W, Zhang S, Li Y, Gao Y (2016) Soil water balance and water use efficiency of dryland wheat in different precipitation years in response to green manure approach. Sci Rep. doi:10.1038/srep26856

Zia MS, Ali A, Aslam M, Baig MB, Saeed Z, Khan MB, Mann RA (1997) Fertility issues and fertilizer management in rice wheat system. Farm Research Notes for Asia and Far East. FAO Publication, Issue No.23, pp 36–50

Zia MS, Baig MB, Tahir MB, Yasin M (1998) Sustaining crop yields through fertilizers: a reality. Sci Vis 4:31–40

Insect-Pests in Dryland Agriculture and their Integrated Management

Ahmad Nawaz, Muhammad Dildar Gogi, and Muhammad Sufyan

1 Introduction

Insects are most adaptable to any condition and form of life. They constitute three-quarters of all the animals and are considered the most diverse of all living species on Earth. Apart from their natural oceanic habitat, they are found in deserts, mountains and also even in very harsh locations such as pools of crude petroleum (Imms 1964). Insects can have direct positive impacts on humans and the environment as predators or parasitoids of harmful pests, pollinators, decomposers or producers (honey or silk). However, only 1 % of all insects is considered pests (Vega and Kaya 2012) and is known to cause damage to crop plants, stored products and buildings, or act as vectors for animal and plant diseases, etc. About one-fifth of the world's total crop production is lost due to herbivorous insects. Several control measures have been developed to minimize the damage caused by insect pests, but the problem still prevails in several agroecosystems. One reason for this failure is the ability of insect pests to evolve new biotypes. They can, thus, adapt to overcome the effect of pesticides or bypass plant resistance, for instance, which further confounds the problem (Roush and McKenzie 1987).

In addition to the water and soil problems in dryland areas, insect pests pose a considerable threat to dryland cropping systems and production. Researchers estimate that food crops produced in the arid and semi-arid tropics are adversely

A. Nawaz (✉) • M.D. Gogi
Integrated Pest Management Laboratory, Department of Entomology, University of Agriculture, Faisalabad, Pakistan
e-mail: nawazrajpoot@yahoo.com

M. Sufyan
Biological Control Laboratory, Department of Entomology, University of Agriculture, Faisalabad, Pakistan

M. Farooq, K.H.M. Siddique (eds.), *Innovations in Dryland Agriculture*,
DOI 10.1007/978-3-319-47928-6_6

affected by more than a thousand species of insect pests, fungi, viruses and weeds (http://exploreit.icrisat.org/page/pests_and_diseases/923). Certain pests flourish in dry environments while others prefer moist conditions. In any monoculture that is often practiced in dryland soils, once a pest settles in a field, serious damage can occur with frequent outbreaks of insect pests (Pimentel 2009). Many insect pests are common in dryland soils with life cycles of up to a year or more. Some dryland pests enter periods of dormancy, unlike wetland pests. Some insect pests such as armyworms, butterflies, locusts and grasshoppers have greater dispersal powers than most wetland insect pests, except for rice plant hoppers and leaf folders (Denno et al. 1991). In addition to insects, dryland pests include birds, rodents, wild pigs, monkeys, squirrels and even elephants and rhinoceroses (Grist and Lever 1969; Fujisaka et al. 1991).

Producers in different dryland regions (for example, USA) are regularly faced with aphid pressure in wheat fields, of which the most prevalent and detrimental are the greenbug (*Schizaphis graminum* R.) and Russian wheat aphid (*Diuraphis noxia* K.) (Kelsey and Mariger 2002; Giles et al. 2003; Momhinweg et al. 2006; Keenan et al. 2007a, b). The greenbug is considered the key pest of wheat in much of the dryland area in the United States due to its frequent occurrence and potential to cause severe crop damage. In the absence of management practices and natural enemies, greenbugs are capable of reproducing quickly in the warmer conditions of the Great Plains and subsequently reducing yields significantly, and sometimes exceeding economic injury levels (Kieckhefer and Kantack 1988; Webster 1995; Kindler et al. 2002, 2003; Giles et al. 2003). Aphids are chronic pests of peas and can transmit several viral diseases. Weevils can also damage peas and lentils in dryland areas.

Residual and other health-related problems regarding synthetic insecticides spur the development of biocontrol and other control practices for insect pests in dryland farming. Biotechnological approaches may provide opportunities to tailor biocontrol agents for the sustainable development of dryland cropping systems. In the long run, genetically-modified crops and integrated pest management (IPM) strategies may be the best options for IPM in dryland regions. This chapter is divided into sections with an emphasis on insect pests in dryland agriculture systems and their economic importance, followed by limitations and problems with management approaches. IPM approaches are discussed along with their advantages and disadvantages in dryland agriculture systems for enhancing productivity and food security.

2 Economic Importance of Insects in Dryland Agriculture Systems

Insects are part of the many components of an agroecosystem and are found in virtually every terrestrial, fresh-water environment. They are abundant on Earth and are involved in many biological processes. Contrary to popular belief, only 5000 insect species are considered harmful to crops, livestock or human beings of the roughly one million described species (Van Lenteren 2006). Insects benefit nature by regulating ecosystem services that are fundamental to the survival of humankind. For instance, insects play a significant role in plant reproduction, and about 70 % of the world's most productive crop species depend on pollinators to some extent, contributing an estimated €153 billion to the global economy and accounting for approximately 9 % of agricultural production (Teeb 2010). Pollinators play an important role in the food production process for several key fruit and certain vegetable crops, and consequently provide ecosystem services beneficial to human health, nutrition and, in turn, food security. In the case of dryland agriculture ecosystems, tight associations between dryland plants and pollinators are known; for example, Agava and its pollinator (Arizaga et al. 2000). The extent to which changes in dryland regions affect pollination services and the dependency of dryland plant species on pollination has not been fully explored. A wide range of plants benefit from pollinators, e.g. *Gloriosa minor* R. (Colchicaceae) is a species of dryland wildflower in Kenya that is pollinated exclusively by butterflies. It relies on pollinators for seed production and survival in arid environmental conditions (Martins 2014). Additionally, seeds of many dryland plant species can be disseminated by fruit-eating birds, often before or after their cross-desert seasonal migration, especially in the Mediterranean basin (Izhaki et al. 1991). Sometimes domestic and wild mammalian herbivores are involved in seed dispersal if seeds become attached to their fur or by consuming seeds and then defecating, which promotes dispersal and enhances the chance of germination, e.g., African acacia trees (Ward 2003). Livestock and other animals may also play a role in seed relocation from improved pasture lands to neighboring non-managed rangelands (CGIAR 1997). Therefore, the services of pollinators in dryland areas has a significant impact; however, changing conditions and cropping biodiversity require more attention for the future of the pollinator's contribution in dryland areas.

In addition to pollination, insects play an equally vital role in waste biodegradation. This service supports soil development and primary production through the breakdown of dead plant parts, enriching the soil with organic matter until it is fit to be consumed by fungi and bacteria and regenerating mineral plant nutrients (Shapiro et al. 2005). A wide range of insects, including beetle larvae, flies, ants and termites, clean up dead plant matter which breaks down the organic matter and nutrients of dead organisms to then be readily available in the soil for plants. Unlike non-dryland areas, where insects and other soil microorganisms are major players in nutrient cycling, invertebrate macro-decomposers play a significant role in dryland areas, and their role increases with aridity. In dry and warmer parts, many termite species contain host nitrogen-fixing organisms in their gut, so they can increase the nitrogen content of soil to benefit crops (Cunningham et al. 2010). Some populations, such as microbes and fungi, decline with aridity due to their stringent moisture dependence. Similarly, the role of large herbivores for nutrient cycling is limited due to the lack of drinking water in arid and hyper-arid areas. However, some macro-decomposers, such as termites, darkling beetles (Tenebrionidae) and other soil dwellers, are less moisture sensitive so become vital for nutrient cycling in dryland areas. These organisms are important in litter preparation for microbial activity and increase the soil infiltration capacity which acts as a key to enhancing crop production (Shapiro et al. 2005).

When dryland areas are used as rangelands, most of the primary production takes place with livestock rather than macro-decomposers. However, livestock may gradually deplete the nitrogen reserves and further exacerbate the nutrient limitations for primary production (Ayal et al. 2005). This depletion may be partially mitigated by biological nitrogen fixation and by urea deposition in the soil (Shachak and Lovett 1998). Conversely, when dryland areas are used for crop production, some cultural practices such as tillage and the excessive use of chemicals to manage insect pests can reduce the role of soil-dwelling macro-decomposers. This, together with the low root biomass of annual crops, can impair nutrient cycling and decrease soil organic carbon and its associated nutrients (Shapiro et al. 2005).

Moreover, natural enemies—insect predators and parasitoids—attack and feed on other insect pests to keep them under economic threshold levels. Thus, they can be manipulated in pest management programs to avoid the use of toxic chemicals (Table 1). Natural enemies help to prevent the outbreak of pest populations and contribute to a type of pest regulation referred to as natural biological control. Natural enemies are responsible for 33 % of natural pest control in cultivated systems (Getanjaly et al. 2015). Natural enemies belong to about 20 insect orders and are characterized as free-living, mobile, larger than their prey, and able to consume several prey throughout their life cycle (DeBach and Rosen 1991). Whereas parasitoids are mainly belonging to the Hymenoptera and Diptera orders with a more specialized host range than predators (Strand and Obrycki 1996). Free-living adult parasitoids parasitize different life stages of their host (i.e. egg, larva, pupa or adult) depending on the parasitoid species. Parasitoids can lay their egg (solitary) or several eggs (gregarious) on or within their host; after hatching, the immature parasitoid(s) feed on their host to complete development by consuming the host and emerging as a free-living adult (DeBach and Rosen 1991). Dryland crops like chickpea (*Cicer arietinum* L.) and pigeon pea (*Cajanus cajan* L.), for example, stand to benefit if natural enemies can be maintained in farmers' fields. However, both species produce trichomes or small hairs that act as a natural defense against predators which can reduce the effectiveness of natural enemies. To increase the effectiveness of these natural enemies, non-trichome-producing genotypes have been developed. Unlike non-dryland environments, hot and dry conditions in dryland areas affect the bionomics and efficiency of natural enemies. For example, the survival of the egg parasitoid, *Trichogramma carverae* O. (Scott et al. 1997) and *Campoletis chlorideae* U. on chickpea pod borer (*Helicoverpa armigera* H.) declines when exposed to abrupt high-temperature changes (Dhillon and Sharma 2009). Similarly, the host searching ability of egg parasitoid *T. carverae* decreases at higher temperatures (Thomson et al. 2001). Reduced fecundity of egg parasitoids, *T. pretiosum* R. and *Trichogrammatoidea bactrae* N. has been recorded at temperatures prevailing above threshold levels (Naranjo 1993). Hot and dry weather conditions in dryland areas reduce parasitism e.g. poor parasitization of egg parasitoid, *Trichogramma* on European corn borer (*Ostrinia nubulalis* H.) affects the natural/biological control of pests in arid areas (Cagan et al. 1998).

Table 1 The natural enemies of important insect pests of dryland crops

Predator/ Parasitoid	Group	Beneficial insect or invertebrate	Target host/prey
Predators	Beetles	Ladybirds (Coccinellidae), red and blue beetles (*Dicranolaius bellulus* G.), green carab beetles (*Calosoma schayeri* E.), green soldier beetles (*Chauliognathus pulchellus* M.)	Aphids, mites, thrips, mealybugs, moth eggs including *Heliothis* spp. and larvae
	Bugs	Assassin bugs (Reduviidae), bigeyed bugs (*Geocoris lubra* K.), brown smudge bugs (*Deraeocoris signatus* D.), damsel bugs (*Nabis kingbergii* R.), glossy shield bug (*Cermatulus nasalis* W.), pirate bug (*Orius spp.*), apple dimple bug (*Campylomma liebknectic* G.), spined predatory shield bug (*Oechalia schellenbergii* G.), broken backed bug (*Taylorilygus pallidulus* B.)	Aphids, diamondback moth, eggs and larvae of *Heliothis* spp., cutworms (*Spodoptera litura* F.), false loopers
	Flies	Hoverfly larvae (Syrphidae)	Aphids
	Mites	Predatory mites from different families (Anystidae, Bdellidae, Erythraeida, Parasitidae and Cunaxidae)	Blue oat mite, lucerne flea, red legged earth mite
	Lacewings	Green lacewing (*Mallada signatus* S.), brown lacewing (*Micromus tasmaniae* W.)	Aphids, moth larvae and eggs, whitefly, thrips, mites and mealybugs
	Spiders	Variety of species including wolf spiders, nights talking spiders, orb weavers, tangle web spiders, flower spiders, jumping spiders and lynx spiders	Predators or a range of insect pests
Parasitoids	Sucking insect parasitoids	*Trioxys complanatus* Q., *Aphidius ervi* H., *Lysiphlebus testaceipes* C., *Aphidius colemani* V.	Aphids
		Eretmocerus spp. and *Encarsia* spp. including *Encarsia Formosa* G.	Whitefly
	Chewing insect parasitoids	Hymenoptera: Numerous parasitic wasps including banded caterpillar parasite (*Ichneumon promissorius* E.), two-toned caterpillar parasite (*Heteropelma scaposum* M.) (Ichneumonidae), *Microplitis demolitor* W., *Cotesia* spp. (Braconidae)	*Heliothis* and other moth larvae
		Sorghum midge parasites (*Eupelmus australiensis* G., *Aprostocetus diplosidis* C., *Tetrastichus* spp.)	Sorghum midge
		Tachinid flies	*Heliothis*, looper, armyworm, grasshopper and other larvae
	Egg parasitoids	Hymenoptera: *Trichogramma* (Trichogrammatidae) and *Telenomus* (Scelionidae) egg parasitoids	*Helicoverpa* and other Lepidopteran insect pests
		Trissolcus basalis W.	Green vegetable bug

Source: Getanjaly et al. (2015), Dixon (2000), and Eggleton and Belshaw (1992)

There are predators of pests that can be affected by dry conditions. Some fungi require humidity to survive, so are less effective during droughts, whereas others perform better in dry conditions. The parasitic wasp lays its eggs in the cereal leaf beetle's pupa, a life-cycle stage between the larva and the adult. After hatching, the wasp larvae eat the pupa's non-vital tissues and finally kill it while emerging from its body. Beetle pupae normally cover themselves in faeces; Sanford Eigenbrode (an entomologist at the University of Idaho in Moscow) suspects that this faecal shield acts as a barrier to keep the wasp at bay (ref?). However, producing faeces requires water and Eigenbrode thinks that in extremely dry conditions, beetle larvae are unable to deploy their shields quickly enough to prevent predation from parasitic wasps (Maxmen 2013).

A better understanding of the beneficial insects in dryland systems will make their conservation easier and their role for pest management. Fields shared by many beneficial insects will positively affect crop yields, so farmers need to be made aware of the beneficial insects in dryland agriculture and decisions should be made carefully as to how to manage insect pests.

3 Insect-Pests in Dryland Agriculture Systems

Insect pests and diseases are serious constraints to crop production and human consumption of these crops is at risk due to their incidence, both in dryland and non-dryland ecosystems. With few exceptions, insect pests in the arid zone do not differ greatly from those encountered in temperate and humid regions; however, the list of major insect pests in dryland crops is provided in Table 2. Dryland areas tend to be hot and insects generally thrive in warmer conditions. As the temperature rises, the excessive heat accelerates an insect's development and provides a favorable thermal environment for growth and development of plant-feeding insects. They eat more, mate more and produce more young (Maxmen 2013). Dry conditions also make plants more attractive and nutritious to insect pests (increased larval weight, survival and reproduction). Water-deficient plants are more susceptible to insects because the production of secondary metabolites or defensive compounds declines under water stress which increases the susceptibility to attack. Unlike non-dryland environments, dry conditions increase insect detoxification systems and insects feeding on water-stressed plants are capable of breaking down certain plant allelochemicals or defensive compounds that would normally have a negative effect on them. Studies suggest that some herbivorous insects specifically target water-stressed plants; however, the effect of drought stress varies depending on the feeding behaviors of insect pests. Insects with piercing-sucking mouthparts (aphids, whiteflies, scales and plant bugs) typically benefit more from dry conditions than those with chewing mouthparts (beetles, caterpillars and sawflies). Water-stressed plants are often susceptible to wood-boring insects such as bronze birch borer (*Agrilus anxius* G.), two lined chestnut borer (*Agrilus bilineatus* W.) and bark beetles due to reduced production of certain compounds (oleoresin) which act to deter

Table 2 Insect pests of dryland crops, host crops and their mode of damage

Common name	Scientific name	Order: Family	Host crop	Mode of damage	Sources
Termites	*Microtermes mycophagous* D.	Isoptera: Termitidae	Wheat, sorghum	Many species are crops with damage caused to stems and roots of seedlings and mature plants which can result in significant yield losses.	Perry and Perry (1989)
Sugarcane beetle	*Holotrichia consanguinea* B.	Coeloptera: Melolonthinae	Pearl millet	Damage soon after sowing and sometimes near maturity. Damaged plants dry up completely and are easily pulled out. Plants damaged at later stages give rise to white ears.	
Locusts	*Chortoicetes terminifera* W.	Orthoptera: Acrididae	Pearl millet, sorghum, barley, oats, etc.	Damage most green plants. Eat a wide range of food and each one eats its weight in food daily.	Maxmen (2013)
Grasshoppers	*Chrotogonus trachypterus* B.	Orthoptera: Pyrgomorphidae	Pasture, grains, forage, vegetables	General feeders on grasses and weeds and often move to cultivated crops. Crop damage is likely to be greatest in years when dry weather accompanies high populations.	Perry and Perry (1989)
Aphids	*Rhopalosiphum maidis* F. *R. padi* L.	Hemiptera: Aphididae	Wheat, maize, barley, forage legumes, sunflower, oilseeds, fruit trees, etc.	Sucking insect pest. Most destructive pest on cultivated plants causing yellowing, mottled leaves, stunted growth, curled leaves, browning, low yields and even death in plants.	Asin and Pons (2001), Irshad (2001) and Taheri et al. (2010)

(continued)

Table 2 (continued)

Common name	Scientific name	Order: Family	Host crop	Mode of damage	Sources
Weevils	*Myllocerus discolor* B.	Coleoptera: Curculionidae	Peas, lentils, pearl millet, sunflower, sorghum, etc.	Feed on plants in the larval stage and as adults. Very destructive to crops. One of the most destructive weevils is the cotton boll weevil.	Butani (1979) and Hill (1987)
Moths	*Achaea janata* L. *Heliocheilus albipunctella* J.	Lepidoptera: Noctuidae	Castor, pearl millet, etc.	Most lepidopterans are moths. Very destructive to crops at larval stages.	Perry and Perry (1989)
Hessian fly	*Mayetiola destructor* S.	Diptera: Cecidomyiidae	Wheat, barley, etc.	Maggots hatch from eggs, and crawl to the crown of seedlings (just above the roots) and feed on plant juices after injecting their unique saliva. Feeding by one larva can permanently stunt plant growth.	Anonymous (1971)
Shoot fly	*Atherigona soccata* R.	Diptera: Muscidae	Sorghum, pearl millet, maize etc.	The larva (maggot) feed on the growing point of the shoot of the seedling and cause "dead heart".	Perry and Perry (1989)
Wheat stem sawfly	*Cephus cinctus* N.	Hymenoptera: Cephidae	Wheat, other cereals	The larvae begin feeding near the oviposition site, eventually feeding up and down the stem, chewing through nodes.	Morrill (1995)

Sugarcane leafhopper	*Pyrilla perpusilla* W.	Hemiptera; Fulgoridae	Sorghum, pearl millet, sugarcane, etc.	Sucks phloem sap from leaves and excretes honeydew onto foliage, leading to fungal diseases. This direct and indirect damage affects sugar yield and quality.	Perry and Perry (1989)
Acacia white fly	*Tetraleurodes acacia* Q.	Hemiptera: Aleyrodidae	Acacia	Feeds by tapping into the phloem of plants, injecting toxic saliva and decreasing the overall turgor pressure of plants.	John et al. (2007) and Srinivasa (2000)
Thrips	*Scirtothrips aurantii* F.	Thysanoptera; Thripidae	Eucalyptus, citrus, etc.	Immature and adult thrips prefer to feed on young leaves in the inner neck of plants and cause severe damage by sucking the cell sap.	Mirab-Balou et al. (2011) and Loomans and Van Lenteren (1995)

feeding. For instance, the mountain pine beetle (*Dendroctonus ponderosae* H.) killed about 750,000 hectares of trees in 2010–2011 in the western United States in an infestation thought to be fueled in part by dry conditions (Chapman et al. 2012). Pine trees normally secrete a sticky resin which suffocates beetles burrowing underneath the bark, but water-stressed trees are unable to produce enough sap. Beetles finding such trees emit chemical signals (pheromones) that attract other beetles. The resulting mass predation further deteriorates the tree and the beetle population multiplies. Additionally, water-stressed plants emit volatile chemicals, e.g. ethanol and alpha pinene that attract these types of insects. Wood-boring insects use these chemical cues which facilitate them to find these plants whose natural defenses are already compromised due to water deficiency. Moreover, the lack of moisture in the upper tree canopy may result in cambial and phloem tissue degradation which is attractive to wood-boring female insects (bronze birch borer) for egg laying. Similarly, bark beetles colonize and weaken the defenses of their target plants. Dry conditions also encourage the development of two spotted spider mite (*Tetranychus urticae* K.) populations as these mites tend to feed more under dry conditions. According to Claudio Gratton (an entomologist at the University of Wisconsin-Madison), the relationship between insect populations and dry conditions is not consistent because sometimes there are more insects and sometimes less (Maxmen 2013). One explanation for this variation might be that insects can respond positively to dry conditions for a period. Eventually, they suffer as the plants they feed on weaken or deteriorate. For example, aphids can flourish during a short dry spell as plant nutrients become more concentrated. However, these benefits cease during prolonged dry seasons due to a drop in fluid pressure within the phloem of water-stressed plants (Huberty and Denno 2004). The dryland environments are associated with long dry seasons, and both insect pests and farmers spring into action upon the resumption of rainfall. Thus, the insects start damaging crops at the seedling stage because aestivation in seedling maggots and white grubs is broken by early heavy rains (Litsinger et al. 2002).

4 Limitations and Problems of Insect-Pest Management Approaches

Integrated management is not a panacea for every insect pest problem. Certain limitations are associated with any insect control program, and IMP is not an exception. With about one million described species, insects are dominant creatures in the world (Vega and Kaya 2012) with their short life cycles, high reproduction rates, variability, and adaptation to the environment. Therefore, every cultivated crop can be attacked by a complex of insect pest species. Modern agriculture practices (monoculture, high rate of plant fertilization and indiscriminate chemical control application, etc.) have decimated natural enemies of insects which formerly played an important role in maintaining insect pest populations at the general equilibrium position.

4.1 Insect Adaptations in Dryland

Insects have adapted to survive in dryland conditions through morphological, physiological, behavioral and ecological modifications (Cloudsley-Thompson 1975). Morphological adaptations of insects in arid environments include the creation of a boundary layer of hairs or scales to reduce the absorption of heat from the environment, and the resistance to water loss thorough spiracles. Some adaptations include modifications of body forms and legs for burrowing and running on hot surfaces. Body colors vary from ochre, brown, sandy grey or other colors, but insects are predominantly black. These colors characterize those insects that are relished by natural enemies. Adaptations to protect the young in harsh desert conditions include the construction of egg pods, larval cases and pupal cocoons. The behavioral adaptations of insects force them to confine themselves to favorable microhabitats. They feed and mate during those hours or seasons when conditions are favorable (Arnon 1992). Some insects in arid regions spend most of their time underground or under stones where the conditions are favorable compared to the open environment. There are many species in arid regions which remain dormant (similar to ephemeral plants) for extensive periods of dryness, heat and hot winds. This may occur in insects of cultivated plants in dryland conditions. For example, *Sesamia* larvae of dryland sorghum become dormant within stalks during hot summer. In contrast, the dormancy of pests occurs much later in irrigated sorghum. In some cases, the eggs remain dormant until sufficient rainfall occurs, e.g., Desert locust (*Schistocerca gregaria* F.). Many carnivorous insects in the desert store food for their progeny.

Physiological adaptations of arid-region insects include tolerance of high temperature, low respiration, and water uptake from the atmosphere, water conservation for metabolic activities, facultative hyperthermia, and resistance to desiccation (Cloudsley-Thompson 1975). These characteristics are not present in all insects, but many have multiple adaptations which help them to survive under adverse conditions. Some insects are so adapted to arid and semi-arid conditions that humid conditions can be harmful, e.g., chinch bug (*Blissus leucopterus* S.) and pale western cutworm (*Porosagrotis orthogonia* Morr.) (Arnon 1992). The fluctuation in temperature affects the life cycle of insects and, if not excessive, can accelerate developmental cycles. As a result, insects have more generations per year. In dryland regions, the rainy season is short but vital for insect breeding while adverse biological and physical conditions cause high mortality in insect populations. Therefore, irrigation in arid regions causes profound changes both in vegetation (cultivated and spontaneous) and insect populations. Irrigation can extend the favorable period for breeding and reduce the unfavorable conditions responsible for checking the increase in the insect population. In the absence of any natural control, epidemics may occur.

4.2 Drylands as Reservoirs for Potential Insect Pests

Indeed, many insect populations are not able to maintain their populations under adverse conditions in dryland areas. They can, however, serve as a reservoir for insect pests which spread to damaged, adjacent irrigated regions. Mckinney (1939) reported that the introduction of irrigation in the isolated Salt River Valley of Arizona increased the abundance and adaptation of certain insect species in cultivated crops that originally subsisted on native wild vegetation. Many moths migrate at night from their arid breeding habitats to irrigated or higher rainfall areas during the dry season, e.g., cotton leafworm (*Alabama argillacea* H.) and black cutworm (*Agrotis ipsilon* H.). The beet leaf hopper (*Circulifer tenellus* B.) overwinters in desert areas of southwestern United States which usually receives winter rainfall. In late spring, the beet leaf hopper migrates to newly-sown irrigated beet crops and causes infection with beet curly-top virus (Siegel and Hari 1980). Similarly, the desert locust is a typical example of desert region insects which become a major pest of cultivated crops far from their native area. They require moist sand for egg laying, and abundant green and tender food for their young. A locust can eat ten times their weight in vegetation. Land development schemes based on irrigation may provide favorable breeding grounds for locusts, making locusts even more dangerous than in the past. In addition to locusts, other insects which feed on the native vegetation of arid and semi-arid regions have become major pests of cultivated crops; for example, wheat thrips (*Haplothrips tritici* K.), sorghum midge (*Contarinia sorghicola* C.), cotton boll weevil (*Anthonomus grandis* B.) and beet leafhopper (*Circulifer tenellus* B.) (Uvarov 1962).

4.3 Injudicious Chemical Use and Insect Resistance

Approximately 45 % of the worldwide consumption of pesticides is in Europe, with 25 % in the USA and 30 % in the rest of the world (De et al. 2014). Estimated crop losses are between 10 and 30 % in developed nations but as high as 75 % in developing countries (Ohayo-Mitoko et al. 1997). Thus, chemical pesticide use is common for increasing crop productivity because pesticides protect crops by eliminating, inhibiting or controlling pests. Pesticides may affect plant development, prevent or kill competitive vegetation, or aid in the management of final products. But many studies have shown that high pesticide use may pose a serious threat to soil and water quality (Arias-Estévez et al. 2008), human health (Athukorala et al. 2012; Nawaz et al. 2014), food safety (Liu et al. 1995), aquatic species (Skevas et al. 2013) and beneficial insects (Mullen et al. 1997). Insecticide resistance is a major constraint in the management of insect pests of agricultural and public health importance (Khan et al. 2011). Cotton is considered a major crop in dryland agriculture, with reported insecticide resistance in major insect pests such as *Helicoverpa armigera* H., *Pectinophora gossypiella* S., *Earias vittella* F., *Spodoptera litura* F. and

Bemisia tabaci G. (Jan et al. 2015; Kranthi et al. 2002; Martin et al. 2000). Similarly, insecticide resistance has been observed in other major pests of dryland agricultural crops, including olive fruit fly (*Bactrocera oleae* R.; Vontas et al. 2001), melon fruit fly (*Bactrocera cucurbitae* C.; Vontas et al. 2011), oriental fruit fly (*Bactrocera dorsalis* H.; Hsu and Feng 2006), codling moth (*Cydia pomonella* L.; Reyes et al. 2015, Reyes et al. 2007).

4.4 Climate Change and Insect Pest Control

Climate change is occurring; the last decade of the twentieth century and the first decade of the twenty-first century have been the warmest periods on record. The global mean surface temperature rose approximately 0.6 ± 0.2 °C during the twentieth century, and climatic models have predicted an average increase in global temperature of 1.8–4 °C over the next 100 years (Collins et al. 2007; Johansen 2002; Karl and Trenbeth 2003). This is the largest increase in temperature in any century in the past 1000 years (Houghton et al. 2001). If temperatures rise about 2 °C in the next 100 years, then the negative effects of global warming would begin to extend worldwide (Griggs and Noguer 2002). Insects are poikilothermic (cold-blooded) organisms, i.e. their body temperatures vary with the surrounding temperatures. They are strongly influenced by changing climatic and weather conditions. Their rate of development, reproduction, migration, adaptation and distribution is directly affected by temperature, humidity, precipitation, wind speed, etc. In addition, host plants, natural enemies and interspecific interactions with other insects indirectly affect insects. Thus, climate change poses a threat to the control of insect pests. Similarly, increasing levels of greenhouse gases in the atmosphere may significantly impact agricultural insect pests. Consequently, existing pests at low densities may spread on a broad spectrum and reach damaging population densities (Bale et al. 2002; Porter et al. 1991).

Population dynamics of insects deal with factors affecting population densities. The rise in temperature positively affects the development of certain pests until it exceeds the optimal requirements of the species. For example, bark beetles profit from accelerated development rates with early completion of life cycles to produce more generations within a season. A rise in temperature above favorable conditions may decrease growth rates and fecundity, and increase mortality rates in many species (Jönsson et al. 2009; Rouault et al. 2006). Many species require a dormancy phase to complete their life cycle. Increased temperatures may benefit those species which actively feed during winter but may have a negative impact on those species which require low temperature for diapause (Bale et al. 2002). Migration and dispersal are essential parameters in the phenology of herbivorous insects for host finding, mating, colonization and brood establishment. Temperature requirements have been described for different phases of flight activities. For instance, black bean aphid (*Aphis fabae* S.) requires 6.5 °C for wing beating, 13 °C for horizontal flight, 15 °C for sustained upward flight and 17 °C for take-off (Cockbain 1961).

Temperature directly affects the survival rate of insects. For example, survival of the brown plant hopper (*Nilapavata lugens* S.) remains unchanged between 25 and 35 °C but significantly declines at 40 °C. The oviposition efficiency of female brown plant hopper (BPH) was relatively higher at a higher temperature (35 and 40 °C) while egg survival declined at 35 °C. The pre-oviposition period also shortened at high temperature (Heong et al. 1995). The viability of eggs and required degree days for hatching of *Helicoverpa ammigera* reduced with increasing temperature (Dhillon and Sharma 2007). Similarly, the effect of temperature on growth rate, voltinism (number of generations per year), and species distribution of various insects has been described (Chakravarthy and Gautam 2002; Régnière et al. 2012; Tobin et al. 2008).

In addition to temperature effects on insects, changes in precipitation and rising CO_2 levels also affect insect life cycles. A number of insects are sensitive to precipitation and heavy rains can kill or remove them from crops. Increased summer rainfall and drought conditions promoted growth in the upper soil of a wireworm population (Staley et al. 2007). The effect of rising levels of CO_2 on levels of herbivory in soybean was tested using FACE (free air gas concentration enrichment) technology. A soybean crop grown in an elevated CO_2 atmosphere had 57 % more damage from insects compared with those grown in the natural atmosphere (Hamilton et al. 2005). This evidence shows that climate change in different agroecosystems and ecological zones might affect the population dynamics of insect pests. The prediction of climate change impacts on insects is somewhat uncertain and may be positive for certain insects and negative for others. The increase in insect outbreaks will increase the use of insecticides which is likely to have a negative impact on the environment. Therefore, the best economic strategy for farmers is to practice integrated pest management.

4.5 Lack of Conservation of Natural Enemies

Paul DeBach defines conservation biological control as "manipulation of the environment to favor natural enemies, either by removing or mitigating adverse factors or by providing lacking requisites" (Naranjo 2001). Natural enemies can play a pivotal role in the management of insect pests and is often credited with being the oldest form of biological control. Conservation of endemic natural enemies has been less successful in field crops and has received little attention as a method of arthropod pest suppression compared with classical and augmentation biological control (Landis et al. 2000).

4.5.1 Insecticides

The factors affecting the conservation natural enemies are frequent application of pesticides, periodic disruption of the soil structure by heavy tillage, frequent planting and rotation, removal of crop residues, and the destruction of plant structures with harvesting. The production of monoculture crops (often practiced in dryland systems) often lacks the alternative food sources—flowers for nectar and pollen as well as shelter—which may natural enemies need. It is evident from both laboratory and field studies that the most significant disrupting factor for the biological control of arthropod pests is the use of toxic insecticides. These insecticides cause species and stage (egg, larvae, pupa and adult) specific toxicity to the predators and parasitoids of different insects (Croft 1990; Gerling and Sinai 1994; Hoseini et al. 2012; Simmons and Jackson 2000; Smith et al. 1999). Some selective insecticides for pest management are available but disruption of biocontrol agents is still likely in some agricultural systems. The presence of multiple key pests is reason to use broad-spectrum insecticides due to the lack of available selective control measures. Additionally, economic consideration is important due to the high cost of IGRs (insect growth regulators); some areas may be forced to opt for cheaper but more destructive pesticides.

4.5.2 Other Factors

Aside from insecticides, other factors contribute to biological control disruption but have received little attention. Intraguild predation (IP) is well known in many cropping systems. The aphelinid heteronomous hyperparasitoids attacking whitefly is possibly the best-known example of IP. They produce males as hyperparasitoids and females as primary parasitoids. This type of behavior can be disruptive to biological control (Mills and Gutierrez 1996). Therefore, IP is possibly common due to the diversity of natural enemies attacking large numbers of insect pests and may play an integral part in determining the role of natural enemy species in affected crops. Characteristics of the host plant, degree of leaf glossiness or levels of nitrogen may affect the biology and behavior of natural enemies (Bentz et al. 1996; Jackson et al. 2000; Wilson and George 1986). These factors highlight the challenges faced in the integration of biological control into economically-sustainable pest management strategies for multiple pest management systems.

In many biological control systems, the efficacy of natural enemies—imported or mass reared—is likely to depend on conservation measures and the suitability of the environment in which they are released. The conservation of natural enemies can occur in dryland agriculture systems by focusing on three overlapping components: (i) survey and identification of extant natural enemies, (ii) elucidation of constraints and manipulation of factors enhancing the abundance of natural enemies, and (iii) evaluation of the biological control efficacy of released natural enemies in that particular system (Fig. 1).

4.6 Impact of Genetically Modified Crops on Insect

Plant breeding history reveals the use of new technologies for improving crop cultivars by manipulating chromosome number, chemical and radiation treatment to induce mutations, and developing addition/substitution lines as well as cell and

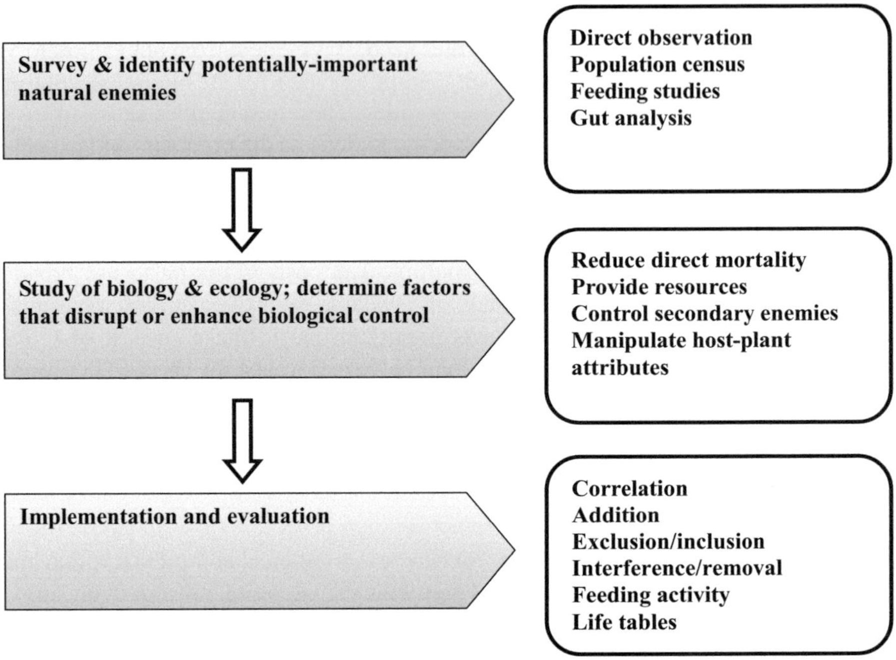

Fig. 1 Components and approaches of conservation biological control

tissue cultures. All have contributed to improved yields, developed resistance to specific diseases and pests, environmental adaptations and improved quality demanded by the food industry and consumers. Cell culture and molecular biology advances have culminated in the genetic modification of crops. Currently, genetic modification of plants is a powerful, widespread but controversial application of biotechnology (Conner et al. 2003; Mendelsohn et al. 2003; Sanahuja et al. 2011). GM crops have reduced, and continue to reduce herbicide, insecticide and overall pesticide use (Benbrook 2012). Despite the potential benefits, environmental and food safety concerns have been raised about GM crops. It is feared that GM crops will harm the human population with undesired impacts on the environment. Concerns regarding the introduction of GM crops into the environment include the effects on biodiversity, becoming agricultural weeds, direct and indirect effects on non-target organisms, and food safety (Conner et al. 2003).

4.6.1 Impact on Insect Pests

The area under *Bacillus thurengiensis* (Bt) crops increased from 1.1 million to 66 million hectares from 1996 to 2011 with a cumulative total of more than 420 million hectares. Insects have a remarkable ability to adapt insecticides and other control tactics. The development of resistance can reduce the effectiveness of insecticidal Bt proteins in transgenic plants. Therefore, the main threat to the continued success of Bt crops is the evolution of resistance in pests. Some reviews have addressed insect pest resistance in GM crops (Carrière et al. 2010; Pardo-Lopez et al. 2013; Tabashnik 1994). Several studies have compared the field outcomes for resistance to Bt crops but with limited sample sizes for both field outcomes and other factors predicted to affect resistance (Carrière et al. 2010; Tabashnik 2008; Tabashnik et al. 2009). A 2012 review summarized pest resistance to Bt crops by analyzing field monitoring data for resistance from 77 different studies, and reported resistance for some populations of 5 of 13 major pest species (lepidopteran and coleopteran) examined when compared with resistant populations of only one pest species in 2005. The number of major pest species with field-evolved resistance and reduced transgenic crop efficacy increased from one to five (three Bt corn and two Bt cotton) from 2005–2010 (Tabashnik et al. 2013). Another study revealed that the insect resistance percentage to GM crops increased from 0.93 % in 2010 to 5.5 % in 2013. Therefore, pest resistance management is necessary to attain the full benefit of GM crops to minimize food security issues. The factors responsible for delaying resistance include the use of abundant non-Bt refuge crops, recessive inheritance of resistance, the low initial frequency of resistant alleles, and separate deployment of two-toxin Bt crops from one-toxin Bt crops. It was predicted that without natural refuge crops, the percentage of resistant insects to GM crops would exceed 98 % in 2013 but would increase by only 1.1 % if natural refuges were as effective as non-Bt cotton refuges. Similarly, the percentage of resistant insects increased from 37 % in 2010 to 84 % in 2013 with the non-recessive inheritance of resistance (Jin et al. 2015; Tabashnik et al. 2013). The integration of two or more toxic genes with other control tactics may further slow increases in resistance.

4.6.2 Impacts on Non-target Insects

Insects play an important role in the ecosystem and human economy as crop pollinators, natural enemies (predators and parasitoids) and detritivores. Honey bees are the best-known pollinators and often improve fruit and seed yields (Crane and Walker 1984) with the additional advantage of honey production. Cotton nectar is attractive to bees, but they are not required for pollination. It is considered that GM crops may have hazardous effects on bees as well as other insect pollinators. Most studies showed no deleterious effects on pollinators, but facts indicate dose-dependent hazardous effects to bees (Malone and Burgess 2009). Similarly, natural enemies (predators and parasitoids) are significant regulators of insect pests. It is evident that the survival of natural enemies depends on the supply of host insects. So, the reduction in insects feeding on GM crops ultimately affects the natural enemies' population. GM crops could have both direct and indirect effects. Many laboratory studies on Bt cotton demonstrated negative tritrophic impacts on predators (*Orius tristicolor* W. and *Geocoris punctipes* S.) (Ponsard et al. 2002) and parasitoids (*Cotesia marginiventris* C. and *Copidosoma floridanum* A.) (Baur and Boethel

2003). Therefore, the impact of GM crops needs to be investigated carefully to develop insect management techniques with minimum effects on non-target organisms.

There is no doubt that an escalated world population has increased the need for intensified food production which will contribute to environmental degradation. Increased food production, while protecting resources, requires an upturn in the implementation of current management practices as well as rapid development of innovative and sustainable ways to mitigate the agricultural impact on the environment and public health.

5 IPM in Dryland A griculture Systems

To feed the ever-increasing world population, agricultural productivity needs to ensure food security in every agroecosystem especially dryland agriculture, as it constitutes a substantial proportion of the global agricultural system. In addition, conservation of environmental quality and lessening of others threats associated with indispensable agroecosystem services are imperative contemplations for the implementation of IPM in dryland agriculture systems (Tilman et al. 2002; Sandhu et al. 2008; Nash and Hoffmann 2012). Regular, unremitting and continuous monocultivation of crops year after year and the introduction of exotic crops to enhance agricultural productivity in dryland cultivation systems has resulted in the precipitous increase in the incidence/level of pest populations and lead to the establishment of invasive pest species, respectively (Royer and Krenzer 2000; Royer et al. 2007). Agricultural pest management systems still depend on broad-spectrum synthetic pesticides; there is an imperative and imperious need for the strategic implementation of IPM tactics to exploit the practical, theoretical and conceptual IPM principles (Samiee et al. 2009; Nash and Hoffmann 2012). The lack of awareness of the operating manpower of dryland agriculture systems and non-implementation of IPM programs by farmers has lead to an erratic and impulsive outbreak of pests, reduced yields and profit margins, and exclusive dependency on pesticides-based approaches (Royer and Krenzer 2000; Royer et al. 2007; Samiee et al. 2009; Nash and Hoffmann 2012). Different fall-on strategies including landscape modifications, host plant resistance (HPR), ecological indicators, reliable predictors, and emergency intervention should be implemented for the manipulation and stability of crop environments or agroecosystems that would not be conducive for the pest population. Despite such strategies, if a pest population outbreak then an operational approach including broad-spectrum multiple chemical control (fall-off strategies) may be a better option to protect productivity (Nash and Hoffmann 2012). The changing scenario of climatic conditions as well as the increasing demand for quality food emphasizes the use of a dynamic approach of IPM. The techniques should be efficient, cost-effective and sustainable, produce quality yields and the least degrading to environmental quality, biodiversity conservation, mitigation of health hazard effects and low rates of development of ecological backlashes in dryland agroecosystems (Way and van Emden 2000; Samiee et al. 2009). In dryland agriculture systems, such comprehensive and flexible IPM approaches should be practiced and implemented that are highly compatible and a best-fit for changing climatic conditions, agronomic/cultural practices, socio-economic factors and landscape usage. Sustainable and successful implementation of such flexible IPM approaches against pests in dryland agriculture systems depend on a comprehensive understanding of the development of simple indicators for system imbalance, host–plant resistance, pest population dynamics, amplified system intricacy and emergency intervention of fall-off strategies (chemical control) (Nash and Hoffmann 2012).

5.1 Characteristics of Sustainable Pest Management

The basic characteristics of sustainable pest management include (1) application of multiple tactics in a highly-compatible manner; (2) reduction of pest numbers or their effects below some economic injury level; (3) conservation of environmental quality (Knipling 1979; Pedigo and Rice 2009). However, some scientists advocate some additional characteristics including (1) highly selective for pest (target specific); (2) comprehensive for a production system (non-phytotoxic and increase yield); (3) compatible with ecological principles, and (4) tolerant of potentially-harmful species but within economically-acceptable limits. An IPM program and strategy constituting these characteristics/elements guarantee its success, efficiency, sustainability, social and economic acceptability, and environmental appropriateness for any pest management system (Dhaliwal et al. 2006; Buurma 2008; Heong et al. 2008; Pedigo and Rice 2009; Alam 2010; Schowalter 2011). Effective and sustainable IPM strategies are also contingent on economic decision levels which are critical for defining the course of action, guaranteeing practical pesticide application, reducing outrageous economic damage, safeguarding producer profits, and conserving environmental quality in any agriculture system and pest situation (Norris et al. 2002; Dhaliwal et al. 2006; Pedigo and Rice 2009; Jha 2010; Schowalter 2011).

5.2 Basic Principles of Pest Management System

Any sustainable agricultural system is characterized by healthier and more productive production and protection cropping systems which demonstrates the least application of highly toxic synthetic pesticides. Such systems depend on a holistic pest management approach based on some basic principles (Joshi 2006; Dhaliwal et al. 2006; Dhaliwal and Koul 2007; Singh 2008; Pedigo and Rice 2009) including

- Pest avoidance/exclusion (a precautionary step which inhibits entry of any pest insect into any ecosystem or agroecosystem using techniques such as hand-picking, bagging, trapping, physical barriers, screening, physical beating, rope dragging, banding, burning, sieving and winnowing, acausting (noise creation etc.);

- Identification of a pest and its status (identification of pest species and life stage is the principal component of any IPM program and identifies whether or not the pest is dangerous) (Dhaliwal and Koul 2007; Pedigo and Rice 2009;
- Understanding the biology and ecology of the pest;
- Understanding the structure/components of the agroecosystem (Pedigo and Rice 2009; Schowalter 2011);
- Economic decision levels (Pedigo and Rice 2009); 6) pest monitoring and pest scouting (Dhaliwal and Arora 2003);
- Selection of single or set control tactics;
- Goal of pest management program [prevention (keeping a pest from becoming a problem), suppression (reducing pest numbers or damage to an acceptable level) and eradication (destroying an entire pest population) (Pedigo and Rice 2009)];
- Study factors causing failure of pest management strategies (incorrect identification of insect pest species, selection of inappropriate control measures, selection of incompatible control measures, selection of inappropriate application technique, improper timing of application of control measures, excessive application of same tactics, development of resistance in insect pest species against control measures, adverse climatic conditions, use of incorrect dosage of pest control measure); and
- Public awareness, long-term commitment, planning and improvement of the IPM tactic and strategy (Pedigo and Rice 2009; Schowalter 2011).

5.3 *Requirements of Integrated Pest Management*

Integrated pest management (IPM) is a holistic dynamic approach involving integrated and strategic implementation of available efficient, effective and highly-compatible IPM tactics using information regarding pest scouting, pest forecasting, survey and surveillance, economic decision levels, knowledge of technologies, and biological knowledge of the pests for suppressing pest populations below the economic threshold level (ETL), conserving environmental quality and biodiversity, and enhancing positive cost-benefit-ratios (CBR) under the acceptable limits of social barriers. This definition reviews the sound pillars required for the foundation of a successful IPM program or strategy against any pest in any agroecosystem (Dhaliwal and Arora 2003; Pedigo 2003; Dhaliwal et al. 2006; Buurma 2008; Heong et al. 2008; Pedigo and Rice 2009). IPM is a poly-strand approach that is established on some prerequisites including knowledge of the pest management technology, biological, morphological and ecological knowledge of the pests, structural and functional components of the ecosystem and their interactions, landscape and habitat management techniques, and biological control conservation techniques (Sandhu et al. 2008; Nash and Hoffmann 2012).

Knowledge of the technology is one of the pillars of IPM. The knowledge of different features of any technology that ensures its proper and effective implementation include its nature and type, mode of application (aerial, foliar, chemigation, baits, traps etc.), bio/shelf life, equipment required for its application, factors affecting its performance, compatibility with other management tactics, target specific or broad spectrum, mode of action etc. (Pedigo and Rice 2009). This information guarantees the successful and effective implementation of any technology.

The devising and efficient implementation of an IPM approach/strategy against any pest in any agroecosystem/farming system also depends on the morphological, biological and ecological knowledge of the pest species. This knowledge lays the foundation for an effective and economical pest management strategy, eliminates the factors that result in the failure of the IPM program, minimizes operational input-costs, enhances profitability, guarantees conservation of environmental quality/stability, and reduces health hazardous effects for mankind (Knipling 1979; Norris et al. 2002; Pedigo and Rice 2009). The knowledge of pests includes information on the types of habitat and food preferred by the pest, its lifespan, longevity of its incubation period, life stages found, breeding places, season and behavior (dispersal, migratory, immigrant etc.) (Knipling 1979; Norris et al. 2002; Sorby et al. 2005, Dhaliwal et al. 2006; Pedigo and Rice 2009; Alam 2010; Jha 2010) and an understanding of the complex effects of insects and their interactions with other organisms on ecosystem services (Schowalter 2011).

5.4 Advantages and Disadvantages of IPM

Economics, environmental conservation and food security are among the indispensable factors emphasized by the philosophy of an IPM approach. Implementation of conventional insecticides exerts a negative impact on the environment (insecticidal pollution in lithosphere, hydrosphere and atmosphere, residual toxic impact on non-target organisms, biomagnification of persistent toxic residues at trophic levels) which results in ecological backlash (resistance, resurgence and replacement) in pests and hazardous effects on human health (carcinogenic, mutagenic, teratogenic, respiratory, eyes, digestive ailments) (Sarfraz et al. 2005; Gogi et al. 2006; Pedigo and Rice 2009). These issues are flagrantly addressed by the strategic integration of various IPM tactics including biorational and ecofriendly insecticides. An IPM approach contributes to various economic profits and benefits to agricultural producers, the environment, pest management professionals and organizations, and the general public. Implementation of an integration of insecticide-free IPM tactics and/or calendar-based application of ecofriendly biorational insecticides when required and at lower application rates can reduce pesticide costs by 30–40 %, enhance the acceptability and marketability of the produce at comparatively higher marketable rates. In addition, it will diminish the probabilities of environmental pollution and health issues in the workforce, and increase the knowledge of pest biology and specific pest management techniques and options used in IPM (Olsen 1997; Cartwright et al. 1989; Collins et al. 1992). The following are details of the advantages and disadvantages of an IPM approach practiced in any farming system.

5.4.1 Advantages

- Reduces the possibility of litho-, hydro- and atmospheric contamination with toxic chemical compounds
- Promotes sound structures and healthy plants
- Guarantees protection and conservation of non-target species through reduced impact of IPM activities
- Encourages sustainable bio-based pest management alternatives
- Reduces environmental risk associated with pest management by encouraging the adoption of more ecologically-sound control tactics
- Diminishes the need for and reliability on pesticides by using several pest management options and methods
- Reduces or abolishes issues of pesticide toxic and lethal residues (MRLs) in consumable plant parts, drinking water and other consumable commodities, and ensures food security and residue-free acceptable and marketable produce in national and international markets
- Reduces or eliminates re-entry interval restrictions
- Reduces the chance of exposure to pesticides by workers, tenants, the public and other stakeholders through direct contact, inhalation, oral, food-chain, etc.
- Assuages and moderates public concern about pest and pesticide-related practices
- Guarantees the maintenance or escalation of cost-effectiveness/cost-benefit-ratio in pest management programs
- In the case of export commodities, removes the issue of MRLs as a barrier of export and consignment are accepted in international markets at high values that would be a source of foreign exchange earnings
- Reduces the chance of developing resistance, resurgence and replacement of pest insects (Dhaliwal et al. 2006; Pedigo and Rice 2009).

5.4.2 Disadvantages

- A successful and sustainable implementation of any IPM program in any farming system requires comprehensive planning which is mostly beyond the capacity of illiterate or less-literate farming communities
- IPM requires more resources as alternatives to pesticides
- Cost-effective implementation of IPM on a sustainable basis requires a more comprehensive knowledge of the pest, pest management techniques, agroecosys-

tems and the types and nature of interactions among the various biotic and abiotic components of ecosystems
- IPM requires the development of economic decision-making tools for the economical application of ecofriendly, biorational and bio-based insecticides (Dhaliwal et al. 2006; Pedigo and Rice 2009).

5.5 IPM Approaches for Enhancing Productivity and Food Security

In dryland farming systems, monocultures or rotated cultivation supports the ephemeral nature of a pest's food resources (host plants) that directly curtails the performance of natural enemies, pollinators and other beneficial insect fauna and indirectly reduces the density of beneficial fauna and crop yields from high pest pressure (Vandermeer 1989; Booij and Noorlander 1992; Way 1988; Altieri and Brewer 2002; Brewer and Elliott 2004; Tscharntke et al. 2005; Clough et al. 2007). In dryland cropping systems, IPM tactics are used in different ways including incorporation into the cropping system, application for near-term future problems and implementation for currently-active problems. Host plant resistance, biological control (importation, release and conservation methods) and cultural control practices including crop rotation, intercropping, relay/trap cropping, planting density, clean cultivation (pest-free materials), sowing time, harvesting time, tillage, sanitation, adjacent land use, fertilizer application management and irrigation management are generally incorporated into cropping system design in dryland farming systems. Pesticides, biological control by augmentation and tillage practices are generally applied while intercropping, cover cropping, sowing and harvesting times, sanitation, soil fertility and irrigation management are occasionally applied for currently-active problems in dryland farming systems. Biological control (BC) by importation and release for future and current problems, by augmentation in cropping system design or for currently-active problems, and cultural practices like crop rotation, plant population/density, pest-free planting materials and field size are not applied for currently-active problems in dryland farming systems (Holtzer et al. 1996). IPM tactics can be categorized into host plant resistance, biological, mechanical, physical, cultural, genetic and chemical control.

5.5.1 Economic Decision Levels

Effective and sustainable insect pest management depends on economic decision levels (EDLs) which are indispensable for determining the course of action, ensuring sensible pesticide application, reducing ludicrous economic damage, safeguarding producer profits, and conserving environmental quality in any pest situation (Dhaliwal et al. 2006; Pedigo and Rice 2009; Alam 2010; Jha 2010). EDLs were developed in the 1950s by entomologists to determine whether the use of insecticides was appropriate and economical. EDLs are also used to determine if other IPM control tactics like HRP, biological, mechanical, physical, cultural and ecological control processes successfully suppress pest populations or fail to reduce it below a tolerable level. If EDLs manifest the failure of other IPM tactics by suppressing pest populations below tolerable levels, then the use of chemical control becomes mandatory to keep the pest population from reaching the economic injury level. EDLs also determine if any active insect pest problems are prevailing in the cropping system which are defined by pest-scouting techniques (Holtzer et al. 1996; Dhaliwal et al. 2006; Pedigo and Rice 2009). The implementation and utilization of EDLs ensures that the deliberate and sensible use of insecticides helps to avoid the indiscriminate use of insecticides, decreases the intensity of insecticide use, increases the producer profit ratio, conserves natural biodiversity and environment quality, provides solutions for some problems like ecological backlash (resistance, resurgence and replacement), health hazard effects, pesticide residues and the negative impacts on non-target organisms. These EDLs include economic injury level, economic threshold level, gain threshold and damage boundary (Knipling 1979; Pedigo and Rice 2009). The decision to implement any of these EDLs is determined using four principles/rules: No-Threshold-Rule, Nominal-Threshold-Rule, Simple-Threshold-Rule and Comprehensive-Threshold-Rule (Knipling 1979) (see Table 3). The development of precise EDLs for specific pests in particular crops is accomplished with years of research and field experiments carried out under controlled biological, economical, agronomic and environmental conditions to predetermine

Table 3 EDL rules and criteria for their selection and application in pest management program

EDL rules	Criteria for application
No-Threshold-Rule	(i) Pest sampling cannot be done economically
	(ii) Practical response to cure a problem cannot be implemented in a timely manner
	(iii) Once detected, the problem cannot be cured
	(iv) ETL is immeasurably low (some quality losses, disease transmission, rapid growth potential)
	(v) Populations are intense with a general level of density always above EIL
Nominal-Threshold-Rule	Based on the experiences and expertise of the entomologist and most frequently used in a pest management program.
Simple-Threshold-Rule	Implies the use of calculated ETLs which are based on market values, management costs, damage done per insect, yield reduction per plant, and the amount of damage avoided.
Comprehensive-Threshold-Rule	Based on interactive effects of biotic and abiotic factors on plant stresses and can be calculated and implemented only if the computer-based information delivery system and acquisition of on-farm computers are ensured.

Knipling (1979)

pest density. EDLs vary with cultivar, growing location, pest-damaging stage, etc. (Holtzer et al. 1996). Along with EDLs, some essential and indispensable additional information (such as resistance, susceptible or tolerance potential, growth stage, yield potential of crop, status and impact of naturally-occurring biocontrol agents, status of other factors such as soil moisture, soil fertility, crop residues etc., and their impact on pest biology and damage potential) are prerequisites and should be obtained for the economical application of insecticides (Knipling 1979; Holtzer et al. 1996; Pedigo and Rice 2009).

5.5.2 Insect Pest Monitoring System

Monitoring and forecasting a pest's resistance, resurgence, replacement and outbreak have reached a crucial and decisive position and the attention of pest management scientists (Maelzer and Zalucki 2000). The prediction of pest population dynamics and its outbreak is determined by understanding the life-history strategies of key pests, including the biotic potential, reproductive and survival potential, mode/rates of reproduction, migratory, trivial and dispersal behavior, diapause, and the growth pattern (r-strategic, k-strategic and a-strategic) of pests (Birch 1948; Greenslade 1983). The implementation of EDLs in a pest management program for crop pests in dryland systems should be emphasized for economical pest suppression. However, EDLs as decision-making tools in dryland pest management require the establishment of ETLs. Farmers should exploit the additional information on monitoring and scouting techniques to determine and implement EDLs in dryland cropping management systems. In such systems, advanced technologies including global positioning system (GPS), remote sensing technology and global information system (GSI) not only make the monitoring and EDL system more effective/efficient but can lay the foundation of precision agriculture. Using these advanced technologies would reduce the costs involved in acquiring pests, host plants, beneficial fauna and soil-related information (pest stage, damage intensity, crop growth stage, soil condition, water requirements, etc.). These technologies would also ensure precise site-specific application of nutrients (fertilizers) for good crop health and pesticides for economical management of pests (insect, weeds, pathogens, etc.) in dryland farming systems. Data maps developed using these technologies will help to detect the precise spatial presence and location of insect pests on crops, and GPS units mounted with pesticide application equipment would help to apply a precise quantity of the pesticides (Holtzer et al. 1996).

5.5.3 Host Plant Resistance (HPR)

In dryland farming systems, HPR is the most commonly-exploited IPM tactic in which resistant cultivars are incorporated into cropping system design. However, HPR may be selected by the farmers in dryland cropping systems if the chances of pest outbreaks are certain during the cropping season (Holtzer et al. 1996). HPR is ecofriendly, easy and simple to apply and targets specific pest management techniques with improved yield potential. It has been practiced widely in dryland and marginal cropping systems (Park et al. 2006; Ortiz et al. 2007; Hillocks 2009). GM crops with toxin-transcribing genes confer resistance to many Lepidopterous, Dipterous, Coleopterous and Hemipterous insects and reduce the application intensity of persistent and lethal pesticides (Cattaneo et al. 2006; Zalucki et al. 2009). HPR technology will guarantee the stability of dryland agroecosystems, but single strategy resistance HPR-techniques may jeopardize ecosystem biodiversity due to the homogenization of the landscape and lead to secondary pest resurgence and replacement (Altieri et al. 2004; Gu et al. 2008; Wang et al. 2008; Hillocks 2009). HPR technology will be effective and acceptable in dryland farming systems if it: (1) demonstrates resistance against the insect pest complex and weeds (multiple pest resistance gene systems); (2) reduces the rate of development of ecological backlash (resistance, resurgence and replacement) in insect pests; (3) eliminates the requirement of chemical control for the pest complex; (4) ameliorates the performance of natural enemies in the system; (5) does not deteriorate yield potential; and (6) does not hamper with anti-herbivory allelochemicals and phyto-allexins synthesis pathways of plants (Glamoustaris and Mithen 1995; Harrington et al. 1998; Horne and Page 2008).

5.5.4 Crop Rotation System

Any modification to the crop rotation system will exert a profound and reflective influence on the agroecosystem, agricultural landscape, interaction and functioning of habitat components, cropping system, biotic potential of pests, performance and activities of natural enemies, soil fertility and intensity of pest problems (Ahern and Brewer 2002; Elliott et al. 2002). Crop rotation systems in dryland areas can also enhance water-use efficiency and confirm the stability or increase in farm profits (Peterson et al. 1996). Growers practicing crop rotations need more frequent insect pest monitoring than those growers who do not. Other than insect pest control, crop rotations benefit dryland growers by impacting weed management, enhancing labor and equipment efficiency, and promoting resistance in crops to insect pests. This rotational system should use cultivars which exhibit potential for resistance to insects and other pests, fewer yield losses and high-yielding potential (Koul and Cuperus 2007).

5.5.5 Ecological Engineering of the Landscape

Ecological engineering of the dryland agricultural landscape ensures an increase in biodiversity, conservation of beneficial fauna and suppression of insect pests as its foremost consequences (Gurr et al. 2004; Schellhorn et al. 2008). Habitat modification will change the composition of the arthropod community, alter pest–parasitoid/pest–predator interactions, promote conservation and augmentation of natural enemies, and ultimately impact the efficacy of any IPM system (Tilman and Knops 1997; Harwood et al. 2009). However, implementation of this strategy will be difficult and intricate in dryland cropping systems on a large scale or including the surrounding landscape such as pasture cropping (Jones 1999; Schellhorn et al. 2008). Undisturbed strips of grassy non-crop intercropped with major crops will provide a habitat conducive to the multiplication of natural enemies and enhance pest control in the crop (Collins et al. 2003; Macleod et al. 2004; Tsitsilas et al. 2011).

5.5.6 Conservation Biological Control

Biological control describes the exploitation of predators, parasites and pathogens for pest control. It is considered ecofriendly, self-perpetuating once established, environmentally safe, target specific and best-fit in an IPM program (Dhaliwal and Arora 2003; Pedigo 2003; van Emden 2003; Dhaliwal and Koul 2007; Jonsson et al. 2008; Pedigo and Rice 2009). It plays an indispensable role in controlling insect pests in economic crops, fruit orchards, vegetables, ornamental plants and fodder/pasture in any cropping system. Dryland cropping systems need to: (1) recognize the values of natural enemies at the farm level; (2) investigate their effectiveness and abundance in different dryland zones and prevailing seasons; (3) explore the limiting biotic/abiotic factors involved in the failure of a biocontrol system; (4) determine and standardize the techniques for the collection, mass propagation, release and conservation of indigenous biocontrol natural enemies, and (5) import and manipulate exotic natural enemies (van Emden 2003; Jonsson et al. 2008). The abundance and performance of natural enemies in any cropping system are regulated by the conservation of floral diversity, modification and manipulation of habitat, maintenance of diversity outside and/or inside the major crop fields planned for biological control, integration and implementation of IPM tactics highly conducive and compatible with biological control, and conservation of the natural enemies/pest ratio through manipulation and augmentation techniques (van Emden 2003; Jonsson et al. 2008). Habitat modification with diversified flowering species or artificial diets supports the abundance and performance of natural enemies, specifically parasitoids. The female parasitoid, *Pimpla examinator* (F.) of the pine shoot moth, *Rhyacionia buoliana* (S.) is not attracted to pine oil fragrance until the female has fed on floral nectar to mature the eggs. Once the female parasitoid has fed and matured the eggs, the female is attracted to the pine oil smell and locates the host moth (Thorpe and Caudle 1938; van Emden 2003; Jonsson et al. 2008). Sometimes, parasitizing and predatory stage(s) of parasitoids and predators, respectively, do not synchronize with the host/prey stage or food is scarce (for host or prey). In these conditions, the provision of obligatory or other alternate host/prey guarantees the conservation of natural enemies (Hardy 1938; van Emden 2003; Jonsson et al. 2008). For example, emergence of the larval parasitoid (*Diadegma fenestralis* H.) of the diamondback moth (*Plutella xylostella* L.) in autumn does not synchronize with the presence of larval stages of the host (available in autumn in pupal stages) and they must, therefore, survive on another host (*Crataegus monogyna* J.) (Hardy 1938; Van Emden 2003). The appearance and outbreak of *Ichneumon dispar* (P.), a parasitoid species of gypsy moth (*Lymantria dispar* L.), occurs when the preferred host is scarce in the forest, but they survive on 45 caterpillar species as alternate hosts (Babaei et al. 2009). Habitat modification also involves practices that make the microclimate highly conducive to the conservation and performance of natural enemies in the cropping system. Taylor (1940) documented that maintaining a conducive microclimate, specifically humidity and temperature, in an abandoned coffee plantation by shading enhanced the abundance and performance of natural enemies which ultimately suppressed antestia bugs (*Antestiopsis* spp.). Conservation of biological control can be accomplished by maintaining diversity outside or inside the cropland by establishing insects/natural-enemies-banks (INEB) or biological-control-conservation-strips (BCCS). An INEB or BCCS system can be established by growing grass/hedgerows/flowering/nectar-plantations banks around the farmland, along roadsides, in the periphery of farmlands and/or stripped cultivation of alternate/trapping plantations in cultivated crops. Such practices/systems can accommodate and accumulate overwintering predators/parasitoids and provide a conducive environment for nectar feeding by the predacious adults of predators (Chrysopids, syrphidflies etc.) and adults of parasitoids. These also help to increase and maintain the natural enemy to pest ratio and enhance the chances of conservation of biological control (Doutt and Nakata 1973; Sotherton 1984; Thomas and Wratten 1988; Boller 1992; Gurr et al. 1998; Murphy et al. 1998; Powell 2000; Van Emden 2003; Jonsson et al. 2008). Growers prefer to exploit insecticides due their knockdown effects and quick control of pests to discourage weeds on their farmland. They are not acquainted with the concepts, complexity and practices of biodiversity conservation and management. It is, therefore, imperative to arrange outreach programs and campaigns at farmers' door steps or through media and distance-learning/online training sessions to create awareness on the implementation and conservation of biological control in dryland cropping systems.

Growers cannot avoid chemical pest control that interrupts the conservation of biological control in any cropping system. The conservation of biological agents in pesticide regimes is a different approach from the concept of conservation biological control by habitat modification. This former approach (conservation of biological control in pesticide regimes) focuses on using selective, ecofriendly and biorational insecticides/approaches suxh as IGRs, Bt crops (GM crops), botanicals, microbial insecticides, spynosins, avermectins, allellochemical, pheromones and plant-incorporated products/poisons (PIP). Conservation of natural enemies in pesticide regimes can also be achieved by modifying pesticide application techniques (seed dressing, chemigation, whorl-application etc.) and applying pesticides at safe pre- or post-application intervals (Morse 1989; van Emden and Peakall 1996; Van Emden 2003; Dhaliwal et al. 2006; Jonsson et al. 2008; Pedigo and Rice 2009).

5.5.7 Biorational and Other Innovative Approaches

'Biorational control' involves using chemicals that suppress insect populations by modifying their behavior, disrupting growth and impeding reproduction. Generally and operationally, biorational pest management involves substances or processes that execute diminutive or no adverse consequences to the environment and non-target organisms (humans, beneficial fauna and flora etc.); but they impose lethal, suppressive or behavior-modifying effects on a target organism and augment the specific control system (Pathak and Dhaliwal 1986; Dhaliwal and Arora 2003). Historically, Carl Djerassi used the term 'biorationals' for the first time for pheromones, insect hormones and hormone antagonists (Dhaliwal and Arora 2003). However, he did not propose any particular definition of biorationals but described their properties such as species specificity, active lethality at low concentrations and low persistency, and toxicity to non-target vertebrates (Djerassi et al. 1974). Biorational approaches include the strategic application of insect growth regulators (IGRs) and semiochemicals (pheromones and allelochemicals) while other innovative approaches include the use of propesticides, light-activated pesticides, avermectins, spinosyns, etc. Insect growth based insecticides include IGRs that interfere with cuticle formation mechanisms (e.g., chitin synthesis and degradation inhibitors, cuticle sclerotization disrupters, etc.) and with the secretion and actions of insect growth hormones (e.g., brain hormones, juvenile hormones, molting hormones, etc.). These are used as foliar sprays on crops to protect them from the attack of both chewing and sucking type insect pests. However, they do not affect the health of human beings (Altstein et al. 2000; Ishaaya 2001; Dhaliwal and Arora 2003; Pedigo 2003; Horowitz and Ishaaya 2004; Dhaliwal et al. 2006). Pheromones are the chemicals that induce any chemical communication between similar species. There are different types including sex, aggregation, alarm, trail and host-marking pheromones (Tschinkel and Close 1973; Verheggen et al. 2010; Heuskin et al. 2011; Chapman 2013) which are used to develop monitoring, male disruption or confusing/decoy, mass trapping and attract-and-kill techniques (Dhaliwal and Arora 2003; Pedigo 2003; Horowitz and Ishaaya 2004; Dhaliwal et al. 2006; Witzgall et al. 2010) (Table 4). Allelochemicals are interspecific semiochemicals which elicit chemical-signal-based communication in some members of different species. Allelochemicals include repellent, attractants, antifeedants and a large group of other compounds/molecules that regulate interspecific behaviors. These phytochemicals induce antixenotic (antifeedant, repellent, anti-oviposition and adverse behavioral effects) and antibiotic effects (growth, development, survival) in insects. They constitute a variety of plant secondary metabolites such as unusual amino acids, sugars, alkaloids, terpenoids, flavonoids, polyacetylenes, etc. (Dhaliwal and Arora 2003; Pedigo 2003; Dhaliwal et al. 2006), and may be of plant (botanicals, phytoalaxins, allomones etc.) or animal origin. The allelochemicals produced by natural enemies such as predators, parasitoids and pathogens are important in pest management programs. For example, delta-endotoxin produced by *Bacillus thuringiensis* is lethal against many Lepidopterous and Coleopterous insects (Dhaliwal and Arora 2003). They are categorized into allomones, kairomones and synomones.

Table 4 List of biorational and innovative insecticides for ecofriendly and sustainable management of insect pests in dryland ecosystems

Categories	Chemical molecules/compounds	Category/type/example of target insect pests of dry ecosystem	Mode of action	References
Biorational insecticides				
Brain hormone (BH)	Proctolin	Insect pest complex	Disrupt calcium influx	Dhaliwal et al. (2006)
Juvenile hormone (JH)	JH-0, JH-I, JH-II, JH-III, JH-B3 and methyl farnesoate	Insect pest complex	Ecdysis disruption	Bowers et al. (1965)
Chitin sysnthesis inhibitor (CSIs)	Diflubenzuron, chlorfluazuron, teflubenzuron, buprofezin, plumbagin	Sucking and chewing insect pest	Ovicidal, chemosterilant activities, molting disruption	Dhaliwal et al. (2006)
Molting hormones (MH)	ecdysone (20-hydroxyecdysone, 26-hydroxyecdysone, 20, 26-dihydroxyecdysone), ecdysteroids and ecdysterone (makisterone-A, 20-deoxymakisterone)	Sucking and chewing insect pest	Disrupt molting, growth and maturation of insects	Dhaliwal et al. (2006)
Sclerotization disruptors	MON-0585 (ditertiary butyl alcohol), DD	Dipteran, Lepidopterans and Coleopterans	Disrupt the metabolism and deposition of phenolic compounds, proteins and other compounds required for cuticular stabilization	Dhaliwal et al. (2006)
Pheromones				
Sex pheromones	Dodecatrienol, dodecadienol, neocembrene, trilinolein, decadienal, dodecanal	Termites	Male attractant	Costa-Leonardo et al. (2009)

	(E,Z,Z)- and (E,E,Z)-4,6,10-hexadecatrienyl acetates	cocoa pod borer, *Conopomorpha cramerella* S. (Lepidoptera: Gracillariidae)	Male attractant	Zhang et al. (2008)
	Alcohol and acetate molecules	Most moth species	Male attractant	Butler and McDonough (1979, 1981)
	(8*E*,10*E*)-tetradecadien-1-ol, and the corresponding aldehyde, acetate and formate ester	Lepidopteran insects	Male attractant	Van-der-Kraan and Ebbers (1990)
	Hexanal, heptanal, octanal, decanal, undecanol and 6,10,14-trimethylpentadecanon-2	Greater wax moth, *Galleria mellonella* L. (Lepidoptera: Pyralidae)	Male attractant	Lebedeva et al. (2002)
	(2Z, 6E)-7-methyl-3-propyl-2,6-decadien-1-ol., blend of *E-8, E*-10-dodecadien-1-ol and *E*-9-dodecen-1-ol and saturated alcohols of 10 to 18 carbons	Codling moth, *Cydia pomonella* L. (Lepidoptera: Tortricidae)	Male attractant	Arn et al. (1985)
	(E)-11-tetradecenyl acetate (E11-14Ac); (E,E)-9,11-tetradecadienyl acetate (E9E11-14Ac); (E)-11-tetradecen-1-ol (E11-14OH); tetradecyl acetate; hexadecanal; (E)-11-hexadecenyl acetate (E11-16Ac), hexadecyl acetate; octadecanal; and octadecyl acetate	Lightbrown apple moth, *Epiphyas postvittana* W. (Lepidoptera: Tortricidae)	Male attractant	El-Sayed et al. (2011)
	(2S,3R,7R)-isomers of the propionates of 3,7-dimethyl-2-tridecanol; 3,7-dimethyl-2-tetradecanol; 3,7-dimethyl-2-pentadecanol	Sawfly, *Gilpinia pallida* K.	Male attractant	Hedenström et al. (2006)

(continued)

Table 4 (continued)

Categories	Chemical molecules/compounds	Category/type/example of target insect pests of dry ecosystem	Mode of action	References
	cis and trans forms of propylure (10-propyl-trans-5,9-tridecadienyl acetate)	Pink bollworm, *Pectinophora gossypiella* S.	Male attractant	Jacobson (1969)
	(Z)-11-hexadecenal, (Z)-9-hexadecenal, (Z)-11-hexadecen-1-ol and hexadecanal	Striped rice stem borer, *Chilo suppressalis* W. and yellow stem borer, *Scirpophaga incertulas* W.	Male attractant	Cork et al. (1985)
	(Z)11-hexadecenal/(Z)9-hexadecenal/(Z)7-hexadecenal (Z11-16:Ald)	*Helicoverpa zea* B.	Male attractant	Lopez et al. (1991)
	(Z)-11-hexadecenal 90–99 % + (Z)-9-hexadecenal 10–1 %	*Helicoverpa armigera* H.	Male attractant	Zhang et al. (2012)
	Phenyl propanoids	Most fruit flies	Male attractant	
	Trimedlure [t-Butyl-2-methyl-4--chlorocyclohexanecarboxylate]	Mediterranean fruit fly, *Ceratitis capitate* W.	Male attractant	Beroza et al. (1961)
	Cuelure[4-(p-acetoxyphenyl)-2-butanone]	Melon fly, *Dacus cucurbitae*	Male attractant	
	Methyl eugenol (ME) (4-allyl-1, 2-dimethoxybenzene-carboxylate) and raspberry ketone (RK) (4-(p-hydroxyphenyl)-2-butanone)	Most fruit flies	Male attractant	Vargas et al. (2010)

Other innovative insecticides				
Light-activated insecticides/phototoxins	Furanocoumarines, alpha-terthienyl, polyecetyle, photooxidative dyes (compound of halogenated fluorescein series)	Mosquitoes, beetles, weevils, house flies	Inhibition of feeding, development of larval–pupal intermediates, failure to extricate from pupal case, deformed wings, reduced fecundity, reduced egg viability, direct mortality	Heitz and Downum (1987)
Propesticides	Carbosulfan, acephate, thiodicarb, cartap	Most sucking and chewing insects	Non-toxic in their actual form but highly toxic when metabolized inside the system of insects. Both contact and systemic, neuromuscular blocking agents, inhibition of synaptic transmission	Dhaliwal et al. (2006)
Avermectins (macrocyclic lactones)	Abamectin, ivermectin, selamectin, doramectin	Mites, insect pests	Acaricidal and insecticidal action, disrupt the action of both ligandated/glutamate-gated (GABA) and voltage-gated chloride channel causing an influx of chloride ions into the cells, leading to hyperpolarisation and subsequent paralysis	Dhaliwal et al. (2006)
Milbemycins (macrocyclic lactones)	Milbemycin	Mites and insect pests	Same as for avermectin	Lasota and Dybas (1990)
Spinosyns	Spinosad (mixture of spynosyn-A & D), spintoram	Coleopteran, Dipterans, Lepidopteran, Thysanoptrans	Contact and stomach poison; disruption of nicotinic acetylcholine receptors	Herbert (2010)

Their application in a pest management program of any cropping system not only guarantees environmental safety, and conservation of biodiversity, biological control and environmental quality but also highly supports the philosophy of IPM approaches (Dhaliwal and Arora 2003; Pedigo 2003; Dhaliwal et al. 2006).

6 Conclusions and Future Research Thrusts

The growing population, overuse, pollution, tidal surges and competing interests are degrading water resources globally. This will increase the area of dryland regions in the coming decades. Therefore, enhancing crop productivity in agroecosystems particularly in dryland areas is essential to feed the future population. Insect pests in dryland agroecosystems will be a major threat to crop production as they modify themselves according to environmental conditions. Sustainable agriculture in dryland areas should be practiced by integrating modern, research-proven technologies that are simple, cheap, easy-to-use and compatible with the respective regions. Climate variability and change have become the main reasons for the increased frequency of drought and its effect on crop planting times, growing season lengths, shifts in crop type or cultivars, pest incidence and crop productivity. Climate change, especially dry conditions, will affect the life-history traits of hosts and natural enemies differently. These effects might be more noticeable on natural enemies as they are at a higher trophic level. A better understanding of the behavioral, physiological and functional adaptations of natural enemies to climate extremes, both at the species and community level, will maximize the extent of natural regulation of insect pests, particularly in dryland areas.

Other IPM options including cultural, mechanical, biorational, genetic and biotechnological approaches with chemical control as a last resort should be tailored to water-stressed conditions. Dryland systems are not in equilibrium, having multiple thresholds and a diverse agricultural landscape, and often exhibit multiple ecological and social ranks. Hence, the extrapolation of laboratory experiments and field-simulating data, especially as they apply to arid environments, must be made with reservation. In dryland cropping systems, multidimensional and multidisciplinary research approaches, scientific/scientists and expert collaboration in long-term and large-scale projects should be executed to tackle/solve the prevailing and emerging plant production and protection issues.

References

Ahern RG, Brewer MJ (2002) Effect of different wheat production systems on the presence of two parasitoids (Hymenoptera: Aphelinidae: Braconidae) of the Russian wheat aphid in the North America Great Plains. Agri Ecosyst Environ 92:201–210

Alam MA (2010) Encyclopaedia of applied entomology. Anmol Publications (Pvt.) Ltd. New Delhi

Altieri MA, Gurr GM, Wratten SD (2004) Genetic engineering and ecological engineering: a clash of paradigms or scope for synergy? In: Gurr GM, Wratten SD, Altieri MA (eds) Ecological engineering for pest management. CSIRO Publishing, Collingwood, Australia

Altstein M, Ben-Aziz O, Schefler I, Zeltser I, Gilon C (2000) Advances in the application of neuropeptides in insect control. Crop Prot 19:547–555

Anonymous (1971) Hessian fly, Mayetiola destructor. In: Insects. N.Z. Deptt Agr Report pp 1–3

Arias-Estévez M, López-Periago E, Martínez-Carballo E, Simal-Ga´ndara J, Mejuto J, L. G´a-R´o (2008) The mobility and degradation of pesticides in soils and the pollution of groundwater resources. Agric Ecosyst Environ 123:247–260

Arizaga S, Ezcurra E, Peters E, de Arellano FR, Vega E (2000) Pollination ecology of Agave Macroacantha (Agavaceae) in a Mexican tropical desert. ii. The role of pollinators. Am J Bot 87:1011–1017

Arn H, Guerin PM, Buser HR, Rauscher S, Mani E (1985) Sex pheromone blend of the codling moth, *Cydia pomonella*: evidence for a behavioral role of dodecan-1-ol. Experientia 41:1482–1484

Arnon I (Eds) (1992) Insects and other plant pests. In: Developments in agricultural and managed-forest ecology agriculture in dry lands, principles and practices. Elsevier, Amsterdam, The Netherlands

Asin L, Pons X (2001) Effect of high temperature on the growth and reproduction of corn aphids (Homoptera: Aphididae) and implications for their population dynamics on the northeastern Iberian peninsula. Environ Entomol 30:1127–1134

Athukorala W, Wilson C, Robinson T (2012) Determinants of health costs due to farmers' exposure to pesticides: an empirical analysis. Agri Econ 63:158–174

Ayal Y, Polis GA, Lubin Y, Goldberg DE (2005) How can high animal biodiversity be supported in low productivity deserts? The role of macrodetrivory and physiognomy. In: Shachak M, Wade R (eds) Biodiversity in drylands. Cambridge University Press, Cambridge, UK

Babaei MR, Barari H, Kara K (2009) Weight differences of male and female pupae of gypsy moth (*Lymantria dispar*) and host-sex preference by two parasitoid species *Lymantrichneumon disparis* and *Exorista larvarum*. Pak Biol Sci 12:443–446

Bale JS, Masters GJ, Hodkinson ID, Awmack C, Bezemer TM, Brown VK, Butterfield J, Buse A, Coulson JC, Farrar J (2002) Herbivory in global climate change research: direct effects of rising temperature on insect herbivores. Global Change Biol 8:1–16

Baur M, Boethel D (2003) Effect of Bt-cotton expressing Cry1A (c) on the survival and fecundity of two hymenopteran parasitoids (Braconidae, Encyrtidae) in the laboratory. Biol Control 26:325–332

Benbrook CM (2012) Impacts of genetically engineered crops on pesticide use in the US–the first sixteen years. Environ Sci Eur 24:2190–4715

Bentz JA, Reeves J, Barbosa P, Francis B (1996) The effect of nitrogen fertilizer applied to *Euphorbia pulcherrima* on the parasitization of *Bemisia argentifolii* by the parasitoid *Encarsia formosa*. Entomol Exp appl 78:105–110

Beroza M, Green N, Gertler SI (1961) New attractants for the Mediterranean fruit fly in. J Acricul Food Chem 9:361–365

Birch LC (1948) The intrinsic rate of natural increase of an insect population. Anim Ecol 17:15–26

Boller EF (1992) The role of integrated pest management in integrated production of viticulture in Europe. Proceedings of the British Crop Protection Conference, Pests and Diseases, Brighton pp 499–506

Booij CH, Noorlander J (1992) Farming systems and insect predators. Agri Ecosyst Environ 40:125–135

Bowers WS, Thompson MJ, Uebel EC (1965) Juvenile and gonadotropic hormone activity of 10, 11-epoxyfarnesoic acid methyl ester. Life Sci 4:2323–2331

Brewer M, Elliott NC (2004) Biological control of cereal aphids in North America and mediating effects of host plant and habitat manipulations. Ann Rev Entomol 49:219–242

Butani DK (1979) Insects and fruits. Periodical expert book agency, D-42, Vivek Vihar, Delhi, India. pp 415
Butler LI, McDonough LM (1979) Insect sex pheromones: evaporation rates of acetates from natural rubber septa. Chem Ecol 5:825–837
Butler LI, McDonough LM (1981) Insect sex pheromones: evaporation rates of alcohols and acetates from natural rubber septa. Chem Ecol 7:627–633
Buurma J (2008) Stakeholder involvement in crop protection policy planning in The Netherlands. ENDURE–RA3.5/SA4.5 Working Paper. LEI Wageningen UR, The Hague, The Netherlands
Cagan L, Tancik J, Hassan SA (1998) Natural parasitism of European corn borer eggs, *Ostrinia nubilalis* Hbn. (Lepidoptera: Pyralidae) by Trichogramma in Slovakia-need for field releases of the natural enemy. Appl Entomol 122:315–318
Carrière Y, Crowder DW, Tabashnik BE (2010) Evolutionary ecology of insect adaptation to Bt crops. Evol Appl 3:561–573
Cartwright B, Edelson JV, Chambers C (1989) Composite action thresholds for the control of Lepidopterous pests on fresh market cabbage in the lower Rio Grande Valley of Texas. Econ Entomol 80:175–181
Cattaneo MG, Yafuso C, Schmidt C, Huang CY, Rahman M, Olson C, Ellers-Kirk C, Orr BJ, Marsh SE, Antilla L, Dutilleul P, Carrière Y (2006) Farm-scale evaluation of the impacts of transgenic cotton on biodiversity, pesticide use, and yield. Proc Natl Acad Sci USA 103:7571–7576
CGIAR (1997) Consultative Group on International Agricultural Research. World's dryland farmers need new agricultural technology – "Green Revolution" never reached them. CGIAR Newsletter 4(4), http://www.worldbank.org/html/cgiar/press/dryland.html
Chakravarthy N, Gautam R (2002) Forewarning mustard aphid, *NATP* Project Report. Division of Agricultural Physics, IARI, New Delhi, India
Chapman RF (2013) The insect: structure and function, 5th edn. Cambridge University Press, New York, USA
Chapman TB, Veblen TT, Schoennagel T (2012) Spatio temporal patterns of mountain pine beetle activity in the southern Rocky Mountains. Ecology 93:2175–2185
Cloudsley-Thompson JL (1975) Adaptations of Arthropoda to arid environments. Ann Rev Entomol 20:261–283
Clough Y, Kruess A, Tschamtke T (2007) Local landscape factors in differently managed arable fields affect the insect herbivore community of a non-crop plant species. Appl Ecol 44:22–28
Cockbain A (1961) Low temperature thresholds for flight in *Aphis fabae* Scop. Entomol Exp Appl 4:211–219
Collins JK, Cuperus GW, Cartwright B, Stark JA, Ebro LL (1992) Consumer attitudes on production systems for fresh produce. Sustain Agri 3:67–77
Collins KL, Boatman ND, Wilcox A, Holland JM (2003) A 5-year comparison of overwintering polyphagous predator densities within a beetle bank and two conventional hedgebanks. Ann Appl Biol 143:63–71
Collins W, Colman R, Haywood J, Manning RR, Mote P (2007) The physical science behind climate change. Sci Am 297:64–73
Conner AJ, Glare TR, Nap JP (2003) The release of genetically modified crops into the environment. Plant J 33:19–46
Cork A, Beevor PS, Hall DR, Nesbitt BF, Arida GS, Mochida O (1985) Components of the female sex pheromone of the yellow stem borer, *Scirpophaga incertulas*. Entomologia Experimentalis et Applicata 37(2):149–155
Costa-Leonardo AM, Casarin FE, Lima JT (2009) Chemical communication in Isoptera. Neotrop Entomol 38(1):1–6
Crane E, Walker P (1984) Pollination directory for world crops. International bee research association
Croft BA (1990) Arthropod biological control agents and pesticides. Wiley, New York, USA

Cunningham SA, Evans TA, Arthur AD, Schellhorn NA, Bianchi FJ (2010) What can insects do to help the farm? Soil improvement, pollination and pest control. In: 25th Annual conference proceedings 28–29 July 2010-Dubbo NSW "Adapting mixed farms to future environments". pp 37–41

De A, Bose R, Kumar A, Mozumdar S (2014) Worldwide pesticide use. In: A D, Bose R, Kumar A, Mozumdar S (eds) Targeted delivery of pesticides using biodegradable polymeric nanoparticles. Springer, New York, USA

DeBach P, Rosen D (eds) (1991) Biological control by natural enemies. Cambridge University Press, Cambridge, MA

Denno RF, Roderick GK, Olmstead KL, Dobel HG (1991) Density-related migration in planthoppers (Homoptera: Delphacidae). The role of habitat persistence. Am Nat 138:1513–1541

Dhaliwal GS, Arora R (2003) Principles of insect pest management, 2nd edn. Kalyani Publishers, Ludhiana, India

Dhaliwal GS, Koul O (2007) Biopesticides and pest management: conventional and biotechnological approaches. Kalyani Publishers, Ludhiana, India

Dhaliwal GS, Singh R, Chhillar BS (2006) Essentials of agricultural entomology. Kalyani Publishers, Ludhiana, India

Dhillon M, Sharma H (2007) Effect of storage temperature and duration on viability of eggs of *Helicoverpa armigera* (Lepidoptera: Noctuidae). Bull Entomol Res 97:55–59

Dhillon MK, Sharma HC (2009) Temperature influences the performance and effectiveness of field and laboratory strains of the ichneumonid parasitoid, *Campoletis chlorideae*. Bio Contr 54:743–750

Dixon AFG (2000) Insect predator-prey dynamics: ladybird beetles and biological control: Cambridge University Press, New York, USA

Djerassi C, Shih-Coleman C, Diekman J (1974) Insect control of the future: operational and policy aspects. Science 186:596–607

Doutt RL, Nakata J (1973) The Rubus leafhopper and its egg parasitoid: an endemic biotic system useful in grape-pest management. Environ Entomol 2:381–386

Eggleton P, Belshaw R (1992) Insect parasitoids: an evolutionary overview. Philos Trans R Soc London B Biol Sci 337(1279):1–20

Elliott NC, Kieckhefer RT, Beck DA (2002) Effect of aphids and the surrounding landscape on abundance of coccinellids in cornfields. Biol Control 24:214–220

El-Sayed AM, Mitchell VJ, Manning LA, Suckling DM (2011) New sex pheromone blend for the light brown apple moth, *Epiphyas postvittana*. Chem Ecol 37(6):640–646

Fujisaka S, Kirk G, Litsinger JA, Moody K, Hosen N, Yusef A, Nurdin F, Naim T, Artati F, Aziz A (1991) Wild pigs, poor soils, and upland rice: a diagnostic survey of Sitiung, Sumatra, Indonesia. IRRI Research Paper Series No. 155. p 9

Gerling D, Sinai P (1994) Buprofezin effects on two parasitoid species of whitefly (Homoptera: Aleyrodidae). Econ Entomol 87:842–846

Getanjaly RVL, Sharma P, Kushwaha R (2015) Beneficial Insects and their Value to Agriculture. Res Agri Forest Sci 3(5):25–30

Giles KL, Jones DB, Royer TA, Elliott NC, Kindler SD (2003) Development of a sampling plan in winter wheat that estimates cereal aphid parasitism levels and predicts population suppression. J Econ Entomol 96:975–982

Glamoustaris A, Mithen R (1995) The effect of modifying the glucosinolate content of leaves of oilseed rape (*Brassica napus* ssp. oleifera) on its interaction with specialist and generalist pests. Ann Appl Biol 126:347–363

Gogi MD, Sarfraz RM, Dosdall LM, Arif MJ, Keddie AB, Ashfaq M (2006) Effectiveness of two insect growth regulators against *Bemisia tabaci* (Gennadius) (Homoptera:Aleyrodidae) and *Helicoverpa armegera* (Hubner) (Lepidoptera: Noctuidae) and their impact on population densities of arthropod predators in cotton in Pakistan. Pest Manag Sci 62:982–990

Greenslade PJM (1983) Adversity selection and the habitat templet. Am Nat 122:352–365

Griggs DJ, Noguer M (2002) Climate change 2001: the scientific basis. Contribution of working group I to the third assessment report of the intergovernmental panel on climate change. Weather 57:267–269

Grist DH, Lever RJAW (1969) Pests of rice. Longmans, London, UK

Gu H, Edwards OR, Hardy AT, Fitt GP (2008) Host plant resistance in grain crops and prospects for invertebrate pest management in Australia: an overview. Aust Exp Agri 48:1543–1548

Gurr GM, van Emden HF, Wratten SD (1998) Habitat manipulation and natural enemy efficiency: implication for the control of pests. In: Barbosa P (ed) Conservation biological control. Academic Press, San Diego

Gurr GM, Wratten SD, Altieri MA (2004) Ecological engineering for pest management: advances in habitat manipulation for arthropods. CSIRO, Collingwood, Australia

Hamilton JG, Dermody O, Aldea M, Zangerl AR, Rogers A, Berenbaum MR, Delucia EH (2005) Anthropogenic changes in tropospheric composition increase susceptibility of soybean to insect herbivory. Environ Entomol 34:479–485

Hardy JE (1938) Plutella maculipennis Curt., its natural and biological control in England. Bull Entomol Res 29:343–372

Harrington R, Burgess AL, Taylor MS, Foster GN, Morrison S, Ward L, Tones SJ, Rogers R, Barker I, Walters KFA, Morgan D (1998) Spread of BYDV-DESSAC module: proceedings of the 6th HGCA R&D conference on cereals and oilseeds. Home Grown Cereals Authority, Cambridge

Harwood JD, Phillips SW, Lello J, Sunderland KD, Glen DM, Bruford MW, Harper GL, Symondson WOC (2009) Invertebrate biodiversity affects predator fitness and hence potential to control pests in crops. Biol Control 51:499–506

Hedenström E, Edlund H, Wassgren AB, Bergström G, Anderbrant O, Ostrand F, Sierpinski A, Auger-Rozenberg MA, Herz A, Heitland W, Varama M (2006) Sex pheromone of the pine sawfly, *Gilpinia pallida*: chemical identification, synthesis, and biological activity. Chem Ecol 32:2525–2541

Heitz JR, Downum KR (1987) Light activated pesticides. Am Chem Soc Symp Ser 339:1–340

Heong KL, Song YH, Pimsamarn S, Zhang R, Bae SD (1995) Global warming and rice arthropod communities. In: Peng S, Ingram KT, Neue HU, Ziska LH (eds) Climate change and rice. International Rice Research Institute/Springer, Berlin/Manila

Heong KL, Escalada MM, Huan NH, Ky-Ba VH, Quynhm PV, Thiet LV, Chien HV (2008) Entertainment-education and rice pest management: a radio soap opera in Vietnam. Crop Prot 27:1392–1397

Herbert AK (2010) The spinosyn family of insecticides: realizing the potential of natural products research. J Antibiot 63:101–111

Heuskin S, Verheggen FJ, Haubruge E, Wathelet JP, Lognay G (2011) The use of semiochemical slow-release devices in integrated pest management strategies. Biotechnol Agron Soc Environ 25:459–470

Hill DS (1987) Agricultural insect pests of temperate regions and their control. Cambridge University Press, Cambridge, UK

Hillocks RJ (2009) GM cotton for Africa. Outlook Agric 38:311–316

Holtzer TO, Anderson RL, McMullen MP, Peairs FB (1996) Integrated pest management of insects, plant pathogens and weeds in dryland cropping systems of the Great Plains. J Prod Agric 9:200–208

Horne PA, Page J (2008) IPM for crops and pastures. CSIRO Landlinks Press, Melbourne, Australia

Horowitz AR, Ishaaya I (2004) Biorational insecticides mechanisms, selectivity and importance in pest management programs. New York, USA. pp 1–28

Hoseini S, Pourmirza A, Ebadollahi A, Jahromi MG (2012) Impacts of two conventional insecticides on different stages of Encarsia inaron Walker parasitizing the whitefly, *Trialeurodes vaporariorum* Westwood under greenhouse condition. Arch Phytopathol Plant Prot 45:268–275

Houghton JT, Ding Y, Griggs DJ, Noguer M, van der Linden PJ, Dai X, Maskell K, Johnson C (eds) (2001) Climate change 2001: the scientific basis. Contribution of working group I to the third assessment report of the intergovernmental panel on climate change. Cambridge University Press, Cambridge, UK

Hsu JC, Feng HT (2006) Development of resistance to spinosad in oriental fruit fly (Diptera: Tephritidae) in laboratory selection and cross-resistance. Econ Entomol 99:931–936

Huberty AF, Denno RF (2004) Plant water stress and its consequences for herbivorous insects: a new synthesis. Ecology 85:1383–1398

Imms AD (1964) Outlines of Entomology (5th eds). Methuen London UK

Irshad M (2001) Aphids and their biological control in Pakistan. Pak Biol Sci 4:537–541

Ishaaya I (2001) Biochemical processes related to insecticide action: an overview. In: Ishaaya I (ed) Biochemical sites of insecticide action and resistance. Springer, Berlin/Heidelberg/New York

Izhaki I, Walton PB, Safriel UN (1991) Seed shadows generated by frugivorous birds in an Eastern Mediterranean scrub. Ecology 79:575–590

Webster JA (1995) Economic Impact of the Greenbug in the Western United States: 1992–1993. Volume Publication No. 155, Great Plains Agricultural Council, Stillwater, Oklahoma.

Jackson DM, Farnham M, Simmons A, Van Giessen W, Elsey K (2000) Effects of planting pattern of collards on resistance to whiteflies (Homoptera: Aleyrodidae) and on parasitoid abundance. Econ entomol 93:1227–1236

Jacobson M (1969) Sex pheromone of the pink bollworm moth: biological masking by its geometrical isomer. Science 163:190–191

Jan MT, Abbas N, Shad SA, Saleem MA (2015) Resistance to organophosphate, pyrethroid and biorational insecticides in populations of spotted bollworm, *Earias vittella* (Fabricius) (Lepidoptera: Noctuidae), in Pakistan. Crop Protect 78:247–252

Jha LK (2010) Applied agricultural entomology. New Central Book Agency (pvt.) Ltd., Kolkata/Pune/Delhi, India

Jin L, Zhang H, Lu Y, Yang Y, Wu K, Tabashnik BE, Wu Y (2015) Large-scale test of the natural refuge strategy for delaying insect resistance to transgenic Bt crops. Nat Biotechnol 33:169–174

Johansen BE (2002) The global warming desk reference. Greenwood Press, Westport

John B, Darryl H, Greg P (2007) Hort Guard TM Initiative AGWEST Spiraling Whitefly Aleurodicus Dispersus Exotic threat to Western Australia. Availabale at (www.agric.wa.gov.au/objtwr/imported_assets/.../pw/.../fs01800.pdf)

Jones C (1999) Cropping native pasture and conserving biodiversity: a potential technique. In: Barlow T, Thorburn R (eds) Balancing conservation and production in grassy landscapes: proceedings of the Bushcare grassy landscapes conference. Environment Australia: Biodiversity Group, Clare

Jonsson M, Wratten SD, Landis DA, Gurr GM (2008) Recent advances in conservation biological control of arthropods by arthropods. Biol Control 45:172–175

Jönsson A, Appelberg G, Harding S, Bärring L (2009) Spatio-temporal impact of climate change on the activity and voltinism of the spruce bark beetle, Ips typographus. Global Change Biol 15:486–499

Joshi SR (2006) Biopesticides: a biotechnological approach. New Age International (pvt.) Ltd. Publishers, New Delhi, India

Karl TR, Trenbeth KE (2003) Modern global climate change. Science 302:1719–1723

Keenan SP, Giles KL, Elliott NC, Royer TA, Porter DR, Burgener PA, Christian DA (2007a) Grower perspectives on areawide wheat integrated pest management in the southern US great plains. In: Koul O, GV C (eds) Ecologically based integrated pest management. CAB International, Vallingford

Keenan SP, Giles KL, Burgener PA, Elliott NC (2007b) Collaborating with wheat growers in demonstrating areawide integrated pest management. J Ext (http://www.joe.org/joe/2007february/a7.shtml)

Kelsey KD, Mariger SC (2002) A survey-based model for setting research, education, and extension priorities at the Land–grant University – a case study of Oklahoma wheat producers: final report. Agricultural Extension, Communications, and 4 -H Youth Development, Oklahoma State University, Stillwater, USA

Khan HAA, Akram W, Shehzad K, Shaalan EA (2011) First report of field evolved resistance to agrochemicals in dengue mosquito, *Aedes albopictus* (Diptera: Culicidae), from Pakistan. Parasit Vectors 4:146

Kieckhefer RW, Kantack BH (1988) Yield losses in winter grains caused by cereal aphids (Homoptera: Aphididae) in South Dakota. J Econ Entomol 81:317–321

Kindler SD, Elliott NC, Giles KL, Royer TA, Fuentes-Grandaos R, Tao F (2002) Effect of greenbug (Homoptera: Aphididae) on yield loss of winter wheat. J Econ Entomol 95:89–95

Kindler SD, Elliott NC, Giles KL, Royer TA (2003) Economic injury levels for the greenbug, Schizaphis graminum, in winter wheat. Southwestern Entomol 28:163–166

Knipling EF (1979) The basic principles of insect population suppression and management, Agriculture handbook No. 512. US Government Printing Office, Department of Agriculture, Washington, DC, USA

Koul O, Cuperus GW (2007) Ecologically based integrated pest management. CABI International, New York, USA

Kranthi K, Jadhav D, Kranthi S, Wanjari R, Ali S, Russell D (2002) Insecticide resistance in five major insect pests of cotton in India. Crop Protect 21:449–460

Landis DA, Wratten SD, Gurr GM (2000) Habitat management to conserve natural enemies of arthropod pests in agriculture. Annu Rev Entomol 45:175–201

Lasota JA, Dybas RA (1990) Abamectin as a pesticide for agricultural use. Acta Leiden 59(1–2):217–225

Lebedeva KV, Vendilo NV, Ponomarev VL, Pletnev VA, Mitroshin DB (2002) Identification of pheromone of the greater wax moth *Galleria mellonella* from the different regions of Russia. IOBC WPRS Bull 25(9):229–232

Litsinger JA, Libetario EM, Barrion AT (2002) Population dynamics of white grubs in the upland rice and maize environment of Northern Mindanao, Philippines. Int Pest Manage 48:239–260

Liu H, Cheng H, Wang X (1995) A general study on Chinese diet: pesticide residue. Health Res 24

Loomans AJM, Van Lenteren JC (1995) Biological control oh thrips pests: a review on thrips parasitoids. Wageningen Agricultural University Papers 95-1, Wageningen, p 237

Lopez JD, Leonhardt BA, Shaver TN (1991) Performance criteria and specifications for laminated plastic sex pheromone dispenser for *Helicoverpa zea* (Lepidoptera: Noctuidae). Chem Ecol 17(11):2293–2305

MacLeod A, Wratten SD, Sotherton NW, Thomas MB (2004) 'Beetle banks' as refuges for beneficial arthropods in farmland: long-term changes in predator communities and habitat. Agri Entomol 6:147–154

Maelzer DA, Zalucki MP (2000) Long range forecasts of the numbers of *Helicoverpa punctigera* and *H. armigera* (Lepidoptera: Noctuidae) in Australia using the southern oscillation index and the sea surface temperature. Bull Entomol Res 90:133–146

Malone LA, Burgess EPJ (2009) Impact of genetically modified crops on pollinators. In: Ferry N, Gatehouse AMR (eds) Environmental impact of genetically modified crops, CAB International, Walligford, pp 199–222

Martin T, Ochou GO, Hala-N'Klo F, Vassal JM, Vaissayre M (2000) Pyrethroid resistance in the cotton bollworm, *Helicoverpa armigera* (Hübner), in West Africa. Pest Manage Sci 56:549–554

Martins DJ (2014) Butterfly pollination of the dryland wildflower gloriosa minor. East Afr Nat Hist 103(1):25–30

Maxmen A (2013) Crop pests: under attack. Nature 501:15–17

McKinney K (1939) Common insects attacking sugar beets and vegetable crops in the Salt River Valley of Arizona. Econ Entomol 32:808–810

Mendelsohn M, Kough J, Vaituzis Z, Matthews K (2003) Are Bt crops safe? Nate Biotechnol 21:1003–1009
Mills N, Gutierrez A (1996) Prospective modelling in biological control: an analysis of the dynamics of heteronomous hyperparasitism in a cotton-whitefly-parasitoid system. Appl Ecol vol:1379–1394
Mirab-Balou M, Tong X, Feng J, Chen X (2011) Thrips (Insecta: Thysanoptera) of China. 7(6): 720–744
Momhinweg DW, Brewer MJ, Porter DR (2006) Effect of Russian wheat aphid on yield and yield components of field grown susceptible and resistant spring barley. Crop Sci 46:36–42
Morrill WL (1995) Insect pests of small grains. APS Press, St. Paul
Morse S (1989) The integration of partial plant resistance with biological control by an indigenous natural enemy complex in affecting populations of cowpea aphid (Aphis craccivora Koch). PhD dissertation, University of Reading, UK
Mullen JD, Norton GW, Reaves DW (1997) Economic analysis of environmental benefits of integrated pest management. Agric Appl Econ 29:243–253
Murphy BC, Rosenheim JA, Granett J, Pickett CH, Dowell RV (1998) Measuring the impact of a natural enemy refuge: the Prune tree/wineyard example. In: Pickett CH, Bugg RL (eds) Enhancing biological control: habitat management to promote natural enemies of agricultural pests. University of California Press, Berkeley, USA
Naranjo SE (1993) The life-history of *Trichogrammatoidea bactrae* (Hymenoptera: Trichogrammatidae) an egg parasitoid of pink bollworm (Lepidoptera: Gelechidae) with emphasis on performance at high temperatures. Environ Entomol 22:1051–1059
Naranjo SE (2001) Conservation and evaluation of natural enemies in IPM systems for *Bemisia tabaci*. Crop Protect 20:835–852
Nash MA, Hoffmann AA (2012) Effective invertebrate pest management in dryland cropping in southern Australia: the challenge of marginality. Crop Protect 42:289–304
Nawaz A, Razpotnik A, Rouimi P, de Sousa G, Cravedi JP, Rahmani R (2014) Cellular impact of combinations of endosulfan, atrazine, and chlorpyrifos on human primary hepatocytes and HepaRG cells after short and chronic exposures. Cell Biol Toxicol 30(1):17–29
Norris RF, Caswell-Chen EP, Kogan M (2002) Concepts in integrated pest management. Prentice-Hall of India Pvt. Ltd., New Delhi, India
Ohayo-Mitoko GJA, Heederik DJJ, Kromhout H, Omondi BEO, Boleij JSM (1997) Acetylcholinesterase inhibition as an indicator of organophosphate and carbamate poisoning in Kenyan agricultural workers. Int J Occup Environ Health 3:210–220
Olsen LG (1997) IPM labeling of food products boosts IPM. The IPM Report 6:4–5
Ortiz R, Trethowan R, Ortiz Ferrara GF, Iwanaga M, Dodds JH, Crouch JH, Crossa J, Braun HJ (2007) High yield potential, shuttle breeding, genetic diversity and new international wheat improvement strategy. Euphytica 157:365–384
Pardo-Lopez L, Soberon M, Bravo A (2013) *Bacillus thuringiensis* insecticidal three-domain Cry toxins: mode of action, insect resistance and consequences for crop protection. FEMS Microbiol Rev 37:3–22
Park RF, Wellings CR, Bariana HS (2006) The Australian cereal rust control program: the challenge posed by changing pathogen populations. Agromeridian 2:76–79
Pathak PK, Dhaliwal GS (1986) Trends and strategies for rice pest problems in tropical Asia. IRRI research paper series, 64 Los Baños, Philippines
Pedigo LP (2003). Plant resistance to insects. In: Entomology and pest management. Prentice Hall of India (Pvt.) Ltd, New Delhi, India, pp 413–424
Pedigo LP, Rice ME (2009) Entomology and pest management, 6th edn. PHI Learning (Pvt.) Ltd, New Delhi, India
Perry AS, Perry RY (1989) Effects in arid regions. In: Bourdeau P, Haines JA, Klein W, Krishna Murti CR (eds) Ecotoxicology and climate with special reference to hot and cold climates. Wiley, New York, India

Peterson GA, Schlegel A, Tanaka DL, Jones OR (1996) Precipitation use efficiency as affected by cropping and tillage systems. Prod Agric 9:180–186

Pimentel D (2009) Pest control in world agriculture. In: Lal R (eds) Agri Sci 2:498

Ponsard S, Gutierrez AP, Mills NJ (2002) Effect of Bt-toxin (Cry1Ac) in transgenic cotton on the adult longevity of four heteropteran predators. Environ Entomol 31:1197–1205

Porter J, Parry M, Carter T (1991) The potential effects of climatic change on agricultural insect pests. Agri Forest Meteorol 57:221–240

Powell W (2000) The use of field margins in the manipulation of parasitoids for aphid control in arable crops. In: Proceedings of British crop protection conference, pests and diseases, Brighton, pp 579–584

Régnière J, St-Amant R, Duval P (2012) Predicting insect distributions under climate change from physiological responses: spruce budworm as an example. Biol Invas 14:1571–1586

Reyes M, Franck P, Charmillot PJ, Ioriatti C, Olivares J, Pasqualini E, Sauphanor B (2007) Diversity of insecticide resistance mechanisms and spectrum in European populations of the codling moth, *Cydia pomonella*. Pest Manage Sci 63:890–902

Reyes M, Barros-Parada W, Ramírez CC, Fuentes-Contreras E (2015) Organophosphate resistance and its main mechanism in populations of Codling moth (Lepidoptera: Tortricidae) from central Chile. Econ Entomol 108:277–285

Rouault G, Candau JN, Lieutier F, Nageleisen LM, Martin JC, Warzée N (2006) Effects of drought and heat on forest insect populations in relation to the 2003 drought in Western Europe. Ann For Sci 63:613–624

Roush DK, McKenzie JA (1987) Ecological genetics of insecticide and acaricide resistance. Annu Rev Entomol 32:361–380

Royer TA, Krenzer EG (2000) Wheat management in Oklahoma: a Handbook Jor Oklahoma's wheat Industry. Publication E-831, Oklahoma Cooperative Extension Service, Oklahoma, USA

Royer TA, Story S, Elliott NC (2007) Cereal aphid expert system and glance n' go sampling education handbook. Oklahoma Cooperative Extension Service, Oklahoma State University Extension, Stillwater

Samiee A, Rezvanfar A, Faham E (2009) Factors influencing the adoption of integrated pest management (IPM) by wheat growers in Varamin County, Iran. Afr Agric Res 4:491–497

Sanahuja G, Banakar R, Twyman RM, Capell T, Christou P (2011) *Bacillus thuringiensis*: a century of research, development and commercial applications. Plant Biotechnol 9:283–300

Sandhu HS, Wratten SD, Cullen R, Case B (2008) The future of farming: the value of ecosystem services in conventional and organic arable land: an experimental approach. Ecol Econ 64:835–848

Sarfraz M, Dosdall LM, Keddie BA (2005) Spinosad: a promising tool for integrated pest management. Outlook Pest Manage 16:78–84

Schellhorn NA, Macfadyen S, Bianchi FJJA, Williams DG, Zalucki PMP (2008) Managing ecosystem services in broadacre landscapes: what are the appropriate spatial scales? Aust Exp Agri 48:1549–1559

Schowalter TD (2011) Insect ecology: an ecosystem approach. Academic Press, London

Scott M, Berrigan D, Hoffmann AA (1997) Costs and benefits of acclimation to elevated temperature in Trichogramma carverae. Entomol Exp Appl 85:211–219

Shachak M, Lovett GM (1998) Atmospheric deposition to a desert ecosystem and its implication for management. Ecol Appl 8:455–463

Shapiro DI, Stuart RJ, McCoy CW (2005) Targeted improvement of Steinernema carpocapsae for control of the pecan weevil, Curculio caryae (Horn) (Coleoptera: Curculionidae) through hybridization and bacterial transfer. Biol Control 34:215–221

Siegel A, Hari V (1980) Plant viruses. In: Carlson PS (eds) The biology of crop productivity. Academic Press, New York, pp 155–201

Simmons A, Jackson D (2000) Evaluation of foliar-applied insecticides on abundance of parasitoids of *Bemisia argentifolii* (Homoptera: Aleyrodidae) in vegetables. Entomol Sci 35:1–8

Singh R (2008) Crop protection by botanical pesticides. CBS Publishers and Distributors, New Delhi, India

Skevas T, Stefanou SE, Lansink AO (2013) Do farmers internalise environmental spillovers of pesticides in production? Agri Econ 64:624–640

Smith K, Smith D, Lisle A (1999) Effect of field-weathered residues of pyriproxyfen on the predatory coccinellids *Chilocorus circumdatus* Gyllenhal and *Cryptolaemus montrouzieri* Mulsant. Anim Prod Sci 39:995–1000

Sorby K, Fleischer G, Pehu E (2005) Integrated pest management in development: review of trends and implementation strategies. In: Agriculture and rural development working paper 5. World Bank, Washington, DC

Sotherton NW (1984) The distribution and abundance of predatory arthropods overwintering in farmland. Ann Appl Biol 105:423–429

Srinivasa MV (2000) Host plants of the spiraling whitefly *Aleurodicus dispersus* Russell (Hemiptera: Aleyrodidae). Pest Manage Hort Ecosyst 6:79–105

Staley JT, Hodgson CJ, Mortimer SR, Morecroft MD, Masters GJ, Brown VK, Taylor ME (2007) Effects of summer rainfall manipulations on the abundance and vertical distribution of herbivorous soil macro-invertebrates. Eur Soil Biol 43:189–198

Strand MR, Obrycki JJ (1996) Host specificity of insect parasitoids and predators. BioScience 46:422–429

Tabashnik BE (1994) Evolution of resistance to *Bacillus thuringiensis*. Annu Rev Entomol 39:47–79

Tabashnik BE (2008) Delaying insect resistance to transgenic crops. Proc Natl Acad Sci, USA 105:19029–19030

Tabashnik BE, Van Rensburg J, Carrière Y (2009) Field-evolved insect resistance to Bt crops: definition, theory, and data. Econ Entomol 102:2011–2025

Tabashnik BE, Brévault T, Carrière Y (2013) Insect resistance to Bt crops: lessons from the first billion acres. Nat Biotechnol 31:510–521

Taheri S, Razmjou J, Rastegari N (2010) Fecundity and development rate of the bird cherry-oat aphid, Rhopalosiphum padi (L) (Homoptera: Aphididae) on six wheat cultivars. Plant Prot Sci 46:72–78

Taylor TCH (1940) Report of the entomologist, Kawande. Report of the Department of Agriculture, Uganda (1938–1939) pp 9–24

TEEB (2010) The economics of ecosystems and biodiversity: mainstreaming the economics of nature: a synthesis of the approach conclusions and recommendations of TEEB. Progress Press, Malta

Thomas MB, Wratten SD (1988) Manipulating the arable crop environment to enhance the activity of predatory insects. Asp Appl Biol 17:57–66

Thomson LJ, Robinson M, Hoffmann AA (2001) Field and laboratory evidence for acclimation without costs in an egg parasitoid. Func Ecol 15:217–221

Thorpe WH, Caudle HB (1938) A study of the olfactory responses of insect parasites to the food plant of their host. Parasitol 30:523–528

Tilman D, Knops J (1997) The influence of functional diversity and composition on ecosystem processes. Science 277(5330):1300–1302

Tilman D, Cassman KG, Matson PA, Naylor R, Polasky S (2002) Agricultural sustainability and intensive production practices. Nature 418:671–677

Tobin PC, Nagarkatti S, Loeb G, Saunders MC (2008) Historical and projected interactions between climate change and insect voltinism in a multivoltine species. Global Change Biol 14:951–957

Tscharntke T, Rand TA, Bianchi FJJA (2005) The landscape context of trophic interactions: insect spillover across the crop-noncrop interface. Ann Zool Fenn 42:421–432

Tschinkel WR, Close PG (1973) The trail pheromone of the termite, Trinervitermes trinervoides. Insect Physiol 19:707–721

Tsitsilas A, Hoffmann AA, Weeks AR, Umina PA (2011) Impact of groundcover manipulations within windbreaks on mite pests and their natural enemies. Aust Entomol 50:37–47
Uvarov B (1962) Development of arid lands and its ecological effects on their insect fauna. Arid Zone Res 18:235–248
van Emden HF (2003) Conservation biological control: from theory to practice. In: Van Driesche R (eds) Proceedings of the international symposium on biological control of Arthropods, Honolulu, Hawaii, 14–18 January 2002. USDA Forest Service, Morgantown, pp 199–208
van Emden HF, Peakall DB (1996) Beyond silent spring: integrated pest management and chemical safety. Chapman and Hall, London
Van Lenteren JC (2006) How not to evaluate augmentative biological control. Biol Contr 39:115–118
Van-der-Kraan C, Ebbers A (1990) Release rates of tetradecen-1-ol acetates from polymeric formulations in relation to temperature and air velocity. Chem Ecol 16(4):1041–1058
Vandermeer J (1989) The ecology of intercropping. Cambridge University Press, Cambridge
Vargas RI, Shelly TE, Leblanc L, Piñero JC (2010) Recent advances in methyl eugenol and cue-lure technologies for fruit fly detection, monitoring, and control in Hawaii. Vitam Horm 83:575–595
Vega FE, Kaya HK (Eds) (2012) Insect pathology, 2nd edn. Academic Press, San Diego, p 490
Verheggen FJ, Haubruge E, Mescher MC (2010) Alarm pheromones. In: Litwack G (ed) Pheromones. Elsevier, Amsterdam, The Netherlands
Vontas JG, Cosmidis N, Loukas M, Tsakas S, Hejazi MJ, Ayoutanti A, Hemingway J (2001) Altered acetylcholinesterase confers organophosphate resistance in the olive fruit fly *Bactrocera oleae*. Pestic Biochem Physiol 71:124–132
Vontas J, Hernández-Crespo P, Margaritopoulos JT, Ortego F, Feng HT, Mathiopoulos KD, Hsu JC (2011) Insecticide resistance in Tephritid flies. Pesticide Biochem and Physiol 100:199–205
Wang S, Just DR, Pinstrup-Anderson P (2008) Bt cotton and secondary pests. Int Biotechnol 10:113–121
Ward KD (2003) Three-way interactions between Acacia, large mammalian herbivores and bruchid beetles-a review. Afr Ecol 41:257–265
Way MJ (1988) Entomology of wheat. In: Harris MK, Rogers CE (eds) The entomology of indigenous and naturalized systems in agriculture. Vestview Press, Boulder, pp 183–206
Way MJ, van Emden HF (2000) Integrated pest management in practice-pathways towards successful application. Crop Prot 19:81–103
Wilson DF, George BW (1986) Smoothleaf and hirsute cottons: response to insect pests and yield in Arizona. Econ Entomol 79:229–232
Witzgall P, Kirsch P, Cork A (2010) Sex pheromones and their impact on pest management. Chem Ecol 36:80–100
Zalucki MP, Adamson D, Furlong MJ (2009) The future of IPM: whither or wither? Aust Entomol 48:85–96
Zhang A, Kuang LF, Maisin N, Karumuru B, Hall DR, Virdiana I, Lambert S, Bin Purung H, Wang S, Hebbar P (2008) Activity evaluation of cocoa pod borer sex pheromone in cacao fields. Environ Entomol 37:719–724
Zhang JP, Christian S, Yu-Ling F, Ruo-Jian Z, Zhong-Ning Z (2012) An overlooked component: (Z)-9-tetradecenal as a sex pheromone in *Helicoverpa armigera*. Insect Physiol 58:1209–1216

Epidemiology and Management of Fungal Diseases in Dry Environments

Abdullah M. Al-Sadi

1 Introduction

Fungi are among the most common microorganisms in nature. They have cell walls consisting of chitin (except for the pseudo-fungi, e.g., Oomycetes) combined with other carbohydrates. Most fungi have hyphae—cylindrical, thread-like structures—that collectively form a mycelium. The mycelium can be very complex; some can colonize the soil and cover several hectares (Smith et al. 1992).

Most fungal species are saprophytic, playing important roles in stabilizing soil, decomposing waste material, producing antibiotics, acting as biocontrol agents, and promoting plant growth and development (Abed et al. 2013; Al-Sadi et al. 2015d, Guo et al. 2015; Matsushita et al. 2015; Van Geel et al. 2015). However, several fungal species are plant pathogens, causing various symptoms and losses in plants, the effects of which vary from limited effects on growth to plant destruction.

Fungal pathogens have long been considered as highly variable organisms, apparent both in field collections (Perkins 1991) and from the behavior of single isolates in the laboratory (Caten and Jinks 1968). Variation in some traits have been reported including aggressiveness, morphology and genetic makeup (Harvey et al. 2000; Harvey et al. 2001; Garzon et al. 2005a, b). Fungal species exhibit different levels of genetic diversity which can vary from species to species (Garzon et al. 2005b), region to region (Harvey et al. 2000), host to host (Harvey et al. 2001) and between different production systems (Al-Sadi et al. 2013).

Fungi exist in tropical, temperate and dry environments. Many fungal pathogens produce resistant survival propagules and have wide host ranges which help them to survive in soil and on plant debris for extended periods of time. Mechanisms by which fungal pathogens spread between different places are important features that

A.M. Al-Sadi (✉)
Department of Crop Sciences, College of Agricultural and Marine Sciences, Sultan Qaboos University, 34, AlKhoud 123, Oman
e-mail: alsadi@squ.edu.om

M. Farooq, K.H.M. Siddique (eds.), *Innovations in Dryland Agriculture*,
DOI 10.1007/978-3-319-47928-6_7

impact the distribution and diversity of these pathogens. Some fungal species complete their life cycle in the soil, with no aerial dispersal of propagules. These fungi usually have limited genetic diversity and show sub-structuring (Al-Sadi et al. 2012). Others complete most of their life cycles on the aerial parts of plants and therefore show higher levels of gene flow and genetic diversity. Their dissemination from location to location is usually a result of human activity including irrigation, tillage practices, movement of equipment in fields and, in many cases, movement of infested soil, potting media, plant residues, or colonized plant material (Al-Sa'di et al. 2008b, 2011a, 2015c, Kazeeroni and Al-Sadi 2016). However, some fungal spores (e.g. zoospores of Oomycetes) can move by themselves when water is available. Infection of plants can lower the quality of propagated plant material, produce poor plant stands, stunt growth, cause wilting and death of affected seedlings and plants, and reduce yield.

Dry regions of the world are usually characterized by relatively higher temperatures and lower precipitation rates. Despite these challenging factors, farmers in these regions grow a variety of crops, including field crops, vegetables and fruit. When the conditions are not suitable for certain high-value crops, farmers turn to protected agriculture in the form of greenhouses. Disease epidemics caused by fungal pathogens are usually affected by environmental conditions, the type of pathogens present, the type of crops grown, and the type of cultural practices employed by farmers. This in turn will affect the cultural, chemical or biological management strategies used for these pathogens.

This chapter focuses on some of the pathogens and diseases threatening crops under dry environments, with specific emphasis on wheat (*Triticum aestivum* L.) diseases. It presents the latest findings in the areas of population structure and dynamics of pathogens in dry environments, the factors affecting disease epidemics, plant–pathogen interactions, case studies of some diseases, and disease management.

2 Epidemiological Aspects of Plant Diseases in Dry Environments

2.1 Population Structure and Dynamics of Fungal Pathogens in Dry Environments

The source and quantity of fungal inoculum represent important determinants of the severity of plant diseases. Fungal propagules can come into contact with seeds, seedlings, and plants from numerous sources including potting mixtures, cultivated soil, irrigation water, wind-driven dust, and insects.

Potting mixtures are frequently used for seed germination before transplanting to farms or greenhouses. Infestation of potting mixtures with fungal propagules, as reported by several workers (Cartwright et al. 1995; Davison et al. 2006; Al-Sadi et al. 2011a, 2015c), can act as the initial inoculum that can kill seedlings in nurseries and infest the soil on farms. A survey done by Davison et al. (2006) recorded the presence of *Pythium* and *Phytophthora* species and some nematodes in consignments of potting media. Also, *Pythium aphanidermatum* and *Fusarium oxysporum* have been isolated from commercial peat moss while farmyard manures have been reported to have high levels of contamination with pathogenic fungi (Al-Sadi et al. 2011a).

Soil previously infested with fungi is often regarded as the most important source of fungal propagules as most incidences of soil-borne diseases occur after sowing/transplanting into infested soil (Stanghellini and Phillips 1975; Al-Sa'di et al. 2008b). Disease severity can vary between and within the same planting row, which is usually attributed to the distribution and different densities of fungal propagules in the soil (Stanghellini and Phillips 1975; Al-Sadi et al. 2012). Wind-driven dust in dry areas of the world (Al-Sa'di et al. 2008b, 2011a), insects (Gardiner et al. 1990) and irrigation water (Pottorff and Panter 1997) are also potential sources of fungal inoculum.

Survival of fungal pathogens in soil is affected by several factors such as the type of spores produced and the characteristics of the soil habitat. Mycelia survive for relatively short periods, usually a few days (Martin and Hancock 1986). Spherical sporangia produced by some *Pythium* species usually survive longer than species producing inflated filamentous or lobulate sporangia (Martin and Loper 1999). Survival of sporangia can be from a few days (Stanghellini and Burr 1973) to a few weeks (Peethambaran and Singh 1977) or several months (Stanghellini and Hancock 1971b). Long-term survival is achieved via the thick-walled resting spores. Depending on the species and the soil characteristics, spore survival can range from months (DeVay et al. 1982) to years (Stanghellini and Nigh 1972). Some spores do not germinate immediately or simultaneously but enter a dormancy period (Hancock 1981) which may be advantageous to escape the prospect of germinating all at once in unfavorable conditions. Upon germination, plant infection results in the colonization of the host, disease development and the production of new spores to complete the disease cycle.

Several airborne fungal pathogens usually survive in infected buds or leaves in the form of mycelium or resting spores. Others can thrive in decaying plant tissues for several years. Disease severity depends on several factors, including the aggressiveness of the isolate and susceptibility of the host (Al-Sadi 2016). The spores of some fungal pathogens, especially rust-inducing fungi, can spread long distances via the wind, which is common in dry places. Short distance spread is usually by rain splashing. *Puccinia*'s aeciospores absorb water, which exerts a force on the fruiting body's wall which ruptures to enable spread around the area. Insects are attracted to sugary and odorous material produced by pycniospores and receptive hyphae of rust fungi. Many airborne fungal pathogens usually cause serious epidemics because they can spread more rapidly and some produce high numbers of spores (e.g. rust and powdery mildew diseases).

The exchange of genetic factors between or within fungal populations, either by interbreeding or migration, can impact on the genetic diversity of a fungal pathogen, and is affected by several factors; most importantly, the ability of a pathogen to move from one area to the other. This occurs faster for airborne fungal pathogens such as rusts, which can travel distances over 12,000 km (Watson and de Sousa 1983) and slower for soil-borne pathogens (Al-Sadi et al. 2012). Limited migration and gene flow may result in a phenomenon known as subdivision or sub-structuring (Loveless and Hamrick 1984), where isolated fungal populations, despite little intra-population diversity, show higher levels of genetic differentiation between populations compared to many foliar pathogens with wind-dispersed spores. This has been documented in seven populations of *Pythium irregulare* obtained from different geographical locations rather than within each population (Harvey et al. 2000).

Genetic drift, defined as random fluctuations in allele frequencies, is another factor affecting the population structure of pathogens, especially in small populations as there is a greater loss of allelic diversity (McDonald and McDermott 1993). Random genetic drift—due to rapid asexual cycles, large fluctuations in population size and man-induced changes to the host—is considered an important factor in shaping populations of plant pathogens (Drenth and Goodwin 1999).

Environmental conditions, host range and management practices can deploy selection pressure on populations of fungal species. The effect of environmental conditions becomes apparent with the occurrence of different fungal species in some geographical areas but not others (Van der Plaats-Niterink 1981). Fungal species exhibiting a wide host range are expected to show higher levels of genetic diversity (Harvey et al. 2001). In addition, the deployment of specific host-resistant genes by man is likely to eliminate pathogen populations without the virulent genes (Drenth and Goodwin 1999). Management strategies, especially the application of fungicides, can lower the level of genetic diversity within fungal populations and selects for fungicide-resistant isolates (Al-Sadi et al. 2012). One example is an increase in metalaxyl-resistant genotypes of *Phytophthora infestans* after the introduction of metalaxyl fungicides (Davidse et al. 1989).

2.2 Factors Affecting the Development of Epidemics

Some physical and biological factors reportedly affect the development of plant disease epidemics and include environmental factors, inoculum level, mode of spread, mode of reproduction, cultivar resistance, uniformity, and crop age.

Temperature affects the composition of fungal species infecting plants under given climatic conditions. Temperature and precipitation influenced the distribution of *Pythium* species infecting wheat in eastern Washington State (Paulitz and Adams 2003). *Pythium aphanidermatum* usually causes severe diseases during summer when temperatures are above 23 °C while *P. dissotocum* is more aggressive during winter when the temperatures are between 17 and 22 °C (Bates and Stanghellini 1984). The development of *Fusarium*-induced wilt diseases is higher when temperatures increase, and is mainly attributed to the stress imposed on plants and

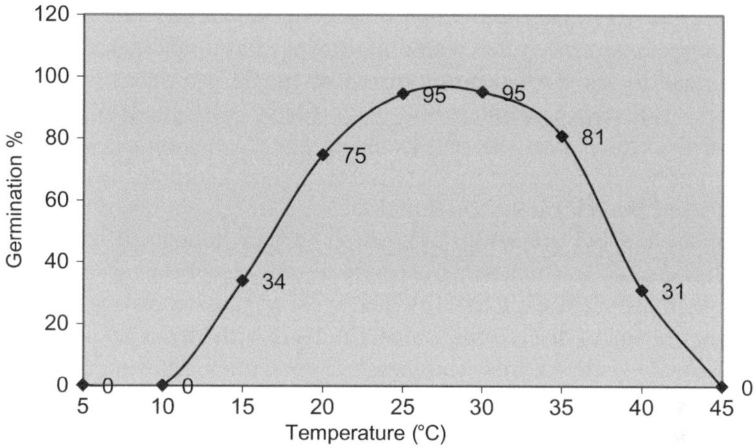

Fig. 1 The effect of temperature on germination percentage of conidia spores from 11 *B. sorokiniana* isolates obtained from wheat and barley (Al-Saadi 2002)

increased activity of the invading pathogens (Al-Mawali et al. 2013). Low air moisture can further increase wilt diseases due to increased transpiration from plants in which water absorption and/or translocation are affected by vascular fungal pathogens. Germination of the dryland root rot pathogen of wheat (*Bipolaris sorokiniana*) is optimum at moderate to high temperatures (Fig. 1).

Moisture is an important determinant of the severity of plant disease, as higher soil moistures usually favor more disease (Hendrix and Campbell 1973). High soil moisture increases the mobility of spores, especially zoospores, saprophytic growth and spore germination (Martin and Loper 1999). However, several plant diseases are favored by drought conditions such as dry rots and charcoal rots caused by *Macrophomina* species (Lodha et al. 2014; Sharma et al. 2015). Increased soil moisture increases the amount of CO_2 within the soil which may offer some fungal species the opportunity to cause more disease while microbial competition decreases as many other microorganisms are sensitive to higher CO_2 concentrations (Gardner and Hendrix 1973).

The conduciveness of soil to pathogen development can have a significant effect on disease; even if the inoculum level varies in different soils, there may not be a strong positive correlation to disease levels due to differences in soil conduciveness (Stasz and Harman 1980). Several workers including Kaiser and Hannan (1983) found higher levels of disease from soil with a low inoculum level than soil with a high inoculum level. Also, the type and amount of propagules and the age of seedlings can affect disease development (Al-Sadi et al. 2011b). A study by Al-Sadi et al. (2010a) showed that increased soil and water salinity stimulates the development of damping-off epidemics.

Disease epidemics are also affected by the inoculum level and the mode of reproduction (Al-Sadi 2016). Some pathogens only have one or few disease cycles in a growing season (e.g. *Fusarium*) while others—those causing powdery mildew—usually complete several cycles within a growing season and subsequently cause greater disease losses. The mode of spread of fungal pathogens is an important determinant of disease epidemics. Soil-borne fungal pathogens, in general, move slowly and therefore cause less and localized disease epidemics, while rusts and powdery mildew are dispersed by the wind to several kilometers which can cause severe disease epidemics in a shorter period.

Older plants are less susceptible to invasion by many fungal species due to thickened and lignified cells which help to restrict the advance of fungal species from the infection point (Al-Sadi et al. 2011b). The level of disease resistance in crops is another determinant of disease epidemics. Cultivars with higher levels of resistance can slow down disease progress and therefore epidemics. All these factors usually interact together and become important determinants of disease levels as well as the efficacy of management strategies implemented in a given area.

3 Plant-Pathogen Interactions

Fungal pathogens interact with their host plants in different ways and are usually influenced by the type of pathogen, the host plant and environmental conditions. An imbibing and germinating seed usually releases significant amounts of exudates into the soil which has the potential to attract mobile pathogen propagules to the infection court and affect spore germination and germ tube growth (Martin 1995). The type and quantity of these exudates vary with seed age and genotype as well as soil moisture content (Stanghellini and Hancock 1971a). Exudates usually consist of polysaccharides, amino acids, fatty acids and carbohydrates. Some may attract propagules while others stimulate germination of fungal spores (Donaldson and Deacon 1993). Zoospores of *P. aphanidermatum* are usually attracted by L-glutamine in a chemotactic response, which has no effect on cyst germination while glucose has the opposite effect. Also, uredospores and teliospores of *Puccinia carthami* are stimulated by some specific volatile polyacetylenes compounds, released by host plants, such as hydrocarbons, fatty acids, aldehydes and alcohols. Germination of fungal spores is usually fast enough to occur within one hour of planting seeds, and it can occur in response to temperature, nutrient availability and moisture in most pathogens (see Martin and Loper 1999). The physical structure and chemical composition of the cuticle wax layer reportedly affects the germination of some fungi such as *Blumeria* spp. (Zabka et al. 2008).

Most rust fungi penetrate the stomata while some penetrate the cuticle. Penetration of *Pythium* species into roots is usually via breaks in the root surface or by direct penetration. Penetration is attained within hours after germination allowing fungal mycelia to grow through the epidermal and cortical cells. Infection of seeds, seedlings or plants by fungal species typically results in seed decay, pre-emergence damping off, post-emergence damping off, wilt, root rot, fruit rot, or other diseases. Although some fungal species can apply mechanical force during infection and invasion (Ravishankar et al. 2001; MacDonald et al. 2002), most of these processes are facilitated by lytic enzyme secretions that help to disintegrate the structural integrity of host cells and tissues (Campion et al. 1997) resulting in tissue maceration and symptom induction. *Colletotrichum* develops a series of infection structures that facilitate colonization of the host tissues. It attacks living plant cells, and then switches to necrotrophic phase resulting in cell death (Kubo et al. 2016).

Plants respond to fungal infection using a hypersensitive response, systemic acquired resistance and other mechanisms. The hypersensitive response is one of the most important plant defense mechanisms against fungi. It can prevent the spread of fungal infection to healthy cells due to the activity of some chemicals such as phenols, sterols and phytoalexins, and the oxidation of other chemicals in the infected cells (Agrios 2005).

Systemic acquired resistance (SAR) is an example of induced resistance. It is used to induce resistance on rust-infected plants using chemicals such as BTH (1, 2, 3-benzothiazole-7-thiocarboxylic acid-S-methyl-ester). This mechanism can also be achieved by inoculating plants with weak strains. Other ways for defense responses include lignification, production of callose papillae, cork layers and phenolic compounds, and detoxification of pathogen toxins.

4 Case Studies of Plant Disease Epidemics in Dry Environments

4.1 Black Point of Wheat

Wheat is subject to various biological stresses, including pathogen stresses (Table 1). The number of wheat diseases exceeds 200. Some of the major pathogens attacking wheat are *Bipolaris sorokiniana, Fusarium graminearum, Fusarium avenaceum, Alternaria alternata, Puccinia graminis, Ustilago tritici, Erysiphe graminis* and *Rhizoctonia solani* (Paulitz and Adams 2003; Smiley et al. 2005; Toklu et al. 2008; Vujanovic et al. 2012; Acharya et al. 2013; Zhu et al. 2015). The losses caused by these pathogens and many others vary from no disease to very severe, depending on the prevailing environmental conditions and the genetics of the host plant and the pathogen.

B. sorokiniana is one of the most prevalent fungi in wheat and barley seeds and is reportedly seed-borne in different temperate, tropical and arid countries (Clarke et al. 2004; Toklu et al. 2008; Al-Sadi and Deadman 2010). Seed infection is favored by warm and humid weather, and may simply discolor seeds or can cause black point disease in barley (*Hordeum vulgare* L.) and wheat. Black point disease affects the kernels of cereals, and exhibits a brown to black tip on the germ end; it can be caused by different fungi, such as *B. sorokiniana, Fusarium graminearum, Alternaria alternata* and *Penicillium* spp. (Toklu et al. 2008; Al-Sadi and Deadman

Table 1 Examples of diseases associated with wheat in dry areas of the world

Disease	Causal organism	Estimated annual yield losses[a]	References
Common root rot	*Bipolaris sorokiniana* and *Fusarium*	5–40 % in Syria	(Wiese (1987), Bailey et al. (1997), and Acharya et al. 2013)
Leaf blight	*Alternaria* spp.	Variable yield loss; important in India, Middle East, Nigeria and other countries	Wiese (1987)
Loose smut	*Ustilago tritici*	Can reach 27 % in different parts of the world	Wiese (1987)
Head blight	*Fusarium* spp.	Variable levels of loss	Wiese (1987) and Wiersma and Motteberg (2005)
Spot blotch	*Bipolaris sorokiniana*	4–43 % in Bangladesh, China, and South Asia	Acharya et al. (2013) and Chowdhury et al. (2013)
Stem rust	*Puccinia graminis* f.sp. *tritici*	5–7 million metric tons worldwide	Pardey et al. (2013)
Cyst nematodes	*Heterodera avenae*	3–49 % in India, 15–20 % in Pakistan, 40–92 % in Saudi Arabia	Bhatti et al. (1981) and Nicol et al. (2011)

[a]The losses are not constant, they depend on the species, cultivar, environment, and virulence of the pathogen isolate (Al-Sadi 2016)

2010). Infection by *B. sorokiniana* in wheat and the black point disease can lower the quality and market value of grains, cause seedling blight, and reduce seed germination and seedling emergence (Al-Sadi and Deadman 2010; Acharya et al. 2013).

Penetration into the seed is achieved through the ovary wall and seed coat. In addition, the spores of *B. sorokiniana* can be carried on seed coat surfaces. *B. sorokiniana* is one of the long living seed-borne pathogens.

4.2 Common Root Rot of Wheat

Common root rot in wheat and barley, caused by *B. sorokiniana*, is an important disease all over the world (Tunali et al. 2008, Al-Sadi and Deadman 2010) and is characterized by symptoms that appear on the roots, sub-crown internodes and crowns. These symptoms include necrosis in the form of dark brown or black colored lesions on the basal stem, crowns, sub-crown internodes and roots (Al-Sadi and Deadman 2010) (Fig. 2). *B. sorokiniana* invades the roots and shoot base causing lignification and death of the epidermal cells. The necrotic symptoms on the sub-crown internodes (Fig. 2) and roots (Fig. 3) are commonly used to evaluate the resistance of different varieties (Liljeroth et al. 1996; Al-Sadi and Deadman 2010).

Fig. 2 Development of necrosis on the shoot base of wheat following infection by *B. sorokiniana* (Photo taken by A.M. Al-Sadi)

Fig. 3 Necrotic lesions on the wheat root as a result of *B. sorokiniana* infection (Photo taken by A.M. Al-Sadi)

Common root rot in wheat is a disease of dry and warm areas of the world, including Oman, Bangladesh, Nepal, Pakistan and others (Iftikhar et al. 2009; Al-Sadi and Deadman 2010; Neupane et al. 2010; Acharya et al. 2013). It is also referred to as dryland root rot (Wiese 1987). Disease severity and incidence are affected by soil moisture, soil temperature, cultural practices, pathogen population in the soil, and time of infection. Disease severity increases when the plant is under stress. Differences in soil surface residue can affect soil moisture and temperature, root growth and nutrient uptake as well as microbial activity, which can modify the conditions for root infections. The severity of the disease in wheat is related to the soil population of *B. sorokiniana* at the time of planting, which is highest in the top 10 cm. This is mainly due to sporulation of the fungus on culm bases, crown roots, stubble, straw, sub-crown internodes and seminal roots near the soil surface (Duczek 1990). Drought and warm temperatures make wheat more susceptible to common root rot disease.

Common root rot can affect the yield and quality of wheat as a result of the effects of the disease on seedling emergence and growth, leaf area and growth, number of tillers per plant, kernel number and weight, shoot and root weight, and size of internodes (Zhang et al. 1999; Tunali et al. 2008). *B. sorokiniana* can also cause root rot disease in rye, oats and other crops (Ovsyankina 2005; Poole et al. 2015).

4.3 Spot Blotch

Spot blotch, caused by *Bipolaris sorokiniana*, is a common disease in wheat and barley in dry environments (Acharya et al. 2013; Agostinetto et al. 2015; Singh et al. 2015; Zhu et al. 2015). It is one of the most important foliar diseases limiting wheat production in warmer non-traditional growing areas. The symptoms of spot blotch disease on wheat appear with distinct, elongated, brown–black lesions that rarely exceed 1 cm in diameter (Agostinetto et al. 2015). Lesions have a distinct margin which varies in size. Enlargement of the spots may form blotches that cover large areas of the leaf, especially on barley cultivars (Fig. 4). Due to sporulation of the fungus on older leaves, the lesions may become olive brown in color. In addition to wheat and barley, *B. sorokiniana* causes leaf spot in maize (*Zea mays* L.), rice (*Oryza sativa* L.), rye (*Secale cereale* L.), and many others (Acharya et al. 2013; Zhu et al. 2015).

Leaf infection by the pathogen may come from the seeds, roots or the air. If the pathogen is in the soil, then the infection occurs through stomata on the hypocotyl and progresses to the root, shoot and coleoptile. In the leaf, penetration of *B. sorokiniana* occurs through stomata and the epidermis. Warm weather favors the pathogen. Leaf surface wetness is also an important factor influencing infection (Kumar et al. 2002; Agostinetto et al. 2015).

Bipolaris sorokiniana produces several phytotoxins (Bach and Kimati 1999; Jahani et al. 2014) which affect many of the physiological and metabolic processes in plants (Bashyal et al. 2012). Toxins of *B. sorokiniana* and other fungal pathogens affect electron transport, ion transport, calcium uptake, chlorophyll and enzyme activities (Liljeroth et al. 1994; Bach and Kimati 1999; Agrios 2005).

A major role of toxins is in pathogenicity. For instance, Liljeroth et al. (1994) suggested that toxins play a role in pathogenicity by killing or weakening plant cells

Fig. 4 Severe necrosis and chlorosis on wheat due to infection by *B. sorokiniana* (Photos taken by A.M. Al-Sadi)

in advance of the growing hyphae, and by facilitating nutrient uptake and further growth of the fungus in plant tissue. The BZR-toxin produced by *B. zeicola* is needed for the initial colonization of *B. zeicola* race 3 in both rice and maize tissues (Xiao et al. 1992). Disease symptoms in wheat and barley which are affected by the spot blotch disease are in the form of necrosis and chlorosis. Helminthosporal and prehelminthosporol are suggested substances responsible for the disease symptoms (Bach and Kimati 1999).

4.4 Wheat Stem Rust

Rust is a serious and widespread disease in wheat (Fetch et al. 2015). Rusts can also attack other crops such as barley and oats (*Avena sativa* L.) (Hernandez et al. 2005; Deadman et al. 2011). Fungi-causing rusts usually attack leaves and stems and produce rusty yellow–orange spots (Chen et al. 2009; Wanyera et al. 2009; Deadman et al. 2011).

Puccinia spp. cause various rust diseases in wheat. They are biotrophic and can complete their life cycle in one, two or more hosts. They are usually dispersed by the wind, insects and rain splashing (Agrios 2005). Germination is stimulated by exudates released from the plant surface and depends on several factors, including water and temperature. Some species grow in different directions on the surface of plants to increase the chance of finding stomata. Penetration into host tissues is usually attained through stomata. Hyphae move between plant cells and produce several haustorium mother cells which produce a penetration peg and a haustorium inside plant cells to absorb nutrients.

Stem rust affects growth, yield and the food value of wheat. Symptoms develop on the stems, leaves, leaf sheath, and neck and glumes of the wheat spike (Szabo et al. 2014). When infections are heavy, especially on stems, the plants may lodge. Although some infections are mild, several studies have reported heavy losses due to stem rust in different parts of the world. Stem rust is favored by hot days and wet leaves. The optimum temperature for stem rust development is around 26 °C (Wiese 1987). There is also evidence for increased aggressiveness of some other rust species at higher temperatures (Gautam et al. 2013).

The disease is caused by various races of the *Puccinia graminis* f.sp. *tritici* (Szabo et al. 2014). The fungus overwinters as thick-walled teliospores that are produced on wheat or other grass hosts. The rust fungi are separated into races by their avirulence or virulence on the different host cultivars (Fetch et al. 2015).

5 Management of Fungal Diseases

The increase in disease incidence in dry environments and other parts of the world has led to the adoption of management strategies including chemical, cultural and biological methods. Many growers focus on applying chemical methods such as fungicides and biofumigation (Deadman et al. 2006). Pre-plant fumigants are effective in reducing soil-borne inoculum. Fungicide-treated seed is an important tool against certain seed and seedling diseases. Although chemical pesticides provide quick, effective and economic management of plant diseases, continuous use of the same pesticides over a long period for the same pathogens may result in the development of resistant strains of the pathogens as well as cause various environmental and health problems. Such pesticides also cause imbalances in the microbial community and inactivate soil-borne antagonists of pathogens (Pose-Juan et al. 2015; Stott and Taylor 2016).

Selection of the best management strategy will depend on some criteria including disease threshold, availability of resources, and applicability to the implementation area. Strategies can be divided into three types: cultural, biological and chemical control which are discussed below.

5.1 *Cultural Practices*

Cultural practices are commonly used to reduce pathogen inoculum and diseases. These include crop rotations, management of irrigation, good sanitation, use of disease-free seeds and seedlings, the use of soil solarization and soil amendments. Crop rotation with non-susceptible crops helps reduce pathogen levels which may build up in the soil when the same crop is grown in the same field year after year (Lamour and Hausbeck 2003).

Solarization of field soils is a widely used method for controlling soil-borne plant diseases (Deadman et al. 2006; Farraq and Fotouh 2010; Carrieri et al. 2013). Soil solarization exposes the soil to heat generated from wavelength sun rays trapped by the polyethylene cover. It is practiced mainly in summer when high ambient temperatures can exceed 40 °C, especially in many dry environments. Soil solarization is a good practice for the management of several soil-borne diseases by reducing the initial pathogen inoculum (Deadman et al. 2006; Farraq and Fotouh 2010; Carrieri et al. 2013). The efficacy of control using solarization can be improved when used in combination with cabbage (*Brassica oleracea* L.), which helps releases gasses that act as a biofumigant to soil (Deadman et al. 2006). Of four chemical and physical treatments, Al-Samarria et al. (1988) found soil solarization to be best control of *P. aphanidermatum* and some other soil pathogens in Iraq. Solarization of potting media effectively eliminated the primary inoculum of many plant pathogenic fungi (Al-Sadi et al. 2015c).

Soil replacement is another cultural practice for managing soil-borne fungal inoculum in some growing systems in arid environments. It is practiced by excluding the top layer (~30–60 cm) of previously-cultivated soil and replacing it with uncultivated soil imported from outside the farm (Al-Sa'di et al. 2008b). In addition, irrigation management is another control tactic for reducing the incidence of soil-borne diseases (Clarke et al. 2004) by reducing the rate of reproduction of pathogens and the movement, germination and growth of fungal spores.

Proper composting is an efficient process for reducing most plant and human pathogen populations which can be transmitted by compost to farms; its success depends on different composting parameters as well as complex microbial interactions. First, temperature–time combinations are the most important factors for eliminating plant pathogens. Temperatures of 60–65 °C during the thermophilic phase for several days should eliminate most plant pathogens and 55 °C for 21 days eliminated most of the tested fungal pathogens (Noble and Roberts 2004). Moisture content is another important factor which can influence the elimination level of pathogens. The occurrence of dry pockets in the composting material is probably the main cause of pathogen survival. The percentage of moisture content should not be lower than 40 % (Bollen et al. 1989).

Creation of unfavorable conditions for disease development by using good soil drainage, optimizing plant spacing to reduce relative humidity around plants, and proper fertilization of the crop are important practices for managing plant diseases.

5.2 *Biological Control*

Biological control of fungal pathogens is an alternative to using chemical pesticides. It involves using an organism or organisms to lower the inoculum density of the pathogen and thus reduce crop losses (Bernard et al. 2014). Disease suppression by biocontrol agents results from interactions between the plant, pathogens, and the microbial community.

Examples of biocontrol organisms that have been widely studied include *Mycorrhizal* fungi, *Trichoderma* spp., *Gliocladium* spp. and *Actinomycetes*. *Mycorrhizal* fungi increase the capacity of plant roots to absorb nutrients which increases the ability of plants to withstand the attack of pathogens. *Trichoderma* spp. are among the most studied fungi and are commercially marketed as bio-pesticides, bio-fertilizers and soil amendments (Guzmán-Valle et al. 2014; Al-Sadi et al. 2015d). They can reduce significant diseases caused by *Phytophthora* spp., *Fusarium* spp., *Pythium* spp. and others (Punja and Yip 2003; Blaya et al. 2013). *Trichoderma* spp. have a high reproductive capability and vigorous aggregativeness against pathogenic fungi (Grondona et al. 1997). Treating soil with *T. harzianum* improved the resistance of bean leaves to diseases caused by *Botrytis cinerea* (Elad et al. 1993).

The mechanisms by which antagonistic microorganisms affect fungal pathogens differ; some attack and kill the mycelium of fungal pathogens while others produce toxins or metabolites that can affect the growth and reproduction of fungal pathogens. Some antagonistic fungi compete for food and space with fungal pathogens while others induce systemic resistance to the invading pathogens (Agrios 2005).

Recent studies have provided evidence that organic materials play an important role in the suppression of soil-borne diseases (Al-Sadi et al. 2015c). Mechanisms of disease suppression in composted materials vary according to the biological and physicochemical characteristics of composts, methods and time of application, as well as pathogen species, densities and their aggressiveness (Hoitink and Boehm

1999). Quality compost can serve as a source of high populations of beneficial microorganisms which are naturally recolonized or artificially inoculated (Al-Sadi et al. 2015c). These organisms suppress pathogenic propagules using different mechanisms such as competition for nutrients, antibiotic production, parasitism against pathogens and activation of disease-resistance genes in plants (Hoitink and Boehm 1999). In addition, they stimulate plant growth by modifying soil conditions. High microbial biomass is responsible for releasing nutrients from organic matter and increasing nutrient availability to crops.

Disease resistance in plants is a cost-effective way to control diseases. It is commonly used for the management of several plant diseases caused by fungal pathogens including resistant cultivars to wheat diseases (van Wyk et al. 2007; Guzmán-Valle et al. 2014; Singh et al. 2015).

Defense responses deployed by host plants in the reaction to pathogen invasion are an important component of the integrated management of plant diseases. They can reduce or eliminate the need to use chemicals and biological formulations and therefore reduce the costs and environmental hazards associated with increased chemical applications.

Increasing the defense response in plants to infection by fungal pathogens has been attempted. For example, by using non-pathogenic microorganisms such as *Fusarium oxysporum* Schltdl. strain Fo47 (Benhamou et al. 2002), compost (Lievens et al. 2001) and silicon (Chérif et al. 1994; Al-Sadi et al. 2010b). Silicon reduced the severity of some cucurbit diseases including powdery mildew (Menzies et al. 1992; Heckman et al. 2003) and wilt (Chérif et al. 1994).

5.3 Chemical Control

Chemical control is a widely used practice for managing diseases in different parts of the world (Wiersma and Motteberg 2005; Chung et al. 2009; Nallathambi et al. 2009; Agostinetto et al. 2015). Several protectant and systemic fungicides are used to control fungal diseases. Fungicides vary in their characteristics and efficacy and are used to treat seeds, soil, foliage and fruits.

Seeds are usually treated with some protectant fungicides to prevent infection of seed in soil (Harman et al. 1989) and to reduce the level of disease in crops. However, soil treatments are typically required for transplanted seedlings and plants, while foliar and fruit applications can help reduce aerial infections and postharvest diseases. Pre-plant fumigants are effective for reducing soil-borne inoculum. However, there are increased restrictions on some fumigants due to their risks. The type of fungicide/fumigant used usually depends on the invading pathogen and other management strategies used on the farm.

Fungicides are more effective for managing plant disease than other strategies. However, there are several environmental and health risks, including negative effects on beneficial microorganisms and risks to wildlife. Some problems have also arisen with the continuous use of fungicides, such as the development of fungicide resistance and the enhanced rate of biodegradation in soil. Fungicide resistance is a genetic adjustment by a microorganism that reduces the sensitivity to a fungicide. It has become more apparent with the introduction of systemic and more selective fungicides such as benomyl (Eckert 1988).

Biodegradation of fungicides and pesticides in soil is considered a driving factor in the failure of many of these chemicals to control pests and diseases. It is characterized by the utilization of these chemicals as the sole or supplemental source of carbon, nitrogen, and other nutrients, or by the creation of specific environments conducive to chemical degradation by soil microflora (Kaufman and Edwards 1983). Biodegradation has also been demonstrated for some pesticides including chlorpropham, dalapon, endothall, metalaxyl and phenoxyalkanoates (Kaufman and Edwards 1983; Al-Sa'di et al. 2008a).

6 Conclusion and Future Research

Crops in dry areas of the world face several biotic and abiotic challenges. Apart from water scarcity and warmer climatic conditions, fungal diseases represent a major threat to several crops (Pilet et al. 2005; de Miranda et al. 2010; Al-Sadi et al. 2011b, Al-Mawali et al. 2013; Al-Sadi et al. 2015a). The challenges faced by crops are not only restricted to current diseases as many fungal pathogens have the potential to evolve and produce races or isolates with higher levels of aggressiveness or tolerance to extreme environmental conditions (Chen et al. 2009; Al-Sadi et al. 2012; Al-Sadi 2013; Al-Sadi et al. 2013, 2015b). This can result in the breakdown of cultivar resistance, the development of fungicide resistance in new fungal generations, or the appearance of new host species for fungal pathogens. Ultimately, the management of plant diseases becomes more challenging. Avoiding monoculture, developing new cultivars with multigene resistance to major plant diseases, and developing integrated disease management strategies are important for reducing the potential for severe disease epidemics (Gilardi et al. 2013; Guo et al. 2015; Neher et al. 2015; Zhu et al. 2015; Ahmad et al. 2016; Kazeeroni and Al-Sadi 2016). Future research on plant diseases under dry environmental conditions should be directed towards characterizing new disease epidemics and their population dynamics, understanding the interactions between hosts, pathogens and environment, and developing durable management strategies for the major diseases.

References

Abed RMM, Al-Sadi AM, Al-Shihi MA, Al-Hinai S, Robinson MD (2013) Diversity of free-living and lichenized fungal communities in biological desert crusts of the Sultanate of Oman and their role in improving soil properties. Soil Biol Biochem 57:695–705

Acharya K, Dutta AK, Pradhan P (2013) Bipolaris sorokiniana (Sacc.) Shoem.: The most destructive wheat fungal pathogen in the warmer areas. Aust J Crop Sci 5:1064–1071

Agostinetto L, Casa RT, Bogo A, Sachs C, Souza CA, Reis EM, Cristina da Cunha I (2015) Barley spot blotch intensity, damage, and control response to foliar fungicide application in southern Brazil. Crop Prot 67:7–12

Agrios GN (2005) Plant pathology. Academic, San Diego

Ahmad A, Thomas GJ, Barker SJ, Macleod WJ (2016) Genotype resistance, inoculum source and environment directly influence development of grey leaf spot (caused by Stemphylium spp.) and yield loss in narrow-leafed lupin (Lupinus angustifolius). Crop Pasture Sci 67:81–90

Al-Mawali QS, Al-Sadi AM, Al-Said FA, Deadman ML (2013) Etiology, development and reaction of muskmelon to vine decline disease under arid conditions of Oman. Phytopathol Mediterr 52:457–465

Al-Sadi AM (2002) Variability in natural host-pathogen interactions; characterization of Cochliobolus sativus (anamorph Bipolaris sorokiniana)-Wheat-Barley Pathosystem in Oman. In: SQU, Muscat.

Al-Sadi AM (2013) Phylogenetic and population genetic analysis of Ceratocystis radicicola infecting date palms. J Plant Pathol 95:47–55

Al-Sadi AM (2016) Variation in resistance to spot blotch and the aggressiveness of Bipolaris sorokiniana on barley and wheat cultivars. J Plant Pathol 98:97–103

Al-Sadi AM, Deadman ML (2010) Influence of seedborne Cochliobolus sativus (anamorph: Bipolaris sorokiniana) on crown rot and root rot of barley and wheat. J Phytopathol 158:683–690

Al-Sa'di AM, Drenth A, Deadman ML, Al-Said FA, Khan I, Aitken EAB (2008a) Association of a second phase of mortality in cucumber seedlings with a rapid rate of metalaxyl biodegradation in greenhouse soils. Crop Prot 27:1110–1117

Al-Sa'di AM, Drenth A, Deadman ML, Al-Said FA, Khan I, Aitken EAB (2008b) Potential sources of Pythium inoculum into greenhouse soils with no previous history of cultivation. J Phytopathol 156:502–505

Al-Sadi AM, Al-Masoudi RS, Al-Habsi N, Al-Said FA, Al-Rawahy SA, Ahmed M, Deadman ML (2010a) Effect of salinity on Pythium damping-off of cucumber and on the tolerance of Pythium aphanidermatum. Plant Pathol 59:112–120

Al-Sadi AM, Al-Said FA, Deadman ML, Drenth A, Aitken EAB (2010b) Can silicon improve defense responses of cucumber seedlings to Pythium-induced damping-off disease? Acta Hortic 871:441–444

Al-Sadi AM, Al-Said FA, Al-Jabri AH, Al-Mahmooli IH, Al-Hinai AH, de Cock AWAM (2011a) Occurrence and characterization of fungi and oomycetes transmitted via potting mixtures and organic manures. Crop Prot 30:38–44

Al-Sadi AM, Al-Said FA, Al-Kiyumi KS, Al-Mahrouqi RS, Al-Mahmooli IH, Deadman ML (2011b) Etiology and characterization of cucumber vine decline in Oman. Crop Prot 30:192–197

Al-Sadi AM, Al-Ghaithi AG, Al-Balushi ZM, Al-Jabri AH (2012) Analysis of diversity in Pythium aphanidermatum populations from a single greenhouse reveals phenotypic and genotypic changes over 2006 to 2011. Plant Dis 96:852–858

Al-Sadi AM, AL-Wehaibi AN, Al-Shariqi RM, Al-Hammadi MS, Al-Hosni IA, Al-Mahmooli IH, Al-Ghaithi AG (2013) Population genetic analysis reveals diversity in Lasiodiplodia species infecting date palm, Citrus, and mango in Oman and the UAE. Plant Dis 97:1363–1369

Al-Sadi, A.M., Z.A. Al-Alawi and A. Patzelt, (2015a) Association of Alternaria alternata and Cladosporium cladosporioides with leaf spot in Cissus quadrangularis and Ficus sycomorus. Plant Pathol J (in press)

Al-Sadi AM, Al-Masoodi RS, Al-Ismaili M, Al-Mahmooli IH (2015b) Population structure and development of resistance to hymexazol among Fusarium solani populations from date palm, Citrus and cucumber. J Phytopathol 163:947–955

Al-Sadi AM, Al-Mazroui SS, Phillips A (2015c) Evaluation of culture-based techniques and 454 pyrosequencing for the analysis of fungal diversity in potting media and organic fertilizers. J Appl Microbiol 119:500–509

Al-Sadi AM, Al-Oweisi F, Edwards S, Al-Nadabi H, Al-Fahdi A (2015d) Genetic analysis reveals diversity and genetic relationship among Trichoderma isolates from potting media, cultivated soil and uncultivated soil. BMC Microbiol 15:147

Al-Samarria FH, El-Bahadli AH, Al-Rawi FA (1988) Comparison between effects of soil disinfestation methods on some pathogens of cucumber. Arab J Plant Protect 6:106–112

Bach EE, Kimati H (1999) Purification and characterization of toxins from wheat isolates of Drechslera tritici-repentis, Bipolaris bicolor, and Bipolaris sorokiniana. J Venomous Anim Toxins 5:184–199

Bailey KL, Duczek LJ, Potts DA (1997) Inoculation of seeds with Bipolaris sorokiniana and soil fumigation methods to determine wheat and barley tolerance and yield losses caused by common root rot. Can J Plant Sci 77:691–698

Bashyal BM, Chand R, Prasad LC, Joshi AK (2012) Influence of growth stage on sensitivity to helminthosporol toxin of Bipolaris sorokiniana of barley (Hordeum vulgare). Indian J Agricult Sci 82:724–726

Bates ML, Stanghellini ME (1984) Root rot of hydroponically grown spinach caused by Pythium aphanidermatum and P. dissotocum. Plant Dis 68:989–991

Benhamou N, Garand C, Goulet A (2002) Ability of nonpathogenic Fusarium oxysporum strain Fo47 to induce resistance against Pythium ultimum infection in cucumber. Appl Environ Microbiol 68:4044–4060

Bernard E, Larkin RP, Tavantzis S, Erich MS, Alyokhin A, Gross SD (2014) Rapeseed rotation, compost and biocontrol amendments reduce soilborne diseases and increase tuber yield in organic and conventional potato production systems. Plant Soil 374:611–627

Bhatti DS, Dalal MR, Malhan I (1981) Estimation of loss in wheat yield due to the cerealcyst nematode heterodera avenae. Trop Pest Manage 27:375–378

Blaya, J., R. López-Mondéjar, E. Lloret, J.A. Pascual, M. Ros, (2013) Changes induced by Trichoderma harzianum in suppressive compost controlling Fusarium wilt Pest Biochem Physiol 107**:**112–119

Bollen GJ, Volker D, Wijnen AP (1989) Inactivation of soil-borne plant pathogens during small-scale composting of crop residues. Neth J Plant Pathol 95:19–30

Campion C, Massiot P, Rouxel F (1997) Aggressiveness and production of cell-wall degrading enzymes by Pythium violae, Pythium sulcatum and Pythium ultimum, responsible for cavity spot on carrots. Eur J Plant Pathol 103:725–735

Carrieri R, Raimo F, Pentangelo A, Lahoz E (2013) Fusarium proliferatum and Fusarium tricinctum as causal agents of pink rot of onion bulbs and the effect of soil solarization combined with compost amendment in controlling their infections in field. Crop Prot 43:31–37

Cartwright DK, Spurr HW Jr, Shew HD (1995) Commercial potting medium as the source of Pythium causing a disease on tobacco transplants. Plant Dis 79:538

Caten CE, Jinks JL (1968) Spontaneous variability of single isolates of Phytophthora infestans. I. Cultural variation. Can J Bot 46:329–348

Chen WQ, Wu LR, Liu TG, Xu SC, Jin SL, Peng YL, Wang BT (2009) Race dynamics, diversity and virulence evolution in Puccinia striiformis f. sp. tritici, the causal agent of wheat stripe rust in China from 2003 to 2007. Plant Dis 93:1093–1101

Chérif M, Asselin A, Bélanger RR (1994) Defense responses induced by soluble silicon in cucumber roots infected by Pythium spp. Phytopathology 84:236–242

Chowdhury AK, Singh G, Tyagi BS, Ojha A, Dhar T, Bhattacharya PM (2013) Spot blotch disease of wheat – a new thrust area for sustaining productivity. J Wheat Res 5:1–11

Chung WH, Chung WC, Ting PF, Ru CC, Huang HC, Huang JW (2009) Nature of resistance to methyl benzimidazole carbamate fungicides in Fusarium oxysporum f.sp. lilii and F. oxysporum f.sp. gladioli in Taiwan. J Phytopathol 157:742–747

Clarke MP, Gooding MJ, Jones SA (2004) The effects of irrigation, nitrogen fertilizer and grain size on Hagberg falling number, specific weight and black point of winter wheat. J Sci Food Agric 84:227–236

Davidse, L.C., J. Henken, A. Dalen, A.B. van Jespers, B.C. Mantel (1989) Nine years of practical experience with phenylamide resistance in Phytophthora infestans in the Netherlands. Netherlands J Plant Pathol 95(Suppl 1):197–213

Davison EM, Drenth A, Kumar S, Mack S, Mackie AE, McKirdy S (2006) Pathogens associated with nursery plants imported into Western Australia. Australas Plant Pathol 35:473–475

de Miranda BEC, Boiteux LS, Reis A (2010) Identification of Solanum (section Lycopersicon) accessions with resistance to Stemphylium solani and S. lycopersici. Hortic Bras 28:178–184

Deadman M, Al-Hasani H, Al-Sa'di AM (2006) Solarization and biofumigation reduce Pythium aphanidermatum induced damping-off and enhance vegetative growth of greenhouse cucumber in Oman. J Plant Pathol 88:333–335

Deadman ML, Al-Sadi AM, Al-Maqbali YM, Farr DF, Aime MC (2011) Additions to the rust fungi (Pucciniales) from northern Oman. Sydowia 63:155–168

DeVay JE, Garber RH, Matheron D (1982) Role of Pythium species in the seedling disease complex of cotton in California. Plant Dis 66:151–154

Donaldson SP, Deacon JW (1993) Effects of amino acids and sugars on zoospore taxis, encystment and cyst germination in Pythium aphanidermatum (Edson) Fitzp., P. catenulatum Matthews and P. dissotocum Drechs. New Phytol 123:289–295

Drenth A, Goodwin SB (1999) Population Structure of Oomycetes. In: Worral J (ed) Structure and dynamics of fungal populations. Kluwer Academic Publishers, Boston, pp. 195–224

Duczek LJ (1990) Sporulation of Cochliobolus sativus on crown and underground parts of spring cereals in relation to weather and host species, cultivar and phenology. Can J Plant Pathol 12:273–278

Eckert JW (1988) Historical development of fungicide resistance in plant pathogens. In: Delp CJ (ed) Fungicide resistance in North America. APS, St. Paul, pp. 1–3

Elad Y, Zimand G, Zaqs Y, Zuriel S, Chet I (1993) Use of Trichoderma harzianum in combination or alternation with fungicides to control cucumber grey mold (Botrytis cinerea) under commercial greenhouse conditions. Plant Pathol 42:324–332

Farraq ESH, Fotouh YO (2010) Solarization as a method for producing fungal-free container soil and controlling wilt and root-rot diseases on cucumber plants under greenhouse conditions. Achieves Phytopathol Plant Protect 43:519–526

Fetch T, Mitchell Fetch J, Xue A (2015) Races of Puccinia graminis on barley, oat, and wheat in Canada in 2007 and 2008. Can J Plant Pathol 37:331–341

Gardiner RB, Jarvis WR, Shipp JL (1990) Ingestion of Pythium spp. by larvae of the fungus gnat Bradysia impatiens (Diptera: Sciaridae). Ann Appl Biol 116:205–212

Gardner DE, Hendrix FF Jr (1973) Carbon dioxide and oxygen concentrations in relation to survival and saprophytic growth of Pythium irregulare and Pythium vexans in soil. Can J Bot 51:1593–1598

Garzon CD, Geiser DM, Moorman GW (2005a) Amplified fragment length polymorphism analysis and internal transcribed spacer and CoxII sequences reveal a species boundary within Pythium irregulare. Phytopathology 95:1489–1498

Garzon CD, Geiser DM, Moorman GW (2005b) Diagnosis and population analysis of Pythium species using AFLP fingerprinting. Plant Dis 89:81–89

Gautam HR, Bhardwaj ML, Kumar R (2013) Climate change and its impact on plant diseases. Curr Sci 105:1685–1691

Gilardi G, Baudino M, Moizio M, Pugliese M, Garibaldi A, Gullino ML (2013) Integrated management of Phytophthora capsici on bell pepper by combining grafting and compost treatment. Crop Prot 53:13–19

Grondona I, Hermosa R, Tejada M, Gomis MD, Mateos PF, Bridge PD, Monte E, Garcia-Acha I (1997) Physiological and biochemical characterization ofTrichoderma harzianum, a biological control agent against soil borne fungal plant pathogens. Appl Environ Microbiol 63:3189–3198

Guo X, Petermann JS, Schittko C, Wurst S (2015) Independent role of belowground organisms and plant cultivar diversity in legume-grass communities. Appl Soil Ecol 95:1–8

Guzmán-Valle P, Bravo-Luna L, Montes-Belmont R, Guigón-López C, Sepúlveda-Jiménez G (2014) Induction of resistance to Sclerotium rolfsii in different varieties of onion by inoculation with Trichoderma asperellum. Eur J Plant Pathol 138:223–229
Hancock JG (1981) Longevity of Pythium ultimum in moist soils. Phytopathology 71:1033–1037
Harman GE, Taylor AG, Stasz TE (1989) Combining effective strains of Trichoderma harzianum and solid matrix priming to improve biological seed treatments. Plant Dis 73:631–637
Harvey PR, Butterworth PJ, Hawke BG, Pankhurst CE (2000) Genetic variation among populations of Pythium irregulare in southern Australia. Plant Pathol 49:619–627
Harvey PR, Butterworth PJ, Hawke BG, Pankhurst CE (2001) Genetic and pathogenic variation among cereal, medic and sub-clover isolates of Pythium irregulare. Mycol Res 105:85–93
Heckman JR, Johnston S, Cowgill W (2003) Pumpkin yield and disease response to amending soil with silicon. Hortscience 38:552–554
Hendrix FF, Campbell WA (1973) Pythium as plant pathogens. Annu Rev Phytopathol 11:77–98
Hernandez JR, Aime MC, Henkel TW (2005) The rust fungi (Uredinales) of Guyana. Sydowia 57:189–222
Hoitink H, Boehm M (1999) Biocontrol within the contex of soil microbial communities: A substrate-dependent phenomenon. Annu Rev Phytopathol 37:427–446
Iftikhar S, Asad S, Kazi AM, Ahmad I (2009) In vitro screening of synthetic hexaploid wheat lines against Bipolaris sorokiniana in Pakistan. Pak J Bot 41:1989–2001
Jahani M, Aggarwal R, Gupta S, Sharma S, Dureja P (2014) Purification and characterization of a novel toxin from Bipolaris sorokiniana, causing spot blotch of wheat and analysis of variability in the pathogen. Cereal Res Commun 42:252–261
Kaiser WJ, Hannan RM (1983) Etiology and control of seed decay and pre-emergence damping-off of chickpea by Pythium ultimum. Plant Dis 67:77–81
Kaufman DD, Edwards DF (1983) Pesticide-microbe interaction effects on persistence of pesticides in soil. In: Proceedings of the 5th international congress of pesticide chemistry, Tokyo, Japan, 29 August to 4 September, 1982. 1983, pp 177–182
Kazeeroni EA, Al-Sadi AM (2016) 454-Pyrosequencing reveals variable fungal diversity across farming systems. Front Plant Sci 7:314
Kubo Y, Harata K, Kodama S, Fukada F (2016) Development of the infection strategy of the hemibiotrophic plant pathogen, Colletotrichum orbiculare, and plant immunity. Physiol Mol Plant Pathol 95(7):32–36
Kumar J, Schafer P, Huckelhoven R, Langen G, Baltruschat H, Stein E, Nagarajan S, Heinzkogel K (2002) Bipolaris sorokiniana, a cereal pathogen of global concern: cytological and molecular approaches towards better control. Mol Plant Pathol 3:185–195
Lamour KH, Hausbeck MK (2003) Effect of crop rotation on the survival of Phytophthora capsici in Michigan. Plant Dis 87:841–845
Lievens B, Vaes K, Coosemans J, Ryckeboer J (2001) Systemic resistance induced in cucumber against Pythium root rot by source separated household waste and yard trimmings composts. Compost Sci Util 9:221–229
Liljeroth E, Franzon AI, Gustafsson M (1994) Effect of prehelminthosporol, a phytotoxin produced by Bipolaris sorokiniana, on barley roots. Can J Bot 75:558–563
Liljeroth E, Franzon AI, Gunnarsson T (1996) Root colonization by Bipolaris sorokiniana in different cereals and relations to lesion development and natural root cortical cell death. J Phytopathol 144:301–307
Lodha, S., R. Mawar and A. Saxena (2014) Compost application for suppression of Macrophomina Phaseolina causing charcoal rot in arid crops. In: Acta Horticulturae, pp 307–312
Loveless MD, Hamrick JL (1984) Ecological determinants of genetic structure in plant populations. Annu Rev Ecol Syst 15:65–95
MacDonald E, Millward L, Ravishankar JP, Money NP (2002) Biomechanical interaction between hyphae of two Pythium species (Oomycota) and host tissues. Fungal Genet Biol 37:245–249

Martin FN (1995) Pythium. In: Kohmoto K, Singh US, Singh RP (eds) Pathogenesis and host specificity in plant diseases; histopathological, biochemical, genetic and molecular bases. Elsevier, Oxford, pp. 17–36

Martin FN, Hancock JG (1986) Association of chemical and biological factors in soils suppressive to Pythium ultimum. Phytopathology 76:1221–1231

Martin FN, Loper JE (1999) Soilborne plant diseases caused by Pythium spp.: ecology, epidemiology, and prospects for biological control. Crit Rev Plant Sci 18:111–181

Matsushita Y, Bao Z, Kurose D, Okada H, Takemoto S, Sawada A, Nagase H, Takano M, Murakami H, Koitabashi M, Yoshida S, Saito M, Sano T, Tsushima S (2015) Community structure, diversity, and species dominance of bacteria, fungi, and nematodes from naturally and conventionally farmed soil: a case study on Japanese apple orchards. Org Agric 5:11–28

McDonald BA, McDermott JM (1993) Population genetics of plant pathogenic fungi. Bioscience 43:311–319

Menzies J, Bowen P, Ehret D, Glass AD (1992) Foliar applications of potassium silicate reduce severity of powdery mildew on cucumber, muskmelon, and zucchini squash. J Am Soc Hortic Sci 117:902–905

Nallathambi P, Umamaheswari C, Thakore BBL, More TA (2009) Post-harvest management of ber (Ziziphus mauritiana Lamk) fruit rot (Alternaria alternata Fr. Keissler) using Trichoderma species, fungicides and their combinations. Crop Prot 28:525–532

Neher DA, Weicht TR, Dunseith P (2015) Compost for management of weed seeds, pathogen, and early blight on brassicas in organic farmer fields. Agroecol Sustain Food Syst 39:3–18

Neupane A, Sharma R, Duveiller E, Shrestha S (2010) Sources of Cochliobolus sativus inoculum causing spot blotch under warm wheat growing conditions in South Asia. Cereal Res Commun 38:541–549

Nicol JM, Turner SJ, Coyne DL, Nijs Ld, Hockland S, Maafi ZT (2011) Current nematode threats to world agriculture. In: Jones J, Gheysen G, Fenoll C (eds) Genomics and molecular genetics of plant-nematode interactions. Springer, Heidelberg, pp. 21–43

Noble R, Roberts SR (2004) Eradication of plant pathogens and nematodes during composting: a review. Plant Pathol 53:548–568

Ovsyankina AV (2005) Species diversity of the rye root rot causative agents in the regions of Russia. Mikol Fitopatol 39:88–91

Pardey PG, Beddow JM, Kriticos DJ, Hurley TM, Park RF, Duveiller E, Sutherst RW, Burdon JJ, Hodson D (2013) Right-sizing stem-rust research. Science 340:147–148

Paulitz TC, Adams K (2003) Composition and distribution of Pythium communities in wheat fields in eastern Washington state. Phytopathology 93:867–873

Peethambaran CK, Singh RS (1977) Survival of different structures of Pythium spp. in soil. Indian Phytopathol 30:347–352

Perkins DD (1991) In praise of diversity. In: Bennett JW, Lasure LL (eds) More gene manipulations in fungi. Academic, San Diego, pp. 3–26

Pilet F, Pelle R, Ellisseche D, Andrivon D (2005) Efficacy of the R2 resistance gene as a component for the durable management of potato late blight in France. Plant Pathol 54:723–732

Poole GJ, Harries M, Hüberli D, Miyan S, MacLeod WJ, Lawes R, McKay A (2015) Predicting cereal root disease in Western Australia using soil DNA and environmental parameters. Phytopathology 105:1069–1079

Pose-Juan E, Sánchez-Martín MJ, Herrero-Hernández E, Rodríguez-Cruz MS (2015) Application of mesotrione at different doses in an amended soil: Dissipation and effect on the soil microbial biomass and activity. Sci Total Environ 536:31–38

Pottorff LP, Panter KL (1997) Survey of Pythium and Phytophthora spp. in irrigation water used by Colorado commercial greenhouses. HortTechnology 7:153–155

Punja ZK, Yip R (2003) Biological control of damping-off and root rot caused by Pythium aphanidermatum on greenhouse cucumbers. Can J Plant Pathol 25:411–417

Ravishankar JP, Davis CM, Davis DJ, MacDonald E, Makselan SD, Millward L, Money NP (2001) Mechanics of solid tissue invasion by the mammalian pathogen Pythium insidiosum. Fungal Genet Biol 34:167–175

Sharma M, Ghosh R, Pande S (2015) Dry root rot (Rhizoctonia bataticola (Taub.) Butler): an emerging disease of chickpea – where do we stand? Arch Phytopathol Plant Protect 48:797–812

Singh S, Singh H, Sharma A, Meeta M, Singh B, Joshi N, Grover P, Al-Yassin A, Kumar S (2015) Inheritance of spot blotch resistance in barley (Hordeum vulgare L.). Can J Plant Sci 94:1203–1209

Smiley RW, Gourlie JA, Easley SA, Patterson LM (2005) Pathogenicity of fungi associated with the wheat crown rot complex in Oregon and Washington. Plant Dis 89:949–957

Smith ML, Bruhn JN, Anderson JB (1992) The fungus Armillaria bulbosa is among the largest and oldest living organisms. Nature 356:428–431

Stanghellini ME, Burr TJ (1973) Germination in vivo of Pythium aphanidermatum oospores and sporangia. Phytopathology 63:1493–1496

Stanghellini ME, Hancock JG (1971a) Radial extent of the bean spermosphere and its relation to the behaviour of Pythium ultimum. Phytopathology 61:165–168

Stanghellini ME, Hancock JG (1971b) Sporangium of Pythium ultimum as a survival structure in soil. Phytopathology 61:157–164

Stanghellini ME, Nigh EL Jr (1972) Occurrence and survival of Pythium aphanidermatum under arid soil conditions in Arizona. Plant Dis Rep 56:507–510

Stanghellini ME, Phillips JM (1975) Pythium aphanidermatum: its occurrence and control with pyroxychlor in the Arabian desert at Abu Dhabi. Plant Dis Rep 59:559–563

Stasz TE, Harman GE (1980) Interactions of Pythium ultimum with germinating resistant or susceptible pea seeds. Phytopathology 70:27–31

Stott MB, Taylor MW (2016) Microbial ecology research in New Zealand. N Z J Ecol 40

Szabo LJ, Cuomo CA, Park RF(2014) Puccinia graminis. In: Genomics of plant-associated fungi: monocot pathogens, pp 177–196

Toklu F, Akgul DS, Bicici M, Karakoy T (2008) The relationship between black point and fungi species and effects of black point on seed germination properties in bread wheat. Turk J Agric For 32:267–272

Tunali B, Nicol JM, Hodson D, Uckun Z, Buyuk O, Erdurmus D, Hekimhan H, Aktas H, Akbudak M, Bagci A (2008) Root and crown rot fungi associated with spring, facultative and winter wheat in Turkey. Plant Dis 92:1299–1306

Van der Plaats-Niterink AJV (1981) Monograph of the genus Pythium. Stud Mycol 21:1–242

Van Geel M, Ceustermans A, Van Hemelrijck W, Lievens B, Honnay O (2015) Decrease in diversity and changes in community composition of arbuscular mycorrhizal fungi in roots of apple trees with increasing orchard management intensity across a regional scale. Mol Ecol 24:941–952

van Wyk M, Al-Adawi AO, Khan IA, Deadman ML, Al-Jahwari AA, Wingfield BD, Ploetz RC, Wingfield MJ (2007) Ceratocystis manginecans sp. nov., causal agent of a destructive mango wilt disease in Oman and Pakistan. Fungal Divers 27:213–230

Vujanovic V, Mavragani D, Hamel C (2012) Fungal communities associated with durum wheat production system: A characterization by growth stage, plant organ and preceding crop. Crop Prot 37:26–34

Wanyera R, MacHaria JK, Kilonzo SM, Kamundia W (2009) Foliar fungicides to control wheat stem rust, race TTKS (Ug99), in Kenya. Plant Dis 93:929–932

Watson IA, de Sousa CNA (1983) Long distance transport of spores of Puccinia graminis tritici in the southern hemisphere. In: Proceedings of the Linnean Society of New South Wales, vol 106, pp 311–321

Wiersma JJ, Motteberg CD (2005) Evaluation of five fungicide application timings for control of leaf-spot diseases and fusarium head blight in hard red spring wheat. Can J Plant Pathol 27:25–37

Wiese MV (1987) Compendium of wheat diseases. APS Press, St. Paul

Xiao JZ, Tsuge T, Doke N (1992) Further evaluation of the significance of BZR-toxin produced by Bipolaris zeicola race 3 in pathogenesis on rice and maize plants. Physiol Mol Plant Pathol 40:359–370

Zabka V, Stangl M, Bringmann G, Vogg G, Riederer M, Hildebrandt U (2008) Host surface properties affect prepenetration processes in the barley powdery mildew fungus. New Phytol 177:251–263

Zhang YH, Peng C, Liu TR, Liu YK, Zhang ZB, Liu XY, Li Y (1999) Loss of spring wheat caused by wheat root rot in Heilongjiang Province. Acta Phytopathol Sin 29:329–332

Zhu Z, Bonnett D, Ellis M, Singh P, Heslot N, Dreisigacker S, Gao C, Mujeeb-Kazi A (2015) Mapping resistance to spot blotch in a CIMMYT synthetic-derived bread wheat. Mol Breed 34:1215–1228

Integrated and Innovative Livestock Production in Drylands

Dominique Blache, Philip E. Vercoe, Graeme B. Martin, and Dean K. Revell

1 Introduction

The need for improvement in the efficiency of food production for humanity has never been greater, with one in seven humans already underfed (FAO 2003). This situation will worsen over the coming decades as our agricultural land, and its ability to meet global demands for food, come under great pressure from climate change, soil degradation, increasing human population, urbanisation, biofuels, and the growing demand for animal protein in developing nations.

In this context, the role of livestock has been controversial, and often the popular view is that ruminant industries cause problems rather than solve them (e.g. (FAO 2006)). The focus of such discussions is invariably the environmental footprint of the livestock sector because it probably accounts for 14–16 % of human-induced greenhouse-gas emissions, much of it due to enteric methane production (Forster et al. 2007). However, such analysis discounts the fact that foraging animals can consume feedstuffs that humans cannot and then produce human food. Moreover, in many parts of the world, ruminants also supply power, fertilize crops, consume post-harvest residues, and play cultural roles as measures of wealth, status, dowry, insurance and long-term economic resilience (Otte et al. 2012). Sustainably managed grazing can also increase biodiversity, maintain ecosystem services and

D. Blache • P.E. Vercoe (✉) • G.B. Martin
School of Agriculture and Environment M085 and The UWA Institute of Agriculture, Faculty of Science, The University of Western Australia, Perth 6001, Australia
e-mail: philip.vercoe@uwa.edu.au

D.K. Revell
School of Agriculture and Environment M085 and The UWA Institute of Agriculture, Faculty of Science, The University of Western Australia, Perth 6001, Australia

Revell Science, 10 Wade Close, Duncraig, WA 6023, Australia

M. Farooq, K.H.M. Siddique (eds.), *Innovations in Dryland Agriculture*,
DOI 10.1007/978-3-319-47928-6_8

improve carbon capture by plants and soil (Garnett 2009). Finally, crop and livestock systems complement each other (Herrero et al. 2010).

If these roles of ruminant livestock are ignored, we could disrupt balanced socioeconomic and ecological systems, exacerbate human dietary problems in poor communities, and magnify problems in sustainability and efficiency (the amount of food produced per unit of land and water). In the end, 'livestock's long shadow'(FAO 2006) would become a self-fulfilling prophecy.

These considerations led Eisler et al. (2014) to use a broader context to plan the contribution of ruminant animals to human food security. Four major needs were identified for livestock management *viz*., (1) better animal health and welfare; (2) selection of appropriate genotypes for the environment; (3) better environmental footprint; (4) better livestock husbandry, nutrition and management.

In addition, Eisler et al. (2014) emphasised the need to reduce the consumption of human food by livestock, an issue that is directly addressed by dryland grazing systems because, in dryland, humans and ruminants are rarely competing for agricultural resources. Many dryland zones cannot reliably produce human-edible grains; even where such crops are possible, there are benefits from associations between cropping and livestock systems (e.g., ecosystem services, pest management, soil management). However, by the nature of their geographical and thus climatic constraints, dryland environments are inherently variable within and among years. This means that, for a production system, dryland systems are risky because the animals will need to cope with unpredictable supplies of feed (quality, quantity, timing) and water (Table 1). Climate change will exacerbate this unpredictability, and also challenge temperature homeostasis in the animals. If not managed correctly, these problems will also raise societal concerns about the ethics of livestock industries.

The solution is *versatile production systems*, a combination of versatile animals, versatile forages and versatile management. The most appropriate combinations of livestock genotypes, forage species and management will depend on local geographical and socio-economic factors, but the principles that underpin the structures of versatile grazing systems will be consistent. The optimal system of husbandry and management will ensure correct livestock nutrition and therefore good health and welfare (reduced susceptibility to parasitism and infectious disease) and reduced environmental impact per unit of product. Achieving these outcomes will maximise productivity within the constraints of the environment.

In this chapter, we have described the risks associated with livestock production in dryland regions and critically analysed some specific management solutions developed to solve particular problems faced by livestock production in difficult environments. In the last section, based on work done with Australian native shrubs and pasture species, we have outlined a silvopastoral system, combined with management of animal behaviour to influence diet and habitat selection that offers innovative and versatile options for the different types of livestock production in dryland regions of the world.

Table 1 A framework for assessing the impact of management interventions on estimated risk levels

Source of risk	Management strategy					
	None[a]	Feed[a]	Water[b]	Heat stress[c]	Stocking density[d]	Integrated (shrubs + behaviour)
Natural environment						
Water						
Availability	High	Serious	Low	Medium	Medium	Medium
Reliability	Serious	Serious	Medium	Serious	Serious	Medium
Quality	Serious	Serious	Medium	Serious	Serious	Medium
Food						
Availability	Serious	Medium	Serious	Serious	Medium	Low
Reliability	Serious	Medium	Serious	Serious	Medium	Low
Diversity	High	High	Serious	Serious	Serious	Low
Quality	Serious	Low	Serious	Serious	Serious	Medium
Heat Stress	High	High	Medium	Low	Medium	Low
Soil degradation	High	High	Serious	Serious	Medium	Medium
Climate predictability	Serious	Serious	Serious	Serious	Serious	Medium
Biota diversity	Serious	Serious	Medium	Medium	Serious	Low
Maladaptation of livestock						
To water supply	Serious	Serious	Medium	Medium	Medium	Medium
To food supply	Serious	Low	Serious	Medium	Medium	Medium
To climate pattern	High	High	Serious	Serious	Medium	Medium
Livestock biodiversity	High	High	Serious	Serious	Serious	Medium
Socio-economic context						
Input costs						
Fertiliser	High	High	High	High	Serious	Medium
Food supplement	Serious	Serious	Serious	Serious	Serious	Low
Use of low productive land	Serious	Medium	Serious	Medium	Medium	Low
Market (outputs)						
Instability	Serious	Medium	Medium	Medium	Medium	Low
Seasonality	Serious	Low	Low	Low	Medium	Low
Demand for "green" production	Serious	Medium	Medium	Medium	Medium	Low
Animal Welfare						
Nutrition	High	Low	Serious	Serious	Medium	Low
Shelter	Serious	Serious	Serious	Medium	Serious	Low
Health	Serious	Low	Serious	Medium	Serious	Low
Behavioural diversity	Serious	Serious	Serious	Serious	Serious	Low

(continued)

Table 1 (continued)

	Management strategy					
Source of risk	None[a]	Feed[a]	Water[b]	Heat stress[c]	Stocking density[d]	Integrated (shrubs + behaviour)
People (producers)						
Adoption	Serious	Medium	Low	Serious	Serious	Serious
Skilled labour	High	Serious	High	Serious	Serious	Medium
Risk aversion	Serious	Medium	Low	Serious	Serious	Serious
Financial						
Borrowing power	Serious	Serious	Serious	Serious	Serious	Medium
Investment	Serious	Serious	Serious	Serious	Serious	Medium
Capital debt	Serious	Serious	Serious	Serious	Serious	Medium
Profit margin	High	Medium	Serious	Medium	Medium	Medium

[a]Improving feedbase
[b]Improving access to water
[c]Reducing heat stress
[d]Adjusting stocking density

2 Risk Associated in Livestock Production in Drylands

Producing livestock in drylands has been described as "living *off* uncertainty" (Kratli and Schareika 2010) because producers are confronted by multiple risks and uncertainties. Before exploring this topic, we first turn to Hardaker et al. (2015) to distinguish between 'risk', when the probabilities of the possible outcomes are known, and 'uncertainty', when the probabilities are not known. However, for practical purposes, it may be more useful to define uncertainty as "imperfect knowledge" and risk as "uncertain and unfavourable consequences". Therefore, the degree of aversion for some of the possible consequences determines the risk profile for a particular production system (Hardaker et al. 2015). For example, a producer might say that he or she is uncertain about the weather in the next few months, which is a value-free statement simply implying imperfect knowledge of the future. However, whether the producer plans to modify his or her flock size is a reflection of their appetite for risk in the face of uncertainty. The concept of uncertainty has to be kept in mind when thinking about animal production systems in dryland areas. While uncertainty cannot be managed, producers do have opportunities for implementing strategies, such as feed supplementation and/or the provision of shelter that can decrease the probability of aversive consequences of uncertain climatic and socio-economic events (risks).

In dryland environments, especially in the context of developing countries and sustainable agriculture, livestock production can be extensive or part of a more intensive mixed farming system based on production of both crops and livestock. Both types of production systems are exposed to the same uncertainties in dryland areas, but the degree of risk (e.g., in profitability or in natural resource management) could be quite different.

The uncertainties can arise from the natural environment, the type of livestock produced and the socio-economic context. Therefore, an integrated approach to livestock production in dryland areas should aim to provide options to minimise multiple risks rather than focussing on a particular risk.

2.1 *Risks from the Natural Environment*

The dryland environment is characterised by an erratic supply of water due to low total rainfall, rainfall patterns that vary within and among years, and high rates of evapo-transpiration from land and plants due to high ambient temperature and radiation. Scarcity of water and high temperature drastically reduce the availability of feed and water, and also increase the risk of heat stress in livestock.

In any farming system, successful livestock production requires a match between the quantity, quality and reliability of feed supply and demand (Martin et al. 2008). Availability of feed resources must match the temporal variations in the animal's requirement for each component of production, especially the high requirement for energy for reproduction (Martin et al. 2008). Production of pasture in drylands is often seasonal, variable and uncertain, even during the so-called 'growing season' (e.g. winter-spring in the Mediterranean and temperate zones of southern Australia; Fig. 1). The variability is such that, depending on the year, the regions could be classified as semi-arid, arid or even sub-humid using the United Nations descriptors (UN 1977). Similarly, ambient temperature is unpredictable, a factor that can disrupt planned relationships between animal production and pasture growth.

Unpredictable natural events related to the weather are the main sources of risk for forage production and, consequently, livestock production. Repeated extreme daily temperatures ('heatwaves') are frequent in dryland regions, and they increase the risk of heat stress in the livestock (Fuquay 1981). Other unpredictable factors, such as flood, pests and diseases, can affect the performance of both crops and livestock. In addition to these relatively short-term events, long-term changes, such as depleted soil quality or climate change, also add risk. For example, a high rate of evaporation combined with loss of vegetation coverage has led to an increase in dryland salinity that has transformed large areas of land to be non-productive (Pannell 2001) with poor livestock productivity. Similarly, climate change can reduce animal production in dryland areas (Nardone et al. 2010) because of the acceleration of desertification and land degradation, outcomes that are exacerbated by a higher frequency of extreme events.

Innovative strategies to improve the resilience of livestock production in dryland areas cannot reduce the unpredictability of the natural conditions, but can minimise the consequences of unpredictable events. In other words, the strategies will not change the uncertainty of adverse events but are expected to reduce the risk by mitigating the consequences of those events for the livestock.

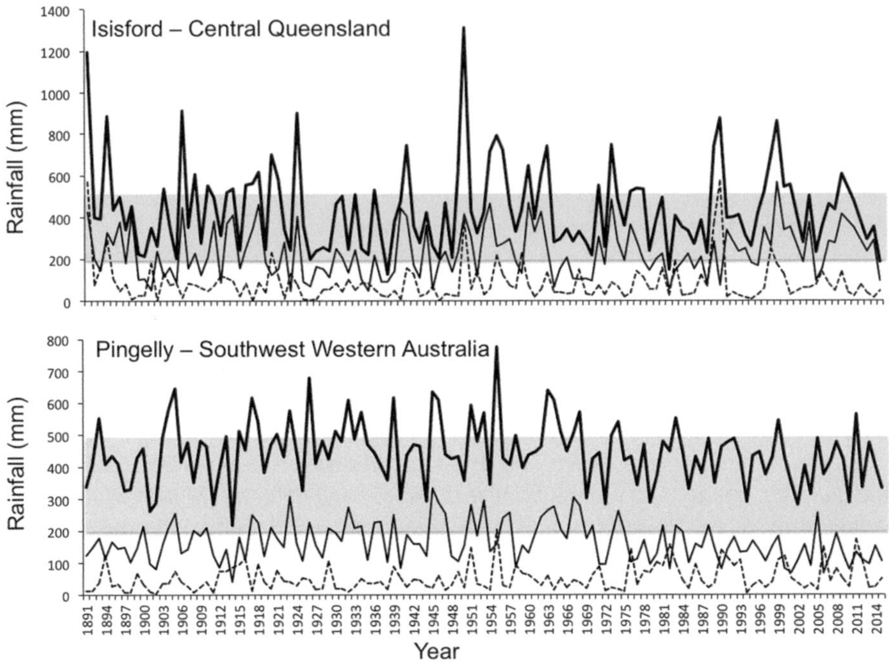

Fig. 1 Illustration of the unpredictability and variability of annual rainfall (bold solid line), and rainfall in the Australian spring (March to May; thin solid line) and summer (June to August; dashed line) in central Queensland (Isisford Post Office: −24.26°S, 144.44°E, 203 m) and in southwest Western Australian (Pingelly: -32.53°S, 117.08°E, 297 m) from 1891 to 2015. The grey box shows the limits of the semi-arid dryland zone as defined by UNESCO (1977). Data from Australian Bureau of Meteorology (http://www.bom.gov.au)

2.2 *Risk from Maladapted Livestock*

In addition to the unpredictability of the environment, the genetics of an animal can influence risk by affecting the capacity of the animal to cope with, and adapt to, the environment and its variability. The need to preserve biodiversity of livestock that are adapted to dryland has been recognised over the last decade (Scherf et al. 2008; Hoffmann 2011). The genetic make-up of productive livestock will favour appropriate levels of production, to drive profitability of the business, and also an ability to adapt to local conditions, including feed and water availability (Revell 2016), climate and diseases (Hoffmann 2011). However, as Provenza (2008) pointed out, genotypes are often selected for high levels of production and therefore high nutritional demand, but that demand is rarely matched by the feed resources available in a dryland landscape. Moreover, exotic target phenotypes are often tested and measured with the animals in confinement, a completely inappropriate environment when the progeny will be raised in dry rangeland (Provenza 2008). Almost inevitably, the outcomes from these mismatches are poor animal performance and a major economic burden when the available pasture does not provide enough feed of the right quality (Provenza 2008). A better option is to conserve the native and traditional breeds and improve production through traditional quantitative genetics, boosted by cutting-edge genomic selection.

About a third of cattle breeds and about half of the breeds of sheep and goat are particularly well adapted to drylands (Scherf et al. 2008), and they clearly provide the base resource for developing versatile production systems and driving profitability in dryland environments. Appropriate livestock genetics will actually reduce eco-systems risks (Doreau et al. 2013) and improve the resilience of the production system (Provenza 2008).

2.3 *Risk from the Socio-economic Factors*

Other critical sources of risk are unpredictable costs of production and profit margins, and the local socio-political environment (Anderson and Dillon 1992). Clearly, these risks are not specific to dryland production, but dryland production is often based in a low-margin system (Scherf et al. 2008) so any change in the level of inputs or outputs could dramatically alter the livelihood of producers. Costs of production can be affected by increases in the costs of inputs such fertiliser, grain or livestock feed supplements. Similarly, demand for the livestock products can also vary dramatically because competitive markets are unpredictable and the global market is complex because it is influenced by seasonal conditions in other parts of the world that affect supply. Additionally, the cost of production as well as the income from selling the product (meat, milk, fibre) can be affected by currency exchange rates and a diverse variety of governmental decisions, such as bans or restrictions on a particular product, or new regulations on waste management in the supply chain.

There has recently been an increase in consumer demand for more sustainable, animal-friendly livestock production (Boland et al. 2013). In fact, this trend is good for dryland livestock production because the natural conditions inherent in the system directly address consumer expectations and demands – for example, the branding of “Uruguay Natural” for beef and lamb meat production (Bervejillo 2015). On the other hand, dryland livestock production may face criticism and lose markets as a consequence of adverse events leading to animal losses or decreases in feed availability (Turner and Dwyer 2007). These risks will not be discussed directly, but the proposed strategies cannot be considered outside of the meat and fibre markets. It is aimed to develop strategies that lower the cost of production, reduce the dependence on supplementation, either in the form of a feed supplement or fertiliser, and thus increase the profit margin of the production system.

Two other sources of risk, arguably the greatest risks of all, are associated with people and financial systems. People risks arise from the attitudes, skills and actions of the managers of the production systems (Hardaker et al. 2015), and can include shortage of skilled labour, illnesses, lack of succession planning, resistance to innovation, and aversion to risk. All of these factors can affect the cost of production and the resilience of the production system (Hardaker et al. 2015). Financial risk includes variable interest rates, limited investment from third parties, or the proportion of capital debt that a business carries (Hardaker et al. 2015). This chapter is focused on the resilience of production systems, but it is worth mentioning that the risks associated with a production system will have direct and indirect consequences on human and financial risks.

In summary, implementation of innovative strategies that reduce one or more risks can improve the resilience of the livestock production systems and buffer the production systems against inevitable uncertainties. Strategies designed to sustain livestock production in dryland areas will need to optimise the interactions between the animals, feed supply, environmental and climatic conditions, and target markets. In the next section, some strategies with potential to address one or two of the interactions listed above are discussed. In the last section of the chapter, a strategy has been proposed that may address most of these interactions. This strategy is being implemented in a range of scenarios but still requires research and, moreover, needs to be adapted to suit localised and specific environments and livestock production systems.

3 Innovative Strategies

3.1 Managing Feed to Avoid Limiting Nutrients

Access to the right quality and quantity of water and feed, including macro and trace elements, is essential for successful animal production. In a dryland environment, where the supply of both feed and water availability can be variable, the problem of feed shortage has been solved in a variety of ways, one of the most common being the provision of supplements of grain, feed blocks or other specific feedstuffs (Blache et al. 2008). For example, in Australia, the use of lupin feeding has long been used to manage sheep during the dry autumn when pasture is sparse and poor, but good nutrition is needed to support reproduction (Blache et al. 2000, Martin et al. 2008). Work done in Tunisia has resulted in the development of a feed block to deliver nutrients and minerals to livestock (Ben Salem and Nefzaoui 2003). When extra nutrients are required, specific plants are included in the diet or in the pasture, such as saltbush (Ben Salem et al. 2010), spineless cactus (Ben Salem et al. 2004), and *Lotus* (Ramirez-Restrepo et al. 2005). In this context, it is critical to consider the capacity of any particular feedstuff or plant species to complement the nutrient supply from all the other options that are available. If an additional feed resource can supply a limiting nutrient at the time of year when it is most needed, then it will add value to the production system and reduce the risks associated with a variable climate and feed supply (Moore et al. 2009). In other words, a supplement, whether provided by hand feeding or by a forage plant on offer to grazing livestock, should not be valued in isolation but on the basis of its impact to the profitability, productivity or risk management within the whole management system.

Supplementation strategies not only have the capacity to boost livestock productivity by reducing feed shortages, but also offer strategic support during periods of high energy demand during the reproductive cycle, from puberty to ovulation and lactation (Blache et al. 2008). Such strategies have great merit because relatively short periods of supplementation ('focus feeding' for a few days; Martin et al. 2008) can have long-term benefits for productivity. However, relying on externally-sourced supplements can add another level of risk to production systems, because of potential uncertainties in availability or price. Therefore, growing and using plants that are suited to the local environment as 'standing supplements' to other feed resources (i.e., plants that nutritionally complement each other) is an attractive proposition, and will be discussed below (Section 4). Moreover, because supplementary feeding can be labour intensive and costly, there is an added advantage in allowing grazing animals to source their own supplements from a mixture of plants on offer. Finally, plant resources used as supplements can also address the sustainability of livestock production in dryland systems (Masters et al. 2006).

3.2 *Managing Water*

The consumption of an adequate amount of water is essential for productivity and welfare of livestock. Water is essential to life and, even if species from semi-arid environments have evolved to be more water-efficient than species living in temperate climates, they still need regular access to quality water. The percentage of water in the body can vary but is kept within a narrow window in which osmolarity (salt water balance) can be tightly regulated. The needs for water repletion depend on the environmental conditions, the physiological status of the animals, and the metabolic activity of the animal (Willmer et al. 2009, Revell 2016). When ambient temperature increases, homeotherms limit a rise in body temperature by using thermoregulatory mechanisms such as sweating or panting, a mechanism that is effective when ambient humidity is not too high. Both mechanisms are based on evaporative water loss and are efficient at maintaining core body temperature within comfortable limits (the 'thermocomfort zone'; Willmer et al. 2009). Consequently, when the temperature rises, animals will necessarily increase their water intake to compensate for their losses (Silanikove 1994). Water needs also depend on the physiological status of the animals. For example, genotype, live weight, sex and reproductive status (pregnancy, lactation), as well as ambient temperature, affect the water requirements of cattle (Winchester and Morris 1956). Livestock will get water from drinking water points (about 10 % of their water need), from the plants they consume (estimated globally at 90 % of consumptive water use), but forage plants contain variable amounts of water and salt, both of which affect their osmoregulation (Hoekstra and Mekonnen 2012; Ran et al. 2016).

With the consumption of dry feeds or senesced forages in dryland systems, a greater proportion of water requirements must be met from drinking water. In fact, feed quality and availability affect the frequency with which animals visit watering points (Squires 1981). Similarly, the quality of the ground water, especially its salt content, can affect their water intake as well as their food intake. Animals fed a high salt content diet can reduce their feed intake by over 35 % (Blache et al. 2007). To mitigate the dependency of ruminants on water, pastoralists must offer water points across the landscape that is to be used for grazing, and then depend on the animals finding and remembering the locations, or else the producers must herd their livestock to the water points (Squires 1981).

A carefully planned distribution of water points is essential because the animals need to balance the distance walked to water and the distance walked to find food. In many situations, feed supply near water points is low because of high grazing pressure in these locations, and because animals choose to camp near water, especially in hot conditions. This behaviour limits use of biomass (Holechek 1997). In rangeland situations, maintenance of water points can be costly and, with free-ranging animals, the monitoring of water points is essential to maintain productivity.

3.3 Management of Heat Stress

Exposure to high temperature or solar radiation can induce heat stress, reducing both welfare and production (Silanikove 2000). In addition to access to enough water, shade should be provided to encourage behaviours that decrease the incidence of heat stress. On hot sunny days, ruminants seek shade from trees, or even from their peers, and will increase their water intake if the water points are not too far from where they spend most of their time. The ability to cope with high temperature varies with breed – for example, the evaporative critical temperature (i.e., the temperature above which the animal starts using evaporative heat loss mechanisms to maintain its body temperature constant) varies between the Jersey (24°C) and the Brahman (35°C; MacFarlane 1968). Regardless of the genotype used in the production system, providing shade and shelter should be an essential component of any integrative and versatile system (discussed further in Section 4).

Management of stocking density is an alternative to managing feed on offer or increasing supplementation when feed demand exceeds feed supply. Management of the stocking density also affects the flow rate required to maintain an adequate supply of water to a group of animals. There are two types of strategy: i) stocking density managed on a long term scheme ('long-term carrying capacity', LTCC); ii) stock numbers adjusted according to the seasonal availability of food on offer ('variable stocking'). Both strategies have been reviewed recently and discussed in the context of northern Australian rangeland (O'Reagain and Scanlan 2013). Briefly, LTCC is a safe strategy designed to use 10–30 % of the annual pasture growth (Scanlan et al. 1994; McKeon et al. 2009). A recent computer simulation suggests that limiting the degree of flexibility in stocking density would enhance sustainability of cattle production in Northern Australia (Pahl et al. 2016). The authors suggest that, in this particular landscape, stocking density should be increased only by 10 % after a good growing pasture season and decreased by 20 % after a poor growing season (Pahl et al. 2016). The impact of grazing on pasture and the need for management are limited, but so is the economic return when the pasture is plentiful (O'Reagain and Scanlan 2013).

By contrast, variable stocking rates, involving a constant adjustment of stocking rate in response to pasture availability, offer a more efficient use of resources. However, variable stocking rate is a risky exercise because, to be efficient, profitable and safe for the environment, the decisions for destocking or restocking need to be made in a very timely fashion, possibly ahead of major changes in pasture availability. In industrialised countries such as Australia, pastoralists have access to computer models (such as in Australia GeoGlam RAAP (Rangelands And Pasture Productivity: http://www.geo-rapp.org) to help them to predict pasture growth, or satellite-based imaging (Hill et al. 2004) that is available now or in the foreseeable future to can help them assess the biomass of vegetation. Nevertheless, in extensive systems, it can still be difficult to change stocking rates quickly and thus make the right decision at the right time. The accuracy and reliability of landscape prediction tools using satellite imaging are progressing rapidly (see http://www.geo-rapp.org/ or www.NRMhub.com.au) but these tools might only be available to producers with the skills or resources to process and interpret the data. They are unlikely to be applicable in less developed countries in the short term.

4 Towards Integrative Innovative Strategies – A Versatile System

To address multiple challenges and opportunities, capitalise on favourable conditions, and cope with the inevitable difficult conditions when they arise, we need livestock systems that function and produce in the face of variability and change. Many simplified agricultural systems are designed for maximum productivity, but often this is only achieved for discrete periods of the year when the supply of water and nutrients is optimal. Of course, in reality, livestock systems need extended periods of predictable nutrient supply. To identify potential options or solutions for a more versatile and robust grazing system, the guidelines proposed by Hobbs and Morton (1999) are valuable, and we have used them in our trans-disciplinary project ('Enrich') in Australia (Revell et al. 2008, Revell et al. 2013). An outline of this approach was used by Revell and Sweeney (2004) to explore grazing systems that can reduce the impact of dryland salinity.

(a) *Identify system functions that are sub-optimal in current managed systems*

The seasonality of feed supply imposes a major constraint on livestock production and, in dryland areas, the resulting feed 'gaps' can be substantial. To reduce the impact of seasonality in feed supply, the periods of high energy requirement in the reproductive process need to coincide with periods of maximum feed quality and quantity. However, even with the best of such strategies, we cannot completely eliminate the constraints. Supplementing animals during periods of feed shortage can be costly, reducing profit margins, so we have considered the possibility of using a mix of forage species that, collectively, can ensure the provision of a predictable and adequate amount of feed throughout the year. In many low-rainfall, dryland environments, the system risks partial failure as a consequence of environmental challenges (Sanderson et al. 2004) or dietary imbalances (Hogan 1982), largely because these grazing systems are typified by monocultures or a limited range of species;

(b) *Identify a suite of species in the natural system that can overcome sub-optimal periods the production system*

The native plant communities of dryland areas of southern Australia are well represented by summer active perennial plants that, collectively, grow actively across the seasons. Importantly, systems based on native plants are stable and resilient to environmental irregularities, and extend the period of water use into periods of high evaporative demand, but the biomass does not necessarily maximise production (Dunin et al. 1999; Johnston et al. 2003; Michalk et al. 2003). To meet production targets, the various native plants need to be combined with other species that are complementary, presenting a suite of species that provides the required biomass and nutrients throughout the year.

(c) *Identify species with key functional roles to mimic more complex native systems*

Perennial shrubs offer particular traits of interest for dryland grazing systems, including deeper roots that confer a capacity to access water and nutrients at depth, an essential factor during prolonged periods without rain. Hydraulic lift can raise water from lower parts of the soil profile so it becomes accessible for surface roots, including those of adjacent, complementary pasture species. In the Enrich project, Australian shrub species were grown with annual pasture legumes, and the annual legumes had higher nitrogen and potassium concentration than when grown in monoculture (J. Emms et al. unpublished data). The shrubs can have modest growth rates and digestibility but, most importantly, they provide green, edible foliage during summer-autumn, the season of serious 'feed gap' in Mediterranean-type environments.

(d) *Identify likely environmental constraints and select an array of species that will confer system resilience*

To mitigate against the environmental uncertainty in dryland systems, it is necessary to choose plant species that are tolerant to the most difficult conditions that are envisaged. In the Enrich project, we targeted regions with less than 350 mm of annual rainfall and with long, hot summers, and included plants that could tolerate moderate salinity, a widespread problem in southern Australia.

(e) *Consider how many species are required for a managed system*

Identifying the required number of species is not straightforward. From a management perspective, fewer rather than more is an attractive proposition but, from the animal's perspective, a diversity of forages can stimulate appetite and feed intake (Agreil et al. 2005, Meuret and Provenza 2015a). In most cases, about two-thirds of the diet of grazing herbivores is made of up 3–6 species, with a large number of species (depending on availability) making up the remainder. Many species might only be consumed in small amounts, but they might still be important to the nutrition and health of the animals, and the health of the landscape (Vercoe et al. 2009). Therefore, focusing on only the 3–6 species that dominate consumption could compromise animal performance.

(f) *Consider the spatial arrangement*

The optimal spatial arrangement of perennial shrubs will depend on the local situation, and could range from high density 'block' plantations to 'alley farming' systems with large inter-row spaces.

(g) *Provide socio-economic instruments that facilitate implementation*

A critical issue is that transformation of a production system is likely to be a big step for managers for several reasons, including social pressures to stay with the norm, potentially high up-front costs, and a lag in profitability whilst the system (plants, animals, people) becomes established.

(h) *Adaptive management, adequate monitoring, and capacity to modify design elements*

A versatile production system must, by definition, be able to respond to external events, so it must not be viewed as 'the solution' but rather as one set of options that can be integrated into current practices, monitored and adapted as required.

A grazing system that incorporates a range of functional plant types can take on many forms, but a common description is *silvopastoralism*, in which shrubs and trees are combined with herbaceous forages. These systems promote vertical and horizontal diversity – ie, diversity by volume, not just area, and span a broad range of environments worldwide. In tropical and equatorial regions of Latin America, such as in Columbia and southern Mexico, innovative silvopastoral systems are being adopted (Murgueitio and Ibrahim 2008; Hall et al. 2011; Balvanera et al. 2012), with a mix of pastures, shrubs, trees and herbage, and the outcome has been an increase in the growth rate in cattle (Paciullo et al. 2011). Silvopastoral systems can be adapted to dryland climates (Guerra and Pinto-Correia 2016), such as Northern Australia, where leguminous *Leucaena* spp shrubs have been planted as extra food for cattle (Dalzell et al. 2006). In addition, cattle and sheep have been supplemented with tagasaste (*Chamaecytisus palmensis* Christ.) in low fertility, sandy soils (Borens and Poppi 1990; Becholie et al. 2005; Assefa et al. 2008) and saltbush (*Atriplex spp.*) in saline land (Norman et al. 2004; Barrett-Lennard and Norman 2009; Masters et al. 2010). These shrub-based strategies give better results if herbaceous species are offered simultaneously (Hopkins and Nicholson 1999; Fancote et al. 2009).

The perennial nature of forage shrubs provides an opportunity to supply green feed at any time of the year, with the most important periods being when the 'traditional' pasture is senescent (Barrett-Lennard and Norman 2009). The shrubs are resilient to harsh conditions but they are also opportunistic because they can respond to rainfall events and present extra feed (Barrett-Lennard and Norman 2009). Producers who adopt these shrubs therefore reduce the risk inherent to both the temporal and quantitative variability in rainfall, because the shrubs also offer valuable forage in early winter when pasture growth is slow (especially in years where there is a late start to the rainy season). Recent studies have shown that a mix of Australian native shrubs (e.g., *Rhagodia preissii* Moq., *Rhagodia candolleana* Moq., *Acacia ligulata* A. Cunn. ex Benth., *Atriplex nummularia* Lindl., *Atriplex amnicola* Paul G.Wilson, *Eremophila glabra* (R.Br) Ostenf., *Maireana brevifolia* (R. Br.) Paul G.Wilson, *Kennedia prostrata* R. Br., *Enchylaena tomentosa* R. Br., *Maireana georgei* (Diels) Paul G.Wilson) planted in rows with 'traditional' pasture between the rows, can sustain livestock production at times when producers would normally need to provide expensive supplements (Lund et al., unpublished; Figs. 2

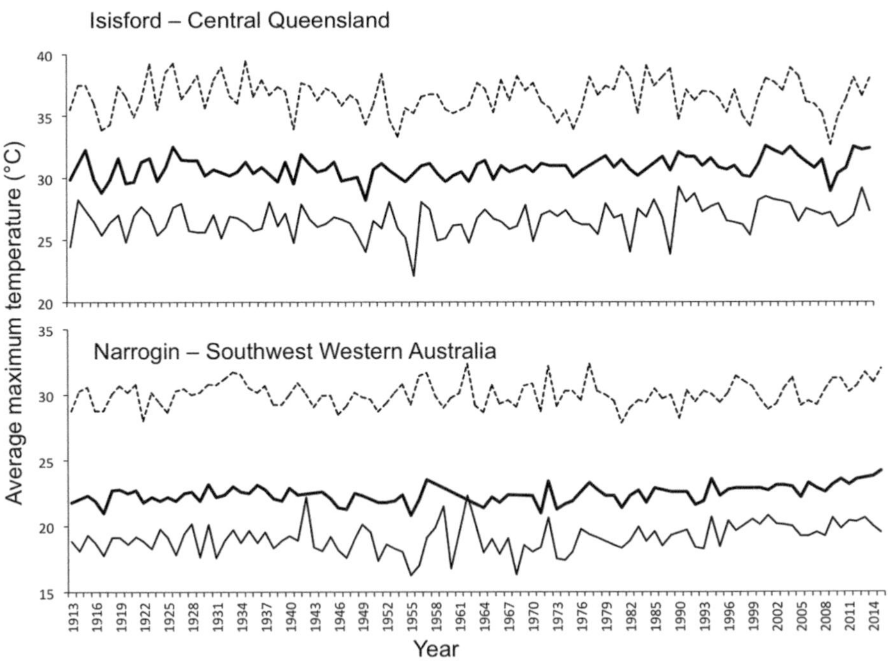

Fig. 2 Illustration of the unpredictability and variability of annual maximum temperature (bold solid line) and maximum temperature in the Australian spring (March to May; thin solid line) and summer (June to August; dashed line) in central Queensland (Isisford Post Office: −24.26°S, 144.44°E, 203 m) and in southwest Western Australian (Narrogin: −32.93°S, 117.18°E, 338 m) from 1913 to 2015 (Data from Australian Bureau of Meteorology (http://www.bom.gov.au))

Fig. 3 Various scenarios illustrating the risks to landscapes in grazing systems, with and without shrubs. (**a**) Effect of an out-of-season (summer) rainfall event in the northern agricultural region of western Australia (photo courtesy of D.K. Revell); (**b**) and (**c**) typical scenarios of overgrazed paddocks during the 'feed gap' in summer and autumn in south western Australia (photos courtesy of P.E. Vercoe and D.K. Revell); (**d**) and (**e**) examples of the 'Enrich' system on two farms in Pingelly, Western Australia, where a mixture of native shrub species has been incorporated into the grazing system. Plate '**d**' was taken in autumn, 8 months after the perennial shrubs had been planted as seedlings (photo courtesy of D.K. Revell) and plate 'e' (Pingelly, Western Australia) was taken in winter (photo courtesy of P.E. Vercoe)

and 3). This is because, when ruminants are offered a diversity of plants, they increase their daily intake and their productivity rises (Baumont et al. 2000). Thus, in Australia, it has been demonstrated that the live weight of sheep is greater if they are offered a diversity of plants rather than a monoculture (Fig. 4).

Complex silvopastoral systems can also improve animal health and welfare (Broom et al. 2013) because, in a complex landscape, feeding behaviour is more rewarding with the animals eating more as well as eating a variety of plants of different flavours (Provenza et al. 1996; Scott and Provenza 1998). Moreover, native shrubs, being wild and unbred, contain an array of natural plant secondary compounds (Tables 2 and 3). Such compounds are typically removed by plant breeding because they can, for example, be responsible for poor palatability and thus intake. However, some of the bioactive secondary compounds can be most valuable – they have antimicrobial and anthelmintic properties, reduce methane emissions, and improve the efficiency of rumen fermentation (Kotze et al. 2009; Vercoe et al. 2009; Durmic et al. 2010; Li 2013). They also have the potential to reduce the risk of lactic acidosis (Hutton et al. 2010) and to modify rumen biohydrogenation, thus driving the fatty acid profiles in meat and milk towards healthier products (Wallace 2004, Durmic et al. 2008).

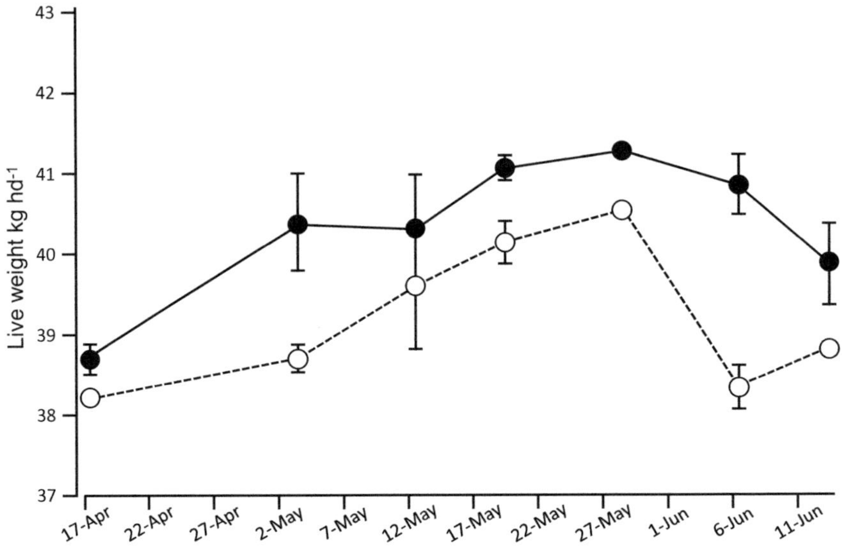

Fig. 4 Live weight change of sheep grazing two shrub-based forage systems during autumn in Monarto, South Australia. There is often a distinct feed gap at this time of year if the pasture consists of annual plant species but, with the inclusion of perennial shrubs, sheep gained weight for about 6 weeks, until feed on offer limited performance. The dashed line represents the growth curve for sheep grazing saltbush (*Atriplex nummularia*) in combination with senesced annual pasture, whereas the solid line represents the growth curve of sheep grazing a mixture of 20 shrub species along senesced annual pasture (Data from J. Emms, South Australia)

Table 2 Examples of shrub species with anthelmintic properties. Larval development score is an *in vitro* method for screening inhibitory effects of compounds on worm parasites

Botanical name and taxa author	Common name	Larval development score (% of control)
Viminaria juncea (Schrad.) Hoffmanns.	Swishbush, golden spray	<1
Eremophila maculata ((Ker Gawel.) F.Muell.	Native fuschia	2
Chenopodium nitrariaceum (F.Muell.) Benth	Nitre goosefoot	3
Acacia pycnantha Benth.	Golden wattle	6
Acacia loderi Maiden	Broken Hill gidgee, nelia	9
Acacia saligna (Labill) H.L.Wendl.	Golden wreath wattle	10
Eremophila longifolia (R. Br.) F.Muell.	Emu bush, Berrigan	10
Medicago sativa L.	Lucerne	10
Kennedia eximia Lindl.	Red coral vine	12
Kennedia prorepens (F. Muell.) F. Muell	Purple flowered pea vine	29
Rhagodia candolleana Moq.	Sea berry saltbush	42
Enchylaena tomentosa R. Br.	Barrier saltbush, ruby saltbush	49

Kotze et al. (2009)

Table 3 Examples of shrub species with anti-methanogenic properties measured using *in vitro* batch culture gas production

Botanical name and taxa author	Common name	Methane (% of control[a])
Acacia loderi Maiden	Broken Hill gidgee, nelia	18
Lotus australis Andrews	Austral trefoil	28
Acacia pycnantha (Benth.)	Golden wattle	28
Acacia saligna (Labill.) H.L. Wendl.	Golden wreath wattle	31
Eremophila glabra (R. Br.) Ostenf.	Tar bush	36
Brachychiton gregorii F. Muell.	Desert kurrajongs	38
Atalaya hemiglauca (F. Muell.) F. Muell. Ex Benth.	Whitewood, cattle bush	38
Acacia neriifolia A. Cunn. ex Benth.	Oleander wattle	38
Acacia iteaphylla F. Muell. ex Benth.	Flinders Range wattle	38
Maireana convexa Paul G. Wilson	Mulga bluebush	47
Kennedia eximia Lindl.	Red coral vine	56
Maireana brevifolia (R. Br.) Paul G. Wilson	Small leaf bluebush, yanga bush	56
Rhagodia candolleana Moq.	Sea berry saltbush	58
Kennedia prorepens (F. Muell.) F. Muell.	Purple flowered pea vine	58
Acacia ligulata A. Cunn. ex Benth.	Umbrella bush, sandhill wattle	62
Chenopodium nitrariaceum (F. Muell.) Benth.	Nitre goosefoot	62
Rhagodia preissii Moq.	Mallee saltbush	71
Chameacytisus palmensis (H. Christ) F.A. Bisby & K.W. Nicholls	Tagasaste	71
Enchylaena tomentosa R. Br.	Barrier saltbush, ruby saltbush	73

[a]Control was oaten chaff

The potential contribution of plant secondary compounds to the control of intestinal parasites, thus reducing our reliance on chemical drenches, is based on evidence *in vitro* that they inhibit the development and motility of the larvae of key parasitic nematodes (Kotze et al. 2009; Kotze et al. 2011). Moreover, livestock grazing shrub-based systems are less likely to ingest parasite larvae because they spend less time grazing near the soil surface when foraging.

Silvopastoral systems can mitigate the risk of stress caused by heat and cold, thus improving productivity (Tucker et al. 2008), because they provide shade and shelter, an essential part of any integrative strategy. The production costs of being outside the zone of thermal comfort arise through multiple mechanisms, all of which effectively increase the maintenance requirements of an animal. A greater portion of ingested energy is spent to maintain body temperature when the ambient temperature is below or above an animal's zone of thermal comfort – shivering at low temperatures or panting at high temperatures. Other production costs are caused by changes in feed intake. Below the animal's lower critical temperature, feed intake increases to generate heat associated with digestion and metabolism (the heat increment of eating) and, at high temperatures, feed intake declines to avoid adding to the body's heat load. Managers of grazing livestock cannot influence climatic conditions but they can modify the microclimate in which animals graze through the provision of shade and shelter, and the benefits to production can be significant. For example, the combination of 30 km/hour wind and an overnight temperature of 0°C can double the maintenance requirement of a recently shorn sheep. Even a sheep with a 30 mm long fleece on a calm night that drops to 5°C will have a maintenance requirement about one-third higher than if it had remained in its zone of thermal comfort. At high temperatures, animals exhibiting rapid, shallow breathing with have an elevated maintenance requirement of about 7 % and, with open-mouthed panting, their maintenance requirement will be elevated by about 25 % at that time. Maintaining animals in their thermal comfort zone is analogous to meeting their nutrient requirements. Either side of meeting requirements, there is a subclinical cost to production and, if the deviations increase further, clinical signs will be apparent. In tropical regions, milk production and profit margin are increased when shade is available (Yamamoto et al. 2007; Murgueitio et al. 2011) and, in semi-arid regions, lamb survival should be improved if shelter were offered since shorn ewes are known to actively seeking shelter at lambing time (Lynch et al. 1980; Mottershead et al. 1982).

In addition, native plants, shrubs and trees benefit natural resource management by reducing the environmental risks of land degradation, ecosystem degradation and carbon emissions that are often a consequence of inappropriate systems and management of grazing livestock. Shrubs can i) protect the companion pasture species from excessive exposure to solar radiation as a secondary mitigation of risk of feed availability and ii) improve soil fertility (Power et al. 2003). The extensive root system of shrubs coupled with their perenniality give them the ability to reduce the risk of dryland salinity by stabilizing the level of the water table and decreasing the risk of desertification, land erosion and salinization of the land (Bienes et al. 2016). Shrubs, especially legume shrubs, also reduce the risk of damage to the ecosystem caused by grazing livestock and can even help the restoration of degraded ecosystems (Emms et al. 2005; Smith et al. 2013). By protecting the topsoil from erosion and compaction (Bienes et al. 2016), as seen in simpler annual pasture systems, shrubs increase the diversity of arthropods, reptiles and birds (Collard and Fisher 2010; Brown et al. 2011; Liu et al. 2016). In addition, forage shrub species host beneficial predatory insects offering potential advantages for integrated pest management in nearby crops and pastures (Taverner et al. 2006).

Shrubs and trees can also improve atmospheric carbon sequestration (Schoeneberger 2009) and mitigate greenhouse gas emissions (Monjardino et al. 2010). For example, *E. glabra* (R. Br.) Ostenf. can decrease the population of methanogens in the rumen and thus reduce methane emissions from grazing livestock (Li et al. 2014). Incorporating shrubs with antimethanogenic effects, such as *E. glabra* (R. Br.) Ostenf., into the pastoral landscape, could have the dual benefits of improving production efficiency and reducing methane emissions, reducing the risk of production of livestock in dryland regions.

Overall, including shrubs in a dryland environment, in either extensive or semi-intensive production systems, will reduce several risks to animal productivity, the environment, and economic outcomes. Bioeconomic modelling has indicated that converting 5–20 % of a mixed farm to shrub-based systems could increase whole-farm profit up to 20 % (Monjardino et al. 2010). Profit is increased by reductions in supplementary feeding and in the grazing of regenerating annual pastures, leading to an increase feed availability throughout the year; moreover, shrubs enable livestock production on a greater area of low productivity land (Monjardino et al. 2010). Investigations are required to determine optimal silvopastoral systems for the world's diverse dryland environments.

5 Managing the Relationship Between Animals and Environment

Compared to the tropics, silvopastoral systems in dryland areas have lower stocking densities and cover a larger area of land, so herds or flocks need to move around more to take advantage of all the food on offer. There are only a few ways to move ruminants over areas of land, the oldest being the skilled shepherd or herdsman. In France, there has been research into the efficacy, importance and complexity of the human-animal interaction in pastoral areas (Meuret and Provenza 2015b, Provenza et al. 2015). Some high-tech solutions have been developed, from virtual fencing technology to location of livestock using Unmanned Aerial Vehicles that allow the remote control of animals (Umstatter 2011, Freeman and Freeland 2015). These technologies seem to be reliable, but expensive, and the virtual fencing technology raises animal welfare questions since the animals are often driven using negative reinforcements (Umstatter et al. 2015). A more promising approach is to make use of natural animal behaviour and leading animals to create a self-managed flock, leading to strategies that positively influence grazing distribution of livestock (Revell et al. 2015). Current projects such as 'Rangelands Self Herding' or 'Rangelands Self Shepherding' (www.selfherding.com) will deliver a method that allows livestock to develop new experiences that consequently shape their grazing range and utilisation of water points. Since feeding behaviour is the result of a complex interaction between reward systems and metabolic status (Ginane et al. 2015), it seems possible that animals can learn about the full array of nutritional and thermal environments in their range if the experience is rewarding. The capacity of animals to learn new feeding behaviours and to adapt to changing circumstances, and the differences in behaviour between experienced and naïve animals when faced with a new feedstuff, shows that ruminants can learn to select one plant from another.

6 Conclusions

The production of ruminants in dryland areas is expected to play an important role in responses to the demands for animal protein production over the coming decades. However, ruminant production in dryland areas is a risky business. There are solutions to manage the risk, but so far they are specific solutions for specific factors causing risk rather than an integrated and innovative approach that simultaneously manages and reduces all of the risk factors associated with the uncertainties of animal production in dryland. To do this requires versatility. An innovative solution to sustainable livestock production in rangelands has arisen through an integrative approach, combining shrubs and native plants in silvopastoral systems with strategies to promote self-herding (or traditional herding) and for careful selection of animal genotypes. This approach would need to be refined for each system and location, underpinned by sound principles, to ensure that the interaction between genotype, environment and management is optimised to maximise productivity and minimise the impact on the environment. However, and perhaps the most essential need, rarely mentioned, is multidisciplinary teams of people –combinations of scientists, livestock producers and communities, with all parties focussed on addressing questions specific for a given environment but with an understanding of the need for production systems.

References

Agreil C, Fritz H, Meuret M (2005) Maintenance of daily intake through bite mass diversity adjustment in sheep grazing on heterogeneous and variable vegetation. Appl Anim Behav Sci 91:35–56

Anderson JR, Dillon JL (1992) Risk analysis in dryland farming systems. FAO, Rome

Assefa G, Kijora C, Kehaliew A, Bediye S, Peters KJ (2008) Evaluation of tagasaste (Chamaecytisus palmensis) forage as a substitute for concentrate in diets of sheep. Livest Sci 114:296–304

Balvanera P, Uriarte M, Almeida-Leñero L, Altesor A, DeClerck F, Gardner T, Hall J, Lara A, Laterra P, Peña-Claros M, Silva Matos DM, Vogl AL, Romero-Duque LP, Arreola LF, Caro-Borrero ÁP, Gallego F, Jain M, Little C, de Oliveira XR, Paruelo JM, Peinado JE, Poorter L, Ascarrunz N, Correa F, Cunha-Santino MB, Hernández-Sánchez AP, Vallejos M (2012) Ecosystem services research in Latin America: The state of the art. Ecosyst Serv 2:56–70

Barrett-Lennard EG, Norman HC (2009) Saltbush for forage production on saltland. In Nuberg I, George B, Reid R (eds) Agroforestry for natural resource management, pp 239–250

Baumont R, Prache S, Meuret M, Morand-Fehr P (2000) How forage characteristics influence behaviour and intake in small ruminants: A review. Livest Prod Sci 64:15–28

Becholie D, Tamir B, Terrill TH, Singh BP, Kassa H (2005) Suitability of tagasaste (Chamaecytisus palmensis L.) as a source of protein supplement to a tropical grass hay fed to lambs. Small Rumin Res 56:55–64

Ben Salem H, Nefzaoui A (2003) Feed blocks as alternative supplements for sheep and goats. Small Rumin Res 49:275–288

Ben Salem H, Nefzaoui A, Ben Salem L (2004) Spineless cactus (Opuntia ficus indica f. inermis) and oldman saltbush (Atriplex nummularia L.) as alternative supplements for growing Barbarine lambs given straw-based diets. Small Rumin Res 51:65–73

Ben Salem H, Norman HC, Nefzaoui A, Mayberry DE, Pearce KL, Revell DK (2010) Potential use of oldman saltbush (Atriplex nummularia Lindl.) in sheep and goat feeding. Small Rumin Res 91:13–28

Bervejillo JE (2015) Uruguay's beef industry. In: Norwegian Institute of International Affairs, Oslo, Norway, p 71

Bienes R, Marques MJ, Sastre B, García-Díaz A, Ruiz-Colmenero M (2016) Eleven years after shrub revegetation in semiarid eroded soils. Influence in soil properties. Geoderma 273:106–114

Blache D, Maloney SK, Revell D (2008) Use and limitations of alternative feed resources to sustain and improve reproductive performance in sheep and goats. Anim Feed Sci Technol 147:140–157

Blache D, Chagas LM, Blackberry MA, Vercoe PE, Martin GB (2000) Metabolic factors affecting the reproductive axis in male sheep. J Reprod Fertil 120:1–11

Blache D, Grandison MJ, Masters DG, Dynes RA, Blackberry MA, Martin GB (2007) Relationships between metabolic endocrine systems and voluntary feed intake in Merino sheep fed a high salt diet. Aust J Agric Res 47:544–550

Boland MJ, Rae AN, Vereijken JM, Meuwissen MPM, Fischer ARH, van Boekel MAJS, Rutherfurd SM, Gruppen H, Moughan PJ, Hendriks WH (2013) The future supply of animal-derived protein for human consumption. Trends Food Sci Technol 29:62–73

Borens FMP, Poppi DP (1990) The nutritive value for ruminants of tagasaste (Chamaecytisus palmensis), a leguminous tree. Anim Feed Sci Technol 28:275–292

Broom DM, Galindo FA, Murgueitio E (2013) Sustainable, efficient livestock production with high biodiversity and good welfare for animals. Proc R Soc B Biol Sci 280:20132025

Brown GW, Dorrough JW, Ramsey DSL (2011) Landscape and local influences on patterns of reptile occurrence in grazed temperate woodlands of southern Australia. Land Urban Plan 103:277–288

Collard SJ, Fisher AM (2010) Shrub-based plantings of woody perennial vegetation in temperate Australian agricultural landscapes: What benefits for native biodiversity? Ecol Manag Restor 11:31–35

Dalzell D, Shelton H, Mullen B, Larsen P, McLaughlin K (2006) Leucaena: a guide to establishment and management. Meat and Livestock Australia Ltd, Sydney

Doreau M, Makkar HPS, Lecomte P (2013) The contribution of animal production to agricultural sustainability. In: Oltjen JW, Kebreab E, Lapierre H (eds) Energy and protein metabolism and nutrition in sustainable animal production. Springer, Berlin, pp. 475–485

Dunin FX, Williams J, Verburg K, Keating BA (1999) Can agricultural management emulate natural ecosystems in recharge control in south eastern Australia? Agrofor Syst 45:343–364

Durmic Z, McSweeney CS, Kemp GW, Hutton P, Wallace RJ, Vercoe PE (2008) Australian plants with potential to inhibit bacteria and processes involved in ruminal biohydrogenation of fatty acids. Anim Feed Sci Technol 145:271–284

Durmic Z, Hutton P, Revell DK, Emms J, Hughes S, Vercoe PE (2010) In vitro fermentative traits of Australian woody perennial plant species that may be considered as potential sources of feed for grazing ruminants. Anim Feed Sci and Technol160: 98–109

Eisler MC, Lee MRF, Tarlton JF, Martin GB, Beddington J, Dungait JAJ, Greathead H, Liu J, Mathew S, Miller H, Misselbrook T, Murray P, Vinod VK, van Saun R, Winter M (2014) Steps to sustainable livestock. Nature (London) 507:32–34

Emms J, Virtue JG, Preston C, Bellotti WD (2005) Legumes in temperate Australia: a survey of naturalisation and impact in natural ecosystems. Biol Conserv 125:323–333

Fancote CR, Norman HC, Williams IH, Masters DG (2009) Cattle performed as well as sheep when grazing a river saltbush (Atriplex amnicola) based pasture. Anim Prod Sci 49:998–1006

FAO (2003) World agriculture: towards 2015/2030: an FAO perspective. FAO, Rome

FAO (2006) Livestock's long shadow: environmental issues and options. FAO, Rome

Forster P, Ramaswamy V, Artaxo P, Berntsen T, Betts R, Fahey DW, Haywood J, Lean J, Lowe DC, Myhre G, Nganga J, Prinn R, Raga G, Schulz M, Van Dorland R (2007) Changes in atmospheric constituents and in radiative forcing. In: Solomon S, Qin D, Manning M, Chen Z, Marquis M, Averyt, KB, Tignor M, Miller HL (eds) Climate change 2007: The physical science basis. Contribution of working group I to the fourth assessment report of the Intergovernmental Panel on Climate Change. Cambridge University Press, Cambridge/New York

Freeman PK, Freeland RS (2015) Agricultural UAVs in the U.S.: potential, policy, and hype. Remote Sens Appl Soc Environ 2:35–43

Fuquay JW (1981) Heat stress as it affects animal production. J Anim Sci 52:164–174

Garnett T (2009) Livestock-related greenhouse gas emissions: impacts and options for policy makers. Environ Sci Pol 12:491–503

Ginane C, Bonnet M, Baumont R, Revell D (2015) Feeding behaviour in ruminants: a consequence of interactions between a reward system and the regulation of metabolic homeostasis. Anim Prod Sci 55:247–260

Guerra CA, Pinto-Correia T (2016) Linking farm management and ecosystem service provision: challenges and opportunities for soil erosion prevention in Mediterranean silvo-pastoral systems. Land Use Policy 51:54–65

Hall JS, Ashton MS, Garen EJ, Jose S (2011) The ecology and ecosystem services of native trees: Implications for reforestation and land restoration in Mesoamerica. Forest Ecol Manag 261:1553–1557

Hardaker JB, Lien G, Anderson JR (2015) Coping with risk in agriculture: applied decision analysis. CABI, Wageningen

Herrero M, Thornton PK, Notenbaert AM, Wood S, Msangi S, Freeman HA, Bossio D, Dixon J, Peters M, van de Steeg J, Lynam J, Rao PP, Macmillan S, Gerard B, McDermott J, Seré C, Rosegrant M (2010) Smart investments in sustainable food production: revisiting mixed crop-livestock systems. Science 327:822–825

Hill MJ, Donald GE, Hyder MW, Smith RCG (2004) Estimation of pasture growth rate in the south west of Western Australia from AVHRR NDVI and climate data. Remote Sens Environ 93:528–545

Hobbs RJ, Morton SR (1999) Moving from descriptive to predictive ecology. Agrofor Syst 45:43–55

Hoekstra AY, Mekonnen MM (2012) The water footprint of humanity. Proc Natl Acad Sci 109:3232–3237

Hoffmann I (2011) Livestock biodiversity and sustainability. Livest Sci 139:69–79

Hogan A (1982) Digestion and utilization of proteins. In Nutritional limits to animal production from pastures (Ed. J Hacker), pp 245–256. Farnham Royal: Commonwealth Agricultural Bureaux, St. Lucia, Queensland, Australia, August 24th–28th

Holechek JL (1997) The effects of rangeland water developments on livestock production and distribution. Proceedings of a Symposium on Environmental, Economic, and Legal Issues Related to Rangeland Water Development. ASU College of Law, Tempe, Arizona, USA, 13–15 November 1997

Hopkins DL, Nicholson A (1999) Meat quality of wether lambs grazed on saltbush (*Atriplex nummularia)* plus supplements or lucerne (*Medicago sativa*). Meat Sci 51:91–95

Hutton PG, Durmic Z, Vercoe PE (2010) Investigating Eremophila glabra as a bioactive agent for preventing lactic acidosis in sheep. Anim Prod Sci 50:449–453

Johnston WH, Garden DL, Rančić A, Koen TB, Dassanayake KB, Langford CM, Ellis NJS, Rab MA, Tuteja NK, Mitchell M, Wadsworth J, Dight D, Holbrook K, LeLievre R, McGeoch SM (2003) The impact of pasture development and grazing on water-yielding catchments in the Murray-Darling Basin in south-eastern Australia. Aust J Exp Agric 43:817–841

Kotze A, O'Grady J, Emms J, Toovey A, Hughes S, Jessop P, Bennell M, Vercoe P, Revell D (2009) Exploring the anthelmintic properties of Australian native shrubs with respect to their potential role in livestock grazing systems. Parasitology 136:1065–1080

Kotze A, Zadow E, Vercoe P, Phillips N, Toovey A, Williams A, Ruffell A, Dinsdale A, Revell D (2011) Animal grazing selectivity and plant chemistry issues impact on the potential of Rhagodia preissii as an anthelmintic shrub. Parasitology 138:628–637

Kratli S, Schareika N (2010) Living off uncertainty: the intelligent animal production of dryland pastoralists. Eur J Dev Res 22:605–622
Li X (2013) Eremophila glabra reduces methane production in sheep. PhD Thesis. University of Western Australia. pp 113
Li X, Durmic Z, Liu S, McSweeney CS, Vercoe PE (2014) *Eremophila glabra* reduces methane production and methanogen populations when fermented in a Rusitec. Anaerobe 29:100–107
Liu R, Zhu F, Steinberger Y (2016) Changes in ground-dwelling arthropod diversity related to the proximity of shrub cover in a desertified system. J Arid Environ 124:172–179
Lynch JJ, Mottershead BE, Alexander G (1980) Sheltering behaviour and lamb mortality amongst shorn Merino ewes lambing in paddocks with a restricted area of shelter or no shelter. Appl Anim Ethol 6:163–174
MacFarlane WV (1968) Adaptation of ruminants to tropics and deserts. In: Hafez ESE (ed) Adaptation of domestic animals. Lea & Febiger, Philadelphia
Martin GB, Blache D, Williams IH (2008) Allocation of resources to reproduction. In: Rauw W (ed) Resource allocation theory applied to farm animal production. Oxford University Press, Oxford, pp. 169–191
Masters D, Edwards N, Sillence M, Avery A, Revell D, Friend M, Sanford P, Saul G, Berverly C, Young J (2006) The role of livestock in the management of dryland salinity. Aust J Exp Agric 46:733–741
Masters D, Tiong M, Vercoe P, Norman H (2010) The nutritive value of river saltbush (Atriplex amnicola) when grown in different concentrations of sodium chloride irrigation solution. Small Rumin Res 91:56–62
McKeon GM, Stone GS, Syktus JI, Carter JO, Flood NR, Ahrens DG, Bruget DN, Chilcott CR, Cobon DH, Cowley RA, Crimp SJ, Fraser GW, Howden SM, Johnston PW, Ryan JG, Stokes CJ, Day KA (2009) Climate change impacts on northern Australian rangeland livestock carrying capacity: a review of issues. Rangel J 31:1–29
Meuret M, Provenza F (2015a) How French shepherds create meal sequences to stimulate intake and optimise use of forage diversity on rangeland. Anim Prod Sci 55:309–318
Meuret M, Provenza FD (2015b) When art and science meet: Integrating knowledge of French herders with science of foraging behavior. Rangel Ecol Manag 68:1–17
Michalk DL, Dowling PM, Kemp DR, King WM, Packer IJ, Holst PJ, Jones RE, Priest SM, Millar GD, Brisbane S, Stanley DF (2003) Sustainable grazing systems for the Central Tablelands, New South Wales. Aust J Exp Agric 43:861–874
Monjardino M, Revell D, Pannell DJ (2010) The potential contribution of forage shrubs to economic returns and environmental management in Australian dryland agricultural systems. Agric Syst 103:187–197
Moore AD, Bell LW, Revell DK (2009) Feed gaps in mixed farming systems: insights from the Grain and Graze program. Anim Prod Sci 49:739–748
Mottershead BE, Alexander G, Lynch JJ (1982) Sheltering behaviour of shorn and unshorn sheep in mixed or separate flocks. Appl Anim Ethol 8:127–136
Murgueitio E, Ibrahim M (2008) Ganadería y medio ambiente en América Latina. In: Murgueitio E, Cuartas C, Naranjo JF (eds) Ganadería del Futuro. Fundación CIPAV, Cali, pp. 19–39
Murgueitio E, Calle Z, Uribe F, Calle A, Solorio B (2011) Native trees and shrubs for the productive rehabilitation of tropical cattle ranching lands. Forest Ecol Manag 261:1654–1663
Nardone A, Ronchi B, Lacetera N, Ranieri MS, Bernabucci U (2010) Effects of climate changes on animal production and sustainability of livestock systems. Livest Sci 130:57–69
Norman HC, Dynes RA, Rintoul AJ, Wilmot MG, Masters DG (2004) Sheep production from saline land - productivity from old man and river saltbush and the value of grain and straw supplements. Proc Aust Soc Anim Prod 25:289–289
O'Reagain PJ, Scanlan JC (2013) Sustainable management for rangelands in a variable climate: evidence and insights from northern Australia. Animal 7:68–78
Otte J, Costales A, Dijkman J, Pica-Ciamarra U, Robinson T, Ahuja V, Ly C, Roland-Holst D (2012) Livestock sector development for poverty reduction: an economic and policy perspective. FAO, Rome

Paciullo DSC, de Castro CRT, Gomide CAM, Maurício RM, Pires MFÁ, Müller MD, Xavier DF (2011) Performance of dairy heifers in a silvopastoral system. Livest Sci 141:166–172
Pahl L, Scanlan J, Whish G, Cowley R, MacLeod N (2016) Comparing fixed and flexible stocking as adaptations to inter-annual rainfall variability in the extensive beef industry of northern Australia. Rangel J 38:85–102
Pannell DJ (2001) Dryland salinity: economic, scientific, social and policy dimensions. Aust J Agric Res Econ 45:517–546
Power IL, Thorrold BS, Balks MR (2003) Soil properties and nitrogen availability in silvopastoral plantings of Acacia melanoxylon in North Island, New Zealand. Agrofor Syst 57:225–237
Provenza FD (2008) What does it mean to be locally adapted and who cares anyway? J Anim Sci 86:E271–E284
Provenza FD, Scott CB, Phy TS, Lynch JJ (1996) Preference of sheep for foods varying in flavors and nutrients. J Anim Sci 74:2355–2361
Provenza FD, Meuret M, Gregorini P (2015) Our landscapes, our livestock, ourselves: Restoring broken linkages among plants, herbivores, and humans with diets that nourish and satiate. Appetite 95:500–519
Ramirez-Restrepo CA, Barry TN, Lopez-Villalobos N, Kemp PD, Harvey TG (2005) Use of Lotus corniculatus containing condensed tannins to increase reproductive efficiency in ewes under commercial dryland farming conditions. Anim Feed Sci Technol 121:23–43
Ran Y, Lannerstad M, Herrero M, Van Middelaar CE, De Boer IJM (2016) Assessing water resource use in livestock production: A review of methods. Livest Sci 187:68–79
Revell D, Durmic Z, Bennell M, Sweeney GC, Vercoe P (2008) The in situ use of plant mixtures including native shrubs in Australian grazing systems: the potential to capitalise on plant diversity for livestock health and productivity. In: Skaife JF, Vercoe P (eds) Harvesting Knowledge, Pharming Opportunities. Cambridge University Press, Cambridge, pp. 36–49
Revell DK (2016) Adaptive responses of rangeland livestock to manage water balance. In: Villalba JJ, Manteca X (eds) Animal welfare in extensive systems. 5m Publishing, Sheffield
Revell DK, Sweeney GC (2004) Aligning profitable grazing systems with reduced water recharge in southern Australia; matching plants, animal grazing behaviour and the environment in mixed forage systems. In Salinity Solutions: Working with Science and Society Eds. A Ridley, P Feikema, S Bennet, M Rogers, R Wilkinson and J Hirth), pp CD-ROM. CRC for Plant-Based Management of Dryland Salinity, Perth, Bendigo, Victoria, 2–5 August 2004
Revell DK, Norman HC, Vercoe PE, Phillips N, Toovey A, Bickell S, Hulm E, Hughes S, Emms J (2013) Australian perennial shrub species add value to the feed base of grazing livestock in low- to medium-rainfall zones. Anim Prod Sci 53:1221–1230
Revell DK, Maynard B, Erkelenz P.A., Thomas D (2015) 'Rangeland Self Herding' – positively influencing grazing distribution to benefit livestock, landscapes and people. In: Proceedings of the Australian Rangeland Society Conference (Alice Springs, 12-17 April 2015)
Sanderson MA, Skinner RH, Barker DJ, Edwards GR (2004) Plant species diversity and management of temperate forage and grazing land ecosystems. Crop Sci 44:1132–1144
Scanlan J, McKeon G, Day KA, Mott JJ, Hinton AW (1994) Estimating safe carrying capacities in extensive cattle grazing properties within tropical semiarid woodlands of north-eastern Australia. Rangel J 16:64–76
Scherf B, Rischkowsky B, Hoffmann I, Wieczorek M, Montironi A, Cardellino R (2008) Livestock genetic diversity in dry rangelands. In: Lee C, Schaaf T (eds) The future of drylands. Springer, Berlin, pp. 89–400
Schoeneberger MM (2009) Agroforestry: working trees for sequestering carbon on agricultural lands. Agrofor Syst 75:27–37
Scott LL, Provenza FD (1998) Variety of foods and flavors affects selection of foraging location by sheep. Appl Anim Behav Sci 61:113–122
Silanikove N (1994) The struggle to maintain hydration and osmoregulation in animals experiencing severe dehydration and rapid rehydration: the story of ruminants. Exp Physiol 79:281–300

Silanikove N (2000) Effects of heat stress on the welfare of extensively managed domestic animals. Livest Prod Sci 67:1–18

Smith FP, Prober SM, House APN, McIntyre S (2013) Maximizing retention of native biodiversity in Australian agricultural landscapes—The 10:20:40:30 guidelines. Agric Ecosyst Environ 166:35–45

Squires V (1981) Livestock management in the arid zone. Inkata Press, Melbourne

Taverner P, Wood G, Jevremov D, Doyle B (2006) 'Revegetation by design' guidebook: a guide to using selected native plants to reduce pests and diseases in the horticulture region of the northern Adelaide Plains. SARDI, Adelaide

Tucker CB, Rogers AR, Schütz KE (2008) Effect of solar radiation on dairy cattle behaviour, use of shade and body temperature in a pasture-based system. Appl Anim Behav Sci 109:141–154

Turner SP, Dwyer CM (2007) Welfare assessment in extensive animal production systems: challenges and opportunities. Anim Welf 16:189–192

Umstatter C (2011) The evolution of virtual fences: A review. Comput Electron Agric 75:10–22

Umstatter C, Morgan-Davies J, Waterhouse T (2015) Cattle responses to a type of virtual fence. Rangel Ecol Manag 68:100–107

UN (1977) Status of dersertification in the arid regions, Climatic aridity index map and experimental world scheme of aridity and drought probability, at a scale of 1:25,000,000. Explanatory note. In UN Conference on desertification A/CONF. 74/31, New York

Vercoe PE, Durmic Z, Revell D (2009) Rumen microbial ecology: Helping to change landscapes. Options Méditérr 85:225–236

Wallace R (2004) Antimicrobial properties of plant secondary metabolites. Proc Nutr Soc 63:621–629

Willmer P, Stone G, Johnston I (2009) Environmental physiology of animals. Blackwell Publishing, Oxford

Winchester CF, Morris MJ (1956) Water Intake Rates of Cattle. J Anim Sci 15:722–740

Yamamoto W, Dewi IA, Ibrahim M (2007) Effects of silvopastoral areas on milk production at dual-purpose cattle farms at the semi-humid old agricultural frontier in central Nicaragua. Agric Syst 94:368–375

Part III
Modeling and Crop Improvement for Dryland Agriculture

Modelling Dryland Agricultural Systems

Daniel Rodriguez, Peter de Voil, and B. Power

1 Introduction

The sustainable intensification of dryland agricultural systems in high- and low-income countries faces different, though equally challenging, and complex problems requiring the application of more integrative and trans-disciplinary science approaches (Rodriguez and Sadras, 2011). On one hand, the limited availability of resources (e.g., land, finance, labour) and the lack of access to inputs, product markets and infrastructure constrain the opportunities and incentives smallholder farmers have to change and improve dryland agricultural systems. On the other hand, in high-income countries, our best farmers are reaching the point where further increases in yield become uneconomical, too risky (Sadras and Rodriguez 2010) or inconsistent with environmental outcomes (CSIRO 2007). This is taking place in a world where the number of hungry people reached record levels in 2009. Despite a slight recovery in 2010, malnutrition among the world's poorest remains higher than that when the 1996 World Food Summit agreed to a hunger-reduction target. The medium-term outlook indicates that agricultural output in the coming decade will not match that of the previous decade, i.e. annual growth will fall from 2 % in 1999–2008 to 1.7 % in 2009–2018 (OECD-FAO 2009) while the expected increase in the world population is 40 % by 2050 and climate changes will become increasingly more evident and serious (Parry 2009). Given the above, the challenge is to

---

There is nothing so practical as a good theory. Emanuel Kant (1724–1804)

D. Rodriguez (✉) • P. de Voil
Queensland Alliance for Agriculture and Food Innovation (QAAFI), The University of Queensland, 102, Toowoomba, QLD 4350, Australia
e-mail: d.rodriguez@uq.edu.au

B. Power
Commonwealth Scientific and Industrial Research Organisation, 102, Toowoomba, QLD 4350, Australia

M. Farooq, K.H.M. Siddique (eds.), *Innovations in Dryland Agriculture*, DOI 10.1007/978-3-319-47928-6_9

increase food production by 70 %, or nearly 100 % in low-income countries, by 2050 to cope with the expected increases in population and food demand.

Dryland agricultural systems are the predominant food production system in the world. Of the total global land area of about 13.3 billion ha, 12 % is used for cultivation of agricultural crops, most of which (80 %, or about 1.3 billion ha) is under dryland conditions (FAO 2015). Rainfed agriculture produces about 60 % of global crop output across the highlands, dry tropics, humid tropics, subtropics and temperate regions of the world (FAO 2011).

Gaps in achievable yield are still significant in dryland agricultural systems; present performance is reportedly up to 50 % below (i.e. yield gap) the achievable yield[1], suggesting that there is ample room for improvement (Lobell et al. 2009). A more detailed analysis, including country and crop-specific yield gaps that take into account the constraints of existing cropping systems, can be found at the Global Yield Gap Atlas website, together with a common methodology for its calculation.

High productivity and consequent small yield and production gaps are the result of optimum combinations of crops, cultivars or hybrids a management variable that best fit a particular environment, as well as how the limited resources e.g. labour, land, finances are allocated across enterprises and fields at the whole-farm level (Calviño and Monzon 2009; Rodriguez et al. 2009, 2011; Power et al. 2011). As a result, focusing on individual crop yields is necessary but insufficient for several reasons. Firstly, large improvements in productivity are likely from interventions at scales beyond the crop, the selection of the cultivar or the management of a particular field. Secondly, changes in individual crop yields may not reflect the fact that farmers manage their farms and resources to satisfy competing objectives: livelihoods, returns, lifestyle and environmental or societal outputs, rather than just increasing crop yields. It is only when the analysis is performed at the whole farm level that changes in one enterprise at any point in time will limit options spatially across the farm e.g., due to land, labour or machinery constraints, and temporally across seasons e.g., due to follow-on implications on soil water and nutrient availability, or the need for pest or disease breaks between successive crops (Sadras et al. 2009; Rodriguez et al. 2011; Power et al. 2011). This chapter aims to scale up the discussion and analysis of farming systems research in dryland systems to the level that farmers think and make decisions, this is the whole farm, or the farm business. The boundaries of the system then become the farm property rather than the field, and key resources are the availability of land of different use, type and fertility, the availability of labour, water for irrigation, cash, machinery and management resources and skills. It is at this level of analysis when agricultural systems modelling tools are most useful, to quantify benefits and trade-offs from alternative

[1] The yield 'achieved' from applying optimum agronomic management under rainfed conditions, also called water-limited yield. Water-limited yield can be calculated using crop simulation models assuming optimum or recommended sowing dates, planting densities and cultivars. An important limitation of this concept is that farmers tend to maximize profits from the entire farm business rather than from individual enterprises.

practices, tactics within alternative strategies and competing objectives, in the background of the need for increasing productivity in highly risky climates.

Since the early 1990s, most participatory farming systems research projects have incorporated systems analysis and a systems modelling tool to support benchmarking and practical decision making. After three decades of 'contestable success' (McCown et al. 2009), the role of farming systems models remains strong at quantifying complex systems interactions, providing ex-ante analyses that identify best bet options for intervention, and quantifying benefits and trade-offs from the adoption of technologies particularly when land, labour and capital limit farmer options.

Clearly agricultural systems modelling techniques have played and will continue to play a significant role in adapting the nature and extent of agriculture to fit the challenges and opportunities from the expected increases in food and energy demand, amid changes in climate and the environment (Rodriguez and Sadras, 2011).

2 Modelling Dryland Agricultural Systems

Modelling in dryland agricultural systems research has two major roles in the design of productive and resilient farming systems: (1) to increase knowledge of the functioning and dynamics of complicated, managed natural systems, where biotic processes interact with climatic, soil and biological drivers at a range of temporal and spatial scales, and (2) to inform (and overcome) the complexities in the system, i.e., those that originate as soon as the manager of the natural system is brought into the analysis – the human dimension which has particular objectives, experiences and perceptions, values and aspirations that also need to be met. Here we are mostly concerned with the second more complex aim, but will also acknowledge the rich contribution from the diversity of approaches used, from simple rules-of-thumb to dynamic crop models, and the analysis of complicated whole-farm models.

Models and frameworks developed and applied to address different questions, usually involve contrasting integration, temporal and spatial scales. Models also vary in their degree of detail with regard to how the processes and relationships are described e.g. they may have a single equation or rule or hundreds of lines of code, they may perform a single calculation i.e. static or multiple calculations over a range of time steps i.e. dynamic models, and they may delve into the minutiae and mechanisms of phenomena or have been developed as learning tools to support the non-specialist farmer or farm advisor.

2.1 Simple Rules-of-Thumb (Models) for Benchmarking Farm Practice

Based on the framework developed by the celebrated Prof. deWitt (1958), the most well-known model for informing practice change and benchmarking dryland agriculture is the water use efficiency (WUE) model by French and Schulz (1984) (Fig. 1).

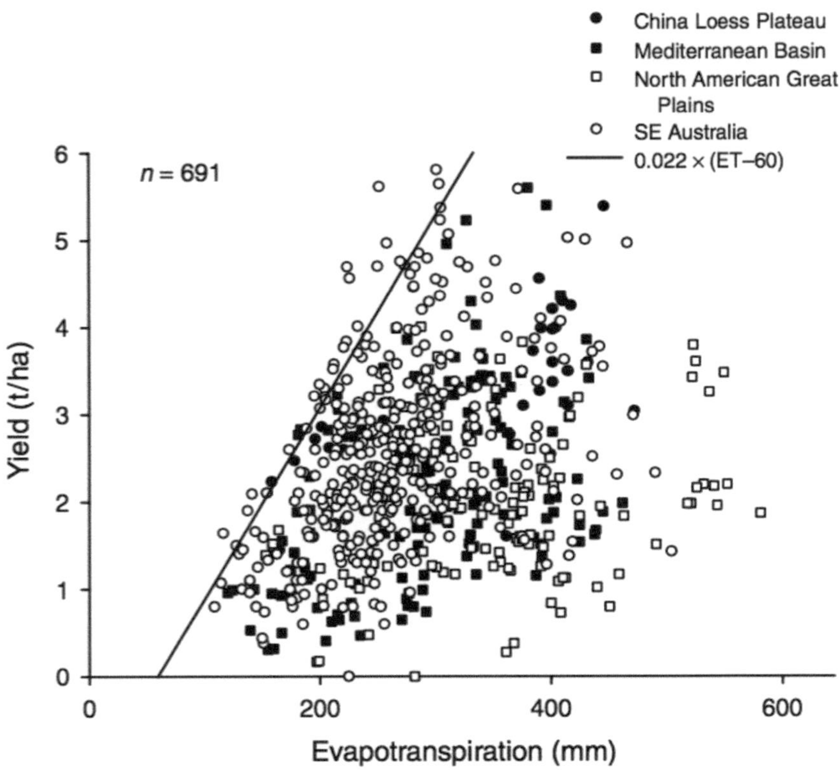

Fig. 1 Scatter plot of grain yield and seasonal evaporation in four mega-environments. The line uses the French and Schultz (1984) frontier concept, with x-intercept = 60 mm and boundary line = 22 kg grain/ha.mm (Source: Sadras and Angus 2006)

The basic principle in this model is that the main determinants of crop production in dryland cropping systems are water availability and water use efficiency (Rodriguez and Sadras, 2007). Figure 1 shows this concept as a single boundary line for the highest value of WUE observed from a range of sites across four continents.

In the dryland cropping systems of Australia, water availability is limited by low and highly-variable rainfall (Nicholls 1986; Stephens and Lyons 1998; Potgieter et al. 2002), often in combination with soils where plant-available water is restricted by physical and chemical constraints to root growth and function (Rengasamy 2002; Nuttall et al. 2003; Sadras et al. 2003; Rodriguez et al. 2006). The influence of latitudinal gradients in vapor pressure deficit and fraction of diffuse radiation on WUE were later recognized by Rodriguez and Sadras (2007) while changes in non-productive evaporative losses in the cropping cycle as a function of latitudinal gradients in the size and structure of rainfall were described by Sadras and Rodriguez (2007). Highlights from that work showed that the size and frequency of rainfall events vary significantly across eastern Australia with important implications for resource availability i.e. water and nitrogen, and for matching crop traits to the environment. In West Asia and Northern Africa, Cooper et al. (1987) promoted management practices favouring rapid canopy growth to reduce soil evaporation and increase transpiration. Currently, the view of breeders in Australia has converged with that of Cooper and his colleagues, to favour plant traits associated with rapid water use thereby reducing soil evaporation in environments with dominant winter rainfall (Richards et al. 1993; Rebetzke and Richards 1999; Richards and Lukacs 2002). In the prevalent climate of south-eastern Australia, saving water through reduced water uptake and slow canopy growth produces little or no benefit, as most of the water is lost through soil evaporation i.e. in environments with a high frequency of small rainfall events (Sadras and Rodriguez 2007). This is not the case in Australia's northern region where winter rainfall is infrequent, and the fraction of soil evaporation relative to total water use is small compared with southern locations (Mitchell et al. 2006; Sadras and Rodriguez 2007). For these reasons, traits favouring slow canopy cover in northern locations and fast canopy cover in southern locations could be adopted as a primary approach to overcome the constraints imposed by rainfall patterns on water use and WUE (Sadras and Rodriguez 2007).

Similar examples from simple models which support the design of dryland agricultural systems in Africa are in Dimes et al. (2015) (Fig. 2). Their work combined

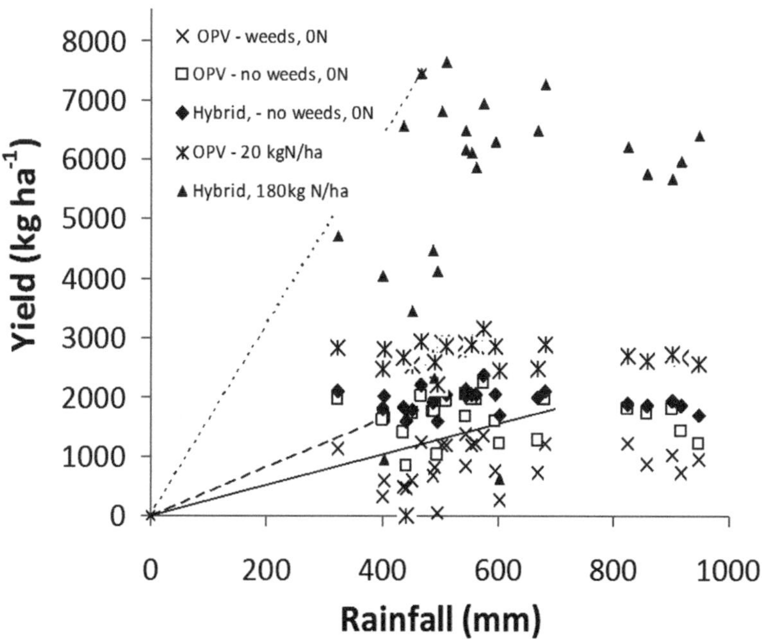

Fig. 2 The relationship between simulated maize yield and seasonal rainfall at Mandela, Tanzania for open pollination varieties and hybrid seed with and without weed competition and varying rates of nitrogen fertiliser (0, 20, and 183 kg N/ha). *Lines* have slopes of 2.5 (*solid*), 4.1 (*dashed*) and 16 (*dotted*) kg/ha.mm (Source: Dimes et al. 2015)

the simple approach from French and Schulz with participatory modelling activities using Agricultural Production Systems Simulator (APSIM) (Holzworth et al. 2014) to design step-wise intensification sequences that increase maize yields in smallholder farming systems. Smallholder agricultural systems from Eastern and Southern Africa are highly vulnerable to biotic and abiotic stresses and price fluctuations, making the intensification of these mostly dryland cropping systems rather challenging. Under these circumstances, increasing the productivity and resource-use efficiency of agriculture is essential to achieve food secured households. Dimes et al. (2015) found that small investments, e.g. low doses of N fertiliser or legume residues, and improved weed control can dramatically increase water use efficiencies and yields by 40–150 % while maintaining yield stability across the seasonal conditions.

These examples show how simple rules-of-thumb or conceptual frameworks such as that from French and Schulz can help to answer questions about how one can manage a crop or improve its genetics to increase yield in dryland cropping systems and water-limited environments (Passioura and Angus 2010). However, the design of farm practices, crop sequences, tactics and strategies that are more resilient to change and better able to profit by emerging opportunities may require more complicated simulation tools.

2.2 Field and Whole-Farm Level Dynamic Models

In the early 1970s, deWitt introduced the state variable approach as the basis for simulation modelling, that the current state determines how the rates of change lead to the next time step in the calculation of crop growth processes (Goudriaan and vanLaar, 1994). This was probably the first step towards dynamic crop modelling. Twenty years later, the Agricultural Production Systems Research Unit (APSRU) in Australia used the existing understanding of eco-physiological processes to develop, in collaboration with practitioners, a range of systems modelling tools with different levels of complexity and scale i.e. field, farm, region, country. The two most-relevant tools to this chapter are APSIM and its multi-field whole-farm configuration (APSFarm).

2.3 Field Level: APSIM

APSIM is a modular modelling framework developed by APSRU in Australia (www.apsim.info) to simulate the biophysical process in farming systems, particularly where there is interest in the economic and ecological outcomes of management practices in the face of climatic risk. APSIM's structure consists of different plant, soil and management modules which include a diverse range of crops, pastures and trees, soil processes including water balance, N and P transformations, soil pH, erosion and a full range of management controls. APSIM has been tested in a diverse range of systems and environments including many small householder systems in Africa (Robertson et al. 2005; Whitbread et al. 2010; Dimes et al. 2015). APSIM has been used in a broad range of applications including support for on-farm decision making, farming systems design for production or resource management objectives, assessment of the value of seasonal climate forecasting, analysis of supply chain issues in agribusiness activities, development of waste management guidelines, risk assessment for government policy-making and as a guide to research and education activities.

2.4 *Risk Management Using APSIM*

One of the most well-known applications of modelling in dryland agricultural systems is to inform risk and uncertainty. Here we define risk as imperfect knowledge where the probabilities of possible outcomes are known – 'known unknowns'. Uncertainty exists when these probabilities are not known – 'unknown unknown'. The distinction is important, as the major source of risk in dryland agricultural systems is usually climate variability, which can be informed using long-term climate records and seasonal climate forecasts – when available. The nature, time frame and intensity of more uncertain events can be difficult to foresee, e.g. price fluctuations, climate change, pest and diseases. In dryland systems, climate change is usually studied using climate projections derived from global circulation models (Rodriguez et al. 2014).

Although discovered by Walker in the 1920s, it was not until 75 years later that the Southern Oscillation Index (SOI) became the major tool to inform climate variability and the state of the El Nino in Australia (Stone et al. 1996). Seasonal climate forecasts based on the SOI have been used, in combination with simulation modelling, probability theory, profit function, and finance techniques, to inform practice change (Meinke and Hochman 2000). A relevant example is the use of SOI and SST-based forecasts in June/July to predict spring rainfall or wheat yields in southeast Australia (Anwar et al. 2008) (Fig. 3). The authors found that the overall predictive skill for spring rainfall and simulated wheat yields was 60–83 % consistent. Levels of predictive skill that should allow farmers use the information to make tactical in crop management interventions, such as top dressing of nitrogen fertilisers.

2.5 *Modelling G×E×M Interactions in Dryland Agriculture*

A more recent application of systems modelling for dryland agricultural systems is in the area of crop design. Crop design uses dynamic models to match crop physiological traits (G) and managements (M) to specific environments (E) in G×E×M studies. For example, APSIM wheat (Chenu et al. 2011), sorghum (Chapman 2008)

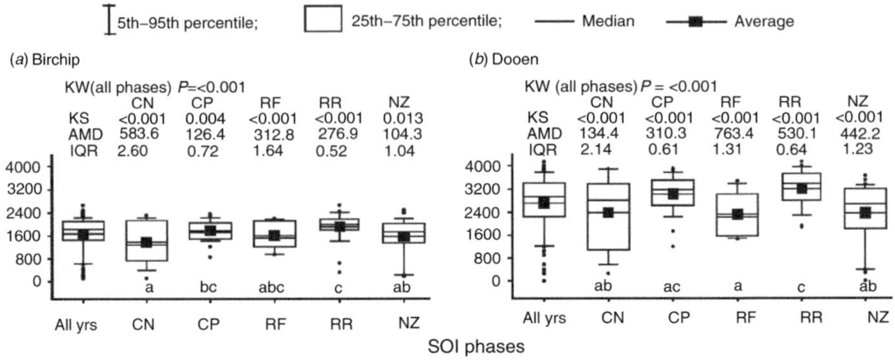

Fig. 3 Simulated distribution of wheat yields at Birchip and Dooen, Australia, for all years and each SOI phase in June/July. Local soils and management conditions were used in the analysis. Means of ENSO phases with common letters below their respective box-and-whisker plot are not significantly different (p≤0.05). The significance of the shift in the median in each phase away from the all-years value was determined using a Kruskal-Wallis test (KW) while the similarity of distributions was determined using the Kolmogorov-Smirnov P-value test (KS). The shift is measured using the absolute median difference (AMD) while changes in the distribution dispersion are measured using the inter-quartile ratio (IQR) (Source: Anwar et al. 2008)

and maize (Chauhan et al. 2013) were used to derive the frequency of occurrence for contrasting water-stress patterns across Australia i.e. in terms of severity and timing with respect to critical stages for yield formation (Fig. 4). This work, together with that from Hammer et al. (2014), shows potential for better matching genotypes and managements to growing environments and seasonal conditions. A common feature of these studies is the use of APSIM to describe contrasting environments based on the ratio between water supply and water demand. By clustering the ratio of simulated supply and demand patterns during the cropping season, major water-stress environment types can be identified. Fig. 4 shows for maize (Chauhan et al. 2013), two environments of little stress (e1) and (e2), two environments of a mild-terminal stress (e3) and (e4), and one environment of severe-terminal stress (e5). The frequency of occurrence of each of these stress patterns was shown to vary across Australia's northern grains region (Fig. 5). This information is valuable in many respects; on the one hand, it supports breeders identify sites with a high frequency of particular stress types to make germplasm screening more effective and efficient; while on the other hand, by understanding the frequency of occurrence of the different stress environments, breeders and agronomists may be able to better match available germplasm and management to particular sites. Clearly, improvements in the skills and lead time of seasonal climate forecasts would add considerable value to this application, as the environments could be predicted and the right combination of genotype and management identified prior to planting.

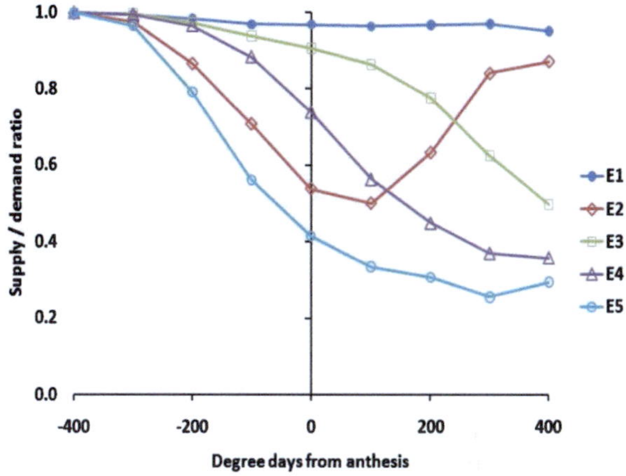

Fig. 4 Five water-stress patterns common in maize growing areas of north-eastern Australia based on clustering of the simulated supply/demand ratio: e1: mid-stress, e2: mild terminal stress, e3: moderate terminal stress, e4: no stress and e5: severe terminal stress (Source: Chauhan et al. 2013)

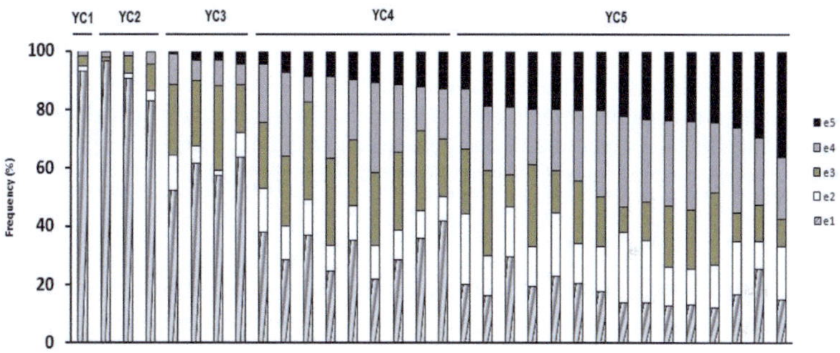

Fig. 5 Simulated frequencies of water-stress environment types (e1 to e5) for maize across sites and soils in eastern and northern Australia. Stress types are e1: no stress, e2: mild terminal stress, e3: mild terminal stress, e4: moderate terminal stress and e5: severe terminal stress (Source: Chauhan et al. 2013)

2.6 Modelling Farms and Farmers

Farms are complex systems (Fig. 6) and modelling dryland agricultural systems is a valuable tool to inform decision making and overcome some of these complexities. Irrespective of their level of endowment, farmers use incomplete or imperfect knowledge to make technical (e.g. agronomic, energy inputs, irrigation scheduling, etc.) and financial (e.g. marketing, loans, off-farm investment) decisions in a context of risk and uncertainty associated with climate variability, market volatility, and political and global change. It is a complicated system dominated by multiple decisions on how to allocate limited resources across a range of competing enterprises and time pressures. Progress in our understanding of crop eco-physiological processes, animal physiology, reproduction and nutrition, and the interactions between different components of the farming system including soil and climate factors (Fig. 6) are

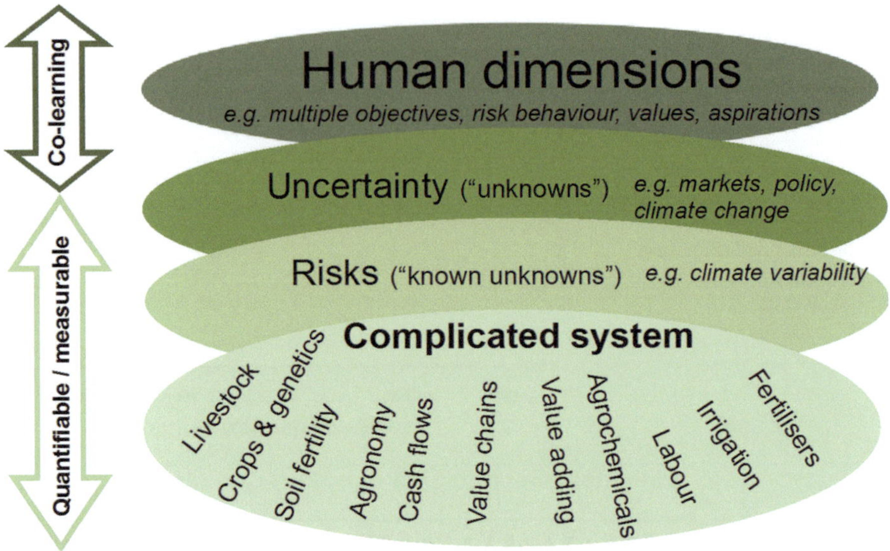

Fig. 6 Conceptualization of the complexities in the management of a farm business and its disaggregation into quantifiable or measurable components, and the social-human dimension accessible via discussion, reflexion and learning (Source: Rodriguez and Sadras 2011)

to a great extent known giving us confidence in the use of more integrative whole-farm modelling tools. In combination with scenarios and sensitivity analyses whole farm models can be used to study the systems' behaviour to internal e.g. resource allocation, and external factors e.g. changes in relative prices, climate change, policy change, market developments, etc.

In contrast to field-level models, whole-farm modelling deals with two main issues: (1) urgent or short-term tactical issues, i.e. more isolated actions that take advantage of opportunities or mitigate impacts within a given farm plan or strategy, and (2) more strategic decisions driven by medium to long-term objectives in the family socioeconomic and institutional context in which they operate. In general, tactical changes mostly deal with incremental changes in response to perceived or expected changes in the operational environment. Examples include seasonal climate or market contingent changes that farmers face every day, season or year. Alternatively, the need to increase preparedness to profit from or mitigate the impacts of time-evolving changes with the availability of resources i.e. over a 5–10-year timeframe, requires the identification of 'best-fit' strategies that allow farm businesses to remain viable and grow in the face of risks and uncertainties. In general, these are the more transformational changes that farmers make in view of the medium to longer-term outlook; decisions that are usually associated with the success or failure of entire farm businesses and potential socioeconomic, environmental and value-chain implications.

The whole-farm model APSFarm (Rodriguez et al. 2011) developed over the last 10 years is an extended configuration of APSIM, which allows users to simulate the impacts (i.e. economic, financial, environmental) of the alternative allocation of limited production resources (e.g. land, labour, time, irrigation water, livestock, machinery, and finance) across alternative farm enterprises at the whole-farm level. Example APSFarm simulation projects were first released in the APSIM 7.2 release, including its application for large-scale commercial rainfed cropping (Rodriguez et al. 2011), irrigated cropping (Power et al. 2011) and mixed grain and grazing farm businesses (Rodriguez et al. 2014) (Fig. 7).

2.7 The APSFarm Technology

Modelling farms (compared to fields) to study resource allocation questions brings further complications, and simulating those complications require a change from fixed 'calendar-driven' management rules to dynamic sets of rules that operate across the whole spectrum of farm activity: land, labour, water, livestock and machinery, to name a few. The introduction of APSIM modules implementing languages that support higher level data manipulation allowed APSFarm developers to simulate farming system management as a set of state and transition networks, or finite state automata, allowing rules for farm management to be studied in detail without affecting higher level function. In APSFarm, each field has a current state (e.g. fallow, wheat, sorghum) and 'rules' that allow transitions between adjacent states (e.g. wheat – fallow – sorghum). These rules represent the capacity (e.g. availability of machinery, land, labour), capabilities (e.g. agronomic and technical skills) and preferences (e.g. farm business strategies, risk attitude of the farm manager). Rules can be expressed as a boolean value (true for feasible, false for otherwise) and can take real values with higher values representing the desirability of a particular action. Each day, the model examines all paths leading from the current state to adjacent states, and if the product of all rules associated with a path is non-zero, it becomes a candidate for action. The highest-ranking path is taken and the process repeats until nothing more can be done that day. Rules can represent farm-level criteria such as sowing windows for each crop, temporal definitions of 'break of the season' such as mm of rainfall over a defined time period, physical limitations of farm components such as machinery operation, and preferential behaviour such as the maximum farm area that could be sown to each crop – farmers' 'risk aversion'. Rules of field-level criteria include minimum extractable soil water (ESW, mm) required to sow a crop, definitions of a 'sowing opportunity', the cropping history of that paddock that may force the inclusion of a break crop, the level of ground cover, etc.

The results from actions having economic implications (e.g. variable costs from fertiliser use, the need for seeds to plant a crop, or profits from crop harvesting) are calculated based on a list of expected costs and prices provided by the participating farmers. Fixed farm operational costs are also obtained from the participating farmer

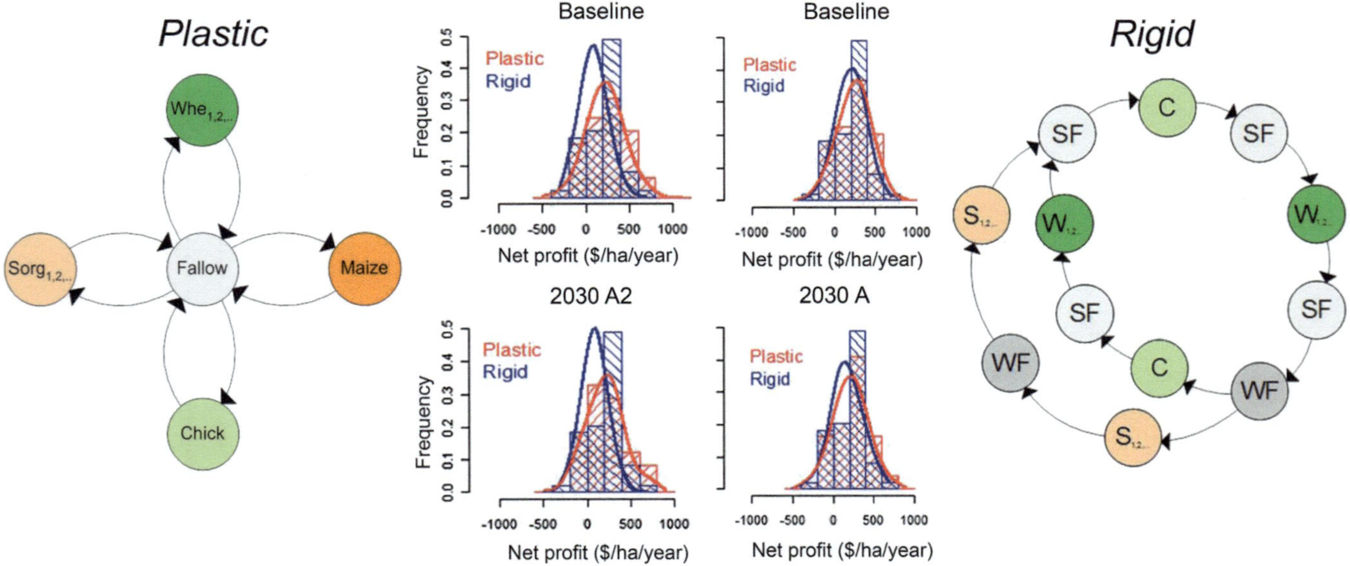

Fig. 7 Sensitivity of two farm businesses characterized by contrasting degrees of plasticity (flexibility in decision making) to a climate change scenario. In APSFarm's user interface, the circles in the graphs represent the state in which a particular field on a farm can be found, while the arrows contain the rules for a change of state to occur (state and flow system). The frequency charts show the likelihood of net profits for the two rotational systems for the baseline and 2030 scenario both at Emerald and Goondiwindi, Queensland, Australia.

and used in the calculation of farm profits (i.e. before tax). Therefore, outputs from APSFarm include, but are not limited to, production measures e.g., yields and crop areas; economic measures such as production costs, crop gross margins, economic risk and farm annual profit; efficiency measures such as crop and whole-farm WUE; and environmental measures such as deep drainage, runoff and erosion.

2.8 Participatory Modelling of Dryland Farms

The need to adjust the scale of economic activity of farm businesses to fit within the boundaries set by shifts in resource availability (e.g. water, climate, finances), environmental change (e.g. emissions, land degradation) and farmer preference required the development and application of more integrative and interdisciplinary modelling tools. In response to this demand, APSFarm uses a participatory framework that captures the major rules and decision-making processes that farmers undergo when managing complicated farm businesses. The framework involves interviews and discussions with farmers and their consultants to identify, quantify and translate rules, preferences, practices and strategies into a whole-farm systems model, becoming a virtual representation capable of mimicking the dynamic operation of their farm business. Once model outputs for the baseline scenario (i.e. climatology) are accepted as realistic by the participating farmer case studies, the model is used to address specific questions from the farmers, researchers, extension officers or agribusinesses via What if? scenario analyses. The final outcome is that farmers, extension officers, researchers and local agribusinesses are more confident that the newly available technologies will successfully achieve their objective and fit in with the existing farming systems, biophysical, social and cultural constraints.

2.9 Systems Characteristics as a Source of Resilience in Dryland Agriculture

Using the whole-farm model APSFarm, Rodriguez et al. (2011) tested the hypothesis that plasticity in farm management introduces resilience to change and allows farm businesses to perform when operating in a highly-variable environment. Two real farm businesses differing in management (plastic vs. rigid rotations) were interviewed and set up in the model. The model was then run for two climatology and two climate change scenarios to introduce stress into the system. Both plastic and rigid locations were run at both locations so that the different systems could be compared to similar baseline and climate projections. The more plastic farming system was located in Emerald, Queensland, Australia and could be characterised by having a management highly contingent on environmental conditions. In the plastic farming system, the farm manager was constantly changing crops and inputs based on the availability of land, water, finances, labour and machinery, and signals from the environment (e.g. climate, markets). In contrast, the more rigid farming system was located in Goondiwindi, Queensland, Australia and could be characterised as more calendar driven following a relatively fixed sequence of crops.

Simulated results showed that in both environments (about 800 km apart in a north–south transect) the more plastic farm management strategy had higher median profits and was less risky for the baseline and 2030 climate change scenarios, suggesting that farming systems with higher levels of plasticity would enable farmers to more effectively respond to climate shifts, thus ensuring economic viability of their farm business. They also found that, in the case studies analysed here, most of the impact from the climate change scenarios on farm profit and economic risk originated from important reductions in cropping intensity and changes in crop mix rather than from changes in the yield of individual crops (Rodriguez et al. 2011).

Changes in cropping intensity and crop mix were explained by the combination of reductions in the number of sowing opportunities around critical times in the cropping calendar, and to operational constraints at the whole-farm level i.e., limited work capacity in an environment having fewer and more concentrated sowing opportunities. This indicates that indirect impacts from shifts in climate on farm operations may be more important than direct impacts from climate on the yield of individual crops. They concluded that, due to the complexity of farm businesses, impact assessments and opportunities for adaptation to climate variability and change may need to be pursued at higher integration levels than the crop or the field i.e. the farm level.

3 Learning by Doing – Learning by Modelling

Irrespective of the levels of wealth/poverty or the scale of the agricultural practice, farmers intuitively adapt their management in response to perceived changes in their operational environment; a process that requires access to relevant experiential information (Schwartz and Sharpe 2006). The decision-making processes that underlie decision making has been described as the combined operation of two systems: a fast, automatic, effortless, unconscious system resembling a neural network and a slow, deliberate, effortful, conscious system better described as being organized by rules (Kahneman 2003). The operation of the former (intuition or practical wisdom – after Schwartz and Sharpe 2006) is mandatory; operation of the latter (conscious, rules-based) is optional. Practical wisdom, requires the right goals, the right motives and builds over time – with practice, as it requires practical knowledge for the decision maker to change old habits – "it takes an enormous amount of practice to change your intuition" (Kahneman 2011). It also requires enough flexibility, autonomy and confidence in the available options e.g. technological or managerial, for the decision maker to respond appropriately to a given situation. An interesting problem arises in the absence of relevant experiential practice (e.g. in the face of unprecedented change such as climate change or when new technologies are presented to traditional farmers). This is because practical wisdom cannot be taught as it is context-sensitive (Schwartz and Sharpe 2006). For example, in the case of adapting to climate change, farmers may have little or no experience to be able to relate to and identify possible actions. This means that medium and long-term farm planning in the face of likely change scenarios will require far greater levels of attention and support. Yet farmers often find long-term climate projections of limited relevance while under pressure to resolve more immediate day-to-day and season-to-season decisions. For traditional farmers from the developing world, new technologies (e.g. conservation agriculture principles or risk management strategies) have the potential to present similar challenges as they may be perceived too difficult, too risky or even culturally wrong. To address some of these issues, participatory discussions and computer-aided farming systems design have proved useful for gaining insight into complex systems, developing intervention strategies and generating awareness on the potential impacts of incremental or transformational changes to adapt to change when applied in real-world situations e.g. commercial farms, or smallholder farming systems.

4 Conclusions and Future Research Thrusts

Modelling tools have shown value in agriculture research from quantifying benefits and trade offs in both short term management practices and tactics that reduce risk in highly-variable climates; and the medium and longer term benefits from changes in strategies, farming systems designs and resource allocations. Advances in people's connectivity, access to internet, developments in big-data, networks of sensors, and computing speed mean that more than ever before, research and practice change in dryland agricultural systems will rely on simulation and prediction. Over the last twenty years we have moved on from modelling plants, crops, cropping systems and farming systems (including livestock), to modelling farms and farmers. A number of challenges though still exist, progress is expected in the near future from modelling multiple farms and farmers in a shared landscape, their interactions and connectivity with markets and rural communities, in addition to the modelling of the multiple functions of agriculture, and the resulting people's health and nutrition. As this chapter was produced we have already started modelling communities of farmers and villages and to answer how alternative innovations, particularly among smallholder farmers, are likely to affect local labour and commodity markets, and how informal economies are likely to develop as we achieve food security and poverty reduction targets in some of the poorest nations in the world.

Acknowledgements The authors thank the Australian Centre for International Agriculture Research (ACIAR) and the Grains Research and Development Corporation for their support with the many studies reported within.

References

Anwar AE, Rodriguez D, Liu C, Power S, O'Leary GJ (2008) Quality and potential utility of ENSO-based forecasts of spring rainfall and wheat yield in south-eastern Australia. Aust J Agri Res 59:112–126

Calvino PA, Monzon JP (2009) Farming systems of Argentina: yield constraints and risk management. In: Sadras VO, Calderini DF (eds) Crop physiology: applications for genetic improvement and agronomy. Elsevier, Amsterdam, pp. 489–510

Chapman SC (2008) Use of crop models to understand genotype x by environment interactions for drought in real-word and simulated plant breeding trials. Euphytica 161:195–208

Chauhan YS, Fekybelu S, Rodriguez D (2013) Characterization of north-eastern Australian environments using APSIM for increasing rainfed maize production. Field Crop Res 144:245–255

Chenu K, Cooper M, Hammer GL, Mathews KL, Dreccer MF, Chapman SC (2011) Environment characterization as an aid to wheat improvement – interpreting genotype – environment interaction by modelling water deficit patterns in northern. Aust J Exp Bot 62:1743–1755

Cooper PJM, Gregory PJ, Tully D, Harris HC (1987) Improving water use efficiency of annual crops in rainfed systems of west Asia and North Africa. Exp Agic 23:113–158

CSIRO Imagery Shows Great Barrier Reef at Risk from Sediment Plumes (2007) CSRIO. February 19, 2007. http://www.clw.csiro.au/new/2007/sedimentplumes.html. Accessed in December 2015

deWitt CT 1958 Transpiration and crop yields. Instituut voor biologisch en scheikundig onderzoek van landbouwgewassen. Wageningen Medeeling 59

Dimes J, Rodriguez D, Potgieter A (2015) Raising productivity of maize based cropping systems in eastern and southern Africa: step-wise intensification options. In Sadras VO, Calderini D (eds) Crop physiology. Applications for genetic improvement and agronomy, 2nd edn. Elsevier, pp 94–110. ISBN: 978-0-12-374431-9

FAO (2011) The state of the world's land and water resources for food and agriculture (SOLAW) – managing systems at risk. Food and Agriculture Organization of the United Nations/Earthscan, Rome/London

FAO, (2015) Food outlook. Biannual report on food markets. FAO. Available on line www.fao.org/publications

French RJ, Schultz JE (1984) Water use efficiency of wheat in a Mediterranean type environment. I. The relationship between yield, water use and climate. Aust J Agri Res 35:743–764

Goudriaan J, van Laar HH (1994) Modelling potential crop growth processes. Current issues in production ecology, vol 2. Kluwer Academic Publishers. doi:10.1007/978-94-011-0750-1

Hammer GL, McLean G, Chapman S, Zheng B, Doherty A, Harrison MT, van Oosterom E, Jordan D (2014) Crop design for specific adaptation in variable dryland production environments. Crop Pasture Sci 65:614–626

Holzworth D, Huth N, deVoil PG, Zurcher EJ, Herrmann NI, McLean G, Chenu K, van Oosterom EJ, Snow V, Murphy C, Moore AD, Brown H, Whish JPH, Verrall S, Fainges J, Bell LW, Peake AS, Poulton PJ, Hochman Z, Thorburn PJ, Gaydon DS, Dalgliesh NP, Rodriguez D, Cox H, Chapman S, Doherty A, Teixeira E, Sharp J, Cichota R, Vogeler I, Li FY, Wang E, Hammer GL, Robertson MJ, Dimes JP, Whitbread AM, Hunt J, van Rees H, McClelland T, Carberry PS, Hargreaves JNG, MacLeod N, McDonald C, Harsdorf J, Wedgwood S, Keating BA (2014) APSIM – Evolution towards a new generation of agricultural systems simulation. Environ Model Softw 62:327–350

Kahneman D (2003) A perspective on judgment and choice. Mapping Bounded Rationality. Am Psychol 58:697–720

Kahneman D (2011) Thinking, fast and slow. Farrar, Strauss, Giroux, New York

Lobell DB, Cassman KG, Field CB (2009) Crop yield gaps: their importance, magnitudes and causes. Annu Rev Environ Resour 34:179–204

McCown RL, Carberry PS, Hochman Z, Dalgliesh NP, Foale MA (2009) Re-inventing model-based decision support with Australian dryland farmers. 1. Changing intervention concepts during 17 years of action research. Crops Pasture Sci 60:1017–1030

Meinke H, Hochman Z (2000) Using seasonal climate forecasts to manage dryland crops in northern Australia – Experiences from the 1997/98 seasons. doi:10.1007/978-94-015-9351-9_11 In: Applications of seasonal climate forecasting in agricultural and natural ecosystems, pp 149–165

Mitchell JH, Chapman SC, Rebetzke GJ, Fukai S 2006 Reduced-tillering wheat lines maintain kernel weight in dry environments. In: Turner NC, Acuna T, Johnson RC (eds) Proceedings of the 13th Australian agronomy conference, Australian Society of Agronomy, Perth. The Regional Institute Ltd, Perth

Nicholls N (1986) Use of southern oscillation to predict Australian sorghum yield. Agric Forest Met 38:9–15

Nuttall JG, Armstrong RD, Connor DJ (2003) Evaluating physicochemical constraints of Calcarosols on wheat yield in the Victorian southern Mallee. Aust J Ag Res 54:487–497

OECD-FAO (2009) OECD-FAO agricultural outlook: 2009–2018, Paris

Parry M, Lowe J, Hanson C (2009) Overshoot, adapt and recover. Nature 458:1102–1103

Passioura JB, Angus JF (2010) Improving productivity of crops in water limited environments. Adv Agron 16:37–75

Potgieter AB, Hammer GL, Butler D (2002) Spatial and temporal patterns in Australian wheat yield and their relationship with ENSO. Aust J Agric Res 53:77–99

Power B, Rodriguez D, deVoil P, Harris G, Payero J (2011) A multi-field bio- economic model of irrigated grain–cotton farming systems. Field Crop Res 124:171–179

Rebetzke GJ, Richards RA (1999) Genetic improvement of early vigour in wheat. Aust J Agric Res 50:291–302

Rengasamy P (2002) Transient salinity and subsoil constraints to dryland farming in Australian sodic soils: an overview. Aust J Exp Agr 42:351–361

Richards RA, Lukacs Z (2002) Seedling vigour in wheat—sources of variation for genetic and agronomic improvement. Aust J Agric Res 53:41–50

Richards RA, Lopez Castaneda C, Gomez-Macpherson H, Condon AG (1993) Improving the efficiency of water use by plant breeding and molecular biology. Irrig Sci 14:93–104

Robertson MJ, Sakala W, Benson T, Shamudzarira Z (2005) Simulating response of maize to previous velvet bean (*Mucuna pruriens*) crop and nitrogen fertiliser in Malawi. Field Crop Res 91(1):91e105

Rodriguez D, Sadras VO (2007) The limit to wheat water use efficiency in eastern Australia. I. Gradients in the radiation environment and atmospheric demand. Aust J Agric Res 58:287–302

Rodriguez D, Sadras VO (2011) Opportunities from integrative approaches in farming systems design. Field Crop Res 124:137–141

Rodriguez D, Nuttall J, Sadras VO, van Rees H, Armstrong R (2006) Impact of subsoil constraints on wheat yield and gross margin on fine-textured soils of the southern Victorian Mallee. Aust J Agric Res 57:355–365

Rodriguez D, Robson AJ, Belford R (2009) Dynamic and functional monitoring technologies for applications in crop management. In: Sadras VO, Calderini D (eds) Crop physiology. Applications for genetic improvement and agronomy. Elsevier, Amsterdam, pp. 489–510

Rodriguez D, deVoil P, Power B, Cox H, Crimp S, Meinke H (2011) The intrinsic plasticity of farm businesses and their resilience to change. Aust Example Field Crops Res 124:157–170

Rodriguez D, Cox H, deVoil P, Power B (2014) A participatory whole-farm modelling approach to understand impacts and increase preparedness to climate change in Australia. Ag Syst 126:50–61

Sadras VO, Angus JF (2006) Benchmarking water use efficiency of rainfed wheat in dry environments. Aust J Agric Res **57**:847–856
Sadras VO, Rodriguez D (2007) The limit to wheat water use efficiency in eastern Australia. II Influence of rainfall patterns. Aust J Agric Res 58:657–669
Sadras VO, Rodriguez D (2010) Modelling the nitrogen-driven trade-off between nitrogen utilisation efficiency and water use efficiency of wheat in eastern Australia. Field Crop Res 118:297–305
Sadras VO, Baldock J, Roget DK, Rodriguez D (2003) Measuring and modelling yield and water budget components of wheat crops in coarse textured soils with chemical constraints. Field Crop Res 84:241–260
Sadras VO, Calderini D, Connor D (2009) Sustainable agriculture and crop physiology. In: Sadras VO, Calderini D (eds) Crop physiology. Applications for genetic improvement and agronomy. Elsevier, Amsterdam, pp. 1–20
Schwartz B, Sharpe KE (2006) Practical wisdom: Aristotle meets positive psychology. J Happiness Stud 7:377–395
Stephens DJ, Lyons TJ (1998) Rainfall–yield relationships across the Australian wheat-belt. Aust J Ag Res 49:211–223
Stone RC, Hammer GL, Marcussen T (1996) Prediction of global rainfall probabilities using phases of the Southern Oscillation Index. Nature 384:252–255
Whitbread AM, Robertson MJ, Carberry PS, Dimes JP (2010) How farming systems simulation can aid the development of more sustainable smallholder farming systems in southern Africa. Eur J Agron 32:51e58

Breeding and Genetic Enhancement of Dryland Crops

Quahir Sohail, Hafsa Naheed, and Reza Mohammadi

1 Introduction

Drylands cover about 41 % of Earth's land area and are inhabited by more than 2 billion people. Of the total area of drylands, dry sub-humid, semiarid, arid and hyper-arid sub-systems cover 21 %, 37 %, 26 % and 16 %, respectively. About 25 % of the world's drylands are croplands, 65 % are rangelands, and the remaining 10 % are other areas such as inland water systems and urban areas. Croplands and rangelands are often interwoven, supporting mixed and integrated livelihoods. Due to increasing populations, a transformation from rangelands to croplands is occurring in many areas.

Water shortage is reaching critical limits in many dry areas and water scarcity has been predicted in some areas where water is abundant at present. Problems like salinity and wind erosion, leading to land degradation, are enhanced by water shortage. Efforts to feed more people will further increase agricultural water use resulting in severity of water scarcity in some areas, and may cause water scarcity even in areas that are relatively well endowed with water resources. It is predicted that the annual maximum temperature by 2050 will increase by 1–3 % (IPCC 2007). The rate of evapotranspiration will also increase with global warming which will lead to changes in agricultural systems and cropping patterns.

Q. Sohail (✉)
International Wheat and Maize Improvement Center (CIMMYT), International Winter Wheat Improvement Program, Ankara, Turkey
e-mail: quahirsohail@gmail.com

H. Naheed
The University of Agriculture, Peshawar, Pakistan

R. Mohammadi (✉)
Dryland Agricultural Research Institute (DARI),
Sararood Branch, AREEO, P O Box 67145-1164, Kermanshah, Iran
e-mail: r.mohammadi@areo.ir

M. Farooq, K.H.M. Siddique (eds.), *Innovations in Dryland Agriculture*,
DOI 10.1007/978-3-319-47928-6_10

Drought and heat are worldwide problems in drylands. What makes planning against drought difficult is that the farmers and researchers don't know when it is going to hit the crop; another problem with drought tolerance is that it is a polygenic trait with low heritability and high genotypic × environmental (G×E) interaction.

Supplying sufficient food for the increasing population is a huge challenge, nonetheless global food security is threatened by climate change (Lal et al. 2005). Climate change refers to changes in the distribution of weather across a period of time that ranges from decades to millions of years. It can be a change in the average weather or a change in the distribution of weather events around an average. Maximum and minimum relative humidity levels have shown increasing trend, climate change is worsening the current situation through its effects on the world wide agricultural systems (Battisti and Naylor 2009; Bloem et al. 2010). Due to climate change, more severe scenarios such as drought in some areas and the increase in frequency and severity of extreme droughts in other areas are expected (IPCC 2007; Carnicer et al. 2010). Climate change is affecting agriculture both directly and indirectly, worldwide; temperature (high and low), emission of greenhouse gases and precipitation patterns directly affect the crops, pathogens, insects, and weeds. Several new diseases, weeds, and insect-pests have started appearing with the changing climate.

Several new diseases such as powdery mildew and foliar blight of wheat have started appearing, and there is a risk of stem rust (black rust), caused by Ug99, which started from Uganda and has now spread up to Iran and further. Likewise, new insect pests such as black aphids and pink stem borer have started appearing on wheat. Crops requiring chilling for flowering, will not be getting enough chilling hours, which may reduce their production. New technologies must be developed to accelerate breeding through improving genotyping and phenotyping methods and by increasing the available genetic diversity in crop breeding programs (Dhillon et al. 2013). Climatic changes in most of the crop growing areas are expected to experience changes in drought, canopy temperature, salinity, etc., For example, wheat is negatively affected by drought in more than 50 % of its cultivated area (Rajaram 2001); its productivity is usually limited by shortage of moisture necessary to maximize biomass and most importantly to complete grain filling (Aprile et al. 2009). The largest producers of pulses (South Asian countries, China and some African countries) having around 70 % of the global production are located in areas that experience shortage of water (Gowda et al. 2009).

As the average temperature is predicted to rise, some regions will be expecting more heat, the greater heat will complicate the situation by increasing agricultural droughts with high evapotranspiration resulting in low soil moisture (Battisti and Naylor 2009). The 1.8 °C rise in average global temperature predicted for 2050 (Meehl and Tebaldi 2004) will force farmers to grow genotypes which are more tolerant, among other factors, to high temperatures (Farooq et al. 2011). Around 9 million hectares of wheat is grown in tropical and subtropical areas of the developing countries with temperatures as high as 17.8 °C in the coolest month of the growing season (Ortiz et al. 2007). Even in the developed countries heat stresses has demonstrated the negative effect on yield. In areas where daily temperature may rise up to 35° C during grain filling, heat stress not only negatively affects yield but also affects protein content, dough quality, and decrease glutenin to gliadin ratio (Blumenthal et al. 1995).

A regional plant genetic resources (PGR) network has recently been established under the auspices of the Association of Agricultural Research Institutions in the Near East and North Africa (AARINENA), as a platform for regional collaboration in setting agriculture priorities to cope with global change. Gene bank in the International Centre for Agricultural Research in the Dry Areas (ICARDA) holds more than 131,000 accessions of cereals, food and feed legumes, and forages including cultivated varieties, landraces and wild relative species, representing more than 50 % of the conserved genetic resources originating from the drylands of Central and West Asia and North Africa. At ICARDA, several promising lines with good tolerance to drought have been derived from inter-specific crosses of barley (*Hordeum vulgare* L.) and wheat (*Triticum aestivum* L.). Wild relatives of different crops have sustained recurrent drought and heat cycles that allowed natural selection to develop resistance to these stresses; these plants therefore may harbor valuable genes for these stresses (FAO 2011).

In this chapter, efficient crops for dryland systems are introduced, and breeding and genetic enhancement strategies for improving the productivity of dryland crops are discussed.

2 Efficient Crops for Dryland System

The choice of crops for the production systems, cultivar, sowing date, plant density, fertilizer management, and control of diseases, insects, and weeds, needs to best suit the local environmental conditions. Given the low and erratic precipitation for crop production in the dry areas, further research should focus on improving water-use efficiency associated with increased production per unit area, and improved production stability. The development of new, well-adapted varieties, efficient crop production and crop production technologies are among the most important factors involved in enhancing crop productivity in dryland agriculture. Among the crops, wheat, barley, millet, sorghum and grain legumes have great potential for drylands.

Wheat and barley are the most important cereal crops in Mediterranean, where high temperatures and water scarcity restricts yield during grain filling (Araus et al. 2007; Francia et al. 2011; Jacobsen et al. 2012). Modern wheat cultivars have been normally selected for improved productivity under high-yielding conditions, whereas barley or traditional wheat were somehow better adopted to poor environments compared to modern high-yielding wheats (Ryan et al. 2008). On the other hand, barley is the dominant cereal crop in the dryland systems of the Mediterranean region (as well as in many other low yielding systems) as farmers assume that barley performs better compared to wheat under these conditions (Cossani et al. 2007; Ryan et al. 2008). Farmers in these systems tend to favor the use of traditional rather than high-yielding modern cultivars (assuming that high-yielding cultivars are more sensitive to stresses than their traditional counterparts (Byerlee 1996).

In Syria, about 23 % of the increase in durum wheat production is due to effects of irrigation, 34 % to the use of improved varieties, 24 % to fertilizer, and 19 % to land and crop management, with 37 % of the impact coming from rainfed areas (Mazid et al. 1998). In Turkey, improved varieties and efficient crop husbandry practices resulted in a three-fold wheat yield in the last 50 years, from 0.8 to 2.4 t ha^{-1} (Avci 1999). This increase is predominantly caused by timely soil management with proper implements (timely tillage, sowing, weeding, etc.), phosphorus application, and improved varieties (Avci et al. 1987).

Barley is the world's fourth most important cereal crop, and drought affects its yields severely. During 2000 in Syria, for example, when rainfall dropped 20 to 30 % below the long-term average, the crop produced little or no grain in some areas. Nonetheless, at four dryland experimental sites, a few barley lines were able to produce grain. Based on these materials, a new drought-tolerant barley variety was developed, with the participation of farmers, through a plant breeding program coordinated by the ICARDA. The economic benefit so far of participatory barley improvement in Algeria, Egypt, Ethiopia, Iraq, Jordan, Morocco, Tunisia and Syria was estimated at about US$91 million in 1997.

Pearl millet (*Pennisetum glaucum* (L.) R. Br.), the most inherently drought-tolerant of all the major staples, together with sorghum (*Sorghum bicolor* (L.) Moench) are key cereal grain crops in the drylands, providing food, feed and, in the case of millet, fuel and construction material as well. Despite formidable obstacles to improvement of these crops for drylands, plant breeders at the International Crops Research Institute for the Semi-Arid Tropics (ICRISAT) and in national partner organizations have made important gains, and farmers are adopting new varieties. In southern Africa, for example, about 34 % of the total millet area is now planted to improved materials and 23 % of the sorghum area (CGIAR 2013).

Grain legume production is increasing worldwide due to their use directly as human food, feed for animals, and industrial demands. Further, grain legumes have the ability to enhance the levels of nitrogen and phosphorus in cropping systems (Sinclair and Vadez 2012). Grain legumes are vital sources of low-cost protein in drylands, and the sale of excess grain generates significant farm income. Grain legumes also help restore soil fertility, since their roots fix nitrogen from the air in forms that can be used by subsequent crops. In addition, the stems and stalks of these crops are valued as livestock feed. Cowpea (*Vigna unguiculata* (L.) Walp.) is the most widely grown grain legume in the dry areas of Africa, while chickpea and pigeon pea (*Cajanus cajan* (L.) Millsp.) predominate in much of the Asian drylands. More than 60 countries have released improved cowpea varieties with support from the International Institute of Tropical Agriculture (IITA). Chickpea (*Cicer arietinum* L.) and pigeon pea varieties resulting from the work of ICRISAT are having a major impact in India, Nepal, Pakistan and increasingly in China (CGIAR 2013).

Plant breeding in general, and cereal breeding in particular, was remarkably successful during the second half of the 20th century, contributing substantially to keep production ahead of population growth. The effect of plant breeding on durum wheat (*Triticum durum* Desf.) yield potential and its physiological determinants has been widely studied in many reports (Motzo et al. 2005; Giunta et al. 2008; Royo et al. 2007) and a genetic gain from 10 to 50 kg $ha^{-1}year^{-1}$ has been recorded over the last century in most countries and often associated with few key genes affecting morpho-phenological traits, mainly *Rht* (Slafer et al. 1994). Some reports have suggested that the improvement for yield potential may have detrimental implications for yield stability in water-limited environments (Ceccarelli et al. 1991; Mohamamdi and Amri 2013), since specific drought tolerant traits (i.e. flowering time, small flag leaves, glaucousness, maintenance of green leaf area, plant height) have been considered inappropriate for achieving high yields under favorable conditions (Acevedo et al. 1991). On the other hand, it has also been suggested that under mild drought conditions, characterized by a wheat/barley grain yield between 2 and 5 t ha^{-1}, selection for high-yield potential had frequently led to some yield improvements under drought (Araus et al. 2002; Rizza et al. 2004). These results were achieved through an empirical selection for high-yield potential and high-yield stability, with the latter being attributed to a minimal genotype × environment (GE) interaction. This implies that traits maximizing productivity normally expressed in the absence of stress can still sustain a significant yield improvement under moderate drought (Slafer et al. 2005; Tambussi et al. 2005).

3 Traits of Importance for Base-Broadening and Stress Resistance

All traits of interest for crop productivity, adaptation and quality are subjected to improvement with the appropriate combination of current and new introgressed alleles. However, due to the genetic bottlenecks and the fact that the most of breeding programs in the past usually exposed their segregating generations during selection to very specific conditions, this can result in genetic diversity of some specific traits may be limited. For instance, International Wheat and Maize Improvement Center (CIMMYT) has based their approach for wide adaptation on selection in environments with high potential (Braun et al. 1997). This approach has been very successful and has led to yield improvements under near-optimal conditions and less limiting environments (Araus et al. 2002; Trethowan et al. 2002). However, due to the intense selection under near-optimal conditions some alleles present in the landraces and cultivars prior to the green revolution that could represent an adaptive advantage under specific environmental conditions may have not been selected and therefore in low frequency or even not present in the modern genetic pool. Moreover, crop landraces and wild relatives have evolved during thousands of years under the most diverse stress scenarios and therefore may carry novel specific genetic combinations due to natural selection and founder effect. Alleles present in landraces and wild relatives providing specific adaptation to environments under drought, heat, salinity, and disease stress pressures are being tackled down and incorporated in the modern gene pool.

Table 1 Traits related to drought and heat

Drought and heat related traits	Function
Leaf angle	To reduce the water loss through transpiration
Leaf area	
Leaf rolling	
Leaf reflectance	
Leaf pubescence	
Root length	To increase the water uptake from soil
Number of lateral roots	
Root diameter	
Root hair density	
Rooting depth	
Root vertical distribution	
Total root size	
Root- shoot ratio	
Large seed and embryo	
Seedling vigor	
Coleoptile length	
Early plant vigor	To decrease evaporation from soil surface and to cover the weeds that compete for limited water resources
Rapid ground cover	
Early flowering	To escape terminal drought
Short life cycle	
Balance phenological development	
Stomatal density	To decrease the rate of transpiration
Stomatal size	
Stomatal conductance	
Cuticle waxes	
Canopy temperature depression	
Osmotic adjustment	Maintenance of tissue water status, and the grain developing rate even under less than optimum conditions.
Stay green character	
Remobilization of assimilate from stem for seed filling	

Many cereals landraces are being tested by several institutions around the world to identify new potential genes/alleles for drought and heat adaptation. The comparison between modern varieties and landraces under near-optimal and sub-optimal conditions has led to the identification of candidates to provide traits of interest for drought and heat adaptation in several countries. In most cases the landraces, as compared to modern varieties, did not show higher grain yield under both near or sub-optimal conditions nor higher yield stability but some candidates and specific traits to improve adaptation to sub-optimal environments have been identified (Denčić et al. 2000; Dodig et al. 2012).

Some important traits related to drought and heat tolerance are given in Table 1. Similarly, canopy temperature, photosynthesis, stomatal conductance, transpiration rate, and intercellular carbon dioxide are associated with drought tolerance (Sohail et al. 2011). On the other hand, wild relatives, synthetic hexaploid wheat (SHW), synthetically derived wheat (SDW) or modern lines with alien translocations have shown higher fitness than their modern counterparts under heat and drought stresses. CIMMYT identified several SHW and SDW with superior adaptation to drought conditions (Trethowan and Van Ginkel 2009). Wheat landraces have also been exploited for tolerance against many abiotic and biotic stresses and are the source of many important genes, such as those conferring drought tolerance, disease resistance and quality traits (Reynolds et al. 2009).

4 Crop Gene Pool, Evolutionary Relationships and Systematics

The concept of the gene pools was proposed by Harlan and de Wet (1971) and later on described by Jiang et al. (1994), on the base of evolutionary distance from each other, their genomic constitution and the idea of the three gene pools i.e. primary, secondary and the tertiary gene pools, this classification is in relation to the cultivated species. The knowledge of the ancestry of cultivated crop species and varieties is important for understanding variation and genetic diversity in their primary and secondary gene pools and the potential for exploiting the valuable genes responsible for disease resistance or stress tolerance, for developing new varieties with resistance and tolerance to various stresses (Smale 1996).

The primary gene pool of a crop is composed of landraces, early domesticates and wild species that hybridize directly with the cultivated species. Bread wheat arose recently (6,000 to 8,000 years ago) from the hybridization of tetraploid (*Triticum turgidum*) and diploid, *Aegilops tauschii* Coss.) so these two species constitute the primary gene pool (Qi et al. 2007) for wheat and the diploid donors of the A and D genome to bread wheat and durum wheat. The primary gene pool is often preferred due the easiness of crossing and gene transfer (Mujeeb-Kazi 2003). The chromosomes of this gene pool are homologous to the cultivated types and can be utilized easily by breeding methods (Feuillet et al. 2008). For barely the progenitor of cultivated barley (*Hordeum vulagare* subsp. *spontaneum*) belongs to this gene pool. Similarly, the primary gene pool of chickpea includes *Cicer arietinum* ssp. *reticulatum* and the variants of the domesticated chickpea.

During last few decades, many useful genes may have been lost in improving varieties for specific environments, some of these lost genes can somehow be recovered from the primary gene pool. The primary gene pool of wheat carries a highly diverse and geographically widespread and sexually compatible germplasm (Feuillet et al. 2008). However, only a small proportion of the existing genetic diversity of the primary gene pool for most crop species has been utilized for crop improvement (Tanksley and McCouch 1997).

The secondary gene pool of wheat contains polyploid species that share at least one homologous genome with the cultivated types (Feuillet et al. 2008). Transferring genetic material from secondary gene pool is comparatively more complex, there are usually problems of hybrid seed death, female sterility of F_1 hybrids, and reduced recombination (Ogobnnaya et al. 2013). To generate F_1 hybrids, embryo rescue is required for crosses of the secondary gene pool with primary gene pool. This gene pool for wheat includes polyploidy *Triticum* and *Aegilops* species, such as *T. timopheevii* (AAGG) and the diploid S-genome (B genome) species from *Aegilops* (Qi et al. 2007; Curtis 2002). In barley, the tetraploid species, *H. bulbosum* comes under this category the closest relative of cultivated barley, after the subsp. *spontaneum,* and shares the 'I genome' with *H. vulgare.* and *H. bulbosum.* The secondary gene pool of chickpea contains *cicer echinospermum* (Ladizinsky et al. 1988).

The tertiary gene pool is composed of more distantly related diploids and polyploids with non-homologous genomes. They have no genome constitutions of the cultivated species. Transfer of genetic material from this gene pool is very complex (Feuillet et al. 2008). Usually special techniques such as, irradiation, gametocidal chromosomes are needed for gene transfer. Embryo rescue is necessary in crossing with species with tertiary gene pool (Jiang et al. 1994). This group for wheat includes mostly germplasm of *Triticeae* that are not within the primary or secondary gene pools. Most of germplasm in this group are perennials (Mujeeb-Kazi 2003; Feuillet et al. 2008). Although tertiary gene pool resources are highly complex to utilize but are important means for developing crop's germplasm. The barley and chickpea tertiary gene pool are composed of all other spices of their respective genus not present in their primary and secondary gene pool. For some crop species like lentils (*Lens culinaris* Medikus) differentiation between the primary, secondary and tertiary gene pools is not straight forward as all the species and sub species can be exploited quite easily for breeding purpose ((Ladizinsky et al. 1988.)

5 Assessment of Gene Flow for Crop Improvement

Gene flow has been one of the main driving forces for crop improvement. Wheat prominent selfing nature has produced a number of genetically differentiated populations around the globe due, among other reasons, to founder effect in combination with local farmer's trait selection. The first human settlers in the Fertile Crescent adopted wheat as one of their main sustaining crops due to its single seed productivity (as bread wheat is believed to have evolved from a natural probably a single cross). However, it was wheat plasticity and adaptive capacity the reasons behind wheat worldwide spread. Wheat, carried by migrating populations, conquered the most diverse environments resulting in populations with specific adaptations to a wide range of environmental conditions. Very few crops can be as useful as wheat to trace migrations and trade relationships of human groups, mostly in Africa and Eurasia but, after 1500s also in America and Oceania. Since the Neolithic, wheat spread-out with the first migrations, reaching Northern Europe by 3000 B.C. and

East Asia around 2000 B.C., North and South America in the 1500s A.C. (Heiser 1990) and Oceania in the late 1700s A.C. (Macindoe and Brown, 1968). Wheat became then a worldwide crop and its adaptive capacity and plasticity resulted in several landraces and cultivars adapted to the most diverse temperature, photoperiod, altitude and cultural practices (Bennet 1970).

The first scientific breeding started in the nineteenth century when the earlier breeders started selecting and crossing landraces and cultivars from different origins and genetic pools and obtained the combinations that led to superior varieties. Especially successful cases of different genepool uses shall include the Italian breeder Strampelli, who in the first decades of the twentieth century include in his crossing block foreign lines like the Japanese variety Akakomugi (Salvi et al. 2013). The crosses between Italian landraces, breeding lines and Akakomugi resulted in the introgression of new alleles for phenological adjustment and suitable agronomical types– such as Rht8c for short straw and Ppd-D1 for photoperiod insensitivity– in the Italian gene pool. The varieties developed by Strampelli, such as Ardito and Mentana became the backbone of most of the new varieties developed in the Mediterranean countries and distant countries like China and some South American countries (Dalrymple 1986; Lupton 1987; Yang and Smale 1996). A similar approach was followed by the Nobel laureate Norman E. Borlaug in the 1950s with the Norin-10/Brevor cross that, with the introduction of the Rht-B1 and Rht-D1 dwarfing alleles, further reduced plant height preventing lodging when the crop was subjected to increased fertilization and irrigation; the efforts of Borlaug and his team led to what is commonly known by "The Green Revolution" (Worland and Snape 2001; Borlaug 2008). These cases could be considered among the most successful human introgression of suitable alleles from a different genepool that led to important increases in crop productivity. Allele introduction from different primary genepools is still nowadays one of the main sources of diversity and traits of interest (Reif et al. 2005). Crosses between winter and spring varieties – two genepools that used to remain isolated from one another due to geographical and physiological barriers – have been widely used by national and international breeding centers like ICARDA and CIMMYT to obtain new high yielding wide adapted varieties.

The usefulness of the primary gene pool in wheat improvement has been and is still nowadays undeniable. However, with the advances in hybridization and biotechnological techniques during the first decades of the 20thcentury, new genepools became available for breeders. The wide genetic diversity of wheat wild relatives that had evolved separately for thousands of years and whose natural hybridization with common wheat normally results in weak progeny and partial or totally sterility is still available. The introgression of new alleles, chromosomes or entire genomes of these relatives has become one of the main sources of diversity and new alleles nowadays. One of the most successful introgressions of wild genomic regions into a modern wheat variety was curiously not the result of a directed cross made by a breeder but a casual hybridization occurred in a farmers' field. The wheat/rye 1RS.1BL translocation allowed CIMMYT wheats – particularly the ones related to the *Veery* group of varieties– to increase their above-ground biomass yield, number of spikes per unit area, grain weight and test weight (Villareal et al. 1992). With this translocation, also novel rust resistance genes were incorporated into the wheat genepool (McIntosh et al. 1995). Perennial ryegrass variety currently being developed by pastoral genomics for improved drought tolerance by inserting H^+ – pyrophosphatase gene isolated from ryegrass (Holme et al. 2013).

Several suitable alleles have been identified in SHW and wild relatives and have been or are being introgressed in modern germplasm. The most successful introgressions of wild relative alleles into released varieties are related to pest and disease resistance (Hajjar and Hodgkin 2007). New leaf and stem rust resistance alleles have been identified in synthetic wheats (Villareal et al. 1992; Aguilar-Rincón et al. 2000) and some of them have already been introgressed in modern released varieties (Hoisington et al. 1999). New tolerance alleles for septoria, tan spot, karnal bunt, fusarium, powdery mildew, Hessian fly, and other pests and diseases have also been identified in SHW (Trethowan and Van Ginkel 2009).

Major genes controlling qualitative traits – for instance most of the disease resistance genes – are more easily introgressed into modern varieties due to their easier genetic control and expression. However, quantitative traits with polygenic control can also be introgressed in modern backgrounds from wild relatives. For instance, grain protein content QTL was successfully introduced in a modern durum wheat background from its diccocoides relative resulting in an average increase of 15 g kg^{-1} in protein content (Chee et al. 2001).

Finally, primary synthetics tend to be better adapted than modern cultivars to extreme abiotic stress conditions (Trethowan and Van Ginkel 2009). The potential of the wild relatives to provide new sources of adaptation, yield potential and disease resistance to modern wheat varieties is confirmed by the high proportion of genes from wild relatives in modern wheat backgrounds. About 20 % of the new CIMMYT and up to 24 % of ICARDA material include synthetic background (Ogbonnaya et al. 2013).

6 Gene Flow Constraints in Plant Breeding

Wheat and rice are the main staple crop in the world while legumes are among the main sources of proteins in the most of developing countries and contributes 33 % human dietary nitrogen requirements. However, to continue feeding the increasing population wheat production has to be increased at least 1.6 % per year (Calderini et al. 1999). Crop landraces and wild relatives have the potential of providing superior alleles that can help in increasing yields (Reif et al. 2005). However, some challenges have to be overcome by the breeders and pre-breeders to be able to fully and efficiently identify useful alleles and make them available for breeding programs.

The main constraint for gene flow in breeding nowadays is the loss of genetic diversity among the modern wheat varieties. Modern lines already adapted to today's cultural practices and ensuring the quality requirements of the wide range of end-uses define the fastest and easiest way to introduce new alleles of interest in a breeding germplasm without important undesirable linked alleles. However, the wide use of Mega-varieties or varieties covering wide extensions and used massively as parents for future varieties may have created a genetic bottleneck for crop diversity.

The use of landraces and other lines non-adapted to the target region for any crop in a breeding program to increase genetic diversity has, however, some inconveniences. The direct use of a landrace in the crossing block will result in a progeny with several unsuitable alleles to select against, increasing the number of F_2 plants needed to identify the ones with acceptable agronomical types. Wheat landraces usually carry alleles for tall straw or photoperiod sensitivity (Worland and Snape 2001) unsuitable for wheat productivity and adaptation while in modern varieties these traits have been already fixed. Especially, adaptation to modern culture or bread-making techniques and processes are significantly different than the ones used when these landraces were selected by farmers. Also, the introduction of foreign material may alter suitable co-adapted gene complexes fixed in the elite material. Normally, backcrosses to elite material will help in overcoming the linkage drag. Even after 20 years of conventional breeding, a single gene transferred by backcrossing may still be linked to a DNA region carrying more than 100 potentially undesirable genes (Young and Tanksley 1989). However, marker assisted backcross, when closely linked markers are available, can greatly simplify this task.

The use of wild relatives, as it was previously stated, can help in improving modern wheat diversity beyond the limits of the landraces germplasm. However, besides the laborious and sometimes expensive process of making synthetic wheat, other complications may reduce the selection efficiency when they are used in a crossing block. Thus, normally primary SHW lines (directly developed from the crosses of wheat with wild relatives normally a tetraploid and *Ae. tauschii*) carry too many undesirable alleles to produce useful F_2 population. Breeders usually prefer to backcross the primary synthetic to increase the chances of obtaining suitable BC_XF_2 plants to select superior SDW. Unfortunately, this implies that two full breeding cycles – the breeding cycle to obtain a suitable synthetic derivative and the breeding cycle to obtain the final variety after the cross of the derivative with a modern adapted variety (SHW>SDW>variety released) – are needed to obtain a variety with superior performance to be released.

Additional complications may occur when crossing SHW and modern varieties: Often the F_1 plants from such crosses have a high frequency of hybrid necrosis, usually lethal or semi lethal, resulting in gradual death or loss of productivity (Tomar et al. 1991; Tomar and Singh 1998). Hybrid necrosis is controlled by two loci, *Ne1* and *Ne2* (Pulkhalky et al. 2000). The presence of at least one copy of these two loci dominant alleles, located in the BB genome (Nishikawa et al. 1974), results in hybrid necrosis. Modern bread and durum wheat varieties vary in the frequency of these alleles (Trethowan and Van Ginkel 2009) and, therefore, the cross between SHW and modern varieties often results in the combination of both dominant alleles and a necrotic F_1. Most of the research in this area has been based on phenotypic characterizations of modern varieties according to the alleles they carry at these loci

(Singh et al. 1989; Pukhalskiy and Bilinskaya 1998; Pukhalskiy et al. 2000). Recently, studies have been conducted for trying to fine-map these loci (Chu et al. 2006) in order to develop diagnostic markers to select the most suitable modern cultivar to cross with SHW avoiding the risks of hybrid necrosis.

In spite of the successful introgression of new alleles from wild relative backgrounds through SHW or SDW, increasing breeder's ability to more effectively identify the best combinations remains a main objective to increase breeding efficiency. For instance, studies have been carried out trying to identify biotic and abiotic resistance traits in *Aegilops tauschii* to increase the chances of expressing them into SHW. The results of these studies showed there is no apparent relationship between *Ae. tauschii* performance and physiological traits and their expression in the synthetic lines obtained (Sohail et al. 2011). In fact, several major rust resistance genes have been identified in *Ae. tauschii* (Villareal et al. 1992; Aguilar-Rincón et al. 2000) but in some cases, the SHW developed with the *Ae. tauschii* lines showed reduced or no resistance (Assefa and Fehrmann 2000). This indicates the importance of assessing the mechanisms of interaction between the three genomes to more effectively introgress genes of interest from wheat wild relatives.

Finally, one of the most important traits for wheat market value, end-use quality, may be negatively affected when introgressing genomic regions from wild relatives. In spite of the number of potential new quality alleles of interest already described in SHW lines (Pflüger et al. 2001; Davies et al. 2006;), suitable wheat quality traits for modern bread-making practices are not often linked to agronomic performance. Since most of the suitable quality alleles identified and fixed in the modern varieties are a result of artificial selection and do not provide actual fitness improvements, natural selection may have negative or no effect on them. Therefore, a number of the SHW selected on the base of other traits of interest will not carry suitable quality alleles needing several backcrossing and selection events to select suitable lines with acceptable quality. For instance, a higher frequency of soft grains has been reported (Lillemo et al. 2006) in SHW as compared to modern lines indicating the need of having this trait in mind when developing SDW. An analogous case would be the 1BL/1RS wheat rye translocation. This translocation, as it was stated before, introduced in the wheat genome several alleles of superior agronomic performance and lines carrying it were widely used as parents in worldwide crossing blocks. However, as a side effect, it also had a negative effect in wheat end-use quality, mostly regarding the modern bread-making techniques (Pena et al. 1990).

7 Breeding Strategies

Plant breeding can be broadly defined as alterations caused in plants as a result of their use by humans, ranging from unintentional changes resulting from the advent of agriculture to the application of molecular tools for precision breeding. The core of plant breeding is the selection of better types among variants, in terms of yield and quality of edible parts; ease of cultivation, harvest, and processing; tolerance to environmental stresses; and resistance to diseases and pests. Each of these aspects of agronomic or food value can be dissected in many specific traits, each presenting its own range of variation. Manipulating a single trait, disregarding all others, is relatively straightforward; however, this is unlikely to result in a useful variety. The vast diversity of breeding methods can be simplified into three categories: (i) plant breeding based on observed variation by selection of plants based on natural variants appearing in nature or within traditional varieties; (ii) plant breeding based on controlled mating by selection of plants presenting recombination of desirable genes from different parents; and (iii) plant breeding based on monitored recombination by selection of specific genes or marker profiles, using molecular tools for tracking within-genome variation. The continuous application of traditional breeding methods in a given species could lead to the narrowing of the gene pool from which cultivars are drawn, rendering crops vulnerable to biotic and abiotic stresses and hampering future progress (Breseghello 2013).

7.1 Screening for Biotic and Abiotic Stresses

Biotic and abiotic stresses are the major limiting factors for high crop productivity in drylands worldwide. In response to abiotic stresses, plants undergo a variety of changes at the molecular level (gene expression) leading to physiological adaptation. Drought, cold and salinity are the major abiotic stresses which severely affect yield and quality in many regions of the world endangering the food security. The situation has become more serious with the global climate change. Therefore, studies on abiotic stress tolerance have become one of the main areas of research worldwide (Patade et al. 2011; Mohammadi et al. 2014, 2015). Indirect selection for stress tolerance may be performed at the laboratory level or in open field conditions. Chlorophyll fluorescence and thermal imaging are well-established, powerful, non-destructive, and rapid techniques for screening plants against abiotic stresses in crop plants (West et al. 2005).

7.2 Genotype × Environment (GE) Interaction

Selection for greater tolerance to abiotic stresses such as drought and heat stresses, via the identification of high-yielding genotypes, is an important element in the development of sustainable agriculture systems in drylands. However, when a genotype is grown in different environments, it frequently shows significant fluctuation in yield performance in comparison to other genotypes. These fluctuations are influenced by environmental conditions (i.e. water availability, temperature) experienced at any given location or year and are often referred to as genotype × environment (GE) interactions (Allard and Bradshaw 1964). GE interaction affects breeding progress because it makes difficult the evaluation and selection of superior genotypes. On the other hand, GE interaction may also offer opportunities, especially in the selection of genotypes showing positive interaction with a specific location and its prevailing environmental conditions (exploitation of specific adaptation) or of genotypes with low frequency of poor yield or crop failure (exploitation of yield stability) (Ceccarelli 1996; Annicchiarico et al. 2005; Mohammadi et al. 2010).

Yan and Kang (2003) described the different types of GE interactions and highlighted the implications of these in plant breeding and crop production. Crossover GE interactions (change in rankings of genotypes across environments) are of greatest interest to breeders as these directly affect genotype selection in specific environments. Consequently, promising selections in one environment may perform poorly in another. Ignoring significant GE interactions in favor of resource savings can have detrimental effects. With regards to genetic gains from selection, large GE interactions, as components of total phenotypic variance, affect heritability (proportion of total phenotypic variance that is due to genetic variance) negatively. The larger the GE interaction, the smaller the heritability estimate; thus, progress from selection would be reduced as well (Yan and Kang 2003). GE interaction has been a focus of plant breeders as early as the 1950s, and there is a wide range of literature outlining examples and methods of dealing with this phenomenon. The primary objective of multi-environment trials (METs) is to identify superior genotypes for a target region, and to determine if the target region can be subdivided into different mega-environments (Yan et al. 2000). A mega-environment may be defined as a portion of a crop species' growing region with a homogeneous environment that causes some genotypes to perform similarly (Gauch and Zobel 1997), and is normally identified through analysis of MET data. Currently, there is a wide range of statistical techniques used for the analysis of yield trials data collected from METs. These models can be linear formulations such as joint-regression (Yates and Cochran 1938; Finlay and Wilkinson 1963; Eberhart and Russell 1966; Tai 1971), multivariate clustering techniques (Lin and Butler 1990), and multiplication approaches such as additive mean effects and multiplicative interaction (AMMI; Zobel et al. 1988; Gauch 1992) and genotype plus GE interaction (GGE; Yan et al. 2000) biplot analyses. Modeling GE interaction in METs helps to determine phenotypic stability of genotypes, but this concept has been defined in different ways and increasing numbers of stability parameters have been developed (Gauch and Zobel 1997).

The effectiveness of genotype evaluation as part of breeding is influenced by (i) understanding of GE interaction and (ii) the degree to which the environments sampled in the MET represent the target population of environments (TPE) (Podlich and Cooper 1998). While statistical approaches to estimate GE interaction have developed over more than 50 years with specific environment variables being included in mixed models (van Eeuwijk et al. 2005) and with environment classifications being identified in principal component analysis (Yan et al. 2000). In parallel with these methodological improvements to analyze GE interactions within a MET, characterization of the complete TPE has been conducted based on (i) climate and soil data (Pollak and Corbett 1993; Hodson and White 2007), (ii) physiological traits that integrate various stresses that plants experience, for example, flowering date or yield (Hernandez-Segundo et al. 2009), and (iii) stress index determined with crop models (Chapman et al. 2000).

7.3 Specific vs. *Wide Adaptation*

There are two possible strategies for a plant breeder interested in developing varieties which show a low GE interaction, namely (i) the subdivision of a heterogeneous area for which the varieties are being bred into smaller regions such that each of them has a more homogeneous environment and its own characteristic varieties, and (ii) the introduction of varieties which show a high degree of stability in performance over a wide range of environments (Annicchiarico et al. 2005). However, the first strategy can also encounter large interactions of genotypes with locations even within a sub-region and even within the same location over seasons. In recent decades, various programs run at international research centres have modified their breeding strategy to produce genotypes suitable for cropping in less favourable areas. A specific adaptation strategy has been pursued for barley improvement at the ICARDA, with selection under conditions similar to those in the target environment (favourable or drought-stressed), and with implications extending also to elements of the breeding strategy, such as the choice of genetic resources and variety type. A "shuttle breeding" procedure- alternate selection in drought-stressed (unfavourable) and irrigated (favourable) environments- was established for the selection of widely adapted germplasm at CIMMYT. However, the possibility of GE interaction between environments of similar ecological potential (based on crop mean yield) may lead to the definition of different sub-regions also within unfavourable and moderately favourable areas. For this reason, barley breeding at ICARDA began producing specific material for different sets of drought-prone countries; rice breeding at the International Rice Research Institute (IRRI) defined specific plant ideotypes for several different ecosystems; and wheat breeding at CIMMYT is attempting a compromise between a wide adaptation prospect and the opportunity to breed specifically for 12 different mega-environments (Braun et al. 1996). The specific adaptation strategy is receiving increasing attention in developing countries, sometimes in combination with participatory plant breeding schemes. The participation of farmers can: i) support the multilocational selection work; ii) allow for exploiting possible specific adaptation effects even within sub region; iii) contribute to enhance the biodiversity of material under cultivation, thereby improving production stability; and iv) facilitate the seed supply to farmers via local seed systems. For small farmers in relatively poor countries, such systems may be far more important than formal seed systems (Annicchiarico et al. 2005; Ceccarelli 2015).

7.4 Classifying Environments and Defining Candidate Sub-Regions

In MET, environments that are similar in terms of genotype response can be grouped by different methods, and each group may identify a cropping area that is relatively uniform because genotype x location (GL) interaction effects are limited or negligible. Such areas have been termed by different authors as subregions, subzones, subareas, macro-environments or mega-environments (Annicchiarico et al. 2005). Based on this, genotypes with specific adaptation can be recommended to specific sub-regions and genotypes with wide adaptation can be recommended to area with different environmental conditions. Different sub-regions may be identified not only within large regions (DeLacy et al. 1994; Mohammadi et al. 2010) but also within relatively small regions, as suggested by results from northern Syria (Ceccarelli 1996); Italy (Annicchiarico 1997) and northern Italy (Annicchiarico 2002); New South Wales (Seif et al. 1979; Basford and Cooper 1998); Queensland (DeLacy et al. 1996); southwest Canada (Saindon and Schaalje 1993); Ontario (Yan et al. 2000); and west of Iran (Mohammadi and Haghparast 2011). Subregions may be defined for variety recommendation: each sub region then coincides with a recommendation domain, grouping those sites with the same best-performing genotype(s) (Gauch and Zobel 1997). The definition of subregions is not just geographical, but may also encompass farming practices (e.g. irrigated or rainfed cropping). Subregions have sometimes been defined on the basis of site similarity for environmental factors that are supposed to be important but are, in fact, chosen without a definite assessment of their impact on GE interaction (Pollak and Corbett 1993). However, additional information on the climatic, soil, biotic and crop management variables closely related to the occurrence of GE interaction may help locate geographic boundaries for subregions, besides contributing to the understanding of causal factors for the interaction (van Eeuwijk et al. 1996, 2005). The zoning process should produce subregions that can be defined on a geographical basis or by other means, such as climatic factors or management practices, in order to be useful for breeding or cultivar recommendation. In this case, recommendations concern the whole region. In other instances, individual locations or small groups of sites may be added to the larger group to form a unique subregion for breeding or recommendation (Annicchiarico 2002).

8 Breeding Methods

Plant breeding has been a key science in improving crop production, with an estimated contribution to productivity increases of around 50 % (Fehr 1984). Over the past decades, it has remained a vibrant science, with continued success in developing and deploying new cultivars on a worldwide basis (Gepts and Hancock 2006).

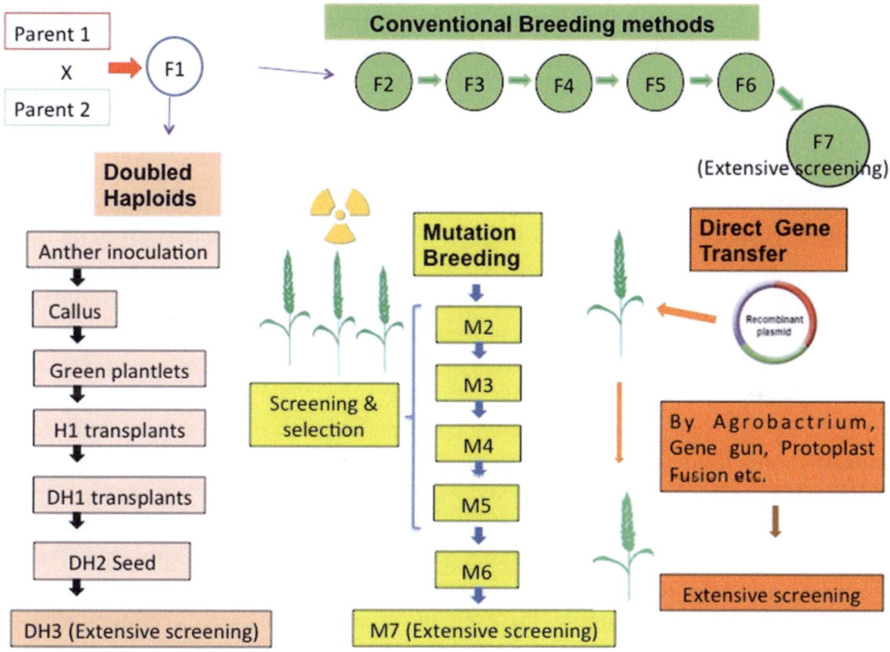

Fig. 1 Comparison of various plant breeding methods

A plant breeding program is a cyclical process aimed at the development of new cultivars, whereby each cycle consists of three major phases (Fig. 1; Gepts 2002; Ceccarelli 2009): (i) generating genetic variability: this includes making crosses, inducing mutation, introducing exotic germplasm, and using genetic engineering techniques; (ii) selection and testing to identify superior recombinants: in self-pollinated, cross-pollinated, and vegetatively propagated crops, which is done with different methods such as marker-assisted selection, introgression of quantitative trait loci (QTL), or use of high-throughput phenotyping platforms, and which terminates with the identification of potential cultivars; (iii) release, distribution, and adoption of new cultivars: the yield testing in MET is either the last step of the second phase or the first step of the third. Two additional steps are often included as essential components of a breeding program, namely, setting the objectives and evaluating the program in terms of reaching those objectives. These two steps come before and after, respectively, the three phases mentioned earlier (Ceccarelli 2015).

8.1 Germplasm Development

Crop genetic diversity, created through natural and human selection over millennia and complemented by the diversity present in wild relatives of crop plants, provides the raw material that can be employed by scientists to improve crop productivity and diversify production systems. But genetic variation, once considered unlimited, is fast eroding as modern breeding lines replace traditional cultivars over large areas (Stamp et al. 2012).

Plant genetic resources need to be preserved to combat evolving and rapidly emerging new strains of pests and diseases, and to produce varieties that are better-adapted to changing climatic and environmental conditions and produce stable yields under such variable conditions. To provide the necessary building blocks for breeders to successfully develop improved and well-adapted varieties, gene bank personnel engage in the collection, assembly, maintenance and conservation, characterization, documentation, and distribution of germplasm for research and development (FAO 2010). Many of the accessions are local landraces, wild relatives of cultivated crops, and indigenous varieties that are being lost as farmers adopt new high yielding varieties of high value vegetable crops. Their preservation and their availability for utilization in research and breeding are of utmost importance to ensure future food and nutrition security of a rapidly growing population (FAO 2010; McCouch 2013). Landraces and crop wild relatives are increasingly valued and exploited for genes that provide increased biotic resistance, tolerance to abiotic stress, and improved yield and quality (Jackson et al. 2007; Frison et al. 2011). These genetic resources, which are threatened by climate change, are the raw materials that are needed to improve the capacity of crops to respond to climate change and other future challenges and to secure nutritious food for a growing world population (Schreinemachers et al. 2014). Genetic variability present in collections and available parental germplasm is important in generating improved germplasm with desired traits that help to increase crop production and thus improve human nutrition. The germplasm of diverse plant species is maintained in gene banks around the world, with collections holding anywhere from hundreds to thousands of accessions. Together with the important role of conserving genetic resources, gene banks also distribute these accessions for use by breeding programs to develop new cultivars of crop and pasture species (Mohammadi and Amri 2013).

8.2 Cultivar Improvement

Increasing CO_2, global mean temperatures, varying rainfall patterns and frequent weather changes are occurring due to climate change. Such factors place a direct impact on the health and wellbeing of crops, thereby affecting small landholders, subsistence agriculture and food security in the developing world (Howden et al. 2007). Crop modeling shows that climate change will likely reduce agricultural production, thus reducing food availability (Lobell and Field 2007) and affecting food security. Plant breeding, appropriate crop husbandry, sound natural resource management and agricultural policy interventions will be needed to ensure food availability and reduce poverty in a world affected by climate change (Howden et al. 2007). Persistent efforts in various research fields have been going on to develop new cultivars that can respond to environments with abiotic stresses (Mohammadi et al. 2014, 2015). A list of crop varieties released by Dryland Agricultural Research Institute, Iran for adaptation to drylands is given in Table 2.

Abiotic stresses aggravated by climate change pose a serious threat to the sustainability of crop yields and account for substantial yield losses. Scientific knowledge of the processes of abiotic stress tolerance in crops continues to develop and guides conventional breeding and genetic engineering of new crop cultivars. The modern tools of cell and molecular biology have shed light on control mechanisms for abiotic stress tolerance, and for engineering stress-tolerant crops based on the expression of specific stress-related genes (Pande et al. 2015).

Allard and Bradshaw (1964) inditaed that variety types where the genetic structure implies high levels of heterozigosity and/or heterogeneity are less sensitive to environmental variation and are, therefore, more stable-yielding. Such types may sometimes offer fewer opportunities for maximizing the yield potential. Within a given variety type, breeding successfully for this trait relies on the adoption of a heritable or repeatable stability measure as a selection criterion. Given the high sampling error, the assessment of yield stability requires numerous test environments (at least eight) to guarantee reliability (Kang 1998). Therefore, direct selection for yield stability may be limited by high costs and can be recommended, even when it has high priority, only for elite material in the final testing stages. The choice of parental germplasm with recognized yield stability and of, if possible, a convenient variety type, can play a major role in breeding for more stable crop yields. In addition, indirect selection for higher yield stability may be attempted by selecting for agro-physiological traits that have proved to be strictly associated with stability (Annicchiarico et al. 2005).

8.3 Participatory Varietal Selection

The centralized plant breeding techniques of the green revolution have yielded good results in more favorable agricultural environments. However, most low-resource farmers in marginal areas have not benefited from these varieties. As an alternative to centralized breeding, farmer participatory approaches using participatory varietal selection (PVS) and participatory plant breeding (PPB) can be used. PPB is an extension of PVS. In PPB, farmers are actively involved in the breeding process, from setting goals to selecting variable, early generation material. In PVS, farmers are given a wide range of new cultivars to test for themselves in their own fields. In PPB programs, the results of PVS were exploited by using identified cultivars as parents of crosses. Participatory research, which allows farmers, research scientists and extension agents to conduct research together, is essential particularly for identifying the traits preferred by farmers. Farmers' fields provide a multitude of diverse environments which allow exploiting GE interaction effect between the farmers' fields and research stations given that in most cases, and particularly under semi-arid conditions, they are different (Ceccarelli and Grando 2007). In addition, farmers may use different selection criteria from those used by the breeders. Conventional

Table 2 Crop varieties released by Dryland Agricultural Research Institute, Iran for adaptation to drylands

Crop	Cultivars released		Mean (kgha^{-1})	Superiority to check (%)	Characteristics	Target areas	Year of release
	Number	Name					
Bread wheat	6	Gahar, Zagros, Niknezhad, Kohdasht, Karim, Aftab	3183–3897	110–119	Earliness, resistance to diseases, drought tolerant	Warm and moderate warm	1996–2015
	3	Azar-2, Baran, Hashtrod	1837–2383	105–114	Earliness, tolerant to cold and drought, high kernel weight, yield stability	Cold and moderate cold	1999–2015
	4	Rasad, Homa, Ohadi, Takab	1932–4436	104–140	Tolerant to cold and drought, high kernel weight, good response to supplemental irrigation	Cold areas	2007–2012
	1	Rijaw	2611	121	Earliness, resistance to lodging, good quality	Moderate cold	2011
	1	Ghaboos	3514	105	Good response to supplemental irrigation, good quality, resistance to diseases	Warm areas	2014
Durum wheat	2	Seimareh, Dehdasht	3190–4015	111–131	Earliness, good quality, resistance to diseases, drought tolerant	Warm and moderate warm	1996–2008
	1	Saji	2669	123	Good quality, good response to supplemental irrigation, resistance to diseases	Moderate cold and moderate warm	2010
Barley	3	Izeh, Mahoor, Khorram	3556–4372	112–131	Earliness, yield stability, resistance to diseases and lodging	Warm and moderate warm	1996–2011
	1	Sahand	2000	–	Moderately resistant to cold, high kernel weight, two-row spike	Cold and moderate cold	1996
	2	Abidar, Ansar	2138–2724	104–105	Earliness, resistance to drought, cold tolerant	Cold and moderate cold	2007–2014

	2	Sararood-1, Nader	3066–3244	105–116	Resistance to diseases, high kernel weight	moderate cold	1998–2012
Chickpea	1	Hashem	2000	–	High plant height, high kernel weight, early cooking, resistant to *Aschochita blight*	Warm and moderate warm	1997
	2	Saral, Samin	600–919	120–132	Resistant to cold, tolerant to *Aschochita blight*, high protein	Cold areas	2012–2015
	3	Arman, Azad, Adel	1472–1651	143–170	High yield, high kernel weight, yield stability, resistant to *Aschochita blight* and Fusarium	Moderate cold and moderate warm	2004–2014
Lentil	1	Gachsaran	914	155	Earliness, tolerant to environmental stresses (drought and heat), resistant to diseases	Warm and moderate warm	1999
	2	Kimia, Bile sawar	951–1133	120–140	High yield, high kernel weight, yield stability, tolerant to Fusarium	Moderate cold and moderate warm	2009–2012
Safflower	1	Sina	1344	242	Earliness, high yield, yield stability, drought tolerant	Moderate cold and warm	2007
	1	Faraman	1380	113	Drought tolerant, spineless, high kernel weight	Moderate cold	2011
Brassica	1	Sadegh	900	125	Earliness, resistant to shattering, drought tolerant	Cold areas	2015
	1	Shiralee	1772	–	High yield, tolerant to environmental stresses (drought and heat)	Warm areas	2009
Forage crops	1	Maragheh	2240	207	Drought tolerant, high stature, relatively earliness, high protein	Cold and moderate cold	2009
	1	Gol sefid	3500	270	High yield, cold tolerant, high protein	Cold areas	2011

plant breeding (CPB) has been successful for farmers in high potential environments. However, it did not make good progress in marginal environments because such environments are highly diverse and farmers are generally poor, hence they cannot afford expensive inputs including certified seed (Morris and Bellon 2004; Ceccarelli and Grando 2007). In CPB, initial selection in large populations of breeding materials is conducted on optimally managed research stations by breeders. PVS has been reported as an efficient approach for disseminating new improved varieties (Ortiz-Ferrara et al. 2001; Thapa et al. 2009). It is capable of better addressing farmers' needs of new varieties that very often are not recognized using conventional non-participatory varietal development approach. PVS could complement ongoing varietal development efforts in the region to help farmers by providing them with a wider option of germplasm to evaluate and adopt under their own conditions (Witcombe et al. 2003). PVS was successfully used in maize in Africa (Snapp 1999; De Groote et al. 2002) and has proven to be an efficient approach to developing and disseminating new varieties through close collaboration with farmers in South Asia as well (Ortiz-Ferrara et al. 2007).

8.4 Indirect Selection Approaches

The genetic diversity of many crops has greatly reduced with the introduction of improved varieties. The choice for selection in modern germplasm is somehow limited now. The diversity of germplasm has a significant impact on the improvement of crop plants. Narrow genetic diversity is a bottleneck in breeding; it is necessary to incorporate more genetically diverse germplasm in the breeding program in order to broaden the gene pools of various crops. Genetic variation among parental lines is necessary to derive superior progeny for crossing and selection. However, crosses are often performed among elite genotypes having similar agronomic traits.

The main goal of most plant scientists is to improve/increase yield. Breeding programs are usually started with the intention of improving some genotype/cultivar etc. Depending on the objective, various approaches are taken while starting the breeding program. The improvement may be thought identifying genotypes tolerant to biotic and abiotic stresses or increase in yield *per se*. Traditionally conventional breeding approaches are taken which are somehow straightforward, however to accelerate the breeding progress advantage of some technologies needs to be taken.

Studies comparing the chlorophyll fluorescence with the conventional techniques indicated that chlorophyll fluorescence can be a useful tool for screening biotic and abiotic stress tolerance in various crops (Matous et al. 2006). Thakur (2004) screened fruit crops for drought tolerance based on indices, namely xylem water potential, relative water content, chlorophyll stability index, drought injury index, and rapid test for drought tolerance. According to Chen et al. (2007), only the chlorophyll fluorescence method has often been the most attractive tool for rapid and sensitive screening with fully automated fluorometers. Positive correlation of grain carbon isotope discrimination under post-anthesis drought stress with economic yield has been established in wheat; therefore, these indices may be used as indirect selection criteria for wheat grown under stress environments (Zhu et al. 2009). Siddiqi et al. (2009) screened safflower accessions for salt tolerance based on biomass (shoot and root dry mass) and other physio-biochemical parameters i.e., photosynthesis, transpiration, stomatal conductance, and chlorophyll a and b at the vegetative stage. Mohammadi et al. (2011) screened durum wheat breeding lines using stress tolerance index (STI) for drought, heat and cold stresses. Mohammadi et al. (2014) identified the agro-physiological traits such as canopy temperature, SPAD reading, plant height, flag-leaf length and heading date as the most important traits for screening drought tolerant durum wheat breeding lines.

8.5 Doubled Haploids

Doubled haploids method accelerates the breeding program in terms of development of homozygous lines from heterozygous plants after crossing two different parents. This procedure saves time by excluding several generations of selfing required before obtaining pure lines. It is also attractive to breeders and genetists because of expression of simple recessive genetic traits or mutated genes. Doubled haploids in wheat have been produced from anther culture (Ouyang et al. 1973), using wide hybridization (Schaeffer et al. 1979) and isolated microspore culture (Wei, 1982). It is noteworthy that the doubled haploids produced by different methods are equally the same (Guzy-Wrobelska and Szarejko 2003), however Guzy-Wrobelska et al. (2007) reported that doubled haploids derived by anther culture had a high recombination frequency.

Anther culture is more cost-effective than other methods of doubled haploid production (Guzy-Wrobelska et al. 2007) such as intergeneric crossing method (Pratap et al. 2006). Anther culture involves the induction of embryoids into plantlets and is known for its highly genotype specific response (Tawkaz 2011). Ouyang et al. (1973) reported the *in vitro* regeneration of plants from pollen. A certain proportion of pollen grains *in situ* are embryogenic. These pollen grains when placed on an artificial media can develop into embryos (Ouyang et al. 1973; Liu et al. 2002).

8.6 Embryo Rescue

Embryo rescue also known as embryo culture refers to subsequent collection of techniques, which enables a weak embryo into viable plant. Embryo rescue is often used after procedures like hybridization (especially incompatible crosses), transformation, doubled haploid production, etc.

The use of embryo rescue in wheat breeding was introduced primarily for the production of interspecific and intergeneric crosses. Since one of the most common reasons in interspecific and intergeneric crosses is embryo abortion, also in case of successful fertilization in wide hybridization the seeds are weak, seed development is not perfect or the seed produced are not viable. Embryo rescue has been very successful in overcoming these main barriers in many plant species (Collins and Grossser 1984).

Artificial media is an important factor in embryo rescue procedure. Young embryos need a more complex media compared to a mature one. Generally, the Murashige and Skoog (Murashige and Skoog 1962) and Gamborgs B-5 (Gamborg et al. 1968) media are used. The concentration of the media depends upon the developmental stage of the embryo. Sometimes growth regulators, vitamins and amino acids are used in the media (Reed 2005).

Embryo rescue is an important procedure utilized in the production of haploids through elimination of chromosomes following wide hybridization. In wide hybridization when the fertilization occurs the pollen parents chromosomes are eliminated afterwards by the seed parent. To obtain a viable haploid the embryo needs to be rescued. The chromosomes of haploid can be doubled to obtain a double haploid (Bridgen 1994). Embryo culture can also avoid dormancy problems if the dormancy is due to seed coat, thus shorting the breeding cycle by rescuing the embryo. Sharma and Gill (1983) reported a reduction of generation time by 40 days.

The success of development of an embryo depend on many factors; along with other factors one of the most important factor is plant genotype, differences were found among different cultivars (Collins and Grosser 1984). In addition to this, light and temperature also effect the development of a reduced embryo.

8.7 *Wide Hybridization*

Genetic diversity of the wheat gene pool has been narrowed down in the last four decades. One reason of this narrowing down is the result of selection of pure line, high yielding and responsive to high input varieties. In ancient times and before the spread of narrow based modern cultivars, there was a lot of genetic diversity in the landraces grown by farmers; sometimes the crops planted by farmers were mixtures of landraces: modern wheat breeders have emphasized pure-line cultivars, narrowing the genetic diversity. Few types of diversity can be measured in the perspective for breeding programs (Smale 1996).

During the Green Revolution, wheat lost a lot more genetic diversity as emphasis was placed on pure lines with higher yields. The improved wheat varieties growing in most of the farmers' fields (in certain area/country) are somehow similar. Despite the fact that these varieties are high yielding and are economically of great importance but mostly have the same mechanism of resistance to most of the wheat diseases. Some disease breaks out or mutation in pest population may cause severe stress resulting devastating crop failure, varieties having similar background will be more vulnerable than varieties having genetically different backgrounds. Breeders should have a stock of resistance; may it even be in the gene banks only. To increase the genetic diversity of any crop species rather broaden the crop gene pool is to take the advantage of available genetic resources present in nature, the easiest way to exploit are the landraces comparatively and the wild relatives.

Variation is the basis for improvement, so we need genetic variation for progress in crop breeding. Currently there is not enough variation to break the yield barrier or the traits/genes have not been discovered yet. Genetic variation in crops can be created and enriched either by gene transformation, or by exploiting variation in the wild wheat relatives through inter specific and inter-generic crosses. Wild relatives of species often the progenitors of that species (Hajjar and Hodgkin 2007). Plant breeders have been continuously exploiting the wild wheat relatives dominantly for biotic stresses (Hajjar and Hodgkin 2007).

Many useful sources of disease and pest resistance are available in the wild relatives of wheat (Zaharieva et al. 2001). Mujeeb-Kazi and Kimber (1985) have reported the usefulness of wild relatives of wheat for wheat improvement. Among the wild species of wheat considerable genetic diversity has been found in *Ae. tauschii* and *T. turgidum* (Xie and Nevo 2008, Sohail et al. 2012). The wild relatives of wheat have adapted to various environmentally harsh conditions for thousands of years, there has been natural selection and only the fittest have survived. There are a total of the 325 wheat wild relatives out of which around 250 are perennials and 75 are annual grasses of the tribe *Triticeae* (Dewey 1984). Very few have been utilized for hybridization with wheat (Mujeeb-Kazi et al. 1996). Although many of the wild relatives of wheat have not been hybridized with wheat, but have been used to transfer useful genes to wheat.

Gene transfer from wild species to cultivated crops have been reported by many researchers; such as resistance to stem rust and loose smut from *Triticum tauschii*, Hessian fly resistance from *T. timopheevi, T. monoccun, and T. turgidum* (McFadden 1930). Goodman et al. (1987) summarized the transfer of important traits and genes for different crops including bread wheat. He reported that high kernel protein from *Aegilops ovata*, stem rust resistance from *Ae. speltiods, T. timopheevi, T. monoccun, and T. turgidum*, leaf rust resistance from *Ae. squarrosa, Ae. umbellulata and Agrophyron elongatum*, yellow rust resistance, powdery mildew resistance, winter hardiness, leaf rust resistance, stem rust resistance from *Secale cereale* have been transferred to bread wheat. Resistance to stem rust, powdery mildew and green bug have been transferred from *Secale cereale* to bread wheat (Friebe et al. 1996; Graybosch, 2001; Yediay et al. 2010).

Nevo et al. (2002) reported a high natural variation of traits in the population of emmer wheat (*Triticum dicoccoides*). The Many of these traits could be useful sources for wheat improvement to stresses such as salinity (Nevo et al. 1992), drought (Peleg et al. 2005), powdery mildew (Moseman et al. 1984), leaf rust, stem rust, stripe rust and tan spot (*Pyrenophora tritici-repentis*) (Chu et al. 2008).

The genetic diversity within the D genome of *Ae. tauschii* (Naghavi and Mardi 2010), which is much higher than within that of wheat (Reif et al. 2005). The D genome is a rich source of resistance to various biotic and abiotic stresses that could contribute to the improvement of wheat (Cox et al. 1994, Assefa and Fehrmann 2000; Colmer et al. 2006).

Wild barley has also shown to be an important source of genes/traits for crop improvement. Various morphological, agronomic and physiological traits of interest have been reported in wild barley, which result in tolerance of barley under drought stress conditions (Ivandic et al. 2000; Grando et al. 2001). Wild barley is also a source of resistance to diseases such as powdery mildew (*Blumeria graminis*f.sp. *hordei*), rust (*Puccinia hordei*) and scald (*Rhynchosporium secalis*) (Ceccarelli and Grando 1987; Eglinton et al. 1999).

Natural hybridization may have occurred for evolution of wheat species and types; the wild diploid wheat *Triticum monococcum* (genome A^mA^m) naturally crossed with the *Aegilops speltoides* (genome BB) millions of years ago which led to evolution of tetraploid emmer wheat, *Triticum dicoccum* (AABB) (Mujeeb-Kazi et al. 1996). The domestication of emmer give rise to durum wheat, *Triticum turgidum* var. *durum*. Somewhere in Fertile Crescent around 8000 to 10,000 years ago the durum wheat ($2n = 28$, AABB) hybridized with a diploid wild grass *Aegilops tauschii* (genome DD) which led to the origin of hexaploid wheat (*Triticum aestivum*) (Mujeeb-Kazi et al. 1996, Feldmann 2001).

Hawkes et al. (2000) reported that the old wheat varieties are now difficult to find in Turkey, Iraq, Afghanistan and Pakistan, which 30 years ago were easily available. Smale et al. (2002) reported that in 1997 the landraces of spring bread wheat were planted on only 3 % of the cropped area. Hirano et al. (2008) reported that landraces have rare alleles. There are many landraces which have not been explored some of them are now only limited to gene banks these genetically diverse germplasm collections in genetic resource centers is of great importance (Hoisington et al. 1999) and the use of this germplasm is vital to raise yield potential and stability and for improvement in tolerance to diseases and pests (Tanksley and McCouch 1997; Van Becelaere et al. 2005).

In grain legumes, important sources of resistance in wild species can be exploited in chickpea, pigeonpea and groundnut. Wild species of chickpea, pigeonpea and groundnut have been extensively studied and quite few of them have demonstrated high level of resistance/tolerance to a number of stresses. Such as wild chickpea species, *C. bijugum, C. judaicum*, and *C. pinnatifidum* have shown important sources resistance/tolerance to multiple stresses (Sharma et al. 2013). Wild pigeonpea species particularly, *C. scarabaeoides, C. acutifolius, C. platycarpus, C. reticulatus, C. sericeus*, and *C. albicans* have resistance to pod borer (Sujana et al. 2008; Sharma et al. 2009). Wild pigeonpea species have resistance to *Alternaria* blight, *Phytophthora* blight, mosaic virus, nematodes and also show tolerance to drought and salinity (Upadhyaya et al. 2013). Wild groundnut species possess resistance/tolerance to many biotic/abiotic stresses (Upadhyaya et al. 2012).

9 Barriers to Interspecific Hybridization

The hybridization procedure starts with pollination of the flower of one species with the pollen of the other. After gametes fusion/fertilization and then cell division, seeds of the hybrid are formed containing the embryo and the endosperm for male and female parents. In the process of producing interspecific hybrids in bread wheat several problems are confronted; these problems are known to be the barriers. The nature of a barrier determines the procedure to be used to overcome the barriers. Producing F_1 hybrids even within the *tritceace* requires special tools for successful hybrid production.

9.1 Pre-fertilization Barriers in Wide Crosses

In a natural cross the pollen of the donor male plant has to germinate on the stigma of the recipient female plant. The pollen and stigma interaction is very specialized in identifying certain genotypic combinations. Protein in the pollen coat, which is originally formed by fraternal parent, plays a crucial role in the recognition events of pollen reception by the stigma on the recipient maternal plant (Baum et al. 1992). The pollen adhesion to the stigma is possible with a cell to cell recognition system and a compatible combination (Heslop-Harrison 1982). Lein in 1943 and later others reported that success of wheat and rye hybridization depends highly on the wheat variety (Lein 1943). He demonstrated that the dominant allele of the genes (*Kr1* on the 5B and *Kr2* on the 5A) were responsible for problems of compatibility of cross between wheat and rye. Snape et al. (1979) reported that the *Kr1* and *Kr2* are also responsible controlling the compatibility of cross between wheat and barley. Majority of the wheat varieties possess the dominant *Kr1* and *Kr2* alleles, the famous variety Chinese spring possess the rare recessive *Kr1* and *Kr2* alleles, as a result this variety can be easily crossed with rye (Baum et al. 1992). Other barriers could be slow pollen tube growth, pollen unable to reach the style, arresting of pollen tube in the style and of course differences in ploidy levels.

9.2 Post-fertilization Barriers

Once the pre-fertilization barriers have been taken care of the fertilization can be restricted by post-fertilization barriers. Post-fertilization barriers are more common in wheat compared to pre-fertilization barriers (Ishii et al. 2012). Usually the more distant the two species are more abnormalities are seen. The most common post fertilization barriers are; (i) sterility of the hybrid, (ii) death of the hybrid embryo as result of endosperm development, (iii) chromosome elimination of one of the parent.

In wheat × sorghum (*Sorghum bicolor* L.) or wheat × maize (*Zea mays* L.) cross, the sorghum and maize chromosomes are eliminated during the early zygotic cell division (Laurie and Bennett 1986). Hybrid necrosis is another problem occurring in wide crosses; it has been reported by Gregory (1980) and explained by Zeven (1981). The dominance of alleles of *Ne1* and *Ne2* depending on the combinations results in various degrees of necrosis (Hermsen 1963). Some interspecific crosses produce seeds with either an abnormal embryo or endosperm or both, which is caused by the imbalance of species polar nuclei activation value or the paternal/maternal genome ratio (Ishii et al. 2012).

Hybridization usually is a simple task when plants from the same species are involved, crossing plants from two different species is not so straight forward; sometimes additional techniques are need to get a successful hybrid. The reason of failure to hybridize may be due to the different mechanisms of the plants that are crossed.

10 Molecular Markers, Genome Mapping and Genomics as an Adjunct to Breeding

There are basically three types of markers; phenotypic markers; biochemical markers and DNA-based markers. Since the discovery of recombinant DNA technology in population genetics in the mid-1980s, a big stock of genetic markers is available for population studies and crop improvement. Plant breeding has changed with the introduction of these molecular techniques.

Molecular markers allow the extension of traditional breeding methods with one important difference, it allows transferring genetic information into new germplasm in a more precise and controlled manner. Marker assisted selection (MAS) for important but complex traits, which are often difficult to select in the routine breeding programs, have enhanced breeding programs in terms of better focused products and saving in time and resources. The use of DNA markers increases our understanding of gene inheritance and action of traits and allows the use of MAS to complement conventional selection procedures of plant breeding.

Biotechnology and especially the molecular markers have great potential for improving plant breeding efficiency and productivity. Plant genetic engineering and DNA marker technologies have now become valuable tools in crop improvement for rapid precision breeding for specific purposes (Dhillon et al. 2013). Molecular markers are present in most breeding programs worldwide since they were first adopted by the Australian Plant Breeding Cooperative Research Centre in 1997. In plant breeding, molecular markers have been used to confirm parent identity, parental characterization for some specific agronomical traits, assess genetic diversity in the crossing block or in specific populations and MAS. Molecular techniques have been and are still being used to monitor the DNA sequence variation in and among the species to create new sources of genetic variation by introducing new and favorable traits from landraces and related grass species. Several types of molecular markers that have been developed and are being used in plants include restriction fragment length polymorphisms (RFLPs), amplified fragment length polymorphisms (AFLPs), randomly amplified polymorphic DNAs (RAPDs), sequence-tagged sites (STS), expressed sequence tags (ESTs), simple sequence repeats (SSRs) or microsatellites, sequence-characterized amplified regions (SCARs), and single nucleotide polymorphisms (SNPs) (Paterson et al. 1991; Hoisington et al. 1998; Joshi et al. 1999). Molecular markers have also proved to be useful for gene pyramiding (Castro et al. 2003). The increased use of crop landraces and wild relatives in modern breeding programs cannot be dissociated from the advances in molecular genetics.

Restriction fragment length polymorphisms (RFLP) and Amplified Fragment Length Polymorphism (AFLP) markers have been the basis for most work in crop plants a decade ago. But with the development of Polymerase Chain Reaction (PCR) based markers a major step forward in the study of the wheat genome has been seen, by providing the tools needed to generate essential knowledge and data in the areas of fingerprinting and genetic conservation (Bennett and Smith 1976).

Valuable markers have been generated from the Simple Sequence Repeats (SSR) or microsatellite markers have been developed in the near past for most of the major crop plants and it is still widely implemented in many breeding programs. The new platforms for genomic sequencing of have greatly reduced the price and time and increased the resolution as compared to the previous AFLP/RFLP based systems. Thus DArT array technologies, SNP platforms or genotype by sequencing (GBS) allows breeders to more easily and cheaply identify new markers and assess wild genome introgressions fastening/accelerating the introduction of new alleles from wild relatives and decreasing the non-desired drag linkage effect. While expecting the results of the wheat genome sequencing, several consensus maps have been developed. In 2004, the union of 4 genetic maps developed out of SSR markers resulted in a more than 1200 loci consensus map (Somers et al. 2004). Plant breeders are now taking advantages of the molecular technologies, which are somehow speeding up the breeding program.

The use of molecular markers in plant breeding programs is relatively recent and its full potential is still being studied and developed. In previous Marker Assisted breeding approaches, the markers used were usually linked to or targeted genes of qualitative or major effect over traits of agronomical or end-use interest. However, these techniques often neglect epistatic effects and small effect genes (Lee et al. 2002). Instead, high-throughput genotyping or genome-wide molecular markers may allow breeders to select based on full genome information, leading to a true genomic selection. Genomic selection will improve breeder's predictions on phenotypic behavior of new breeding lines according to the genetic information obtained from genome-wide molecular markers (Meuwissen et al. 2001). Efforts are being made to exploit traits that are expected to play an important role in drought avoidance under receding soil moisture conditions by improving water availability to the plant through more efficient extraction of available soil moisture.

Drought tolerance is a complex phenomenon involving many known and unknown pathways. To improve drought tolerance, QTLs have been identified for stomatal conductance (Sanguineti et al. 1999; Juenger et al. 2005), transpiration efficiency (Kholova et al. 2010), osmotic adjustment (Diab et al. 2004; Saranga et al. 2004; Robin et al. 2003), relative water content (Diab et al. 2004; Sanguineti et al. 1999), canopy temperature (Diab et al. 2004; Saranga et al. 2004), drought sensitivity index (Sanguineti et al. 1999), leaf ABA (Kholova et al. 2010), chlorophyll content (Shen et al. 2007), root traits (Chandra Babu et al. 2003; Li et al. 2005) and some yield-related traits (Moreau et al. 2004; Xu et al. 2005; Xiao et al. 2005).

Selection of germplasm with reproductive drought-tolerance has been based on grain yield-related traits, with selection often being conducted under field conditions. Several QTLs with widely varying contribution to the grain yield phenotype under drought conditions have identified in wheat and rice (Lanceras et al. 2004; Wang et al. 2005; Bernier et al. 2007; Kirigwi et al. 2007). A major problem with drought tolerance selection is interference with avoidance/escape mechanisms (e.g. early flowering), especially under field conditions, where occurrence, timing, severity and length of water stress conditions cannot be controlled (Yue et al. 2006; Dolferus et al. 2011).

11 Conclusion

Although dryland agriculture has great importance and above 40 % of the world agriculture is in dry areas not much attention has been given to in the past. It is worthy to note that with the current trend of global warming and climate change dry areas may increase and the severity of water scarcity will also increase. Most of the varieties developed in the past for drylands are the spillovers from breeding programs for irrigated areas. More breeding programs using high- throughput technologies specifically targeting dry areas needs to be developed. Some of areas having enough water for crop growth may have water shortage in critical crop growth stages. Development of water efficient or drought tolerant will be a big advantage for farmers, it will not only avoid crop failure but also increase crop yield. For the global food security in the future, dryland agriculture is going to play a major role. More investment and research needs to be done to insure the development and stability of agricultural system in drylands.

References

Acevedo E, Craufurd PQ, Austin RB, Perezmarco P (1991) Traits associated with high-yield in barley in low-rainfall environments. J Agric Sci 116:23–36

Aguilar-Rincón VH, Singh PR, Huerta-Espino J (2000) Inheritance of resistance to leaf rust in four synthetic hexaploid wheats. Agrociencia. 354:235–246

Allard RW, Bradshaw AD (1964) Implications of genotype environment interactions. Crop Sci 4:503–508

Annicchiarico P (1997) Joint regression vs AMMI analysis of genotype–environment interactions for cereals in Italy. Euphytica 94:53–62

Annicchiarico P (2002) Genotype × environment interactions: challenges and opportunities for plant breeding and cultivar recommendations. FAO Plant Production and Protection Paper No. 174. FAO, Rome

Annicchiarico P, Bellah F, Chiari T (2005) Defining subregions and estimating benefits for a specific-adaptation strategy by breeding programs: a case study. Crop Sci 45:1741–1749

Aprile A, Mastrangelo AM, De Leonardis AM, Galiba G, Roncaglia E, Ferrari F, De Bellis L, Turchi L, Giuliano G, Cattivelli L (2009) Transcriptional profiling in response to terminal drought stress reveals differential responses along the wheat genome. BMC Genomics 10:279

Araus JL, Ferrio JP, Buxo R, Voltas J (2007) The historical perspective of dryland agriculture: lessons learned from 10,000 years of wheat cultivation. J Exp Bot 58:131–145

Araus JL, Slafer GA, Reynolds MP, Royo C (2002) Plant breeding and drought in C-3 cereals: what should we breed for? Ann Bot 89:925–940

Assefa S, Fehrmann H (2000) Resistance to wheat leaf rust in Aegilops tauschii Coss. and inheritance of resistance in hexaploid wheat. Genet Resour Crop Evol 47:135–140

Avci M (1999) Impact of water use effieicnecy research and OSWU, and use of information tools and methodologies in Turkey. In: Van Duivenbooden N, Pala M, Studer C, Bielders CL (eds) Efficient soil water use: the key to sustainable crop production in the dry areas of West Asia, and North and Sub-Saharan Africa. ICARDA/ICRISAT, Aleppo/Patancheru, pp. 405–412

Avci M, Guler M, Pala M, Durutan N, Karaca M, Eyuboglu H (1987) The effects of the different components of package of cultural practices on wheat yield under central Anatolian conditions (in Turkish). TUBITAK, U.U. Faculty of Agriculture, Bursa, pp 147–157

Basford KE, Cooper M (1998) Genotype x environment interactions and some considerations of their implications for wheat breeding in Australia. Aust J Agri Res 49:153–174

Battisti DS, Naylor RL (2009) Historical warnings of future food insecurity with unprecedented seasonal heat. Science 9:240–244

Baum M, Lagudah ES, Apples R (1992) Wide crosses in cereals. Annu Rev Plant Physiol Plant Mol Biol 43:117–143

Bennett E (1970) Adaptation in wild and cultivated plant populations. In: Frankel OH, Bennett E (eds) Genetic resources in plants: their exploration and conservation. Blackwell Scientific Publications, Oxford

Bennett MD, Smith JB (1976) Nuclear DNA amounts in angiosperms. Philos Trans R Soc Lond B 274:227–274

Bernier J, Kumar A, Ramaiah V, Spaner D, Atlin G (2007) A large-effect QTL for grain yield under reproductive-stage drought stress in upland rice. Crop Sci 47:507–516

Bloem MW, Semba RD, Kraemer K (2010) Castel Gandolfo workshop: an 18 introduction to the impact of climate change, the economic crisis and the increase in the food prices on malnutrition. J Nutr 140:132–135

Blumenthal C, Bekes F, Gras PW, Barlow EWR, Wrigley CW (1995) Influence of wheat genotypes tolerant to the effects of heat stress on grain quality. Cereal Chem 72:539–544

Borlaug NE (2008) Feeding a world of 10 billion people: Our 21st century challenge. In: Scanes CG, Miranowski JA (eds) Perspectives in world food and agriculture 2004. Iowa State Press, Ames

Braun HJ, Rajaram S, van Ginkel M (1996) CIMMYT's approach to breeding for wide adaptation. Euphytica 92:175–183

Braun HJ, Rajaram S, van Ginkel M (1997) CIMMYT's approach to breeding for wide adaptation. In: Tigerstedt PMA (ed) Adaptation in plant breeding. Kluwer Academic Publishers, Dordrecht

Breseghello F (2013) Traditional and modern plant breeding methods with examples in rice (*Oryza sativa* L). J Agric Food Chem 61:8277–8286

Bridgen MP (1994) A review of plant embryo culture. HortSci 29:1243–1246

Byerlee D (1996) Modem varieties, productivity, and sustainability: recent experience and emerging challenges. World Dev 24:697–718

Calderini DF, Reynolds MP, Slafer GA (1999) Genetic gains in wheat yield and main physiological changes associated with them during the 20th century. In: Satorre EH, Slafer GA (eds) Wheat: ecology and physiology of yield determination. Food Product Press, New York, pp. 351–377

Carnicer J, Coll M, Ninyerola M, Pons X, Sánchez G, Peñuelas J (2010) Wide spread crown condition decline, food web disruption, and amplified tree mortality with increased climate change-type drought. Proc Natl Acad Sci USA 108:1474–1478

Castro AJ, Chen X, Hayes PM, Johnston M (2003) Pyramiding Quantitative Trait Locus (QTL) alleles determining resistance to barley stripe rust: effects on resistance at the seedling stage. Crop Sci 43:651–659

Ceccarelli S (1996) Positive interpretation of genotype by environment interactions in relation to sustainability and biodiversity. In Cooper M, Hammer GL (eds) Plant adaptation and crop improvement. CAB International, Wallingford; ICRISAT, Patancheru; IRRI, Manila, pp 467–486

Ceccarelli S (2009) Main stages of a plant breeding programme. In: Ceccarelli S, Guimaraes EP, Weltzien E (eds) Plant breeding and farmer participation. Rome, FAO, pp. 63–74

Ceccarelli S (2015) Efficiency of Plant Breeding. Crop Sci 55:87–97

Ceccarelli S, Acevedo E, Grando S (1991) Breeding for yield stability in unpredictable environments: Single traits, interaction between traits, and architecture of genotypes. Euphytica 6:169–185

Ceccarelli S, Grando S (1987) Diversity for morphological and agronomic characters in Hordeum vulgare ssp. spontaneum C. Koch. Genet Agr 41:131–142

Ceccarelli S, Grando S (2007) Decentralized-participatory plant breeding: an example of demand driven research. Euphytica 155:349–360

CGIAR (2013) New research approaches to improve drylands agriculture to deliver a more prosperous future. http://www.cgiar.org/web-archives/www-cgiar-org-impact-global-des_fact2-html. Accessed on 12 July 2015

Chandra-Babu R, Nguyen BD, Chamarerk V, Shanmugasundaram P, Chezhian P, Jeyaprakash P, Ganesh SK, Palchamy A, Sadasivam S, Sarkarung S, Wade LJ, Nguyen HT (2003) Genetic analysis of drought resistance in rice by molecular markers: association between secondary traits and field performance. Crop Sci 43:1457–1469

Chapman SC, Hammer GL, Butler DG, Cooper M (2000) Genotype by environment interactions affecting grain sorghum. III. Temporal sequences and spatial patterns in the target population of environments. Aust J Agric Res 51:223–233

Chee PW, Elias EM, Anderson JA, Kianian SF (2001) Evaluation of a high grain protein QTL from Triticum turgidum L. var. dicoccoides in an adapted durum wheat back- ground. Crop Sci 41:295–301

Chen ZH, Zhou MX, Newman IA, Mendham NJ, Zhang GP, Shabala S (2007) Potassium and sodium relations in salinised barley tissues as a basis of differential salt tolerance. Funct Plant Biol 34:150–162

Chu C-G, Faris JD, Friesen TL, Xu SS (2006) Molecular mapping of hybrid necrosis genes Ne1 and Ne2 in hexaploid wheat using microsatellite markers. Theor Appl Genet 112:1374–1381

Chu C-G, Friesen TL, Xu SS, Faris JD (2008) Identification of novel tan spot resistance loci beyond the known host-selective toxin insensitivity genes in wheat. Theor Appl Genet 117:873–881

Collins GB, Grosser JW (1984) Culture of embryos. In: Vasil IK (ed) Cell culture and somatic cell genetics of plants. Laboratory procedures and their applications. Academic, New York, pp. 241–257

Colmer TD, Flowers J, Munns R (2006) Use of wild relatives to improve salt tolerance in wheat. J Exp Bot 57:1059–1078

Cossani CM, Savin R, Slafer GA (2007) Contrasting performance of barley and wheat in a wide range of conditions in Mediterranean Catalonia (Spain). Ann Appl Biol 151:167–173

Cox TS, Sears GR, Gill BS, Jellen EN (1994) Registration of KS91WGRC11, KS92WGRC15, and KS92WGRC23 leaf rust-resistant hard red winter wheat germ- plasms. Crop Sci 34:546

Curtis BC (2002) Wheat in the world. In: Curtis BC, Rajaram S, Go´mez Macpherson H (eds) Bread wheat. FAO, Rome

Dalrymple DG (1986) Development and spread of high-yielding wheat varieties in developing countries. Bureau for Science and Technology, U.S. Agency for International Development, Washington, DC

Davies J, Berzonsky WA, Leach GD (2006) A comparison of marker-assisted and phenotypic selection for high grain protein content in spring wheat. Euphytica 152:117–134

De Groote H, Siambi M, Friesen D, Diallo A (2002) Identifying farmers' preferences for new maize varieties in eastern Africa. In: Bellon MR, Reeves J (eds) Quantitative analysis of data from participatory methods in plant breeding. CIM MYT, Mexico, pp. 82–102

DeLacy IH, Basford KE, Cooper M, Bull JK, McClaren CG (1996) Analysis of multi-environment trials–an historical perspective. In: Cooper M, Hammer GL (eds) Plant adaptation and crop improvement. CAB International, Wallingford, pp. 39–124

DeLacy IH, Fox PN, Corbett JD, Crossa J, Rajaram S, Fischer RA, van Ginkel M (1994) Long-term association of locations for testing spring bread wheat. Euphytica 72:95–106

Denčić S, Kastori R, Kobiljski B, Duggan B (2000) Evaluation of grain yield and its components in wheat cultivars and landraces under near optimal and drought conditions. Euphytica 113:43–52

Dewey DR (1984) The genomic system of classification as a guide to intergeneric hybridization with the perennial Triticeae. In: Gustafsom GP (ed) Gene manipulation in plant improvement. Plenuim Press, New York, pp. 209–279

Dhillon NK, Gosal SS, Kang MS (2013) Improving crop productivity under changing environment. In: Tuteja N, Sarvajeet SG, Tuteja R (eds) Improving crop productivity in sustainable agriculture, pp 23–48

Diab AA, Teulat-Merah B, This D, Ozturk NZ, Benscher D, Sorrells ME (2004) Identification of drought inducible genes and differentially expressed sequence tags in barley. Theor Appl Genet 109:1417–1425

Dodig D, Zorić M, Kandić V, Perović D, Šurlan-Momirović G (2012) Comparison of responses to drought stress of 100 wheat accessions and landraces to identify opportunities for improving wheat drought resistance. Plant Breeding 131:369–379

Dolferus R, Ji X, Richards RA (2011) Abiotic stress and control of grain number in cereals. Plant Sci 181:331–341

Eberhart SA, Russell WA (1966) Stability parameters for comparing varieties. Crop Sci 6:36–40

Eglinton Jk, Evans DE, Brown AHD, Langridge P, McDonald G, Jefferies SP, Barr AR (1999) The use of wild barley (*Hordeum vulgare* ssp. spontaneum) in breeding for quality and adaptation. 9 Australian barley technical symposium, Melbourne

FAO (2010) The second report on the state of the world's plant genetic resources for food and agriculture. Food and Agriculture Organization, Rome

FAO (2011) Highlands and drylands – mountains, a source of resilience in arid regions. Published by FAO, UNCCD, Mountain Partnership, Swiss Agency for Development and Cooperation, and CDE, with the support of an international group of experts. Food and Agriculture Organization, Italy, Rome

Farooq M, Bramley H, Palta JA, Siddique KHM (2011) Heat stress in wheat during reproductive and grain filling phases. Crit Rev Plant Sci 30:491–507

Fehr WR (1984) Genetic contributions to yield gains of five major crop plants. CSSA Special Publication, 7. ASA and CSSA, Madison

Feldmann M (2001) Origin of cultivated wheat. In: Bonjean AP, Angus WJ (eds) The world wheat book. A history of wheat breeding, Lavoiser Publishing, France

Feuillet C, Langridge P, Waugh R (2008) Cereal breeding takes a walk on the wild side. Trends Genet 24:24–32

Finlay KW, Wilkinson GN (1963) The analysis of adaptation in a plant-breeding programme. Aus J Agric Res 14:742–754

Francia E, Tondelli A, Rizza F, Badeck FW, Nicosia OLD, Akar T, Grando S, Al-Yassing A, Benbelkacem A, Thomas WTB, van Eeuwijk F, Romagosa I, Stanca AM, Pecchioni N (2011)

Determinants of barley grain yield in a wide range of Mediterranean environments. Field Crops Res 120:169–178
Friebe B, Jiang J, Raupp WJ, McIntosh RA, Gill BS (1996) Characterization of wheat-alien translocations conferring resistance to diseases and pests: current status. Euphytica 91:59–87
Frison EA, Cherfas J, Hodgkin T (2011) Agricultural biodiversity is essential for a sustainable improvement in food and nutrition security. Sustainability 3:238–253
Gamborg OL, Miller RA, Ojima K (1968) Nutrient requirements of suspension cultures of soybean root cells. Expt Cell Res 50:151–158
Gauch HG (1992) AMMI analysis of yield trials. In: Kang MS, Gauch HG (eds) Genotype-by-environment interaction. CRC Press, Boca Raton, pp. 1–40
Gauch HG, Zobel RW (1997) Identifying mega-environments and targeting genotypes. Crop Sci 37:311–326
Gepts P (2002) A comparison between crop domestication, classical plant breeding, and genetic engineering. Crop Sci 42:1780–1790
Gepts P, Hancock J (2006) The future of plant breeding. Crop Sci 46:1630–1634
Giunta F, Motzo R, Pruneddu G (2008) Has long term selection for yield in durum wheat also induced changes in leaf and canopy traits? Field Crops Res 106:68–76
Goodman RM, Hauptli H, Croosway A, Knauf VC (1987) Gene transfer in crop improvement. Science 236:48–54
Gowda CLL, Parthasarathy Rao P, Bhagavatula S (2009) Global trends in production and trade of major grain legumes. International Conference on Grain Legumes: Quality Improvement, Value Addition and Trade; February 14–16; Indian Institute of Pulses Research, Kanpur, India: Indian Society of Pulses Research and Development; 2009, pp 282–301
Grando S, von Bothmer R, Ceccarelli S (2001) Genetic diversity of barley: use of local adapted germplasm to enhance yield and yield stability of barley in dry areas In: Cooper HD, Hodgkin T, Spillane C (eds) Broadening the genetic base of crop production. CABI, New York/AO, Rome/IPRI, Rome, pp 351-372
Graybosch RA (2001) Uneasy unions: Quality effects of rye chromatin transfers to wheat. J Cereal Sci 33:3–16
Gregory RS (1980) Hybrid necrosis as a problem in handling hexaploid × tetraploid wheat crosses. J Agric Sci 94:377–382
Guzy-Wrobelska J, Labocha-Pawlowska A, Kwasniewski M, Szarejko I (2007) Different recombination frequencies in wheat doubled haploid populations obtained through maize pollination and anther culture. Euphytica 156:173–183
Guzy-Wrobelska J, Szarejko I (2003) Molecular and agronomic evaluation of wheat doubled haploid lines obtained through maize pollination and anther culture methods. Plant Breed 122:305–313
Hajjar R, Hodgkin T (2007) The use of wild relatives in crop improvement: a survey of developments over the last 20 years. Euphytica 156:1–13
Harlan JR, de Wet JMJ (1971) Toward a rational classification of cultivated plants. Taxon 20:509–517
Hawkes JG, Maxted N, Ford-Lloyd BV (2000) The genetic resources of plants and their value to mankind: the ex situ conservation of plant genetic resources. Kluwer Academic Publishers, London, pp. 1–18
Heiser CB (1990) Seed to cultivation: the story of food. Harvard University Press, Cambridge
Hermsen JG (1963) Hybrid necrosis as a problem for the wheat breeder. Euphytica 12:1–16
Hernandez-Segundo E, Capettini F, Trethowan R, van Ginkel M, Mejia A, Carballo A, Crossa J, Vargas M, Balbuena MA (2009) Mega-environment identification for barley based on twenty-seven years of global grain yield data. Crop Sci 49:1705–1718
Heslop-Harrison J (1982) Pollen-stigma interaction and cross-incompatibility in the grasses. Science 215:1358–1364
Hirano R, Kikuchi A, Kawase M, Watanabe KN (2008) Evaluation of genetic diversity of bread wheat landraces from Pakistan by AFLP and implications for a future collection strategy. Genet Res Crop Evol 55:1007–1015
Hodson DP, White JW (2007) Use of spatial analyses for global characterization of wheat-based production systems. J Agri Sci 145:115–125

Hoisington D, Listman GM, Morris ML (1998) Varietal development: applied biotechnology. In: Morris ML (ed) Maize seed industries in developing countries. Lynne Rienner Publishers/ CIMMYT, Boulder, pp. 77–102
Hoisington D, Kahairallah M, Reeves T, Ribaut JM, Skovmand B, Taba S, Warburton M (1999) Plant genetic resources: What can they contribute toward increased crop productivity. Proc Natl Acad Sci 96:5937–5943
Holme IB, Wendt T, Holm PB (2013) Current developments of intragenic and cisgenic crops. ISB news report
Howden SM, Soussana JF, Tubiello FN, Chhetri N, Dunlop M, Meinke H (2007) Adapting agriculture to climate change. Proc Natl Acad Sci 104:19691–19696
IPCC (2007) Climate change 2007: the physical science basis. Contribution of working group I to the fourth assessment report of the intergovernmental panel on climate Change. In: Salomon S, Quin D, Manning M, Chen Z, Marquis M, Averyt KB, Tignor M, Miller HD (eds) Cambridge University Press, Cambridge/New York, pp 996
Ishii T, Tanaka H, Eltayeb AE, Tsujimoto H (2012) Wide hybridization between oat and pearl millet belonging to different subfamilies of Poaceae. Plant Rep 26:25–32
Ivandic V, Hackett CA, Zhang ZJ, Staub JE, Nevo E, Thomas WTB, Forster BP (2000) Phenotypic responses of wild barley to experimentally imposed water stress. J Exp Bot 51:2021–2029
Jackson LE, Pascual U, Hodgkin T (2007) Utilizing and conserving agrobiodiversity in agricultural landscapes. Agri Ecosys Environ 121:196–210
Jacobsen SE, Jensen CR, Liu F (2012) Improving crop production in the arid Mediterranean climate. Field Crops Res 128:34–47
Jiang J, Friebe B, Gill BS (1994) Recent advances in alien gene transfer in wheat. Euphytica 73:199–212
Joshi SP, Ranjekar PK, Gupta VS (1999) Molecular markers in plant genome analysis. Curr Sci 77:230–240
Juenger TE, McKay J, Hausmann N, Keurenties J, Sen S, Stowe K, Dawson T, Simms E, Richards J (2005) Identification and characterization of QTL underlying whole-plant physiology in Arabidopsis thaliana II. QTL analysis of new mapping population, Kas-1 Tsu-1. Evolution 62:3014–3026
Kang MS (1998) Using genotype-by-environment interaction for crop cultivar development. Adv Agron 62:199–252
Kholova J, Hash CT, Lava-Kumar P, Yadav RS, Kocova M, Vadez V (2010) Terminal drought-tolerant pearl millet [*Pennisetum glaucum* (L.) R. Br.] have high leaf ABA and limit transpiration at high vapor pressure deficit. J Exp Bot 61:1431–1440
Kirigwi FM, Van Ginkel M, Brown-Guedira G, Gill BS, Paulsen GM, Fritz AK (2007) Markers associated with a QTL for grain yield in wheat under drought. Mol Breed 20:401–413
Ladizinsky G, Pickersgill B, Yamamoto K (1988) Exploitation of wild relatives of the food legumes. In: Summer field RJ (eds) Word crops: cool season food legumes, Kluwer Academic Publishers, Dordrecht, pp 967–978
Lal R, Stewart B, Uphoff N (2005) Climate change, soil carbon dynamics, and global food security climate change and global food security. CRC Press, Boca Raton, pp. 113–143
Lanceras JC, Pantuwan G, Jongdee B, Toojinda T (2004) Quantitative trait loci associated with drought tolerance at reproductive stage in rice. Plant Physiol 135:384–399
Laurie DA, Bennett MD (1986) Wheat × maize hybridization. Can J Genet Cytol 28:313–316
Lein A (1943) Die genetische Grundlage der Kreuzbarkeit zwischen Weizen und Roggen. Z. Indukt. Abstamm. Vererbungsl 8:28–61
Li Z, Mu P, Li C, Zhang H, Li Z, Gao Y, Wang X (2005) QTL mapping of root traits in a doubled haploid population from a cross between upland and lowland japonica rice in three environments. Theor Appl Genet 110:1244–1252
Lillemo M, Chen F, Xia XC, William M, Pena RJ, Trethowan R, He ZH (2006) Puroindoline grain hardness alleles in CIMMYT bread wheat germplasm. J Cereal Sci 44:86–92
Lin CS, Butler G (1990) Cluster analyses for analyzing two way classification data. Agron J 82:344–348
Liu W, Ming YZ, Polle AE, Konzak CF (2002) Highly efficient doubled-haploid production in wheat (*Triticum aestivum* L.) via induced microspore embryogenesis. Crop Sci 42:686–692

Lobell DB, Field CB (2007) Global scale climate crop yield relationships and the impacts of recent warming. Environ Res Let 2:1–7
Lupton FGH (1987) History of wheat breeding. In: Lupton FGH (ed) Wheat Breeding: Its Scientific Basis. Chapman and Hall, London
Macindoe SL, Brown CW (1968) Wheat breeding and varieties in Australia. Science bulletin No. 76, 3rd edn. New South Wales Department of Agriculture, Sydney
Matous K, Benediktyova Z, Berger S, Roitsch T, Nedbal L (2006) Case study of combinatorial imaging: what protocol and what chlorophyll fluorescence image to use when visualizing infection of *Arabidopsis thaliana* by *Pseudomonas syringae*? Photosynth Res 90:243–253
Mazid A, Tutwiler R, Al-Ahmad H (1998) Impact of modern technologies on durum wheat production in Syria. In: Nachit MM, Baum M, Porceddu E, Monneveaux P, Picard E (eds) Proceedings of the SEWANA (South Europe, West Asia and North Africa) Durum Research Network Workshop. ICARDA, Aleppo, pp 88–102
McCouch S (2013) Feeding the future. Nature 499:23–24
McFadden ES (1930) A successful transfer of emmer characters to vulgare wheat. J Am Soc Agron 22:1020–1034
McIntosh RA, Wellings CR, Park RF (1995) Wheat rusts: an atlas of resistance genes. CSIRO Publications, East Melbourne
Meehl GA, Tebaldi C (2004) More intense, more frequent, and longer-lasting heat waves in the 21st century. Science 305:994–997
Meuwissen THE, Hayes BJ, Goddard ME (2001) Prediction of total genetic value using genome-wide dense marker maps. Genetics 157:1819–1829
Mohammadi R, Amri A (2013) Genotype × environment interaction and genetic improvement for yield and yield stability of rainfed durum wheat in Iran. Euphytica 192:227–249
Mohammadi R, Amri A, Ahmadi H, Jafarzadeh J (2015) Characterization of tetraploid wheat landraces for cold tolerance and agronomic traits under rainfed conditions of Iran. J Agric Sci 153:631–645
Mohammadi R, Haghparast R (2011) Evaluation of rainfed promising wheat breeding lines on farmers' fields in west of Iran. Int J Plant Breed 5:30–36
Mohammadi R, Haghparast R, Amri A, Ceccarelli S (2010) Yield stability of rain-fed durum wheat and GGE biplot analysis of multi-environment trials. Crop Pasture Sci 61:92–101
Mohammadi R, Haghparast R, Sadeghzadeh B, Ahmadi H, Solimani K, Amri A (2014) Adaptation patterns and yield stability of durum wheat landraces to highland cold rainfed areas of Iran. Crop Sci 54:944–954
Mohammadi R, Sadeghzadeh D, Armion M, Amri A (2011) Evaluation of durum wheat experimental lines under different climate and water regime conditions of Iran. Crop Pasture Sci 62:137–151
Moreau L, Charcosset A, Gallais A (2004) Use of trial clustering to study QTL environment effects for grain yield and related traits in maize. Theor Appl Genet 110:92–105
Morris ML, Bellon MR (2004) Participatory plant breeding research: opportunities and challenges for the international crop improvement system. Euphytica 136:21–34
Moseman JG, Nevo E, El-Morshidy MA, Zohary D (1984) Resistance of *Triticum dicoccoides* to infection with *Erysiphe graminis tritici*. Euphytica 33:41–47
Motzo R, Fois S, Giunta F (2005) Relationship between grain yield and quality of durum wheats from different eras of breeding. Euphytica 140:147–154
Mujeeb-Kazi A (2003) Wheat improvement facilitated by novel genetic diversity and in vitro technology. Plant Cell Tissue Organ Cult 13:179–210
Mujeeb-Kazi A, Kimber G (1985) The production, cytology and practicality of wide hybrids in the Triticeae. Cereal Res Commun 13:111–124
Mujeeb-Kazi A, Rosas V, Roldan S (1996) Conservation of the genetic variation of *Triticum tauschii* (Coss.) Schmalh. (*Aegilops squarrosa* auct. non L.) in synthetic hexaploid wheats (*T. turgidum* L. s.lat. *T. tauschii*; 2n1⁄46x 1⁄442, AABBDD) and its potential utilization for wheat improvement. Genet Resour Crop Evol 43:129–134
Murashige T, Skoog F (1962) A revised medium for rapid growth and bioassays with tobacco tissue cultures. Physiol Plant 15:473–497

Naghavi MR, Mardi M (2010) Characterization of genetic variation among accessions of *Aegilops tauschii*, AsPac. J Mol Biol Biotechnol 18:93–96

Nevo E, Gorham J, Beiles A (1992) Variation for 22Na uptake in wild emmer wheat, Triticum dicoccoides in Israel: salt tolerance resources for wheat improvement. J Exp Bot 43:511–518

Nevo E, Korol AB, Beiles A, Fahima T (2002) Evolution of Wild Emmer and Wheat Improvement. Population Genetics, Genetic Resources, and Genome Organization of Wheats Progenitor, *Triticum dicoccoides*. Springer, Berlin

Nishikawa K, Mori T, Takami N, Furuta Y (1974) Mapping of progressive necrosis genes Ne1 and Ne2 of common wheat by the telocentric method. JPN J Breed 24:277–281

Ogbonnaya FC, Abdalla O, Mujeeb-Kazi A, Kazi AG, Gosnian N, Lagudah ES (2013) Synthetic hexaploids: harnessing species of the primary gene pool for wheat improvement. Plant Breed Rev 37:35–122

Ortiz R, Iwanaga M, Reynolds MP, Wu H, Crouch J (2007) Overview on crop genetic engineering for drought prone environments. J Semi-Arid Trop Agric Res 4:1–30

Ortiz-Ferrara G, Bhatta MR, Pokharel TP, Mudwari A, Thapa DB, Joshi AK, Chand R, Muhammad D, Duveiller R, Rajaram S (2001) Farmer participatory variety selection in South Asia. In: Research Highlights of the Wheat Program 1999–2000. Centro Internacional de Mejoramiento de Mai´zy Trigo, Mexico, pp. 33–37

Ortiz-Ferrara G, Joshi AK, Chand R, Bhatta MR, Mudwari A, Thapa DB, Sufian MA, Saikia TP, Chatrath R, Witcombe JR, Virk DS, Sharma RC (2007) Partnering with farmers to accelerate adoption of new technologies in South Asia to improve wheat productivity. Euphytica 157:399–407

Ouyang YW, Hu CC, Chuang CC, Tseng CC (1973) Induction of pollen plants from anthers of Triticum aestivum L. cultured in vitro. Sci Sin 16:79–95

Pande A, Rawat M, Nayan R, Guru SK, Arora S (2015) Nanotechnological interventions for mitigating global warming. In: Rakesh SS, Sengar K (eds) Climate change effect on crop productivity. CRC Press, Taylor & Francis Group, Boca Raton/London/New York, pp. 143–158

Patade VY, Bhargava S, Suprasanna P (2011) Transcript expression profiling of stress responsive genes in response to short-term salt or PEG stress in sugarcane leaves. Mol Biol Rep 39:3311–3318

Paterson AH, Tanksley SD, Sorrels ME (1991) DNA markers in plant improvement. Adv Agron 46:39–90

Peleg Z, Fahima T, Abbo S, Krugman T, Nevo E, Yakir D, Saranga Y (2005) Genetic diversity for drought resistance in wild emmer wheat and its ecogeographical associations. Plant Cell Environ 28:176–191

Pena RJ, Amaya A, Rajaram S, Mujeeb-Kazi A (1990) Variation in quality characteristics associated with some spring 1B/1R translocation wheats. J Cereal Sci 12:105–112

Pflüger LA, D'Ovidio R, Margiotta B, Peña R, Mujeeb-Kazi A, Lafiandra D (2001) Characterisation of high- and low-molecular weight glutenin subunits associated to the D genome of *Aegilops tauschii* in a collection of synthetic hexaploid wheats. Theor Appl Genet 103:1293–1301

Podlich DW, Cooper M (1998) QU-GENE: a simulation platform for quantitative analysis of genetic models. Bioinformatics 14:632–653

Pollak LM, Corbett JD (1993) Using GIS datasets to classify maize growing regions in Mexico and Central America. Agron J 85:1133–1139

Pratap A, Gurdeep SS, Harinder K, Chaudhary HK (2006) Relative efficiency of anther culture and chromosome elimination techniques for haploid induction in triticale × wheat and triticale × triticale hybrids. Euphytica 150:339–345

Pukhalskiy VA, Bilinskaya EN (1998) Distribution of hybrid necrosis genes in winter wheat genotypes. Annu Wheat Newsl 44:191–193

Pukhalskiy VA, Martynov SP, Dobrotvorskaya TV (2000) Analysis of geographical and breeding-related distribution of hybrid necrosis genes in bread wheat (*Triticum aestivum* L.). Euphytica 114:233–240

Qi LL, Friebe B, Zhang P, Gill BS (2007) Homoeologous recombination, chromosome engineering and crop improvement. Chromosome Res 15:3–19

Rajaram S (2001) Prospects and promise of wheat breeding in the 21st century. Euphytica119:3–15

Reed SM (2005) Embryo rescue. In: Trigiano Robert N, Gray DJ (eds) Plant development and biotechnology. CRC Press, Boca Raton

Reif JC, Zhang P, Dreisigacker S, Warburton ML, van Ginkel M, Hoisington D, Bohn M, Melchinger AE (2005) Wheat genetic diversity trends during domestication and breeding. Theor Appl Genet 110:859–864

Reynolds M, Foulkes MJ, Slafer GA, Berry P, Parry MAJ, Snape JW, Angus WJ (2009) Raising yield potential in wheat. J Exp Bot 60:1899–1918

Rizza F, Badeck FW, Cattivelli L, Li Destri Nicosia O, Di Fonzo N, Stanca AM (2004) Diversity of yield potential and actual yield in barley genotypes under rainfed and irrigated conditions. Crop Sci 44:2127–2137

Robin S, Pathan MS, Courtois B, Lafitte R, Carandang S, Lnceraas S, Amante M, Nguyen HT, Li Z (2003) Mapping osmotic adjustment in an advanced back-cross inbred population of rice. Theor Appl Genet 107:1288–1296

Royo C, Alvaro F, Martos V, Ramdani A, Isidro J, Villegas D, Garcia del Moral LF (2007) Genetic changes in durum wheat yield components and associated traits in Italian and Spanish varieties during the 20th century. Euphytica 155:259–270

Ryan J, Singh M, Pala M (2008) Long-term cereal-based rotation trials in the Mediterranean region: implications for cropping sustainability. Adv Agron 97:273–319

Saindon G, Schaalje GB (1993) Evaluation of locations for testing dry bean cultivars in western Canada using statistical procedures, biological interpretation and multiple traits. Can J Plant Sci 73:985–994

Salvi S, Porfiri O, Ceccarelli S (2013) Nazareno Strampelli, the 'Prophet' of the Green Revolution. J Agri Sci Cambridge 151:1–5

Sanguineti MC, Tuberosa R, Landi P, Salvi S, Macaferri M, Casarini E, Conti S (1999) QTL analysis of drought related traits and grain yield in relation to genetic variation for leaf abscisic acid concentration in field-grown maize. J Exp Bot 50:1289–1297

Saranga Y, Jiang C-X, Wright RJ, Yakir D, Paterson AH (2004) Genetic dissection of cotton physiological responses to arid conditions and their inter-relationships with productivity. Plant Cell Environ 27:263–277

Schaeffer GW, Baenziger PS, Worley J (1979) Haploid plant development from anthers and in vitro embryo culture of wheat. Crop Sci 19:697–702

Schreinemachers P, Wu MH, Ebert AW (2014) Costing the ex situ conservation of plant genetic resources at AVRDC – The World Vegetable Center. Genet Res Crop Evol 61:757–773

Seif E, Evans JC, Balaam LN (1979) A multivariate procedure for classifying environments according to their interaction with genotypes. Aus J Agric Res 30:1021–1026

Sharma HC, Sujana G, Rao DM (2009) Morphological and chemical components of resistance to pod borer, *Helicoverpa armigera* in wild relatives of pigeonpea. Springer, SBM, Dordrecht, Netherlands. Arthropod Plant Interact 3:151–161

Sharma HC, Gill BS (1983) New hybrids between Agropyron and wheat. 2. Production, morphology and cytogenetic analysis of F1 hybrids and backcross derivatives. Theor Appl Genet 66:111–121

Sharma S, Upadhyaya HD, Roorkiwal M, Varshney RK, Gowda CLL (2013) Chickpea. In: Singh M, Upadhyaya HD, Bisht IS (eds) Genetic and genomic resources of grain legume improvement. Elsevier Inc, London/Waltham, pp. 81–104

Shen B, Zhang JY, Zhang KQ, Dai WM, Lu Y, Fu LQ, Ding JM, Zheng KL (2007) QTL mapping of chlorophyll contents in rice. Agri Sci China 6:17–24

Siddiqi EH, Ashraf M, Hussain M, Jamil A (2009) Assessment of inter-cultivar variation for salt tolerance in safflower (*Carthamus tinctorius* L.) using gas exchange characteristics as selection criteria. Pak J Bot 41:2251–2259

Sinclair TR, Vadez V (2012) The future of grain legumes in cropping systems. Crop Pasture Sci 63:501–512

Singh RP, Singh I, Chowdhury RK (1989) Hybrid necrosis in bread wheat. Wheat Inf Serv 68:6–8

Slafer GA, Laraus J, Royo C, Garcia del Moral LF (2005) Promising ecophysiological traits for genetic improvement of cereal yields in Mediterranean environments. Ann Appl Biol 146:61–70

Slafer GA, Satorre EH, Andrade FH (1994) Increases in grain yield in bread wheat from breeding and associated physiological changes. In: Slafer GA (ed) Genetic Improvement of Field Crops. Marcel Dekker, Inc., New York

Smale M (1996) Understanding Global Trends in the Use of Wheat Diversity and International Flows of Wheat Genetic Resources. Economics Working Paper 96-02. Mexico, D.F.: CIMMYT

Smale M, Reynolds MP, Warburton M, Skovmand B, Trethowan R, Singh RP, Ortiz-Monasterio I, Crossa J (2002) Dimensions of diversity in modern spring bread wheat in developing countries from 1965. Crop Sci 42:1766–1779

Snape JW, Chapman V, Moss JP, Blanchard CE, Miller TE (1979) The crossabilities of wheat varieties with *Hordeum bulbosum*. Heredity 42:291–298

Snapp S (1999) Mother and baby trials: a novel trial design being tried out in Malawi. In TARGET. The Newsletter of the Soil Fertility Research Network for Maize-Based Cropping Systems in Malawi and Zimbabwe, CIMMYT, Harare, Zimbabwe (January 1999 issue)

Sohail Q, Inoue T, Tanaka H, Eltayeb AE, Matsuoka Y, Tsujimoto H (2011) Applicability of *Aegilops tauschii* drought tolerance traits to breeding of hexaploid wheat. Breed Sci 61:347–357

Sohail Q, Shehzad T, Kilian A, Eltayeb AE, Tanaka H, Tsujimoto H (2012) Development of diversity array technology (DArT) markers for assessment of population structure and diversity in *Aegilops tauschii*. Breed Sci 62:38–45

Somers DJ, Isaac P, Edwards K (2004) A high-density microsatellite consensus map for bread wheat (*Triticum aestivum* L.). Theor Appl Genet 109:1105–1114

Stamp P, Messmer R, Walter A (2012) Competitive underutilized crops will depend on the state funding of breeding programmes: an opinion on the example of Europe. Plant Breed 131:461–464

Sujana G, Sharma HC, Rao DM (2008) Antixenosis and antibio- sis components of resistance to pod borer *Helicoverpa armigera* in wild relatives of pigeonpea. Int J Trop Insect Sci 28:191–200

Tai GCC (1971) Genotypic stability analysis and its application to potato regional trials. Crop Sci 11:184–190

Tambussi E, Nogues S, Ferrio P, Voltas J, Araus JL (2005) Does higher yield potential improve barley performance in Mediterranean conditions? A case study. Field Crops Res 91:149–160

Tanksley SD, McCouch SR (1997) Seed banks and molecular maps: unlocking genetic potential from the wild. Science 277:1063–1066

Tawkaz S (2011) Response of Ssme wheat genotypes to anther culture technique for doubled haploid production. M.Sc thesis. Sudan Academy of Science, Kartum

Thakur A (2004) Use of easy and less expensive methodology to rapidly screen fruit crops for drought tolerance. Acta Hort 662:231–235

Thapa DB, Sharma RC, Mudwari A, Ortiz-Ferrara G, Sharma S, Basnet RK, Witcombe JR, Virk DS, Joshi KD (2009) Identifying superior wheat cultivars in participatory research on resource poor farms. Field Crops Res 112:124–130

Tomar SMS, Kochumadhavan M, Nambisan PNN (1991) Hybrid weakness in Triticum dicoccum Schubl. Wheat Inf Serv 72:9–11

Tomar SMS, Singh B (1998) Hybrid chlorosis in wheat × rye crosses. Euphytica 99:1–4

Trethowan RM, van Ginkel M (2009) Synthetic wheat-An Emerging Genetic Resource. In: Carver BF (ed) Wheat science and trade. Wiley-Blackwell, Ames

Trethowan RM, van Ginkel M, Rajaram S (2002) Progress in breeding wheat for yield and adaptation in global drought affected environments. Crop Sci 42:1441–1446

Upadhyaya HD, Sharma S, Dwivedi SL (2012) Arachis. In: Kole C (ed) Wild crop relatives: genomic and breeding resources, legume crops and forages. Springer, Berlin/Heidelberg, pp 1–19

Upadhyaya HD, Sharma S, Reddy KN, Saxena R, Varshney RK, Gowda CLL (2013) "Pigeonpea," In: Singh M, Upadhyaya HD, Bisht IS (eds) Genetic and Genomic Resources of Grain Legume Improvement. London; Waltham, MA: Elsevier Inc, pp 181–198

Van Becelaere G, Lubbers EL, Paterson AH, Chee PW (2005) Pedigree- vs. DNA marker-based genetic similarity estimates in cotton. Crop Sci 45:2281–2287 van Eeuwijk FA, Denis JB, Kang MS (1996) Incorporating additional information on genotypes and environments in models for two-way genotype by environment tables. In: Kang MS, Gauch HG (eds) Genotype-by-environment interaction. CRC Press, Boca Raton, pp. 15–49

van Eeuwijk FA, Malosetti M, Yin X, Struik PC, Stam P (2005) Statistical models for genotype by environment data: from conventional ANOVA models to eco-physiological QTL models. Aust J Agric Res 56:1–12

Villareal RL, Singh RP, Mujeeb-Kazi A (1992) Expression of resistance to *Puccinia recondita* f.sp. tritici in synthetic hexaploid wheats. Vortr Pflanzenzuchtg 24:253–255

Wang XS, Zhu J, Mansueto L, Bruskiewich R (2005) Identification of candidate genes for drought stress to lerance in rice by the integration of a genetic (QTL) map with the rice genome physical map. J Zhejiang Univ Sci 6:382–388

Wei ZM (1982) Pollen callus culture in *Triticum aestivum*. Theor Appl Genet 67:71–73

West JD, Peak D, Peterson JQ, Mott KA (2005) Dynamics of stomatal patches for a single surface of *Xanthium strumarium* L. leaves observed with fluorescence and thermal images. Plant Cell Environ 28:633–641

Witcombe JR, Joshi A, Goyal SN (2003) Participatory plant breeding in maize: A case study from Gujarat, India. Euphytica 130:413–422

Worland AJ, Snape JW (2001) Genetic basis of worldwide wheat varietal improvement. In: Bonjean AP, Angus WJ (eds) The World Wheat Book: a History of Wheat Breeding. Lavoisier Publishing, Paris

Xiao YN, Li XH, George ML, Li MS, Zhang SH, Zheng YL (2005) Quantitative trait locus analysis of drought tolerance and yield in maize in China. Plant Mol Biol Rep 23:155–165

Xie WL, Nevo E (2008) Wild emmer: genetic resources, gene mapping and potential for wheat improvement. Euphytica 164:603–614

Xu JL, Lafitte HR, Gao YM, Fu BY, Torres R, Li ZK (2005) QTLs for drought escape and tolerance identified in a set of random introgression lines of rice. Theor Appl Genet 111:1642–1650

Yan W, Hunt LA, Sheng Q, Szlavnics Z (2000) Cultivar evaluation and mega-environment investigation based on GGE biplot. Crop Sci 40:597–605

Yan W, Kang MS (2003) GGE biplot analysis: a graphical tool for breeders, geneticists, and agronomists. CRC Press, Boca Raton

Yang N, Smale M (1996) Indicators of Wheat Genetic Diversity and Germplasm Use in the People's Republic of China. NRG Paper 96-04. CIMMYT, Mexico, D.F.

Yates F, Cochran WG (1938) The analysis of groups of experiments. J Agri Sci 28:556–580

Yediay FE, Baloch FS, Kilian B, Ozkan H (2010) Testing of rye-specific markers located on 1RS chromosome and distribution of 1AL. RS and 1BL. RS translocations in Turkish wheat (*Triticum aestivum* L., T. durum Desf.)

Young ND, Tanksley SD (1989) RFLP analysis of the size of chromosomal segments retained around the Tm-2 locus of tomato during backcross breeding. Theoret Appl Genet 77:353–359

Yue B, Xue WJ, Xiong LZ, Yu XQ, Luo LJ, Cui KH, Jin DM, Xing YZ, Zhang QF (2006) Genetic basis of drought resistance at reproductive stage in rice: separation of drought tolerance from drought avoidance. Genetics 172:1215–1228

Zaharieva M, Monneveux P, Henry M, Rivoal R, Valkoun J, Nachit MM (2001) Evaluation of a collection of wild wheat relative *Aegilops geniculata* Roth and identification of potential sources for useful traits. Euphytica 119:33–38

Zeven AC (1981) Eighth supplementary list of wheat varieties classified according to their genotypes for hybrid necrosis. Euphytica 30:521–539

Zhu L, Liang ZS, Xu X, Li SH, Monneveux P (2009) Evidences for the association between carbon isotope discrimination and grain yield—Ash content and stem carbohydrate in spring wheat grown in Ningxia (Northwest China). Plant Sci 176:758–767

Zobel RW, Wright MG, Gauch HG (1988) Statistical analysis of yield trial. Agron J 80:388–393

Part IV
Dryland Agriculture: Some Case Studies

Dryland Agriculture in Australia: Experiences and Innovations

Walter K. Anderson, David Stephens, and Kadambot H.M. Siddique

1 Introduction

The vast majority of crop and livestock production in Australia is conducted under dryland or rainfed conditions. The terms 'dryland' and 'rainfed' are used interchangeably to refer to production using only natural rainfall without any form of irrigation. Australia is the driest, inhabited continent in the world and agricultural production is entirely absent from a large part of it, except for extensive grazing, mostly of cattle or sheep. In 2007–08, only 54 % of the land surface was managed for agricultural businesses and, of that area, 87 % was under grazing management and only 8 % was cropped (ABS 2009a).

As might be expected under such low rainfall conditions, the average size of agricultural holdings is large, the population density is low and the distances between settlements are large. Agriculture is highly mechanised, driven by the size of the holdings and the high cost of labour. The average farm size in Australia increased from about 1700 ha to more than 2200 ha from 1990 to 2010 (ABARES 2011). In most of the cropping zones however, the average size of dryland cropping farms has doubled over the last 20 years. In addition, Australian farmers experience greater volatility in yield and price than most other farmers in the world (Australian Farm Institute 2012). Despite these apparent disadvantages for productivity, Australian farmers have managed to keep the cost of production of most agricultural products at relatively low levels. In turn, this has enabled Australian farmers to

W.K. Anderson (✉) • K.H.M. Siddique
The University of Western Australia Institute of Agriculture, The University of Western Australia, MO82, LB 5005, Perth, WA 6001, Australia
e-mail: wmanderson@bordernet.com.au

D. Stephens
Australian Export Grains Innovation Centre, 3 Baron-Hay Court, South Perth, WA 6151, Australia

M. Farooq, K.H.M. Siddique (eds.), *Innovations in Dryland Agriculture*, DOI 10.1007/978-3-319-47928-6_11

remain competitive, an essential requirement for a country that relies substantially on production for export.

Another feature of Australian dryland agriculture is the very low, and declining, level of government subsidies. The major direct government subsidy comes from matching some of the funds that producers contribute for research into their respective industries (see section on research below). Public research, funded at both state and federal levels, is declining in favour of partnerships with private producer organisations. Research into aspects of the Australian, broad-acre agricultural industries nevertheless has been highly profitable (Mullen 2007).

There have been a series of previous reviews on dryland agriculture in Australia, especially with respect to the cropping industries (Anderson and Impiglia 2002; Freebairn et al. 2006; Passioura and Angus 2010; Anderson and Angus 2011; Stephens et al. 2011; Fischer et al. 2014; Anderson et al. 2016) and the pasture/forage industries (Wolf 2009). These reviews have given detailed descriptions of the situations at their relative times of publication. This chapter is focussed on some general characteristics of Australian agriculture, recent experiences and innovations, with comparatively less discussion of specific agricultural industries.

2 Brief History of Agriculture in Australia

2.1 Pre-European Settlement

There is a long history of human habitation on the Australian continent dating back at least 40,000 years. Evidence has been accumulating that the aboriginal peoples practised some form of settled agriculture through harvesting and processing grains and root crops well before European settlers arrived in the late eighteenth century. This early agriculture included management of pastures, and thus native animals, through judicious use of fire (Mitchell 1839; Gamage 2011; Pascoe 2014).

The introduction of hard-footed animals such as sheep, cattle and horses and the widespread clearing of natural vegetation changed the agricultural environment in ways that modern farmers have been striving to remedy ever since. Despite the long-term use of Australian native plants prior to European settlement, no native grain crops have been developed for widespread human use until the present time.

The Last 200+ Years

The Australian winter crop yields have gone through four phases (See yield of wheat in Fig. 1). In the first phase, European settlers brought with them crop types and cultivars that proved unsuitable for the local conditions with the result that early settlement on the eastern seaboard almost failed. Subsequently a phase of nutrient depletion prior to about 1900 lead to a steady decline in yield. It was not until James Farrer improved drought resistance, disease resistance and grain quality in wheat (Farrer 1898) that the early colonies began to thrive. The history of wheat yields and the influence of sown legume pastures, as described by Donald (1965), largely

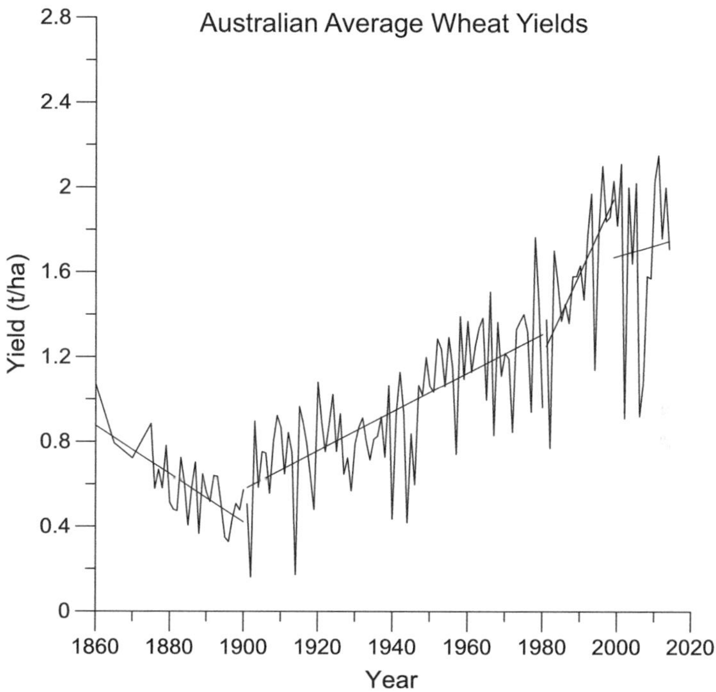

Fig. 1 Average Australian wheat yields with fitted linear trends for the four main epochs of productivity: 1860–1999 (declining), 1901–1980 (steady increase), 1981–1999 (rapid increase), 2000–2014 (mixed and highly variable)

reflect the history of the development of the wider agricultural industry. The use of superphosphate fertiliser, trace elements, fallowing and new cultivars became widespread, increasing the average yield of wheat above earlier levels (second phase, Fig. 1). The third stage of accelerated yield increase occurred in the 1980s and 1990s, which was accompanied by earlier sowing, the switch to minimum tillage, improved weed control using chemical herbicides, more nitrogen fertiliser and the use of rotation crops, pasture legumes, and eventually semi-dwarf cultivars. All these changes benefited both crops and pastures which were largely integrated into mixed systems. Finally, average yields levelled out after about 2000 as yield variability increased, farmers reduced the use of nitrogen fertilisers and grain legumes, and a severe drought restricted irrigation in south-eastern Australia between 2006 and 2009 (Stephens et al. 2011).

Australian farmers have been relatively quick to adopt innovations that have improved their productivity. This is likely a function of economic necessity (see Pannell et al. 2006). A large proportion of output is exported so it has to compete favourably for both price and quality. The poor soils and low rainfall conditions have also focused the minds of producers on reducing costs to remain viable.

Reducing costs, often at the same time as improving product yield and quality, has only been possible through maintaining soil and environmental quality, a challenging task that would have been impossible without consistent innovation. Reduced or zero tillage, stubble retention, improved herbicides and cultivars and adoption of precision agricultural techniques have proved pivotal in these respects in recent times (Anderson et al. 2005; Derpsch and Friedrich 2009; Kingwell and Fuchsbichler 2011; Robertson et al. 2012; Scott et al. 2013).

Landholdings in Australia have always been relatively large by world standards, often exceeding 2000 ha in the crop/pasture zone. Even when the land was first divided and cleared it was necessary to have larger holdings to support the farming family. Some early subdivisions proved too small to be viable, leading to farm amalgamations; the trend towards increasing farm size has continued from the mid-twentieth century until the present. This has been accompanied by the adoption of larger power units and agricultural machines, driven by the high costs of farm labour.

2.2 The Current Situation

Dryland agriculture in Australia is currently in a period of great change. In addition to the trend towards larger farms and increasing mechanisation, which continues, there is a growing recognition of the need to farm more sustainably to protect the natural resource base (Ogilvy et al. 2015). In this transition, there are many questions yet to be answered such as 'Does this mean reverting to entirely natural or organic methods?', 'What place will there be for genetically-modified plants and animals?', 'Do we need to abandon chemical herbicides, pesticides and fungicides altogether?', 'Can we restore degraded lands to profitable production?', 'How do we adapt to more variable climates in the future?' and 'What will be our future sources of energy to work the land?'

Further complicating these questions is the current trend towards decreasing public funds for agricultural research. It appears that the support to address the questions above, and many others, will need to come increasingly from farmers themselves although a case can be made for continued public support given that many of the demands for greater knowledge about production methods is coming from consumers. There is already a small demand from consumers for greater, measurable evidence of the use of sustainable practices in the production of our food and fibre (Ogilvy et al. 2015).

Australian dryland agriculture is also under worldwide pressures from trends such as the increasing demand for food in developing countries and for better quality food from emerging middle classes, from the threat of climate change and from new technologies, in both the genetic and digital areas (Hajkowiczs and Eady 2015).

3 Resource Management

3.1 Climates and Soils

Dryland cropping zones in Australia mainly occur between the 300 and 600 mm rainfall isohyets (Fig. 2). In the northern part of the dryland agricultural areas, summer rainfall dominates such that production of winter crops and pastures depends on conserving some of the rainfall. This accounts for the occurrence of production on the soils in those areas that can store substantial amounts of water and the widespread use of summer fallow. In the southern and western parts of the agricultural area, winter rainfall predominates so that soil water storage becomes less important for plant production and sandy soils that store less water can be used.

Australian soils used for agriculture are deeply weathered and depleted of plant nutrients. They have been derived from the oldest geology on the earth. Soil types ranging from deep sands (largely in Western Australia) to self-mulching, cracking clay loams (Queensland and northern New South Wales) are used for agriculture in various parts of the continent (Leeper 1964; McGarity 1975; Hamblin and Kyneur 1993; Freebairn et al. 2006).

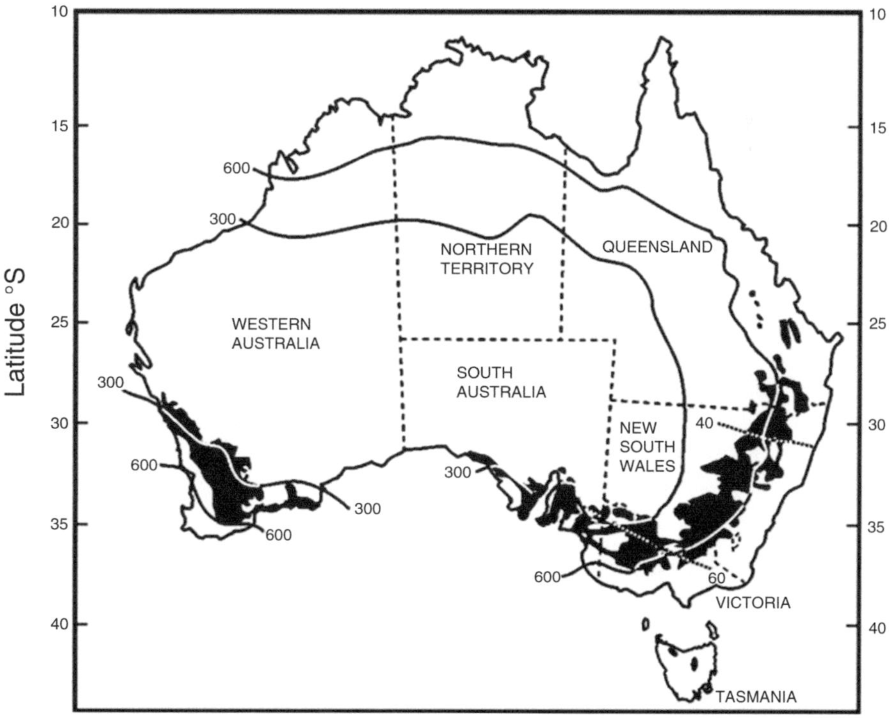

Fig. 2 Dryland cropping zones in Australia between the 300 and 600mm rainfall isohyets. Dotted lines show average percentage of winter rainfall. All cropping zones in Western Australia receive >60 % of rainfall in the winter months (After Anderson and Angus 2011)

3.2 Landcare

Concern among both rural and urban communities in Australia has led to the organisation of local groups of landholders and other groups into a Landcare movement consisting initially of volunteers convened to address issues of natural resource management (NRM), environmental improvement and preservation. The network has spread across both rural and peri-urban areas and has joined the NRM agencies funded by the state and federal governments.

The following is a quote from the Australian Agriculture website: "Landcare is a unique grass-roots movement that started in the 1980s through initiatives to tackle degradation of farmland, public land and waterways. The movement has expanded and evolved significantly since then, and is achieving results Australia-wide. Individuals and groups .... are focused on best-practice sustainable agriculture and expert management of natural assets such as soil, water and native vegetation.

Caring for the land includes a range of activities such as:

- sustainable farm practices
- restoring native habitats and revegetation
- controlling weeds and pests
- developing and sharing local natural resource management skills and knowledge". (landcareaustralia.com.au)

The movement includes Landcare groups, farming systems and farm improvement groups, 'Friends of' groups and Indigenous land management groups. It is estimated there are 6000 groups and over 100,000 volunteers across Australia caring for the land. Many farmers and landholders also undertake this important work but are not always affiliated with a particular Landcare group.

3.3 Natural Resource Management

The understanding that the natural resource base, especially the soil, has suffered long-term decline is leading to some gradually emerging trends towards restoring and even improving the agricultural landscape and farming practices (Ogilvy et al. 2015). For example, in the 2006–2007 season over 94 % of Australian farmers reported using some aspect of NRM including control of weeds, pests and diseases, land and soil management (Australian Bureau of Statistics 2009b). Given the naturally-poor soil fertility and fragile landscapes (McKenzie et al. 2004), it is probably unsurprising that concern for the environment is relatively high in Australia. After the initial degradation to the land caused by extensive clearing of native forests following European settlement, much work has been devoted to restoring soil fertility and reducing erosion (e.g., Freebairn and Wockner 1986; Sallaway et al. 1990). A comprehensive review of the status of Australian soils and nutrient balances was summarised in the National Land and Water Resources Audit (NLWRA 2001).

3.4 Introduced Pests, Diseases and Weeds

Many animals that were introduced intentionally or accidentally as a result of European settlement have become feral pests (Bomford and Hart 2002). These include rabbits, foxes, cats, camels, goats, donkeys, horses and pigs. Considerable resources are expended annually to keep these pests at manageable levels and overall productivity has clearly been reduced as a consequence (Canyon et al. 2002). Many introduced weeds and diseases that are found in similar dryland areas around the world also occur in Australia and these have contributed to substantial costs to industry (Groves 2002). However, a rigorous quarantine and biosecurity system has succeeded in excluding others (Nairn et al. 1996).

4 Cropping Systems

4.1 Yield Improvement – Management and Breeding

The average yield of crops has not increased at a uniform rate since European settlement but rather at different rates in different periods (Fig. 1). The differing adoption of various practices and improved cultivars of wheat is illustrated in Fig. 3 as an example. However, the adoption of practices and cultivars at given times can only infer causation. Comparisons based on the yield progress of cultivars in variety trials with the yield increases of average farm yields can give an indication of the relative contributions of management and genotypes to overall yield progress (e.g. Byerlee 1994). Such comparisons suggest that about 70–80 % of past yield advances have been due to changes in management practices (Anderson et al. 2005). The term 'management' can be further divided into 'tactical' or in-season practices and 'strategic' or practices that have a long-term impact, largely soil improvement. It is proposed that past yield improvements can be equally attributed to genetic, tactical and strategic factors.

However, data from field trials of wheat over a number of locations and seasons that include both genotypes and agronomic treatments have shown that the environ-

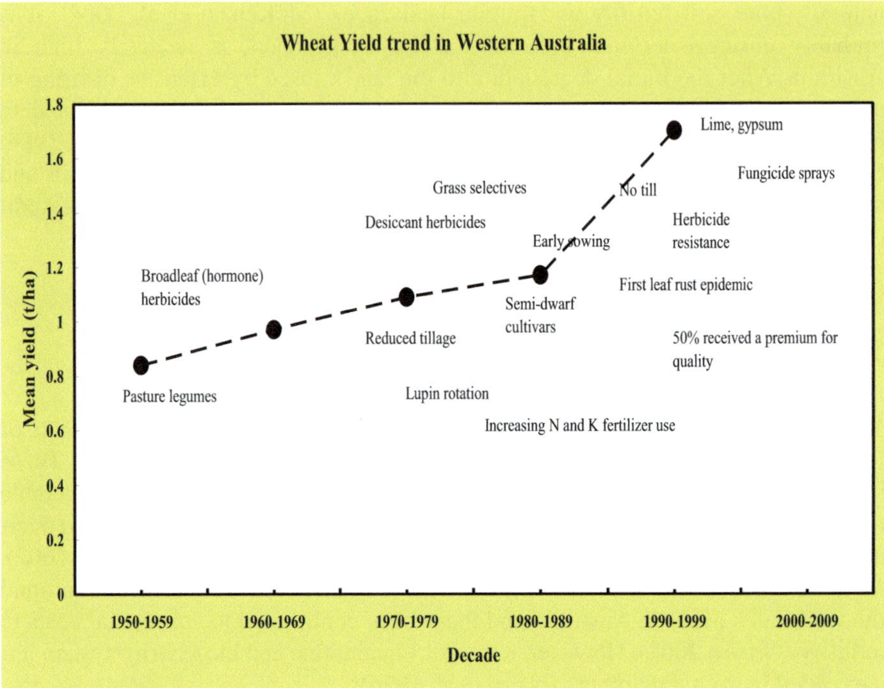

Fig. 3 Some practices associated with average wheat yield improvement in Western Australia, 1950–2000 (After Anderson et al. 2005)

ment (site × season) can dominate the responses but management (tactical agronomy) and interactions with the environment can account for much more of the yield variance than genotype or its interactions with either environment or management (Anderson 2010).

In the Australian dryland environments since about 2000, average grain yields have either reached a plateau or become much more variable from season to season (Fig. 1). For example, in Western Australia, the season-to-season range of average wheat yield was about 0.8 t/ha before 2000 compared to about 1.7 t ha^{-1} in the new millennium. The reasons proposed for this increased variability range from increased variability in rainfall (climate change) to inappropriate application of inputs, especially N fertilisers, according to seasonal conditions (tactical management).

Breeding advances, often underpinned by physiological research, have made major contributions to yield improvement in the Australian wheat crop (Fischer and Wall 1976; Siddique et al. 1989; Loss and Siddique 1994; Evans 1987; Passioura 2006). Further improvements have been attributed to the interactions that exist between breeding and agronomy or management (Anderson and Impiglia 2002; Hochman et al. 2009; Cooper et al. 2001; Passioura and Angus 2010; Sadras and Lawson 2011; Richards et al. 2014). Improved transpiration efficiency (Evans 1987; Passioura and Angus 2010), competitive ability against weeds (Palta and Peltzer

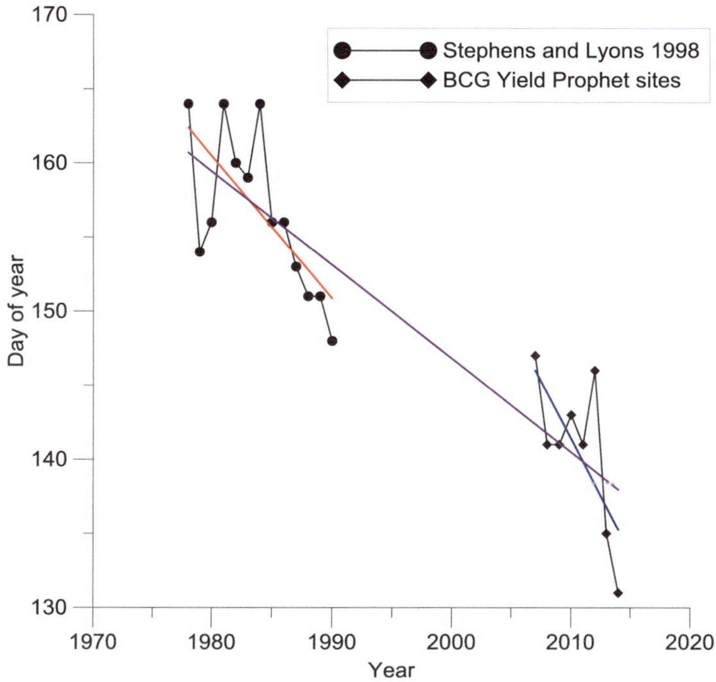

Fig. 4 The mean midpoint of wheat sowing for Australia from 1978–1990 (Stephens and Lyons sowing date survey 1998), 2007–2014 (Source: Birchip Cropping Group, BCG- sites were spread across Australia)

2001; Lemerle et al. 2004), nutrient use efficiency (Anderson and Hoyle 1999) and suitability for dual-purpose use (grazing and grain recovery (e.g., Anderson 1985; Virgona et al. 2006) are also considered traits likely to contribute to future yield increases.

4.2 Adoption of the Components of Conservation Agriculture

Australian dryland farmers have adopted zero tillage over more than 80 % of the cropped area (ABS 2009b; Llewellyn and D'Emden 2010; Ward and Siddique 2014). Some of the main reasons that farmers adopted zero tillage were - reduced fuel costs from fewer cultivations, quicker seeding programmes, reduced wind erosion, better water retention and better weed management from herbicides (Scott et al. 2013). One of the biggest benefits of this new technology has been that farmers have been able to sow crops much earlier. From 1978 to 1990, the mean midpoint of wheat sowing progressed a day earlier per year on a national scale (Stephens and Lyons 1998). When the recent midpoint of wheat sowing was calculated, this trend to earlier sowing has continued with farmers now planting in late April and early May (Fig. 4).

The main benefit of this change is that crops are now growing in a period of lower average evaporative demand and better pre-seeding weed management with herbicides.

Adoption of stubble retention has been much less complete and much slower according to a review of Scott et al. (2010). In a further review Scott et al. (2013) again concluded that, in the dominant cropping systems of southern Australia, there is little compelling evidence that retention of crop residues has led reliably to economic benefits.

The practices of zero tillage and residue retention have fitted into the long-standing systems of rotation of cereals either with grain legumes such as field peas and lupins, oilseed crops such as canola (ABS 2009a) or sunflower (*Helianthus anuus* L.), or with pasture legumes such as subterranean clover (*Trifolium subterraneum* L.) and medics (*Medicago* spp., Reeves and Ewing 1993). Crop and pasture sequences are seldom fixed over long periods but vary according to grain prices, requirements for disease, weed and soil management, and demand for animal products (wool, meat, dairy).

Evidence in Australian rainfed crops using direct drilling with residue retention, has indicated that there may be no increase in soil organic matter in a range of soil types even after 10 years unless annual rainfall exceeds about 500 mm (data summarised by Chan et al. 2003). This is likely due to the lower levels of crop yield and residue produced under lower rainfall conditions, or to the likelihood of higher soil temperatures in low rainfall areas which can prevent accumulation of soil organic matter (Hamza and Anderson 2010). This variability in response to the various components of the CA system has likely led to partial adoption by farmers in the various Australian environments as discussed by Kirkegaard et al. (2014).

In higher rainfall areas (>500 mm annual rainfall) and where perennial pastures are part of the dominant farming system, soil organic matter tends to accumulate more across a range of soil types than where continuous cropping is practised (Hoyle et al. 2014). In any case, organic matter largely accumulates in the top 10 cm of soil in a zero tillage system such that, even if the topsoil is saturated with respect to the SOC level, the content below that depth may still be low.

Adoption of precision agriculture methods such as controlled traffic, yield mapping and variable rate technology has been slower but is gaining momentum (ABS 2009a; Kearns and Umbers 2010). Problems with different machinery configurations and the complexity of the technology have slowed uptake. However, farmers who have adopted this technology are finding improved yields from less soil compaction and a more efficient allocation of resources on soils that yield the greatest return per unit of input.

4.3 Trends in Water Use Efficiency

All these varietal and management improvements result in a higher yield per unit of water from water supplied from rainfall. Water use efficiency is related to the amount of grain produced per millimetre of total water used in the crop growing period. To assess water use (or production) efficiency of wheat yields across Australia, state yields and statistical sub-division yields were divided by potential yields (Y_{pot}) defined by French and Schultz (1984). However, instead of using a fixed 110 mm of rainfall as soil evaporation, as used by French and Schultz (1984), we assumed that a third of growing season rainfall was lost to soil evaporation. This assumption prevents the actual yield from exceeding the potential yield in drought conditions. Therefore,

$$WUE = \mathrm{Y_a} / \mathrm{Y}_{pot}$$

$$\mathrm{Ypot} = \mathrm{W} \times WUE_{\mathrm{p}}$$

$$= (GSR + SM - \mathrm{E}) \times WUE_{\mathrm{p}}$$

$$= (GSR + SM - (0.33 \times GSR) \times WUE_{\mathrm{p}}$$

where:

WUE = water use efficiency
Ya = actual yield
Yp = potential yield
$\mathrm{WUE_p}$ = potential water use efficiency (20 kg/ha/mm)
W = water used
GSR = growing season rainfall (between sowing and rainfall 'ending date' in STIN)
SM = soil moisture at sowing determined by STIN
E = soil water losses to evaporation.

A modelling analysis used this approach to examine regional and state yields with the Australian Export Grains Innovation Centre's STIN wheat forecasting model (Stephens et al. 1989). This model calculates a daily water balance up to a district midpoint of sowing and then determines a maturity date when rainfall stops contributing to yield. Excess heavy rain in the last three weeks of the season is also removed. This approach has a number of simplifications, like assuming a dry soil profile at the beginning of the soil moisture accumulation on 1 October in the year before sowing. However, since we were more concerned with the changes in WUE over time, this method gave similar results to more complex model calculations. When the water supply at 800 rainfall stations across the Australian grain belt was added and weighted to the proportion of grain planted near each station, state and national potential wheat yields were determined. When the actual yields were divided by the potential yield through time, the increasing water use efficiency for different parts of the country and the variability of WUE with season type is highlighted, as are longer term trends (Fig. 5).

Figure 5 shows that lower WUEs occur in severe drought years which are typically El Niño years where many paddocks were not harvested or were used to feed

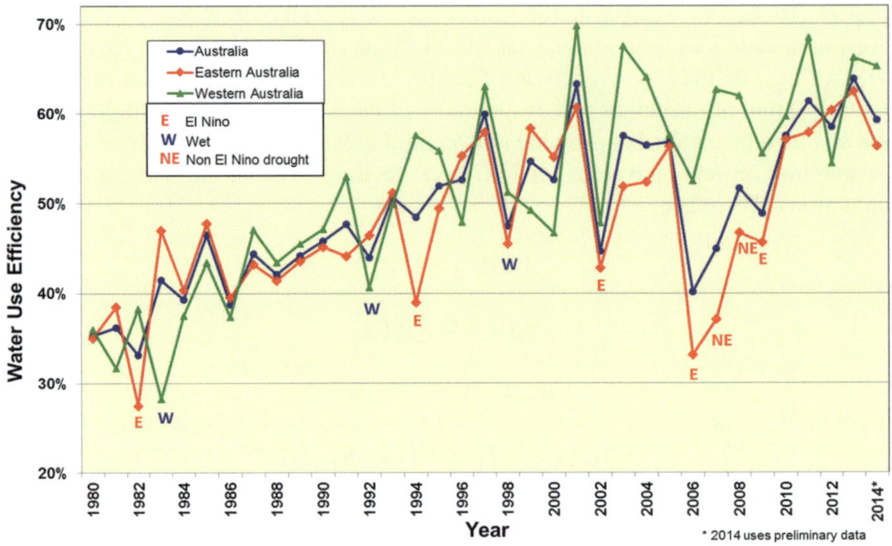

Fig. 5 Average Australian water use efficiency for wheat 1980–2013, where E is an El Nino drought year, NE is a non-El Nino drought year and W is a wet year

livestock. This is particularly the case in eastern Australia where ENSO (El Niño – Southern Oscillation) events dominate rainfall variability. Lower WUEs also occurred in very wet years in 1983 and 1992 where waterlogging was prevalent in Western Australia and in 1998 in eastern Australia.

In terms of WUE, Fig. 5 shows that the largest increases occurred in Western Australia where almost a doubling in efficiency occurred from about 33 to 64 % over the last three decades. Significant increases appear related to:

- a more steady and significant increase in nitrogen fertiliser, the highest proportion of farmers matching their fertiliser use to soil testing (60–80 %)
- the highest rates of adoption of zero tillage
- spraying out summer weeds
- up to 80 % stubble retention
- a major uptake of dry sowing which enables optimal plant emergence in favourable moisture conditions (Kearns and Umbers 2010; Stephens et al. 2011)
- a doubling of lime application (to over a million tonnes a year) to address acidic soils
- fewer wet winters which contributed to waterlogging on shallow soils (Stephens and Lyons 1998; Ludwig et al. 2009, Stephens et al. 2011).

In eastern Australia, the WUE has also increased by about 70 % since the 2006–2008 drought based on the adoption of much the same practices as in the west.

4.4 *Product Quality – The Case for Wheat*

There has been an increasing understanding across the supply chain of Australian wheat that improving grain quality for specific products can make a major contribution to the profitability of the crop. Wheat grain that is exported to various countries from Australia can qualify for a price premium if the quality is suitable for various specific end uses. These include white, salted noodles, yellow, alkaline noodles, biscuits, pasta and high-quality bread. The amount of the premium varies according to market conditions of supply and demand, but grain samples that conform to specifications of kernel size and hectolitre weight (related to milling yield), protein percentage and varietal characteristics including flour colour and starch quality may qualify. The premiums vary from about 3 to 20 % of the base price but may mean the difference between profit and loss in some cases. Plant breeders have produced a range of high-yielding cultivars that are acceptable for the various export markets and local research has revealed the appropriate combinations of soil type selection, rotation and agronomy that maximise the probability of achieving quality that will attract premiums (Anderson et al. 1995; Anderson et al. 1997; Anderson and Sawkins 1997; Miyan et al. 2011). The adjustments to crop management that have been suggested by this research can often be made at a very low cost to the farmer and have resulted in major increases in the percentage of the crop that receives a premium (see Fig. 6).

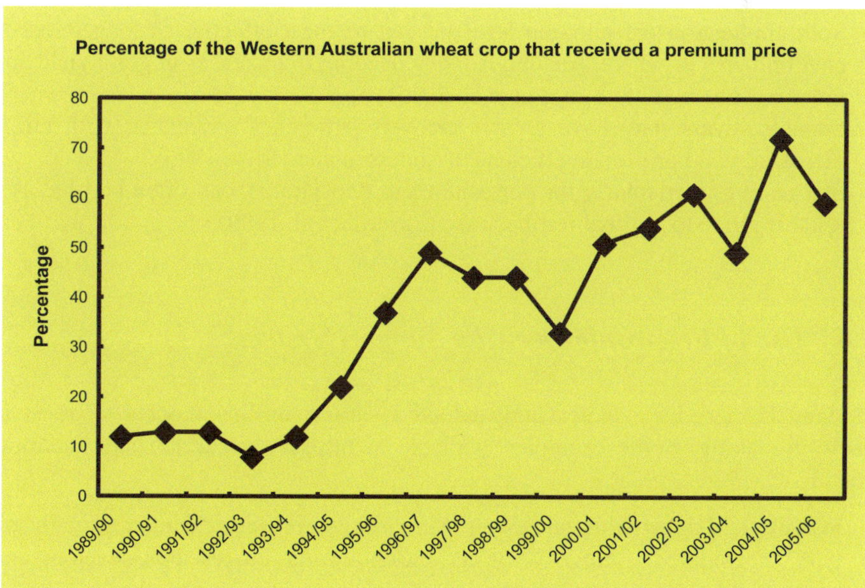

Fig. 6 Percentage of the Western Australian wheat crop that qualified for a premium price (After Anderson et al. 2005)

4.5 Nutrient Efficiency

More precise management of fertiliser application to wheat crops is desirable to increase profits and to reduce soil acidification and nutrient losses from farms. Systems that estimate all components of the nutrient balance and assess the spatial variability of nutrients will ultimately be useful for decision makers. In practice, factors such as the availability of fertilisers and cash, the relative return from other inputs, interactions with factors such as weeds, variable rainfall and the perceived risk to the environment will all modify fertiliser decisions.

Grain yield per unit of applied nutrient (economic efficiency) as a function of uptake per unit of applied nutrient (uptake or recovery efficiency) and yield per unit of nutrient taken up (physiological efficiency) is receiving increased attention from researchers and farmers alike (e.g., Ladha et al. 2005).

Recovery of applied fertilisers by the wheat crop is often low. For example, recovery of applied nitrogen seldom exceeds 50 % in rainfed wheat crops (Fillery and McInnes 1992). Improving the efficiency of fertiliser use, and probably of fertiliser recovery, can be influenced by several management practices:

- Fertilisers applied to early sown crops are often used more efficiently (Anderson et al. 1995, 1997)
- Banding of nitrogen and potassium fertilisers below and at some distance from the seed often achieves better results than broadcasting (Jarvis and Bolland 1990)
- Deep placement can have advantages over shallow placement (Jarvis and Bolland 1990)
- Split applications of nitrogen fertiliser can be most effective in longer season environments or where leaching is likely, leading to increases in grain yield and protein (Mason 1975; Simpson et al. 2015)
- Some cultivars may have greater recovery efficiency associated with either increased yield or protein efficiency (Anderson and Hoyle 1999)
- Soil testing, particularly for potassium and phosphorus, can often be used as a reliable guide to optimal fertiliser use (Peverill et al. 1999).

4.6 Use of Electronic Decision Support Systems

Dryland farmers have increasingly adopted various on-line models to assist in decision-making. Some examples available at https://www.agric.wa.gov.au/tools are:

- **Wheat yield constraint calculator**: Simple water balance calculations with soil type considered. https://www.agric.wa.gov.au/grains-research-development/wheat-yield-constraint-calculator

- **MyPestGuide**: To identify pests. https://www.agric.wa.gov.au/plant-biosecurity/mypestguide-app
- **Wheat diagnostic tool**: A systematic tool to assess factors likely to limit yield. https://www.agric.wa.gov.au/wheat-diagnostic-tool
- **WA crop sequence calculator**: Rotation effects considered. https://www.agric.wa.gov.au/sowing/wa-crop-sequence-calculator
- **Flower power**: Wheat phenology; a tool to estimate when wheat varieties will flower from given sowing dates at a range of locations. Used to manage the risk of frost damage. https://www.agric.wa.gov.au/frost/flower-power
- **My paddock**: Compares the effect of different crop problems on the level of grain yield loss. https://www.agric.wa.gov.au/mypaddock
- **Seasonal climate information**: To assess the seasonal outlook during the year.https://www.agric.wa.gov.au/agseasons/seasonal-climate-information
- **Weather stations**: Contains current weather records for a range of stations in cropping areas. https://www.agric.wa.gov.au/weather-stations
- **Rainfall to date**: In-season. https://www.agric.wa.gov.au/climate-weather/rainfall-date

In addition to the crop models under development and testing, some models have been developed and used in pasture management (Bell and Allen 2000; Clark et al. 2000).

4.7 Diagnostic Research and Development

There is recognition that crop and pasture yields that do not approach the rainfall-limited potential yield are likely limited by factors that are not easily identified by less than an objective trial-and-error process. The gap between actual or average grain yields and the rainfall-limited potential is likely greater in the wetter seasons (Anderson 1985, 2010).

Field diagnosis of the factors likely to limit production, as practised by advisors, agronomists and farmers at an informal and practical level, has been a part of system development. However, more formal and systematic diagnosis is developing. Objective field diagnosis using measurements and observations in a systematic manner have been developed by agronomists in Western Australia (www.agric.wa.gov.au/objtwr/imported assets/content/live/land/bn detect doug.pdf). Further, an experimental approach based on objective soil, plant and management measurements followed by field experiments and on-farm verification has been successfully trialled (Sharma and Anderson 2014; Anderson et al. 2014, 2016). Extension of the results from such field experiments can be achieved by either verification on similar soil types and/or the use of crop models.

5 Grazing and Pastoral Systems

Grazing animals, largely sheep and cattle, have comprised a major part of the dryland agricultural systems in Australia over the last 200+ years. Sheep numbers have declined from the peak of more than 170 million in 1970–1971 to less than 100 million in the mid-2000s in response to the declining price for wool and sheep meats. Cattle numbers continue to increase, reaching 28.5 million in 2005–2006 (data from Wolf 2009 derived from the Australian Bureau of Statistics). However, a mixture of crops and pastures in some form has been used to reduce the risks of both the weather and the markets. The proportion of each has varied according to economic conditions, but concerns for the long-term viability, or sustainability, of the systems has increased recently.

Dryland farming has been based on ley farming, a system that incorporates a short phase of self-regenerating, legume-based, annual pastures with cereal crops. However, the system has come under increasing pressure for change as the prices for animal products (largely wool and meat) have fallen (Reeves and Ewing 1993). This has resulted in much longer cropping phases ('phase farming') which has raised concerns regarding reduced soil fertility and perceptions of increased economic risk, even if the current profitability of crops exceeds that of pastures (Bell et al. 2014).

Recent innovations in the pastures and grazing area have included the revival of the use of dual-purpose crops (cereals and canola) for grazing and later recovery for grain harvest, and the introduction of perennial species to the crop/pasture sequence (Bell et al. 2014). Some of the concerns that confront farmers, and that perennial pastures may address, include dryland salinity associated with rising water tables, waterlogging and herbicide resistance.

In terms of the livestock component of the dryland system, economic pressures have resulted in an increase in the production of fat lambs for meat production from traditional wool (Merino) flocks (Kopke et al. 2008). This has come both from the use of meat-type sires over Merino ewes and from the development of dual-purpose Merino breeds.

6 Key Conclusions

6.1 Research Funding – Legislated, Voluntary and Private

Various research and development corporations (RDCs) have been pivotal in supporting agricultural research for Australian dryland agricultural systems. Essentially, farmers pay a levy equivalent to a percentage (mostly 1 %) of the net farm gate value of their production. The resulting funds are matched by the federal government (fewer administration costs) and the combined funds are available for competitive grants for research, development and extension by government, universities and private organisations. In addition, various groups have engaged in research and development with more specific goals based on voluntary contributions.

Much, if not all, of the private research conducted by companies with interests in chemicals, plant breeding and fertilisers is conducted in other countries, with local verification.

As discussed above, there has been a partial withdrawal of public funds for agricultural research over the past decade or so. Various public–private research partnerships have been established in areas considered to have some component of public benefit.

6.2 Genetic Modification

Genetically-modified crops have been introduced to Australian dryland systems over the past two decades. The main type of modification has been for herbicide resistance, in addition to insect-resistant cotton cultivars. The use of herbicide-resistant crops has been controversial in some cases. Nevertheless, adoption of glyphosate-resistant canola, for example, has been widespread, even if problems with herbicide resistance in weeds such as annual ryegrass (*Lolium rigidum* Gaud.) have been encountered, requiring the development of integrated weed management systems.

The convenience and effectiveness of genetically-modified crops for weed or pest management have largely outweighed the possible hazards. However, there is some concern about the safety and efficacy of these cultivars as summarised from worldwide data (e.g. Antoniou et al. 2012).

6.3 Closing the Yield Gap

There is still a gap between average farm yields and the theoretical potential yield as limited by seasonal rainfall (Anderson et al. 2016). There is a need for objective, field investigations to determine the appropriate measures to address this yield deficit and to estimate the profitability of the various measures that may be required.

Acknowledgements We acknowledge the assistance of Clinton Revell, Cora Castens and Philip George in the preparation of this chapter.

References

Anderson WK (1985) Actual and potential wheat yield in Western Australia. Western Australian Department of Agriculture technical note no. 11/85

Anderson WK (2010) Closing the gap between actual and potential yield of rainfed wheat. The impacts of environment, management and cultivar. Field Crop Res 116:14–22

Anderson WK, Angus JF (2011) Agronomy and cropping practices in semi-arid conditions in Australia. In: Bonjean AP, Angus WJ, van Ginkel M (eds) The world wheat book, vol 2. Lavoisier Publishing 11, rue Lavoisier F-75384 Paris cedex 08 France, pp 563–605

Anderson WK, Hoyle FC (1999) Nitrogen efficiency of wheat genotypes in a Mediterranean environment. Aust J Exp Agric 39:957–965

Anderson WK, Impiglia A (2002). Management of dryland wheat. In: Curtis BC, Rajaram S, Gomez Macpherson H (eds) Bread wheat improvement and production. FAO, Rome, pp 407–432

Anderson WK, Sawkins D (1997) Production practices for improved grain yield and quality of soft wheats in Western Australia. Aust J Exp Agric **37**:173–180

Anderson WK, Crosbie GB, Lemsom K (1995) Production practices for high protein, hard wheat in Western Australia. Aust J Exp Agric 35:589–595

Anderson WK, Crosbie GB, Lambe WJ (1997). Production practices in Western Australia for wheats suitable for white, salted noodles. Aust J Agric Res 48, 49–58.doi:10.1071/A95133

Anderson WK, Hamza MA, Sharma DL, D'Antuono MF, Hoyle FC, Hill N, Shackley BJ, Amjad M, Zaicou-Kunesch C (2005)The role of management in yield improvement of the wheat crop – a review with special emphasis on Western Australia. Aust J Agric Res 56:1137–1149

Anderson WK, McTaggart RM, McQuade NC, Carter D, Overheu T, Bakker D, Peltzer S (2014). An approach to crop yield improvement through diagnostic systems research in a winter-dominant rainfall environment. Crop Past Sci, 2014, 65, 922–933. http://dx.doi.org/10.1071/CP14065

Anderson WK, Johansen C, Siddique KHM (2016) Addressing the yield gap in rainfed crops: a review. Agronomy for sustainable development and environment. Agron Sustain Dev 36(18). doi:10.1007/s13593-015-0341-y

Antoniou M, Robinson C, Fagan J (2012) GMO myths and truths. An evidence-based examination of the claims made for the safety and efficacy of genetically modified crops. www.earthopensource.org

Australian Bureau of Agriculture and Resource Economics (ABARES) (2011) Productivity and farm size in Australian agriculture: reinvesting the returns to scale. Conference paper 11.06. Available at www.agriculture.gov.au/abares/publications

Australian Bureau of Statistics (ABS (2009a) Land management and farming in Australia 2007–08 Available at www.abs.gov.au

Australian Bureau of Statistics (ABS (2009b) Natural resource management on Australian farms, 2006/07. Publication 4620.0. Available at www.abs.gov.au

Australian Farm Institute (2012) Managing uncertainty in the world's riskiest business' Farm Policy Journal. Autumn Q 9:1

Bell AK, Allan CJ (2000) PROGRAZE – an extension package in grazing and pasture management. Aust J Exp Agric 40:325–330

Bell LW, Moore AD Kirkegaard JA (2014) Evolution in crop-livestock integration systems that improve farm productivity and environmental performance in Australia. Eur J Agron 57:10-20.

Bomford M and Hart Q (2002) Non-indigenous vertebrates in Australia. In: Pimentel D (ed) Biological invasions: economic and environmental costs of alien plant, Animal and microbe species. CRC Press, New York, pp 25–44

Byerlee D (1994) Technology transfer systems for improved crop management: lessons for the future. In: Anderson JR (ed) Agricultural technology: policy issues for the international community. Commonwealth Agricultural Bureaux International, Cambridge, pp. 208–230

Canyon D, Speare R, Naumann I, Winkel K (2002) Environmental and economic costs of invertebrate invasions in Australia. In: Pimentel D (ed) Biological invasions: economic and environmental costs of alien plant, Animal and microbe species. New York, CRC Press, pp. 45–66

Chan KY, Heenan DP, So HB (2003) Sequestration of carbon and changes in soil quality under conservation tillage in light-textured soils in Australia: a review. Aust J Exp Agric 43:325–334

Clark SG, Donnelly JR, Moore AD (2000) The GrassGro decision support tool: its effectiveness in simulating pasture and animal production and value in determining research priorities. Aust J Exp Agric 40:247–256

Cooper M, Woodruff DR, Phillips IG, Basford KE, Gilmour AR (2001) Genotype-by-management interactions for grain yield and grain protein concentration of wheat. Field Crop Res 69:47–67

Derpsch R, Fredreich T (2009) Development and current status of no-till adoption in the world. In: Proceedings on CD, 18th triennial conference of the International Soil Tillage Research Organization (ISTRO), June 15–19, 2009, Izmir, Turkey

Donald CM (1965) The progress of Australian agriculture and the role of pastures in environmental change. Aust J Sci 27:187–198

Evans LT (1987) Opportunities for increasing yield potential in wheat. In: The future development of maize and wheat in the third world. CIMMYT, Mexico. 79-93. ISBN 968-6127-16-X

Farrer W (1898) The making and improvement of wheats for Australian conditions. Agric Gaz NSW 9:131–167

Fillery IR, McInnes KJ (1992) Components of the fertilizer nitrogen balance for wheat production on duplex soils. Aust J Exp Agric 32:887–899

Fischer RA, Wall PC (1976) Wheat breeding in Mexico and yield increases. J Aust Inst Agric Sci 42:139–148

Fischer RA, Byerlee D, Edmeades GO (2014) Crop yields and global food security: will yield increase continue to feed the world?Canberra: Australian Centre for International Agricultural Research. Retrieved from http://aciar.gov.au/publication/mn158

Freebairn DM, Wockner GH (1986) A study of soil erosion on vertisols of the eastern Darling Downs, Queensland. 1. The effect of surface conditions on soil movement within contour bay catchments. Aust J Soil Res 24:135–158

Freebairn DM, Cornish PS, Anderson WK, Walker SR, Robinson JB, Beswick AR (2006) Management systems in climate regions of the world—Australia. In: Dryland agriculture monograph. ASA,CSSA, SSSA, Madison, pp 837–878

French RS, Schultz JE (1984) Water use efficiency of wheat in a Mediterranean-type environment. Aust J Agric Sci 35:765–775

Gamage B (2011) The biggest estate on earth. Allen and Unwin, Sydney

Groves R (2002) The impacts of alien plants in Australia. In: Pimentel D (ed) Biological Invasions: Economic and Environmental Costs of Alien Plant, Animal and microbe species. CRC Press, New York, pp. 11–23

Hajkowiczs S, Eady S (2015) Megatrends impacting Australian agriculture over the coming twenty years. Rural Industries Development Corporation, Publication No. 15/065

Hamblin A, Kyneur G (1993) Trends in wheat yields and soil fertility in Australia. Department of Primary Industries and Energy and Bureau of Resource Sciences, Australian Government Publishing Service, Canberra, 141 p

Hamza MA, Anderson WK (2010) Potential and limitations of soil organic matter build-up in dry areas. Afr J Agric Res 5:2850–2861

Hochman Z, Holzworth D, Hunt JR (2009) Potential to improve on-farm wheat yield and WUE in Australia. Crop Pasture Sci. 60:708–716

Hoyle FC, D'Antuono MF, Overheu T, Murphy DV (2014) Capacity for increasing soil organic carbon in dryland agricultural systems. Soil Res 51:657–667

Jarvis RJ, Bolland MDA (1990) Placement of fertilizers for cereal crops. Fertil Res 22:97–107

Kearns S and Umbers, A (2010) GRDC farm practices baseline report. GRDC final report, 215 pp

Kingwell R, Fuchsbichler A (2011) The whole-farm benefits of controlled traffic farming: an Australian appraisal. Agric Syst 104:513–521

Kirkegaard JA, Conyers MK, Hunt JR, Kirkby CA, Watt M, Rebetzke GJ (2014) Sense and nonsense in conservation agriculture: principles, pragmatism and productivity in Australian mixed farming systems. Agric Ecosyst Environ 187:133–145

Kopke E, Young J, Kingwell R (2008) The relative profitability and environmental impacts of different sheep systems in a Mediterranean environment. Agric Syst 96:85–94

Ladha JK, Pathak H, Krupnik TJ, Six J, van Kessel C (2005) Efficiency of fertilizer nitrogen in cereal production: retrospects and prospects. Adv Agron 87:85–156. doi:10.1016/S0065-2113(05)87003-8

Leeper GW (1964) Introduction to soil science. Melbourne University Press, Parkville, p. 253
Lemerle D, Cousens RD, Gill GS, Peltzer SJ, Moerkerk M, Murphy CE, Collins D, Cullis BR (2004) Reliability of higher seeding rates of wheat for increased competitiveness with weeds in low rainfall environments. J Agric Sci 142:395–409
Loss SP, Siddique KHM (1994) Morphological and physiological traits associated with wheat yield increases in Mediterranean environments. Adv Agron 52:229–276
Llewellyn RS and D'Emden, F. (2010). Adoption of no-till cropping practices in Australian grain growing regions. CSIRO report for GRDC, SA No-till Farmers Association and CAAANZ. 32 pp. www.grdc.com.au/notill_adoption
Ludwig F, Milroy SP, Asseng S (2009) Impacts of recent climate change on wheat production systems in Western Australia. Climate Change 92:495–517
Mason MG (1975) Nitrogenous fertilizers for cereal production. J Agric WA 4th Ser 16:103–110
McGarity JW (1975) Soils of the Australian wheat-growing areas. In: Lazenby A, Matheson EM (eds) Australian field crops 1. Angus and Robertson, Sydney, pp. 227–278
McKenzie N, Jacquier D, Isbell R Brown K (2004) Australian soils and landscapes. an illustrated compendium. CSIRO Publishing, Melbourne. www.publish.csiro.au/pid/3821.htm
Mitchell TL (1839) Three expeditions into the interior of Eastern Australia, 2 volumes. T. and W Boone, London
Miyan MS, Impiglia A, Anderson WK (2011) Agronomic practices for durum wheat in an area new to the crop. Commun Biometry Crop Sci 6(2):64–79
Mullen J (2007) Productivity growth and the returns from public investment in R&D in Australian broadacre agriculture. Aust J Agric Res Econs 51:359–384
Nairn ME, Allen PG, Inglis AR, Tanner C (1996) Australian quarantine: a shared responsibility. Department of Primary Industries and Energy, Canberra, p. 284
NLWRA (2001) Project 4D: nutrient balance in regional farming systems and soil nutrient status. National Land and Water Resources Audit, Canberra 120 pp
Ogilvy S, Kulkarni A, Hurley S (2015). From vicious to virtuous cycles: a sustainable future for Australian agriculture. Centre for Policy Development, Discussion paper. Available at www.cpd.org.au
Palta JA, Peltzer S (2001) Annual ryegrass (*Lolium rigidum*) reduces the uptake and utilisation of fertilizer nitrogen by wheat. Aust J Agric Res 52:573–581. doi:10.1071/AR00085
Pannell DJ, Marshall GR, Barr N, Curtis A, Vanclay F, Wilkinson R (2006) Understanding and promoting adoption of conservation practices by rural landholders. Aust J Exp Agric 46:1407–1424
Pascoe B (2014) Dark emu black seeds: agriculture or accident? Magabala Books Aboriginal Corporation, Broome
Passioura J (2006) Increasing crop productivity when water is scarce – from breeding to field management. Agric Water Manag 80:176–196. doi:10.1016/j.agwat.2005.07.012
Passioura JB, Angus JF (2010) Improving productivity of crops in water limited environments. Adv Agron 106:37–75
Peverill KI, Sparrow LA, Reuter DJ (1999) Soil analysis – an interpretation manual. CSIRO Publishing, Melbourne.
Reeves TG and Ewing MA (1993) Is ley farming in Mediterranean zones just a passing phase? In: Proceedings of the XVII international grasslands Congress 1992, pp 2169–2177
Richards RA, Hunt JR, Kirkegaard JA, Passioura JB (2014) Yield improvement and adaptation of wheat to water-limited environments in Australia – a case study. Crop Pasture Sci 65:676–689
Robertson MJ, Llewellyn RS, Mandel R, Lawes R, Bramley RGV, Swift L, Metz N, O'Callaghan C (2012) Adoption of variable rate fertiliser application in the Australian grains industry: status, issues and prospects. Precision Agricult. doi:10.1007/s11119-011-9236-3
Sadras VO, Lawson C (2011) Genetic gain in yield and associated changes in phenotype, trait plasticity and competitive ability of South Australian wheat varieties released between 1958 and 2007. Crop Past Sci 62:533–549

Sallaway MM, Yule DF, Mayer D, Burger PW (1990) Effects of surface management on the hydrology of a Vertisol in semi-arid Australia. Soil Tillage Res 15:227–245

Scott BJ, Eberbach PL, Evans J, Wade LJ (eds) (2010) Stubble retention in cropping systems in Southern Australia: benefits and challenges. EH Graham Centre monograph no. 1. Graham Centre for Agricultural Innovation, Wagga Wagga, NSW, Australia. Available at: http://www.csu.edu.au/research/grahamcentre/ ISBN:978-1-74256-020-5

Scott BJ, Podmore CM, Burns HM, Bowden PI, McMaster CL (2013) Developments in stubble retention in cropping systems in southern Australia. Report to GRDC on Project DAN 00170. (Ed. C .Nicholls and E.C. (Ted) Wolfe).Department of Primary Industries, Orange NSW. 103 pp. Available at:http://www.csu.edu.au/research/grahamcentre/research/Pub_downloads/Developments-in-Stubble-Retention-in-Cropping-Systems-in-Southern-Australia-11Dec2013.pdf

Sharma DL, Anderson WK (2014) Success of diagnostic approach to rainfed, wheat-based cropping systems in Western Australia. Agric Syst 123:22–33

Siddique KHM, Belford RK, Perry MW, Tennant D (1989) Growth, development and light interception of old and modern wheat varieties in a Mediterranean-type environment. Aust J Agric Res 40:473–487

Simpson NL, Brennan RF, Anderson WK (2015) Grain yield increases in wheat and barley to nitrogen applied after transient waterlogging in the high rainfall cropping zone of Western Australia. J Plant Nutr. doi:10.1080/01904167.2014.993475

Stephens DJ, Lyons TJ (1998) Variability and trends in sowing dates across the Australian wheatbelt. Aust J Agric Res 49:1111–1118

Stephens DJ, Lyons TJ, Lamond MH (1989) A simple model to forecast wheat yield in Western Australia. J Roy Soc Western Aust 71:77–81

Stephens D, Anderson W, Nunweek M, Potgeiter A, Walcott J (2011) GRDC planning for investment based on agro-ecological zones – second phase. Department of Agriculture and Food Western Australia. Available at www.agric.wa.gov.au

Virgona JM, Gummer F, Angus JF (2006) Effects of grazing on wheat growth, yield, development, water use and nitrogen use. Aust J Agric Res 57:1307–1319

Ward P and Siddique KHM (2014). Conservation agriculture in Australia and New Zealand. In: Farooq M, Siddique KHM (eds) Conservation agriculture. Springer International Publishing Switzerland, pp 335–355

Wolf E (2009) Country pasture/forage resource profiles – Australia http://www.fao.org/ag/agp/AGPC/doc/Counprof/Australia/australia.htm

Pastures in Australia's Dryland Agriculture Regions

Ann Hamblin

1 Introduction

The focus of this chapter is the Intensive Land Use Zone (ILZ) of Australia, from which much of the original, native vegetation was cleared or modified between 200 and 30 years ago (Graetz et al. 1995). This corresponds to the agricultural belt (Fig. 1) which lies between the latitudes 23° and 41°S—excluding the arid and tropical rangelands (the Extensive Landuse Zone, or ELZ) where inter-annual rainfall variability exceeds the long-term average annual rainfall[1] which is insufficient for dryland cropping. The rangelands support 5 % of Australia's sheep and 45 % cattle. The division between the ELZ and the ILZ (Fig. 1) is an accepted boundary between farmed and pastoral regions, and distinguishes freehold from leasehold tenures.

Sheep were a mainstay of rainfed Australian agriculture until recent decades, reaching an all-time high of 170 million at the end of the 1980s as a result of inappropriate interventionist wool pricing policies that encouraged farmers to keep large flocks for wool (Massey 2011). After the collapse of the export wool market, this number dropped to 72 million (90 % in the ILZ) in 2014 (the lowest since 1910) with a concomitant shift to other animal products and an increase in cropping.

The persistently low returns from wool and relatively better returns to grains during the past 25 years have been a game-changer, resulting in substantial changes in land use and farming systems (Nossal and Sheng 2010; Walcott et al. 2013). There has been an expansion in cropping into wetter areas, while the total area of dryland grain cropping has fluctuated between 20 and 23 Mha. Cattle numbers have increased; by >5 % in medium to low rainfall parts of New South Wales (NSW) and

[1] Index of annual rainfall variability defined as [90p-–10p]/50p > 0.5-–0.75; p = 90th, 50th and 10th percentiles: http://www.bom.gov.au/jsp/ncc/climate_averages/rainfall-variability/index.jsp

A. Hamblin (✉)
The UWA Institute of Agriculture, The University of Western Australia,
LB-5005, Perth, WA 6001, Australia
e-mail: ahamblin@optusnet.com.au

M. Farooq, K.H.M. Siddique (eds.), *Innovations in Dryland Agriculture*,
DOI 10.1007/978-3-319-47928-6_12

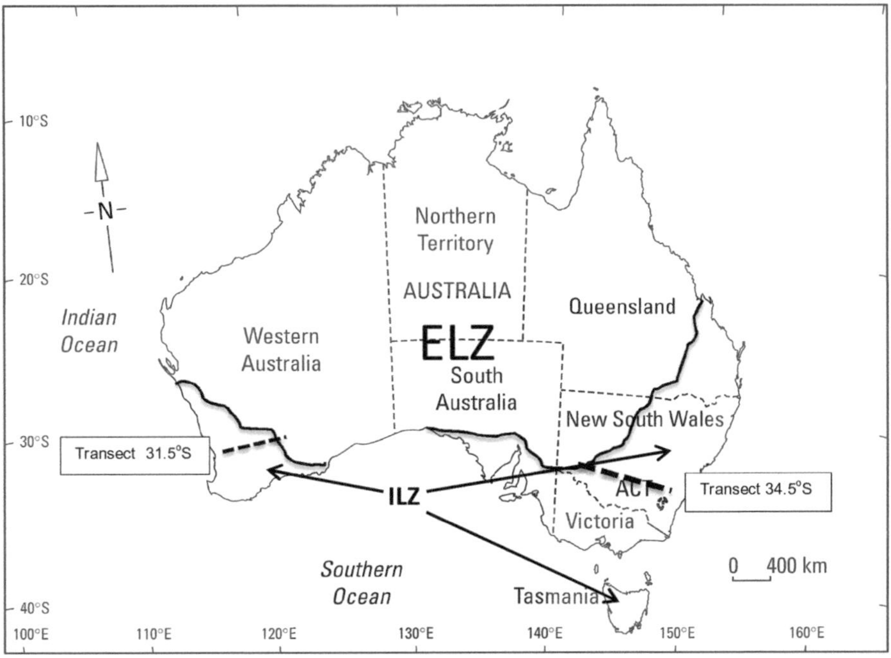

Fig. 1 Location of the Intensive Land Use Zone (ILZ) and Extensive Land Use Zone (ELZ) in Australia showing the transects at latitudes 31.5°S Western Australia and 34.5°S New South Wales referred to in this chapter

southern Queensland, and throughout the high rainfall coastal zone (Bell et al. 2014; Walcott et al. 2013). In southern Queensland and northern NSW, in particular, the increase in cattle numbers has been associated with the development of hundreds of feedlots that source animal feed from surrounding grain-growing areas (Queensland Transport Logistics Council 2014). In 2011, the high rainfall region of the ILZ had 8.5 million cattle (excluding 1.8 million dairy cows) compared with 6.6 million in the mixed-farming zone, and a national total of 28.5 million (Meat and Livestock Australia 2012a). After peaking at 29 million in 2013, cattle numbers are forecast to decline from 2016–2019 due to the persistent droughts in northern farming areas causing increased sell-off. Cattle numbers were fairly stable in the ILZ from 1990 to 2000 but increased to >50 % of the national herd between 2001 and 2012 (Fig. 2).

Despite the expansion of the red meat industry, income from crops has outperformed livestock across most regions in temperate Australia, while meat has replaced wool as the most profitable animal product (Dahl et al. 2014; Robertson 2010). Most rainfed farms have reduced their total flock size, increased their proportion of ewes, focussed on cross-bred lamb rather than wool production, or changed to beef while some have completely destocked (Curtis 2009).

There have been substantial changes in the methods of sale, processing and product development in the livestock industry. The reduction in the number of sale yards and processing plants has been rapid; for instance in 2003, 25 large processors were

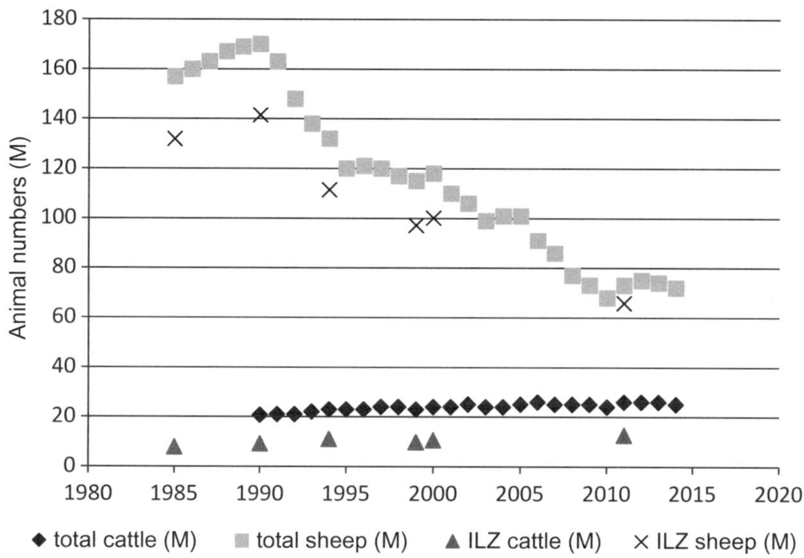

Fig. 2 Total Australian sheep and cattle (millions) and numbers in the ILZ from 1985 to 2014 (Sources: ABS 2014, ABS 1982–1996, National Land and Water Resources Audit 2001 and Meat and Livestock Australia 2012a, b)

responsible for 60 % of production, but by 2015 these had merged into two large, vertically-integrated processing companies (Teys-Cargill and JBS) responsible for 50 % of the red meat processed in Australia. As Asian, and particularly Chinese, demand for agricultural products is projected to increase rapidly over the next 30 years, there is potential for significant expansion in red meat products in future (Linehan et al. 2012).

2 Extent of Pastures and Grazing in the ILZ

Grazing land occurs throughout the ILZ, extending in an arc around the continent, across winter, equi-seasonal and summer-dominant rainfall regimes. Permanent pastures of perennial, introduced grasses and legumes support dairy and beef cattle in wetter coastal regions and eastern slopes of the Great Dividing Range. On the inland lower rainfall slopes and plains, annual and perennial species are rotated with crops in phase farming. Native perennial grasses occur on the western hinterlands of the Great Dividing Range, fenced and later improved with fertiliser and/or introduced species (Hill et al. 1999), variously described as 'native pastures', 'modified native pastures' or 'improved pastures' (Benson 1996; Donald 2012). In south-west Western Australia (WA) and South Australia (SA), introduced annual grasses and legumes predominate. A comprehensive description of Australian pastures has been given by Wolfe (2009).

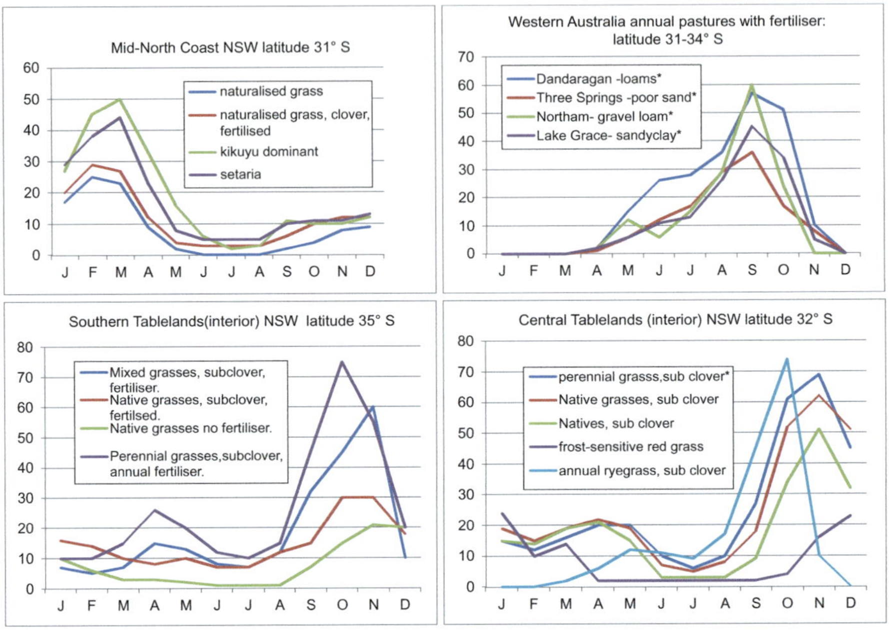

Fig. 3 Pasture growth rates (kg ha^{-1} d^{-1}) for perennial and annual, exotic and native grass species from summer, equi-seasonal and winter-dominant rainfall regions (Source: Meat and Livestock Australia (MLA) Sustainable Grazing Systems (www.mla.com.au/tipsandtools8.2). 'kikuyu'- Pennisetum clasestrinum (Hoscht ex. Chicov), 'setaria' – Setaria shacelata (Schumach), 'sub-clover'- subterranean clover, WA= all sites fertilised pastures of annual ryegrass, subterranean clover and exotic weeds

Australian livestock producers must manage marked seasonality in rainfall incidence, except in some coastal regions and those with a more equi-seasonal rainfall distribution (Fig. 3).

In the Mediterranean climates of WA and SA, there may be seven or more months without effective rain, restricting the use of perennial grasses and legume species. Fig. 3 demonstrates that native grasslands (Central Tablelands NSW examples) do not achieve the same level of production as sown pastures, even with introduced legumes and fertiliser, and deficiencies in soil fertility or physical properties (Western Australian examples) cannot be overcome by fertiliser and legumes. Native grasslands have undergone many transformations since the European occupation two centuries ago including grazing by introduced herbivores, the addition of exotic grasses, legumes and forbs, and the use of phosphate fertilisers. Sown pastures have been the major focus of research and development, but native pastures extend over a much larger area (Wolfe and Dear 2001). Of all the introduced species, perennial and annual ryegrass (Lolium perenne L. and L. rigidum Gaud.) and subterranean clover (Trifolium subterrraneum L.) are the most extensively used, but all legumes play an important role in maintaining adequate animal nutrition and contributing to soil fertility. Table 1 summarises the geographical range of the most commonly-used legumes.

Table 1 The principal geographical zones and soils where introduced pasture legumes are sown in rotation with crops (*brackets* indicate subdominant legumes)

Latitude	Winter rainfall (%)	Geographical location	Principal soil types	Length of pasture phase	Principal legumes
25–29°S	<25	Southern Queensland	Medium to heavy loams and clays, neutral pH	Long	Lucerne (alfalfa)
				Variable	(annual medics)
29–32°S	25–40	northern NSW	Medium to heavy loams, clays, neutral pH, some acid sands	Long	Lucerne
				Variable	(annual medics)
32–38°S	40–65	Central-southern NSW, Victoria	Acid light to medium sands or clay at neutral pH	Long	Subclover, annual medics, lucerne
				Short	(balansa clover)
				Long	
				Variable	
38–44°S	70–80	Western Victoria, SA, Tasmania	Neutral-alkaline sandy loams	Short	Annual medics, lucerne, subclover, (balansa clover)
			neutral loams	Long	
				Short to variable	
29–37°S	70–80	South-west WA	Acidic sands	Short	Serradella
			Acidic sands	Short	Subclover
			Sandy loams	Short	Annual medics

Lucerne (Medicago sativa L.), Subclover (Trifolium subterraneum L.), Serradella (Ornithopus compressus L.), Annual medics (Medicago polymorpha L., Medicago truncatula Gaetnr), Balansa clover (Trifolium michelianum Savi)
Adapted from Wolfe (2009)

In mixed-farming regions, pasture legumes traditionally supplied much of the nitrogen (N) demand for crops (Hill 1996) and enhanced the value of animal feed into the dry season. Together with phosphate (P) fertilisers, legumes were the foundation for increasing animal production from native as well as sown pastures (Doyle et al. 1993), but over the past 20 years, crop N demand has been met by increasing rates of fertiliser, and pasture phases have shortened or been replaced by oilseeds and pulses as break crops, apart from the specific use of long-phase lucerne (alfalfa) (see Anderson et al. Chap. 11, this volume).

The total area of land grazed by domestic stock in the ILZ from 1990 to 2015 is difficult to establish because pasture definitions have varied in census questionnaires. Farm land that is grazed but does not qualify as sown pasture has been reported by the Australian Bureau of Statistics (ABS) as 'native, volunteer pasture' (1991), 'all improved pastures' (1996) and 'all grazed land: improved plus other' (2011). Land use areas have been estimated from the Australian Collaborative National Land Use Mapping Program (ACLUMP) since 1992/1993, with

Table 2 Area (Mha) of sown pastures and all sown + modified grazing

	ABS censuses	NLUM	ABS resource	ABARES	ABARES
Date	*sown pastures ** all sown + modified	all modified grazing land	Management practices surveys[##]	average % of 300–600 mm zone farm not cropped	average % of >600 mm zone farm not cropped
1989–1990	74.0**	19.9[#3]	–	72	96
1996–1997	70.0**	25.8[#3]	–	80	94
2000–2001	25.8*[–]**	23.8	–	71	94
2001–2002	22.9*[–]**	23.8[#3]	–	75	94
2005–2006	24.5*[§][–]**	72.0[#4]	–	78	92
2007–2008	–	–	66.7	70	93
2010–2011	19.1*[55.0**]	70.0[§#5]	60.5	73	93
2012–2013	–	–	48.5	70	92

Sources Australian Bureau of Statistics agricultural censuses, National Land Use Mapping (NLUM), ABS Resource Management Practices surveys, and Australian Bureau of Agricultural and Resource Economics and Sciences (ABARES) farm surveys for high rainfall (>600 mm) and mixed farm (300–600 mm rainfall) categories (ABARES 2014)
[§]estimate only; [#3–5]AgStats (NLUM) data for Versions 3,4,and 5; significantly under-reported in Version 3 which assumed no grazing where native woodland had a crown-cover >50 %; [##]ABS (2013, 2014) Catalogue numbers: 46270DO001_2007-13 'improved' grazing areas'

remote-sensed NDVI[2] (ABARE-BRS 2010), but methods have varied, with grassy woodlands sometimes included but sometimes not (Mewett et al. 2013), so this measure of total grazed areas has fluctuated widely (Table 2).

Fertilised, sown pastures in the mixed-farming and high-rainfall regions have occupied between 19 and 26 Mha since 1990 but the total area of all grazed land on farms is much larger despite an overall reduction of ~15 Mha since the 1980s, due to increased cropland, reduced farm areas from urban expansion, and farm forestry (ABS 1991–2011). The total pasture area on farms can be estimated by subtracting the cropped area from the total operational area of farm holdings. By this calculation, grazing land accounts for 70 % of mixed farms and >90 % of high rainfall farms (ABARES 2014).

As grazing continues to be the largest single land use in the ILZ, domestic grazing animals impact at least two-thirds of the agricultural landscape, except where fencing and watering points have been deliberately removed, for example in north-eastern WA and SA–Victorian mallee (Norris 2009; van Rees et al. 2011). Elsewhere, farmers have retained the infrastructure needed for animal production to provide flexibility in the face of fluctuating seasons and prices. Hutchings and Nordblom (2011) showed that mixed farming with livestock in NSW significantly reduced the risk of financial loss compared with farms devoted solely to cropping. Robertson et al. (2014) found that mixed crop–livestock systems in south-east Australia were financially most robust with an optimum distribution of 45 % pasture

[2]NDVI= Normalised Difference Vegetation Index, the ratio of near infra-red to visible wavelength reflectance spectra for green vegetation received by satellite sensors.

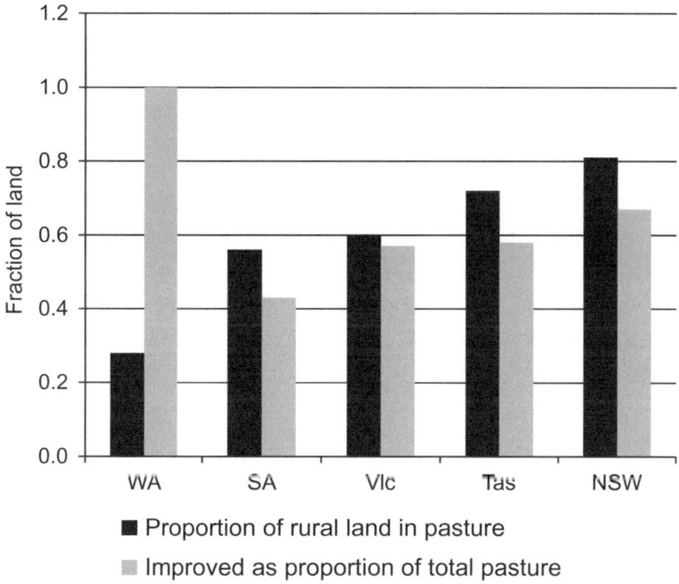

Fig. 4 The proportion of rural land in pasture and in improved pasture by state in southern Australia, after Donald (2012). Values for NSW exclude the Western Division Pastoral Lands which fall outside the ILZ

and 55 % cropped. Recent benchmarking studies undertaken in the mixed-farming zone have shown that most cropping farmers have between 30 and 60 % of their land in pasture depending on rainfall, some of which may be a sown pasture within the main crop rotation, but more is in perennial pasture in longer phase rotation (Burns and Norton 2015; Harries et al. 2015; Llewellyn and D'Emden 2010). Grazing land may not always be fully used, but forms an insurance against crop loss risk, particularly during drought years.

2.1 Regional Differences

The distribution of native grasslands that existed across Australia before white settlement still influences the extent and type of grazing land on farms. Figure 4 draws on data from the 2011 pasture audit of southern Australia (Donald 2012) to compare the extent and type of pastures in each state. In WA there is no native pasture, and sown pasture occupied only 28 % of farmland in 2011, whereas in NSW the total extent of pasture was 80 % of the farmland area, of which only 42 % was sown to introduced species.

In northern NSW and south-eastern Queensland, where open scattered woodland occurs with native grasses, sporadic clearing of remnant woody vegetation continues under licence. This has increased the total extent of 'managed' or 'modified' pasture categories in agricultural censuses. Between 1990 and 1995, clearing increased grazing of native vegetation areas in Queensland and NSW on average by 0.5 Mha year^{-1}. The rate of clearing in Queensland was most rapid, increasing to 654,000 ha year^{-1} in the last two years of the twentieth century, with 2.2 Mha cleared by 2012 (Accad et al. 2013). The greatest extent of recent clearing has been south of latitude 27° in more fertile brigalow (Acacia harpophylla F. Muell. ex Benth.), Mitchell grass (Astrebla lappacera Lindl.(Domin) and other Astrebla species) and mulga (Acacia aneura Benth.) subtropical bioregions where leguminous native shrubs and palatable grasses are traditional cattle-rearing areas. Such clearing maintains 'improved pastures' for producers by counteracting woody regrowth, with increased productivity as the competition for light, water and nutrients from trees declines (Scanlan 1991).

2.2 Responses to Changing Climates and Other Stressors

Independent of changes arising from market conditions, some land use changes have occurred in response to changing climates. Since 1990, there has been a 15–20 % decline in winter rainfall and a 25 % reduction in spring rain in south-eastern Australia. In the south-west, there has been a permanent reduction of 20–25 % in winter rainfall since the mid-1970s (Bureau of Meteorology 2014). Cropping has therefore expanded into southern high rainfall pastures (>600 mm) where waterlogging and frost are now lesser problems than in the past. In the interior margins of the wheatbelt in WA (Northampton and Yilgarn statistical local areas (SLAs)), SA (West Coast and Flinders Ranges SLAs) and throughout southern Queensland, the area sown to crops has declined and more grazing now occurs due to the increased variability in growing season rainfall (Mewett et al. 2013).

In the Mediterranean-type environments, deep-rooted perennial pasture legumes and grasses have been introduced with some success to take advantage of increased out-of-season summer rainfall (Wolfe 2009). Warmer average temperatures in the past two decades have increased the opportunity to introduce subtropical grasses and legumes (Bell et al. 2013; Descheemaeker et al. 2013) and increase planting of new lucerne varieties for forage (Moore et al. 2009). Novel systems have been developed for sheep to graze young cereal and canola crops during the winter feed gap in south-western and south-eastern regions, providing autumn feed and reducing grazing pressure during periods of low growth (Virgona et al. 2006; Kirkegaard et al. 2008).

The development of herbicide-resistant weeds such as multiple-resistance annual ryegrass has reduced the reliance on annual grass-based pastures in crop rotation sequences (Powles et al. 1997), and price signals have shifted rotations more toward oilseeds than pulses as break crops from cereals (Harries et al. 2015). These changes have tended to reduce the reliance on traditional legume-based ley pasture rotation systems (Seymour et al. 2012) and contributed to the reduced extent and poorer condition of many pastures in the ILZ.

3 Managing and Maintaining Pastures

Two pasture audits have been carried out across the ILZ since 1990. The first occurred from June to October 1994, across 544 local government areas (LGAs) from southern Queensland to WA using SLA map boundaries (Pearson et al. 1997). An edited database provided information on the proportion of ~2500 plant types across the country (Hill and Donald 1998). Most pastures originally sown to self-regenerating legumes and improved grasses in southern WA, much of SA, and northern and central parts of NSW were reported as weedy, with weeds comprising up to 70–80 % of the ground cover. These weeds, of low nutritional value, included barley grass (Hordeum leporinum Link.), silver grass (Vulpia bromoides L. and V. myuros L.) and capeweed (Arctotheca calendula L. Levyns). Exotic temperate perennial grasses such as cocksfoot, phalaris and fescue (Festuca rubra. L.) in the higher rainfall areas were in good condition in the south and east where there was a wide spectrum of introduced pasture legumes.

The second pasture audit occurred in 2011 across the southern ILZ, omitting south-eastern Queensland, using 404 SLAs as the reporting base (Donald 2012). This survey categorised pastures by type, dominant species and varieties, use in crop rotation and carrying capacity, and estimated pasture condition. Condition was assessed as stable, declining or improving, and given a score of 1–10 on the basis of legume and weed content (Table 3). Of the pastures surveyed, 35 % (17 of 48 Mha) were scored as in decline, which is similar to the results of the earlier audit which reported 32 % of SLAs having the majority of pastures in poor condition.

Table 3 Total grazed area (ha) in ILZ by state, % area in decline and estimated area (ha) of dominant sown species in the 2011 pasture audit

Area (ha)	NSW	Victoria	Tasmania	SA	WA
Total surveyed	22,500,00*	7,391,400	1,186414	7,576,000	9,462,335
'in decline'	7,016,250	805,340	444,450	4,348,600	3,186,550
Decline as % total	31 %	11 %	38 %	57 %	37 %
Subterranean clover	1,127,082	495,415	69,332	208,203	1,455,555
Annual rye grass	533,960	511,150	48,016	160,086	1,528,430
Medics**	1,401,157	0	0	1,213,500	296,050
Perennial rye grass	305,288	934,521	265,939	47,949	37,742
All other clovers	713,203	261,957	37,839	40,620	51,300
Pasture area with legume	22 %	13 %	11 %	23 %	19 %
Lucerne	1,756,799	208,221	11,598	266,908	33,369
Cocksfoot	372,727	349,068	103,822	23,408	25,424
Phalaris	746,512	912,337	11,227	266,900	21,020

After Donald (2012)

*Excluding Western Division NSW pastoral region; **Principally burr medic (Medicago polymorpha L.) and barrel medic, Cocksfoot = Dactylis glomerata L., Phalaris = Phalaris aquatica L.

Between the two surveys, the area of forage lucerne (alfalfa) increased to 2.3 Mha, and specific legumes adapted to acidic or alkaline conditions, such as *Serradella* spp. in WA and burr medic in SA, were successfully introduced or expanded, with more subtropical perennial grasses and legumes planted in southern localities in response to changing climatic conditions (Nichols et al. 2012). The Millennium drought (2002–2009) contributed to the reduced legume content and seed set in NSW (Donald 2012), but the low status of pastures in Tasmania, WA and SA that were less affected by prolonged drought indicates a deeper underlying problem affecting many sown pastures.

Deterioration of legume-based pastures in rainfed mixed-farming regions has been reported since the late 1980s. Some of the factors listed by Wolfe and Dear (2001), such as insect pests, herbicide damage, effects of drought, acidification and suboptimal fertiliser use, occur today. National initiatives have been mounted by research and development agencies to address the low productivity of pastures across different rainfall zones. Between 1994 and 2001 the Sustainable Grazing Systems Program (SGS) involved 23,700 producers in high rainfall regions, and between 2003 and 2008 the Grain and Graze program involved 6800 mixed farms in the wheatbelt. These programs resulted in topical research findings and practical guidelines based on a network of experimental sites and farmer participation groups (Hacker et al. 2009; Johnson et al. 2003). SGS promoted the extension of rotational grazing to avoid selective overgrazing of more palatable species, rather than set stocking which had been the traditional method of rearing stock in many areas. By 2012, ~30 % of sheep farmers had converted to rotational grazing (Barson et al. 2012) compared with <15 % a decade earlier.

Many native grasslands in the interior plateaux (the Tablelands) of NSW, Victoria, southern Queensland and central Tasmania became severely degraded and invaded by exotic grasses of low herbage value after decades of set stocking. Remedial management systems were developed, where stock are grazed at low densities in winter then at high levels in spring to suppress flowering of annual exotic species and allow later-flowering native grasses[3] to flower and seed (Kemp and Dowling 2000). Native species, adapted to nutritionally-poor soil conditions, can then compete more effectively against introduced grasses. This system, however, requires active paddock monitoring, subdivision of large paddocks into smaller cells, additional fencing and watering points, and frequent movement of sheep (Evergraze 2014). Existing land degradation (bare scalds and gullies) and a preponderance of unpalatable grasses such as serrated tussock (Nassella tricotoma *(*Nees) Hack ex. Arechav) can make restoration of these degraded native pastures a long and costly process.

Stocking density—the number of dry sheep equivalents (DSE) per unit area—is the simplest indicator of pasture productivity. A rough calculation demonstrates a reduction rather than an increase in stocking density since 1990. At that time, the ILZ had 142 million sheep—213 million DSE (assuming a half ewe, half wether

[3] such as Microlaena stipoides Labil (stipa), Themeda australis .(kangaroo grass) and Austrodanthonia bipartita (Link) H.P. Linder (wallaby grass).

national flock)—and 9.2 million cattle—74 million DSE (using a rate of one cattle animal to eight sheep)—totalling 278 million DSE. In 2011, cattle numbers had increased to 10 million (80 million DSE) while sheep had declined to 66 million (110 million DSE for a 70 % ewe flock composition), totalling 190 million DSE. During this time (1990–2011), the estimated total area of grazed land declined from 74 million to ~65 million hectares (Table 2) and the overall stocking density fell from 3.8 to 2.9 DSE ha^{-1}. Such a generalised calculation does not reflect the actual variations in different regions (Walcott et al. 2013; Mewett et al. 2013), but it does suggest no overall improvement in productivity. In the mixed-farming zone, Angus and Peoples (2012) calculated that the stocking rate had decreased from 1990 to 2010 by ~20 % and Bell and Moore (2012) reported a decline from 2.8 to 2.1 DSE ha^{-1} from 2002 to 2010. The only areas where regional stocking rates have either remained the same or increased are the temperate, medium-to-high rainfall regions where perennial legumes and grasses of high nutritional quality dominate and grow for 6–8 months of the year, such as the lucerne–phalaris pastures in Central NSW which are estimated at 9–11 DSE ha^{-1} (Bell and Moore 2012) and in western Victoria–south SA at 11–12 DSE ha^{-1} (Donald 2012).

3.1 Managing and Maintaining Pasture Productivity

One of the challenges in maintaining or improving pasture is that many are located in areas of the farm that are too steep or too stony where soils are nutritionally poor, or too acidic or alkaline for profitable crop production. Such land can be difficult to access and/or expensive to fertilise, lime or reseed, with the net result that dry matter production per mm rainfall is less than that on more fertile, flatter land on the same farm. In addition, most annual pasture plants, which dominate pastures in southern Australia, have shallow rooting systems that deplete soil water only in the top 50 cm compared with deeper-rooted perennial grasses, such as phalaris, cocksfoot, native grasses (e.g. stipa and kangaroo grass) and lucerne, which can extract water to >2.5 m (Dolling et al. 2005; Singh et al. 2001). On most Australian rainfed farms, pastures with lower water use efficiencies (WUE) can become water-limited earlier in the growing season than cereal crops. Careful grazing management is needed to maximise WUE by maintaining the production of young shoots without overgrazing which results in premature senescence. While 20 kg ha^{-1} mm^{-1} is used as a potential WUE for crops (Anderson et al. Chap. 11, this volume), pasture growth models use an average potential WUE of 15 kg ha^{-1} mm^{-1} rainfall for improved pastures and 10 kg ha^{-1}mm^{-1} for native pastures (Section 4.1 this chapter). Water use of shallow-rooted annual ryegrass pasture and deep-rooted perennial phalaris grazed pastures in >600 mm regions can typically differ by +40 mm year^{-1}, with more water extracted under the perennial pasture (Heng et al. 2001). On undulating and sloping terrain, the planting of deep-rooted fodder shrubs, such as tagasaste or tree lucerne (Chamaecytisus palmensis (H.Christ)) and saltbush (e.g., Atriplex nummilaria Lindl), has improved grazed land WUE as well as helped to control secondary salinity and waterlogging (Lefroy 2002).

3.2 Soil Nutrient Status and Management

The N status of Australian pastures has changed substantially since the 1980s when the pasture phase in mixed-farming regions was longer, and crops obtained up to half their nitrogen from the carry-over of N mineralised by legume fixation (Ellington 1986). Legume-dominated pastures that produce 3–6 t ha^{-1} will fix between 90 and 160 kg N ha^{-1} annually (Unkovich et al. 2010), but this requires active rhizobia, a soil pH >5.5 and <8.4, and adequate available soil P. In equi-seasonal and winter-rainfall environments, the proportion of atmospheric N_2 fixed by all legumes is high, ranging between 65 and 94 %, and is regulated by biomass production, with 20–25 kg of shoot N fixed for every tonne of shoot dry matter produced (Peoples et al. 2001). However, in northern, summer-dominant rainfall regions the higher rainfall variability, fluctuations in soil stored water, and irregularity of crop–pasture phases produces large variations in N_2 fixation with a less reliable supply to subsequent crops.

Biological N_2 fixation in pastures has declined sharply since 1990 (Angus and Peoples 2012), even in permanent pastures due to a decline in legume content, often to <10 %, so that N_2-fixation supplies only 15 % of requirements (Spiers et al. 2013). However, the overall use of N fertilisers on mixed farms has doubled since 1990 as the benefit of additional nitrogen to crop water use efficiency became widely appreciated (see Anderson et al. Chap. 11, this volume), compensating for the reduced biological-N_2 input from the loss of pasture legumes in rotations (Lake 2012). In 2011–2012, N fertiliser was applied to 20.6 Mha farmland, but only 1.9 Mha of this was applied to pastures (ABS 2013). Nevertheless, the loss of the slow, steady supply of biological-N_2 from soils on mixed farms is a cause for concern for the long-term sustainability of mixed-farming systems (Moore 2014).

Historically, pastures were regularly dressed with subsidised single superphosphate fertiliser—because P is deficient across many Australian soils—but subsidies were phased out in 1988 and P use has been influenced by world prices. P-fertiliser consumption continued to increase, from 580 kt P_2O_5 in 1990 to 1059 kt in 1999, but dropped from 2006 when the price doubled (Ryan 2010). P-fertiliser use has risen since 2010 but is applied mainly to crops in mixed-farming regions; in 2010, an estimated 455 kt P_2O_5 was applied to crops and 290 kt to permanent pastures (White et al. 2010). The use of single superphosphate (SSP), traditionally applied to permanent pastures by topdressing, has declined steadily since 1990 due to the low value per tonne carted (Blair 2008). A minority of farmers test the nutrient staus of their pasture soils each year and these numbers declined from 20 % in 2007–2008 to 16 % in 2010–2011 (Barson et al. 2012).

In 1999–2000, the National Land and Water Resources Audit mapped the nutrient status of agricultural soils across Australia from large archived data sets (National Land and Water Resources Audit 2001). Most soils in the mixed-farming zone (300–600 mm rainfall) had Colwell[4] P values between 5 and 30 mg P kg^{-1}

[4] A test for the amount of P available to plants in soils where soil pH in water <7.4, suitable for most Australian pasture soils. Soils require a Colwell P kg^{-1} >30 mg P to reach their equivalent 'critical value' (CV) that provides 90–95 % of maximum plant production.

indicating P deficiencies (Wong et al. 2012). In permanent pasture regions receiving >600 mm rainfall, Colwell-P values were often >40 mg P kg^{-1}, but most soils had a pH_{Ca} <5.0. These soils had adequate or even excess P, but their high acidity inhibited good plant growth. A decade later, Weaver and Wong (2011) analysed soil and farm records from 2007 to 2010, creating P balance sheets for different farm industries. Soil P levels on 78 % of cropping farms and 63 % of animal farms were higher than their critical values (CV) and more than half of all farms had a 0–10 cm pH_{Ca} <5.5. Many moderately-acidic soils with Colwell-P values of <40 mg P kg^{1} in 2001 (National Land and Water Resources Audit 2001) had values of >40 mg P kg^{-1} by 2010, indicating continuing build up of P beyond the capacity for plant uptake, which was constrained by low pH and other nutrient deficiencies, such as potassium (49 % of soils deficient), and sulphur (61 % deficient). In the north eastern part of the ILZ (latitude 31° northwards) long-farmed soils were P-deficient because, although originally of high natural fertility (Vertisols, Chromosols and Sodosols) they had received no fertiliser since clearance.

Many pasture soils are therefure currently nutritionally imbalanced and too acid for high levels of production. There is substantial scope for improving the productivity of many Australian pastures if producers invested more in soil testing and balanced fertiliser regimes to overcome the run-down in soil fertility that occurred over the past two decades when drought and low prices restricted on-farm inputs.

3.3 Acidity and Liming

Legume-based pastures in higher rainfall environments in Australia have been acidifying since the 1950s as a result of excess NO_3^- ions leaching into subsoils, thus increasing their natural acidity (Helyar 1976). Between 1957 and 1995, measured acidification rates ranged from 0.001 to 0.06 units per year depending on soil type, farming system and climate (Porter et al. 1995). In 2001, approximately 50 Mha of farmland soils in the ILZ had a surface pH_{Ca} <5.0 (National Land and Water Resources Audit 2001). A decade later, further acidification had occurred in most farming regions (Australian State of the Environment Committee 2011). Table 4 shows that a higher proportion of both permanent and rotational pasture soils continue to have lower pHs than associated cropped land, even in medium rainfall regions.

Since 2000, only 8–12 % of livestock properties needing amelioration have spread lime, gypsum or dolomite in any one year; Barson et al. (2012) estimated that 21 % of pastures were at high risk of further acidification. In WA, where the extent of acid soils is greatest, 0.8–1.0 Mt of lime have been spread each year over the past decade, which is only half the calculated need (Gazey and Gartner 2009). One strategy (Section 4.2 in this chapter) that has successfully overcome the problems of soil acidity, particularly for subsoils that are only slowly amenable to liming, is to select for acid-tolerant legume species.

Table 4 Percentage of soil samples in each pH increment with a soil pH_{Ca} between <4.5 and 8.5 (0–10 cm), sampled from 2006 to 2015

pH_{Ca}	Landuse	<4.5	4.5–4.8	4.9–5.5	5.6–7.5	7.6–8.5	No. sites
S inland NSW	C	3	20	51	25	1	75
	C+P	6	19	44	30	1	141
	PP	16	24	37	23	–	107
SW Central WA	C	4	18	50	28	–	64
	C+P	10	35	44	11	–	119
	PP	65	16	18	1	–	135

Source: www.soilquality.org.au open-source website
C cropped, *C* + *P* crop and pasture rotation, *PP* permanent pasture

4 Supporting Technologies

Two different strands of advanced technologies underpin efficient and profitable animal industries. One is information and electronic management systems. The vast pool of data gained from decades of animal and agronomic research and development is today captured by a wide range of electronic tools, in software packages and mobile devices. On-farm microchip ear tags and barcodes are used in automatic stock monitoring, herd/flock management and animal tracking, used in conjunction with the National Livestock Identification System and Property Identification Code to trace stock movements and disease-free status. Novel applications such as drones and video cameras point to future solutions to the growing problem of a labour shortage in the animal industries (Doole et al. 2009). One area that urgently requires further development is the design and implementation of cheap, virtual fencing systems to assist in grazing management and the exclusion of livestock from conservation areas (Umstatter 2011). The other strand is animal breeding which lies outside the scope of this chapter but is a critical component in the development of successful animal breeds to withstand heat stress, maintain disease resistance, and provide superior energy conversion rates and desirable market traits.

4.1 Decision Support and Remote Sensing

Agronomic experiments conducted on grass and legume species across the ILZ in the 1970–1990s formed the basis of simulation growth models used in pasture, animal and farm management today. These models have passed through many iterations to become user-friendly decision-making tools. For example, simulation models were developed from the SGS program in 2001–2003 as the SGS Model with GrazeMod, DairyMod and EcoMod modules (Johnson 2013). Public sector and livestock industry agencies maintain and update software derived from such models online, ranging from spreadsheet calculators such as Meat and Livestock Australia's 'More Beef from Pastures' (www.mla.com.au) to CSIRO's GrazPlan©

(www.grazplan.csiro.au) which contains modules such as Grazfeed© (feed calculator) and Grazgro© (pasture growth, Clark et al. 2000). Specific software programs include Prograze for livestock production (Bell and Allan 2000) and AusFarm©, a whole-farm support tool for mixed crop and stock farming. These decision-support models calculate production, feed requirements, management of stock attributes and numbers, and farm budgeting, with location-specific input variables. They are used collaboratively by producers and consultants to manage physical and business components of farms.

Remote sensing has also been harnessed to pasture production. An application of multi-spectral MODIS-NDVI satellite imagery has operated since 2003 using the ratio of red and near-infrared wavelength detection of photosynthetic activity to derive aboveground dry matter (DM) and feed-on-offer (FOO) via empirical modelling of growth estimates with daily climate data (Donald et al. 2010; Hill et al. 2004). Products are available from paddock (90 m^2 resolution) to regional scale on an open-source platform *Pastures-from-Space* (PGR®, www.pasturesfromspace.csiro.au). This provides farmers with real-time information on the amount and variation over large expanses of pasture more easily and accurately than can be obtained on the ground as well as synoptic assessments of regional production. Figure 5 gives an example using the *Pastures-from-Space* (PGR®) data to assess the inter-annual variations from 2003–2014 in total DM ha^{-1} $year^{-1}$ across the transects identified in Fig. 1.

The localities along each 600 × 200 km transect have progressively lower annual rainfall toward the interior, and nearly three-fold variation in production over the period, which included both a severe and prolonged drought (2003–2009 in NSW) with annual rainfall in the lowest tenth percentile of long-term averages and two exceptionally wet years (2010–2011). One-third of the annual aboveground DM values in southern NSW and 50 % of those in WA were ≤2.5 t ha^{-1} $year^{-1}$. These low pasture growth values are typical in large parts of the mixed-farming belt and illustrate the observed low levels of pasture production.

4.2 *Plant Improvement*

Over the past 70 years, an estimated 8,200 exotic pasture plants have been collected and introduced intentionally for use in Australian pastures by scientists, predominantly from public sector organisations (Cook and Dias 2003). In the past 15 years, public breeding programs have focussed principally on pasture legumes of Mediterranean origin with 58 new annual and short-lived perennial pasture legumes released, principally for southern Australia (Nichols et al. 2006). Newer varieties are adapted to a wider range of edaphic conditions, such as fluctuating rainfall regimes, and tolerance to acidity, drought, waterlogging and disease.

Traditionally, funding for pasture plant breeding was shared between public agencies and the research and development (R&D) levy organisations, but the role of public research and extension organisations has diminished in the past two

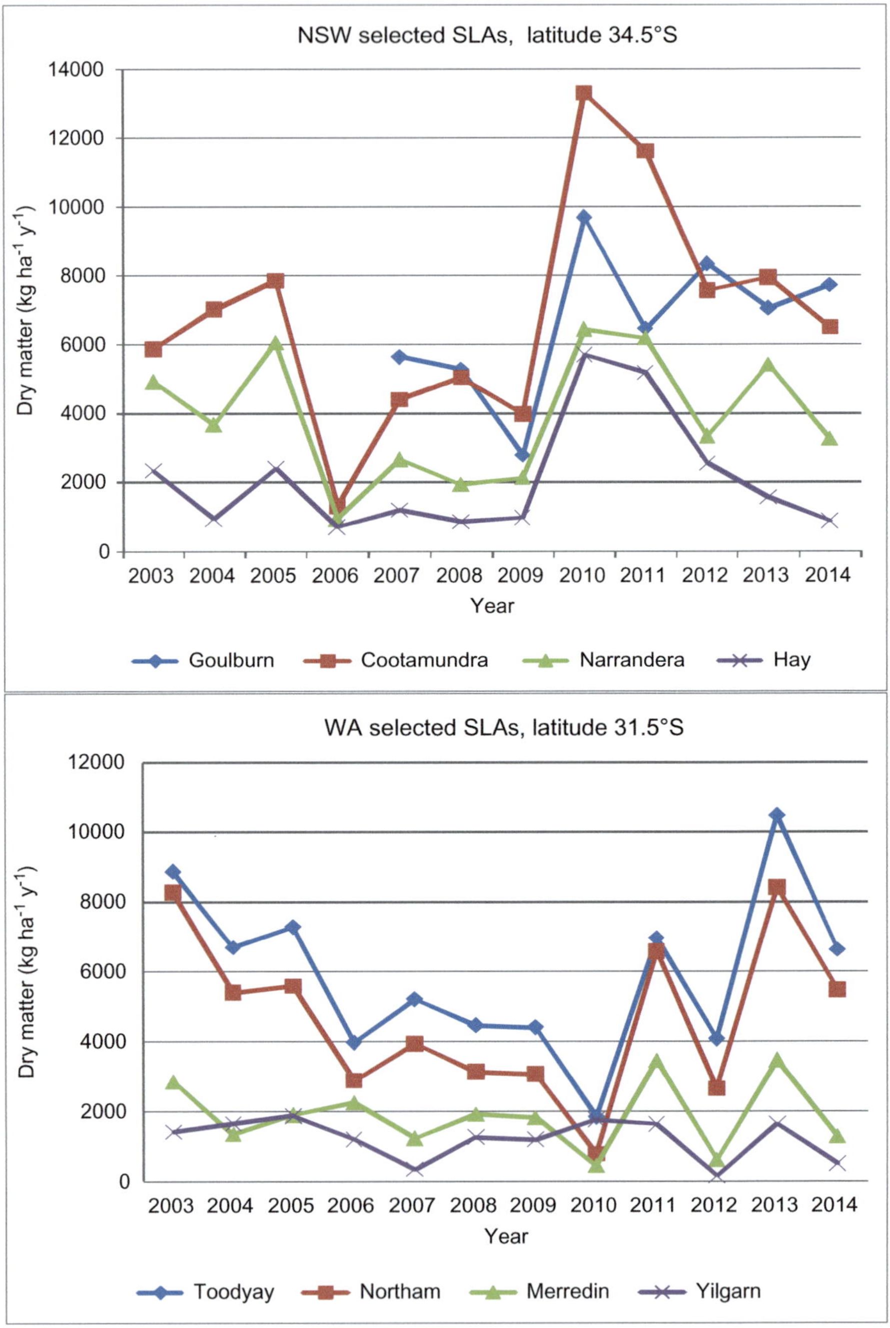

Fig. 5 Remote-monitored total pasture growth (kg ha^{-1} $year^{-1}$) from 2003 to 2014 averaged across statistical local areas (SLAs) in eastern Australia at latitude 34.5°S and western Australia at latitude 31.5°S. Long-term annual rainfall (mm): Goulburn 628, Cootamundra 650, Narrandera 433, Hay 433. Toodyay 521, Northam 428, Merredin 326, Yilgarn 303

decades and, since 2007, an integrated service has been formed called Pastures Australia. Pasture breeding achievements include the selection of species and breeding of varieties adapted to low and high pH with their species-specific symbiotic rhizobia (Howieson et al. 2000) and the introduction of new annual legumes with high growth rates and seed set that can revitalise old pastures (Wolfe 2009). The adoption of new varieties has been variable, and the benefits to animal and crop production need further quantification to build confidence in producers. Unlike dryland crops, pasture breeding has not been attractive to commercialisation due to the long-lasting nature of pastures, small opportunity of annual revenues to breeding companies, high costs to farmers in ensuring establishment or persistence in some environments (Burns and Norton 2015) and the difficulty of demonstrating an immediate benefit (Whitbread et al. 2005). Seed companies report that many recent varieties enter the market for only a short time, and many old varieties retain a strong market share (Grains Research and Development Corporation 2007). The loss of most specialist pasture agronomists with a wide local knowledge of soils, climates and plant growth habits over the past 20 years has made many farmers cautious of attempting to introduce new cultivars or species in periods of financial uncertainty (Bell et al. 2014; Nichols et al. 2012).

5 Current Restrictions and Future Opportunities

Farm performance studies consistently show a large difference in profitability between the top 10–20 % of producers and the industry or regional average. In a comprehensive benchmarking study Hooper and Levantis (2011) found the top 10 to 20 % most profitable farmers of mixed farms across 13 agro-ecological zones in Australia used higher inputs compared with the average grower in each zone. Intensification is greatest in most profitable groups. For example in the high rainfall (Gippsland) region of Victoria the top 20 % of profitable farms ran 30 % more stock than other farms, irrespective of the enterprise (beef, lamb, wool or a mix of these), translating into 50-100 % higher gross margins per hectare (Webb Ware 2014). It is more difficult for small to medium sized livestock-dominant farms to achieve profitability than for larger enterprises; their unit costs of production (c/kg live weight gain) are fifty to a hundred percent higher than for large farms, and for some, getting bigger has increased their debt in the past decade as a result of land purchases (Thompson and Martin 2014).

As demonstrated by research and advisory programs, the key to greater profitability is good management, a substantial investment in pasture production and farming by soil type. This was the main principle of the extensive SGS and Grain and Graze programs (Price and Hacker 2008). Despite the success of those popular programs, much grazing land is currently underused or deteriorating. Both financial and social reasons are implicated. Labour costs and scarcity have been identified as restricting animal industry productivity more than cropping in the past decade (Doole et al. 2009; Rose and Kingwell 2009). Comparisons of different broadacre industries show that many farms have maintained positive total factor productivity (TFP) only by reducing input costs when output values decline, with specialised sheep farms having had the lowest TFP (ABARES 2014; Nossal and Gooday 2009). In parts of more-densely settled NSW and Victoria, many smaller grazing properties have been subdivided as hobby farms close to towns, and greater investment can be made from capital appreciation than from on-farm income. Behrendt and Eppelston (2011) found that the returns from capital appreciation were three- to ten-fold that of conventional grazing returns for farms within one hour's drive from regional centres. Across NSW, Eaves (2010) found that between 1990 and 2008, the returns on rural land devoted to mixed farming outperformed the returns from traditional high rainfall grazing farms, but with higher risk from price volatility. On smaller properties, social factors are also influential. In a survey of Victorian farmers, Wilkinson et al. (2011) found that many livestock farms were small and run by older farmers, two-thirds of whom had gross annual incomes of less than half the industry average. Of these, 33 % of sheep and 45 % of beef farmers used no agency or other advisory services, and farmers were essentially semi-retired and marking time. Small to medium-sized farms are also inhibited from intensifying their livestock systems because of the increase in complexity of the farming operation, exposing the farmer to higher risk and uncertainty (Kingwell 2011).

6 Conclusions and the Way Ahead

While grazing still forms the largest land use in the Australian farming belt, many pastures are in no better or poorer condition than 20 years ago. Red meat production has risen but at a much slower rate than grain production. The initial trigger for the lower performance of the animal industries came from the negative returns in wool production in the early 1990s, but pressure on input costs from the constant cost–price squeeze during that decade also shifted the production focus to more immediate returns from crops than the longer lifecycle of meat industries (Nossal and Sheng 2010). Cash receipts on animal-dominated farms have been low as a result of the Millennium drought, a high Australian dollar and export market competition, so mixed farms have concentrated their efforts on cropping with low expenditure on pasture renovation, soil nutrition and liming, and negative effects on pasture productivity. The upturn in prices since 2013 has increased cattle and sheep sales by 12–23 % in 2014–2015 and returns to farm business profit (Martin 2015). If such conditions continue, confidence may return to livestock producers with more investment on-farm.

Changes to regional climates have also interacted with farm financial conditions to reduce the focus on pastures with the reduction in rainfall in southern Australia, thereby shortening the growing season and increasing variability in summer rainfall regions with adverse effects on legume persistence. These trends are projected to continue, and future research directions need to incorporate changing climatic parameters into all aspects of production research (Bell et al. 2014; International

Panel on Climate Change 2007). Research on intercropping, the greater persistence of legumes within grass–legume pastures, and the introduction winter-dominant lucerne and summer-dormant perennial grasses into southern Australia are continuing priorities. The mismatch between pasture production and animal demand often lowers the effective use of rainfed pasture, resulting in conservative stocking rates. Trends in diversification of the feed base, such as grazing winter cereals, production of on-farm forage crops, increased use of no-till stubble for grazing, and more feedlotting, are useful strategies to increase livestock production, but the decline in soil fertility from loss and deterioration of pastures requires more research and extension if most farms are to achieve higher overall long-term productivity.

There is considerable scope for increasing pasture productivity in higher rainfall districts where more reliable rainfall regimes reduce investment risk, and where higher winter temperatures are more favourable to grass growth. Most farms retain livestock-related infrastructure and the capacity for intensification of animal production systems so reversing the run-down in soil fertility and pasture condition may be less costly than maintaining the high levels of soil nutrition, weed and pathogen-free soil environments needed for high-yielding continuous cropping. This would be a fruitful area for economic modelling. The key driver that is often lacking for improving pasture production, and hence higher profits from animals, is the incentive of clear information that links the prices paid for quality meat products to the quality of the pasture. The red meat industry recognises that this requires a shift from a supply-chain mentality to whole-of-value-chain approach, but it also implies a further shift in market power from producers to retailers, and from smaller to larger producers (AMPC 2013; Jie and Parton 2009).

In general, producers on small farms, or farms where livestock forms a minor part of the business, are vulnerable as price-takers; their business profits are less and rates of return on capital lower than for large-scale enterprises (Martin 2015). Alternative financing systems and farming structures are needed, such as cooperatives and partnerships among neighbours, with clearer information flows through the whole-of-value chain to provide incentives for large-scale reinvestment of pastures. Many animal-dominant farms are too small to gain the economies of scale enjoyed by large farms which are also mainly run as family businesses (Thompson and Martin 2014). Research into alternative business structures would be valuable; for example, individual adjacent farms might form associations to lease out larger areas of adjoining land on long leases to animal management specialists (or vertically-integrated meat processing firms) thus providing a more regular and consistent type of animal aimed at specific market requirements (Kingwell 2005). Opportunities exist in expanding exports to Asian markets that demand higher value products such as red meat of high quality, where Australia's guaranteed food security is advantageous. Together with the recent signing of several international free trade agreements (2015–2016), this may provide the needed incentives to greater investment in underperforming parts of the livestock industry.

References

ABARE-BRS (2010) Australian collaborative land use and management program– national land use mapping. Land use Australia, Version 4. Australian Bureau of Agricultural and Resource Economics. Bureau of Resource Sciences. Canberra. Available at: http://apps.daff.gov.au/AGSURF/. Last checked 20-03-16

ABARES (2014) Australian farm surveys: all broadacre industries selected physical and financial estimates, 1990–2014. Australian Bureau of Agricultural and Resource Economics and Sciences, Department of Primary Industries. Commonwealth of Australia. Available at: http://apps.daff.gov.au/AGSURF/ last checked 20-10-15

Accad A, Neldner VJ, Wilson BA, Niehus RE (2013) Remnant vegetation in Queensland. Analysis of remnant vegetation 1997–2011, including regional ecosystem information. Queensland Department of Science Information Technology, Innovation and the Arts, Brisbane

AMPC (2013) AMPC Strategic Plan 2013–2017. Australian Meat Processing Corporation, St Leonards, NSW, p 48

Angus JF, Peoples MB (2012) Nitrogen from Australian dryland pastures. Crop Pasture Sci 63:746–758

Australian Bureau of Statistics (ABS) (1982–1991, 1996, 2001, 2006, 2011) Agricultural census records concorded, consistent sets, courtesy of ABARES

Australian Bureau of Statistics (ABS) (2013) Agricultural Resource Management Practices, Australia 2011–2012, Catalogue No. 4630.0. Issued 28/06/2013

Australian Bureau of Statistics (ABS) (2014) Land management and farming in Australia 2012–13, Catalogue number 4627.0 (46270DO001_2007-13). Data collected from the rural environment and agricultural commodities survey 2013. Commonwealth of Australia Available at http://www.abs.gov.au/ausstats/abs@.nsf/mf/7121.0 last checked 14-03-16

Australian State of the Environment Committee (2011) The state of the environment report 2011. Chapter 5, Land. pp. 287–293, and Figures 10a and 10b. Department of Environment, Sport, Arts and Communities. Commonwealth of Australia

Barson M, Mewett J, Paplinska J (2012) Land management practice trends in Australia's grazing (beef cattle/sheep) industries. Caring for our Country Sustainable Practices, fact sheet 2. Department of Agriculture, Fisheries and Forestry, Canberra

Behrendt K, Eppleston J (2011) Threats, realities and opportunities of grassland farming in the Central Highlands. In: Proceedings of the 26th annual conference of the Grassland Society, New South Wales

Bell AK, Allan CJ (2000) PROGRAZE – an extension package in grazing and pasture management. Aus J Exp Agric 40:325–330

Bell LW, Moore AD (2012) Integrated crop-livestock systems in Australian agriculture: trends, drivers and implications. Agric Syst 111:–12

Bell MJ, Eckard RJ, Harrison MT, Neal JS, Cullen BR (2013) Effect of warming on the productivity of perennial ryegrass and kikuyu pastures in south eastern Australia. Crop Pasture Sci 64(1):61–70

Bell LW, Hayes RC, Pembleton KG, Waters CM (2014) Opportunities and challenges in Australian grasslands: pathways to achieve future sustainability and productivity imperatives. Crop Pasture Sci 65:489–507

Benson J (1996) What is a native grassland? In: Proceedings of the 11th annual conference of the Grassland Society New South Wales, 1996, pp 92–96

Blair G. (2008) SSP: single superphosphate, a scenario slowly passing. In: Unkovich M (ed) Global issues. Paddock Action. Proceedings of the 14th Australian Agron conference 21–25 September 2008. Adelaide, South Australia. Available at: http://www.regional.org.au/au/asa/2008/concurrent/plant-nutrition/5950_blairg.htm last checked 20-01-16

Bureau of Meteorology (2014) State of the Climate 2014. Full Report. Commonwealth of Australia, p 16. Available at: http://www.bom.gov.au/state-of-the-climate/documents /state-of-the-climate/2014/ last checked 2-02-16

Burns H, Norton M (2015) Constraints to greater use of pulses and forage legumes on acid soils of the high rainfall zone of south-eastern Australia- focus groups and a farmer survey. In: Proceedings of the 17th Australian Society Agron conference 20–24 September, Hobart. Australia

Clark SG, Donnelly JR, Moore AD (2000) The GrassGro decision support tool: its effectiveness in simulating pasture and animal production and value in determining research priorities. Aus J Exp Agric 40:247–256

Cook GD, Dias L (2003) It was no accident: deliberate plant introductions by Australian government agencies during the twentieth century. Aus J Bot 54:601–625

Curtis K (2009) Recent changes in the Australian sheep industry (the disappearing flock). Issue No. 012. Department of Agriculture and Food. Government of Western Australia

Dahl A, Martin P, Gray E (2014) Australian lamb: financial performance of slaughter lamb producing farms 2011–12 to 2013–14. Australian Bureau of Agricultural and Resource Economics and Sciences Res Rep, 14.9. Meat and Livestock Australia, Canberra, August

Descheemaeker K, Llewellyn R, Moore A, Whitbread A (2013) Experimental trials and simulation modelling indicate that summer-growing perennial grasses are a potential new feed source in the Mallee region of southern Australia. In: Michalk DL, Millar GD, Badgery WB, Broadfoot KM (eds) Proceedings of the 22nd international Grasslands Congress 15–19 Sept 2013, pp 479–482

Dolling PJ, Latta RA, Ward PR, Robertson MJ, Asseng S (2005) Soil water extraction and biomass production by lucerne in the south of Western Australia. Aus J Agric Res 56:389–404

Donald G (2012) Final Report: analysis of feed-base audit. Project number: B.PAS 0297. Meat and Livestock Australia, North Sydney

Donald GE, Gherardi ASG, Edirisinghe BEA, Gittins CSP, Henry D, Mata G (2010) Using MODIS imagery, climate and soil data to estimate pasture growth rates on farms in the south-west of Western Australia. Anim Prod Sci 50:611–615

Doole GJ, Bathgate AD, Robertson MJ (2009) Labour scarcity restricts the potential scale of grazed perennial plants in the Western Australian wheatbelt. Anim Prod Sci 46:860–893

Doyle PT, Grimm M, Thompson AN (1993) Grazing for pasture and sheep management in the annual pasture zone. In: Kemp DR, Michalk DL (eds) Pasture management technology for the 21st century. CSIRO, Melbourne, pp. 71–90

Eaves C (2010) NSW rural land performance: 1990–2008. Aust Agribus Rev 18:85–102

Ellington A (1986) Nitrogen inputs and utilisation in leguminous pasture: a review of recent Australian literature. Tech Rep No. 128. Victorian Department of Agriculture and Rural Affairs, Melbourne

Evergraze (2014) More livestock from perennials. Available at: http://www.wool.com/on-farm-research-and-development/production-systems-eco/whole-farm-systems/evergraze/ last checked 7-03-16

Gazey C, Gartner D (2009) Survey of West Australian agricultural lime sources. Bulletin 4760, Department of Agriculture and Food. Government of Western Australia

Graetz RD, Wilson MA, Campbell SK (1995) Landcover disturbance over the Australian continent: a contemporary assessment. Biodiversity series paper 7. Department of Environment, Sports and Territories, Commonwealth of Australia. Canberra

Grains Research and Development Corporation (2007). The value of pastures. Available at: http://www.grdc.com.au/Research-andDevelopment/Major-Initiatives/ last checked 22-02-16

Hacker RB, Robertson MJ, Price RJ, Bowman AM (2009) Evolution of mixed farming systems for delivery of triple bottom line outcomes; a synthesis of the Grain and Graze program. Anim Prod Sci 49:966–974

Harries M, Anderson GC, Hüberli D (2015) Crop sequences in Western Australia: what are they and are they sustainable? Findings of a four-year survey. Crop Pas Sci 66:634–647

Helyar KR (1976) Nitrogen cycling and soil acidification. J Aus Inst Sci 42:217–221

Heng LK, White RE, Helyar KR, Fisher R, Chen D (2001) Seasonal differences in the soil water balance under perennial and annual pastures on an acid sodosol in south eastern Australia. Eur J Soil Sci 52:227–236

Hill MJ (1996) Potential adaptation zones for temperate pasture species as constrained by climate: a knowledge-based logical modelling approach. Aust J Agric Res 47:1095–1117

Hill MJ, Donald GE (1998) Australian Temperate Pastures Database [electronic resource]. Compiled for National Pastures Improvement Coordinating Committee, Hill MJ, Donald GE. CSIRO Animal Production Division 1998, Perth, Western Australia

Hill MJ, Vickery PJ, Furnival EP, Donald GE (1999) Pasture land cover in eastern Australia from NOAA-AVHRR NDVI and classified Landsat TM. Remote Sens Environ 67:32–50

Hill MJ, Donald GE, Hyder MW, Smith RCG (2004) Estimation of pasture growth rate in the south west of Western Australia from AVHRR NDVI and climate data. Remote Sens Environ 93:528–545

Hooper S, Levantis C (2011) Physical and financial performance benchmarks for grain producing farms for the six Southern agro-ecological regions, and seven Western and Northern agro-ecological regions. Australian Bureau of Agricultural and Resource Economics and Sciences reports prepared for the Grains Research and Development Corporation, February 2011

Howieson JG, O'Hara GW, Carr SJ (2000) Changing role for legumes in Mediterranean agriculture; developments from an Australian perspective. Field Crop Res 65:107–122

Hutchings TR, Nordblom TL (2011) A financial analysis of the effect of the mix of cropping and sheep enterprises on the risk profile of dryland farms in south-east Australia. Aus Agric Res Econ Soc 55th Ann Conf, Melbourne, Victoria

International Panel on Climate Change (2007) Fourth assessment report 2007. The physical science basis. Section 11.7 Australia and New Zealand. 11.7–11.7.3.6. Working Group 1

Jie F, Parton K A (2009). Analysing beef supply chain strategy in Australia, the United States and the United Kingdom. In: Proceedings of the Australian NZ Marketing Academy conference, Melbourne, pp 1–6

Johnson IR (2013) DairyMod and the SGS Pasture Model: a mathematical description of the biophysical model structure. IMJ Consultants, Dorrigo

Johnson IR, Lodge GM, White RE (2003) The sustainable grazing systems pasture model: description, philosophy and application to the SGS national experiment. Aus J Exp Agric 43:711–728

Kemp DR, Dowling PM (2000) Towards sustainable temperate perennial pastures. Aust J Exp Agric 40:135–132

Kingwell R (2011) Managing complexity in modern farming. Aus J Agric Econ 55:12–34

Kingwell R (2005) Future broadacre farming systems in Australia: a personal view Available at http://www.agrifood.info/Agrifood/connections/autumn_2005/Kingwell.htm last checked 16-03-16

Kirkegaard JA, Sprague SJ, Dove H, Kelman WM, Marcroft SJ, Lieschke A (2008) Dual-purpose canola, a new opportunity in mixed farming systems. Aus J Agric Res 59:291–302

Lake A (2012) Australia's declining crop yield trends: II. The role of nitrogen nutrition. 16th Australian Agronomy Society conference, Armidale, New South Wales

Lefroy EC (2002) Forage trees and shrubs in Australia; their current use and future potential. A report for the Joint Venture Agroforestry Program. Rural Industries Research and Development Corporation Publication 02/039. Available at: https://rirdc.infoservices.com.au/items/02-039 last checked 27-02-16

Linehan V, Thorpe T, Andrews N, Kim Y, Beaini F (2012) Food Demand to 2050: opportunities for Australian agriculture. 42nd Australian Bureau of Agriculture and Resource Economics and Sciences Outlook conference, 6–7 March 2012, Canberra

Llewellyn R, D'Emden FH (2010) Adoption of no-till cropping practices in Australian grain growing regions. Grains Research and Development Corporation and CSIRO, Kingston, p. 31

Martin P (2015) Australian beef. Financial performance of beef cattle producing farms 2012–3 to 2014–5. Australian Bureau of Agricultural Economics and Sciences Res Rep 15.5, Department of Agriculture, Canberra.

Massey C (2011) Breaking the sheep's back. pp. 436. University of Queensland Press.

Meat and Livestock Australia (2012a) Cattle numbers as at June 2011- natural resource management region. Published by Meat and Livestock Australia Ltd

Meat and Livestock Australia (2012b) Sheep numbers as at June 2011- natural resource management region Published by Meat and Livestock Australia Ltd

Mewett J, Paplinska J, Kelly G, Lesslie R, Pritchard P, Atyeo C (2013) Towards national reporting on agricultural landuse in Australia. Australian Bureau of Agricultural and Resource Economics and Sciences Technical report 13.06

Moore AD (2014) Integrated farming systems for a sustainable future-what is the role of mixed farming in sustainable agriculture? Available at: http://www.grdc.com.au /Research-and-Development/GRDC-Update-Papers/2014/09/Integrated-farming-systems-for-a-sustainable-future last checked 4-03-16

Moore AD, Bell LW, Revell DK (2009) Feed gaps in mixed-farming systems: insights from the Grain and Graze program. Anim Prod Sci 49:736–748

National Land and Water Resources Audit (2001) Australian agricultural assessment volume 2. Natural Heritage Trust, Canberra

Nichols PGH, Revell CK, Humphries AW, Howie JH, Hall EJ, Sandral GA, Ghamkhar K, Harris CA (2012) Temperate pasture legumes in Australia- their history, current use and future prospects. Crop Pasture Sci 63:691–725

Nichols PGH, Loi A, Nutt BJ, Evans PM, Craig AD, Pengelly BC, Dear BS, Lloyd DL, Revell CK, Nair RM, Ewing MA, Howieson JG, Auricht GA, Howie JH, Sandral GA, Carr SJ, de Koning CT, Hackney BF, Crocer GJ, Snowball R, Hughes SJ, Hall EJ, Foster KJ, Skinner PW, Barbetti MJ, You MP (2006). New annual and short-lived perennial pasture legumes for Australian agriculture – 15 years of revolution. Field Crops Res 104: 10-23

Norris P (2009) 'No till', no sheep and 'knowledge' farming system. pp. 36–38. Aus Grain, May–June 2009

Nossal K, Sheng Y (2010) Productivity growth; trends, drivers and opportunities for broadacre and dairy industries. Australian Bureau of Agricultural and Resource Economics, Australian Commodities 17: 217–229 Australian Government, Canberra

Nossal K, Gooday P (2009) Raising productivity growth in Australian agriculture. Technical report 09.7, Australian Bureau of Agricultural and Resource Economics, Australian Government, Canberra

Pearson CJ, Brown R, Collins WJ, Archer KA, Wood MS, Petersen C, Booth B (1997) An Australian temperate pasture database. Aus J Agric Res 48:453–465

Peoples MB, Bowman AM, Gault RR, Herridge DF, McCallum MH, McCormick KM, Norton RM, Rochester IJ, Scammell GJ, Schwenke GD (2001) Factors regulating the contributions of fixed nitrogen by pasture and crop legumes to different farming systems in eastern Australia. Plant Soil 228:29–41

Porter WA, McLay CDA, Dolling PJ (1995) Rates and sources of acidification in agricultural systems in southern Australia. In: Date RA, Grundon NJ, Rayment GE (eds) Plant soil interactions at low pH. Kluwer, Amsterdam, pp. 75–83

Powles SB, Preston C, Jutsum AR, Bryan IB (1997) Herbicide resistance: impact and management. Adv Agron 58:5–93

Price RJ, Hacker RB (2008) Grain and graze: an innovative triple bottom line approach to collaborative and multi-disciplinary mixed-farming systems research, development and extension. Anim Prod Sci 49:729–735

QTLC (Queensland Transport and Logistics Council) (2014) Supply chain perspective; livestock/ meat. Available at: http://www.qtlc.com.au/wp-content/uploads/2013/01/QTLC-Supply-Chain-Perspective_Livestock.pdf last checked 03/04/16

Robertson M, Bathgate N, Reynolds M, Hayes R, McCormick J, Li G, Casburn G (2014) EverCrop uniform rainfall zone – assessing the impact of input costs on enterprise mix on crop-livestock farms in southern NSW. Technical report no. 11. Future Farms Cooperative Research Centre

Robertson M (2010) Australian productivity in Australia and New Zealand; constraints and opportunities. Plenary paper, 15th Australian Society Agron conference, Lincoln, New Zealand

Rose G, Kingwell R (2009) Seasonal labour is the most profitable use of labour in broadacre crop dominant farms. Final report, Land and Water Australia, Australian Government

Ryan T (2010) The Australian fertiliser industry—values and issues. Australian Fertilizer Industry conference, Fertiliser Industry Federation of Australia. Available at: http://www.iama.org.au/sites/default/files/Australian%20Fertilizer%20Industry%20Value%20and%20Issues%20August%202010.pdf. Last checked 12-02-16

Scanlan JC (1991) Woody overstorey and herbaceous understorey biomass in Acacia harpophylla (brigalow) woodlands. Aus J Ecol 16:521–529

Seymour M, Kirkegaard JA, Peoples MB, White PF, French RJ (2012) Break crop benefits – insights from three decades of research. Crop Pas Sci 63:1–16

Singh DK, Bird PR, Saul GR (2001) Maximising the use of soil water by herbaceous species in the high rainfall zone of southern Australia: a review. Aus J Agric Res 54:677–691

Spiers SD, Reuter DJ, Peverell KI, Brennan RF (2013) Making better fertiliser decisions for cropping systems in Australia: an overview. Crop Pas Sci 64:412–423

Thompson T, Martin P (2014) Australian beef. Financial performance of beef production farms 2011–12 to 2013–14. Australian Bureau of Agricultural and Resource Economics and Sciences Res Rep 14.7. Australian Government, Department of Agriculture.

Umstatter C (2011) The evolution of virtual fences; a review. Comput Electron Agric 75:10–22

Unkovich M, Baldock G, Peoples M (2010) Prospects and problems of simple linear models for estimating symbiotic N_2 fixation by crop and pasture legumes. Plant Soil 329:75–89

van Rees H,White B, Laidlaw J, McKinley D (2011) Farming during a period of extreme climate variability: consequences and lessons. Final Report December 2011, prepared for the Birchip Cropping Group

Virgona JM, Gummer FAL, Angus JF (2006) Effects of grazing on wheat growth, yield, development, water use and nitrogen use. Aus J Agric Res 57:1307–1309

Walcott JJ, Zuo H, Loch AD, Smart RV (2013) Patterns and trends in Australian agriculture: a consistent set of agricultural statistics at small areas for analysing regional changes. J Land Use Sci 9:453–473

Weaver DW, Wong MTF (2011) Scope to improve phosphorus (P) management and balance efficiency of crop and pasture soils with contrasting P status and buffering indices. Plant Soil 349:37–54

Webb Ware J (2014) Opportunities to maximise livestock profit in mixed-farming enterprises. Mackinnon Project, University of Melbourne. Grains and Development Research Corporation. Australian Government. Available at: https://www.grdc.com.au/Research-and-Development/GRDC-Update-Papers/2014/03/Opportunities-to-maximise-livestock-profit-in-mixed-farming-enterprises. Last checked 13-13-16

Whitbread AM, Pengelly BC, Smith BR (2005) An evaluation of three tropical legumes for use in mixed farming systems on clay soils in southern inland Queensland. Australia Trop Grass 39:9–21

White S, Cordell D, Moore D (2010) Securing a sustainable phosphorus future for Australia; implications of global phosphorus scarcity and possible solutions. Prepared for CSIRO Sustainable Agriculture Flagship. Available at: http://cfsites1.uts.edu.au/find/isf/publications/whitecordell-moore2010phosphorusfuture.pdf. Last checked 11-01-16

Wilkinson R, Barr N, Hollier C (2011) Segmenting Victoria's farmers. Department of Primary Industries, State Government of Victoria, p 63

Wolfe E (2009) Country pasture/forage resource profiles. Australia. Food and Agriculture Organisation, p 45. www.fao.org/ag/agp/AGPC/doc/Counprof/PDF%20files/Australia.pdf. Last checked 05-02-16

Wolfe E, Dear BS (2001) The population dynamics of pastures with particular reference to southern Australia. In: Tow PG, Lazenby A (eds) Competition and succession in pastures. CABI, Wallingford, pp. 119–148

Wong MTF, Grundy M, Barson M, Walcott J (2012) A strategic framework to improve phosphorus management in the Australian Grains industry. CSIRO, Canberra

Dryland Agriculture in South Asia: Experiences, Challenges and Opportunities

Cherukumalli Srinivasa Rao, K.A. Gopinath, C.A. Rama Rao, B.M.K. Raju, R. Rejani, G. Venkatesh, and V. Visha Kumari

1 Introduction

South Asia—comprising eight countries *viz.* Afghanistan, Bangladesh, Bhutan, India, the Maldives, Nepal, Pakistan and Sri Lanka—has a population of about 1.5 billion (22 % of the world's population) is the most populated region in the world but only 4.8 % of the world's total land area (Lal 2006). Among different south Asian countries, India is the largest with about two-thirds of the geographical area and coastline, and nearly three-quarters of the population (Table 1). Its topography includes a variety of mountains, plateaus, dry regions, intervening structural basins and beaches. The elevation varies from the world's highest point, Mount Everest, to the world's lowest, the sea beach. It has about 10,000 km of coastline. The region has a largely tropical monsoon climate with two monsoon systems: the southwest monsoon (June–September) and the northeast monsoon (December–April). The region features large year-to-year variations in rainfall which frequently cause severe floods and droughts over large areas. South Asia has some of the world's largest river systems: the River Indus flows from China to Pakistan, the Ganga stretches for about 2525 km, and the Brahmaputra flows for about 2900 km through Tibet, India and Bangladesh (Sharda 2011). Soil and water are considered the principal natural resources of the South Asian region and the ultimate source of people's livelihood. However, the sustainability of these resources poses a challenge due to land degradation. Soil erosion and landslides are critical environmental hazards in the region.

C. Srinivasa Rao (✉) • K.A. Gopinath • C.A. Rama Rao • B.M.K. Raju • R. Rejani
G. Venkatesh • V. Visha Kumari
Indian Council of Agricultural Research-Central Research Institute
for Dryland Agriculture, Hyderabad, India
e-mail: cherukumalli2011@gmail.com

M. Farooq, K.H.M. Siddique (eds.), *Innovations in Dryland Agriculture*,
DOI 10.1007/978-3-319-47928-6_13

Table 1 Profile of South Asia

Particular	Afghanistan	Bangladesh	Bhutan	India	Maldives	Nepal	Pakistan	Sri Lanka
Geographical area (Mha)	65.22	14.85	3.84	328.73	0.03	14.72	79.61	6.56
Land area (Mha)	65.22	13.02	3.84	297.32	0.03	14.34	77.09	6.46
Population (millions)	32.56	168.96	0.74	1251.70	0.39	31.55	199.09	22.05
Coastline (km)	0	580	0	7000	644	0	1046	1340

Source: Central Intelligence Agency (2015)

The region is characterized by diverse climates, and equally-diverse soil and water resources. South Asian economies are agriculture based, so the land constitutes a valuable resource. The region shows extraordinarily diverse landforms due to the diverse climatic regimes, latitudes, altitudes and topography. Afghanistan and Bhutan are mostly rugged with mountains. Bangladesh is mainly flat alluvial plains. India has an upland plain (Deccan Plateau) in the south, a flat to rolling plain along the Ganges, deserts in the west, and the mountainous Himalayas in the north. The topography in Maldives is flat with white sandy beaches. Nepal has the flat river plain of the Ganges in the south, a central hill region and the rugged Himalayas in the north. Pakistan has the flat Indus plain in the east, mountains in the north and northwest, and the Baluchistan plateau in the west. The terrain of Sri Lanka is mostly low, flat to rolling plains with mountains in the south-central interior. Land degradation is one of the biggest problems in South Asia due to water erosion resulting from the steep topography coupled with high-intensity rainfall. Modern methods of agriculture further aggravate the situation, with practices such as overuse of fertilizers and pesticides, excessive irrigation of saline lands, and shifting cultivation. About 50 % of the total land area in South Asia is used for agriculture. Due to the high population pressure on the land, the percentage of agricultural to total land area is much higher in the region than the global average. In South Asia, agricultural land occupies more than 50 % of the land area in Afghanistan, Bangladesh and India and less than 50 % in the other countries (Table 2).

Most of the South Asian region is under rainfed agriculture. Afghanistan, Bhutan and Sri Lanka are predominantly rainfed ($\geq$80 %), as are India and Nepal (60–70 %). Irrigated agriculture predominates in Pakistan (26 %) and Bangladesh (45 %). India, Bangladesh, Pakistan, Nepal and Sri Lanka produce a wide range of agricultural and animal husbandry products. The forest area compared to land area is less than 30 % in all countries except Bhutan where 86 % of the land area is forest. Of the South Asian countries, Pakistan and Afghanistan have the least forest (2.1 %).

The climate in Afghanistan is arid to semiarid. Mountains in Afghanistan cause many variations in climate. More than three-quarters of the annual precipitation (327 mm) is received as snow in the mountain ranges of central Afghanistan. Bangladesh is located in the deltaic plains of river basins and the sea shore. This

Table 2 Natural resources and land use in South Asia

Country	Afghanistan	Bangladesh	Bhutan	India	Maldives	Nepal	Pakistan	Sri Lanka
Agricultural land (% of land area)	58.1	70.1	13.6	60.5	23.3	28.8	35.2	43.5
Rainfed land (% of agricultural land)	94	45	94	63	–	72	26	80
Forest (% of land area)	2.1	11.1	85.5	23.1	3	25.4	2.1	29.4
Average precipitation (mm) per year	327	2666	2200	1083	1972	1500	494	1712
Natural hazards	Earthquakes, floods, droughts	Droughts, cyclones, floods	Violent storms, landslides during rainy season	Droughts, floods, severe thunderstorms, earthquakes	Tsunami, rising sea level	Severe thunderstorms, floods, landslides, drought	Earthquakes, floods	Occasional cyclones and tornadoes
Per capita total renewable water resources (m^3)	2006	7262	105133	1527	–	6662	1240	2394
Electricity production per capita (kWh)	26	251	10176	779	674	109	451	535
Crude oil production bbl/day/million people	60	25	0	607	0	0	299	0

Source: Central Intelligence Agency (2015)
Note: Reference year ranges from 2011–2013 except for rainfed land (2003–2011)
bbl billion barrels, *kWh* kilo watt hour

country has a tropical climate—summer (March to June) is hot and humid while winter (October to March) is mild—the rainy season is (June to October) warm and humid. Annual rainfall is more than 2500 mm. Bhutan has a tropical climate in the southern plains, cool winters and hot summers in the central valleys, and severe winters and cool summers in the Himalayas. Annual precipitation is 2200 mm. The Indian climate varies from tropical in the south to temperate in the north. The annual average precipitation in India is 1083 mm with about 85 % of this rainfall received in 100–120 days (southwest monsoon). The climate in the Maldives is tropical; hot and humid. Annual precipitation is about 2000 mm. The climate in Nepal varies from cool summers and severe winters in the north to subtropical summers and mild winters in the south. Annual precipitation is about 1500 mm. Pakistan's climate is mostly hot, desert with a temperate northwest and arctic north. Annual precipitation is about 500 mm. Sri Lanka is an island nation with a tropical monsoon. Average annual rainfall is around 1700 mm. The northeast monsoon occurs between December and March and the southwest monsoon between June and October.

Most of the South Asian region is under rainfed agriculture, but some regions have been facing challenges regarding drought. Farmers in India, Afghanistan, Bangladesh and Sri Lanka have incurred large losses due to drought. Drought is affecting the economies in general and threatening food security in particular. Floods occur in India, Bangladesh, Pakistan, Nepal and Afghanistan and can cause substantial damage to standing crops. Cyclones and severe thunderstorms cause irrevocable damage to the general life of public as well as to agriculture and livestock. Bangladesh is extremely prone to floods and cyclones. The westernmost and easternmost parts of Bangladesh are prone to drought. About 33 % of India receives less than 750 mm rainfall, and 68 % of the sown area is subject to drought in varying degrees. Floods and cyclones are also frequent in India. The east coast of the country is hit by more cyclones. Sri Lanka has been experiencing drought since the ancient times. An average of 11,000 hectares of paddy land is destroyed every year due to the lack of water in sufficient quantities (Bhaskara Rao 2011). Soil erosion, deforestation, limited freshwater resources, and water pollution are among the major environmental concerns in South Asian countries.

Per capita renewable water resources are highest in Bhutan due to its low population. Bangladesh and Nepal have reasonably good water resources on a per capita basis. Per capita water resources in the remaining five countries range from 1000–2500 m^3 per year. Per capita electricity production (in kWh) is highest in Bhutan on account of the rich water resources, and ranges from 600–800 kWh in India and Maldives, 100–500 kWh in Bangladesh, Nepal, Pakistan and Sri Lanka, and 26 kWh in Afghanistan. Crude oil production per day per million people is highest in India (607 billion barrels (bbl) followed by Pakistan (299 bbl), with the remaining countries at less than 100 bbl.

The South Asian region sustains the world's poorest people; the contribution of this region (by 22 % of world population) to the world's GDP (Gross Domestic Product) is less than 5 %. Agriculture is the main source of livelihood in this region. High agricultural population densities in the south Asia have reduced the amount. The region has a high concentration of poverty and hunger. The Asian Development

Bank estimated that 451 million people in the subcontinent live below their respective national poverty lines (Chand 2010). The region has a 2.62 % share in global income, which is very low. The region is home to 40 % of the world's poor, with 29.5 % of its population living on less than $1 a day (ADB 2009).

The per capita gross national income (GNI) in these countries in 2014 ranged from US$690–5600. The lowest per capita income is in Afghanistan and the highest is in the Maldives. Per capita GNI in India and Pakistan ranged from US$1300–1600. The population density in 2014 was more than 1000 persons per km^2 in the Maldives and Bangladesh, but less than 400 persons per km^2 in the other countries with Bhutan at 19 and Afghanistan at 43. The population growth rate ranged from 0.08 % (Maldives) to 2.32 % (Afghanistan) per annum. The undernourished population was as low as 17 % in Nepal to as high as 30 % in Bangladesh. In 2013, the percentage of the population in poverty ranged from 6.7 % in Sri Lanka to 36.5 % in Afghanistan, with more than 60 % in rural areas in all countries except the Maldives. The infant mortality rate (deaths per 1000 live births) ranged from 20–60 in all countries except Afghanistan (115) and Sri Lanka (8.8). The life expectancy ranged from 67–77 years in all countries except Afghanistan (51 years). The literacy rate is very high in the Maldives (99 %) and Sri Lanka (93 %) but very low in Afghanistan (38 %). The average size of landholding is more than 3 ha in Pakistan but less than 1 ha in Nepal, Sri Lanka and Bangladesh. Unemployment in South Asia ranges from 4 % in Sri Lanka to 46 % in Nepal. The percentage of the labor force engaged in agriculture is highest in Afghanistan (78.6 %) and lowest in the Maldives (15 %). Per capita electricity consumption is less than 100 kWh in Afghanistan and Nepal. Bhutan has the highest per capita consumption of electricity with 2488 kWh. Emissions of carbon dioxide (ton/capita), a greenhouse gas causing climate change and global warming is highest in India (1.7) and lowest in Nepal (0.2).

The dependency on land for livelihoods is rising in South Asian countries (Chand 2010). The study compared agricultural workers per 100 ha of arable land between 1989–90 and 2003–04 and observed a rise in the number of workers per hectare of land in all South Asian countries except Sri Lanka (data not available for Bhutan and the Maldives). Other serious issues in South Asia's agriculture are the heavy dependence on rainfed agriculture and the small size of operational holdings (except Pakistan). The percentage of cropland area under dryland agriculture is 70 % in Afghanistan, 46 % in Bangladesh, 80 % in Bhutan, 66 % in India, 65 % in Nepal, 17 % in Pakistan and 33 % in Sri Lanka. About 60 % of the total arable land in South Asia is under dryland agriculture. About 83 % of the total dryland agricultural area of 129 Mha in South Asia is in India (Lal 2006). Cereals are cultivated in all countries except the Maldives. Rice is a major crop in all countries except the Maldives and Afghanistan.

The area under cereals is about 100 Mha in India, and less than 15 Mha in the other countries. As per the triennium ending (TE) 2013, the cereal yield was highest in Bangladesh (4.38 t/ha) followed by Sri Lanka (3.73 t/ha) and may be explained to a large extent by high fertilizer use. The productivity of cereals was lowest and less than 2 t/ha in Afghanistan. Compared with TE 1993, cereal yield doubled in

Bhutan and the Maldives and increased by 25–70 % in the other countries. In TE 2013, the yield of pulses was more than 1 t/ha in Sri Lanka and Bangladesh and increased from TE 1993 in all countries except Afghanistan by 5–65 %. Access to electricity in rural areas is lowest in Afghanistan (32 %), Bangladesh (49 %) and Bhutan (53 %), but is 100 % in the Maldives. The number of tractors per 100 km^2 of arable land is regarded as an indicator of mechanization in agriculture. India, Pakistan and Nepal have more than 100 tractors per 100 km^2 but Afghanistan, Bangladesh and Bhutan have less than 12. Fertilizer consumption in Afghanistan, Bhutan and Nepal is less than 30 kg/ha of arable land but ranges from 140 to 280 kg/ha in other countries. Livestock plays a key role in agriculture, especially in rainfed agriculture, in minimizing the risk of livelihood and coping with crop failures due to natural hazards. The cattle population per km^2 of land is highest in India (64), Nepal (51) and Pakistan (48) but less than 20 in Afghanistan, Bhutan and Sri Lanka. Similarly, the buffalo population ranges from 35–45 in India, Nepal and Pakistan, but is only 1.4 in Sri Lanka; the goat population is very high in Nepal (315) but very low in Bhutan, Afghanistan and Sri Lanka (<15); and the sheep population is >20 in India and Pakistan and <15 in Bhutan, Afghanistan and Nepal.

The South Asian region has the highest rate of irrigated agriculture (40 % of the cultivated area), but the water resources are becoming stressed due to the increasing population coupled with poor management practices. The mean annual precipitation—1083 mm in India, 2666 mm in Bangladesh, 280 mm in Pakistan, 1500 mm in Nepal, 1712 mm in Sri Lanka, 300 mm in Afghanistan, and 2091 mm in Myanmar—is characterized by high temporal and spatial variability resulting in excess surface water during summer and water shortfalls during winter. Groundwater and surface storage along with efficient utilization of available water resources are of utmost importance for agriculture in South Asian countries (Hasanain et al. 2012). In the last three decades, large irrigation projects have not been viable financially or environmentally, which has led to increased exploitation of groundwater, and increased the share of groundwater in total irrigated area from 50 % two decades ago to 75 %. This has serious consequences for the declining groundwater table.

The eight countries within South Asia are characterized by low volumes of intra-regional trade in goods and services. While a quarter of the world's population lives in the region, South Asia accounts for only 3 % of the global gross domestic product (GDP), 1.9 % of world exports, and 1.7 % of the world's foreign direct investment (ADB 2009). Nevertheless, South Asia's economy has grown by an annual average of 6 % in the last ten years. Dominance of small holder (small and marginal farmers) agriculture in South Asia lead to very low marketed surplus ratio (proportion of the produce available with farmer to market after meeting requirements such as family consumption, payment of wages in kind, feed, seed and wastage). Marketed surplus ratio is bound to be negligible with the livelihood options the farmer is left with.. The export performance of the region was credible during the pre- World Trade Organization (WTO) period. Growth rates during the implementation period, 1995–2000, declined sharply (−0.54 %). The entire post-WTO period, 1995–2003, had an average growth of 1.82 % compared with almost 8.00 % pre-WTO export. Imports to South Asia as a whole increased by 11 % in the pre-WTO period. The imports growth rate in WTO implementation period (1995–2000) was 2.38 % whereas entire post-WTO period (1995–2003) recorded a growth rate of 4.73 %. The 'lack of trade balance, therefore, hurt the region due to the unanticipated and extraordinary decline in commodity prices. Consequently, exports declined and imports spiraled, which adversely affected farmer incomes (George 2005).

2 Technological Developments: Experiences

Rural livelihood systems in dryland areas have, by persisting over several decades, demonstrated a resilience which runs counter to some predictions of imminent, irreversible degradation or collapse (Mortimore et al. 2000). The potential for technically-based interventions varies across South Asian countries. For example, in India, a diverse bank of proven technologies for dryland farming and conservation has grown over eight decades due to a massive investment in research. Research in dryland agriculture in India began in the 1930s. During the 1950s, the research focused on soil management and water conservation including bunding, terracing, gully plugging and check dams along with improved agronomic practices such as deep plowing, early sowing, improved varieties and crop rotations, optimum crop stands and weed control. Location-specific technologies—including *in-situ* moisture conservation, rainwater harvesting in farm ponds and its efficient utilization, integrated nutrient management modules, foliar sprays for drought mitigation, resilient crops and cropping systems, seed priming, improved sowing methods and contingency crop plans—have been developed by different research organizations including the Central Research Institute for Dryland Agriculture (CRIDA) to improve the productivity and profitability of dryland systems (Srinivasarao et al. 2014d).

2.1 Natural Resources Management

The management of natural resources in dryland areas is important not only because the livelihoods of millions of rural poor (>500 million) are directly connected to these areas but also because these areas will continue to play a crucial role in determining food security for the growing population and in reducing poverty in the coming decades (Rockstrom et al. 2007). Enhancing the efficiency and sustainability of natural resource management (NRM) projects in these areas is a universal challenge faced by concerned stakeholders.

2.1.1 Water

Rainwater management is a critical component of rainfed farming; the successful production of crops largely depends on how efficiently soil moisture is conserved *in-situ* and how the surplus runoff is harvested, recharged, stored and reused for supplemental irrigation (Rao et al. 2010; Srinivasarao et al. 2013a). Dryland areas receive an annual rainfall of less than 750 mm and experience more frequent water scarcity events during summer, in years with deficient monsoon rainfall, and during drought years. In these regions, agriculture is the prime source of income for local inhabitants and the major constraints to agricultural production is the availability of water during dry spells and a shortage of drinking water due to the declining groundwater table. The seasonal distribution of rainfall and temperature affects crop water requirements and hence the soil and water conservation interventions needed (Murty and Jha 2011). The adoption of *in-situ* and *ex-situ* soil and water conservation techniques is essential for arid, semiarid and rainfed regions due to the erratic nature of monsoon rainfall (Rejani et al. 2015b). These interventions need to be based on the runoff potential and resulting soil loss.

In-situ soil and water conservation techniques based on soil loss (Reddy et al. 2005; Rejani et al. 2016a); soil, rainfall and slope of the land (Reddy et al. 2005; Pathak et al. 2009; Srivastava et al. 2010); slope and soil depth (Kalgapurkar et al. 2012); and precipitation, slope, soil depth, texture, salinity, land use, land cover and geological information (De Pauw et al. 2008) have been reported. The major *in-situ* soil and water conservations interventions planned for dryland regions are agronomic measures such as contour cultivation, strip cropping, proper crop rotations, tillage practices, mulching, planting of grasses for stabilizing bunds, and deep plowing in black soils once every three years to reduce soil losses (Table 3).

An important strategy to enhance the infiltration rate of water into the soil during the 1970s was deep tillage because traditional tillage using the wooden plow (noninverting plow) was usually less than 10-cm deep (Vittal et al. 1983). In addition to crop yield, deep plowing improved porosity, infiltration and available water capacity, and reduced runoff and erosion. In dryland areas, water harvesting and storage in farm ponds, which is then used for supplementary irrigation of crops using efficient water application methods like drip and sprinkler irrigation, can substantially increase crop productivity (Murty and Jha 2011; Srinivasarao et al. 2014a). In nonarable lands with black soils, graded bunds with waterways, farm ponds, gully stabilization structures like check dams, gabion structures and horticultural crops such as pomegranate (*Punica granatum* L.), amla (*Phyllanthus emblica* L.) and guava (*Psidium guajava* L.) are recommended (Reddy et al. 2005). In non-arable areas, soil conservation measures such as contours or staggered trenching on foothills, plugging of stream courses, gabion structures and check dams are preferred (Reddy et al. 2005). The selection of suitable structures mentioned above for a specific location and its optimal spacing for drainage line treatments are key factors for the effective and economic control of sedimentation and runoff (Kadam et al. 2012; Rejani et al. 2016b). Since the implementation of drainage line treatments is expensive, site selection and construction need precision. The literature on site selection procedures for water harvesting structures considers slope, runoff, watershed area, stream order and socioeconomic aspects (IMSD 1995; Geetha et al. 2007). Researchers have used remote sensing and geographical information systems (GIS) to find suitable locations for rainwater harvesting structures (Chowdary et al. 2009; Ramakrishnan et al. 2008, 2009; Shanwad et al. 2011; Rejani et al. 2016b).

Table 3 Location-specific *in-situ* moisture conservations practices in different countries

Country	*In-situ* moisture conservation practices	References
Afghanistan	Organic and inorganic mulches for three ecoregions of Afghanistan: lowland (900–1300 m), upland (1300–2400 m) and mountains (above 2400 m)	Bhuchar et al. (2016)
	Pit composting suitable for all three ecoregions: pit composting, conservation tillage.	Virgo et al. (2006)
	Low and upland regions: vermicomposting.	
	Lower slopes: water harvesting bunds.	
	Simple contour plowing would reduce erosion and retain moisture.	
Bangladesh	Plantation along contours, mulching, zero tillage with surface mulching	Uddin and Saheed (2016)
Bhutan	*In-situ* management practices include multiple cropping, cover crops, intercropping, strip cropping, mulching using crop residue and organic matter, terracing and planting of fodder trees and grasses, terraced wetland with bunds for rice cultivation, stone bunds along contour lines	Katwal (2010)
India	Arid regions (rainfall <500 mm): contour farming / cultivation, conservation furrows, mulching, deep plowing and inter-row water conservation systems.	NRAA (2009)
	Semiarid regions (500–1000 mm): conservation furrows, contour farming, compartmental bunding, runoff strips, tied ridges, graded ridging, mulching, live hedges, ridge and furrow system, off-season tillage on conserved soil moisture, broad beds and furrows, graded border strips.	
	Subhumid regions (>1000 mm): field bunds, graded bunds, vegetative bunds, level/graded terraces, contour trenches, inter-plot water harvesting, raised bed and sunken system.	
Maldives	Traditional farming systems are based on shifting agriculture, polycultural home gardens, agroforestry and taro pits. Improved agricultural practices include crop rotation, intercropping, composting, irrigation.	FAO (2016)
Nepal	Contour bunding	Tamang (2016)
	Crop rotation using legumes, traditional plowing	Pokhrel and Pokhrel (2013)
	Zero/minimum tillage for rice–wheat system	Hobbs and Giri (1997)
	Application of farmyard manure and rice stubble left in field	
Pakistan	Terracing, contouring, strip cropping, construction of soil and water conservation structures, contour planting, hedgerows, living fences and barriers.	Baig et al. (2013)
	Improved tillage practices include conservation tillage, mulches, addition of crop residues such as wheat straw, cover crops including nitrogen-fixing legumes	
	Zero tillage for rice–wheat and cotton–wheat systems, gully land management for degraded lands, conservation tillage, stubble mulch, trash farming, strip tillage.	Zia et al. (2004)
	On steep sloping lands, stone bench terraces in cultivated areas, and installation of water disposal system with grass waterways and water drop structures.	

(continued)

Table 3 (continued)

Country	*In-situ* moisture conservation practices	References
Sri Lanka	Conservation bunds and drains, vegetative measures such as glyricidia, vetiver or citronella hedges to control soil erosion and restore degraded lands	Dharmasena (2003)
	Orchard or eyebrow terraces and sand pits (modified terrace system) for shallow soils with steep slopes	Wijayaratna and Weerakoon (1996)
	Graded bunds, drains and stabilization of bunds in undulated or rolling dry zones; bund stabilization by vegetative means (Vetiver grass).	Somasiri et al. (1990)
	Application of organic matter increases the waterholding capacity of soil.	
	Shade management using glyricida (hedge row cultivation) and roof water harvesting.	

Under the Technology Demonstration Component (TDC) of the National Innovations on Climate Resilient Agriculture (NICRA), farm ponds are considered a key intervention to cope with climate variability (Fig. 1). Various cropping system modules have been developed using harvested water. Most farmers opted to cultivate vegetables with harvested water in a ratio of 1:10 (command to catchment area) with sustained profits (Prasad et al. 2014).

Watershed management could be a key strategy to unlock rainfed production potential. An integrated watershed management approach shows promise in the sustainable development of land and water resources. Watershed development projects are designed to harmonize the use of water, soil, forest and pasture resources while raising agricultural productivity by conserving moisture in the soil and increasing irrigation through tank- and aquifer-based water harvesting. Of the rainfed cropped area in India, it is estimated that 15 Mha is in arid regions with less than 500 mm of annual rainfall, 15 Mha is in the 500–700 mm rainfall zone, 42 Mha is in the 750–1100 mm rainfall zone and 20 Mha receives more than 1150 mm. A single supplemental irrigation of 100 mm in a rainfed area of 27.5 Mha increased annual production of food grains by about 9.3 Mt (Sharma et al. 2010). Significant production improvements could be realized in cotton (*Gossypium* spp.), sesame *(Sesamum indicum* L.), groundnut (*Arachis hypogaea* L.), soybean (*Glycine max* L. Merrill) and chickpea (*Cicer arietinum* L.) (Sharma 2011). On a regional basis, collecting small amounts of runoff using macro-catchments during the rainy season for supplementary irrigation can improve agricultural production in rainfed areas (Molden 2007) by more than 50 % (Sharma 2010). The theory of water pricing, and improved water use efficiency, are also better technical solutions. In many northern states of India like Uttar Pradesh, Punjab and Haryana, the conjunctive use of surface water and groundwater has been practiced using canal systems and tube or dug wells to increase crop yields and the efficiency of the water system (Frenken 2011). In recent years, water-saving technologies like sprinkler and drip irrigation have been used.

In Pakistan, 90 % of the country's food grain production comes from rice (*Oryza sativa* L.) and wheat (*Triticum* spp.). Key resource conservation technologies

Fig. 1 HDPE-lined farm pond at Namakkal, Tamil Nadu

include zero tillage, direct seeding, parachute transplanting, bed planting, laser land leveling and crop residue management (PARC-RWC 2003; Ahmad et al. 2007). Resource conservation technologies (RCTs) in Pakistan have improved the field irrigation efficiency (Gupta et al. 2002; Humphreys et al. 2005) and saved water. However, the water-saving impacts of RCTs beyond the field level are not well documented. It is possible that real water savings are much lower than assumed when field-level calculations are extrapolated to broader scales (Ahmad et al. 2007).

In Sri Lanka, policymakers have focused on alleviating seasonal water scarcity in the dry zone using large-scale storage tanks and inter-basin transfers (Ariyabandu 2008). An unlined rooftop rainwater harvesting (RRWH) pond concept is practiced by the poorest rainfed farmers living in the more vulnerable and marginal areas of Sri Lanka. Farmers realized that household food security increased using RRWH and that there was an indirect impact on the local microenvironment around the system, particularly the survival of vegetation during dry spells. In general, a sensitivity analysis under various scenarios indicated that RRWH pond investment was economically viable under the given circumstances (Bandara and Aheeyar 2010).

In Nepal, community management of watersheds and water systems has been popular (Pretty 2003). Many irrigation systems use surface irrigation methods such as basins and furrows, and limited areas in the hills and mountains use sprinkler irrigation. In Nepal, there are public irrigation systems and farmer-managed irrigation systems (FMIS); in 2008, 70 % of the irrigated area was under FMIS. In non-FMIS areas, some systems are managed by Water User Associations (WUAs), while others are jointly managed by the government and WUAs. Farmer- and community-managed systems are more efficiently managed than government-managed systems (Frenken 2011); however, the government plays a crucial role in research and development, extension services, and other regulatory fiscal and non-fiscal mechanisms. Government assistance in the rehabilitation and repair of irrigation systems is essential to sustain farmer-managed systems (MOIR 2005).

2.1.2 Soils

As with the climates, the soils of South Asia are equally diverse (Lal 2006). The predominant soils are Alfisols and Vertisols in the semiarid regions, Inceptisols and Entisols in the alluvial plains of the main river systems, and Aridisols in the arid regions or desert climates. In terms of land area, Entisols (169 Mha) > Aridisols (122 Mha) > Inceptisols (95 Mha) > Alfisols (79 Mha) > Vertisols (60 Mha) > Ultisols (42 Mha), Mollisols (19 Mha), and others. Challenges persist in the alleviation of soil physical constraints such as crusting, compaction and hard setting which lead to high runoff, erosion, frequent drought stress and low soil fertility.

Soil organic carbon, which is the seat of major soil processes and functions, is <5 g/kg in rainfed soils of India, while the desired level is 11 g/kg. Maintaining or improving soil organic matter is a prerequisite for ensuring soil quality, productivity and sustainability. Srinivasarao et al. (2013b) summarized the results of several long-term manure experiments conducted under rainfed conditions. In a groundnut-based system, the highest soil organic carbon (SOC) stock (t/ha) was found with applicastion of 50 % recommended dose of fertilizer (RDF) + 4 t/ha groundnut shells (47.2) followed by 100 % RDF (36.2) and the control (32.2). In finger millet (*Eleusine coracana* L.) monocropping, the profile SOC stock (t/ha) was the highest in the Farmyard manure (FYM) 10 t/ha + 100 % NPK (85.7) followed by 100 % NPK and control (63.5) treatments. In the groundnut–finger millet rotation, the SOC stock (t/ha) was the highest in the FYM + 100 % NPK (73.0) and lowest in control (51.7) treatments. Sorghum (*Sorghum bicolor* L. Moench) had the highest SOC stock (t/ha) in the 25 kg N (crop residue) + 25 kg N Subabul (*Leucaena*) (68.5) followed by 25 kg N (crop residue) + 25 kg N (urea) N (65.8), and the control (49.0). In the pearl millet (*Pennisetum glaucum* (L.) R.Br.) system, the profile SOC stock (t/ha) was the highest in 50 % recommended dose of nitrogen (RDN) (fertilizer) + 50 % RDN (FYM) (25.5) followed by 50 % RDN (FYM) (23.4) and the control (17.9). On-farm generation of organic matter with appropriate policy support needs to be promoted to maintain soil health and crop productivity (Srinivasarao et al. 2014b).

Soils in most parts of India are not only deficient in NPK but also in secondary nutrients (S, Ca). Magnesium (Mg) deficiency is also prevalent in many rainfed areas. A study done by CRIDA across diverse agroecological regions highlighted the extent of Mg levels in major Indian soil types and recommended further attention on Mg nutrition in current intensive agriculture (Srinivasarao et al. 2015a) and micronutrients (e.g., B, Zn, Cu, Fe, Mn). Balanced nutrient application to crops based on the nutrient requirement to produce a unit quantity of yield and the native nutrient supplying capacity of the soil, improves crop yields while minimizing nutrient losses and cultivation costs. In Andhra Pradesh and Telangana (states of India), eight districts addressed nutrient deficiencies within farmers' fields using balanced nutrition, with promising results. For example, in Warangal, balanced nutrition significantly improved cotton yields in many farmers' fields by 10–25 %, reaching up to 1.6 t/ha (Srinivasarao 2011).

Community tanks in peninsular India collect and store rainwater, as well as the nutrient-rich topsoil, eroded from catchment areas. Analysis of tank silt collected from several tanks in 100 districts identified the potential of tank silt for supplying organic carbon and several nutrients to improve soil health. Tank silt application in degraded Alfisols of Telangana improved yields in maize (*Zea mays* L.) and castor (*Ricinus communis* L.) as well as soil physical, chemical and biological properties.

Sodic soils occupy 3.6 Mha in India (Bhargava 1989). The establishment of permanent tree cover with suitable salt-tolerant species is an important option for the reclamation of such soils (Gill and Abrol 1991; Garg and Jain 1992). Of the fuelwood species evaluated, mesquite (*Prosopis juliflora* Sw. DC.) is reportedly the most widely-adapted to alkaline soils and produces the most biomass (Singh 1989). Egyptian pea (*Sesbania sesban* L. Merr.) and salt cedar (*Tamarix dioica* Roxb. ex Roth.) also exhibit good adaptability. Growing salt-tolerant woody species like *Prosopis juliflora*, babul (*Acacia nilotica* L. Del.), casuarina (*Casuarina equisetifolia* J. R. & G. Forst.), athel tamarisk (*Tamarix articulata* Vahl.) and sprangle top (*Leptochloa fusca* L. Kunth.) improve the quality of salt-affected soils (Singh et al. 1994). Some grasses are also useful for reclaiming sodic soils (Gupta et al. 1990), especially when grown in association with trees. Salt-tolerant trees and forage species are valuable for reclaiming sodic soils in Pakistan (Qadir et al. 1996). In addition to improving structural properties, trees affect the salt balance by increasing the depth of the water table leading to a net downward leaching. Similarly, bioremediation of sodic soils by the silvopastoral system is economically and ecologically feasible (Singh et al. 1998; Kaur et al. 2002).

Soil erosion is a major problem on the 82 Mha affected by water erosion and 59 Mha affected by wind erosion in South Asia. Afforestation, reforestation, reducing deforestation, and controlling grazing on steep land prone to erosion are needed to reverse the degradation trends of erosion. Establishing forest plantations of *Eucalyptus* spp., subabul (*Leucaena leucocephala* Lam. de Wit), casuarina (*Casuarina equisetifolia* J. R. & G. Forst.), *Prosopis* spp., teak (*Tectona grandis* L. f.), sissoo (*Dalbergia sissoo* Roxb. ex DC.), etc has numerous benefits including reductions in soil erosion and nonpoint-source pollution. Other important biological measures include establishing contour hedges of perennial grasses (e.g., vetiver or *Vetiveria zizanioides* L.) Nash ex Small; NRC 1993; Prasad et al. 2003) and shrubs, buffer strips, riparian zones (Ranada et al. 1997), and converting marginal lands to restorative uses through afforestation and nature reserves.

2.2 *Farming Systems*

The term 'farming system' refers to a particular arrangement of farming enterprises that is managed in response to physical, biological and socioeconomic environments in accordance with farmers' goals, preferences and resources (Shaner et al.

1982). "The household, its resources and the resource flows and interactions at the individual farm levels are together referred to as a farm system" (FAO 2001). A common characteristic of integrated farming systems is that they invariably have a combination of crop and livestock enterprises and, in some cases, may include combinations of poultry, agroforestry, horticulture, apiary, etc. Further, there are synergies and complementarities between different enterprises that form the basis of the concept of Integrated farming system (IFS) (Lightfoot and Minnick 1991; Jitsanguan 2001; Radhammani et al. 2003). Integration usually occurs when outputs (usually by-products) of one enterprise are used as inputs by another within the context of the farming system. The difference between mixed farming and integrated farming is that enterprises in the integrated farming system are mutually supportive and depend on each other (Csavas 1992). The synergy between enterprises increases with on-farm diversity and is fundamental to the IFS concept. Diversification of farming activities improves the utilization of labor, reduces unemployment in areas where there is a surplus of underutilized labor, and provides a source of living for those households that operate their farms as a full-time occupation.

In South Asia, annual crops, perennial tree crops, ruminants and non-ruminants are maintained in integrated farming systems. However, on the small farms, ruminants are more widely reared than non-ruminants. More than 90 % of the total population of large and small ruminants is kept on mixed farms in the South Asian region (Devendra 1983). In rainfed annual cropping systems, ruminants graze native grasses and weeds on roadside verges, on common property resources, or in stubble after the crop harvest. In India, in regions with 500–700 mm of rainfall, farming systems should be based on livestock with the promotion of low-water-requiring grasses, trees and bushes to meet fodder, fuel and timber requirements of the farmers (Vittal et al. 2003). In 700–1100 mm rainfall regions, crop, horticulture and livestock-based farming systems can be adopted depending on the soil type and the marketability factors. Runoff harvesting is a major component in this region in the watershed-based farming system. In areas where the rainfall is more than 1100 mm, the IFS module integrating paddy with fisheries is ideal (Fig. 2). There are several modules of rainfed rice cultivation, along with fisheries, in the medium to lowlands of rainfed rice-growing regions in the eastern states of India.

There are few examples of improved pastures being used in these systems; in Sri Lanka, integrated perennial-tree-crop–animal systems include coconut/fruits/cattle/goats, where the ruminants graze the understorey of native vegetation or leguminous cover crops. However, these systems can evolve into more intensive production systems depending on the availability of feed, markets, and the development of cooperative movements. This is evident in South Asia, e.g. Bangladesh, where root crops are produced, and pig production is based on cassava and sweet potato (Devendra and Thomas 2002). Rearing livestock such as small and large ruminants at home by transitory vulnerable-category communities for different livestock products has been a traditional activity in rainfed Pothwar areas of Pakistan; also an important means to fulfill the livelihood requirements of people below the poverty line (Zahra et al. 2014). Range-based small ruminant production is the major activity in the area coupled with rainfed agriculture. Sheep and goats are the main live-

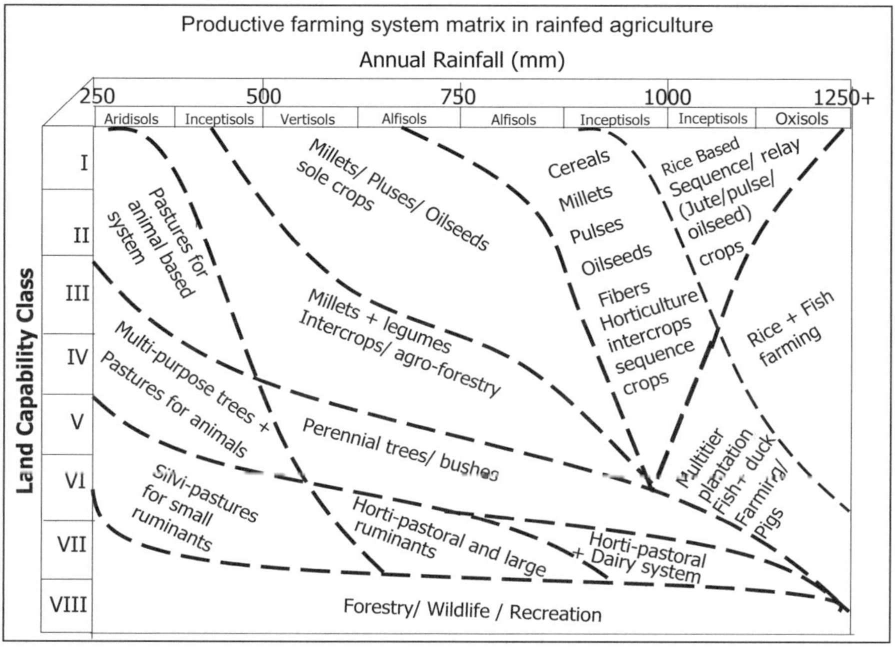

Fig. 2 Productive farming system matrix in rainfed agriculture in India

stock of the province. In Baluchistan approximately 87 % of the people directly or indirectly drive their livelihood from livestock rearing (Heymell 1989).

Farmers in the drought-prone areas of the Gulmi District of Nepal grow drought-resistant crops like elephant foot yam (*Amorphophallus paeoniifolius (*Dennst.) Nicolson), taro (*Colocasia spp* (L.) Schott), cassava (*Manihot esculenta* Crantz), winter bean (*Vicia faba* L.), air potato (*Dioscorea bulbifera* L.), cush-cush yam (*Dioscorea trifida* L.f.) and brinjal (*Solanum melongena* L.). Similarly, in the marginal areas in the dryland Gulmi District, farmers grow cassava, winter bean, elephant foot yam, taro, sugarcane (*Saccharum officinarum* L.), turmeric (*Curcuma longa* L.), ivy gourd (*Coccinia grandis* (L.) Voigt) and legumes (Regmi et al. 2009).

Based on an ex-post facto study involving 120 farmers, Desai et al. (2009) identified the most economical rainfed farming system models for Andhra Pradesh, Karnataka and Tamil Nadu states in India (Table 4). They observed the prevalence of different farming systems across farm sizes and states. The major components of the farming systems were cereal crops, oilseeds, vegetables, fruits, bovines and caprines.

The inclusion of perennial components like trees and grasses in dryland farming systems imparts stability to farming by reducing the effect of yearly variations in rainfall on these components, protecting the crops from water and wind erosion, and improving soil fertility. The returns in arid/dryland ecosystems are much higher when trees are associated with crops/grasses in silvi-pasture, agri-horti and agroforestry systems. In addition to the economic benefits, tree-based systems improve soil

Table 4 Farming systems providing higher economic returns in different rainfed regions of India

	Farm size			
State	Marginal	Small	Medium	Large
Andhra Pradesh	Maize–paddy–caprine (Rs. 14,334)	Castor–maize–bovine (Rs. 18,625)	Castor–paddy–bovine (Rs. 28,581)	Maize–paddy–pulses (Rs. 18,886)
Karnataka	Pulses–bovine (Rs. 13,180)	Bajra–pulses–groundnut–bovine (Rs. 17,690)	Sorghum–pulses–sugarcane–bovine (Rs. 16,280)	Pulses–banana (*Musa* spp. L.)–sugarcane–bovine (Rs. 1,74,105)
Tamil Nadu	Paddy–sorghum–onion (*Allium cepa* L.)–bovine–poultry (Rs. 30,082)	Groundnut–sesame–onion–bovine–caprine (Rs. 19,490)	Paddy–sesame–vegetables–bovine–caprine (Rs. 23,500)	Paddy–sesame–groundnut–bovine–caprine (Rs. 29,058)

Source: Adapted from Desai et al. (2009)

fertility through the build-up of organic matter and nutrients in the soil. Studies have found that multiple-use species such as bamboo (*Bambusa nutans* Wall.ex Munro) have the potential to help bind soil nutrients during the restoration of abandoned shifting agricultural lands (*jhum* fallows) in northeastern India (Arunachalam and Arunachalam 2002). Shelterbelts or windbreaks with *Eucalyptus* spp. in farm bunds or the borders of farm land can save crops from the desiccating effects of the wind in dry and sandy areas of dryland Mastung of Balochistan (Mohammad and Ehrenreich 1993). Planting shelter belts and windbreaks can enhance agricultural crop yields in these areas (Mohammad and Ehrenreich 1993). In this area, Rehman (1978) reported increased wheat yields of 8, 15 and 14 % with one row of French Tamarisk (*Tamarix gallica* L.), two rows of French Tamarisk + Giant Reed (*Arundo donax* L.) and three row shelterbelt of French Tamarisk + Giant Reed + Phog (*Colligonum polygonoides* L.), respectively. Soil moisture in this study in the 0–150 mm layer in the plots protected by the belts was consistently 26 % higher than in the unprotected plots.

Perennial grass components, besides imparting stability to farming systems in dryland areas, also act as vegetative filter strips to prevent wind and water erosion. Some important tree species which are compatible with the grass component for silvi-pasture are siris (*Albizia lebbeck* L. Benth.), desert teak (*Tecomella undulate* L.), mopane (*Colophospermum mopane* J. Kirk ex Benth. J. Leonard), gum Arabic tree (*Acacia senegal* L. Willd), umbrella thorn (*Acacia tortilis* Forssk.), jujube (*Zizyphus nummularia* Burm. F. Wight & Arn.) and wild jujube (*Zizyphus rotundifolia* Mill.). Of the pasture legumes, blue pea (*Clitorea ternatea* L.) and Indian bean (*Lablab purpureus* L.) are compatibile with sewan grass (*Lasiurus sindicus* Henr.) and buffel grass (*Cenchrus ciliaris* Linn.) (Samra 2004). Promising multipurpose trees, fruits, crops and grasses for various agroforestry systems in dryland areas of arid and semiarid regions in India are summarized in Table 5.

Table 5 Promising plant species for dryland systems in India

Country	Zone	System	Promising species		
			Forestry plants	Fruit plants	Crops/grasses/shrubs
India	Arid	Agri-silviculture[a]	Khejri (*Prosopis cineraria* L. Druce.), Desert Teak, Anjan (*Hardiwickia binata* Roxb.) and Wild Jujube	Indian Plum (*Ziziphus mauritiana* Lamk.) and Date Palm (*Phoenix dactylifera* L.)	Crops: Mung bean (*Vigna radiata*), Moth bean (*Vigna aconitifolia* Jacq) Marechal), Cowpea (*Vigna unguiculata* L. Walp), Cluster bean (*Cyamopsis tetragonoloba* L. Taub), Pearl millet and Sesame
		Silvi-pasture[a]	Mopane, Jujube and Anjan	Caper (*Capparis decidua* Forssk. Edgew), Indian Plum and Khejri	Grasses: Buffel grass,irdwood grass (*Cenchrus setigerus* Vahl.) Sewan grass and Marvel grass (*Dicanthium annulatum* Forsk. Stapf.)
		Shelterbelts[a]	Umbrella thorn, Kassod tree (*Cassia siamea* Lamarck Irwin et Barneby), Mesquite, Siris and Neem (*Azadirachta indica* A. Juss.)	–	–
	Semiarid	Agri-silviculture[a]	Babul, Tree of Heaven (*Ailanthus excelsa* Roxb.), Sissoo, Khejri, Cottonwood (*Populus deltoides* Bartr.) and Anjan	Indian plum Mango (*Mangifera indica* L.), Guava, Citrus, Amla and Bael (*Aegle marmelos* L. Correa)	Crops: Pearl millet, Sorghum, Cluster bean, Pigeon pea, Cowpea, Mung bean, Sesame and Groundnut
		Silvi-pasture[a]	Babul, Sissoo, Khejri and White Bark Acacia (*Acacia leucophloea* Roxb. Willd.)	–	Seasonal grasses: Rat's tail grass (*Sehima nervosum* Rottl. Stapf.), Blue Panic grass (*Panicum antidotale* Retz.) and Buffel grass
		Farm boundary[a]	Babul, *Eucalyptus* spp. Cottonwood, Butter tree (*Madhuca latifolia* Roxb.) and Sissoo	–	–

(continued)

Table 5 (continued)

Country	Zone	System	Promising species		
			Forestry plants	Fruit plants	Crops/grasses/shrubs
Pakistan	Arid	Silvi-pasture[b,c]	Khejri, Toothbrush tree (*Salvadora oleoides* Decne.), Tamarisk (*Tamarix aphylla* L. Karsten.), Desert teak Babul, gum Arabic tree, Baonli (*Acacia jacquemontii*). Mulga (*Acacia aneura* F. Muell. ex. Benth.), *Acacia victoriae* Benth. sens lat. and Meswak (*Salvadora persica* L.)	–	Shrubs: Phog, Giant milkweed (*Calotropis procera* Ait. Ait. f.), Khereit (*Salsola foetida* Del.ex Spreng.) and *Haloxylon* spp. Grasses: Lemongrass (*Cymbopogon* spp.) and Sabth grass (*Pennisetum divisum*)
	Arid to semiarid	Silvi-pasture[b,c]	Babul, Khejri, Toothbrush tree, Tamarisk, Indian plum, Jujube, Indian Mulberry (*Morus alba* L.), Siris, Phulai (*Acacia modesta*), *Acacia victoriae* Benth. sens lat., Mulga, Desert teak and Baonli		Shrubs: Phog, Salt cedar, Giant milkweed and Jujube Grasses: Wiregrass (*Eleusine compressa*), Sewan grass, Sugarcane and Blue Panic grass
		Rainfed farm boundary[d]	*Populus euramericana* and River Red Gum (*Eucalyptus camaldulensis* Dehnh.)		
		Agri-silviculture[e]	*Populus euramericana*, Cottonwood, Varnish tree (*Ailanthus altisima* Mill. Swingle) and River red gum		Crops: Wheat, Maize, Sugarcane and different vegetables
		Agrosilvopastoral system[f]	Phulai, Mulberry (*Morus nigra* L.) and Sissoo	Jujube	Shrubs: Egyptian pea, Subabul and *Acacia* spp. Grasses: Native species

Sri Lanka	Low country dry zone	Live fences[g]	Hill mango (*Commiphora caudata* Wight & Arn.), Portia tree (*Thespesia populnea* L. Soland. ex Correa), White gul mohur (*Delonix elata* L. Gamble), Gliricidia, Aal (*Morinda coreia* Buch. Ham), Drumstick (*Moringa oleifera* Lam.), Subabul, Teak, Neem, Indian balm of gilead (*Commiphora berryi* Arn.) and Indian Labernum (*Erythrina indica* Lam.)	Custard apple (*Annona squamosa* L.), Betelnut palm (*Areca catechu* L.), Jack fruit (*Artocarpus heterophyllus* Lam.), Palmyra (*Borassus flabellifer* L.), Papaya (*Carica papaya* L.), Coconut (*Cocos nucifera* L.), Indian wood-apple (*Limonia acidissima* L.), Mango, Cassava, Wild date palm (*Phoenix sylvestris* Roxb.), Guava, Indian blackberry (*Syzygium cumini* L.) and Tamarind tree (*Tamarindus indica* L.)	Shrubs: China rose (*Hibiscus rosa-sinensis* L.), Gunja (*Lannea grandis* (Dennst.) Engl.) Wild sage (*Lantana camara* L.), Castor (*Ricinus communis* L.), Crepe jasmine (*Tabernaemontana divaricata* R.Br. Ex Roem. & Schult.) and Yellow oleander (*Thevetia peruviana* (Pers). K Shum)
		Home gardens[h]	Neem, Cashew (*Anacardium occidentale* L.) and Coconut		
Bang-ladesh	Dryland	Traditional home gardens[i]	Neem, Big leaf mahogany (*Swietenia macrophylla* King.), River red gum and Teak	*Ficus* spp., Monkey jack (*Artocarpus lakoocha* Roxb.), Jack fruit, Drumstick, Indian blackberry, Guava, Wild date palm, Palmyra, Coconut, Pomegranate, Indian plum. Kadam (*Anthocephclus cadamba* Roxb. Miq.) and Bael	

Source: [a]Venkateswarlu (2004); [b]Rahim and Hasnain (2010); [c]Gintings and Lait (1994; [d]Ali et al. (2011); [e]Subhan (1990); [f]Amin (1987); [g]Jayavanan et al. (2014); [h]Mattsson et al. (2015); [i]Mustafa and Haruni (2002)

The hilly terrain of the North-Eastern Hill (NEH) region of India is suitable for sustainable multi-enterprise systems. As an alternative to shifting cultivation in NEH, Satapathy (2003) suggested using watershed-based farming systems that involve appropriate soil conservation measures, mixed land use of agri-horti-silvi-pastoral systems, a subsidiary source of income through livestock rearing, and the creation of water harvesting and silt retention structures at lower reaches. Economic analysis of different micro-watershed-based farming systems namely dairy farming, agro-pastoral and agri-horti-silvi-pastoral systems have shown the economic viability of these systems as an alternative to shifting cultivation.

Coastal areas offer tremendous opportunities if proper strategies are adopted using well-focused action plans under an integrated farming systems approach so that agriculture will be more productive, profitable, sustainable, competitive and ecofriendly. In these regions, land holdings are small and fragmented, operational resources are scarce, and rice continues to be the major crop. The rice-based farming system is the single most important source of income and employment for most of the people in this tract and is a viable option for meeting their livelihood requirements. Rice-based farming systems such as rice + fish, rice + duck, rice + fish + duck, rice + fish + *Azolla,* rice + fish + poultry/duck + vegetables + fruits + agroforestry are practiced particularly in the rainfed lowlands of coastal areas (Nanda and Garnayak 2010).

2.3 Institutional Mechanisms

Institutions play a major role in agricultural development along with other resources like technology, capital and enterprise. In small producer-dominated situations like Indian agriculture, the role of institutions becomes more crucial as there are structural- and enterprise-specific constraints like high transaction costs, lack of market integration, and interlocking of factor and output markets which only institutions and organizations can tackle effectively. Institutions help small farmers by reducing transaction costs, managing or reducing risk, building social capital, enabling collective action, and/or readdressing missing markets (Singh 2013a, b). Institutional innovation occurs when there is a "change of policies, standards, regulations, processes, agreements, models, ways of organizing, institutional practices, or relationships with other organizations" so as to "create a more dynamic environment that encourages improvements in the performance of an institutions or system to make it more interactive and competitive" (IICA 2014). Institutional innovations needed for agricultural transformation in South Asia are distinct due in part to the unique nature of the sector. Most farmers are resource-poor, face diverse challenges particularly under dryland ecosystems, and have specific needs that must be addressed.

Timely access to farm machinery for sowing, harvesting, etc. is a major component of any adaptation strategy to deal with climatic variability. The sowing window in rainfed areas is mostly very short and access to farm machinery is poor for smallholder farmers. As a result, many farmers are not able to sow the crop in a timely fashion and incur significant yield losses. In India, an innovative institutional arrangement in the form of a Custom Hiring Center (CHC) for farm machinery has been created in 100 selected villages in 100 districts. Farm implements and machines available to hire include ferti-seed drills, zero-till drills, power weeders, harvesters, threshers, power tillers, sprayers, rotavators for residue incorporation, sprinklers, chaff cutting machines, and weighing machines. Some of the implements are common across districts, but there are district-specific items at each center depending on local needs. The village committee decides the price to hire each farm implement on a consensus basis which is displayed at the CHC. The income generated by the CHCs goes to a common account. Of the 100 CHCs, most are performing well (Srinivasarao et al. 2013b).

Producer companies are another legal institutional innovation providing more business-like entities to primary producers for organizing and conducting business without any bureaucratic/government control or interference (Singh 2008). In India and many other developing countries, traditional cooperatives were mostly organized under the cooperative structure, e.g. the State Cooperative Societies Acts in India. However, for several reasons, the cooperatives lost their vibrancy and became known for their poor efficiency and loss-making ways. In light of these experiences of traditional cooperatives in India, it was felt that more freedom should be given to cooperatives to operate as business entities in a competitive market. This led to the amendment of the Companies Act, 1956 in 2003, which provided provision for incorporation of producer companies. At present, there are more than 130 producer companies in India, promoting agencies, crops and products, and types of primary producers.

3 Challenges in Dryland Systems

The climate in South Asia varies widely, from the warm humid and subhumid tropical climates predominating in the south-eastern and southern-most regions to the semiarid and arid subtropical regions of western India and Pakistan. These variations in climate are largely determined by latitude and altitude gradients. The variability in rainfall is greatest in drier zones, particularly in Pakistan and western parts of southern India, making them among the high-risk areas for rainfed agriculture. Most of the Himalayan mountain region, which covers part of Nepal and northern India, is climatically unsuited to agriculture (Pender 2008). The main constraints to rainfed agriculture include frequent drought, soil degradation, low SOC content, multiple nutrient deficiencies, low external inputs, low investment capacity, and poor market linkages (Srinivasarao et al. 2015b). Furthermore, agriculture options available in these regions depend on the prevailing socioeconomic conditions. As developing and underdeveloped nations, many hiccups underly such as declining natural resource base, climate change, and food and nutritional insecurity etc . Any improvement in or adaptation to these hiccups, may bring a change.

3.1 Declining Natural Resource Base

The scarcity of soil and water resources, exacerbated by soil degradation and environmental pollution, are the principal challenges to enhancing dryland agriculture production in South Asia. High population densities and high growth rates have accentuated the demands on soil, water, vegetation, and other natural resources (Lal 2006). The per capita renewable freshwater resources are likely to be a severe constraint by 2050 in Afghanistan, Iran, India, and Pakistan (Table 6). The shortfall in water availability in Pakistan was 40 million acre feet (MAF) in 2000, which was projected to 150 MAF by 2025 (Afzal 2001). Consequently, some 30 million Pakistanis face food insecurity due to water shortages (Afzal 2001). Furthermore, water pollution is a serious issue since industrial and urban affluents are often discharged into rivers. Surface and groundwater resources are also prone to nonpoint-source pollution by runoff from agricultural lands, especially in densely-populated regions of India (Pachauri and Sridharan 1999).

About 70 % of the Indian population lives in rural areas, and most depend on rainfed agriculture and fragile forests for their livelihoods (World Bank 2011). The mean annual rainfall varies from less than 100 mm in the western part of Rajasthan to around 11,700 mm in Chirapunji in Meghalaya (Kumar 2011). The per capita annual water availability in India decreased from 5177 m^3 in 1951 to 1654 m^3 in 2007 (MOWR 2008). According to the Central Ground Water Board, 15 % of districts are overexploited and growing at a rate of 5.5 % per year (Rejani et al. 2015a). The groundwater level in the 16 states of India has dropped by more than 4 m from 1981–2000, with the most substantial decline in north-western India. In many parts of Gujarat and Rajasthan, the groundwater level declined by more than 16 m (Sheetal Sekhri 2012).

Similarly, there is significant spatial variation in rainfall in Pakistan, with most of the country receiving very low rainfall. The whole of Sindh, parts of Baluchistan, most of Punjab, and central parts of northern areas receive less than 250 mm of rainfall while northern Sindh, southern Punjab, and north-western Baluchistan receive less than 125 mm of rainfall (Kumar 2011). Soil salinity, waterlogging and

Table 6 Per capita renewable fresh water resources (m^3/person) in South Asia

Country	1955	1990	1995	2025	2050
Afghanistan	5137	3020	2543	1091	815
Bangladesh	56,411	–	19936	14153	10,803
Bhutan	12,9428	–	53,672	26,056	18,326
India	5227	2464	2244	1496	1360
Iran	6203	2025	1719	816	690
Nepal	19,596	8686	7923	4244	3170
Pakistan	10,590	3962	3435	1803	1310
Sri Lanka	4930	2498	2410	1738	1600

Sources: Engelman and LeRoy (1993), Gardner-Outlaw and Engelman (1997) and Lal (2006)

soil erosion are the prevailing land degradation issues in Pakistan (Sharda 2011). Irrigation systems in Pakistan are now underperforming, and groundwater use is increasing, leading to soil salinity. The major challenge is to improve productivity levels by managing water resources in an integrated way to benefit the people and the environment (IWMI 2015).

Sri Lanka has per capita water resources availability of 2400 m^3 with an average annual rainfall of 1712 mm (Ariyabandu 2008). Rainfall ranges from less than 1250 mm in the northwest and southeast to more than 5080 mm in the southwest. There are two rainfall zones in Sri Lanka namely, dry zone and wet zone (Kumar 2011); some districts of Sri Lanka experience prolonged dry periods due to the high temporal and spatial variability of rainfall. Sri Lanka has the second highest annual variability of rainfall of 22 Asian and Pacific countries. Even though the per capita availability of water is higher than many other countries, the scarcity of water is starting to threaten Sri Lanka's development (Ariyabandu 2008). Demand for water has increased and, at the same time, its availability has been affected by prolonged dry spells and droughts, and the pollution of sources. The Ministry of Environment and Natural Resources (2002) in Sri Lanka states that 85 % of developed water is used for irrigation, 6 % is for domestic use and 5 % is for industry. The number of wells and irrigation pumps has increased resulting in the indiscriminate withdrawal of groundwater from aquifers, which seriously threatens the sustainability of groundwater. Nearly 30 % of the population lives in coastal areas and depends on the shallow lens of freshwater saddled on saline waters for their livelihood, and the precious and limited water resources are threatened by over extraction, pollution, saltwater intrusion and rising sea levels due to climate change. A systematic research program is urgently needed to guide the policy, legislation and institutions that include adaptation to climate change (IWMI 2010).

Rainfall in Nepal varies spatially from more than 1300 mm in Kathmandu to more than 6000 mm along the southern slopes of the Annapurna Range in central Nepal to less than the 250 mm in the north central part near the Tibetan plateau. Rainfall amounts ranging from 1500 mm to 2500 mm are predominant over most parts of the country. High mountains cover nearly 35 % of the geographical area, middle mountains cover nearly 42 %, and the Tarai region covers nearly 23 % (Kumar 2011). Major problems include soil erosion, deterioration of soil quality, waterlogging along canal systems, quarrying which denudes hill slopes, excessive erosion, and the accumulation of debris in rivers (Sharda 2011). The government has adopted the principles of Integrated Water Resources Management (IWRM) and River Basin Management for planning and managing water resources (IWMI 2015). Global warming is expected to cause changes in the timing of monsoonal rains and the release of snow and glacier melt. In Nepal, nearly 30 million people are vulnerable to recurrent floods, landslides and droughts, and more than 80 % of the population depends on agriculture, which consumes 99 % of water withdrawn, but only 24 % of arable land is irrigated. The poorly-managed watersheds increase the stress and decrease the sustainability of water resources. The drying up of water sources, reduced surface and groundwater flows, and pollution are adversely affecting the domestic water supply demands for the increasing population.

Myanmar has abundant water resources, but the problems in this country are related to the uneven spatial and temporal distribution of rainfall. Soil and river bank erosion are the two land degradation issues in Myanmar (Sharda 2011). About 992.1 km^3 of surface water is produced internally per year, of which 453.7 km^3 is groundwater and about 443 km^3 is the base flow of rivers and surface runoff. Rest of the water is annual inflow from other countries About 89 % of the surface water is withdrawn for agriculture. Due to the non-uniform rainfall distribution, the need for irrigation is highest in the central dry zone, but there is concern about drainage and flood protection (Frenken 2011). Despite the availability of groundwater aquifers in Myanmar, their exploitation has been limited to municipal water supplies and intensive irrigation of vegetables and other high-value crops from hand-dug wells.

The mean annual precipitation in Afghanistan is less than 300 mm and ranges from 50 mm in the southwest to 700 mm in the region of Salang. Towards the eastern part of the country, the total annual precipitation is about 100 mm (Kumar 2011). Almost 90 % of irrigation water in Afghanistan comes from karez, springs, wells and rivers, etc. Major water losses are related to the low efficiency of irrigation systems and mismanagement of water distribution (Palau 2012). Areas with an unreliable river discharge and the largest drought deficits are in need of drought mitigation measures on river flows (Eriyagama et al. 2009). There is widespread natural resource degradation including drying of wetlands, with drought compounding the problem caused by improper management of the river basins, irrigation projects and dams. Soil erosion, irrigation and the spread of sand dunes into settlements, agricultural areas and roads has resulted in a scarcity of water potentials. During past years, floods have significantly increased due to deforestation and vegetation losses which have, in turn, decreased the waterholding capacity of the land.

Soils hold the key to productivity and resilience to climate vagaries including drought in rainfed agriculture. The loss of fertile soil due to erosion, depletion of soil organic matter, emerging secondary and micronutrient deficiencies, soil compaction, surface crusting, and loss of soil biodiversity are potential limiting factors for productivity enhancement in these regions (Srinivasarao et al. 2012a, 2014c). The per capita land area is progressively declining, even in southern India, which has a low rate of population growth (Lal 2006). At a medium rate of population growth, the projected per capita arable land area by 2025 will be 0.05 ha in Bhutan, 0.07 ha in Nepal and Pakistan, 0.08 ha in Sri Lanka, 0.12 ha in India and Iran, and 0.17 ha in Afghanistan.

Most soils in dryland regions are prone to human-induced degradation and desertification, which are a serious problem in South Asia (Lal 2006). Soil erosion by water is a principal constraint in mountainous regions and undulating terrains. The land area affected by water erosion is estimated to be 82 Mha, of which 33 Mha are in India and 26 Mha in Iran. In comparison, the land area affected by wind erosion is 59 Mha, of which 11 Mha each are in India and Pakistan, and 35 Mha are in Iran. Closely related to wind erosion is the issue of desertification or the degradation of soil and vegetation in arid regions. Desertification affects 67 Mha of Asia's drylands, of which 60 Mha are in India, 3 Mha in Pakistan and 2 Mha in Iran. Poor soil structure, slaking and excessive tillage lead to crusting in Aridisols (Hemmat and

Khashoei 2003; Barzegar et al. 2002a, b), Inceptisols (Ishaq et al. 2001) and Alfisols (Smith et al. 1992). In addition to crusting, Alfisols in central India are also prone to hard setting (Smith et al. 1992).

The principal constraints to enhancing crop production on Alfisols, which occur extensively in southern Asia, are: (i) low SOC pool, (ii) poor soil structure, (iii) crusting and mechanical impedance, (iv) high runoff and erosion, (v) high soil temperatures, (vi) severe drought stress, and (vii) low soil fertility (Lal 2006). In India, Alfisols occur mainly in southern parts of the country and constitute about 30 % of the soils under rainfed farming. Being light-textured and shallow, their available water capacity (AWC) is low. Furthermore, hard setting and structural instability exacerbate surface sealing and crusting. These soils in semiarid climates support a single rainy season crop (*Kharif* or summer) with productivity levels of 0.7–0.8 Mg/ha. These soils are characterized by low SOC and N stocks despite large variations in the cropping system, soil type, rainfall, temperature and supplementary management practices such as manuring and fertilization (Srinivasarao et al. 2012b). The high clay content, high waterholding capacity and favorable moisture release characteristics are important attributes of Vertisols which reduce the severity and duration of drought under conditions of low and erratic rainfall. These soils have numerous constraints especially low infiltration rates, poor internal drainage, inundation, high runoff and erosion, poor trafficability, narrow workable water content range, high evaporation, shrinkage cracks and risks of salinization. Similarly, alluvial soils (Inceptisols and Entisols), dominant in the Indo-Gangetic Plains, are light-textured (sandy and sandy loams), have low SOC concentrations and low inherent soil fertility. Crusting and high soil strength are serious problems on alluvial soils in arid regions (140–180 mm rainfall), which adversely affect seedling emergence and the crop stand. High crust strength reduces the emergence of cotton in Pakistan (Nabi et al. 2001) and elsewhere. However, Inceptisols respond to inputs (e.g., irrigation, fertilizers) and are highly productive under irrigated conditions. Arid climate prevails over 32 Mha of land area in seven states of India alone. Surface crusting, low water and nutrient retention capacity, drought stress, subsoil salinity and wind erosion are severe constraints to achieving high yields.

3.2 Climate Change

South Asia is home to more than one-fifth of the world's population and is the most natural disaster-prone region in the world. The high rates of population growth and natural resource degradation, with continuing high rates of poverty and food insecurity, make South Asia one of the most vulnerable regions to the impacts of climate change. In general, past and present climate trends and variability in South Asia can be characterized by increasing air temperatures, which are more pronounced during winter than in summer. There has been an increasing trend in the intensity and frequency of extreme events such as heat waves, cold waves, untimely and high-intensity rainfall, hailstorms and frost in South Asia over the last century (Sivakumar and Stefanski 2011; Srinivasarao et al. 2015c)

During recent decades, observed increases in temperature in some parts of Asia have ranged from 1–3 °C per century. Across all of Asia, inter-seasonal, inter-annual and spatial variability in rainfall has been observed in the past few decades. Decreasing trends in mean annual rainfall have been observed in the coastal belts and arid plains of Pakistan and parts of northeast India with increasing trends in Bangladesh (Cruz et al. 2007). The intensity and frequency of these events in South Asia have tended to increase in the last century. There have been significantly longer heat waves in many countries of South Asia with several cases of severe heat waves. In general, the frequency of more intense rainfall events in many parts of Asia has increased, causing severe floods, landslides and debris/mud flows. It is interesting that, at the same time, the number of rainy days and the total annual amount of precipitation has decreased, but the rain has been concentrated over fewer days. A long-term analysis of rainfall trends in India (1901–2004) by CRIDA, India indicated a significant increase in rainfall trends in West Bengal, central India, coastal regions, south-western Andhra Pradesh and central Tamil Nadu. Rainfall is likely to decline by 5–10 % in southern parts of India while a 10–20 % increase is likely in other regions (Venkateswarlu 2010). The number of rainy days in most of the country is likely to decrease, which points to a likely increase of extreme events (Venkateswarlu 2010). Recent models projected that the frequency of extreme precipitation days (e.g. 40 mm/day) are likely to rise (Venkateswarlu 2010). Changes in average annual temperatures are expected to increase by 2–2.5 °C. Warming is likely to be more in northern parts of India. A rise in night temperatures is also likely in India except for some small pockets in the peninsular region. This is concerning for agriculture as increased night temperatures increase the crop water requirements, accelerate respiration, hasten crop maturity and reduce yields.

Many parts of South Asia have been experiencing an increasing frequency and intensity of droughts, especially in the tropics and subtropics, since the 1970s (IPCC 2007). Rainfall has also been decreasing, with a 7.5 % reduction from 1900 to 2005 (significant at <1 %). The number of cyclones originating from the Bay of Bengal and Arabian Sea has decreased since 1970 but the intensity of these storms has increased, and the damage caused by intense cyclones has risen significantly in India and the Tibetan Plateau (Sivakumar and Stefanski 2011; Table 7). Melting of the Himalayan glaciers could increase flooding and affect water resources within the next two to three decades. Crop yields could decrease by up to 30 % in South Asia by the mid-twenty-first century.

Projections indicate that climate variations in South Asia will be varied and heterogeneous, with some regions experiencing more intense precipitation and increased flood risks, while others will encounter more sparse rainfall and prolonged droughts (Sivakumar and Stefanski 2011). Both the extent and severity of drought in rainfed areas have increased, and the consequences are poverty, food insecurity and hunger. With climate change, the dry regions of Sri Lanka (northern and eastern provinces), are expected to lose large portions of revenue from dryland agriculture (Seo et al. 2005).

Table 7 Summary of observed changes in extreme events and severe climate anomalies in South Asia

Climatic event	Observed change
Droughts	About 50 % of droughts associated with El Niño; consecutive droughts in 1999 and 2000 in Pakistan and northwest India led to sharp decline in water tables; consecutive droughts between 2000 and 2002 caused crop failures, mass starvation and affected ~11 million people in Odisha, India; droughts in northeast India during summer monsoon of 2006; about 400 million people affected by drought in 2009 in India.
Intense rains and floods	Serious and recurrent floods in Bangladesh, Nepal and northeast states of India during 2002, 2003 and 2004; floods in Surat, Barmer and Srinagar of India during summer monsoon season of 2006; 17 May 2003 floods in the southern province of Sri Lanka triggered by 730 mm rain.
Cyclones/typhoons	Frequency of monsoon depressions and cyclone formation in Bay of Bengal and Arabian Sea on the decline since 1970 but the intensity is increasing causing severe floods with damage to life and property.
Heat waves	Frequency of hot days and multiple-day heatwaves has increased in past century in India with an increase in deaths due to heat stress in recent years.

Source: Adapted from Cruz et al. (2007)

The projected impacts of climate change will vary across sectors, locations and populations. With most of the land area (approximately three-fifths of the cropped area) in these countries rainfed, the economy of South Asia hinges critically on the annual success of monsoons. In the event of a failure, the worst affected will be the landless and the poor whose sole source of income is from agriculture and allied activities. Water scarcity and water stress are big issues that will negatively impact the farmers in South Asia regarding water availability, soil moisture, and insect pest and disease incidence. Again, the worst hit would be the farmers in rainfed areas with small and marginal holdings, and poor financial capacity to cope with climate variability. There is a need to mainstream climate-resilient practices for adaptation to climate change into sustainable development planning in the region. An improved understanding of climate change impacts, vulnerability and adaptation practices to cope with climate change would help this process.

3.3 *Food and Nutritional Insecurity*

Reducing hunger by half is the first of the Millennium Development Goals (MDGs), and there have been some spectacular success stories in the Asia-Pacific region. However, success in reducing poverty and hunger varies among subregions and is closely associated with economic performance and investment in social capital. Poverty and hunger are particularly serious in South Asia and small islands in the

Pacific (Mukherjee 2008). For some countries, like Bangladesh and India, about 30 % of children are born underweight and run the risk of dying in infancy, being stunted physically and cognitively during childhood, having reduced working capacity and earnings as adults, and giving birth (females) to low-weight babies. The impact of hunger is more prevalent in women and girl children, who eat last, eat the least and eat leftovers. Further, the economic cost of hunger in terms of lost productivity, earnings and consumption run into billions of dollars, apart from the direct cost of dealing with the damage it causes.

The future requirements of food grains should be based on population growth, the composition of rural and urban populations, growth in per capita income in rural and urban areas, and changes in taste and preferences. According to demand projections prepared by Paroda and Kumar (2000) under low- and high-income scenarios, demand in the region will grow by 1 % for cereals and by 1.7 % for pulses. Demand for edible oils is projected to grow by more than 1.6 %. Demand for fruits, vegetables and livestock would rise by around 3 % or more under the high-income scenario. These growth rates can be used to work out feasible levels of agricultural diversification.

The long-term trend in the consumption pattern at the household level shows that the per capita direct consumption of food grains has been declining, and that of livestock products and fruits and vegetables has been increasing in most The South Asian Association for Regional Cooperation (SAARC) countries. Despite this shift in dietary pattern, food grains are paramount for household food and nutritional security because: (1) cereals and pulses are staple foods, and there is no perfect substitute; (2) due to the inadequate intake of almost all foods, increased consumption of other foods, in most cases, will fill any dietary deficiency; (3) food grains are the major and cheapest source of energy and protein compared with other foods, and are vital for food and nutritional security of low-income masses; and (4) increased production and consumption of livestock products resulting from increased per capita income requires high growth in the use of grain as feed for livestock. For these reasons, food grains continue to be the pillars of food security in South Asia and any slack in their production translates into persistent price shocks and adverse impacts on common people (Chand 2010).

3.4 Socioeconomic Issues

The ultimate goal of all development efforts is to improve socioeconomic conditions and, therefore, the wellbeing of people. South Asia is an important region in the world that needs attention as the bulk of the poor live in these nations. The widespread poverty is reflected in the relatively higher prevalence of malnutrition in these countries. Addressing the issue of malnutrition should, therefore, be an explicit objective of any program or intervention intended to tackle food insecurity.

Improving the situation with respect to food security, poverty and nutrition is a recognized goal of all policy interventions, but certain elements in the existing socioeconomic setting can impede these goals. Among them, an important challenge is improving the situation with high population pressure with smaller size of farms. Poor infrastructure, both hard and soft, is another determinant and determined aspect of socioeconomic conditions. At the regional level, the lack of political trust also results in suboptimal policy making which is reflected in poor adherence to or inadequate implementation of various regional agreements.

4 Opportunities for Resilient Dryland Agriculture

The elimination of poverty, improvement in nutrition and the supply of safe food are of prime importance for most SAARC countries. There is renewed emphasis on agriculture for development and on addressing poverty and hunger. Agricultural growth not only offers a pathway out of poverty but it promotes employment in non-farm rural activities and facilitates the migration to non-agricultural avenues without distress. Based on this, growth in agriculture and overall rural development are considered essential for a sustainable exit from poverty (FAO 2002). Agricultural growth helps to reduce poverty and hunger, not only by raising the income of the poor but by keeping food prices in check. Recognition of the importance of agriculture has shifted the emphasis from growth *per se* to inclusive growth (Chand 2010). Growth in output and farm income depends on many factors, such as input and output prices and non-price factors. Increasing growth requires remunerative price for output, access to improved technology, application of quality inputs, and the use of modern machinery. There is vast potential for enhancing agricultural production in dryland agriculture, especially in South Asia. The potential of rainfed agriculture needs to be unlocked through the efficient management of natural resources to increase farm productivity and profitability in South Asian countries. Soil and water management will play a critical role in this.

4.1 Technological Approaches

Technology is the prime mover for agricultural growth. Considering the costs and constraints of resources such as water, nutrients and energy, the genetic enhancement of productivity should be coupled with input-use efficiency. This will be possible only by using the existing improved technology and by developing new technologies (Chand 2010). Agricultural research systems have developed some promising technologies for overcoming the barriers, outlining the needs of farmers to achieve high growth and promoting farming systems to improve the natural resource base. Current approaches in development are aimed at enhancing the resilience to climate change and contributing to mitigation. This is a clear pointer to the potential for raising output through the effective dissemination of agricultural technologies. With advanced management and high input levels, yields could increase by 2–5-fold (Venkateswarlu and Prasad 2012). Specialized dryland management practices such as water harvesting and reducing soil moisture loss can increase yields by an additional 5–15 % on average across the SAT regions and reduce yield variability from year-to-year, producing a more reliable yield (Fischer et al. 2009).

4.1.1 Rainwater Management

Soil and water need special attention for their sustainable use as degradation is evident in all SAARC countries. The optimal use of water resources has three distinct adoption strategies for the sustainable development of dryland agriculture during the 21st century: (i) cropping strategy based on rainfall analysis and moisture availability, (ii) *in-situ* moisture conservation technologies, and (iii) rainwater harvesting and use for crops at critical stress periods (Singh et al. 1999). The rainwater management strategy in arid and semiarid regions involves the selection of short duration and low-water-requiring crops and the conservation of as much rainwater as possible so that crops can escape moisture stress during the growing period. In addition, in relatively high rainfall regions, the surplus water can be harvested for lifesaving irrigation, enhancing cropping intensity and maximizing returns. Apart from enhancing the availability of water, increasing the water use efficiency by arresting various kinds of losses should be the focus. The frequent occurrence of midseason and terminal droughts lasting 1–3 weeks is the main reason for crop failure and low yield (Srinivasarao et al. 2015c). The provision of critical irrigation during this period has the potential to improve yields by 29–114 % for different crops (Prasad et al. 2014). Critical irrigation to high-value vegetables, fruit and flower crops would contribute a higher benefit:cost ratio and higher rainwater use efficiency.

The watershed development approach adopted in India and elsewhere since the 1970s is hydrologically ideal. However, within the large hydrologic unit or watersheds there are many soil types, landscape units, land uses and other social, economic, cultural, and ethnic factors to be considered (Lal 2006). Only a small portion of the entire watershed may be prone to excessive runoff and severe erosion. Rather than treating the entire watershed, it is important to treat those landscape units which are most prone to degradation using the partial area concept. Watershed development programs are silently revolutionizing dryland areas to become a growth engine for inclusive and sustainable development in vast tracts of dryland areas in South Asia. Meta-analysis of 311 watershed case studies from different agroecoregions in India revealed that watershed programs benefitted farmers with enhanced irrigation areas by 33.5 %, increased the cropping intensity by 63 %, reduced soil losses to 0.8 t/ha and runoff to 13 %, and improved groundwater availability. Economically, the watershed programs are beneficial and viable with a benefit:cost ratio of 1:2.14 and an internal rate of return of 22 % (Joshi et al. 2005).

4.1.2 Soil Management

Soils hold the key to productivity and resilience to climate vagaries in dryland agriculture in South Asia. The potential limiting factors for productivity enhancement in these regions are the loss of fertile soil as erosion, depletion of soil organic matter, emerging secondary and micronutrient deficiencies, soil compaction, surface crusting, and loss of soil biodiversity. Soil management systems should not only take into account the risks inherent in the farm and field locations but also in the choice of crops, cultivation methods and/or stocking levels. Soil health restoration can be addressed through better management practices, which include (i) timely tillage at suitable moisture levels to prevent bringing up clods which require more tillage, (ii) reducing secondary tillage, no-till, or ridge–tillage systems to leave crop residues on the soil surface, (iii) using crop rotations which include grasses and legumes where possible, (iv) using cover crops, (v) using manure to build soil organic matter, and (vi) if a crust has formed before the crop emerges, using a rotary hoe to break up the crust to assist with crop emergence. The main emphasis in rainfed farming systems is to build soil organic matter (SOM) for soil health restoration. Management to improve and restore soil health involves a combination of practices that enhance the soil's biological, chemical, and physical suitability for crop production, including erosion control, correction of nutrient deficiencies, reclamation of problematic soils, reducing compaction by decreasing heavy equipment traffic, and using best nutrient management practices such as integrated nutrient–water management (Fig. 3).

Fig. 3 Components of recommended management practices (RMPs) for soil health restoration (Source: Srinivasarao et al. 2015a or b?)

Low SOM in tropical soils, particularly those under the influence of arid, semi-arid, and subhumid climates, is a major factor contributing to poor productivity (Katyal et al. 2001). Proper management of SOM is important for sustaining soil productivity and ensuring food security and protection on marginal lands. Under dryland conditions, the process of organic matter decomposition is faster than irrigated condition; organic matter disappears rapidly due to its rapid oxidation under prevailing high temperatures, low rainfall and high potential evapotranspiration (PET). Therefore, frequent and large quantities of added organic manure are essential to maintaining SOC concentrations. An important general strategy is to add as much organic matter as possible through management practices (*viz.* crop and cover crop residues, manures, the inclusion of legumes in the cropping sequence or as intercrops, green manure crops, green-leaf manuring, tank silt addition, farmyard manure, biofertilizers and vermicompost) (Srinivasarao et al. 2011).

Green manuring is a viable option to increase SOM. The incorporation of Gliricidia (*Gliricidia sepium* Jacq. Walp.) green-leaf manuring technology in the light soils of rainfed tribal and backward districts of Andhra Pradesh and the All India Coordinated Research Project for Dryland Agriculture (AICRPDA) and Operational Research Project (ORP) villages in different regions of India had a significant and positive effect on increasing SOM and macro- and micro-nutrients (Srinivasarao 2011).

The effective management of residues, roots, stubbles and weed biomass can have beneficial effects on soil fertility through the addition of organic matter, plant nutrition and improved soil condition. Agricultural waste is usually handled as a liability, often because the means to transform it into an asset is lacking. Crop residues in fields can result in crop management problems as they accumulate. In India, the availability of biomass (2010–2011) was estimated at 500 million tons/year (MNRE 2009). Studies sponsored by the Ministry of New and Renewable Energy (MNRE) in India have estimated surplus biomass availability at about 120–150 million tons/year (MNRE 2009). Of this, about 93 million tons are burned each year. The lack of availability of proper chipping and soil incorporation equipment is one reason for the colossal wastage of agricultural biomass, and the increased cost of labor and transport is another. Many technologies exist such as briquetting, anaerobic digestion, vermicomposting, biochar, but they have not been commercially exploited.

The correction of nutrient deficiencies can be achieved through site-specific nutrient management (SSNM) and integrated nutrient management (INM). SSNM takes into account all nutrient deficiencies to ensure that crop demands are met, and soil fertility is improved, which in turn ensures higher nutrient use efficiency, crop productivity and economic returns (Dobermann 2004). The results of on-farm demonstrations across crops and soils in India showed that S application increased grain yield by 650 kg/ha (+24 % over NPK) in cereals, 570 kg/ha (+32 % over NPK) in oilseeds, and 375 kg/ha (+20 % over NPK) in pulses (Singh 2001). Cotton yields increased in response to balanced nutrition in the Warangal, Adilabad and Khammam districts of India by 20, 60 and 30 %, respectively, compared with traditional farmer practices (Fig. 4).

Fig. 4 Enhanced growth of Bt-cotton (left) with balanced nutrition compared to farmers' practice (right) in Adilabad and Khammam districts of Andhra Pradesh, India

4.1.3 Conservation Agriculture

Conservation agriculture (CA) is gaining momentum as an alternative strategy to sustain agricultural production due to the growing resource degradation problems, particularly under dryland conditions. Sufficient information is available to indicate that CA practices save time, reduce production costs and contribute substantially towards profitability and sustainability of carbon in soils (Srinivasarao et al. 2013a). However, in dryland ecosystems, having sufficient crop residue to leave on the surface is not feasible as part of CA, nor is the conservation of first rains by deep tillage before the rainy season to allow sufficient infiltration. Hence, typical CA with three principles (minimum tillage/soil disturbance, permanent soil cover and crop rotation/intercropping) needs to be investigated further to include *in-situ* water conservation of rainwater to recharge the soil profile. A modified CA with four principles for dryland agriculture is presented in Fig. 5.

4.1.4 Agricultural Intensification

The goal of agricultural intensification can be realized through the adoption of land-saving technologies which enhance crop yields per unit area per unit time and unit input of off-farm resources (Lal 2006). With available dryland technologies such as rainwater management, crop options, short-duration varieties, and various agronomical practices, more dryland areas can be used for intensive cropping including relay cropping and double cropping. Double cropping is possible in areas with sufficient rainfall (usually more than 750 mm) that have a soil storage capacity of more than 150 mm of available soil moisture, or with rainwater harvested in farm ponds which can be used to establish winter crops. There is plenty of scope for raising farm outputs and income with diversification to high-value crops and by harnessing niche areas particularly in mountainous regions. This will require markets for those

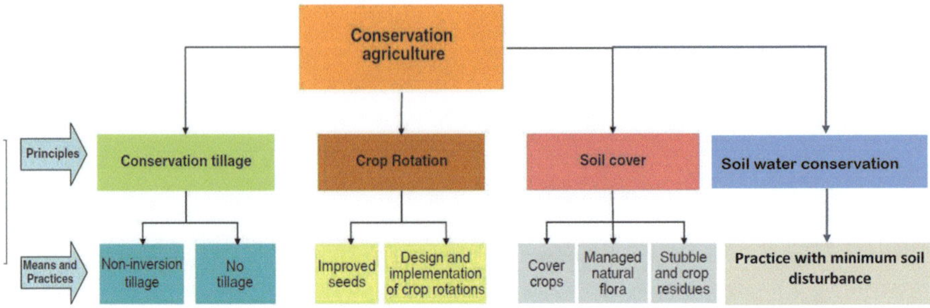

Fig. 5 Modified CA for tropical dryland ecosystems

products, which depend on the private sector, cooperatives or joint ventures of public and private agencies.

Studies on intercropping over time and across locations in India have identified useful and productive crop combinations with matching production technologies. For example, pearl millet and pigeon pea (*Cajanus cajan* (L.) Millsp) (Umrani and Subba Reddy 1999; Srinivasarao et al. 2014d) which has done extremely well in regions where late rains are likely to occur in September/October (Bijapur and Solapur) or where deep, moisture-retentive soils exist (Akola). Cotton and mung bean (*Vigna radiata* (L.) Wilczek), black gram (*Vigna mungo* (L.) Hepper) or soybean are other successful intercropping combinations. Although intercropping has not been recommended for post-rainy or winter season cropping, the system is gaining importance as a major adaptation strategy to cope with the vagaries of climate.

4.1.5 Adaptation to Climate Change

While climate change impacts the agricultural sector in general, dryland agriculture is expected to be more vulnerable given its high dependency on monsoons, and the likelihood of increased extreme weather events due to the aberrant behavior of monsoons. Given South Asia's geographical location and high incidence of poverty, the region is suffering from climate change. Increased temperatures have not only affected cropping seasons but also melted the Himalayan glaciers at an alarming rate. These changes have increased flooding and raised sea levels, severely impacting rural livelihoods in the region. Recurrent natural hazards such as drought and heavy flooding have affected the region's poor population disproportionately. Farmers have reported shortened rainy seasons and increased temperatures over the years (Chatterjee and Khadka 2011). A wide variety of adaptive actions may be taken to face the adverse effects of climate change on dryland agriculture. The Indian Council of Agricultural Research (ICAR) responded to the challenge of climate change and launched NICRA in 2011 with a major focus on infrastructure development for strategic research, and demonstration of best practices on farmers' fields to cope with climate variability. The major achievements of the project include micro-level agrometeorological advisories to cope with seasonal climate variability, tolerant genotypes including a few advanced lines for climatic stresses in wheat, rice and pulses in multi-location testing stage, a strengthened database of greenhouse gas fluxes in various crops and agroecosystems, an assessment of water and carbon footprints, and the identification of climate-resilient technologies with co-benefits of low global warming potential. The TDC of the project, implemented through 121 *Krishi Vigyan Kendras* (KVKs) across the country, identified appropriate location-specific technologies that can enhance resilience to climatic variability. Successful demonstrations to cope with deficit rainfall situations, floods and cyclones were possible through the implementation of climate-focused action plans. On-farm adjustments to climate change would require crop varieties suitable for late/early sowing, new cropping sequences, the supply of seed and inputs on demand, water conservation, diversified production, etc. All of which would require major investments.

Farmers, in particular, and society, in general, have attempted to adapt to climatic stresses by resorting to practices such as mixed cropping, by changing varieties and planting times, and by diversifying their sources of income (Jat et al. 2012). In future, such adaptation strategies would need to be considered along with innovations to cope with climate change. For dryland agriculture to successfully adapt to climate changes and variability, climate-resilient crops and cultivars for different regions need to be identified. Maheswari et al. (2015) have documented the available crop varieties that are suitable for cultivation under stresses like drought, heat, cold, salinity and flooding with details on agroclimatic zones and the possible sources of seed availability in India. Other technical adaptation measures range from changes in production systems such as adjusting planting or fishing dates; rotations; multiple cropping/species diversification; crop–livestock pisciculture systems; agroforestry; soil, water and biodiversity conservation and development by building soil biomass; restoring degraded lands; rehabilitating rangelands; harvesting and recycling water; planting trees; developing adapted cultivars and breeds; and protecting aquatic ecosystems to maintain long-term productivity. Adaptation measures also take into account the establishment of disaster risk management plans and risk transfer mechanisms, such as crop insurance and diversified livelihood systems.

4.2 Institutional Innovations

Institutional innovations are required both at the domestic and regional level in South Asia. Some fundamental aspects of agricultural production are similar between countries. For example, smaller farms lead to smaller marketable surpluses, resulting in less bargaining power for producers and exposure to reduced economies of scale. There are many institutional innovations emerging in these countries that are successfully bringing farmers together to aggregate their small marketable surpluses. There are also efforts to integrate producers with consumers through supply and value chains to increase the producers' share of what the consumer pays for the commodity at the end of the value chain. These institutional innovations vary in that the nature and participation of public, private and civil society sector players varies. Self-help groups—the initiatives of private sectors in linking farmers to the markets (e.g. e-choupal in India)—are one example. There are also community-based initiatives to enhance smallholder access to some key inputs, for example community seed banks, 'seed villages' and CHCs to access farm mechanization. In India, Bangladesh and Nepal, there are several community-based interventions for the management of water resources, the lessons from which could be replicated or adapted to local conditions in other countries. The diversification of agriculture towards high-value crops has been a driver of growth in these south Asian countries. For such a strategy to be effective for smallholders, institutional arrangements such as contract farming and producer supply organizations are needed so that farmers have access to improved technologies and better, more stable markets.

At the regional level, despite some improvement in intra-regional trade, progress has been slower than desired. To further trade within the region, institutions and their protocols should be strengthened, especially in food and agricultural trade, so that the food security concerns are well addressed. The agricultural innovations generated from respective agricultural research systems should be shared as these countries share similar agro-economic settings. There can be a mechanism whereby certain outputs of the national agricultural research system can be identified as regional public goods so that other countries within the region can have. This will require suitable modification and changes to laws and regulations related to intellectual property rights.

The timely completion of agricultural operations has significant benefits both on research farms and farmers' fields. In the recent past, due to a high degree of weather aberrations, the timeliness of agricultural operations has come into focus. Often, the ideal conditions for an agricultural operation such as sowing or intercultural operation only exist for a short period. If the farmer fails to complete the operation within this timeframe, then the output will be compromised. This problem can be tackled by using appropriate agricultural implements to carry out the operations. However, smallholders often cannot afford such equipment. This calls for sharing the cost of implements by innovative institutional arrangements. Recently, the custom hiring of agricultural machinery has been seen as an appropriate institutional arrangement which can promote the mechanization of agricultural operations on small farms.

Weather-based agro-advisory services (AAS) in farming activities are important for accessing real-time weather information, timely agricultural operations, improved crop yields, reduced costs of cultivation, need-based changes in cropping patterns and improved livelihoods. The district level weather forecast could be used along with current crop and weather conditions to prepare district level advisories by respective KVKs or scientists of Agricultural Universities (AU). A pilot methodology for preparing and issuing agrometeorological advisories at the block level has been tested in the Belgaum district of Karnataka, India. The main innovation of this project was to set up a framework involving KVKs, state-line departments and field information facilitation to collect real-time crop data, formulate an appropriate advisory and disseminate what?. Field Information Facilitators (FIF) were appointed in 10 blocks of the district to collect information on weather, crops, disease and pest incidence. They supplied information by phone or email to contact staff at KVK who in turn developed a qualitative Agromet Advisory specific to the village/farmers, in consultation with an agrometeorologist of AU and scientists of KVK.

The increasing frequency and severity of droughts, storms and other extreme weather events associated with climate change reduce the livelihood options for millions of small-scale farmers in South Asia. Weather index-based insurance is an attractive approach to managing weather and climate risk because it uses a weather index, such as rainfall, to determine payouts and these can be made more quickly and with fewer arguments than is typical for conventional crop insurance. The underlying premise of weather insurance is that weather parameters can be reliable proxy indicators for the actual losses incurred by farmers (Singh 2013a). Weather index-based insurance was formally introduced to Indian farmers in 2003 through a program initially supported by the World Bank. The first weather index insurance in India was a rainfall insurance contract underwritten in 2003 by the ICICI-Lombard General Insurance Company for groundnut and castor farmers of the BASIX water user association in the Mahabubnagar district of Andhra Pradesh. By 2007, the national government had adopted it as an alternative to the existing crop-yield-index insurance. Moreover, by 2012, up to 12 million farmers growing 40 different crops on more than 15 million hectares were insured against weather-related losses (CCAFS 2013). There is still gap in these payout structures that need to be refined by the insurance companies in consultation with agrometeorologists and other scientists.

4.3 Agricultural Markets and Trade

Intra-regional trade can enhance food security in South Asia. If South Asian countries trade agricultural produce extensively among themselves, then the region's food security concerns can be addressed in a sustainable manner. Improvement in the performance of agricultural markets, domestic and international, is a means to enhance farmer income in developing countries, and South Asia is no exception. As a region, South Asia is one of the least integrated regions in terms of trade (Weerahewa 2009), and domestic agricultural markets are less developed and integrated. These countries trade more with countries outside the region than with those within the region. The contribution of trade to agricultural GDP is relatively more in the Maldives and Sri Lanka than in other countries. USA, EU and UAE are among the more important countries in terms of the value of exports and imports to these South Asian countries. Within the region, India is the hub of agricultural trade because of its size, demographic, geography and economy. Import substitution has been the driver of development regimes in South Asia and, as a result, trade is characterized by restrictions and interventions by governments. Policies related to trade, especially in India, have changed. Non-tariff barriers, rather than tariff barriers, were hindering intra-regional trade. The lack of variation in comparative advantage and poor infrastructure along with supply-side barriers due to lower productivity levels are the major reasons for relatively lower trade within and outside the region (Nanda 2012).

4.4 Policy Needs

Despite some progress, poverty, food insecurity, hunger and malnutrition continue to be major development goals of the economies of South Asia. Countries in the region should plan and implement strategies that optimize the resources and efforts in a coherent and efficient manner so that the desired progress is achieved. Within agriculture, much of cropping remains rainfed, except in Pakistan, so investing in enhancing productivity and profitability of rainfed agriculture is needed. Some threats facing the region are land degradation, rising input costs, climate change and higher dependence on agriculture for livelihoods. Despite broad similarities within the agricultural frame work, considerable diversity exists between of these South Asian countries. Hence should be adequately recognized by agricultural policy makers to devise policies and interventions for technology generation and transfer in dryland agriculture. Public investment in agricultural research needs to be increased, especially in natural resource management, without which the potential of genetic enhancement and other productivity enhancing technologies cannot be realized. Enhancing the use efficiency of inputs such as water and fertilizer nutrients should attract the attention of researchers and policymakers. Other areas that need attention are land reform and market infrastructure. The former will positively affect the investment decisions of farmers, and the latter will help farmers to obtain better prices for their products. Investments in road and communication infrastructure and processing facilities are needed. Only then, will farmers in South Asia be able to take advantage of expanding global trade. Such investments are needed to encourage diversification towards high-value commodities, which have been identified as a source of income growth in these countries (Joshi et al. 2003). Diversification is more common in regions with less irrigation, but all countries should have regionally-differentiated approaches to technology generation and transfer, with policies and programs flexible enough to accommodate location-specific requirements.

The pursuance of domestic policies for addressing location-specific issues within these countries should be accompanied by better coordination in dealing with issues that have cross-border relevance. At present, trade is hampered by a lack of standardization of procedures and protocols and by inadequacies in soft and hard infrastructure in this regard. Information from research systems and policy making should be shared between countries within the region. Arrangements for conservation and the use of diverse genetic resources in these countries should be strengthened further. The National Agricultural Research System of India is stronger than its counterparts in other countries, and outputs should be shared with other countries where relevant. Some crop varieties developed in India are finding their way into other countries. Formal and informal trade in agricultural inputs such as seed, fertilizers has occurred between India and countries like Bangladesh, which could be optimized by streamlining the procedures.

At the regional level, relevant SAARC Declarations should be implemented with vigor. The SAARC Food Bank initiative must be implemented with greater emphasis and coordination to help poor consumers during food crises. The SAARC countries must collaborate in sharing knowledge and transferring technology as a regional mitigation strategy. Similarly, the concept of the SAARC Seed Bank must be promoted. Poor and marginal farmers should have access to better quality seeds that can resist the vagaries of climate change.

On the whole, policies are needed to put in place programmes and resources that make dryland agriculture more sustainable in the face of emerging threats such as climate change, and that dovetail domestic policies with regional policies for a better-coordinated effort to deal with larger issues such as food insecurity, climate change and globalization of agricultural trade.

5 Conclusions and Future Research Thrusts

The South Asian region sustains the world's poorest people, contributing less than 5 % to the world's GDP. Agriculture is the main source of livelihoods in this region. There is potential to enhance agricultural production in dryland agriculture with the adoption of recommended management practices. These practices are location-specific and involve knowledge-based management of natural resources including soils and rainwater, agricultural intensification, INM, rainwater harvesting and efficient use, and conservation agriculture. The efficient use of water, soil and farm management practices in an integrated approach is both essential and a prerequisite to making dryland farming more economical and sustainable under the increasing frequency of droughts, reduced number of rainy days, and extreme and untimely rainfall. Location-specific needs of soil and water conservation measures *vis-a-vis* changing rainfall scenarios will better address water issues. Public–Private–Partnerships (PPP) that create market links have proved successful in several watersheds sites and created win–win situations for all stakeholders involved, particularly in India. Therefore, it is necessary to formulate a coherent set of guidelines to enable governments and consortium partners to approach the private sector and begin fruitful collaborations in PPPs. These partnerships need to strengthen market linkages and value chains and increase investments by the private sector in watershed development.

Field burning of crop residues must be stopped to minimize environmental pollution; surplus crop residues should be moved into the soil system either as manure or surface cover. Residue management must be taught at all levels, including extension workers and land managers. In addition, a region-specific, needs-based crop residue management plan should be developed. The adoption of various location-specific INM practices in dryland systems should be promoted further and implemented with the help of various government programs. Weather-based AAS in farming activities are needed for access to real-time weather information, timely agricultural operations, improved crop yields, reduced cultivation costs, needs-based changes in cropping patterns, and improved livelihoods. Extension departments should increase exposure visits, training and demonstrations of location-specific farming system models, to increase awareness and capacity building of the farming community towards upscaling relevant dryland technologies.

Despite the broad similarities within the agricultural settings of these countries, the diversity within these countries should be recognized by those concerned with agricultural policy to devise policies and interventions for technology generation and transfer in dryland agriculture. Public investment in agricultural research should be increased, especially in natural resource management, without which genetic enhancement and other productivity enhancing technologies cannot be realized. Researchers and policymakers should focus on enhancing the use efficiency of inputs such as water and fertilizer nutrients. Pursuance of domestic policies to address location-specific issues within these countries should be accompanied by better coordination of cross-border issues. For example, there is potential to address food insecurity at a regional level by improving the functioning of regional institutions, and coordinating and streamlining trade mechanisms.

References

ADB (2009) Study on intraregional trade and investment in South Asia, Development Partnership Program for South Asia-Aus AID: Australian Government, Asian Development Bank, Manila, Philippines

Afzal M (2001) Water resources development and management: Pakistan's vision. Pak J Agric Econ 4:67–81

Ahmad MD, Turral H, Masih I, Giordano M, Masood Z (2007) Water saving technologies: myths and realities revealed in Pakistan's rice–wheat systems. IWMI Research Report, vol 108. International Water Management Institute, Colombo, 44 p

Ali CA, Khan A, Hakim Shah H (2011) Agroforestry in Khyber Pakhtunkhwa: current situation and future prospects. Pak J Forest 61(1):1–11

Amin AE (1987) Interaction of climatic components on agroforestry systems by management practices in Pakistan. In: Reifsnyder, WS, Darnhofer TO (eds) Meteorology and agroforestry. Proceedings of an international workshop on the application of meteorology to agroforestry systems planning and management, Nairobi

Ariyabandu R (2008) Swings and round abouts: a narrative on water policy development in Sri Lanka. Working paper 296. Overseas Development Institute. 111 Westminster Bridge Road, London SE1 7JD

Arunachalam A, Arunachalam K (2002) Evaluation of bamboos in eco-restoration of 'jhum' fallows in Arunachal Pradesh: ground vegetation, soil and microbial biomass. For Ecol Manag 159:231–239

Baig MB, Shahid SA, Straquadine GS (2013) Making rainfed agriculture sustainable through environmental friendly technologies in Pakistan: a review. Int Soil Water Conserv Res 1(2):36–52

Bandara MACS, Aheeyar MMM (2010) Runoff Rainwater Harvesting Interventions in Sri Lanka. HARTI research report no: 133. Hector Kobbekaduwa Agrarian Research and Training Institute, Colombo, Sri Lanka
Barzegar AR, Asoodar MA, Ansari M (2002a) Effectiveness of sugarcane residue incorporation at different water contents and the proctor compaction loads in reducing soil compactibility. Soil Tillage Res 57:167–172
Barzegar AR, Yousefi A, Daryashenas A (2002b) Changes in soil physical properties and wheat yield upon addition of different types and amounts of organic material. Plant Soil 247:295–301
Bhargava GP (1989) Salt-affected soils of India: a source book. New Delhi, Oxford, IBH
Bhaskara Rao M (2011) Drought scenario in South Asia – an overview of integrated water resources development & management techniques for drought management in South Asia. In: Mishra PK, Osman M, Satendra, Venkateswarlu B (eds) Techniques of water conservation & rainwater harvesting for drought management, SAARC Training Program, 18–29 July, 2011, Central Research Institute for Dryland Agriculture, Hyderabad, India, pp 413–427
Bhuchar S, Billingsley A, Dhakal M, Providoli I, Sherpa SL, Siddiqi AS (2016) Soil moisture retention and soil fertility options for agricultural and degraded lands. Mountain Development Resource Book for Afghanistan, pp 69–81. Accessed on March 2016. http://lib.icimod.org/record/26742/files/c_attachment_648_5826.pdf
CCAFS (2013) Weather index-based insurance: a tool for managing climate risk, Research program on climate change, agriculture and food security. University of Copenhagen, Copenhagen
Central Intelligence Agency (2015) https://www.cia.gov/library/publications/the-world-factbook/index.html
Chand R (2010) SAARC agricultural vision 2020. Agric Econ Res Rev 23:197–208
Chatterjee B, Khadka M (2011) Climate change and food security in South Asia. CUTS International, Jaipur, 119 p
Chowdary VM, Ramakrishnan D, Srivastava YK, Chandran V, Jeyaram A (2009) Integrated water resource development plan for sustainable management of Mayurakshi Watershed, India using Remote Sensing and GIS. Water Resour Manage 23:1581–1602
Cruz RV, Harasawa H, Lal M, Wu S, Anokhin Y, Punsalmaa B, Honda Y, Jafari M, Li C, Huu N (2007) Asia. Climate change, impacts, adaptation and vulnerability. In: Parry ML, Canziani OF, Palutikof JP, van der Linden, Hanson CE (eds) Contribution of Working Group II to the Fourth Assessment Report of the Intergovernmental Panel on Climate Change. Cambridge University Press, Cambridge, pp 469–506
Csavas I (1992) Regional review on livestock–fish production systems in Asia. In: Mukherjee TK, Moi PS, Panandam JM, Yang YS (eds) Proceedings of the FAO/IPT Workshop on integrated livestock–fish production systems, Institute of Advance Studies, University of Malaya, Kuala Lumpur pp 16–20
De Pauw E, Oweis T, Youssef J (2008) Integrating expert knowledgein GIS to locate biophysical potential for water harvesting: methodology and a case study for Syria. ICARDA, Aleppo, Syria 59 p
Desai GR, Manohari PL, Ramana Rao SV (2009) Performance of rainfed farming systems in selected districts of Southern India. J Agric Ext Manage 10(2):35–60
Devendra C (1983) Small Farm Systems Combining Animals with Mixed Cropping. In: Proceedings of the world conference of animal production, Tokyo, Japan 1, pp 173–191
Devendra C, Thomas D (2002) Smallholder farming systems in Asia. Agric Syst 71:17–25
Dharmasena PB (2003) Soil erosion and conservation – A case study in the dry zone of Sri Lanka. Trop Agricult Res Ext 6:128–131
Dobermann A (2004) Increasing productivity of intensive rice systems through site-specific nutrient management. Science Publishers and IRRI, Enfield 410 p
Engelman R, LeRoy P (1993) Sustaining water: Population and the future of renewable water supplies. Population Action International, Washington, DC
Eriyagama N, Smakhtin V, Gamage N (2009) Mapping drought patterns and impacts: a global perspective. IWMI Research Report, vol 133. International Water Management Institute, Colombo, 31 p

FAO (2001) Farming systems and poverty: improving farmers' livelihoods in a changing world. Food and Agriculture Organization of the United Nations, Rome 412 p

FAO (2002) Reducing poverty and hunger: the critical role of financing for food, Agriculture and Rural Development, FAO, Rome, Paper prepared for the international conference on financing for development, Monterrey, Mexico, 18–22 March 2002

FAO (2016) Biological diversity of the Maldives. FAO Corporate document repository. Accessed on March 2016. www.fao.org/docrep/005/ac792e/ac792e02.html

Fischer G, Velthuizen HV, Hizsnyik E, Wiberg D (2009) Potentially obtainable yields in the semi-arid tropics. Global theme on agroecosystems report no. 54 Patancheru 502 324, India. International Crops Research Institute for the Semi-Arid Tropics, 68 p

Frenken K (2011) Irrigation in Southern and Eastern Asia in figures AQUASTAT Survey, FAO water reports, vol 37. http://www.fao.org/nr/water/aquastat/countries_regions/mmr/index.stm

Gardner-Outlaw T, Engelman R (1997) Sustaining water, easing scarcity: a second update. Population Action International, Washington, DC

Garg VK, Jain RK (1992) Influence of fuel wood trees on sodic soils. Can J For Res 22:729–735

Geetha K, Mishra SK, Eldho TI, Rastogi AK, Pandey RP (2007) Modifications to SCS-CN method for long-term hydrologic simulation. J Irrig Drain Eng 133(5):475–486

George J (2005) Agriculture trade negotiations after the July package – the way forward for South Asia, briefing paper No. 1/2005. CUTS International– Centre for International Trade, Economics & Environment (CUTS-CITEE), Jaipur, India

Gill HS, Abrol IP (1991) Salt-affected soils, their afforestation and its ameliorating influence. Int Tree Crops J 6:239–260

Gintings FAN, Lait CK (1994) Agroforestry in Asia and the Pacific: with Special Reference to Silvopasture Systems, In: Copland JW, Djajanegra A, Sabrani M (eds) Agroforestry and animal production for human welfare: Proceedings of an international symposium held in association with the 7th AAAP Animal Science Congress, Bali. Indonesia, 11–16 July 1994. ACIAR proceedings vol 55, pp 33–38

Gupta RK, Naresh RK, Hobbs PR, Ladha JK (2002) Adopting conservation agriculture in the rice–wheat system of the Indo-Gangetic Plains: new opportunities for saving water. In Bouman BAM, Hengsdijk H, Hardy B, Bindraban PS, Tuong TP, Ladha JK (eds) Water wise rice production. Proceedings of the international workshop on water wise rice production, April 8–11, 2002, International Rice Research Institute, Los Banos, Philippines

Gupta SR, Sinha A, Rana RS (1990) Biomass dynamics and nutrient cycling in a sodic grassland. Int J Ecol Environ Sci 16:57–70

Hasanain A, Ahmad S, Mehmood MZ, Majeed S, Zinabou G (2012) Irrigation and Water use efficiency in South Asia. Policy research paper 9, supporting policy research to inform agricultural policy in Sub-Saharan Africa and South Asia. Global Development Network(GDN). http://www.gdn.int/admin/uploads/editor/files/SA_4_ResearchPaper_Irrigation_Use.pdf

Hemmat A, Khashoei AA (2003) Emergence of irrigated cotton in flatland planting in relation to furrow opener type and crust-breaking treatments for Cambisols in Central Iran. Soil Tillage Res 70:153–162

Heymell V (1989) Evaluation of pasture conditions in Nichara Union Council in relation to conservation of water. Studies 09/89. Pak-German self help project for rural development. Balochistan, Quetta, Pakistan

Hobbs PR, Giri GS (1997) Reduced and zero-tillage options for establishment of wheat after rice in South Asia. In: Braun HJ, Altay F, Kronstad WE, Beniwal SPS, McNab A (eds) Wheat: prospects for global improvement. Springer, Dordrecht, pp 455–465

Humphreys E, Meisner C, Gupta R, Timsina J, Beecher HG, Lu TY, Singh Y, Gill MA, Masih I, Guo ZJ, Thompson JA (2005) Water savings in rice–wheat systems. Plant Prod Sci 8:242–258

IMSD (1995) Integrated mission for sustainable agriculture: technical guidelines. National Remote Sensing Agency, Hyderabad, pp. 1–127

IICA (2014) Inter-American Institute for Cooperation on Agriculture Innovation in Agriculture (IICA): a key process for sustainable development. Institutional position paper. Inter-American Institute for Cooperation on Agriculture, San Jose

IPCC (2007) Climate change 2007: the physical science basis. In: Solomon S, Quin D, Manning M, Chen X, Marquis M, Averyt KB, Tignor HL, Miller M (eds) Contribution of Working Group I to the Fourth Assessment Report of the Intergovernmental Panel on Climate Change. Cambridge University Press, Cambridge, pp. 1–996

Ishaq M, Hassan A, Saeed M, Ibrahim M, Lal R (2001) Sub-soil compaction effects on crops in Punjab, Pakistan. I. Soil physical properties and crop yield. Soil Tillage Res 59:57–66

IWMI (2010) Water issue brief– Sri Lanka: issues and opportunities for investment. http://www.iwmi.cgiar.org/Publications/Water_Issue_Briefs/PDF/Water_Issue_Brief_7.pdf.

IWMI (2015) http://www.iwmi.cgiar.org/regions/asia/south-asia-region/pakistan/key-water-issues. Accessed on 14 Oct 2015

Jat RA, Craufurd PQ, Sahrawat KL, Wani SP (2012) Climate change and resilient dryland systems: experiences of ICRISAT in Asia and Africa. Curr Sci 102(12):1650–1659

Jayavanan GK, Pushpakumara DKNG, Sivachandran S (2014) Role of live fence agroforestry in Jaffna peninsula. Tropical Agriculturist 162:25–44

Jitsanguan T (2001) Sustainable agricultural systems for small-scale farmers in Thailand: implications for the environment. Food and Fertilizer Technology Center (FFTC), Taipei

Joshi PK, Gulati Ashok, Birthal PS, Tewari L (2003) Agricultural diversification in South Asia: Patterns, determinants and policy implications. MSSD discussion paper no. 57, International Food Policy Research Institute, Washington, DC. p 51

Joshi PK Jha AK, Wani SP, Joshi L, Shiyani RL (2005) Meta analysis to assess impact of watershed program and people's participation. Comprehensive assessment research report, vol 8. Comprehensive Assessment Secretariat, IWMI, Colombo, Sri Lanka, p 18

Kadam AK, Kale SS, Pande NN, Pawar NJ, Sankhua RN (2012) Identifying potential rainwater harvesting sites of a semi-arid, basaltic region of western India, using SCS CN method. Water Resour Manage 26:2537–2554

Kalgapurkar A, Mishra PK, Tripathi KP (2012) Application of RS and GIS in soil and water conservation measures. Indian J Soil Conser 40:190–196

Katwal T (2010) Multiple cropping in Bhutanese agriculture: Present status and opportunities. Paper presented during the regional consultative meeting on popularizing multiple cropping innovations as a means to raise productivity and farm income in SAARC countries, Peradeniya, Kandy, Srilanka. http://www.nbc.gov.bt/wp-content/uploads/2010/06/Multiple-Cropping-Paper-Bhutan-for-SAC-1.pdf

Katyal JC, Rao NH, Reddy MN (2001) Critical aspects of organic matter management in the Tropics: the example of India. Nutr Cycl Agroecosyst 61:77–88

Kaur B, Gupta SR, Singh G (2002) Bioamelioration of a sodic soil by silvopastoral systems in northwestern India. Agrofor Syst 54:13–20

Kumar MD (2011) Potentials of rainwater harvesting in different hydrological regimes: practical and policy implications for South Asian countries. In: Mishra PK, Osman M, Satendra, Venkateswarlu B (eds) Techniques of water conservation & rainwater harvesting for drought management. SAARC training program, 18–29 July, 2011, Central Research Institute for Dryland Agriculture, Hyderabad, India, 714 p

Lal R (2006) Dryland farming in South Asia. In Peterson GA, Unger PW, Payne WA (eds) Dryland agriculture, agronomy monograph no. 23, 2nd edn, American Society of Agronomy, Crop Science Society of America, Soil Science Society of America, 677 S. Segoe Rd., Madison, WI 53711, USA, pp 527–576

Lightfoot C, Minnick DR (1991) Farmer-first qualitative methods: farmers diagrams for improving methods of experimental design in integrated farming systems. J Farming Syst Res Ext 2:11–34

Maheswari M, Sarkar B, Vanaja M, Srinivasa Rao M, Srinivasa Rao Ch, Venkateswarlu B, Sikka AK (2015) Climate resilient crop varieties for sustainable food production under aberrant weather conditions. Central Research Institute for Dryland Agriculture (ICAR), Hyderabad, India, 47 p

Mattsson E, Ostwald M, Nissanka SP, Pushpakumara DKNG (2015) Quantification of carbon stock and tree diversity of home gardens in a dry zone area of Moneragala district. Sri Lanka Agroforest Syst 89:435–445

Ministry of Environment and Natural Resources (2002) State of the environment in Sri Lanka: a national report prepared for South Asian Association for Regional Cooperation, Sri Lanka

MNRE (2009) Ministry of New and Renewable Energy Resources, Government of India, New Delhi. www.mnre.gov.in/biomassresources

Mohammad T, Ehrenreich JH (1993) Scope of agroforestry research and development in Pakistan. In: Proceedings of the third North American agroforestry conference, Ames, Iowa, pp 377–379

MOIR (2005) National water plan, Ministry of Irrigation. Available at: http://www.moir.gov.np/index.php

Molden D (2007) Water for food, water for life: a comprehensive assessment of water management in agriculture. International Water Management Institute IWMI. Earthscan, London/Colombo, 40 p

Mortimore M, Adams B, Harris F (2000) Poverty and systems research in the drylands. Gatekeeper series no. 94. International Institute for Environment and Development, UK, 27 p

MOWR (2008) Report of Sub-Committee on Policy and Institutional Framework. National Water Mission under National Action Plan on Climate Change. Comprehensive Mission Document, Volume – II, Ministry of Water Resources, Government of India, New Delhi. http://wrmin.nic.in/writereaddata/nwm28756944786.pdf

Mukherjee A (2008) Food insecurity: a growing threat in Asia, Accessed online at http://www.unapcaem.org/publication/FoodInsecurityInAsia.pdf

Murty VVN, Jha MK (2011) Land and water management engineering, 6th edn. Kalyani Publishers, Ludhiana 604 p

Mustafa MDM, Haruni AKO (2002) Vegetation characteristics of Bangladesh home gardens. Schweiz Z Forstwes 12:454–461

Nabi G, Mullins CE, Montemayor MB, Akhtar MS (2001) Germination and emergence of irrigated cotton in Pakistan in relation to sowing depth and physical properties of the seedbed. Soil Till Res 59:33–44

Nanda SS, Garnayak LM (2010) Integrated farming systems for livelihood security in coastal ecosystems. In: Extended summaries: XIX national symposium on resource management approaches towards livelihood security, Indian Society of Agronomy, India, pp 160–167.

Nanda, Nitya (2012) Agricultural trade in South Asia: barriers and prospects. SAWTEE Working paper No.3/12. South Asia Watch on Trade, Economics and Environment (SAWTEE), Khatmandu, p 31

National Research Council (1993) Vetiver grass: a thin green line against erosion. NRC, Washington, DC

NRAA (2009) Drought management strategies–2009, National Rainfed Area Authority, Ministry of Agriculture, Government of India, New Delhi

Pachauri RK, Sridharan PV (1999) Looking back to think ahead: Green India 2047. Tata Energy Resources Institute, New Delhi

Pakistan Agricultural Research Council – Rice–Wheat Consortium for the Indo-Gangetic Plains. (PARC-RWC) (2003) Proceedings of the national workshop on Rice–Wheat Systems, Islamabad, Pakistan, Dec 11–12, 2002. Pakistan Agricultural Research Council, Islamabad, Pakistan and Rice–Wheat Consortium for the Indo-Gangetic Plains, New Delhi, India, 118 p

Palau RG (2012) The decision to plant poppies: irrigation, profits & alternatives crops in Afghanistan. Part5 of a 5-Part Series on Counter-Narcotics Issues in Afghanistan, http://reliefweb.int/sites/reliefweb.int/files/resources/Full_Report_4263.pdf

Paroda RS, Kumar P (2000) Food production and demand in South Asia. Agric Econ Res Rev 13:1–24

Pathak P, Mishra PK, Rao KV, Wani SP, Sudi R (2009) Best-bet options on soil and water conservation. In: Wani SP, Venkateswarlu B, Sahrawat KL, Rao KV, Ramakrishna YS (eds) Best-bet options for integrated watershed management proceedings of the comprehensive assessment of watershed programs in India, 25–27 July 2007. ICRISAT Patancheru, Andhra Pradesh, India http://oar.icrisat.org/3916/

Pender J (2008) Agricultural technology choices for poor farmers in less-favoured Areas of South and East Asia. International Fund for Agricultural Development (IFAD), 81 p

Pokhrel S, Pokhrel S (2013) Legumes crop rotation can improve food and nutrition security in Nepal. Agronomy J Nepal 3:123–127

Prasad A, Singh AK, Katiyar VS, Singh KD (2003) The effect of vetiver vegetative barrier in association with bund on runoff, soil loss and yield of soybean and pigeon pea intercropping. Indian J Soil Conserv 31:199–202

Prasad YG, Maheswari M, Dixit S, Srinivasarao Ch, Sikka AK, Venkateswarlu B, Sudhakar N, Prabhu Kumar S, Singh AK, Gogoi AK, Singh AK, Singh YV, Mishra A (2014) Smart practices and technologies for climate resilient agriculture. Central Research Institute for Dryland Agriculture (ICAR), Hyderabad, 76 p

Pretty J (2003) Social capital and the collective management of resources. Science 302(5652):1912–1914

Qadir M, Qureshi RH, Ahmad N, Ilyas M (1996) Salt-tolerant forage cultivation on a saline-sodic field for biomass production and soil reclamation. Land Degrad Dev 7:11–18

Radhammani S, Balasubramanian A, Ramamoorthy K, Geethalakshmi V (2003) Sustainable integrated farming systems for dry lands: a review. Agricult Rev 24:204–210

Rahim SMA, Hasnain S (2010) Agroforestry trends in Punjab, Pakistan. Afr J Environ Sci Technol 4(10):639–650

Ramakrishnan D, Durga Rao KHV, Tiwari KC (2008) Delineation of potential sites for water harvesting structures through remote sensing and GIS techniques: a case study of Kali watershed, Gujarat, India. Geocarto Int 23(2):95–108

Ramakrishnan D, Bandyopadhyay A, Kusuma KN (2009) SCS-CN and GIS-based approach for identifying potential water harvesting sites in the Kali Watershed, Mahi River Basin, India. J Earth Syst Sci 118(4):355–368

Ranada DH, Jain LK, Gupta RK, McCool DK (1997) Comparative performance of mechanical and vegetative soil conservation measures in Vertisols. Indian J Soil Conserv 25:182–185

Rao KV, Venkateswarlu B, Sahrawath KL, Wani SP, Mishra PK, Dixit S, Reddy KS, Kumar M, Saikia US (2010). Proceedings of national workshop-cum-brain storming on rainwater harvesting and reuse through farm ponds: experiences, issues and strategies, ICAR-CRIDA and ICRISAT, Hyderabad, 242 p

Reddy RS, Nalatwadmath SK, Krishnan P (2005) Soil erosion in Andhra Pradesh. NBSS Publication, no. 114, NBSS&LUP, Nagpur, India, 76 p

Regmi BR, Suwal R, Shresha G, Sharma GB, Thapa L, Manandhar S (2009) Community resilience in Nepal. Tiempo. Issue 73

Rehman S (1978) Effect of shelterbelts on yield of wheat crop in Mastung valley (Balochistan). Pak J For 28(1):4–6

Rejani R, Rao KV, Osman M, Chary GR, Puspanjali, Sammi Reddy K, Srinivasa Rao Ch (2015a) Location specific identification of *in-situ* soil and water conservation interventions for sustainable management of drylands. J Agrometeorol 17(1):55–60.

Rejani R, Rao KV, Osman M, Chary GR, Pushpanjali SRK, Srinivasa Rao C (2015b) Spatial and temporal estimation of runoff in a semi-arid micro-watershed of Southern India. Environ Monit Assess 187(8):1–16

Rejani R, Rao KV, Osman M, Srinivasa Rao C, Sammi Reddy K, Chary GR, Samuel J, Pushpanjali (2016a) Spatial and temporal estimation of soil loss for the sustainable management of a wet semi-arid watershed cluster. Environ Monit Assess 188(3):1–16

Rejani R, Rao KV, Srinivasa Rao C, Osman M, Sammi Reddy K, George BA, Chary GR, Pratyusha Kranthi GS, GR C, MV S, Rao PJ (2016b) Irrigation and drainage. Wiley, Chichester

Rockstrom J, Hatibu N, Oweis T, Wani SP (2007) Managing water in rainfed agriculture. In: Molden D (ed) Water for food, water for life: a comprehensive assessment of water manage-

ment in agriculture. Earth Scan and Colombo/International Water Management Institute, London/Sri Lanka, pp 315–348
Samra JS (2004) Resource management options in multi-enterprise agriculture. In: Acharya CL, Gupta RK, Rao DLN, Rao AS (eds) Multi-enterprise systems for viable agriculture. NAAS, New Delhi, pp. 19–39
Satapathy KK (2003) Watershed based farming systems as alternative to shifting cultivation in north eastern region. J Agric Eng 40(1):41–48
Seo SN, Mendelsohn R, Munasinghe M (2005) Climate change and agriculture in Sri Lanka: a Ricardian valuation. Environ Develop Econ 10:581–596
Shaner WW, Phillip PF, Schmehl WR (1982) Farming systems research and development: guidelines for developing countries. West View Press, Boulder
Shanwad UK, Patil VC, Gowda HH, Dasog GS, Shashidhar KS (2011) Generation of water resources action plan for Medak Nala Watershed in India using Remote sensing and GIS technologies. Aust J Basic Appl Sci 5(11):2209–2218
Sharda VN (2011) Soil and water conservation measures for checking land degradation in India. In: Mishra PK, Osman M, Satendra, Venkateswarlu B (eds) Techniques of water conservation & rainwater harvesting for drought management, SAARC Training Program, 18–29 July, 2011, Central Research Institute for Dryland Agriculture, Hyderabad, India, pp 1–20
Sharma B, Rao KV, Vittal KPR, Ramakrishna YS, Amarasinghe U (2010) Estimating the potential of rainfed agriculture in India: prospects for water productivity improvements. Agric Water Manage 97(1):23–30
Sharma KP (2010) Climate change trends and instances of socio-economic effects in Nepal. Vedams eBooks, New Delhi
Sharma KD (2011) Rain-fed agriculture could meet the challenges of food security in India. Curr Sci 100:1615–1616
Sheetal Sekhri (2012) Sustaining groundwater: role of policy reforms in promoting conservation in India. Report of Indian policy forum 2012. India International Centre, New Delhi
Singh B (1989) Rehabilitation of alkaline wasteland on the Gangetic Alluvial plains of Uttar Pradesh (U.P.), India through afforestation. Land Degrad Rehab 1:305–310
Singh MV (2001) Evaluation of current micronutrient stocks in different agroecological zones of India for sustainable crop production. Fertil News 46(2):25–42
Singh S (2008) Producer companies as new generation Co-operatives. Econ Polit Wkly 43(20):22–24
Singh R (2013a) Agricultural livelihoods and crop insurance in India: Situation analysis and assessment, Deutsche Gesellschaft fürInternationale Zusammenarbeit (GIZ) GmbH, Germany
Singh S (2013b) Institutional innovations for agricultural development in India: experience and prospects. Accessed online at http://www.punjabiuniversity.ac.in/cdeiswebsite/papers/24%20Sukhpal%20Singh%20institutionalinnovationspupatiala.pdf
Singh G, Singh NT, Abrol IP (1994) Agroforestry techniques for the reclamation of degraded salt-affected lands in India. Land Degrad Rehab 5:223–242
Singh HP, Mouat DA, McGinty HK (1998) Sustainable development of the Indian desert: the relevance of the farming systems approach. J Arid Environ 39:279–284
Singh RP, Farod AS, Joshi NL, Kathju S, Kar A (1999) Water resources and optimal water use for sustainable agriculture. In: Management of arid ecosystems. In: Proceedings of the Symposium, March 1997. Arid Zone Research Association, Jodhpur, India, pp 409–426
Sivakumar MVK, Stefanski R (2011) Climate change in South Asia. In: Lal R, Sivakumar MVK, Faiz MA, Mustafizur Rahman AHM, Islam KR (eds) Climate change and food security in South Asia. Springer, Dordrecht, pp. 13–30
Smith GD, Coughlan KJ, Yule DF, Laryea KB, Srivastava KL, Thomas NP, Cogle AL (1992) Soil management options to reduce runoff and erosion on a hard-setting Alfisol in the semi-arid tropics. Soil Tillage Res 25:195–215
Somasiri S, Handawela J, Weerakoon WL, Dharmasena PB, Jayawardena SN (1990) Rainfed upland farming for the dry zone, agro-technical bulletin for presidential task force on land alienation, Department of Agriculture, Peradeniya, 17 p

Srinivasarao C (2011) Nutrient management strategies in rainfed agriculture: constraints and opportunities. Indian J Fertil 7(4):12–25

Srinivasarao C, Venkateswarlu B, Dixit S, Kundu S, Devi KG (2011) Livelihood impacts of soil health improvement in backward and tribal districts of Andhra Pradesh. Central Research Institute for Dryland Agriculture, Hyderabad 119 p

Srinivasarao C, Venkateswarlu B, Lal R, Singh AK, Kundu S, Vittal KPR, Balaguruvaiah G, Vijaya Shankar Babu M, Ravindra Chary G, Prasadbabu MBB, Reddy Y (2012a) Soil carbon sequestration and agronomic productivity of an Alfisol for a groundnut based system in a semi arid environment in South India. European J Agron 43:40–48

Srinivasarao C, Venkateswarlu B, Lal R, Singh AK, Vittal KPR, Kundu S, Singh SR, Singh SP (2012b) Long-term effects of soil fertility management on carbon sequestration in a rice–lentil cropping system of the Indo-Gangetic Plains. Soil Sci Soc Am J 76(1):168–178

Srinivasarao C, Venkateswarlu B, Lal R, Singh AK, Kundu S (2013a) Sustainable management of soils of dryland ecosystems of India for enhancing agronomic productivity and sequestering carbon. Adv Agron 121:253–329

Srinivasarao C, Dixit S, Srinivas I, Reddy BS, Adake RV, Borkar S (2013b) Operationalization of custom hiring centres on farm implements in hundred villages in India. Central Research Institute for Dryland Agriculture, Hyderabad 151 p

Srinivasarao Ch, Gopinath KA, Venkatesh G and Jain MP (2014a) Climate change adaptation and mitigation strategies in rainfed agriculture. In: Souvenir, national seminar on technologies for sustainable production through climate resilient agriculture, Jabalpur, 8–9 August, 2014, 59–73 p

Srinivasarao Ch, Rejani R, Kumar P (2014b) Climate resilient agriculture – NICRA experiences to improve the adaptive capacity through sustainable water management. In: Proceedings of the national workshop on climate change and water: improving WUE. Water and Land Management and Research Institute (WALAMTARI), Hyderabad, pp 9–30

Srinivasarao C, Venkateswarlu B, Lal R, Singh AK, Kundu S, Vittal KPR, Patel JJ, Patel MM (2014c) Long-term manuring and fertilizer effects on depletion of soil organic carbon stocks under pearl millet-cluster bean-castor rotation in western India. Land Degrad Develop 25:173–183

Srinivasarao Ch, Chary GR, Mishra PK, Subba Rddy G, Sankar GRM, Venkateswarlu B, Sikka AK (2014d) Rainfed farming – a compendium of doable technologies, All India Coordinated Research Project for Dryland Agriculture, CRIDA, Indian Council of Agricultural Research, Hyderabad, 194 p

Srinivasarao C, Kundu S, Sharma KL, Reddy S, Pharande AL, Vijayasankarbabu M, Satish A, Singh RP, Singh SR, Ravindra Chary G, Osman M, Gopinath KA, Yasmin C (2015a) Magnesium balance in four permanent manurial experiments under rainfed agro-ecosystems of India. Crop Pasture Sci 66:1230–1240

Srinivasarao C, Lal R, Prasad JVNS, Gopinath KA, Singh R, Jakkula VS, Sahrawat KL, Venkateswarlu B, Sikka AK, Virmani SM (2015b) Potential and challenges of rainfed farming in India. Adv Agron 133:113–181

Srinivasarao Ch, Venkateswarlu B, Sikka AK, Prasad YG, Chary GR, Rao KV, Gopinath KA, Osman M,. Ramana DBV, Maheswari M and Rao VUM (2015c). District agriculture contingency plans to address weather aberrations and for sustainable food security in India. ICAR-Central Research Institute for Dryland Agriculture, Natural Resource Management Division, Hyderabad-500 059, India, 22 p

Srivastava RK, Sharma HC, Raina AK (2010) Suitability of soil and water conservation measures for watershed management using geographical information system. J Soil Water Conserv 9(3):148–153

Subhan FD (1990) Financial analysis of selected shelterbelts systems in Pakistan. Pak J Forest 44(1):247–252

Tamang BB (2016) Contour bunding. A traditional low-cost method of soil conservation suitable for sloping land. Accessed on March 2016. http://www.icimod.org/?q=10412

Uddin MM, Saheed SM (2016) Soil and water conservation and dryland farming in the hilly areas of Bangladesh. Bangladesh Agricultural Research Council New Airport Road, Farmgate, Dhaka-1215, Bangladesh. Accessed on March 2016. Available at http://natres.psu.ac.th/Link/SoilCongress/bdd/symp36/2324-t.pdf

Umrani NK, Reddy GS (1999) Cropping systems research in drylands – a review. In: Singh HP, Ramakrishna YS, Sharma KL, Venkateswarlu B (eds) Fifty years of dryland agricultural research in India. Central Research Institute for Dryland Agriculture, Hyderabad, pp. 105–114

Venkateswarlu J (2004) Rain-fed agriculture in India Research and Development Scenario, published by ICAR, New Delhi, pp 1–508

Venkateswarlu B (2010) Climate change: adaptation and mitigation strategies in rainfed agriculture. Dr SP Raychaudhuri Memorial Lecture, Birsa Agricultural University, Ranchi, India, 11 p

Venkateswarlu B, Prasad JVNS (2012) Carrying capacity of Indian agriculture: issues related to rainfed agriculture. Curr Sci 102:882–888

Virgo KJ, Aslami MH, Ahmed B (2006) Participatory watershed management: examples from Herat, Western Afghanistan. P World Assoc Soil Water Conserv P1:65–81

Vittal KPR, Vijayalakshmi K, Rao UMB (1983) Effect of deep tillage on dryland crop production in red soils of India. Soil Till Res 3:377–384

Vittal, KPR, Singh HP, Prasad JVNS, Rao KV, Victor US, Maruthi Sankar GR, Ravindra Chary G, Gurbachan Singh, Samra JS (2003) Bio-diverse farming system models for dryland agriculture. All India Coordinated Research Project, Central Research Institute for Dryland Agriculture, Hyderabad 500 059, Andhra Pradesh, India, 58 p

Weerahewa J (2009) Impact of trade facilitation measures and regional trade agreements on food and agricultural trade in South Asia. Asia-Pacific Research and Training Network on Trade. Working paper series, no 69. Available at http://www.unescap.org/sites/default/files/AWP%20No.%2069.pdf. Accessed on 1/11/2015

Wijayaratna CM, Weerakoon WL (1996) Water harvesting and soil and water conservation practices in the dry zone of Sri Lanka. Consulation on small water harvesting schemes, Society for promotion of waste lands and development, New Delhi, India, March 1996

World Bank (2011) Agriculture in South Asia. Thimphu Statement on Climate Change. 10th Feb 2012. http://www.saarc-sec.org/userfiles/ThimphuStatementonClimateChange-9April2010.pdf

www.adb.org/publications/basic-statistics-2015

Zahra N, Shah H, Khan AM, Anwar MZ (2014) Livelihood assets and poverty nexus: a case study from rainfed Pothwar area of Pakistan. Pak J Agric Res 27(3):255–264

Zia MS, Muhmood T, Baig MB, Aslam M (2004) Land and environmental degradation and its amelioration for sustainable agriculture in Pakistan. Science Vision 9(1–3):21–25

Integrated Dryland Agriculture Sustainable Management in Northwest China

Feng-Min Li, You-Cai Xiong, Xiao-Gang Li, Feng Zhang, and Yu Guan

1 Introduction

China dryland agriculture is developed in the region with an average annual rainfall of 250–550mm, at longitude 73°50′–135°05′, latitude 33°20′~53°30′. The total area of this region is up to 5.42 million km^2, accounting for 56% of total land area of Chinese territory. The area of regional arable land is 51 million hm^2, i.e. about 51 % of total arable land area of China. Sixty five percentage of which belongs to the area of dryland without irrigation. Local population in this region accounts for 32 % of total population of the country. Dryland agriculture is mainly distributed in the hilly area and few are distributed in the plain area for the region. As for local weather condition, winter is cold and dry, and summer is warm and rainy. Monthly average temperature ranges from −10 to −5 °C in January, and from 20 to 25 °C in July. Seasonal precipitation amount during the period from June to September occupies 60–70 % of annual precipitation amount. In most areas, average annual temperature is almost 6–10 °C, and the frost-free period reaches up to 100–170 days (Li 2007). As affected by southeast and west monsoon, local rainfall is at a low level but with high fluctuation, resulting in low and unsteady agricultural productivity, which forms the basic characteristics of dryland agriculture (Li et al. 2003).

China's dryland agriculture is mainly distributed in hilly area, which favors the formation of rich biodiversity and the maintenance of ecosystem stability. Supposing the reclamation intensity is lower, fine biodiversity would be maintained and turn to be beneficial for the stability and sustainable development of agricultural ecosystems. However, due to low yield of crops in this region, a vicious circle has been

F.-M. Li (✉) • Y.-C. Xiong • X.-G. Li • F. Zhang • Y. Guan
State Key Laboratory of Grassland Agroecosystem, Institute of Arid Agroecology, School of Life Sciences, Lanzhou University, Lanzhou 730000, China
e-mail: fmli@lzu.edu.cn

M. Farooq, K.H.M. Siddique (eds.), *Innovations in Dryland Agriculture*,
DOI 10.1007/978-3-319-47928-6_14

formed between extensive cultivation and poor production since ancient times. In order to survive from food shortage, local farmers tended to enlarge planting area using the method of setting on fire and burning mountains, i.e. so-called "slash and burn cultivation". At the initial few years following reclamation cultivation, a huge amount of topsoil was lost, and soil water and fertility conditions were increasingly worsened, which further resulted in the decline of crop yield. Facing this dilemma, local farmers tended to abandon the low-yield farmland, and choose another uncultivated land for clearing and farming. This vicious circle was going round and further accelerated the shrink and degradation of vegetation cover.

Under weak population pressure before, self-rehabilitating capacity of ecosystem could to some extent compensate the decline of soil quality caused by continuous cropping, and maintain the sustainability of regional ecosystem. With the increase in population pressure, the sustainability of ecosystem would decline gradually till the collapse of ecosystem at the threshold of pressure. For more food production, farmers chose to reclaim the lands which was previously used for natural restoration. The newly reclaimed lands were used for grazing and cutting down trees excessively, resulting in continuous deterioration of natural vegetation and serious soil erosion. With further degradation of soil system, raw black loam in topsoil layer, accompanying with native vegetation, was quickly lost (Guo and Li 2014). Instead, the loess soil was remained at topsoil layer in a large area, with a very low content of organic matter up to only about 1 % (Li et al. 1998, 2003). In this case, natural ecosystem suffered much from intensive grazing activities, bringing about a vicious cycle between low agricultural productivity and land degradation. Therefore, extensive reclamation and decentralized management are major social-economic characteristics in the society of small household farmer in the historical period, and tended to become the major reasons to cause the degradation of agricultural ecological environment (Zhang 1989). Over 60 years since the foundation of new China, the government enforced several mega projects to focus on soil and water conservation and agricultural development in the region, which produced positive impacts on agricultural production, soil erosion, land degradation, and vegetation restoration. However, owing to the rapid increase in regional population and adverse conditions, the shortage of food production could not be fundamentally resolved over a long period. Vegetation coverage and soil quality were still extremely deteriorated, which was difficult to reverse fundamentally. The protection of biodiversity resources and its cushioning effects on agricultural ecosystem were not exhibited. At the beginning of this century, the situation of ecological environment in this region was shown as "more damages than protection, partial improvement but overall deterioration"(Yang et al. 2001).

From the perspective of ecosystem, the problems above mentioned were originated from strongly weakening assimilation and increasingly activating catabolism within the system. Those factors including expanding reclamation area under the condition with destructing vegetation, low agricultural productivity, serious soil erosion by water and wind, and continuously declining soil fertility, have led to more active catabolism in the system. The signals of weakened assimilation are the decline of vegetation cover and the low productivity of agricultural ecosystem and natural ecosystem. The uncoordinated changes in the functions of two systems are called the disorders of ecological balance. The key strategy to solve the problem is striving to improve the assimilation scale and productivity of the system. Improving productivity of people-centered agriculture (in a broad sense) is the key to improve ecological environment and develop sustainability of ecosystem (Li 1999).

Over last decades, great achievements have been made in the basic construction of farmland including land leveling and terrace construction programs in the dryland agricultural areas of Loess Plateau. In recent ten years, innovative cultivation system of dryland crops, mainly featured by ridge-furrow covering farming technology, led to a great increase in the yield of crops, and accordingly produced a large amount of straw resources. This help provides rich sources of crude forage for the development of animal husbandry, and therefore gradually updated the people's understanding on the mode of animal husbandry development. This also resulted in the rapid development of the animal husbandry, which has been in long delay before, and provided an opportunity for the integrated development of dryland ecosystem. In this chapter, we provide a brief introduction to the key points of these developments, and summarize the main principles and practices involved. These will provide references for harmonizing agricultural development, ecological protection and regional sustainable development in those similar areas in the world.

2 Terraces and Fertility Betterment

Dryland farming in China is mainly distributed in the hilly area where the terrace is a primary measure to control soil erosion and stabilize the development of agricultural production. Terrace construction has a history of about 2000 years in China, and plays an important role in the development of dryland agriculture. Terrace establishment can help intercept runoff and sediment, increase precipitation infiltration, and improve agricultural productivity while reduce soil erosion (Fig. 2).

In general, the terrace can be categorized into four basic types (Guo and Li 2014): sloping terrace, interval terrace, bench terrace and back terrace. Sloping terrace is relatively primitive and was generally established under the pervasive availability of preliminary tool of production. It consists of several ridges along the contour to intercept sediment from runoff on sloping land, which is easy to build and abolish. Interval terrace is more efficient than sloping terrace, which remains a slope allocated between the upper and lower terraces. This kind of terrace looks flat to reduce the risk of runoff. Flat terrace is shaped by shoveling and flatting the planting surface, and then constructing a dyke at the outer edge of the terrace. There exists a relatively steep slope land between the upper and lower terraces, which enables land use more efficiently and crop productivity to be significantly improved. On the other hand, back-slope terrace is built as reverse gentle slope, which intercepts sediment from runoff more effectively. The establishment requires more labor input, yet the capability of controlling soil erosion is significantly increased. This sort of terrace is suitable for the areas with high rainfall amount. Since the founding of new China, the government has taken large-scale actions to build terraces. The terraces construction during the period from 1950s to 1970s had laid a strong base for food security of newly established new China. When it came to the 1980s at the beginning of the Policy of Reform and Opening Up, people's enthusiasm of terrace construction declined slightly. Yet in the 1990s, terrace construction revived once again. Since entering the twenty-first century, the scale and speed of construction were further accelerated. Efficiency and quality of terrace building have been greatly improved by the application of machines (Xie et al. 2001; Wang 2006).

Numerous studies showed the effect of terraces on receiving the infiltration of precipitation and increasing production (Ye 1984). In the hilly region of semi-arid Loess Plateau, 92.4 % of rainfall in the summer effectively was trapped by terraces and then infiltrated into soil system; the sediment loss was also greatly reduced correspondingly. The soil water storage in the depth of 200 cm with terraces was 110–180 mm more than that of slope farmland in a normal year. The water use efficiency of crops in terraces built over 3 years was 1.01–3.38 kg/(mm hm^2) more than that of farmland with a slope more than 10° (Yang 2006). In the semi-arid region of Ningxia, compared to sloping farmland, terraces decreased soil bulk density in the depth of 0–100 cm, and increased the percentage of capillary porosity. During the period from April to November, terraces increased the soil moisture in the depth of 0~100 cm, suggesting that terraces were more suitable for the infiltration and preservation of rainfall, and maintaining high and stable yield. (Cai et al. 2007).

The major features of newly built terraces include low soil fertility, biological activity and crop yields. To seek the ways to effectively improve soil fertility and crop yield is an important issue. Many agricultural practices, including farming, irrigation, film mulching, fertilizer application, and crop rotation system and others could increase crop yield and water use efficiency (Ye 1984; Xie et al. 2001; Agbede and Ojeniyi 2009; Behera and Panda 2009; Kunzová and Hejcman 2009). Among all these measures mentioned above, fertilization was the most direct and effective way to achieve higher crop production and maintain soil quality (Hong et al. 1998; Xie et al. 2001; Bhattacharyya et al. 2007;). Our group had conducted a field experiment in newly built terraces located in the region with an annual average rainfall of 300 mm for seven consecutive years. It was found that crop yield and water use efficiency with the mixed application of nitrogen and phosphorus fertilizer were significantly greater than those with the sole application of organic fertilizer and the control of no fertilizer application. Yet, with the year extending, the magnitude of the increased yield by nitrogen and phosphorus fertilizer application tended to decrease. In contrast, the effect of organic fertilizer application on yield turned to have a significant increase (Fig. 1). After 4 years, the yields of peas and spring wheat with organic fertilizer were higher than those of nitrogen and phosphorus fertilizer treatments (Fig. 3). Meanwhile, the amount of macro-aggregates particles (> 0.25 mm) with the application of organic fertilizer was significantly higher than those of nitrogen and phosphorus fertilizer and no fertilizer control; the water holding capacity of soil was also significantly improved. Thus, application of organic fertilizer was the key measure to promote soil aging of newly built terraces and sustainable increase of soil fertility and land productivity (Liu et al. 2013).

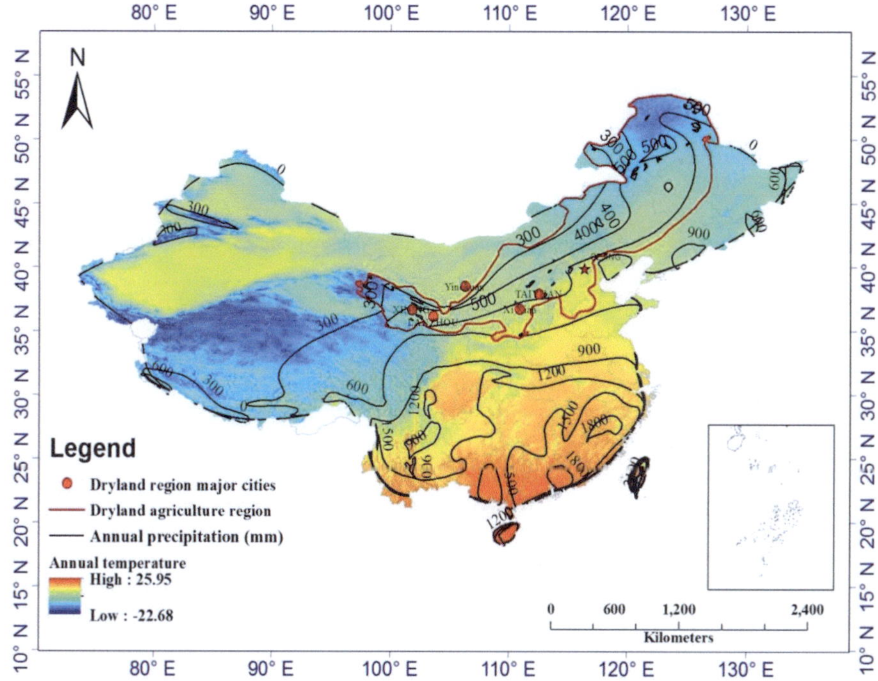

Fig. 1 Dryland agricultural areas in China (Data (1950–2000) of air temperature and precipitation are from China Meteorological DADA Sharing Service System)

Fig. 2 Terrace system in the dryland agricultural areas in China (Photo by Feng Zhang)

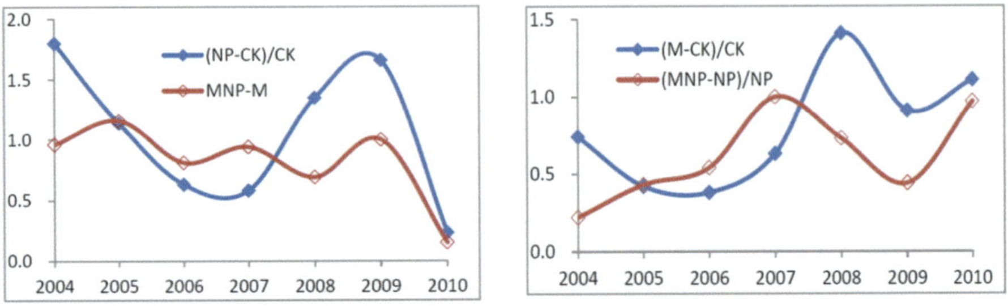

Fig. 3 Grain yield effect by application of nitrogen and phosphorus chemical fertilizer (*Left*) and organic fertilizer of manure (*Right*). CK, NP, M, MNP represented no any fertilizer application, NP (Nitrogen and phosphorus) fertilizer application, manure fertilizer application, and both manure and NP chemical fertilizers application, respectively. NP yield effects (*Left*) are fluctuating very much and trend to decline but manure application yield effects (*Right*) are trending to increase as the terrace is developing (Data from Liu et al. 2013)

3 Plastic Film Mulched Ridge-Furrow Cultivation

In many countries around the world except China, plastic film mulch is widely used to grow vegetable plants, horticulture and other cash crops rather that grain crops. In China, plastic film mulch was firstly also used for vegetables production in green house. In 1990s, scientists had studied the role of plastic film mulch in field grain crops' production and the tested crops included wheat, potato, and corn etc. (Wan 2013). In the past 20 years, dryland agriculture has developed rapidly based on the mulched ridge-furrow cultivation.

Historically, wheat is the most important food crop and the source of local people's staple food in dryland agricultural areas of China. However, wheat growth only covers about 1/3–1/2 of the rainy season. As a result, drought is a major obstacle to wheat production, and the yield is normally low and unstable. Naturally, wheat was chosen for the first crop to research its response in the productivity to the plastic film mulch technology. In Lanzhou, the yield of spring wheat with plastic mulch was increased by about 50 % relative to no plastic mulch (Li et al. 1999b). The invention of seeders (dibblers) to sow wheat seeds underneath the plastic film has brought an easy use of plastic film mulching in wheat cropping and thus a popularization of the technology in the wheat production in dryland areas in the 1990s. In 2000, a promotion of plastic film mulch in wheat production was proposed, in which furrows were dug and gaps (ridges) between the furrows were mulched with plastic film. Wheat crops were planted beside the edge of film in furrows. This new technology of plastic film mulch largely reduced management procedures in whet growth. By using this mulched ridge-furrow cropping, wheat yield reached to 7311 kg ha^{-1} at a site with a precipitation of 685 mm, 40 % higher than under non-mulched fields, 31% higher than crushed-stubble mulched no-till cropping, and 20 % higher than stubble-mulched sub-tilled treatment (Xue et al. 2006). Prior to wheat planting, covering a layer of soil on mulching-film significantly reduced

Table 1 Potato tuber yield of rainfall-harvest cultivation of ridge-furrow mulched with plastic film

Site/Year	Growth season	Rainfall (mm)	Un-mulch Flat (kg/ha)	Mulched R-F(kg/ha)	Sources
Gaolan/2001	April 29–Sept 5	162.1	995 D	3602 D	Wang et al. (2005)
C Yuzhong/2002	April 29–Sept 5	181.6	671 D	2084 D	
Dingxi/2009	April 30–Oct 8	245.5	2739 D	3993 D	Zhao et al. (2012)
Dingxi/2010	April 25–Oct 5	243.0	2973 D	5288 D	
Dingxi/2010	April 12–Sept 24	227.8	26,590 F	49,657 F	Qin et al. (2014)
Dingxi/2011	April 15–Sept 26	213.2	28,181 F	45,227 F	

Note: D and F mean dry weight and fresh weight, respectively

weed growth and saved labor; with this practice, the land needed no-tillage for three to four years. The grain yield was also increased by 35–37 % (Han 2013) and 12 % (Ma 2010) compared with conventional uncovered cultivation and film side planting technology and film side planting technology, respectively, producing a significant economic benefit than other farming methods.

Potato is one of dominant crops in the most dryland agricultural regions in China. The water demand of potato during growth largely matches the distributions of rainfall and temperature in the season. Entering the twenty-first century, the area of potato growing was quickly enlarged due to increasing demand of tubers in the market. The plastic film mulched ridge-furrow technologies improved potato yield by 20–50 %, even 100 % than non-mulched flat or ridge-furrow cultivation (Table 1) (Tian et al. 2003; Wang et al. 2005; Li 2011; Qin et al. 2014; Zhao et al. 2014). In the potato production, black (opaque) plastic film had a better increasing effect on yield than transparent plastic film (He et al. 2015). Removing off the film in the middle of the growing season not only avoided the risk of plastic residue pollution, but also maintained the same yield of 5280 kg ha^{-1}, as was achieved from season long mulching (Fan et al. 2012; Zhao et al. 2012).

Maize is a thermophilic crop with higher consumption of water and fertilizer. In the past, farmers in dryland areas had little enthusiasm to plant maize. In recent ten years, along with the increase demand of the global market and livestock feeding in China, the price of maize continuously increased, almost up to the same level as wheat. As a high-yielding crop, maize had more obvious advantages to produce economic returns than wheat. The straw of maize can be used as forage and biogas, and increasing soil fertility when applied back to the field. With so many advantages, maize drew many attentions of the government, farmers and scientists.

Plastic film mulched ridge-furrow is currently a major cultivation pattern of maize in dryland agriculture (Table 2). Big and small ridges are made and then completely covered with transparent polyethylene film (Fig. 4). Maize is planted in

Table 2 Maize grain yield of rainfall-harvest cultivation of R-F (ridge-furrow) mulched with plastic film

Site/Year	Growth season	Rainfall (mm)	Un-mulch Flat (kg ha^{-1})	Mulched R-F(kg ha^{-1})	References
N Yuzhong/2006	March 29–Oct 5	171	170	1150	Zhou et al. (2009)
N Yuzhong/2007	April 3–Sept 29	259	536	6130	
C Yuzhong/2008	Early April	314	5350	8230	Liu et al. (2014b)
C Yuzhong/2009	to	257	3460	7770	
C Yuzhong/2010	Early Oct	291	4030	9120	
C Yuzhong/2011	April 27–Oct 8	215	2320	4880	
C Yuzhong/2012	April 27–Oct 8	336	3810	8470	
Heyang/2007	April 28–Sept 15	398	7674	9402	Li et al. (2012)
Heyang/2008	April 15–Sept 5	330	8844	11,792	
Heyang/2009	April 26–Sept 18	379	7136	10,103	
Heyang/2010	April 24—Sept17	390	7524	10,501	

Rainfall is within the growing period of maize; N Yuzhong means Northern mountain area of Yuzhong County; C Yuzhong means Central plain area of Yuzhong County

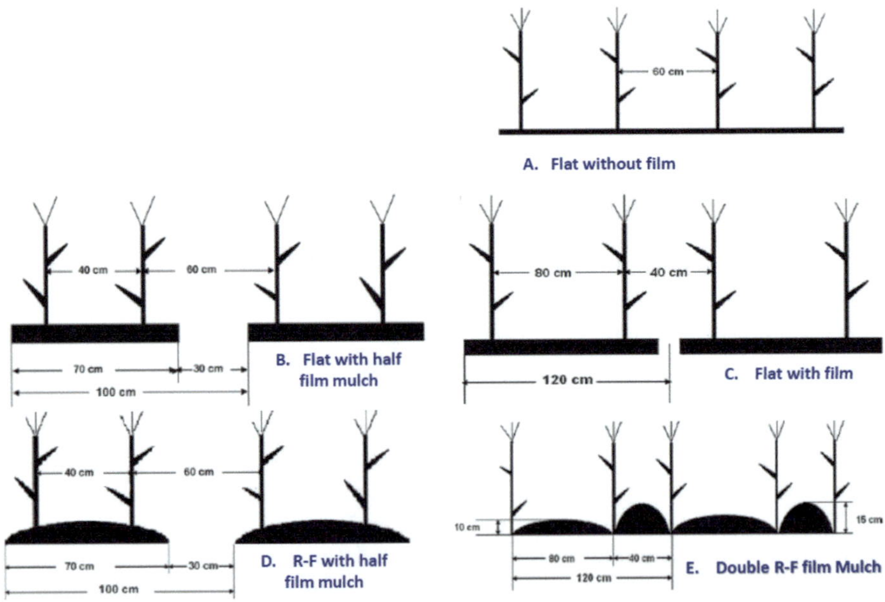

Fig. 4 Plastic film mulching patterns are in succession for maze production, which lead to yield increasing significantly. Pattern e is now popular and highest yield

the furrows between ridges. The width of the ridge and furrow is dependent on the specific water and heat conditions. In the area with an annual rainfall of 300–400 mm, V-shaped furrow is often used due to insufficient hydrothermal conditions. A V-shaped furrow is a naturally formed furrow of two adjacent ridges, which has a width of zero. The main function of ridge-furrow with film-mulching is harvesting rainwater. The ridge and furrow itself has a capability to collect precipitation from surface to bottom of the ridge. With film-mulching, the precipitation on the ridge surface is collected to the bottom of furrows, and then is infiltrated into soil through small holes on the film. Thus, useless precipitation during light rains in the non-mulched fields is efficiently overlapped to maize root zones in the film mulched ridge-furrow cropping (Jiang and Li 2015; Li et al. 1999a).

Plastic film mulch on ridge-furrows can effectively reduce water evaporation from soil. Compared to un-mulched cultivation, plastic film mulch prepared before the previous winter increases the water storage in 2 m-depth soil profiles by 43 mm at maize sowing time (Liu et al. 2009). Plastic film mulch also significantly increased soil moisture during the early growth of maize. But at the peak time of maize growing, this effect of increasing soil moisture is usually not significant. In some cases, plastic mulch even reduces soil moisture compared with no mulch (Li et al. 1998, 2005; Liu et al. 2014b; Wang et al. 2014). Liu et al. (2014b) once assessed the effect of plastic film mulched ridge-furrow copping to conseve rainwater in soil under non-maize-cropped condition. They found that in one growing season plastic film mulched ridge-furrow cropping increased soil water contents in the 170 cm-depth to the field capacity, and conserved 100 mm more rainwater in soil than un-mulched ridge-furrow cropping.

Plastic film mulch can effectively enhance temperature in the topsoil. This function of the plastic film mulch expands maize planting to the areas where previously maize was not a profitable crop due to low yield under insufficient rainfall and low temperature conditions. The effect of plastic film mulch in increasing soil temperature mainly occurs in the early stage of crop growth (seedling and jointing) (Liu et al. 2009, 2014b; Zhou et al. 2009; Wang et al. 2014; Hai et al. 2015). At a site with annual temperature of about 6.6 °C and precipitation of about 300 mm, found that daytime soil temperature in the 0–15 cm was 3.0 °C higher in mulched than un-mulched plots throughout maize growing season in the absence of maize growing (Hai et al. 2015). However, in the presence of maize growing, daytime soil temperature in the same depth increased by 2.8 °C in mulched relative to non-mulched plots, only in the early growth (before jointing) but not in the late growth (Hai et al. 2015). This indicates that during the late growth when soil surface is mostly shaded by canopy, the plastic mulch effect in increasing soil temperature is not significant. The increasing effect of film-mulching on soil temperature is the key to plant maize in high altitude areas with a lack of heat. In the areas with an average annual rainfall of only 300 mm, average annual temperature of 6.5 °C, and altitude of 2400 m above sea level, maize was completely unable to grow without film-mulching; the application of ridge-furrow with film-mulching made maize yield reach to 6–7 t ha^{-1}, 5–10 times more than conventional uncovered cultivation (Liu et al. 2009; Zhou et al. 2009). In a continuous four-year experiment, it was found that the lower limit of the average annual rainfall was 273 mm, which was appropriate for the use of ridge-furrow with film-mulching (Liu et al. 2014a). With the application of ridge-furrow with film-mulching, maize yield could reach to 6–9 t ha^{-1}, 80–100 % more than conventional uncovered cultivation in the region with an annual precipitation of 300–400 mm; and reach to 9–12 t ha^{-1}, 30–80 % more than conventional uncovered cultivation in the region with an annual precipitation of 400–600 mm (Li et al. 2009; Ye and Liu 2012). With the increase of precipitation and temperature, the yield-increasing rate of film-mulching decreased gradually, reaching to 10–30 % in the region with an annual precipitation of 550–600 mm (Ye and Liu 2012). In the region with better conditions of water and heat, the black film could avoid unwanted increase in temperature, resulting in better increasing effect of maize yield than transparent film (Wang and Li 2011; Lu et al. 2016).

4 Grassland Productivity and Development of Animal Husbandry

China dryland agriculture lies mainly in hilly regions, where rainfall is concentrated mainly in summer and autumn, frequently with heavy rains, which is prone to serious soil erosion. Historically, with the increasing of population, in order to solve the food shortage, natural vegetation clearance had been continued for thousands of years. Overgrazing and firewood collection were also among the major direct anthropogenic factors responsible for destruction of vegetation cover and causing soil erosion. In order to maintain and increasing the vegetation coverage and to control soil erosion, governments and scientists are advocating to promote grassland animal husbandry (Ren and Chang 2009; Zhang 2010). Through the establishment of artificial grasslands and restoration of native vegetation, grassland productivity is extended to be improved. It has been hoped that with development in livestock sector, it may increase the income of local people, and thereby may improve the living standards of the local people with increased vegetation, soil and water conservation at the same time (Ren and Lin 2009). Since the 1980s, the Chinese government and the scientific communities have made great efforts to achieve this objective. Breeding new grass varieties, improving grassland managements and reseeding grassland were applied in the vast of grassland in China. Meanwhile, international collaborations, foreign experts' technical support and the World Bank special funding were supplied to reach this goal. However, the objective is still hard to achieve with nearly 30 years' implementation of the project, and lots of effort with little success, since the 1980s to the first decade of the twenty-first century (Yang et al. 2001; Shan and Xu 2009).

Raising cattle and sheep is one major part of animal husbandry in the dryland agriculture area of China. To increase vegetation cover and animal husbandry development, people have been trying to develop artificial grassland. There is a consensus that grazing domestic animals on natural grasslands in hilly areas leads to the

Table 3 Aboveground biomass comparison between alfalfa and wheat in areas with various annual mean precipitations

AMP (mm)	Survey site	Alfalfa BPY (kg ha^{-1}, year)	Alfalfa AAABY (kg ha^{-1}, year)	Sources	Wheat B (kg ha^{-1})	References
300	N Yuzhong	5527, 7	4471, 10 years	Jia et al. (2009)	5390	Jia et al. (2009)
450	Guyuan	6368, 6	4533, 8 years	Du et al. (1999)	6179	EV
580	Changwu	7284, 5	4852, 6 years	Hao et al. (2004)	7500	Guo et al. (2008)

Note: AMP is Annual Mean Precipitation; Alfalfa BPY is Alfalfa Biomass Peak and occurred Year after sowing; Alfalfa AAAB is Averaged Annual Accumulated Biomass and occurred Year after sowing; Wheat B is wheat biomass including straw and grain at the same site as alfalfa. EV is an estimated value: (5390 / 300 + 7500 / 580) / 2 * 450 = 6179

soil erosion, and thus barn feeding must be the dominate measures for husbandry developing, with supplement of grazing animals in areas around the croplands (Zhang et al. 2012; Li and Shen 2013). The grazing is not allowed on natural vegetation. Grassland and crop residues are the major sources of forage for barn feeding sheep. Productivity of artificially cultivated grasslands is much higher than natural grasslands, and it has become the dominated source of forage grass. As the alfalfa productivity is significantly higher than other forage species and it also can fix nitrogen from the atmosphere as legumes, it becomes one of most important legume forage for artificial grasslands (Zeng 2002; Shan et al. 2008).

Biomass productivity of sown alfalfa grassland is generally lower than that of wheat at the same type of region (Table 3) (Du et al. 1999; Hao et al. 2004; Guo et al. 2008; Jia et al. 2009; Jia and Li 2011), even if furrow-ridge with plastic film mulching is used for alfalfa (Jia et al. 2006, 2009). In the area with 300 mm average annual rainfall, although grass yield may be higher than local wheat aboveground biomass, it was still smaller than the half of maize aboveground biomass under the ridge-furrow with plastic film mulching. Undoubtedly, alfalfa grassland is inefficient in providing the forage grass for the husbandry developing compared to maize planting, or even it's less efficient than wheat planting. Therefore, with dependence only on the planting of alfalfa grassland to develop the animal husbandry, it is hard to solve the food problem and feed the local people, and in improving people's income. This could be an important reason of the unwillingness of local people in planting alfalfa for a long time.

The household survey further confirmed this view (Zhang et al. 2012). In the North Mountainous area of Yuzhong County in Gansu (average annual rainfall of about 300 mm). The study pointed out that the feeding sheep numbers or willingness have significant negative correlation with the number of sold potatoes, and it had also significant positive correlation with the maize dominated cereal production, and no significant correlation with the alfalfa production (Fig. 5). This result showed that planting alfalfa is insufficient to support the development of sheep hus-

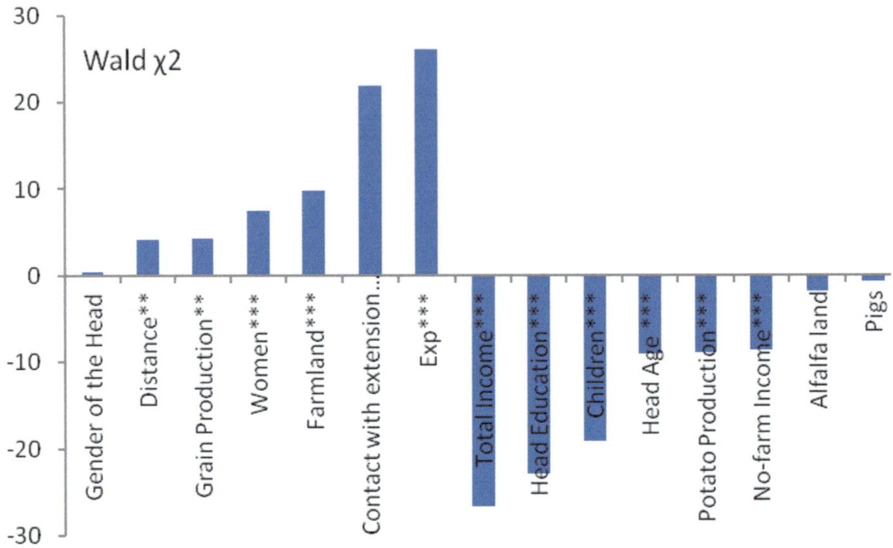

Fig. 5 Logistic regression analysis of influencing factors of local farmer household adopting raising sheep indoor in North mountain area (annual precipitation 300 mm, annual mean temperature 6.5 °C) of Yuzhong County, Gansu Province. Wald X^2 indicated that farmland and grain production influenced positively on the adoption but the influence of alfalfa land has negatively influence trend to the adoption although this influence is not statistically significant. (Data from Zhang et al. 2012)

bandry industry, and maize as one of main food crops. However, the production capacity determines the size of feeding livestock. As potato stalks and tubers cannot be directly used as feeding forage, so the most of the area under potato will induce the smaller scale feeding sheep.

Mowing is the most common form to use alfalfa pasture. In another aspect, the soil fertility restoring ability of mowing alfalfa land is an important reflection of the productivity of alfalfa grassland. Our research indicates that soil organic carbon (SOC), soil total nitrogen (STN) and soil SOC/STN ratio showed a slow downward trend in a conventional mowing alfalfa land. It implied that the mowing alfalfa was unable to improve soil quality, so that it cannot restore the farmland fertility through rotation between crop land with this type of mowing alfalfa grassland (Jiang et al. 2006; Wang et al. 2006, 2009).

Vegetation restoration is an important way to improve the ecological environment of the semi-arid areas. Returning farmland to grassland or forest is the main vegetation restoration policy. According to 6 years' continuous field observation of vegetation restoration process with three methods: planting alfalfa (*Medicago sativa* Linn.); biennial sweet clover (*Melilotus officianalis* Linn.) and natural recovered abandoned land (Yuan et al. 2015b), it was found that abandoned land biomass is only 1000 kg ha^{-1} within the whole experiment period; sweet clover biomass above 2000 kg ha^{-1} in the majority of the years, except decreased significantly in the third year. The alfalfa biomass more than 2000 kg ha^{-1} in the third year, furthermore, it

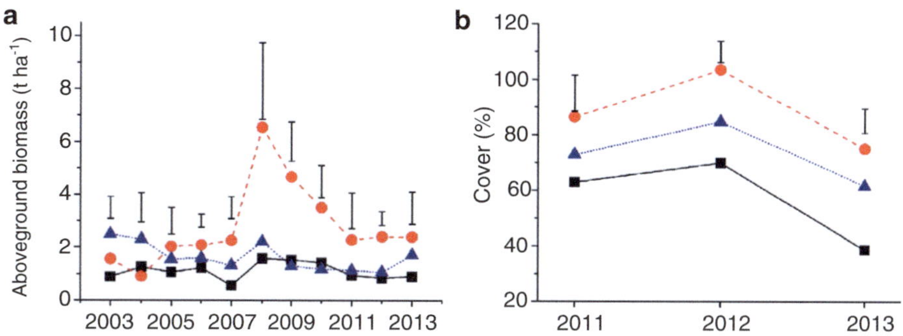

Fig. 6 Perennial legume species alfalfa (*Medicago sativa* Linn. ●) revegetation has higher aboveground biomass and coverage than biennial legume species sweetclover (*Melilotus officianalis* Linn. ▲) revegetation and natural restoration vegetation (■) after cropland abandonment. (From Yuan et al. 2015b)

grows up gradually and as high as 6000 kg ha^{-1} at the end of 6 years' experiments (Fig. 6). With the plantation of alfalfa and sweet clover as an artificial vegetation restoration measure, the biomass was significantly higher than the natural recovered abandoned land. In the perspective of ecological benefits from carbon sequestration, solid soil, reduce soil erosion the alfalfa clearly has more advantages (Fig. 6). Alfalfa as the preferred species for vegetation recovery, cutting once a year in summer also helps to maintain vegetation vigor (Guo et al. 2010; Yuan et al. 2015a, c).

There is a concern that alfalfa could exhaust deep soil water. Our research indicated that for a cropland plowed after 9-year-old alfalfa pasture, planting annual crops (millet grass) first year and then with different crop rotations for four years, 0–5 m soil moisture can restore up to about 90 % of soil moisture in normal annual crop farmland (Wang et al. 2008). Several researchers have obtained similar results indicating that the deeper soil drier problems led by alfalfa planting can be alleviated (Wang et al. 2009). Alfalfa pasture is as a production system in all the researches. And, if we use alfalfa pasture as a re-vegetation for ecological protection purpose, and alfalfa and other species biomass are not harvested, alfalfa vegetation will lead to weak soil water utilization, which soil moisture should not be lower than the production system.

From the analysis above, the alfalfa pasture productivity remained lower than the main crops (for example corn, potatoes, sometime in wheat), and it is difficult to be the region's major source of forage for large-scale animal husbandry development. Natural vegetation, which was restored from abandoned land, have very low productivity, which also cannot support large-scale grassland animal husbandry. The local farmer's adoption of barn feeding sheep also clearly illustrates the alfalfa pasture has disadvantages in forage yield compare to rapidly expanded maize planting. The annual crops, with maize as the representative, are becoming the major forage source of livestock feeding. Meanwhile, the alfalfa can only be used as auxiliary forage jointly support the development of animal husbandry mixed with the grain and grassland production. On the other hand, due to higher productivity and vegetation coverage, the alfalfa plantation gained the significant advantage in ecological vegetation restoration compare to vegetation restoration in abandoned land and sweet clover plantation land.

5 Sustainability of Dryland Agriculture Ecosystem

To control soil erosion, accumulate rainwater and improve water use efficiency via large-scale land leveling is the fundamental of rapid development of dryland agriculture over the last 20 years in China (Li et al. 1999a, 2003; Li and Xu 2002). Land leveling and rainwater accumulation programs have been basically accomplished through the establishment of various terraces across different periods of past few decades. In addition to ridge-furrow plastic-mulching system used for improving water use efficiency as mentioned above, another key technology is the construction of underground rainwater storage cellar, with the aim to provide the necessary supplementary irrigation water for the plants with high economic value. The rainwater-harvesting cellar can provide a variety of combinations with water, fertilization, and soil thermal supply, which offered the conditions for the introduction of high-value economic plants. Greater diurnal temperature difference and better lighting resource were conducive to the accumulation of photosynthetic products in semi-arid areas. This help creates the most favorable conditions to increase crop quality and yield. Therefore, with the development of high-value economic plants, the economic efficiency of agricultural production process was increased. The new dryland agriculture system was expected to alter the hopeless situation of traditional dryland agriculture, and to become a promising agriculture development system. The efforts have injected new power into the entire agriculture and rural development (Cui and Wang 2002; Wang et al. 2001; Cui et al. 2005).

Livestock industry in the semi-arid region is endowed with a certain advantage. A large number of studies suggested that this advantage should play more important roles (Ren 2007; Ren and Lin 2009; Shan and Xu 2009; Zhang 2010). However, as the population is increasing with time, the contradiction between food production and livestock production has become increasingly prominent. For the past 20 years, under the support of rain-harvesting farming techniques, grain yields per unit area were substantially increased, and the area of farmland used for self-consumption crops was gradually reduced. Moreover, due to the introduction and cultivation of cash crops, economic benefits have been improved, which has provided a more flexible preconditions to return farmland to grassland for feeding livestock, adjust and optimize industrial structure. Furthermore, farmers could optimize the production structure interactively with the market (Li and Xu 2002; Li et al. 2003).

Based on the previous analysis, combined with characteristics of the hilly landscape in dryland agriculture region, we proposed a management model of dryland agricultural landscape system (Fig. 7). The resource conditions gradients tended to be worse from the foot to the top of low hill. This can also be seen as an ecological band spectrum (Li et al. 2003).

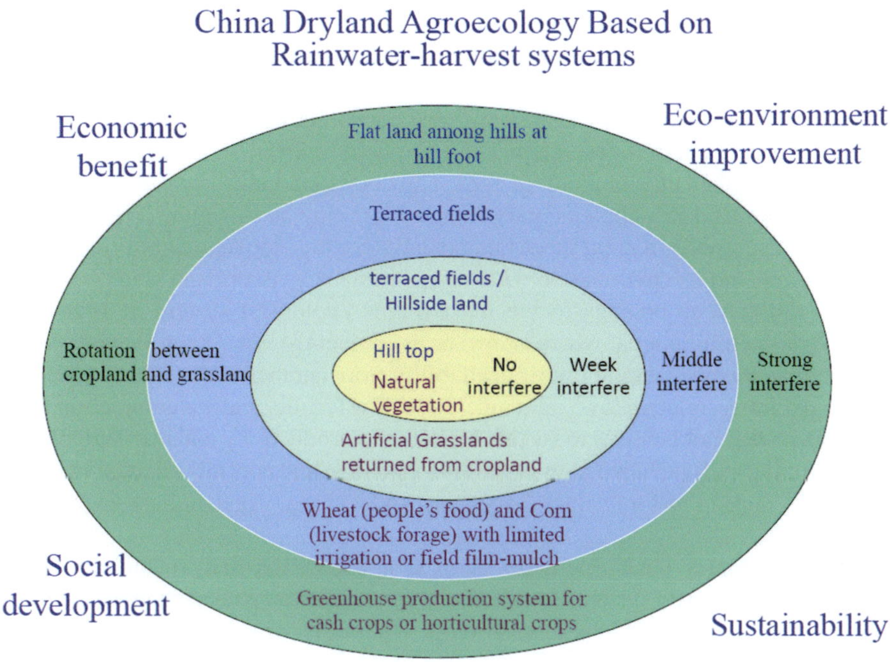

Fig. 7 Dryland landscape agroecological model based on rainwater harvesting technology. This system returned from cropland almost all in the hillside before to diverse ecosystems distributed in a hillside, which is beneficial to increase in local farmer household income, to promote land use/land coverage optimization, social development and regional sustainable development (Modified from Li et al. 2003)

In this model diagram, the flat land (first zone) at the foot of hill has the best ecological conditions; it could only grow a variety of dryland cash crops after basic food supply was guaranteed in the past. Rainwater-harvesting system provides a certain amount of available water resource, which can be used for introducing the cash crops with high additional value in flat lands, such as off-season greenhouse vegetables, flowers, fruits and so on. However, these lands only consisted of a small proportion of total lands, i.e. generally less than 5 % in hilly and gully region.

The second zone comprises most of the remaining flat fields and the lower terrace lands. This lands have the advantages as fine soil quality, excellent production conditions, and high suitability for developing food production. This zone can take the advantage of favorable terrains to introduce ridge-furrow system with film mulching and also develop supplemental irrigation agriculture in the flat areas under the support of rainwater storage system. Moreover, it can ensure self-sufficient supply of food production, and provide a certain degree of crop products for the market and accordingly achieve economic benefits.

The third zone is allocated at general terraces and the lands with a gentle slope. It is the major component of land use. In the past, food production is in a very low level for this zone, where for the lands are much influenced by human activities, and mostly suffered from strong disturbance including constant reclamation, repeated abandon, and subsequent fallow-rotation practice. Furthermore, this zone is also confronted with high soil erosion risk. Critically, farmers need to produce sufficient food and clothing, and meanwhile get stable economic income resource from first and second zones. In this case, farmers are required to first give up inefficient farming practice at the lands of the third zone, and correspondingly convert the lands to artificial grassland for supplementary forage to develop barn-feeding livestock husbandry. This provides a great space and potential for the innovation of dryland cropping system, and its combination with animal husbandry in dryland area.

The fourth zone belongs to the most fragile ecological system. It is usually located at the hilltop area, where rainwater is frequently lost. Other characteristics of this zone include lacking runoff availability from another area, strong wind, low air temperatures, relative low soil temperature, low fertility and water retention ability. This zone is not possible to sustain higher crop productivity and is unsuitable for agricultural use. To achieve high vegetation coverage and control soil erosion, fencing can be done to restore natural vegetation and increase vegetation coverage (Sun et al. 2015).

In this complex landscape ecosystem, as a whole, the first, second and third zones are in an elastic system, which can be closely linked with each other via rotation system. The proportions of cash crop, grain crop, grassland/pasture for livestock husbandry will be adjusted according to market quotations. Importantly, this strategy can help maintain soil fertility equilibrium, and thereby control soil erosion, and ultimately achieve the more balanced development between vegetation restoration, soil quality improvement and economic boost.

Predictably, in the program of this agroecosystem model, the efficient rainwater use based intensive dryland agriculture productivity will be achieved within the first and second zones. Under such a base, the economic benefits will also be significantly improved. In the third and fourth zones, the grasslands with the aim to develop animal husbandry is considered as the main vegetation cover landscape and land use pattern. This design was made in accordance with the principles of ecological zone distribution of natural vegetation and agricultural system. In this integrated landscape model, the diversity of ecosystems and the development of agricultural industry were maintained simultaneously. In the meantime, biodiversity and industrial diversity were increased synchronously. As a whole, the ecosystem can be kept at a stable state, in which agricultural system, livestock husbandry system and economic system would obtain stronger flexibility. Therefore, this model diagram of dryland agro-ecosystem as an innovative theoretical framework for regional development acts as a general methodology to guide restore the degraded dryland ecosystem (Sun et al. 2015), and coordinate the development of economic, social and ecological systems in dryland agricultural area.

6 Conclusion and Future Prospective

Dryland agriculture of China is located in semi-arid and semi humid but drought-prone areas. Despite the low rainfall, it is still able to support a certain crop production. In the absence of external water resources for irrigation, it can still support a certain level of social economic development. Owing to insufficient rainfall with large fluctuation, however, agricultural productivity has been remained at a low and unstable state in a long run. Also due to low vegetation coverage and soft soil texture, the soil has been seriously eroded. Furthermore, because of social and economic development was lagged behind in the region, there existed a widespread poverty issue. Water and soil erosion, over-reclamation and overgrazing resulted in serious land degradation. There existed a complex interaction between extensive poverty and serious land degradation, which brought about a very prominent contradiction between agricultural production and ecological protection. Following the exploration and efforts for decades, significant progress has been achieved in recent 10 years. In the presence of large-scale construction of terraced fields, the ridge-furrow covering farming technology has enhanced the efficient use of rainwater in dry farmland, and agricultural productivity has continued to improve, which provided plenty of crude forage sources for the development of animal husbandry. In this case, natural grassland and artificial grassland are no longer the main support of the development of animal husbandry. And instead, they turned to play more roles of ecological protection. Under such circumstance, the pattern of coordinated development of farming, animal husbandry and grassland industries has formed. As a result of remarkable improvement of land productivity, the organic matter input to the soil was accordingly increased. What's more, soil erosion from farmland and grassland has been effectively controlled. Land degradation has been suppressed and turned to improve the quality of soil as future trend of development. Under the global change of decreasing rainfall and increasing air temperature, dryland ecological system did not further degrade, but instead got a continuous improvement and development. This provides a successful case for adaptive management of dryland ecosystem under the background of global change.

However, optimal management of dryland ecosystems is still facing a lot of challenges. Continuous improvement of land productivity and soil quality is the base for sustainable development of ecosystem. It is an inevitable requirement to ensure the sustainability of ecological system via the explorations on scientific basis and technical means to continuously improve agricultural productivity and soil quality. The structure of agricultural production is frequently driven by the market process. It is the key of determining the direction of dryland ecosystem development, to keep seeking and forming environment-friendly production process of agricultural products. In the context of global change, we also have to face the increasing challenges of drought and extreme weather.

Acknowledgements The research was funded by International Cooperation Program of Ministry of Science and Technology of China (2015DFG31840 and 2012DFG31450), State Technology Support Program (2015BAD22B04) and Fundamental Research Funds for the Central Universities of China (lzujbky-2015-ct02 and lzujbky-2015-br02) and "111" Program (2007B051) and Innovative Team Program of Ministry of Education (IRT_13R26).

References

Agbede TM, Ojeniyi SO (2009) Tillage and poultry manure effects on soil fertility and sorghum yield in southwestern Nigeria. Soil Tillage Res 104:74–81

Behera SK, Panda RK (2009) Effect of fertilization and irrigation schedule on water and fertilizer solute transport for wheat crop in a sub-humid sub-tropical region. Agric Ecosyst Environ 130:141–155

Bhattacharyya R, Chandra S, Singh R, Kundu S, Srivastva A, Gupta H (2007) Long-term farmyard manure application effects on properties of a silty clay loam soil under irrigated wheat–soybean rotation. Soil Tillage Res 94:386–396

Cai J-J, Jiang J, Zhang Y-R, Wang Y-L, Dong L-G, Ji B, Li S-B (2007) The soil characteristic analysis of model dry farmlands in the south mountainous area of Ningxia. Acta Agri Boreali-Occiden Sin 16:75–79 (In Chinese with English abstract)

Cui Q-F, Wang J (2002) Temperature and energy-saving effects of applying the mobile double layers thermal screen in a greenhouse. Trans CSAE 18:111–114 (in Chinese)

Cui Q-F, Cao C-H, Hu X-J, Yi L-X, JIng-Wang (2005) Practice and theoretic analysis of supplementing illumination in the solar greenhouse with reflecting film. Chin J Eco-Agric 13:82–84 (In Chinese with English abstract)

Du S-P, Wang L-F, Long M-X (1999) _Study on soil moisture and yield dynamic of arid alfalfa grassland in mountain area of southern Ningxia. Prataculturae Sinica 16:12–16 (In Chinese with English abstract)

Fan S-J, Wang D, Zhang J-J, Bai J-P, Song J-x, Ma Z-X (2012) Effects of tillage strategies on the topsoil water content and the yield of potato. Acta Prataculturae Sinica 21:271–279 (In Chinese with English abstract)

Guo R-Y, Li F-M (2014) Agroecosystem management in arid areas under climate change: experiences from the Semiarid Loess Plateau, China. World Agri 4:19–29

Guo S-L, Wu J-S, Dang T-H (2008) Effects of crop rotation and fertilization on aboveground biomass and soil organic C in semi-arid Region. Sci Agric Sin 41:744–751 (In Chinese with English abstract)

Guo ZB, Yan GJ, Zhang RH, Li FM, Zeng ZX, Liu H (2010) Improvement of Soil Physical Properties and Aggregate-Associated C, N, and P After Cropland was Converted to Grassland in Semiarid Loess Plateau. Soil Sci 175:99–104

Hai L, Li XG, Liu XE, Jiang XJ, Guo RY, Jing GB, Rengel Z, Li FM (2015) Plastic Mulch Stimulates Nitrogen Mineralization in Urea-Amended Soils in a Semiarid Environment. Agron J 107:921–930

Han X-J (2013) Thin soil covering plastic film mulching cultivation and hole sowing for dryland winter wheat. Agri Technol Equip 2:8–10 (in Chinese)

Hao M-D, Zhang C-X, Wei X-R, Wang X-G, Gao C-Q (2004) Effect of rotation and fertilization to alfalfa productivity on the Loess Plateau. Acta Agron Sin 12:195–198 (In Chinese with English abstract)

He E, Jie W, Lv T, Guo T, Wang P, Li F, Qi X (2015) Plastic film colors on potato yield of cultivation with film mulching ridge-furrow. Gansu Agri Sci Technol 7:55–57 (in Chinese)

Hong S-q, Zhu Y-c, Jiang H, Li H (1998) Effects and ways of soil fertility improvement for new terrace field in mountain areas of Qinghai Province. Agric Res Arid Areas 16:80–83 (In Chinese with English abstract)

Jia Y, Li F-M (2011) Production and ecological benefits of alfalfa fields in the Loess Plateau. Ed Jianguo Wu & Feng-Min Li, *Lectures in Modern Ecology (V)--Large scale Ecology and Sustainability Science* 262–288. (In Chinese with English abstract)

Jia Y, Li FM, Wang XL, Yang SM (2006) Soil water and alfalfa yields as affected by alternating ridges and furrows in rainfall harvest in a semiarid environment. Field Crop Res 97:167–175

Jia Y, Li F-M, Zhang Z-h, Wang X-L, Guo R, Siddique KHM (2009) Productivity and water use of alfalfa and subsequent crops in the semiarid Loess Plateau with different stand ages of alfalfa and crop sequences. Field Crop Res 114:58–65

Jiang X, Li XG (2015) Assessing the effects of plastic film fully mulched ridge–furrow on rainwater distribution in soil using dye tracer and simulated rainfall. Soil Tillage Res 152:67–73

Jiang H-M, Jiang J-P, Jia Y, Li F-M, Xu J-Z (2006) Soil carbon pool and effects of soil fertility in seeded alfalfa fields on the semi-arid Loess Plateau in China. Soil Biol Biochem 38:2350–2358

Kunzová E, Hejcman M (2009) Yield development of winter wheat over 50 years of FYM, N, P and K fertilizer application on black earth soil in the Czech Republic. Field Crop Res 111:226–234

Li F-M (1999) On agricultural productivity and ecosystem's sustainability in semiarid areas of china. *Res Sci* 21:25–30 (In Chinese with English abstract)

Li SX (2007) Dryland Agriculture in China. Science Press, Beijing/New York, pp 22–52

Li J-M (2011) Experimentation on various mulching methods for rain water harvest and soil moisture conservation in the potato production of Anding. Chinese Potato J 25:275–278 (In Chinese with English abstract)

Li J, Shen Z (2013) Influencing factors analyses of local farmer willing to feeding animal indoor under the policy of banning grazing in hillsides – case study in Yulin, Shaanxi Province. China Rural Econ:78–86 (in Chinese)

Li F-M, Xu J-Z (2002) Rainwater-collecting eco-agriculture in semi-arid region of Loess Plateau. Chin J Eco-Agric 10:105–107 (In Chinese with English abstract)

Li FM, Wang TC, Cao J (1998) Effect of organic matter on total amount and availability of nitrogen and phosphorus in loess soil of Northwest China. Commun Soil Sci Plant Anal 29:947–953

Li F-M, Wang J, Zhao S-L (1999a) The rainwater harvesting technology approach for dryland agriculture in semi-arid Loess plateau of China. Acta Ecol Sin 19:259–264 (In Chinese with English abstract)

Li FM, Guo AH, Wei H (1999b) Effects of clear plastic film mulch on yield of spring wheat. Field Crop Res 63:79–86

Li F-M, Xu J-Z, Sun G-J (2003) Restoration of degraded ecosystems and development of water-harvesting ecological agriculture in the semi-arid Loess Plateau of China. Acta Ecol Sin 23(9):1901–1909 (In Chinese with English abstract)

Li FM, Wang J, Xu JZ (2005) Plastic film mulch effect on spring wheat in a semiarid region. J Sustain Agric 25:5–17

Li L-X, Liu G-C, Yang Q-F, Zhao X-W, Zhu Y-Y (2009) Research and application development for the techniques of whole plastic film mulching on double ridges and planting in catchment furrows in dryland. Agric Res Arid Areas 27:114–118 (In Chinese with English abstract)

Li R, Wang M, Jia ZK, Hou QX, Yang BP, Han QF, Nie JF, Zhang R (2012) Effects of different mulching patterns on soil temperature, moisture water and yield of spring maize in Weibei Highland. Transactions of the CSAE 28(2):106–113 (In Chinese with English abstract)

Liu CA, Jin SL, Zhou LM, Jia Y, Li FM, Xiong YC, Li XG (2009) Effects of plastic film mulch and tillage on maize productivity and soil parameters. Eur J Agron 31:241–249

Liu CA, Li FR, Zhou LM, Zhang RH, Yu J, Lin SL, Wang LJ, Siddique KHM, Li FM (2013) Effect of organic manure and fertilizer on soil water and crop yields in newly-built terraces with loess soils in a semi-arid environment. Agric Water Manag 117:123–132

Liu CA, Zhou LM, Jia JJ, Wang LJ, Si JT, Li X, Pan CC, Siddique KHM, Li FM (2014a) Maize yield and water balance is affected by nitrogen application in a film-mulching ridge-furrow system in a semiarid region of China. Eur J Agron 52:103–111

Liu XE, Li XG, Hai L, Wang YP, Li FM (2014b) How efficient is film fully-mulched ridge-furrow cropping to conserve rainfall in soil at a rainfed site? Field Crop Res 169:107–115
Lu H, Xue J, Hao Y, Gao J (2016) Effects of black film mulching on soil environment and maize growth in dryland. Acta Ecol Sin 36(7) (In Chinese with English abstract)
Ma Q-Q (2010) Plastic film mulching winter wheat cultivation expreiment on the tarrace of Zhuanglang County. Gansu Agri Sci Technol 11:20–22 (in Chinese)
Qin S, Zhang J, Dai H, Wang D, Li D (2014) Effect of ridge–furrow and plastic-mulching planting patterns on yield formation and water movement of potato in a semi-arid area. Agric Water Manag 131:87–94
Ren J-Z (2007) Call for the wisdom return of "Reverse Playing Pipa". Prataculturae Sinica 24:68 (In Chinese with English abstract)
Ren J-Z, Chang S-H (2009) Using grassland agricultural systems to ensure the food security. Chinese J Grassland 31:3–6 (In Chinese with English abstract)
Ren J-Z, Lin H-L (2009) Promoting prataculture development in arable region to ameliorate the farming system and insure food security in China. Acta Prataculturae Sinica 18:1–9 (In Chinese with English abstract)
Shan L, Xu B-C (2009) Discussion on how to establish stable artificial grassland in the semiarid Loess Plateau. Prataculturae Sinica 18:1–2 (In Chinese with English abstract)
Shan L, Zhang S-Q, Li W-R (2008) Productivity and drought resistance of alfalfa. J Agric Sci Technol 10:12–17 (In Chinese with English abstract)
Sun W, Song X, Mu X, Gao P, Wang F, Zhao G (2015) Spatiotemporal vegetation cover variations associated with climate change and ecological restoration in the Loess Plateau. Agric For Meteorol 209-210:87–99
Tian Y, Su DR, Li FM, Li XL (2003) Effect of rainwater harvesting with ridge and furrow on yield of potato in semiarid areas. Field Crop Res 84:385–391
Wan BR (2013) Grain production will be more increasing by reducing the dependence on the weather – an investigation of dryland faring development in Gansu Province and other areas. Qiushi 2013(19):25–27 (in Chinese)
Wang Y (2006) Exploration of Tarrace resource status and development in Gansu Province. Soil Water Conserv China:37–38 (In Chinese with English abstract)
Wang J, Li J (2011) Effect of Black/clear films and mulching initiating time on maize yield with ridge-furrow system. Gansu Agric Sci Technol 2:18–20 (in Chinese)
Wang J, Li Z-X, Wang Z-Z, Liu H-S, Nan D-C (2001) Combined construction of rainwater catchment techniques with methane pool and greenhouse. Chin J Appl Ecol 12:51–54 (in Chinese)
Wang XL, Li FM, Ha Y, Shi WQ (2005) Increasing potato yields with additional water and increased soil temperature. Agric Water Manag 78:181–194
Wang J, Li FM, Jia Y (2006) Responses of soil water, nitrogen, and organic matter to the alfalfa crop rotation in semiarid Loess area of China. J Sustain Agric 28:117–130
Wang XL, Sun GJ, Jia Y, Li FM, Xu JZ (2008) Crop yield and soil water restoration on 9-year-old alfalfa pasture in the semiarid Loess Plateau of China. Agric Water Manag 95:190–198
Wang X-L, Jia Y, Li X-G, Long R-J, Ma Q, Li F-M, Song Y-J (2009) Effects of land use on soil total and light fraction organic, and microbial biomass C and N in a semi-arid ecosystem of northwest China. Geoderma 153:285–290
Wang YP, Li XG, Hai L, Siddique KHM, Gan YT, Li FM (2014) Film fully-mulched ridge-furrow cropping affects soil biochemical properties and maize nutrient uptake in a rainfed semi-arid environment. Soil Sci Plant Nutr 60:486–498
Xie Y-H, Hong J-P, Zhou S-Q, Chang H-Q (2001) Approach of fertility betterment and ripeness and high yield techniques on newly-built fields in loess hilly region. J Soil Water Conserv 15:126–128 (In Chinese with English abstract)
Xue S-P, Zhu R-X, Yang Q, Han S-M, Han W-T (2006) Rainfall mechanical utilization techniques for winter wheat in Loess Plateau dryland. J Northwest Sci-Tech Uni Agri Forestry Nat Sci Edn 34:1–8 (In Chinese with English abstract)

Yang F-K (2006) Research on rainwater harvesting and yield increasing effect of terrace in semi-arid area of Loess Hilly Region. J Soil Water Conserv 20:158–161 (In Chinese with English abstract)
Yang Q-Y, Jing K, Shen Y-C (2001) Eco-environment construction and disaster reduction in Loess Plateau. Dis Reduc China 11:19–22 (In Chinese with English abstract)
Ye Z-O (1984) Experimentation for high and stable yield of dry terraces by fertilizing and preserving soil water. Soil Water Conserv China:24–28 (In Chinese with English abstract)
Ye JS, Liu CA (2012) Suitability of Mulch and Ridge-furrow Techniques for Maize across the Precipitation Gradient on the Chinese Loess Plateau. J Agric Sci 4
Yuan Z-Q, Epstein H, Li F-M (2015a) Factors affecting the recovery of abandoned semi-arid fields after legume introduction on the Loess Plateau. Ecol Eng 79:86–93
Yuan Z-Q, Yu KL, Epstein H, Fang C, Li JT, Liu QQ, Liu XW, Gao WJ, Li FM (2015b) Effects of legume species introduction on vegetation and soil nutrient development on abandoned croplands in a semi-arid environment on the Loess Plateau, China. Sci Total Environ 541:692–700
Yuan Z-Q, Yu K-L, Wang B-X, Zhang W-Y, Zhang X-L, Siddique KHM, Stefanova K, Turner NC, Li F-M (2015c) Cutting improves the productivity of lucerne-rich stands used in the revegetation of degraded arable land in a semi-arid environment. Sci Rep 5:12130
Zeng Y-M (2002) Regulating plantation structure and promoting alfalfa pasture so as to develop the confined animal husbandry. Inner Mongolia Prataculture 14:17–19 (in Chinese)
Zhang W (1989) Why the ecological environment of the Loess Plateau is destroyed thoroughly. Bull Soil Water Conserv 9:21–27 (In Chinese with English abstract)
Zhang X-S (2010) China has to go the way to grassland agriculture by developing artificial grasslands. Impact of Science on Society 18–21. (in Chinese)
Zhang W-S, Li FM, Xiong YC, Xia Q (2012) Econometric analysis of the determinants of adoption of raising sheep in folds by farmers in the semiarid Loess Plateau of China. Ecol Econ 74:145–152
Zhao H, Xiong YC, Li FM, Wang RY, Qiang SC, Yao TF, Mo F (2012) Plastic film mulch for half growing-season maximized WUE and yield of potato via moisture-temperature improvement in a semi-arid agroecosystem. Agric Water Manag 104:68–78
Zhao H, Wang RY, Ma BL, Xiong YC, Qiang SC, Wang CL, Liu CA, Li FM (2014) Ridge-furrow with full plastic film mulching improves water use efficiency and tuber yields of potato in a semiarid rainfed ecosystem. Field Crop Res 161:137–148
Zhou LM, Li FM, Jin SL, Song YJ (2009) How two ridges and the furrow mulched with plastic film affect soil water, soil temperature and yield of maize on the semiarid Loess Plateau of China. Field Crop Res 113:41–47

Dryland Agriculture in North America

Neil C. Hansen, Brett L. Allen, Saseendran Anapalli, Robert E. Blackshaw, Drew J. Lyon, and Stephen Machado

1 Introduction

Dryland crop production is found in semi-arid regions of North America including much of Mexico, the Western United States, and the Canadian Prairies (Fig. 1). These lands are classified as semi-arid based on an aridity index, with values of P/ETP between 0.2 and 0.5, where P is annual precipitation and ETP is annual potential evapotranspiration. The map (Fig. 1) was created using data from WorldClim Global Climate Data (http://WorldClim.org) as modelled in the CGIAR-CSI Global-Aridity Database (Zomer et al. 2007, 2008). Nearly 40 % of North America is classified as an arid or semi-arid region. To illustrate the extent of dryland farming within the semi-arid zones of North America, a map was created that shows areas where wheat (*Tritcum aestivum* L.) is produced (Fig. 2). Wheat in semi-arid zones was chosen as a means of illustrating locations of dryland farming because it is the

N.C. Hansen (✉)
Department of Plant and Wildlife Sciences, Brigham Young University, Provo, UT, USA
e-mail: neil_hansen@byu.edu

B.L. Allen
United States Department of Agriculture-Agricultural Research Service, Sidney, MT, USA

S. Anapalli
United States Department of Agriculture-Agricultural Research Service, Stoneville, MS, USA

R.E. Blackshaw
Agriculture and Agri-Food Canada, Lethbridge, AB, Canada

D.J. Lyon
Department of Crop and Soil Sciences, Washington State University, Pullman, WA, USA

S. Machado
Columbia Basin Agricultural Research Center, Oregon State University, Pendleton, OR, USA

M. Farooq, K.H.M. Siddique (eds.), *Innovations in Dryland Agriculture*, DOI 10.1007/978-3-319-47928-6_15

Fig. 1 Semi-arid lands in North America as defined by an aridity index. The index classifies land as semi-arid when P/ETP is between 0.2 and 0.5 (*P* annual precipitation, *ETP* annual potential evapotranspiration). The map was created using data from the CGIAR-CSI Global-Aridity Database (Zomer et al. 2007, 2008)

most common dryland crop in North America. Omitted from the map are significant areas of rainfed wheat production in more humid regions to the east of the semi-arid zone. The map shows variation in harvested wheat area within 5-min by 5-min latitude/longitude grid cells and is based on production census data (Monfreda et al. 2008). Because some of the semi-arid wheat is produced under irrigation, the map also shows areas where a majority of cropland is equipped for irrigation (Siebert et al. 2013). The remainder of the harvested wheat area is representative of the extent and importance of dryland farming in North America. The map shows that the Northern and Central Great Plains of the U.S. and the Canadian Prairies are areas with very high density dryland farming and that the harvested area of dryland wheat production far outweighs irrigated wheat. The map also shows an area of high density dryland farming in the inland pacific northwest (IPNW) of the U.S. The southern Great Plains of the U.S. and Mexico are also areas with high density dryland farming, but the map shows only moderate density of harvested wheat because much of this area is in production of other crops such as sorghum (*Sorghum bicolor* L.) (Baumhardt and Salines-Garcia 2006). Dryland crop production in the high

Fig. 2 Harvested wheat area within 5 min by 5 min longitude/latitude grid cells was used to show the dryland farming lands of North America. Only wheat produced in semi-arid zones is show. Wheat produced under irrigation is shown in blue to separate these from the drylands. Wheat data from (Monfreda et al. 2008) and irrigation data from (Siebert et al. 2013)

density land areas shown on this map have been discussed previously (Baumhardt and Salines-Garcia 2006; Cochran et al. 2006; Hansen et al. 2012; Schillinger et al. 2006). However, previous maps depicting the areas of dryland agriculture in North America have not illustrated the extent of lower density dryland crop production. Nearly every state in the western U.S and in northern and central Mexico has significant land areas where some dryland farming occurs. In addition to wheat, North American dryland farming is important for the production of maize (*Zea maize* L.), sorghum, pulses, and oilseeds.

The Great Plains region of the United States and Canada is an area of widespread dryland crop production, with wheat being the dominant crop. This region is characterized by a continental climate, with precipitation in the region ranging from 300 to 500 mm annually. Wide fluctuation in annual precipitation and periods of extended drought are common. There is a strong, increasing west to east precipitation gradient. Winters become increasing long and cold from south to north. The majority of the annual precipitation falls during summer months, but winter snow is important, especially in the central and northern areas. The Great Plains are roughly divided into the Canadian Prairies and U.S. Northern Great Plains, The Central Great Plains, and the Southern Great Plains.

The Canadian Prairies and the U.S. Northern Great Plains have long, cold winters and short, warm summers. In this area, spring wheat is the most commonly produced crop. The traditional rotation of spring wheat-summer fallow has largely been replaced with rotations that substitute fallow with cool season, annual crops that mature early in summer and are well adapted for the region. For instance, in NE Montana and NW North Dakota fallow land area decreased by nearly fourfold (1.2 to 0.3 million ha) from 1990 to 2015, with only 12 % of the total cropped area in fallow during 2015. Pulses, such as lentil (*Lens culinaris* Medik.), field pea, chickpea (*Cicer arietinum* L.), and faba-bean (*Vicia faba* L.), and oilseed crops, such as canola (*Brassica napus* L.) and sunflower (*Helianthus annuus* L.), are commonly grown in rotation with spring wheat. No-till is practiced on more than half of the dryland area (Hansen et al. 2012). Compared to the Central and Southern Great Plains, the Northern areas have more diverse crop rotations and higher adoption of no-till, largely due to lower potential evapotranspiration in summer months.

In the central Great Plains, the prevailing cropping system is a two-year rotation of winter wheat and summer fallow. In this traditional practice, shallow tillage is used during fallow periods to control weeds and help store moisture in the soil. Sustainability of this practice is limited by soil degradation and erosion and poor water use efficiency. The adoption of no-till practices on more than 25 % of the drylands has resulted in greater precipitation storage and more diverse crop rotations. For example, a no-till rotation of winter wheat-maize-fallow increased total annualized grain yield by 75 % compared to winter wheat – summer fallow. Soil erosion was reduced to just 25 % of that from a conventional tillage wheat-summer fallow system (Peterson et al. 1996, 1998). Other crops rotated with wheat in the Central Great Plains include sorghum, sunflower, and proso millet (*Panicum mileaceum* L.).

In the Southern Great Plains, the hotter climate and longer growing seasons compared to the Northern and Central Great Plains result in much greater ETP and more arid conditions. Winter wheat is commonly grown in rotation with grain sorghum (Baumhardt and Salines-Garcia 2006). Annual small grains are often produced as dual purpose crops for both grazing and grain production. Other dryland crops include maize, cotton (*Gossypium hirsutum* L.), and proso millet. No-till adoption is limited due to lack of residue accumulation and residue removal from grazing.

The dryland farming area of the Inland Pacific Northwest (IPNW) is distinct from the Great Plains. It has a semi-arid Mediterranean-like climate with cool wet winters and warm dry summers. Three quarters of the precipitation in this region is received from September to April. Inland PNW is broadly classified into three precipitation zones and dryland winter-wheat (*Triticum aestivum* L.)-based production systems are the predominant cropping systems (Schillinger and Papendick 2008). Low precipitation zone (< 300 mm annual precipitation) covers 1.56×10^6 ha and more than 90 % of this zone is grown mostly to winter wheat under a winter wheat-summer fallow system in which one crop is grown every 2 years. In the intermediate precipitation zone with 300–450 mm annual precipitation (0.97×10^6 ha) the traditional wheat-summer fallow system and 3-year rotation of winter wheat – spring barley (*Hordeum vulgare* L.) – fallow in drier and wetter regions, respectively, are prevalent in this zone. In addition, several new crops such as spring legume crops (pea (*Pisum sativum L.*), lentil, chickpea), and sunflower have been introduced in this zone (Schillinger and Papendick 2008). In the high precipitation zone (> 450 mm annual precipitation), annual cropping is practiced and wheat is rotated with grain-legume and oilseed crops.

Much has been written about dryland agriculture in North America (Baumhardt and Salines-Garcia 2006; Cochran et al. 2006; Hansen et al. 2012; Schillinger et al. 2006). This chapter seeks to add to the established literature by highlighting some new understanding, issues, and innovations that have not been widely reviewed. In this chapter, we have discussed (1) the role of integrated pest management for herbicide resistant weeds, (2) diversification of crop rotations, including oilseed crops produced for biofuels, (3) understanding effects of residue management on soil carbon dynamics, and (4) the application of models to aid decision making.

2 Integrated Weed Management as an Approach to Deal with Herbicide Resistant Weeds

Retaining crop residues at the soil surface with no-till production systems has proven to be critical to the success of modern dryland agriculture in North America. The adoption of no-till production systems by many dryland farmers has greatly reduced soil erosion by wind and water, preventing a repeat of the infamous decade of severe wind erosion in the 1930s, despite subsequent drought conditions that were more severe and longer in duration than in the 1930s. Additionally, no-till production systems have increased soil water storage and retention compared to tilled production systems. This increase in stored soil water has allowed dryland farmers to increase cropping intensity and diversity, which has accrued many benefits including increases in grain and biomass yields on an annual basis (Peterson et al. 1996; Norwood 1994; Jones and Popham 1997), net returns to producers (Peterson and Westfall 2004), and potentially active surface soil organic C and N (Peterson et al. 1998). Diverse cropping systems also effectively controlled winter annual grass weeds in winter wheat (Daugovish et al. 1999) and reduced yield loss in wheat resulting from soil borne disease (Krupinsky et al. 2002). No-till dryland cropping systems are heavily reliant on herbicides for weed control. This reliance on herbicides makes no-till production systems particularly vulnerable to the rapidly increasing problem of herbicide resistance in weeds.

2.1 Extent and Occurrences of Herbicide Resistant Weeds

We have been selecting for weed biotypes with resistance to herbicides ever since the commercial introduction of the synthetic organic herbicides in the late 1940s. The first documented cases of herbicide resistant weeds in North American dryland wheat production were reported in 1987. Kochia [*Kochia scoparia* (L.) Schrad.], prickly lettuce (*Lactuca serriola* L.), and Russian-thistle (*Salsola tragus* L.) biotypes resistant to chlorsulfuron were reported from the U.S. Great Plains and the Western U.S.A. (Heap 2016). Chlorsulfuron inhibits acetolactase synthase (ALS) and this site of action is common for many of the herbicides used in wheat production in North America. Consequently, weed biotypes resistant to this site of action are plentiful throughout the Western U.S.A. and Prairie Provinces of Canada (Table 1).

As of 2016, there were 72 resistant weed biotypes reported in wheat from the western U.S.A. and Canada (Heap 2016). Amongst these biotypes, resistance to seven different sites of action were reported (Table 1). Eight of the biotypes were resistant to two or more different sites action, i.e., they exhibited multiple resistance. Varanasi et al. (2015) confirmed the first case of resistance to four herbicide sites of action (ALS inhibitors, synthetic auxins, photosystem II inhibitors, and EPSP synthase inhibitors) in a single kochia population from western Kansas. Multiple resistance in weeds is of growing concern to weed scientists and growers.

No-till production systems in North America are heavily reliant on herbicides and no herbicide has played a larger role in no-till production than glyphosate. Monsanto's patent for glyphosate ended in 2000 and generic glyphosate quickly followed. Generic glyphosate played a major role in the adoption of no-till practices throughout the dryland farming regions of North America because it was cheap and effective, particularly at higher use rates that were unaffordable for many prior to generic glyphosate. However, after more than a decade of reliance on this single active ingredient, weed resistance to glyphosate is threatening to stall, and possibly reverse, the gains in no-till management adoption (CAST 2012). Glyphosate-resistant kochia was first documented in several fallow fields in Kansas in 2007 (Waite et al. 2013). By 2014, more than five million hectares from Texas into the Prairie Provinces of Canada were infested with glyphosate-resistant kochia (Stahlman, personal communication). In 2015, glyphosate-resistant Russian-thistle was documented in Montana and in 2016 another population was identified in Washington (Heap 2016). Kochia and Russian-thistle are troublesome weeds in dryland agriculture and both species have a history of developing herbicide-resistant biotypes (Table 1).

2.2 *Integrated Weed Management concepts*

In order for dryland agriculture to remain economically and environmentally sustainable, production systems will need to rely less on herbicides and more on an integrated approach to weed management. Integrated weed management (IWM) involves the sustainable use of all available methods to reduce weed pressure, including sanitation, mechanical, cultural, chemical, and biological methods. The three general principles of integrated weed management are: (1) use agronomic practices that limit the introduction and spread of weeds, (2) help the crop compete with weeds, and (3) use a diverse set of practices, altered over time, to prevent weed

Table 1 Weed species with identified biotypes resistant to herbicides in dryland wheat production systems from the U.S. Great Plains, Canadian Prairie Provinces, and Western U.S.A. as reported by Heap (2016)

Site of action (herbicide group no.[a])	Weed species with resistant biotypes
Accase inhibitors (A/1)	Green foxtail, *Setaria viridis* (L.) Beauv.[b]
	Italian ryegrass, *Lolium perenne* L. ssp. *multiflorum* (Lam.) Husnot[b]
	Persian darnel, *Lolium persicum* Boiss. & Hohen. Ex. Boiss
	Wild oat, *Avena fatua* L.[b]
ALS inhibitors (B/2)	Ball mustard, *Neslia paniculata* (L.) Desv.
	Cheat, *Bromus secalinus* L.
	Common chickweed, *Stellaria media* (L.) Vill.
	Common lambsquarters, *Chenopodium album* L.
	Cowcockle, *Vaccaria hispanica* (P. Mill.) Rauschert
	False cleavers, *Galium spurium* L.[b]
	Field pennycress, *Thlaspi arvense* L.
	Horseweed, *Conyza canadensis* (L.) Cronq.
	Italian ryegrass, *Lolium perenne* L. ssp. *multiflorum* (Lam.) Husnot[b]
	Kochia, *Kochia scoparia* (L.) Schrad.[b]
	Mayweed chamomile, *Anthemis cotula* L.
	Pale smartweed, *Polygonum lapathifolium* L.
	Prickly lettuce, *Lactuca serriola* L.
	Redroot pigweed, *Amaranthus retroflexus* L.
	Russian-thistle, *Salsola tragus* L.
	Shepherd's-purse, *Capsella bursa-pastoris* (L.) Medik.
	Smallseed falseflax, *Camelina microcarpa* Andrz. ex DC.
	Spiny sowthistle, *Sonchus asper* (L.) Hill
	Wild buckwheat, *Polygonum convolvulus* L.
	Wild oat, *Avena fatua* L.[b]
Microtubule assembly inhibitors (K/3)	Green foxtail, *Setaria viridis* (L.) Beauv.[b]
Synthetic auxins (O/4)	Common hempnettle, *Galeopsis tetrahit* L.
	False cleavers, *Galium spurium* L.[b]
	Kochia, *Kochia scoparia* (L.) Schrad.
Lipid synthesis inhibitors but not ACCase inhibitors (N/8)	Wild oat, *Avena fatua* L.[b]
EPSP synthase inhibitors (G/9)	Kochia, *Kochia scoparia* (L.) Schrad.[b]
	Russian-thistle, *Salsola tragus* L.
Inhibitors of very-long-chain fatty acid synthesis (K/15)	Italian ryegrass, *Lolium perenne* L. ssp. *multiflorum* (Lam.) Husnot[b]

[a]The uppercase letter is the Herbicide Resistance Action Committee (HRAC) classification and the number is the Weed Science Society of America classification

[b]One or more biotypes of this species exhibit multiple herbicide resistance, i.e., resistance to two or more herbicides with different sites of action

species or biotypes from adapting to production systems. The following are specific examples of integrated weed management systems used in dryland wheat production systems in North America.

2.2.1 Winter Wheat-Summer Crop-Fallow in the U.S. Central Great Plains

In the U.S. Central Great Plains, dryland agriculture has developed around winter wheat production. Summer fallow, the practice of controlling all plant growth during the non-crop season, was quickly adopted to stabilize winter wheat production in the region. Winter wheat—fallow was the predominant crop rotation in the Central Great Plains during most of the twentieth century (Baumhardt and Anderson 2006). Downy brome (*Bromus tectorum* L.), jointed goatgrass (*Aegilops cylindrica* Host), and feral rye (*Secale cereale* L.) are weeds that cause significant economic loss in winter wheat—fallow production regions of the Great Plains and western USA, particularly where conservation tillage is used (Lyon and Baltensperger 1995). In addition to wheat yield loss, jointed goatgrass and feral rye seed frequently contaminate winter wheat grain, resulting in economic loss from dockage and grade reduction. These three winter annual grass weeds have a similar life cycle and physiology to winter wheat, which limits effective control methods. While plowing with a moldboard plow can effectively control these weeds (Donald and Ogg 1991; Kettler et al. 2000; Stump and Westra 2000), tillage that buries nearly all surface crop residues defeats the goal of no-till systems, i.e., maintaining crop residues on the soil surface (Unger et al. 2006). An IWM approach must balance crop production, weed control, and soil and water conservation.

Several herbicides including sulfosulfuron, propoxycarbazone, pyroxsulam, and imazamox can provide selective control of one or more of these grass weeds. Imazamox can only be used with imazamox-tolerant wheat cultivars or serious crop injury will occur. Concerns with these products include high cost, long soil residual that restricts rotation flexibility, and development of weed resistance with frequent use.

Crop rotation with late spring-planted crops effectively controls these winter annual grass weeds in winter wheat (Daugovish et al. 1999). Growing a winter wheat crop every three or four years rather than every other year promotes depletion of the soil seed bank as long as no plants are allowed to produce seed during the non-wheat portion of the rotation. This is easily accomplished if the rotational crops are not growing during the early spring when the use of nonselective herbicides or tillage can be used to kill emerged plants.

Seed production of winter annual grass weeds can be reduced by combining cultural practices. Feral rye and jointed goatgrass seed production was reduced by applying nitrogen fertilizer five months before wheat seeding, increasing the wheat seeding rate, and planting a standard height cultivar (Anderson 1997). Standard height cultivars frequently yield less than many semi-dwarf cultivars. If a semi-dwarf cultivar is used, row spacing can be reduced to help compensate for the loss in weed competitiveness.

By combining crop rotation, cultural practices, and herbicides, winter annual grass weeds are effectively controlled in winter wheat and the entire cropping system is made more sustainable. The winter wheat—summer crop—fallow system is an example of an effective IWM system with implications beyond weed control.

2.2.2 Zero-till Spring Wheat in North America

Zero tillage (no-till) has become a widely adopted agronomic practice in the spring wheat production areas of Canada and the northern USA. Research has shown that some weed species may become more prevalent with zero tillage, but overall weed densities decline with time (Derksen et al. 2002; Anderson 2003; Blackshaw 2005). Weed seed mortality tends to be greater when weed seeds are left on the soil surface compared to when buried in the soil with tillage. Additionally, crop residues on the soil surface may inhibit weed germination and growth through physical suppression and/or allelopathic interactions. Thus, zero tillage has contributed greatly to improved weed management as well as higher spring wheat yields.

Improved soil moisture conservation with zero tillage has allowed a greater variety of crops to be grown in recent years in the semi-arid Canadian Prairies. Wheat-based rotations now include more oilseeds [e.g., canola (*Brassica napus* L.), flax (*Linum usitatissimum* L.)] and pulses [e.g., field pea, lentil, and faba bean]. Inclusion of forages such as alfalfa (*Medicago sativa* L.) or red clover (*Trifolium pretense* L.) in rotation with spring wheat, with the main goal of managing weeds, is gaining acceptance in areas where forage demand is high. Survey results indicate that 83 % of farmers had lower weed densities after two to four years of forage production (Entz et al. 1995). Diverse crop rotations have resulted in lower weed populations in spring wheat.

Weed management in spring wheat can be improved by including fall-seeded crops in rotation (O'Donovan et al. 2007; Beres et al. 2010). Many spring-germinating weeds emerge after canopy closure of fall-seeded crops, which makes them noncompetitive. Winter wheat, winter rye, and winter triticale (*Triticosecale* spp.) are being more widely grown on the Canadian Prairies. Systematically changing planting dates and crop species prevents any one weed species from developing into a major problem (Derksen et al. 2002; Harker et al. 2016).

Spring wheat farmers in Canada are slowly but surely adopting IWM systems. Foxtail barley (*Hordeum jubatum* L.) is an example of a weed species that became a greater problem with zero tillage (Blackshaw 2005). However, Blackshaw et al. (1999) determined that good control of this weed could be attained by combining crop rotation, higher wheat seeding rates, banded nitrogen fertilizer, and timely herbicide use in a multi-year approach. Farmer adoption of an IWM system for foxtail barley was one of the first success stories, and it occurred in part because the need was so great. Farmers will readily adopt new practices when they perceive a need for change and when those practices are effective and affordable.

Another multi-year study conducted at three locations assessed the merits of combining several crop production practices to manage weeds in the context of full or reduced herbicide rates in spring wheat and other major field crops of the Canadian Prairies (Blackshaw et al. 2005a, b). Factors included in the study were crop rotation, seeding date, seeding rate, fertilizer timing, and herbicide rate. The combination of earlier seeding date (3 weeks earlier), higher crop seeding rate (50 % higher), and spring-applied subsurface-banded fertilizer resulted in the most competitive cropping system. Weeds were controlled with this IWM approach and it is notable that the weed seed bank was not greater after four continuous years of using 50 % herbicide rates in a competitive cropping system at two of three sites. Farmers were impressed with the level and consistency of weed control in this study but were only truly convinced of the merits of these IWM systems when they were shown to be economically viable (Smith et al. 2006).

These examples illustrate that weed populations adapt to changes in farm management. Over-reliance on herbicides for weed control has led to weed resistance and weed shifts that complicate weed management. With a limited array of available herbicide classes, weed resistance will remain a challenge to wheat production in the future. Farmers must strive to alter management practices and maximize the competitive advantage of the crop at the expense of weeds. The ubiquitous use of glyphosate jeopardizes the sustainability of no-till systems that have proven so effective for diversifying and intensifying wheat production systems in semi-arid environments. Proper glyphosate stewardship is critical to maintaining these effective production systems. Sustainable weed management in dryland wheat production will best be achieved through continued development and adoption of integrated weed management and crop production practices.

3 Diversifying Rotations to Include Forages and Short Season Crops

Throughout the dryland regions of North America, a 2-year rotation of wheat-summer fallow was commonplace during the previous century and continues to be widely practiced. A two-year rotation with one crop functions in drylands because (1) the early maturing wheat avoids the late-season heat and drought common in North American dryland regions, and (2) the fallow period stabilizes yields and reduces the incidence of crop failure (Greb et al. 1970; Baumhardt and Anderson 2006). While the rotation can function in the short term, many long-term problems have been documented with the wheat-summer fallow rotation, including increased soil erosion, saline seep formation, reduced precipitation use efficiency, decreased soil organic C and N, and reduced annualized crop yield and economic returns (Black et al. 1981; Janzen 1987; Campbell et al. 1990; Wienhold et al. 2006; Sainju et al. 2015a). Wheat-summer fallow cropping systems are unlikely to maintain soil quality, much less improve the quality of degraded soils. For example, in a 19-year study in the Northern Great Plains, Allen et al. (2011a) concluded that a conventionally tilled wheat-fallow cropping system did not restore soils damaged by erosion even if yields were maintained with fertilizer application. In a different 30-year study, Sainju et al. (2015a) reported 18 % lower soil organic C and 20 % greater bulk density in the surface 7.5 cm soil of a spring wheat-tilled fallow rotation compared to continuous wheat in spring tilled or no-till treatments. In the same experiment, annualized crop yield was 23–30 % lower, and P, K, Zn and CEC were 23–60 % lower in a spring wheat-tilled fallow rotation than for continuously cropped treatments with or without tillage (Sainju et al. 2015b). These studies highlight the need for diversifying and intensifying the traditional wheat-summer fallow rotation in North America dryland cropping. Improved precipitation storage efficiency during fallow through adoption of reduced-till or no-till systems is key to successfully diversifying and intensifying crop rotations (Black and Power 1965; Tanaka and Aase 1987; Peterson et al. 1996).

Among the dryland cropping areas of North America, the Northern Great Plains has been the most successful in adoption of diversified crop rotations. Cropping systems in the Northern Great Plains have evolved to include fallow replacement crops ranging from green manure crops to continuously cropped systems that include forage, oilseed, and pulse crops. Obtaining the advantage of diversified crop rotations can require time for the system to evolve and for the producer to adapt. In a simple example, the fallow phase of a 2-year spring wheat-summer fallow rotation was replaced with lentil grown as a green manure crop in a 12-year study (Allen et al. 2011b). In that study, low soil nitrate in the green manure treatment during the first five years resulted in 33 % reduction in spring wheat yield compared to wheat grown after fallow. However, beginning with the sixth year, wheat yield after the lentil green manure crop differed by less than 2 % compared to wheat after summer fallow due to 26 % greater spring soil nitrate and a two-fold increase in overall nitrogen cycling. The study demonstrates that over time, adopting the diversified rotation increased sustainability of the system, as evidenced by enhanced nitrogen cycling in the soil. The study also illustrated the importance of careful management to achieve success in the diversified rotation. Soil water at the time of wheat planting was similar for fallow and green manure cropping systems if the lentil green manure crop was terminated at full bloom. However, if the lentil green manure crop was allowed to mature, there was a 20 % reduction in soil water at the time of planting the subsequent wheat crop.

Several other studies document the advantages of crop diversification in the NGP. Inclusion of an annual forage crop in rotation with wheat has been shown to improve sustainability. For example, Lenssen et al. (2010) showed that replacing fallow in a wheat-fallow rotation with forage barley enhanced water use efficiency and led to greater profit. Annual forages can be beneficial in dryland systems because they require less water than grain crops, compete with weeds, and protect the soil from erosion. A drawback to fallow replacement crops is the possibility that the subsequent wheat crop will have reduced soil water at planting and reduced yield. Reduced yields, however, are offset by the overall benefit to the system (Lyon et al. 2004; Lenssen et al. 2010). Management practices becomes more critical

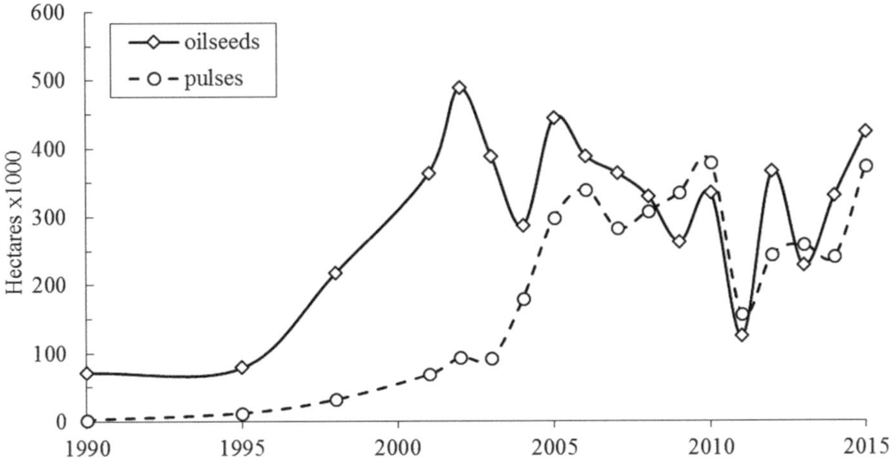

Fig. 3 Harvested areas of oilseed crops (camelina, canola, flax, mustard, safflower, sunflower) and pulse crops (chickpea, dry bean, lentil, field pea) in 25 counties from NE Montana and NW North Dakota 1990–2015. (Data from: USDA-Farm Service Agency)

when dryland cropping systems are intensified and diversified. A study by Lenssen et al. (2015) showed that including ecologically managed forage barley (33 % greater seeding rate and banded N fertilizer) in dryland crop rotation produced more forage and used water more efficiently than barley with conventional seeding rate and broadcast N. The same study also showed greater forage yield and water use efficiency in the ecologically managed 4-year than 3-year rotation, illustrating the advantages of increasingly diversified cropping rotations in the NGP.

Oilseeds and pulses are important crops for diversification of dryland cropping systems in North America, with greatest adoption in the NGP and IPNW (Fig. 3). Among the oilseed crops, canola is the most widely produced dryland crop, with lesser production of camelina, flax, mustard, safflower, and sunflower. The vegetable oils are marketed for both food and industrial uses. Further opportunity for diversifying dryland rotations is being driven by interest in agriculturally produced oils as feedstocks for biofuels (Lenssen et al. 2012; Allen et al. 2014; Gesch et al. 2015; Long et al. 2016). Pulse crops produced in North American dryland cropping include chickpea, dry bean, lentil, and field pea. While adoption of dryland cropping systems with oilseeds and pulses is increasing in North America, more information is needed about the sustainability of these systems. Lenssen et al. 2012 concluded from a 4-year study in NE Montana that 2-year durum-oilseed rotations could be used for grain production and biofuel feedstock, but that production system sustainability would require additional research on soil quality, no-till oilseed stand establishment, and pest management. Allen et al. (2014) reported after the fifth year of the same experiment that durum yield following the oilseed B. juncea was similar to that of durum following chem fallow, though yield of durum following camelina and crambe was less than that of durum following fallow.

Table 2 Summary of the general properties of three categories of tillage practices

Tillage system[a]	Tillage depth (cm)	Residue management cover	Soil organic carbon Loss/gain (%)	Soil erosion+/−	Literatures
Conventional tillage	20 cm	< 15 %	Loss – 0.22–0.42 Mg ha^{-1} $year^{-1}$	−	Machado (2011) and Williams and Wuest (2011)
Reduced tillage	< 20 cm	15–30 %		+	Schillinger and Young (2014)
No-tillage	Surface/ None	>30 %	Predicted gain – 0.12–0.21 Mg C ha^{-1} $year^{-1}$ in 10–12 years	+	Brown and Huggins (2012)

[a]Tillage system

4 Effects of Crop Residue and Nutrient Management Practices on Soil Organic Carbon

Decisions on crop residue management practices in dryland cropping systems are mainly influenced by agroecological and agronomic considerations. Crop residues are managed by burning, harvest and removal, or incorporation by tillage. Burning is practices on about 1.24×10^6 ha of cropland in the contiguous U.S. annually. Although most of the burned acreage is under grassland production some dryland farmers use burning. For example, burning is used on about 30,000 ha of dryland farms in the IPNW (McCarty et al. 2009). Residue management practices such as burning residue and intensive tillage are not sustainable in dryland areas. In these areas, soils are highly vulnerable to wind and water erosion and loss of soil organic matter, especially in a wheat-summer fallow rotation where only one crop is produced in two years.

Concern about erosion and soil degradation have led to an evolution of tillage practices (Table 2) in North American dryland that were designed to leave crop residues in the field during fallow periods and cultivate soil in spring using disc or sweep plows. Reduced tillage practices such as stubble mulching leaves about 10–20 % surface cover, whereas no-till can leave as much as 50 % cover. While this cover is highly beneficial for soil conservation, there are some challenges. Under these high residue systems seed emergence is reduced due to lack of good seed-soil contact and increased disease incidences such as *Rhizoctonia* and *Pythium* (Paulitz 2006). Emergence and early season growth can be slow due to cooler temperatures and N deficiency caused by N tie-up due to high C:N ratio of the wheat straw (80:1). Thus, tillage decisions in dryland cropping systems must weigh the advantages of soil and water conservation against the potential direct and indirect costs of the systems.

Results of a long term study can shed light on the sustainability of various crop residue management practices. The crop residue long-term experiment (CR-LTE)

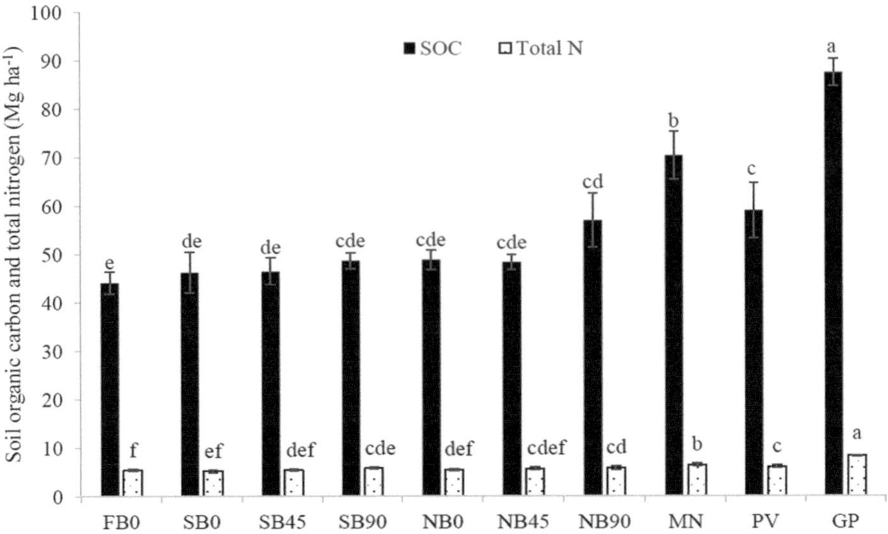

Fig. 4 Soil organic carbon (SOC) and nitrogen (TN) stock (0–60 cm) after 80 years (1931–2010) of different crop residue and nitrogen management treatments in a conventionally tilled (plow) WF system and undisturbed grassland at the Columbia Basin Agricultural Research Center (CBARC) near Pendleton, OR. *FB* fall burn, *SB* spring burn, *NB* no burn, *MN* manure application, *PV* pea vine, and *GP* undisturbed grassland. 0, 45, 90 accompanied with FB, SB, and NB indicate amount of N (kg ha^{-1}) applied from chemical fertilizer (Source: Ghimire et al. (2015))

was established in 1931 at the Oregon State University's Columbia Basin Agricultural Research Center near Pendleton, OR. The on-going experiment evaluates residue and N treatments that included no N addition with fall burning (FB0), spring burning (SB0), and no burning (NB0), 45 kg N ha^{-1} with SB (SB45) and NB (NB45), 90 kg N ha^{-1} with SB (SB90) and NB (NB90), manure (MN, 5.32 Mg dry mass ha^{-1} year^{-1}), and pea vines (PV, 0.99 Mg dry mass ha^{-1} year^{-1}) under a WF system receiving 405 mm of mean annual precipitation. A nearby grassland (GP) was used as a reference. After 80 years of experimentation all WF treatments lost SOC compared to the grassland (Fig. 4). The MN and PV treatments maintained higher SOC stocks than other treatments. All burn treatments had lower SOC stocks than corresponding no-burn treatments. The lowest SOC stocks were observed under treatments where residues were burned in the fall soon after harvest. Wheat – summer fallow systems have lost more than 60 % of the original SOC (Machado 2011; Ghimire et al. 2015). Using the same data set, Machado (2011) observed that 30 years (1976–2005) of fall burning of wheat residue depleted about 17.3 % SOC (0.34 Mg SOC ha^{-1} year^{-1}) in the top 60 cm soil and the depletion of SOC was significantly higher than a reduction of 10.4–11.9 % SOC (0.22–0.28 Mg SOC ha^{-1} year^{-1}) with no residue burning treatments. A similar experiment at the same location also revealed lower SOC and N with fall burning (9.9 g C kg^{-1} and 0.73 g N kg^{-1}) of wheat residues compared to spring burning (11.4 g C kg^{-1} and 0.88 g N kg^{-1}) and no-burning (10.9 g C kg^{-1} and

0.86 g C kg^{-1}) treatments without any N addition in a winter wheat – summer fallow rotation over seven decades (Wuest et al. 2005). Furthermore, burning of residues can reduce arbuscular mycorrhizal fungal (basidiomycetes and total glomalin) and earthworm activities, that are responsible for stabilizing soil aggregates and enhance soil water infiltration (Wuest et al. 2005).

Carbon (C) sequestration in soil occurs when C in plant residues, root exudates, and microbial biomass is incorporated into soil organic matter (SOM) and not lost to the atmosphere as CO_2. A dynamic balance exists in the soil between the additions of plant residue, the conversion of plant residues into SOM, and loss of SOC to the atmosphere as CO_2. Increasing SOM and hence C sequestration occurs when the production of SOM is greater than its decomposition by soil microbes (and loss by erosion). Based on this premise, SOM accumulation would be unattainable in conventional tillage WF systems predominant in the IPNW and Great Plains because of insufficient residues (one crop in 2 years) and rapid SOM oxidation of buried residues. Crop residue burning exacerbates SOM depletion by further reducing the amount of residues returned to the soil under conventional tillage. Annual cropping systems, even under conventional tillage, tend to maintain SOM although most of it is located in the depth of incorporation (Machado et al. 2006). Under no-till systems residues remain at the surface and are protected from decomposition and consequently do not significantly add to SOM buildup as has been shown in other studies (Soon 1999; Hooker et al. 2005). There were no differences in SOM content between treatments with residues returned and removed (Soon 1999; Hooker et al. (2005) or burned (Rumpel 2008). Roots play an important role in SOM accretion under no-till systems (Gale and Cambardella 2000). Under no-till, increasing cropping intensity increased SOM accretion due to root biomass production (Franzluebbers et al. 1994). The little or no increase in SOM when conventional tillage WF systems were converted to no-till WF systems in the INPNW and Great Plains was probably attributed to less root biomass produced as a result of fallow. Cropping intensity has been shown to increase SOM accretion in no-till cropping systems (West and Post 2002; Luo et al. 2010; Hansen et al. 2012).

There is an increasing body of knowledge, however, that indicates SOM buildup is not solely dependent on C input but also on crop residue quality. Kirkby et al. (2013, 2014) observed that SOM buildup was influenced by humification efficiency, the conversion of C inputs to humus or the stable or fine fraction of SOM. The conversion of C inputs into SOM was influenced by the carbon-nutrient stoichiometry and more stable SOM was formed when elements (N, P, S) were added to crop residues to match the stoichiometry of these nutrients in stable SOM. Soils that were supplemented with nutrients to match this ratio sequestered more carbon than soils without additional nutrients (Kirkby et al. 2013, 2014). Crop residue quality, therefore, play an important role in SOC dynamics. In the CR-LTE the MN and PV treatments had higher SOC than treatments receiving artificial N only (Fig. 4). The MN and PV added 3.68 and 2.54 MG C ha^{-1} $year^{-1}$, and supplied about 110 and 45 kg N ha^{-1} $year^{-1}$, respectively. Crop quality effects were clearly demonstrated by comparing SOC under PV and other N treatments.

Despite supplying only 45 kg N ha, the PV treatment produced more SOC than a treatment receiving 90 kg N ha (Fig. 4).

5 Modelling Management Practices for Water Productivity in Dryland Cropping Systems

In the U.S. Great Plains, reduced tillage system have allowed for replacement of the summer fallow period with a summer annual crop, a dryland practice that improves productivity and precipitation use efficiency (Farahani et al. 1998; Nielsen et al. 2005). The summer crop uses water for growth (evapotranspiration) that would otherwise have been lost through soil evaporation during the summer fallow period. While this practice has been widely adopted, the selection of the most advantageous summer crop in the rotation is challenging and costly due to variation in geographic and temporal conditions. The choice of summer crop must consider its economic return but also not compromise the subsequent wheat crop due to excessive water use. The summer crop selected should also help with weed and pest control and soil conservation needs. Research has documented conditions suitable for summer crops including maize, grain sorghum, proso millet, and annual forages (Dhuyvetter et al. 1996; Nielsen et al. 2006; Shanahan et al. 1988; Smika and Unger 1986). However, results of field research can only be applied within the spatial and temporal limits that it was conducted in. This limitation highlights the need for tools to extrapolate research information beyond its limits. The use and application of cropping system models is one approach, which has been successful in expanding knowledge from field research to answer important production question over a broader range of locations and years.

Several cropping systems models have been developed and applied to dryland cropping systems including The Root Zone Water Quality Model (RZWQM; Ahuja et al. 2000; Ma et al. 2009), APSIM (McCown et al. 1996), CropSyst (Stockle et al. 2003), DSSAT-CSM (Jones et al. 2003; Jame and Cutforth 1996; Saseendran et al. 2004, 2005a; Elliott and Cole 1989; Mathews et al. 2002). Here we highlight the use of the Root Zone Water Quality Model (RZWQM2). RZWQM2 is a process-oriented model that uses algorithms to integrate biological, physical, and chemical processes for integration, synthesis, and extrapolation across soils and climates of effects at the system level. The model includes a wide spectrum of cultural practices to allow for simulations of varying tillage, residue management, fertilizers, and crop rotation. Driving factors such as weather, soil, and crop parameters can be modified and outputs include crop production and water quality parameters (Ahuja et al. 2000; Ma et al. 2009). RZWQM2 has been successfully used to inform management in dryland cropping systems for the U.S. Great Plains (Ma et al. 2003; Saseendran et al. 2004, 2005a, b, 2008, 2009).

One application of the use of RZWQM2 is development of a decision-aid tool to guide the choice of a summer crop to replace summer fallow based on knowledge of soil water at planting for use in the U.S. Central Great Plains. The model was parameterized and validated based on published field research from two locations, Akron, CO and Sidney, NE, in the U.S. Central Great Plains (Saseendran et al. 2008, 2009, 2010a). The model was used to simulate long-term yields and output was used to calculate net returns for maize, canola, and proso millet and two annual forage crops, foxtail millet and spring triticale, in a wheat-summer crop-fallow rotation. The simulations were performed with varying plant available water (PAW) at the time of planting (25, 50, 75 and 100 % PAW). By simulating multiple crop years using historic weather but variable PAW at planting as model inputs, yield probability curves were generated (Fig. 5). With this information, growers can forecast probable yields of different crops based on the PAW at planting. Yield increased for all crops as initial PAW increased, but the response was greatest for forage crops and least for maize (Fig. 5). Probable yields can then be used with current production costs and commodity prices to select the most advantageous summer crop. Under the conditions used in the study, forage crops had greater net economic returns than any of the grain crops (Fig. 6). Among the grain crops, proso millet was the most profitable crop at Akron, while canola was the most profitable crop at Sidney, illustrating the value of the model for guiding site specific decisions. Modelling allows users to evaluate many alternatives, including different environmental conditions, climate scenarios, production costs, and commodity prices.

In addition to guiding management decisions, models can provide mechanistic understanding of how management practices or environmental conditions control observed cropping system outcomes. This understanding can lead to development of new ideas and technology for improving resource use. To illustrate, Saseendran et al. (2005a) evaluated RZWQM for its ability to simulate a 2-year winter wheat-summer fallow rotation under tilled and no-till conditions on a Weld silt loam soil in semi-arid northeastern Colorado. Field observations of increased water storage during no-till fallow were explained by model predictions of crop residue dynamics and the effects of residue on soil evaporation (Fig. 7). In the tilled wheat-summer fallow rotation, residue mass increased at harvest but decreased shortly thereafter due to tillage. In the no-till wheat-summer fallow rotation, residue remaining at the soil surface during the summer fallow period increased evaporative resistance and decreased water lost from evaporation relative to the tilled system. The model output of residue dynamics explained why increased soil moisture was observed in no-till fallow and gives details about the temporal dynamics. This information can be useful in developing different cropping practices. For example, the model showed that the increased soil moisture during fallow does not always translate to greater yield in a wheat-summer fallow rotation because the is often adequate water for a single wheat crop even in the tilled system. This gave rise to intensified wheat-summer crop-fallow rotations to take advantage of the increased water retention in no-till systems.

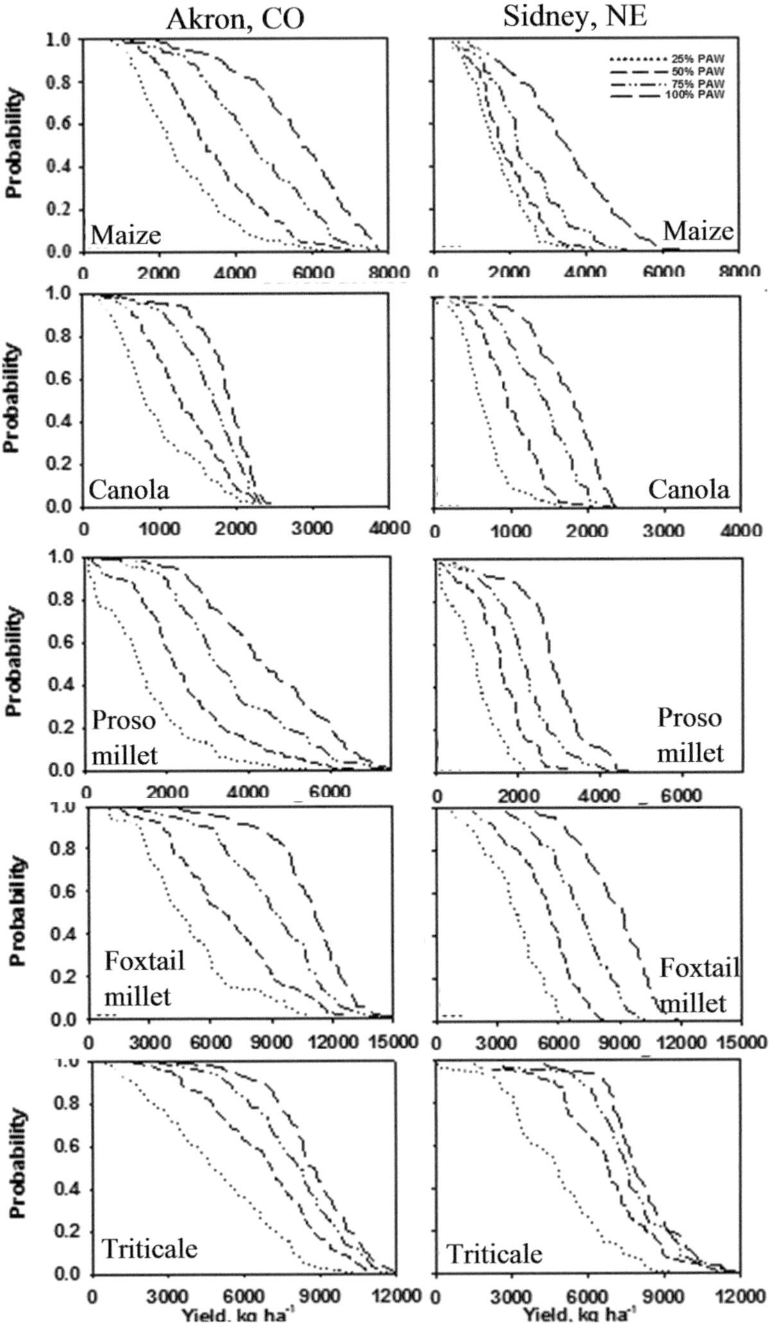

Fig. 5 The RZWQM2 model was used to predict the probability of obtaining a minimum yield for three grain crops (maize, canola, proso millet) and two forage crops (foxtain millet, triticale) based on the plant available water (PAW) at planting for at Akron, Colorado and Sidney, Nebraska

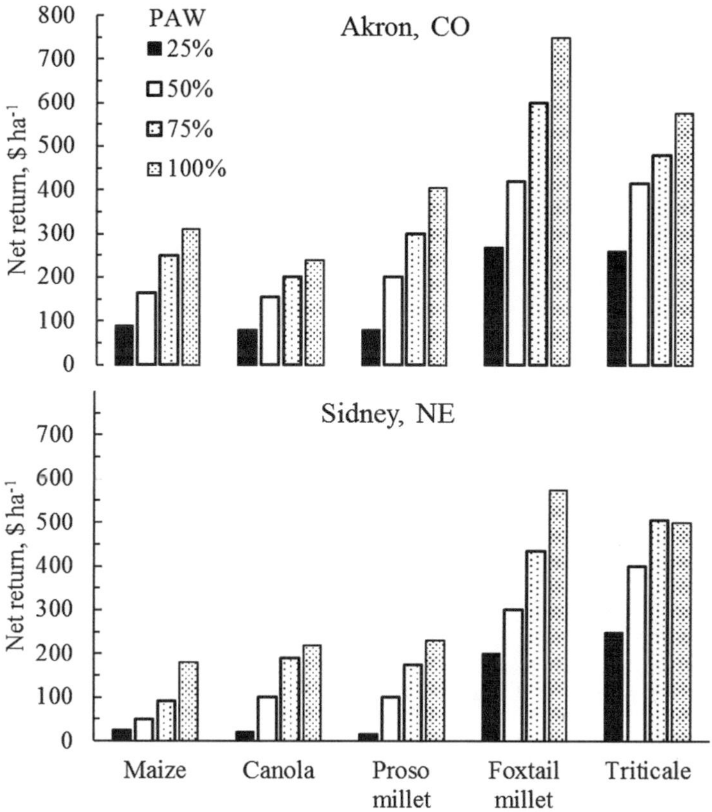

Fig. 6 Mean net returns of summer crops in a wheat-summer crop-fallow rotation based on output from the RZWQM2 model in response to 25, 50, 75 and 100 % plant available water (PAW) at planting. Summer grain crops were maize, canola, and proso millet, and summer forage crops were foxtail millet and triticale. The model was based on two locations in the U.S. Great Plains, Akron, Colorado and Sydney, Nebraska

6 Future Direction and Research Needs

Attaining food security for an increasing population while minimizing soil degradation is a major global challenge facing agriculture. Dryland cropping systems play a very important role in food, feed, and fiber production in North America. Improving soil health is critical for sustaining dryland crop production. Soil organic matter, through the provision of ecosystems services such as increased water and nutrient holding capacity, increased microbial diversity and function, and increased soil aggregation and improved soil structure, underpins agricultural production. Management practices that maintain or improve SOM improve soil health and agricultural sustainability. The WF system in the dryland regions of the INPW and Great Plains have lost more than 60 % of SOC from topsoil in the last century

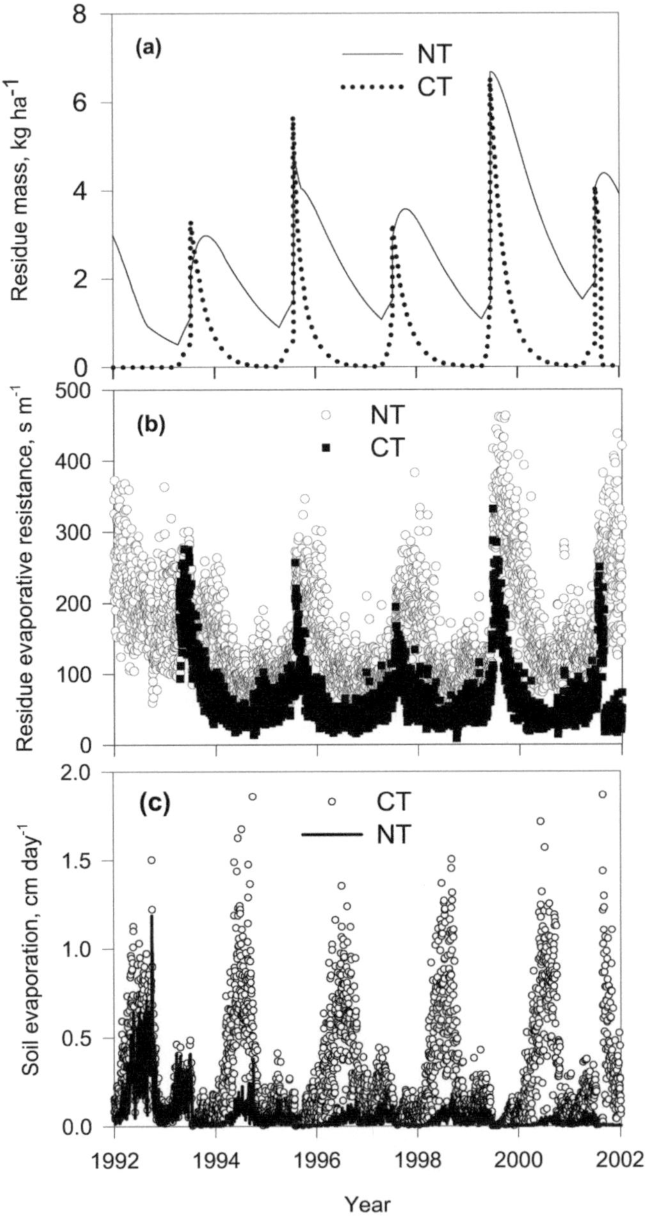

Fig. 7 RZWQM simulations of residue dynamics and residue effects on (**a**) flat residue mass on the soil surface, (**b**) residue evaporative resistance, and (**c**) potential soil evaporation in conventional till (CT) and no-till (NT) wheat-fallow cropping systems

(Machado 2011; Brown and Huggins 2012), decreased soil health and crop productivity. The same trend has been observed in other agricultural systems worldwide (Lal 2004, 2006). Restoring SOM that has been lost due to years of fallow and intensive tillage is among the greatest sustainability challenges for dryland cropping in North America and is especially challenging because of the low biomass production in dryland areas (Aase and Pikul 1995; Brown and Huggins 2012; Liebig et al. 2004; Lloyd et al. 2013). Application of current research and new research must address soil health as affected by management systems such reduced tillage, no-tillage, increased cropping intensity, cover cropping, and soil amendments. Application of manures, biosolids, and biochar have tremendous potential to increase SOM and ecosystems services to improve soil health in dryland crop land. In addition to field research, use of cropping systems models can provide direction and guidance on how to improve management practices.

Increasing cropping system diversity is a high priority for dryland cropping systems in North America. While dryland cropping systems in the Northern Great Plains have been diversified in recent decades, much of the rest of the dryland cropping region in North America is very limited in cropping system diversity. The traditional wheat-summer fallow rotation is widely practiced, resulting in declining soil health, wind and water erosion, low precipitation use efficiency, and low economic diversity. Work is needed for genetic improvement and variety development of alternative crops and how to fit alternative crops into dryland crop rotations for various dryland cropping regions of North America. Among the improvements to be develop are crops that produce quality residues that reduce soil loss and increase potential for sequestration of soil carbon. Both experimental and modelling research can be used to advance cropping system diversification. There is especially a need for viable pulse crops adapted to the conditions of the Central and Southern Great Plains. This effort must include development of markets for alternative crops and overcoming barriers limiting producers from diversifying crop rotations.

Gains in the diversification and intensification of the traditional wheat-summer fallow rotations have been made largely by improving precipitation storage and use through adoption of reduced tillage or zero-till systems (Aase and Siddoway 1980; Aase and Reitz 1989; Cochran et al. 2006). Maintaining the benefits of these reduced and no-till systems is being threatened by weeds, insect pests, and diseases that are harbored in the crop residues. For example, crop damage from the wheat stem sawfly, *Cephus cinctus* Norton (Hymenoptera: Cephidae), to wheat has increased with increasing adoption of no-till (Weaver et al. 2009). Similar observations have been made for crop diseases and weeds that have a competitive advantage in no-till environments. Understanding and adoption of integrated pest management practices to manage pests and diseases as a result of increasing no-till production is of high priority.

7 Conclusions

Dryland crop production is of major importance in North America, with large areas of dryland farming in Canada, Mexico, and the United States. The Canadian Prairie and the Great Plains and Inland Pacific Northwest of the U.S. are the areas with the highest density of dryland farming. In addition, there are less dense areas of dryland production in nearly every state in the western U.S and in northern and central Mexico. The traditional production system is a wheat-summer fallow rotation with conventional or stubble mulch tillage. Sustainability of this practice is limited by soil degradation and erosion and poor water use efficiency. Where adopted, no-till practices improve precipitation storage and use efficiency, which has led to crop intensification and diversification and improvements in soil properties. While wheat is the most common dryland crop in North America, dryland farming is also important for the production of maize, sorghum, pulses, and oilseeds. No-till systems are being adopted together with more intensive crop rotations that reduce fallow frequency, increase precipitation use efficiency, reduce erosion, and improve soil properties. No-till adoption is greatest in the northern region of the Great Plains, where climate conditions are favorable, and least in the Southern Great Plains.

As adoption of reduced tillage practices has grown, dependence on herbicides for weed control has also increased and has led to weed resistance and weed shifts that complicate weed management. With a limited array of available herbicide classes, weed resistance will remain a challenge to wheat production in the future. Herbicide resistance to glyphosate is creating unique challenges as a result of its ubiquitous use. Sustainable weed management in dryland wheat production will best be achieved through continued development and adoption of integrated weed management and crop production practices.

When reduced or no-till practices are implemented, crop rotations can be intensified and diversified. In the Northern Great Plains, inclusion of oilseed crops in continuous crop rotations without fallow is common. Annual forage or pulse crops have also been successfully integrated into dryland rotations, improving soil health such as soil carbon levels and nutrient cycling. Gaining the soil improvements associated with intensified rotations requires a period of time for the system to adapt and also requires careful management. Choice of a fallow replacement crop is a major decision for dryland farmers in the Great Plains. Use of a cropping system model has been demonstrated as a way to test alternative crops and crop rotations with different environmental conditions and climate.

Sustainability of dryland cropping systems must consider practices that maintain soil organic matter and restore soils degraded by past practices. Practices such as residue burning and fallow period tillage have reduced soil organic carbon levels by as much as 60 %. Current no-till dryland systems with intensified crop rotations have been shown to stabilize soil carbon and increase soil organic matter accretion. Future research using experimentation and modelling should continue to develop alternative no-till crops and crop rotations that increase precipitation use efficiency, improve soil properties, reduced dependence on N fertilizers, adapt to climate change, and develop alternative markets.

References

Aase JK, Pikul JL (1995) Crop and soil response to long-term tillage practices in the northern Great Plains. Agron J 87:652–656

Aase JK Reitz LL (1989) Annual cropping may be practical and profitable. Montana Farmer-Stockman. pp 20.daw

Aase JK, Siddoway FH (1980) Stubble height effects on seasonal microclimate, water balance, and plant development of no-till winter wheat. Agric Meteorol 21:1–20

Ahuja LR, Rojas KW, Hanson JD, Shafer MJ, Ma L (eds) (2000) Root zone water quality model. Modeling management effects on water quality and crop production. Water Resources Publications, LLC, Highlands Ranch

Allen BL, Cochran VL, Caesar TC, Tanaka DL (2011a) Long-term effects of topsoil removal on soil productivity factors, wheat yield and protein content. Arch Agron Soil Sci 57:293–303

Allen BL, Pikul JL, Waddell JT, Cochran VL (2011b) Long-term lentil green-manure replacement for fallow in the semiarid northern Great Plains. Agron J 103:1292–1298

Allen BL, Lenssen AW, Sainju UM, Caesar TC, Evans RG (2014) Nitrogen use in durum and selected brassicaceae oilseeds in two-year rotations. Agron J 106:821–830

Anderson RL (1997) Cultural systems can reduce reproductive potential of winter annual grasses. Weed Technol 11:608–613

Anderson RL (2003) An ecological approach to strengthen weed management in the semi-arid Great Plains. Adv Agron 80:33–62

Baumhardt RL, Anderson RL (2006) Crop choices and rotation principles. In: Peterson GA, Unger PW, Payne WA (eds) Dryland agriculture. ASA, CSSA, and SSSA, Madison, pp. 113–139

Baumhardt RL, Salines-Garcia J (2006) Dryalnd agriculture ini Mexico and the U.S. Southern Great Plains. In: Peterson GA, Unger PW, Payne WA (eds) Dryland agriculture. ASA, CSSA, and SSSA, Madison, pp. 341–364

Beres BL, Harker KN, Clayton GW, Bremer E, Blackshaw RE, Graf RJ (2010) Weed competitive ability of spring and winter cereals in the Northern Great Plains. Weed Technol 24:108–116

Black AL, Brown PL, Halvorson AD, Siddoway FH (1981) Dryland cropping strategies for efficient water-use to control saline seeps in the northern Great Plains, U.S.A. Agric Water Manag 4:295–311

Black AL, Power JF (1965) Effect of chemical and mechanical fallow methods on soil moisture storage wheat yields, and soil erodibility. Soil Sci Soc Am Proc 29:465–468

Blackshaw RE (2005) Tillage intensity affects weed communities in agroecosystems. In: Injerjit S (ed) Invasive plants: ecological and agricultural aspects. Birkhauser Verlag, Basel, pp. 209–221

Blackshaw RE, Beckie HJ, Molnar LJ, Entz T, Moyer JR (2005a) Combining agronomic practices and herbicides improves weed management in wheat-canola rotations within zero-tillage production systems. Weed Sci 53:528–535

Blackshaw RE, Moyer JR, Harker KN, Clayton GW (2005b) Integration of agronomic practices and herbicides for sustainable weed management in zero-till barley field pea rotation. Weed Technol 19:190–196

Blackshaw RE, Semach G, Li X, O'Donovan JT, Harker KN (1999) An integrated weed management approach to managing foxtail barley (*Hordeum jubatum*) in conservation tillage systems. Weed Technol 13:347–353

Brown TT, Huggins DR (2012) Soil carbon sequestration in the dryland cropping region of the Pacific Northwest. J Soil Water Conserv 67:406–415

Campbell CA, Zentner RP, Janzen HH, Bowren KE (1990) Crop rotation studies on the Canadian prairies. Research Branch, Agriculture Canada, Ottawa. 1841/E, Canada

Cochran V, Danielson J, Kolber R, Miller P (2006) Dryalnd cropping in the Canadian Prairies and the U.S. Northern Great Plains. In: Peterson GA, Unger PW, Payne WA (eds) Dryland agriculture. ASA, CSSA, and SSSA, Madison, pp. 293–339

Council for Agricultural Science and Technology (CAST) 2012 Herbicide-resistant weeds threaten soil conservation gains: finding a balance for soil and farm sustainability. Issue Paper 49. CAST. Ames

Daugovish O, Lyon DJ, Baltensperger DD (1999) Cropping systems to control winter annual grasses in winter wheat (*Triticum aestivum*). Weed Technol 13:120–126

Derksen DA, Anderson RL, Blackshaw RE, Maxwell B (2002) Weed dynamics and management strategies for cropping systems in the northern Great Plains. Agron J 94:174–185

Donald WW, Ogg AG Jr (1991) Biology and control of jointed goatgrass (*Aegilops cylindrica*), a review. Weed Technol 5:3–17

Dhuyvetter KC, Thompson CR, Norwood CA, Halvorson AD (1996) Economics of dryland cropping systems in the Great Plains: a review. J Prod Agric 9:216–222

Elliott ET, Cole CV (1989) A perspective on agroecosystem science. Ecology 70:1597–1602

Entz MH, Bullied WJ, Katepa F (1995) Rotational benefits of forage crops in Canadian prairie cropping systems. J Prod Agric 8:521–529

Farahani HJ, Peterson GA, Westfall DG, Sherrod LA, Ahuja LR (1998) Soil water storage in dryland cropping systems: significance of cropping intensification. Soil Sci Soc Am J 62:984–991

Franzluebbers AJ, Hons FM, Zuberer DA (1994) Long-term changes in soil carbon and nitrogen pools in wheat management-systems. Soil Sci Soc Am J 58:1639–1645 Available at ISI:A1994PT17400009

Gale WJ, Cambardella CA (2000) Carbon dynamics of surface residue- and root-derived organic matter under simulated no-till. Soil Sci Soc Am J 64(1):190–195 Available at <Go to ISI>://WOS:000089446000025

Gesch RW, Isbell TA, Oblath EA, Allen BL, Archer DW, Brown J, Hatfield JL, Jabro JD, Kiniry JR, Long DS, Vigil MF (2015) Comparison of several brassica species in the north central U.S. for potential jet fuel feedstock. Ind Crops Prod 75:2–7

Ghimire R, Machado S, Rhinhart K (2015) Long-term crop residue and nitrogen management effects on soil profile carbon and nitrogen in wheat–fallow systems. Agron J 107:2230 Available at https://dl.sciencesocieties.org/publications/aj/abstracts/107/6/2230

Greb BW, Smika DE, Woodruff NP, Whitfield CJ (1970) Summer fallow in the Central Great Plains. In: Haas JJ, Willis WO, Bond JJ (eds) Summer fallow in the western United States, Conservation Res. Rep. No. 17. USDA-ARS, U.S. Government Printing Office, Washington, DC, pp. 51–85

Hansen NC, Allen BL, Baumhardt RL, Lyon DJ (2012) Research achievements and adoption of no-till, dryland cropping in the semi-arid U.S. Great Plains. F Crop Res 132:196–203

Harker KN, O'Donovan JT, Turkington TK, Blackshaw RE, Lupwayi NZ, Smith EG, Johnson EN, Pageau D, Shirtliffe SJ, Gulden RH, Rowsell J, Hall LM, Willenborg CJ (2016) Diverse rotations and optimal cultural practices control wild oat (*Avena fatua*). Weed Sci 64:170–180

Heap I (2016) The international survey of herbicide resistant weeds. Online. Internet. Friday, April 29, 2016. Available at: www.weedscience.org.

Hooker BA, Morris TF, Peters R, Cardon ZG (2005) Long-term effects of tillage and corn stalk return on soil carbon dynamics. Soil Sci Soc Am J 69:188–196 Available at http://www.scopus.com/inward/record.url?eid=2-s2.0-12944286379&partnerID=40&md5=daa554449b1ffc1d2e89f7afdba97a60

Jame YW, Cutforth HW (1996) Crop growth models for decision support systems. Can J Plant Sci 76:9–19

Janzen HH (1987) Soil organic matter changes after long-term cropping to various spring wheat rotations. Can J Soil Sci 67:845–856

Jones JW, Hoogenboom G, Porter CH, Boote KJ, Batchelor WD, Hunt LA, Wilkens PW, Singh U, Gijsman AJ, Ritchie JT (2003) The DSSAT cropping system model. Eur J Agron 18:235–265

Jones OR, Popham TW (1997) Cropping and tillage system for dryland grain production in the Southern High Plains. Agron J 89:222–232

Kettler TA, Lyon DJ, Doran JW, Powers WL, Stroup WW (2000) Soil quality assessment after weed-control tillage in a no-till wheat – fallow cropping system. Soil Sci Soc Am J 64:339–346
Kirkby CA, Richardson AE, Wade LJ, Batten GD, Blanchard C, Kirkegaard JA (2013) Carbon-nutrient stoichiometry to increase soil carbon sequestration. Soil Biol Biochem 60:77–86
Kirkby CA, Richardson AR, Wade LJ, Passioura JB, Batten GD, Blanchard C, Kirkegaard JA (2014) Nutrient availability limits carbon sequestration in arable soils. Soil Biol Biochem 68:402–409
Krupinsky JM, Bailey KL, McMullen MP, Gossen BD, Turkington TK (2002) Managing plant disease risk in diversified cropping systems. Agron J 94:198–209
Lal R (2004) Soil carbon sequestration to mitigate climate change. Geoderma 123:1–22
Lal R (2006) Enhancing crop yields in the developing countries through restoration of the soil organic carbon pool in agricultural lands. Land Degrad Dev 17:197–209
Lenssen AW, Iversen WM, Sainju UM, Caesar TC, Blodgett SL, Allen BL, Evans RG (2012) Yield, pests, and water use of durum and selected crucifer oilseeds in two-year rotations. Agron J 104:1295–1304
Lenssen AW, Cash SC, Hatfield PG, Sainu UR, Grey WR, Blodgett SL, Goosey HB, Griffith DA, Johnson GD (2010) Yield, quality, and water and nitrogen use of durum and annual forages in two-year rotations. Agron J 102:1261–1268
Lenssen AW, Sainju UM, Jabro JD, Allen BL, Evans RG (2015) Management and tillage influence barley forage productivity and water use in dryland cropping. Agron J 107:551–557
Liebig MA, Tanaka DL, Wienhold BJ (2004) Tillage and cropping effects on soil quality indicators in the northern Great Plains. Soil Tillage Res 78:131–141
Lloyd G, Hansen NC, Sherrod L, Inman D, Peterson GA (2013) Constraints and capabilities of no-till dryland agroecosystems as bioenergy production systems. Agron J 105:364–376
Long DS, Young FL, Schillinger WF, Reardon CL, Williams JD, Allen BL, Pan WL, Wysocki DJ (2016) Development of dryland oilseed production systems in northwestern region of the USA. Bioenergy Res 9:412–429
Luo Z, Wang E, Sun OJ (2010) Can no-tillage stimulate carbon sequestration in agricultural soils? A meta-analysis of paired experiments. Agric Ecosyst Environ 139:224–231
Lyon DJ, Baltensperger DD (1995) Cropping systems control winter annual grass weeds in winter wheat. J Prod Agric 8:535–539
Lyon DJ, Bruce S, Vyn T (2004) Achievements and future challenges in conservation tillage. In: Proceedings of the 4th international crop science congress, 26 September–1 October 2004, Brisbane
Ma L, Hoogenboom G, Saseendran SA, Bartling PNS, Ahuja LR, Green TR (2009) Estimates of soil hydraulic properties and root growth factor on soil water balance and crop production. Agron J 101:572–583
Ma L, Nielsen DC, Ahuja LR, Malone RW, Saseendran SA, Rojas KW, Hanson JD, Benjamin JG (2003) Evaluation of RZWQM under varying irrigation levels in eastern Colorado. Trans ASAE 46:39–49
Machado S (2011) Soil organic carbon dynamics in the Pendleton long-term experiments: implications for biofuel production in pacific Northwest. Agron J 103:253–260
Machado S, Rhinhart K, Petrie S (2006) Long-term cropping system effects on carbon sequestration in eastern Oregon. J Environ Qual 35:1548–1553
Mathews R, Stephens W, Hess T, Middleton T, Graves A (2002) Applications of crop/soil simulation models in tropical agricultural systems. Adv Agron 76:31–24
McCarty JL, Korontzi S, Justice CO, Loboda T (2009) The spatial and temporal distribution of crop residue burning in the contiguous United States. Sci Total Environ 407:5701–5712
McCown RL, Hammer GL, Hargreaves JNG, Holzworth DL, Freebairn DM (1996) APSIM: a novel software system for model development, model testing, and simulation in agricultural systems research. Agric Syst 50:255–271

Monfreda C, Ramankutty N, Foley JA (2008) Farming the planet. Part 2: geographic distribution of crop areas, yields, physiological types, and net primary production in the year 2000. Global Biogeochemical Cycles 22, GB1022, doi:10.1029/2007GB002947.

Nielsen DC, Unger PW, Miller PR (2005) Efficient water use in dryland cropping systems in the Great Plains. Agron J 97:364–372

Nielsen DC, Vigil MF, Benjamin JG (2006) Forage yield response to water use for dryland corn, millet, and triticale in the Great Plains. Agron J 98:992–998

Norwood CA (1994) Profile water distribution and grain yield as affected by cropping system and tillage. Agron J 86:558–563

O'Donovan JT, Blackshaw RE, Harker KN, Clayton GW, Moyer JR, Dosdall LM, Maurice DC, Turkington TK (2007) Integrated approaches to managing weeds in spring-sown crops in western Canada. Crop Prot 26:390–398

Paulitz TC (2006) Low input no-till cereal production in the Pacific Northwest of the U.S.: the challenges of root diseases. Eur J Plant Pathol 115:271–281

Peterson GA, Halvorson AD, Havlin JL, Jones OR, Lyon DJ, Tanaka DL (1998) Reduced tillage and increasing cropping intensity in the Great Plains conserves soil carbon. Soil Tillage Res 47:207–218

Peterson GA, Schlegel AJ, Tanaka DL, Jones OR (1996) Precipitation use efficiency as affected by cropping and tillage systems. J Prod Agric 9:180–186

Peterson GA, Westfall DG (2004) Managing precipitation use in sustainable dryland agroecosystems. Ann Appl Biol 144:127–138

Rumpel C (2008) Does burning of harvesting residues increase soil carbon storage. J Soil Sci Plant Nutr 8(2):44–51

Sainju UM, Allen BL, Caesar TC, Lenssen AW (2015a) Dryland soil carbon and nitrogen after thirty years of tillage and cropping sequence combination. Agron J 107:1822–1830

Sainju UM, Allen BL, Caesar TC, Lenssen AW (2015b) Dryland soil chemical properties and crop yields affected by long-term tillage and cropping sequence. SpringerPlus 4:320

Saseendran SA, Nielsen DC, Ma L, Ahuja LR (2010a) RZWQM2 simulations of alternative cropping systems with and without summer crops in the Central Great Plains. Agron J 102:1521–1534

Saseendran SA, Nielsen DC, Ma L, Ahuja LR, Halvorson AD (2004) Modeling nitrogen management effects on winter wheat production using RZWQM and CERES-wheat. Agron J 96:615–630

Saseendran SA, Nielsen DC, Ma L, Ahuja LR, Vigil MF, Halvorson AD (2005b) Effectiveness of RZWQM for simulating alternative Great Plains cropping systems. Agron J 97:1183–1193

Saseendran SA, Ma L, Nielsen DC, Vigil MF, Ahuja LR (2005a) Simulating planting date effects on corn production using RZWQM and CERES-Maize. Agron J 97:58–71

Saseendran SA, Nielsen DC, Lyon DJ, Ma L, Felter DG, Baltensperger DD, Hoogenboom G, Ahuja LR (2009) Modeling responses of dryland spring triticale, proso millet and foxtail millet to initial soil water in the High Plains. Field Crop Res 10:48–63

Saseendran SA, Ahuja LR, Nielsen DC, Trout TJ, Ma L (2008) Use of crop simulation models to evaluate limited irrigation management options for corn in a semiarid environment. Water Resour Res 44:W00E02. doi:10.1029/2007WR006181

Schillinger WF, Papendick R (2008) Then and now: 125 years of dryland wheat farming in the Inland Pacific Northwest. Agron J 100(3):S166–S182 Available at https://dl.sciencesocieties.org/publications/aj/articles/100/Supplement_3/S-166

Schillinger WF, Papendick R, Guy SO, Rasmussen PE, VanKessel C (2006) Dryland cropping in the Western United States. In: Peterson GA, Unger PW, Payne WA (eds) Dryland agriculture. ASA, CSSA, and SSSA, Madison, pp. 365–393

Schillinger WF, Young DL (2014) Best management practices for summer fallow in the world's driest rainfed wheat region. Soil Sci Soc Am J 78:1707–1715

Shanahan JF, Anderson RL, Greb BW (1988) Productivity and water use of proso millet grown under three cropping rotations in the Great Central Plains. Agron J 80:487–492

Siebert, S., Henrich, V., Frenken, K., and Burke, J. 2013. Global map of irrigation areas version 5. Rheinische Friedrich-Wilhelms-University, Bonn/Food and Agriculture Organization of the United Nations, Rome

Smika DE, Unger PW (1986) Effect of surface residues on soil water storage. In: Stewart BA (ed) Advances in soil science, vol 5. Springer, New York, pp. 111–138

Smith EG, Upadhyay BM, Blackshaw RE, Beckie HJ, Harker KN, Clayton GW (2006) Economic benefits of integrated weed management systems in field crops of western Canada. Can J Plant Sci 86:1273–1279

Soon YK (1999) Crop residue and fertilizer management effects on nutrient use and barley production. Can J Soil Sci 79(2):389–394 Available at <Go to ISI>://WOS:000081215100018

Stockle CO, Donatelli M, Nelson R (2003) CropSyst, a cropping systems simulation model. Eur J Agron 18:289–307

Stump WL, Westra P (2000) The seedbank dynamics of feral rye (*Secale cereale*). Weed Technol 14:7–14

Tanaka DL, Aase JK (1987) Fallow method influences on soil water and precipitation storage efficiency. Soil Tillage Res 9:307–316

Unger PW, Payne WA, Peterson GA (2006) Water conservation and efficient use. In: Peterson GA, Unger PW, Payne WA (eds) Dryland agriculture. ASA, CSSA, and SSSA, Madison, pp. 39–85

Varanasi VK, Godar AS, Currie RS, Dille JA, Thompson CR, Stahlman PW, Jugulam M (2015) Field evolved resistance to four modes of action of herbicides in a single kochia (*Kochia scoparia* L. Schrad.) population. Pest Manag Sci 71:1207–1212

Waite J, Thompson CR, Peterson DE, Currie RS, Olson BLS, Stahlman PW, Al-Khatib K (2013) Differential kochia (*Kochia scoparia*) populations response to glyphosate. Weed Sci 61:193–200

Weaver DK, Buteler M, Hofland ML, Runyon JB, Nansen C, Talbert LE, Lamb P, Carlson GR (2009) Cultivar preferences of ovipositing wheat stem sawflies as influenced by the amount of volatile attractant. J Econ Entomol 102:1009–1017

West TO, Post WM (2002) Soil organic carbon sequestration rates by tillage and crop rotation: a global data analysis. Soil Sci Soc Am J 66(6):1930–1946

Wienhold BJ, Pikul JL, Liebig MA, Mikha MM, Varvel GE, Doran JW, Andrews SS (2006) Cropping system effects on soil quality in the Great Plains synthesis from a regional project. Renewable Agric Food Syst 21:49–59

Williams JD, Wuest SB (2011) Tillage and no-tillage conservation effectiveness in the intermediate precipitation zone of the inland Pacific Northwest, United States. J Soil Water Conserv 66:242–249

Wuest SB, Caesar-TonThat TC, Wright SF, Williams JD (2005) Organic matter addition, N, and residue burning effects on infiltration, biological, and physical properties of an insensitively tilled silt-loam soil. Soil Tillage Res 84:154–167

Zomer RJ, Trabucco A, Bossio DA, van Straaten O, Verchot LV (2008) Climate change mitigation: a spatial analysis of global land suitability for clean development mechanism afforestation and reforestation. Agric Ecosyst Environ 126:67–80

Zomer RJ, Bossio DA, Trabucco A, Yuanjie L, Gupta DC, Singh VP (2007) Trees and water: smallholder agroforestry on irrigated lands in Northern India, IWMI Research Report 122. International Water Management Institute, Colombo, p. 45

Nurturing Agricultural Productivity and Resilience in Drylands of Sub-Saharan Africa

Tilahun Amede and Admassu Tsegaye

1 Introduction

In most African countries, agriculture accounts for about 70 % of the labour force, 25 % of the Gross Domestic Product (GDP) and 20 % of agribusiness (UNECA 2009). However, the natural resources base in SSA has not been exploited to improve food security and poverty. The region remains one of the most food-insecure parts of the world, where poverty is prevalent, food shortages are common, and food aid is the major coping strategy to feed people during food-deficit months. There has been some progress in Sub-Saharan Africa towards halving the proportion of its population that suffers from food insecurity, partly by helping farmer to access agricultural technologies and expand market opportunities. The prevalence of hunger in the region declined by 31 % between the base period (1990–1992) and 2015 (FAO 2015). For example, between 1990–1992 and 2012–2014, food availability in Ethiopia and Mozambique increased by 41 and 31 %, respectively (FAO 2015). However, much of Eastern and Southern Africa has been affected by unfavourable climatic and drought conditions which have undermined any progress toward improving food security and nutrition. For instance, in 2015–2016, Ethiopia and Zimbabwe have been suffering the worst drought in 50 years due to El Nino-associated extreme events. Maize stocks in Malawi—a net exporter of maize just a few years ago—declined to a quarter of its annual average after the worst harvest in seven years in 2012–2013. Meanwhile, maize prices have more than doubled recently (https://www.gov.uk/government/news/southern-africa-facing-disaster-as-food-crisis-looms).

T. Amede (✉)
International Crops Research Institute for the Semiarid Tropics (ICRISAT), Addis Ababa, Ethiopia
e-mail: t.amede@cgiar.org

A. Tsegaye
College of Development Studies, Addis Ababa University, Addis Ababa, Ethiopia

M. Farooq, K.H.M. Siddique (eds.), *Innovations in Dryland Agriculture*,
DOI 10.1007/978-3-319-47928-6_16

Food insecurity in the region is generally caused by low and stagnant agricultural productivity as expressed by low crop yields (Amede et al. 2014b) and rural poverty. For instance, about 55 % of smallholder farms in Malawi are less than 1.0 ha, with 25 % of these less than 0.25 ha (Dorward 1999). It is becoming increasingly difficult for these farmers to satisfy their basic household food requirements with the existing low-input, low-output production practices. The land resources base is no different for Ethiopia. Food production on such small plots commonly supplies six to nine months of the household food demand per year, but this can vary with weather conditions (drought or wet years). For the remainder of the year, resource-poor smallholder farmers depend on off-farm activities to raise the necessary cash for their household needs or rely on food aid. Some communities, sell their livestock, mainly goats and sheep, to cover the food deficit periods as is the case in Zimbabwe. Recent government initiatives for improving food security and tackling poverty in SSA region have had mixed results, and the potential effects of these policies on rural livelihoods are yet to be evaluated. For instance, land tenure policies in Mozambique are a major concern for farmers as they affect long-term investments (Amede unpublished data).

In general, rural poverty has decreased in most SSA countries—mainly those that have increased food availability and experienced economic growth (FAO 2015). For example, poverty rates in South Africa declined by 64 %, from 26 % in 2000 to 9 % in 2011, while Ethiopia declined by 33 % from 1999 to 2014 (World Economic Forum 2015). Investment in the agricultural sector in SSA was found to be more effective than other sectors in reducing hunger and poverty and contributing to economic growth. The agricultural sector has the potential to generate capital surplus, release labour for other sectors, and provide a stable food supply at affordable prices, thus contributing to the competitiveness of the economy as a whole and acting as a major stimulus for the demand of goods and services in other sectors (FAO 2015).

Key problems facing dryland countries include:

(i) Food production systems are highly fragile.
(ii) Some 16 % of the population lives in poverty.
(iii) Food imports are untenably high.
(iv) Water scarcity is a constant and growing problem.
(v) Adverse climate events (extreme heat and cold; drought and flooding) are aggravating vulnerability (Dryland Systems 2012).

This chapter describes the various dryland farming systems in Sub-Saharan Africa and reviews the different interventions required to increase productivity and curb the adverse effects of recurrent drought in the region. The various entry points to facilitate sustainable intensification of dryland systems at farm and landscape scales are also outlined.

2 Dryland Farming Systems

SSA farming systems can be classified into 13 broad farming systems, where each has a unique core concept or central tendency, and contains a substantial degree of subsystem heterogeneity (Garrity et al. 2012). Dry area production systems include a diverse mix of food, fodder and fiber crops; vegetables, rangeland and pasture species; fruit and fuel-wood trees; medicinal plants; and livestock and fish. They are found where precipitation is low and erratic, and water supply is often the most limiting factor to agricultural production (Dryland Systems 2012). The dryland systems in Eastern and Southern Africa (ESA) extend from north to south across the various countries, as presented in the aridity index (Fig. 1).

Maize (*Zea mays* L.) is the dominant crop in the dryland systems of ESA, with some 91 million ha cultivated. The maize mixed farming system has a higher agricultural population (just under 91 million in 2010) and more poverty than any other farming system in Africa. This farming system is the major food basket of the region, as well as the driver of agricultural growth and food security, though its production has peculiar characteristics with important distinctions across countries. Maize-based food constitutes about 50 % of the daily calorie intake in Zimbabwe, Kenya, Malawi, Zambia and Mozambique (Haggblade et al. 2009), which is much higher than other parts of Africa. Maize accounts for 60–70 % of the total cropped area in Zambia and Zimbabwe and > 90 % of the total cereal production (Mukanda and Moono 1999). However, about 40 % of Africa's maize growing area faces occasional drought stress which reduces yields by 10–15 % (Fisher et al. 2013), and about 25 % of the area suffers frequent drought with yield losses reaching 50 % (Abate et al. 2013). Because most of the farmland is allocated to maize—from 45.9 % in Mozambique to 69.8 % in Malawi—the risk of crop failure due to drought cannot be ignored. Moreover, most of the maize land in Southern Africa is covered with local landraces, which are commonly long maturing and low yielding; except in Zambia and Zimbabwe where mostly hybrid maize is grown (Kassie et al. 2012).

Maize is also an important crop in Eastern Africa, with annual plantings on 7.3 million ha (corresponding to 21 % of the arable area and 41 % of land under cereals). However, there are some marked regional variations in growing maize, the largest area allocated to maize compared to all other cereals being in Kenya and lowest in Ethiopia where maize comes second after teff (*Eragrostis teff*) (Erenstein et al. 2011). Maize yields in East Africa average only 1.6 t ha^{-1}. Ethiopia has substantial areas sown to sorghum (*Sorghum bicolor*), wheat (*Triticum aestivum*), barley (*Hordeum vulgare*), teff, chickpea (*Cicer arietinum*), faba bean (*Vicia faba*), enset (*Ensete ventricosum*) and coffee (*Coffee Arabica, Coffee robusta*).

Most of the farmed land in ESA is rainfed, though there is an increasing trend for irrigation agriculture. Crop yields are generally low and below global averages, though productivity varies between countries. For instance, maize yields are close to 3 t ha^{-1} in Ethiopia but < 2 t ha^{-1} in Mozambique (Abate et al. 2013). Despite the availability of virgin land, yield is generally low in Mozambique due to poor agronomic practices. For example, the planting density of maize is very low in small-scale

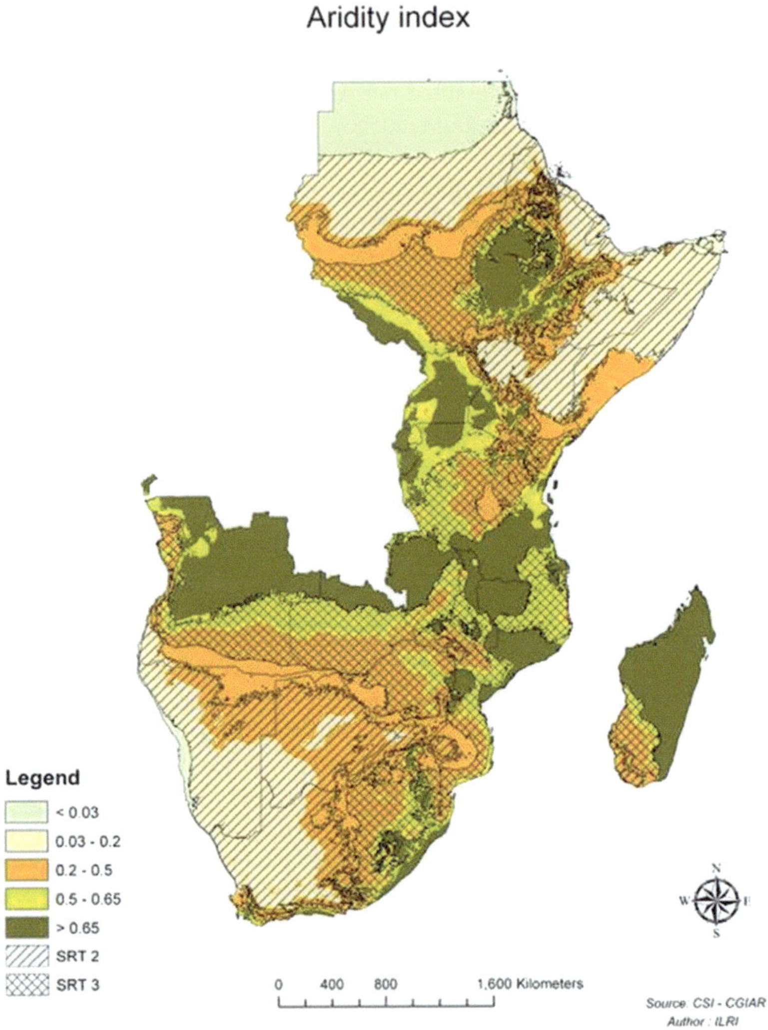

Fig. 1 The distribution of dryland systems in Eastern and Southern Africa (ESA). Values between 0.2 and 0.5 are considered semiarid. (Source: http://geoagro.icarda.org/en/cms/category/maps/15/regional)

farmers' fields in Mozambique, often as low as 40 % of the recommended density, which significantly reduces farm productivity. Most farmers practice low-input low-output (low-risk) agriculture, with very low economic and agronomic yield per unit of land and water, which could be stated as low water productivity.

Other important food crops in ESA are cassava (*Manyhot esculenta*), millet (*Panicum miliaceum*), beans (*Phaseolus vulgaris*), sweet potato (*Ipomoea batates*), Irish potato (*Solanum tubersum*), groundnut (*Arachis hypogaea*) and pigeon pea

(*Cajanus cajan*). Cotton (*Gossypium hirsutum*), tea (*Camellia sinensis*), macadamia nuts (*Macadamia*), tobacco (*Nicotiana tabacum*), sesame (*Sesamum indicum*) and grain legumes are also grown as cash crops in various countries.

Livestock is an integral part of the farming systems in Eastern and Southern African regions. The savannah pasturelands support a large number of animals, mainly cattle and goats, that represent an important livelihood strategy given the high risk of crop production due to the relatively low and unreliable rains and high evapotranspiration rates.

The frequency of drought, decline in soil fertility, and changing market opportunities became major drivers for farmers to change from traditional, commonly long-maturing varieties to early-maturing varieties which can be grown in degraded soils. Drought is the most significant challenge for the people in Malawi, Zambia and Zimbabwe (Kassie et al. 2012) creating food insecurity and affecting rural livelihoods.

Although various reasons have been given for rural poverty and food insecurity in Southern and Eastern Africa, possible solutions/intervention areas to achieve food security and sustainable agriculture are presented below.

3 Adopting Climate-Smart Agriculture

Africa, in general, and the dryland systems of Southern and Eastern African regions, in particular, are considered the most vulnerable to climate variability and change compared to the rest of the world, due in part to the lack of financial, institutional and technological capacity to respond to these changes. Climate change, whether due to natural variability or as a result of human activity (http://www.gao.gov/new.items/d07285.pdf), is threatening livelihoods due to increased greenhouse gas emissions and the subsequent warming of the Earth's surface. Long- and short-term climate data in ESA shows that temperatures have increased over the last 40 years (IPCC 2007), consistent with the global trend of temperature rise. Between 1988 and 1992, more than 15 drought events were reported in various parts of Southern Africa, with an increase in the frequency and intensity of El Niño and La Niña episodes (IPCC 2007).

In Eastern Africa, particularly in Ethiopia, decadal climate variability has been related to ENSO (El Niño/Southern Oscillation, the interaction between the atmosphere and ocean in the tropical Pacific) and the Southwest monsoon over the Arabian Sea (Camberlin et al. 2001). Deep convection over India establishes a temperature gradient that drives a tropical upper level easterly jet over the Arabian Sea. When the jet surges, boreal summer rainfall over Ethiopia tends to increase. Ethiopian rainfall is suppressed following an active tropical cyclone season in the south-west Indian Ocean. However, the rainfall varies with altitude, location and regional monsoon situations. Spatially drought frequency and magnitude also varies considerably from one region to the other. Understanding and predicting inter-annual, inter-decadal and multi-decadal variations in climate is important for planners and development actors. However, it is difficult to realistically predict mainly because there is a lack of historical data in the region, particularly at district level. This variability is however likely to increase with the onset of climate change, and some trends are already observable. The implication is that more and more communities will be exposed to climate change related extreme events, including drought and flood.

A wide variety of weather systems may bring extreme weather to the ESA region—including tropical cyclones and cut-off lows (low pressure centers aloft) that bring widespread flooding to countries including Mozambique, Malawi and Zambia (Davis 2011)—which destroy agricultural enterprises and livelihoods. Climate, especially rainfall, varies from intra-seasonal, through inter-annual to decadal and multi-decadal regimes (Kandji et al. 2006) with annual rainfall variability reaching 40 %. Climate variability and associated drought is the most frequently-recurring cause of food insecurity in the region. Of the 24 El Niño events recorded between 1875 and 1978, 17 corresponded to rainfall decline in the region (Rasmussen 1987) and the 1991–1992 El Niño caused a severe drought, putting millions of people on the brink of famine. The recent floods in downstream parts of the Zambezi River, particularly in Malawi, Zambia and Mozambique, which were mainly caused by La Niña, affected humans and livestock through drowning and landslides, reduced crop production, displaced people, and damage to assets and infrastructures (Kandji et al. 2006). Recently, in early 2013, flooding displaced about 200,000 people from their homes in the Mozambique lowlands. In contrast, periods of sustained anti-cyclonic circulation and subsidence can cause heat waves and prolonged dry spells over the Southern African region, which is expected to worsen in the future (Davis 2011). The impacts will likely include increases in surface and ocean temperatures, a rise in sea level, glacial melting, and more extreme weather events, such as droughts and floods, and less precipitation in some areas (Freimuth et al. 2007)—resulting in reduced agricultural productivity over time and space.

The capacity of small-scale farmers to adapt to climate change is strongly linked to their ability to change to water-efficient agronomic practices and drought-resistant crop types, diversify their crop choices, and improve land and water management at the farm and landscape scales. Watershed management—which is an integrated strategy for increasing vegetative cover, improving water yield, reducing erosion effects, and efficiently using available resources—is becoming an important intervention to enhance the resilience of systems and minimising climatic shocks. Strategies that aim to increase production and income can help to reduce the impact of climate shocks on rural communities. It is assumed in this case that increased production/productivity will lead to increased income, which will be used to support food security, enable investment to protect farms and landscapes, and allow households to acquire productive assets, shelter and safety nets during climate shocks (Amede et al. 2014b). Increasing production usually requires an expansion of the area under cultivation, provision of irrigation to expand the cropping season and/or application of critical inputs (fertilisers, seeds and pesticides) along with agronomic improvements. Increasing farm productivity entails producing more per unit of land, labour and inputs such as water, which implies maximizing efficient use of scarce natural resources. Careful balancing of how this maximization is achieved can help increase the productivity of ecological services while minimising the risk of their depletion. Approaches that combine both production and conservation objectives, for example conservation agriculture (or climate-smart agriculture) and other conservation-based approaches, provide win–win scenarios for communities.

In countries where agricultural resource efficiency is very high, the carbon sequestration capacity of the landscape also tends to be relatively high. The global progress in increasing cereal yields per hectare and producing more meat and milk per animal, and more farm outputs per unit of labour through agricultural intensification, has reduced the encroachment into forests and grasslands, which are critical for mitigating climate change effects. However, improved natural resource management practices for mitigating climate change are commonly adopted by farmers when farmers get short-term benefits from investments interms of increased yield and income (Wichelns 2006).

The key investments required by small-scale farmers to promote drought resilience are summarized below.

3.1 *Improved Water Management in Dryland Environments*

The threat of water scarcity in SSA is real, due to the expanding agricultural needs, and is exacerbated by the increase in climate variability and inappropriate land use (Amede et al. 2009). Competition for water between different uses and users is increasing at global, national and community scales although agriculture will remain the largest water user. Up to 70 % of the water from rivers and groundwater globally goes into irrigation (http://www.lenntech.com/water-food-agriculture.htm). Irrigation in SSA is the lowest of all countries worldwide despite the increasing need to improve food security and minimise climatic variability. In SSA, the water needs for the future for food production and livelihoods will triple by 2025 compared to the year 2000 (Rockström et al. 2004). About 75 % of the additional food required over the coming decades could be met by increasing the production levels of the subsistence farmers' up to 80 % of those of high-yield farmers, which could be achieved mainly through improved water management (CA 2007). However, major trade-offs are forecast between agriculture and ecosystem services, including trade-offs between increasing food security and safeguarding ecosystems (de Fraiture et al. 2007; Bossio 2009). These demands will include water allocated to ecosystem services. Moreover, focusing solely on irrigation and agricultural production could result in freshwater shortages for wetlands and other aquatic ecosystems (Postel 2000), and degraded water quality, with serious impacts on terrestrial and aquatic ecosystems.

Irrigated agriculture is becoming an increasingly important intervention in response to the increasing food demand in dryland systems, managing climate variability, farm employment and reducing poverty. Irrigation, along with improved agricultural water management practices, could provide opportunities to cope with the impacts of increasing climatic variability and enhance the productivity per unit of land, which would significantly increase the annual production volume of crops (Awulachew et al. 2005). A substantial yield gap still exists between achievable and actual yields both in terms of yield per unit of land but also yield per unit volume of water that should be exploited to ensure food security. Current yields from rainfed crops are only about 50 % of those on irrigated land when all other inputs remain the same. If ESA countries are to achieve their stated aims of food self-sufficiency and food security, the current production shortfalls call for drastic measures to improve water productivity in both irrigated and rainfed systems.

With irrigation, non-productive water depletion could be reduced by improved irrigation water management, which includes choice of water-efficient enterprises, minimising conveyance, drainage losses, and multiple uses of water for household use, fishing and irrigation (Amede et al. 2014a). Irrigation farming is becoming a necessity in the drought-stricken regions of SSA to: (i) reduce the vulnerability of farmers to annual rainfall variability and the associated crop and livestock risks; (ii) increase agricultural production per unit of land, water and labour investments, thereby reducing the expansion of farming to less productive hillsides and valley bottom wetlands; (iii) enable communities to produce high-value enterprises in homesteads and selected plots thereby enhancing the capacity of communities to reinvest on their farms, demand better services and production inputs, and strengthen collective action for broader land and water management; and iv) become an incentive to mobilise communities to better manage upper watershed and command areas. In Ethiopia, there is a strong association between small-scale irrigation (SSI) and the protection of upper slopes from erosion, landslide and gully formation. In the last two decades, irrigation has become an incentive to rehabilitate catchments through area enclosure, soil and water conservation, and to enrich the natural vegetation.

There are huge opportunities for developing SSI in various river basins from the Nile in the east to Limpopo and Zambezi basins in Southern African regions. In Mozambique, it is estimated that about 1.7 million ha of land has irrigation potential within the Zambezi basin (FAO 1997), while in Ethiopia this potential is expected to be 5.1 million ha (Awulachew et al. 2005).

However, irrigation development in SSA is still in its infancy (Fig. 2) and is only likely to be exploited if policy incentives are in place that would improve water access, reduce irrigation costs and increase farm returns. Despite irrigated agriculture starting in the region during the colonial era to produce cash crops (e.g. sugar cane, tobacco, tea), by the 1960s it remained highly localised, contributing to <5 % of food production, much lower than the global average. About 5.2 million ha of land are irrigated in SSA, representing 3.3 % of cultivated land, which is much lower than the irrigated share of crop lands in other continents. In Ethiopia, for example, 4.3 % of its estimated potential area is irrigated which contributes about 3 % of total food crop production in that country. Moreover, existing irrigation schemes are not giving the expected returns, due to excessive siltation, poor agronomic and water management practices, and the failure of local institutions to

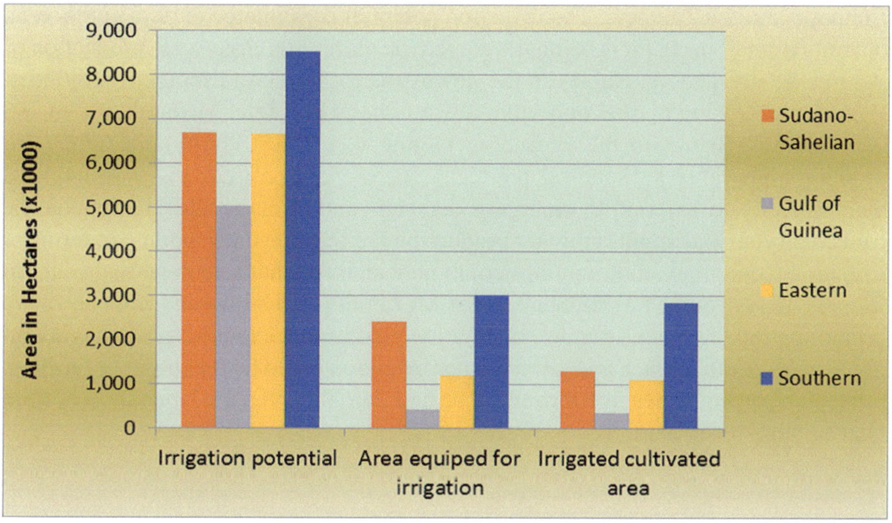

Fig. 2 Potential, equipped and actual irrigated areas in the SSA dryland countries. (Source: Ward and Ringler, IFPRI, undated)

sustainably manage them (IFAD 2004). However, there is increasing government interest and investment to develop the water resources for irrigation and other uses.

Zambia has enormous irrigation potential with estimates of about 2.75 million ha of land (FAO 2006) and a large portion of it in the Zambezi basin. However, the total area currently under irrigation is only about 10 % of this potential and is mostly practiced by large-scale and medium-scale farmers using surface water irrigation (Evans et al. 2012). About 15,000 ha of land is irrigated with motor pumps despite the region having high ground water potential that could be exploited by increasing access to motor pumps (Amede et al. 2014a). Most of the smallholder irrigation schemes produce food crops, such as rice, maize and horticultural crops, but they usually underperform due to poor scheme infrastructure, inadequate water supply, and inefficient use of available water (MoAFS 2011). The high rainfall from December to March commonly saturates the soil and creates seasonally water-logged low-lying dambos. There is potential to develop and expand small reservoirs in the region for multiple uses, namely livestock drinking, fishery, household irrigation and other domestic uses (Evans et al. 2012).

Mozambique has multiple river basins, with most of the rivers having highly seasonal, torrential flow regimes, with high flows for 3–4 months and low flows for the remainder of the year, corresponding to the distinct wet and dry seasons (FAO 2006). Of these basins, the Zambezi basin is the most important as it accounts for about 50 % of the surface water resources in the country and about 80 % of its hydropower potential, including the Cahora Bassa Dam (FAO 2006) which is the second largest dam in Africa. The main source of water for irrigation in Mozambique is surface water. Irrigation in Mozambique is in its infancy despite being a downstream country with large seasonal flows from the region's big rivers, including the

Limpopo and the Zambezi. The irrigation potential is estimated to be >3 million ha but only a small part of it is being developed, primarily for large-scale production of sugarcane, rice and vegetables in the downstream central and southern provinces. With the increasing market opportunities in regions like Tete, Mozambique and its surroundings, thanks to the expanding mining sector, SSI could help farmers to produce high-value agricultural products, access regional markets, and improve their capacity to respond to emerging demands and climatic shocks. Vegetables, fruits, dairying and small ruminant production are feasible entry points. In terms of irrigation technology, motor pumps could play an important role in getting water to farmers' fields yet the governments in the Mozambique, Malawi and Zimbabwe are promoting treadle pumps because of their low maintenance requirements and operational costs. Both surface and groundwater irrigation could be the future of farming, particularly in countries like Mozambique and Zambia given the large areas of farm land suitable for irrigation.

3.2 Rainwater Management

The majority of farming populations in ESA region rely mainly on rainwater. On an annual basis, there is generally sufficient rainfall to support full-season crops, but variability in temporal and spatial distribution calls for improved rainwater management (RWM). RWM is an integrated strategy to systematically map, capture, store and efficiently use runoff and surface water emerging from farms and watershed in a sustainable way for both productive purposes and ecosystem services (Amede et al. 2011). It has three major components, namely water storage, water management and water productivity. Interventions aim to reduce unproductive water losses (runoff, evaporation, conveyance losses, deep percolation) as well as improve the water productivity of respective enterprises to increase returns per unit of water investment (Amede et al. 2014b).

Small-scale farms in the region often occupy fragmented, marginal and rainfall-dependent lands that are commonly prone to erosion, droughts, floods and fluctuating market prices. Improving soil and water conservation is the first action to improve the water supply for agriculture, i.e. to make more rainwater available for plants (Rockström 2000). Hence, strategies to reduce rural poverty will depend largely on improved RWM across space and time. Interventions are required not only to minimise risk but also to improve water storage and productivity for increased water access for food production and environmental services.

Access to groundwater is beyond the reach of most farmers, mainly due to financial constraints. However, RWM has the potential to provide enough water to supplement rainfall thereby increasing crop yields, reducing the risk of crop failure (Oweis et al. 2001) and providing a water supply for livestock. Enhancing and stabilising crop yields and livestock production for farmers in these crop–livestock systems will encourage farmers to invest in rainwater harvesting and the accompanied nutrient management at the plot, farm and landscape scales. The choice of a certain agricultural enterprise or management would also influence water productivity as it affects the quantity and quality of water used to grow crops, forage, and pasture. Improved vegetative soil cover and strategic selection of cropping pattern (e.g. close row spacing), cropping system (e.g. intercropping and agroforestry) and crop (variety) (e.g. crops with early development of a closed canopy) could reduce unproductive water losses such as evaporation and runoff and increase productive transpiration of rainfed systems (Bouman 2007).

A promising RWM intervention to minimise drought effects is conservation agriculture (CA), a crop management system using three basic principles in a mutually-reinforcing manner (Thierfelder et al. 2013), namely: i) minimum soil disturbance, i.e. no soil inversion with the plough or hoe; ii) surface crop residue retention as mulch with living or dead plants; and iii) crop rotations and associations of different crop species over time. Initial research from the 1990s in the region largely focused on the effects of CA on soil quality, soil erosion, carbon, weeds and water dynamics. These studies highlighted that reduced tillage and mulch cover reduced erosion and increased soil moisture, which led to overall greater yields, especially in dry years (Thierfelder et al. 2013). CA is becoming an attractive intervention in Southern Africa (e.g. Zambia), particularly in areas where there is limited competition for biomass between livestock feed and CA, and where there are large amounts of crop residue produced per unit area, particularly on mechanised farms. CA has been practiced for generations in Southern Ethiopian enset–coffee based systems, particularly for high-value crops grown around homesteads. The application of CA principles may vary from system to system and farmer to farmer.

3.3 Improved Soil Fertility Management for Sustainable Productivity

Soil fertility has declined in SSA, partly due to nutrient mining for generations, which has been aggravated by soil erosion and poor agronomic management. The other major causes include the high cost of fertilisers and the failure of traditional methods for maintaining soil fertility (e.g. fallowing), which is almost non-existent given the scarcity of land and high population pressure. The soils in SSA are also mostly unstable, fragile and prone to erosion.

The two most important nutrients for crop production in the region are nitrogen and phosphorus, although crops also respond to the application of potassium, particularly in Nitisols located in high rainfall regions. However, fertiliser use in the SSA is very low compared to other continents (Fig. 3).

In most dryland systems in SSA, fertiliser application may not be an option given the market distortions (e.g. middle men, government monopoly) and low financial capacity of farmers. ICRISAT and its partners have been promoting microdose fertiliser application since the early 1990s (Twomlow et al. 2008). Microdose is about reducing costs but improving fertiliser use efficiency by applying about 25 % of the recommended NP fertilisers close to the plant roots. While the application of

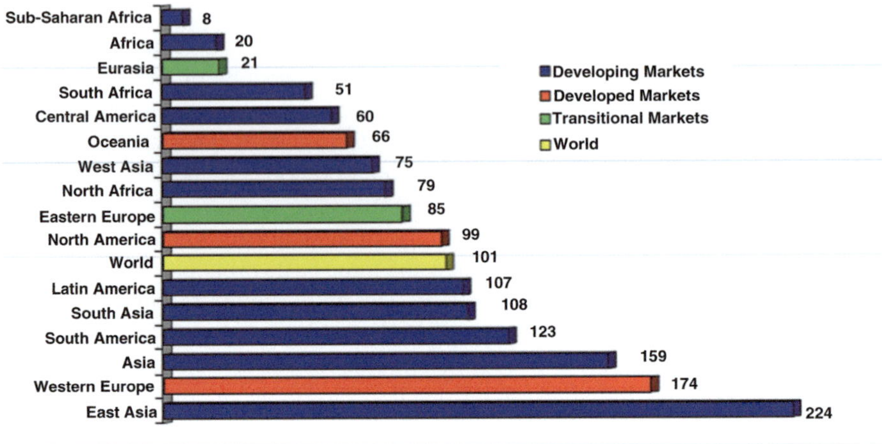

Fig. 3 Fertilizer use per ha by markets and sub-continents (2004–2005) (Roy 2007)

judicious amounts of fertilisers and improved nutrient management are the major components of soil fertility management, investment in chemical fertilisers is limited in SSA for various reasons: (1) chemical fertilisers are expensive while agricultural products are sold relatively cheaply such that it may not cover the cost of investment; (2) most of the nutrient movement is facilitated by erosion, which commonly emerges from upstream, communal lands. However, there is rarely an institution or a community strategy committed to manage these resources as there is limited incentive for individuals to invest labour and money to manage them; (3) farmers manage multiple enterprises of crops, livestock and/or woodlots which are interdependent on each other. Hence priority is given to those which have a direct impact on farmer income rather than those contributing to longterm food security. In some cases, the trade-offs among components is much stronger than the benefits in terms of environmental services (e.g. effect on water, nutrients and so on) and yet they could be economically attractive in the longer term. In this case, farmers single out enterprises with better economic benefits regardless of the negative impact on other system components and future production scenarios.

Moreover, fertiliser efficiency is very low due to soil erosion, recurrent drought and inappropriate choice of crop varieties. Farmers are also reluctant to apply chemical fertilisers for food crops due to high fertiliser costs, which are higher in landlocked African countries (e.g. Ethiopia, Zambia) than coastal countries. Most importantly, there is limited market regulation to ensure the quality of inputs (Amede et al. 2014a). There is a general concern that the chemical fertilisers currently available could be adulterated. There is also a lack of expert knowledge on what fertiliser mix would best fit where within the different agroecological zones and landscapes. For instance, ammonium sulphate could aggravate acidity if applied to soils with low pH while it may be appropriate on saline or sodic soils. Moreover, the major fertiliser inputs that have been applied to date are predominantly nitrogen and phosphorus which create long-term nutrient imbalances. Costly ameliorative measures such as liming and corrective fertiliser compounds are currently in demand though small-scale farmers cannot easily cover the costs of these measures. Application of conventional fertilisers may not necessarily improve crop yield unless accompanied by supplementary plant nutrients and expert advice on fertiliser use on a case-by-case basis.

In an IFPRI review on the rate of returns from agriculture-related investments in the region, fertiliser returns did not make it to the list of priorities, while improved crop varieties had a return of 35 to 70 % (Alston et al. 2000). The low returns could be due to other yield-limiting factors. Low soil water holding capacity accompanied by high evapotranspiration could reduce nutrient uptake and yield. In some soils, e.g. calcareous savannah soils, the most important yield determinants could be micronutrients (e.g. zinc) while in high rainfall areas aluminium toxicity and P-fixation is to be expected. Moreover, the drier parts of SSA experience recurrent soil water deficits that reduce crop yields, when the drought period coincides with flowering (maize) and key tuber extension (root crops).

Fertilizer subsidies (e.g. Malawi and Ethiopia) have made a huge difference in terms of increasing crop yields and improving food security in the region however these subsidies have been phased out. In 2006–2007, maize yields in Malawi improved significantly, and some was even exported (Amede et al. 2014a). Other African governments were interested in copying the 'Malawi green revolution' but enthusiasm quickly faded when the direct financial support of donors to the Ministry of Agriculture and farmer organizations dwindled, and the government removed fertilizer subsidies. Moreover, investing in fertilizer and seed alone will not solve the food crisis in the long term unless a parallel investment in complementary services, including market infrastructure and marketing system, is established that will allow producers to connect to wider markets.

There are ongoing continent-wide initiatives to characterise soil quality and develop fertiliser recommendation domains for the whole SSA region. The Africa Soil Information Service (AfSIS) Online Map Tool is an interactive mapping application that can display more than 30 maps of soil and related environmental characteristics for Africa. http://blogs.ei.columbia.edu/2012/12/12/new-understanding-of-soil-quality-throughout-africa/. The tool creates an online map that allows users to examine soil characteristics from existing, legacy soil maps and data as well as a new collection of soil samples gathered by AfSIS in the past four years. The Ethiosis, a localised tool for Ethiopia, can also create maps indicating soil nutrient availability in major regions. These maps should be converted to usable tools to guide farmers toward agronomically-efficient and economically-viable fertiliser application schemes.

3.4 Integrated Watershed Management

Resource degradation is a serious problem facing the agricultural systems of SSA. Given the extent of land degradation (Fig. 4), particularly in the Eastern African highlands, reducing poverty and improving livelihoods may not be achieved

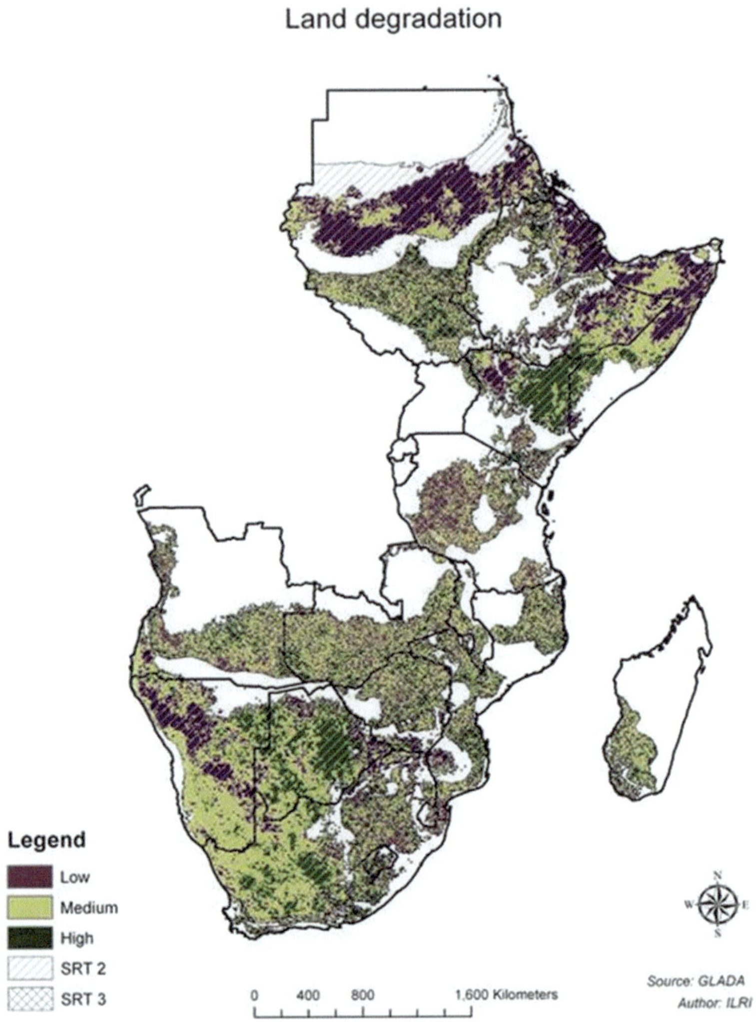

Fig. 4 The level of land degradation in Eastern and Southern Africa (Source: http://geoagro.icarda.org/en/cms/category/maps/15/regional)

without addressing this major system constraint. The proximate drivers of land degradation include forest clearance and surface soil exposure (high removal of vegetative cover), detrimental cultivation practices with emphasis on small-seeded crops that require fine tillage, and overgrazing (Gebreselassie et al. 2015). Due to land shortages and the lack of alternative livelihoods, farmers cultivate land that slopes more than 60 % with shallow and stony soils prone to erosion.

Water erosion is considered the major source of nutrient depletion, with N, P and organic carbon nutrient losses estimated to be 74, 5.5 and 539 kg ha^{-1} $year^{-1}$, respectively, with N and P losses estimated at more than $US 300 million per year

(Saka et al. 1999), based on fertiliser prices prevailing in the 1990s. A study by Nakhumwa (2004) showed that the severity of soil erosion and its increasing trend equated to gross annual losses of $US 6.6–19.0 million. In Mozambique, average nutrient losses from agricultural fields is estimated to be 34, 6 and 25 kg ha^{-1} $year^{-1}$ of N, P, K, respectively (Folmer et al. 1998), which is threatening agricultural production given the very low application rates of chemical fertilisers. This value could be even higher in the mountainous parts of Eastern Africa, where rainfall is higher, erosion is more severe, and farmers rarely adopt soil and water conservation practices.

Oldeman (1998) estimated that about 10 % of cumulative productivity losses in the last 60 years in Africa resulted from human-induced soil degradation, associated losses of soil carbon, and accelerated water depletion. About 65 % of agricultural cropland and 31 % of permanent grazing land in Africa were degraded in the same period (Scherr 1999). Gross fertiliser use also fell by 9 % in the late 1990s (Wichelns 2006). These low-input agricultural practices in the region have led communities into a downward spiral of poverty, deforestation, resource degradation and local conflicts, all of which affect the adaptation capacity of communities to climate change. For example, cultivation with low-input methods (no fertiliser) in the humid savannah zones of SSA induced a 30 % loss of soil organic matter after 12 years and 66 % after 46 years, with rice yields declining from 1 t ha^{-1} to 300 kg ha^{-1} (Barrett et al. 2000), substantially decreasing the soil carbon stock due to loss in soil organic matter. The return of crop residues to the soil alone is not sufficient to offset these losses. Soil carbon decline and recurrent spatial and temporal climate variations, aggravated by the lack of water storage capacity, hinder the effective use of water and nutrients by plants, leading to frequent crop failures which, in turn, contribute to substantial poverty and resource degradation.

One area of intervention to minimize nutrient losses and enhance sustainable and climate-smart agriculture is improved watershed management. Integrated watershed management (IWSM) reduces erosion, regulates runoff, reduces unproductive water losses (runoff, evaporation, conveyance losses, deep percolation) from a system, and increases the water use efficiency of respective enterprises (Amede et al. 2011). It capitalises on rainwater harvesting principles, by storing and efficiently using water in the soils, farms, landscapes, reservoirs and other facilities. Experiences from the Eastern African highlands (German et al. 2012) showed that watershed management is an effective strategy to improve vegetation cover on hillsides, reduce the negative effects on downstream farms and water facilities, and manage the consequences of climate change (e.g. floods and drought) by combining water management with land and vegetation management at landscape scales. Research results from Tigray, Ethiopia showed that IWSM decreased soil erosion, increased soil moisture, reduced sedimentation and runoff, set the scene for some positive knock-on effects such as stabilisation of gullies and riverbanks, and rehabilitated degraded lands. IWSM also increased the recharge in subsurface water (Alemayehu et al. 2009). IWSM approaches would increase the resilience of systems by capturing, storing and efficiently using runoff and surface water emerging from farms and landscapes for production and ecosystem services. This is particularly critical for

SSA, where about 70 % of the land falls within drought-prone arid, semiarid zones. Unlike conventional approaches, it focuses more on the institutions and policies than on the technologies. On the other hand, there has been limited experiences and institutional arrangements to date to employ watershed management practices in the region.

IWSM could be used efficiently for fostering sustainable land management practices, which need a suitable policy framework to consider the interests of present and future generations. Moreover, watershed management interventions should be selected based on context-specific and intensification levels of the farming systems (German et al. 2012). The proximate factors that significantly determine the likelihood of adopting watershed management technology include climate and agroecological zonal characteristics. Communities residing in warm humid/subhumid, cool, arid semiarid agroecological zones or cool humid/subhumid are less likely to adopt watershed management technologies compared to those located in warm, arid/semiarid agroecology (Gebreselassie et al. 2015). This is partly because of the urgency of the system to produce food and fodder in the drylands, while in warm, humid areas the probability of getting alternative food and feed sources is usually high.

3.5 Improved Livestock Management Under Changing Climates

The farming systems in SSA are largely crop–livestock systems, whereby livestock production is integrated into crop production in both rainfed and irrigated systems. The only exception, where livestock predominates as a livelihood strategy is the pastoral and agropastoral systems, which is beyond the scope of this review. In the semiarid parts of the region, livestock provide around 45 % of the families' income for the poorest and nearly 60 % for the better-off households (World Bank 2006). Poor livestock keepers are overwhelmingly most abundant (only next to Nigeria) in Ethiopia and Tanzania with 12.4 and 10 million poor livestock keepers in dryland mixed systems, respectively (Dryland Systems 2012). On the other hand, in areas with cereal mixed systems, farmers keep some livestock mainly for draught power, sale in times of need, and as a reserve for bad times. There has been a significant increase in the number of livestock in Southern Africa; for instance, in Mozambique (by 28 %) between 1999 and 2009, following recovery from long war or drought. Poultry has also been growing at a rate of 52 % annually for five consecutive years (FAO 2013), facilitated by government policy for import replacement (Technoserve 2011). In Malawi, 57 % of households owned or kept livestock or poultry, with male-headed households' owning more than female-headed households. In general, households in Southern Africa are more likely to have kept fewer livestock than households in other African dryland systems, partly due to resource scarcity (NSO 2010). The very poor households commonly keep some chickens and pigs, while those with medium resource status can add goats and cattle. The better-off households can afford dozens of cattle and goats and large numbers of chicken (Amede et al. 2014a).

Livestock numbers in SSA are projected to increase by 2.5- to 5-fold, from 200 M head in 2005 to 500–970 M head in 2050 (Cork et al. 2005), which will put huge pressure on water and land resources unless productivity per unit of water investment increases significantly (Amede et al. 2009). Although the livestock revolution offers a chance for smallholders to benefit from the rapidly growing market and raise their incomes, it may have negative environmental, social and health impacts if not managed well (Steinfeld et al. 2006). How livestock production triggers and aggravates water resource degradation in the drylands includes:

1. By satisfying the increasing feed demands, pastures and arable land for growing feed expand into protected and natural ecosystems.
2. Overstocking and inadequate watering points degrades rangelands.
3. In peri-urban environments, soils and water resources become contaminated from manure and wastewater mismanagement.
4. Growing feed crops demand intensification, which may lead to resource mining and soil degradation (Steinfeld et al. 2006).

Several factors undermine the potential contribution of livestock systems to rural livelihoods, of which shortage of feed and veterinary services are the major ones. Livestock mortality is commonly caused by feed shortages during drought years, lack of drinking water, and the prevalence of animal diseases. Newcastle disease in poultry, African swine fever in pigs and Trypanosomiasis in cattle are the most prevalent livestock diseases in the region (World Bank 2006). These diseases commonly cause mortality but also reduce meat and milk productivity, reduce animal traction power, and affect overall productivity and profitability of livestock systems. Most of the feed is obtained from the natural pasture and crop residues. Quality feed is usually allotted to draught oxen, mainly in the peak farming months, when land preparation and planting operations are commonly practiced. In systems where oxen plough is common (e.g Ethiopia), crop residues are the major feed source. Crop residue from pulses is considered a quality feed resource and it is fed mainly to oxen and milking cows mixed with cereal straw. Crop residues from cereal fields are low in metabolisable energy and protein content (Blummel et al. 2014). This problem can be addressed to some extent by mixing crop residues with various forage legumes, which can enhance rumen fermentation and the availability of energy from the total diet. Improved forages provide a good source of energy throughout most of the year. Despite recognition by farmers of the potential contribution of forage legumes to crop–livestock farming systems, their integration is relatively slow. Growing feed is a new concept for most farmers; they are used to collecting natural forages from roadsides, weeding crops, fallow lands or forests. Some farmers also fear that forages will become weeds. For farmers who are convinced of the value of improved forages, the lack of availability of seeds and planting materials often forms a bottleneck. Steinbach (1997) indicated that six other factors affect the integration of forage legumes into subsistence farming systems:

- Available arable land per capita
- Number of crops that can be grown per year
- Market access to animal products
- Labour availability
- Farmer's perceptions of the risks and
- Rewards of investing in their livestock enterprises

Extensive grasslands in pastoral and agropastoral systems have multiple uses in addition to being an important source of livestock for stock raisers and herders. Most grasslands are important catchment areas, and the management of their vegetation is of prime importance for the water resources of downstream lands (FAO 2009). For instance, in the Nile basin, about 70 % of the water is depleted through grassland pastoral and agropastoral systems. Grassland management, which encompasses erosion control, controlled grazing, availability of strategic watering points for livestock drinking, and different forms of water harvesting structures could be effective adaptation strategies to minimize drought effects. Minahi et al. (1993) stated that grasslands are almost as important as forests in the recycling of greenhouse gases and that soil organic matter under grassland is of the same magnitude as in tree biomass; while the carbon storage capacity under grasslands could be increased by avoiding overgrazing. Improved grazing management can increase soil carbon stocks by an average of 0.35 t C ha^{-1} $year^{-1}$ but under good climate and soil conditions improved pasture and silvopastoral systems can sequester 1–3 t C ha^{-1} $year^{-1}$ (FAO 2009). It is estimated that 5–10 % of global grazing lands could be placed under C sequestration management by 2020 (FAO 2009).

There is evidence that developing multiple watering points in the various niches of dryland systems, thereby reducing long walks for livestock, would enhance livestock productivity and reduce land degradation caused by livestock free movement. In Northern Ethiopia, North Wollo Zone, reducing livestock walking from 9 km to 2 km per day reduced the energy spent for walking from 1956 to 584 MJ ME^{-1} TLU^{-1}, which is equivalent to 343 litres of additional milk per lactation period (Descheemaeker et al. 2010).

There is increasing conflict between livestock keepers and crop producers, particularly in sorghum–millet based subsystems, concerning access to pasture land and watering points, which is getting worse in seasons of drought and feed scarcity. In general, the livestock sector has potential for growth to improve the livelihoods of rural communities in the region. However, the sector receives limited policy attention regarding access to markets, veterinary services, watering points, household credit, and overall marketing and processing infrastructure. On the other hand, there is increasing opportunity to invest in livestock systems due to increasing demand for livestock products to feed the growing middle-class population, particularly in Southern Africa, where mining has become a major economic activity.

4 Land Tenure and Use

There are three major categories of land tenure system in the dryland regions of SSA, namely public land, private land and customary land (Amede et al. 2014a). These tenure systems vary from country to country depending on their historical perspectives. In Ethiopia, the land belongs to the state while farmers have unrestricted user rights. In some other countries, land belongs to the government, but the land under customary tenure could represent up to 85 % of the total land holdings (e.g. Mozambique) (Nabhan et al. 1999). In Zambia, the land tenure system is both customary and state land/leasehold tenure. State land tenure is defined as reserved or gazetted land (national forests, local forests and parks), towns and permanent commercial farms, while customary land means traditional land or "open land" (non-gazetted) where traditional chiefs and their village headmen decide on how the land is to be used (Olson 2007). National and foreign investors can obtain concessions (effectively leases, known as DUATs) for unused land for 100 years, subject to community consultations. Communities and individuals have permanent occupation rights. The Land Law recognizes customary rights and gives them formal legal rights, while encouraging the growth of private sector in the regions (De Wit and Norfolk 2010). Land that is not under any form of use is considered community property, which is under the jurisdiction of the local chief (Saka et al. 1999). However, the governments can declare customary land as public land as deemed necessary and allocate it to investments when the need arises.

This land policy creates insecurity in local communities, putting pressure on the farming systems, reducing fallow periods and the time required for soil fertility replenishment, and squeezing crop and livestock farmers to increasingly smaller landholdings. Moreover, the system of land inheritance varies, whereby the patrilineal or matrilineal system of inheritance is practiced depending on the cultural setups of the respective communities. The consequence is increasing land scarcity, the major cause of local conflicts. For instance, in 2006–2007 in Malawi, 47 % of villages had conflicts over land; 29 % between family groups and households, 20 % between villages, and 5 % between villages and estates (NSO 2010). These conflicts are partly due to the weak institutional capacity to enforce land laws in the respective countries. The general trend is that state ownership has been increasing in the region with the view to expand investments and public ownership of resources. The implication is that there are limited incentives for farmers and investors to invest in their farm unless there is certified land security.

5 Capacitating Local Institutions

The extension system in both regions is generally weak and disorganized in terms of reach and effective service supply. Although there are differences in the extension capacity among countries, the capacity is weak and rarely supported by the required infrastructure. Besides the fact that most of the districts are far from the centre of powers of the respective countries, with poor road infrastructure, the reach of small-scale farmers is aggravated by the limited number of extension staff on the ground (except in Ethiopia where there are about 65,000 extensionists in the field), limited financial capacity, and limited access to farm inputs. In Southern Africa, access to technologies and good practices is limited mainly to large- and medium-scale farmers. Moreover, the common perception that there will be transfer of knowledge from large- and medium-scale to resource-poor farmers is unrealistic due to the fact that better-off farmers rely more on high-value commercial crops with different levels of input-output farming while small-scale farmers are primarily growing subsistence food crops with low-input/low-output scenarios.

Given the weak public institutions in the region to facilitate dissemination of technologies and best practices to wider communities, there is a need to build strong local institutions. Various local institutions are filling the gaps and are engaged in input distribution, marketing and collective action at various levels in all countries in the region. The traditional authorities, sub-traditional authorities, group village chiefs and village chiefs play an important role in the agricultural sector, particularly in organizing communities, disseminating agricultural intervention and guiding farmer organizations (Amede et al. 2014a). The paramount chief is the highest order of the traditional institutions, which has a very strong influence on both policies and local investment flows to the localities.

The current farmer associations in the respective countries need to be organized and facilitated to ensure that their engagement enables local action and creates a wider movement to improve land, water and vegetation management at farm and watershed scales. The current institutional setup rarely entertains community priorities in the planning and implementation of development projects and programmes. Moreover, most programs in the region are run and managed by large NGOs, with top-down approaches, without creating local capacity and institutional innovation. Although the NGOs play a vital role in organizing small-scale farmers, there is still a significant risk that if these international NGOs leave the scene, there will be little capacity left to carry on the development process. There appears to be multiple and parallel initiatives in the region, sometimes with conflicting approaches. Various donors and NGOs promote different philosophies, objectives and activities on the ground, which do not necessarily align with government development directions. There is also poor linkage within the government structure between ministries and local governments. For instance, in Mozambique, beyond the fact that the government extension service was understaffed, there appears to be little effort to create linkages between the various officers at different levels. The major hurdles across the different hierarchy seem to be poor communication, lack of a joint forum for learning and planning, and weak monitoring systems. The formation of local and national forums would serve as a platform for sharing knowledge, identifying gaps and providing comprehensive policy recommendations that would help to avoid past mistakes.

6 Conclusions and Future Research Thrusts

Dryland systems in SSA have been affected by recurrent drought, which has been aggravated by climate change, the decline in soil fertility, and poor agronomic management of production systems. There are proven interventions that should be promoted to help communities to minimise the drought effects and develop sustainable and resilient dryland systems. However, these interventions are either not reaching the intended farmers on time, in the required amount and knowledge detail, or the inputs are beyond the financial reach of poor, smallholder farmers. There is also limited access to markets for farmers to invest in inputs and produce more than what is required for their household consumption. The livestock sector is also receiving limited policy attention regarding access to markets, veterinary services, watering points, household credit, and the overall marketing and processing infrastructure. As the research capacity to develop context-specific agricultural technologies is limited, there is a need to develop the innovation capacity of the local communities by providing various capacity building courses and on-the-job training. Key development investments are required to curb the recurrent effects of drought in the regions, including strengthening the extension system in the respective countries, developing agriculture-friendly policies, developing high-yielding drought-resistant crop varieties, developing SSI facilities using surface and groundwater options, and creating alternative water sources to satisfy feed and food demands. In all these interventions, strong policy support is needed regarding subsidies, improving market infrastructure, capacity building, and the development of function extension systems.

References

Abate T, Menkir A, MacRobert JF, Tesfahun G, Abdoulaye T, Setimela P, Badu-Apraku B, Makumbi D, Magorokosho C, Tarekegne A (eds) (2013) DTMA highlights for 2012/13. CIMMYT-Kenya, 120 pp

Alemayehu F, Taha N, Nyssen J, Girma G, Zenebe A, Behailu M, Deckers J, Poesen J (2009) The impacts of watershed management on land use and land cover dynamics in Eastern Tigray (Ethiopia). Resources Conservation and Recycling 53:192–198

Alston JM, Chang-Kang C, Marra M, Pardey P, Wyatt JJ (2000) A meta analysis of rate of return to agricultural R&D. Ex pede herculem? IFPRI Research Report 113. Washington, DC

Amede T, Descheemaeker K, Peden D, van Rooyen A (2009) Harnessing benefits from improved livestock water productivity in crop–livestock systems of sub-Saharan Africa: synthesis. Rangeland J 31:169–178

Amede T, Tarawali S, Peden D (2011) Improving water productivity in crop livestock systems of drought prone regions. Exp Agri 47:1–6

Amede T, Tamene Desta L, Harris D, Kizito F, Xueliang C (2014a) The Chinyanja Triangle in the Zambezi River Basin, Southern Africa: Status of, and Prospects for, Agriculture, Natural Resources Management and Rural Development. Research for Development Learning Series 1. WLE. International Water Management Institute, Colombo, SriLanka. 41 pages.

Amede T, Awlachew S, Matti B, Yitayew M (2014b) Chapter 26: Managing rainwater for resilient dryland systems. In: Melesse A, Abtew W, Setegn S (eds) Nile River Basin: ecohydrological challenges, climate change and hydropolitics. Springer, Cham

Awulachew SB, Merrey DJ, Kamara AB, Van Koppen B, Penning de Vries F, Boelee E with editorial assistance from G. Makombe (2005). Experiences and opportunities for promoting small-scale/micro irrigation and rainwater harvesting for food security in Ethiopia. IWMI Working Paper 98. IWMI, Colombo

Barrett C, Place F, Aboud A, Brown D (2000) The challenge of improved natural resource management practices adoption in African agriculture: A social science perspective. Paper prepared for the workshop on "Understanding Adoption Processes for Natural Resource Management Practices for Sustainable Agricultural Production in Sub-Saharan Africa," Nairobi

Blummel M, Haileselassie A, Samiteddypalle A, Notembaert AM (2014) Livestock water productivity: feed resourcing, feeding and coupled feed-water resource data bases. Animal Prod Sci 54:1584–1593

Bossio D (2009) Livestock and water: understanding the context based on the comprehensive assessment of water management in agriculture. Rangeland J 31:179–186

Bouman B (2007) A conceptual framework for the improvement of crop water productivity at different spatial scales. Agri Syst 93:43–60

CA (Comprehensive assessment of water management in agriculture) (2007) Water for food, water for life. London, Earth scan/Colombo, IWMI

Camberlin P, Janicot S, Poccard I (2001) Seasonality and atmospheric dynamics of the teleconnection between African rainfall and tropical sea surface temperature: Atlantic vs. ENSO. Int J Climatol 21:973–1005

Cork S, Peterson G, Petschel-Held G, Alcamo J, Alder J, Bennett E, Carr E, Deane D, Nelson G, Ribeiro T, Butler C, Mendiondo E, Oluoch-Kosura W, Zurek M (2005) Four scenarios. In: Carpenter SR, Pingali PL, Bennett EM, Zurek MB (eds) Millennium ecosystem assessment. Ecosystems and human well-being, vol 2. Island Press, Washington, DC

Davis CL 2011 Climate risk and vulnerability: a handbook for Southern Africa. Council for Scientific and Industrial Research, Pretoria, pp 92. http://www.sarva.org.za/sadc/download/sadc_handbook.pdf. Accessed on May 21 2013

de Fraiture C, Wicheins D, Rockstrom J, Kemp-Benedict E (2007) Looking ahead to 2050: scenarios of alternative investments approaches. In: Molden D (ed) Water for food, water for life: comprehensive assessment on water in agriculture. International Water Management Institute/Earthscan, London/Colombo

Descheemaeker K, Mapedza E, Amede T, Ayalneh W (2010) Effects of integrated watershed management on livestock water productivity in water scarce areas in Ethiopia. Phys Chem Earth Parts A/B/C 35(13–14):723–729

De Wit P, Norfolk S (2010) Recognizing rights to natural resources in Mozambique. Rights and Resources Initiatives, Washington, DC

Dorward A (1999) Farm size and productivity in Malawian smallholder agriculture. J Dev Stud 35(5):141–161

Dryland Systems (2012) Strategies for combating climate change in drylands agriculture. Synthesis of dialogues and evidence presented at the International conference on food security in dry lands, Doha, November, 2012. http://drylandsystems.cgiar.org/sites/default/files/Agriculture%20and%20Climate%20Change_%20Input%20to%20COP%20%288%29.pdf

Erenstein O, Kassie GT, Langyintuo A, Mwangi W (2011) Characterization of maize producing households in drought prone regions of Eastern Africa. Socioeconomics Working Paper 1. Mexico: CIMMYT

Evans A, Giordano M, Clayton T (2012) Investing in agricultural water management to benefit smallholder farmers in Burkina Faso. AgWater Solutions Project country synthesis report. (IWMI working paper 149). Colombo: International Water Management Institute (IWMI). 22 p

FAO (1997) Irrigation potential in Africa: a basin approach. FAO Land and water bulletin 4. ISBN 92-5-103966-6. http://www.fao.org/docrep/w4347e/w4347e00.htm#Contents

FAO (2006) AQUASTAT. Malawi, Zambia and Mozambique. http://www.fao.org/nr/water/aquastat/countries_regions/malawi/index.stm
FAO (2009) A review of evidence on dryland pastoral systems and climate change: implications and opportunities for mitigation and adaptation land and water discussion paper no. 7, FAO, Rome
FAO (2013) Poultry sector Mozambique. FAO animal production and health livestock country reviews no. 5 Rome
FAO (2015) Regional overview of food insecurity of Africa. Accra. 21 pp
Fisher M, Abate T, Rodney, W, Lunduka RW, Asnake W, Alemayehu Y, Madul R (2013) Drought tolerant maize for farmer adaptation to drought in Sub-Saharan Africa: determinants of adoption in eastern and southern Africa. Climatic Change. DOI 10.1007/s10584-015-1459-2
Folmer ECR, Geurts PMH, Francisco JR (1998) Assessment of soil fertility depletion in Mozambique. Agri Ecosyst Environ 71:159–167
Freimuth L, Bromberg G, Mehyar M, Al Khateeb N (2007). Climate change: a new threat to Middle East security prepared for the United Nations climate change conference Bali. EcoPeace/Friends of the Earth Middle East Amman, Bethlehem, and Tel-Aviv Available from http://www.foeme.org/index_images/dinamicas/publications/publ78_1.pdf
Garrity D, Dixon J, Boffa JM (2012) Understanding African farming systems. A paper presented at a conference on 'Food Security in Africa'. Sydney, November, 29–30
Gebreselassie S, Kirui OK, Mirzabaev A (2015) Economics of land degradation and improvement in Ethiopia. In Nkonya E, et al (eds), Economics of land degradation and improvement – a global assessment for sustainable development. DOI 10.1007/978-3-319-19168-3_14
German L, Mowo J, Amede T, Masuki K (eds) (2012) Integrated natural resources management in the highlands of Eastern Africa: from concept to practice. World Agroforestry Centre (ICRAF), Nairobi and International Development Research Centre (IDRC), Ottawa. Earth Scan, London/ New York, 350 pp
Haggblade S, Longabaugh S, Tshirley D (2009) Spatial patterns of food staple production and marketing in South East Africa: implication for trade policy and emergency response. Michigan State University, ISBN 0731-3483
IFAD (2004) Effects of small scale irrigation on system productivity, Natural resource management and community innovation in Ethiopia: IFADs' interventions. SCPII Evaluation Report. Rome
IPCC (Intergovernmental panel on Climate Change) (2007) Climate change 2007. The physical science basis; summary for policy makers. http://www.pnud.cl/recientes/IPCC-Report.pdf
Kandji ST, Verchot L, Mackensen J (2006) Climate change and variability in Southern Africa: impacts and adaptation in the agricultural sector. World Agroforestry Centre (ICRAF) and United Nations Environment Programme (UNEP). Nairobi
Kassie GT, Erenstein O, Mwangi W, La Rovere R, Setimela P, Langyintuo A (2012) Characterization of maize production in Southern Africa: synthesis of CIMMYT/ DTMA household level farming system surveys in Angola, Malawi, Mozambique, Zambia and Zimbabwe. Socio-economics program working paper 4. CIMMYT, Mexico
Minahi K, Goudriaan J, Lantinga EA, Kimura T (1993) Significance of grasslands in emission and absorption of greenhouse grasses. In: Barker MJ (ed) Grasslands for our world. SIR Publishing, Wellington
MoAFS (Ministry of Agriculture and Food Security) (2011) Malawi agricultural sector wide approach. A prioritized and harmonized agricultural development agenda, 2011–2015
Mukanda N, Moono D (1999) Zambia. 337–353 pp. In Nebhan H, Mashali AM, Mermut AR (eds), Integrated soil management for sustainable Agriculture and food security in Southern and East Africa. Proceedings of the expert consultation, Harare, December, 1997. FAO, Rome. 415 pp
Nabhan H, Mashali AM, Mermut AR (1999) Integrated soil management for sustainable agriculture and food security in Southern and East Africa. Proceedings of the expert consultation, Harare, 8–12 December 1997. FAO, Rome
Nakhumwa TO (2004) Dynamic costs of soil degradation and determinants of adoption of soil conservation technologies by smallholder farmers in Southern Malawi. PhD thesis. University of Pretoria, Pretoria

NSO (National Statistical Office) (2010) National census of agricultural and livestock 2006/2007. Main report
Oldeman LR (1998) Soil degradation: a threat to food security? Wageningen, International Soil Reference and Information Centre
Olson GK (2007) Forest and farming: an analysis of rural livelihood programmes for poverty reduction in Eastern Zambia. MSc thesis, University of Montana, Missole
Oweis T, Prinz D, Hachum A (2001) Water harvesting: indigenous knowledge for the future of the drier environments. ICARDA, Aleppo
Postel S (2000) Entering an era of water scarcity: the challenges ahead. Ecolog Appl 10:941–948
Rasmussen EM (1987) Global climate change and variability: effects on drought and desertification in Africa. In: Glantz MH (ed) Drought and hunger in Africa: denying famine a future. Cambridge University Press, Cambridge, MA
Rockström J (2000) Water resources management in smallholder farms in Eastern and Southern Africa: an overview. Phys Chem Earth 25(3):278–288
Rockström J, Folke C, Gordon L, Hatibu N, Jewitt G, Penning de Vries F, Rwehumbiza F, Sally H, SavenijeH SR (2004) A watershed approach to upgrade rainfed agriculture in water scarce regions through water system innovations: an integrated research initiative on water for food and rural livelihoods in balance with ecosystem functions. Phys Chem Earth 29:1109–1118
Roy A (2007) Fertilzers: key to achieving hunger and poverty goals. Paper presented at 'The fertilizer outlook and technology conference', Tempa. Nov 5–7, 2007. IFDC
Saka AR, Bunderson WT, Lowole MW, Kumwenda JDT (1999) Malawi. 231–246 pp. In Nebhan H, Mashali AM, Mermut AR (eds), Integrated soil management for sustainable Agriculture and food security in Southern and East Africa. Proceedings of the expert consultation, Harare. FAO, Rome. 415 pp
Scherr SJ (1999) Soil degradation: a threat to developing-country food security by 2020? International Food Policy Research Institute, Washington, DC
Steinbach J (1997) Alternatives to crop residue as feed sources in mixed farming systems. In: Renard C (ed) Crop residue in sustainable mixed crop-livestock farming systems CABI International
Steinfeld H, Gerber P, Wassenaar T, Castel V, Rosales M, de Haan C (2006) Livestock's long shadow. Environmental issues and options. FAO, Rome
Technoserve (2011) The emergency of the poultry industry & the decline of poultry imports in Mozambique. http://www.technoserve.org/files/downloads/poultry-presentation.pdf
Thierfelder C, Rusinamhodz L, Ngwira AR, Mupangwa W, Nyagumbo I, Kassie G, Cairns JE (2013) Conservation agriculture in Southern Africa: advances in knowledge. Renew Agri Food Syst:1–21
Twomlow S, Mugabe FT, Mwale M, Delve R, Nanja D, Carberry P, Howden M (2008) Building adaptive capacity to cope with increasing vulnerability due to climatic change in Africa: a new approach. Phys Chem Earth 33(8–13):780–787
UNECA (2009) Economic report on Africa 2009: developing African agriculture through regional value chains. UNECA, Addis Ababa
Ward C, Ringler C (undated) Improving resilience in SSA drylands: the contribution of agricultural water management. http://www.terrafrica.org/Drylands/African_Drylands_SS_Ward_Ringler_Feb5.pdf
Wichelns D (2006) Improving water and fertilizer use in Africa: challenges, opportunities and policy recommendations. Background paper. Africa Fertilizer Summit, 9–13, June 2006. NEPAD and IFDC, Abuja
World Bank (2006) Mozambique agricultural development strategy; stimulating small holder agricultural growth. AFTSI, Agriculture, Environment, and Social Development Unit, Africa Region. Report No. 32416-MZ
World Economic Forum (2015) The Africa competitiveness report 2015. Geneva: World Economic Forum. © World Economic Forum. https://openknowledge.worldbank.org/handle/10986/22014 License: CC BY-NC-ND 3.0 IGO

Part V
Innovations in Dryland Agriculture

Soil Carbon Sequestration in Dryland Agriculture

Muhammad Shakeel Arshad, Muhammad Sanaullah, and Muhammad Farooq

1 Introduction

Carbon sequestration is the transfer of CO_2 from atmosphere into the enduring pools and keeping it stable, so that it cannot directly remitted. Hence, the soil carbon (C) sequestration is the enhancement of soil organic carbon (SOC), and soil inorganic carbon (SIC) pools by wisely land usage and suitable management techniques. The well-managed ecosystems have the possible soil sink capability of almost equals to the collective historic C loss projected at 55–78 Gt (Lal 2004a). The achievable soil C sink capability is only about 50–66 % of the total possible volume. This strategy is quite cost-effective and eco-friendly. The areas where proportion of total annual precipitation to the potential evapotranspiration (PET), the aridity index (AI), ranges from 0.05 to 0.65 are called as drylands, and consist of dry sub-humid areas from 0.50 to 0.65, semi-arid from 0.20 to 0.50, arid from 0.05 to 0.20, and hyper-arid with <0.05 AI, covering 9.9, 17.7, 12.0 and 7.5 % of the world's total land area, respectively (Reynolds and Smith 2002). Drylands cover approximately 6.15 billion hectares, primarily in the Southwestern and Northern Africa, Central and Southwestern Asia, Northwestern Pakistan and India, Australia, Southwestern Mexico and United States (Noin and Clarke 1997; Hillel and Rosenzweig 2002).

M.S. Arshad (✉)
Department of Agronomy, University of Agriculture, Faisalabad 38040, Pakistan
e-mail: mshakeel1264@gmail.com

M. Sanaullah
Institute of Soil and Environmental Sciences, University of Agriculture, Faisalabad 38040, Pakistan

M. Farooq
Department of Agronomy, University of Agriculture, Faisalabad 38040, Pakistan

The UWA Institute of Agriculture, The University of Western Australia, Perth, WA 6001, Australia

M. Farooq, K.H.M. Siddique (eds.), *Innovations in Dryland Agriculture*,
DOI 10.1007/978-3-319-47928-6_17

Globally, drylands hold 241 Pg of SOC (Eswaran et al. 2000), almost 40 times higher C as compared to the addition into atmosphere by anthropogenic actions, during 1990s, which was 6.3 Pg C year^{-1} (Schimel et al. 2001). Drylands contain higher levels of SIC as compared to the SOC (Eswaran et al. 2000).

In drylands, mutual management of SIC and SOC can play a key role in the reduction of CO_2 concentration in the atmosphere (Lal 2002b). Global C cycle is strongly affected by drylands due to vast areas and significance of these soil C pools. On the other hand, in these areas, desertification and land degradation are prevalent, which often consequence in CO_2 emission into the atmosphere (Lal 2004a). The historic C loss from the ecosystem was projected due to losses from vegetation/biotic pool (10–15 Pg) and by desertification (9–14 Pg of SOC pools) (Lal et al. 1999). Likewise, global C (13–24 Pg) has been lost due to desertification of drylands and grasslands (Ojima et al. 1995). Whereas all the ecosystem C loss estimates through desertification are hypothetical, the figures are higher at about 20–30 Pg (Lal 2004a). If two third from this would be sequestered by controlling desertification, and with the implementation of recommended soil management techniques and land use practices, in 50 years this amount would be at 12–20 Pg (Cole et al. 1997). Technical options vary among diverse soil types and systems of land utilization (Hinman and Hinman 1992).

Although some studies have been conducted on C sequestration in dry lands (Lal 2002b), however, these were conducted on different and individual aspects e.g., C sequestration, effect of climate change, and biophysical aspects etc., This chapter describes the potential of soil C sequestration, mitigation of accelerated greenhouse effect, biophysical aspects of C sequestration, impact of C sequestration on climate change and food security, land use management practices and vegetation management options to combat land degradation in dryland agriculture (Tables 1, 2, and 3).

2 Biophysical Aspects of Carbon Sequestration in Drylands

Grasslands, forests and cultivated lands constitute the major potential C sinks in the drylands. In the followings lines, biophysical aspects of carbon sequestration in drylands are described.

2.1 Grasslands

Natural biome in many drylands developed from grasslands is inadequate for plantation due to low rainfall. Both C sequestration and grassland productivity are thought provoking issues. Because increase in the grassland production, enhances the C sequestration process (Scurlock and Hall 1998), about 70 t ha^{-1} C is stored under grassland areas. However, most of the grasslands are prone to degradation resulting in less sequestration of C. Nonetheless the contribution of grasslands is

Table 1 Improved management systems for soil organic carbon sequestration and soil quality enhancement

Farming system	Improved management	Carbon sequestered	Country/region	References
1. Cropland				
(a) Tillage methods	Time and method of seedbed preparation		Syria	Ryan (1997) and Schomberg and Jones (1999)
(b) Supplemental irrigation	Increasing wheat yield		Syria	Oweis et al. (1998)
(c) N fertilization	50–100 kg N ha^{-1}		Syria	Oweis et al. (1998)
(d) N and P management	Enhancing biomass yield		Sudan, WANA	Matar et al. (1992)
(e) Water/nutrient recycling	Sewage irrigation		Egypt	El-Naim et al. (1987)
(f) Crop rotations	Legume-based system	0.6 t ha^{-1} $year^{-1}$	Syria, Midemanian region	Jenkinson et al. (1999)
2. Grazing/ rangelands				
(a) Soil P level	Residual phosphate and grassland restoration		Syria, WANA	Matar et al. (1992)
(b) Controlled grazing	Extending growing period of shrubs		Egypt	Duivenbooden (1993)
(c) Seed dispersal	Using sheep to disperse leguminous seeds		Syria	Ghassali et al. (1998)
(f) Fodder trees	Nutrient cycling		–	Le Houreou (2000)
(e) Crop/livestock integration	Improving carrying capacity		WANA	Thomson and Bahhady (1995)

higher (1–2 t ha^{-1}) than the croplands (Jenkinson and Rayner 1977). Grasses also sequester more C than leguminous cover crop (Lal et al. 1999). However, in grasslands, controlled grazing led to more C in soil than in conservation tillage systems of the grasslands (Franzluebbers et al. 2000). Plant roots have more potential to add organic matter in soil, because of their physical protection, chemical recalcitrance and physico-chemical complexes of rhizodeposits (Rasse et al. 2005) and as a result more C is stored in the grasslands (Woomer et al. 1994). Garten and Wullschleger (2000) reported that in degraded lands, switch grass (*Panicum virgatum* L.) improved the SOC by 12 %. Through proper grazing management, US rangelands can increase

Table 2 Strategies of soil management in dryland ecosystems for carbon sequestration

Strategy	Practice	Carbon sequestered	Location/Region	References
Erosion control/water conservation	(A) No-till farming		Bushland, TX, USA	Jones et al. (1997)
			Northern CO, USA	Potter et al. (1997)
			Queensland, Australia	Dalal et al. (2000)
		4 Mg ha^{-1}	West Africa Sahel	Bationo et al. (2000)
	(B) Conservation tillage	1.34 % in 0–5 cm	Southern Spain	Murillo et al. (1998)
	(C) Mulching			
	(a) Stone cover		Negev Desert	Lahav and Steinberger (2001)
	(b) Residue mulch		Chihuahuan Desert	Rostagno and Sosebal (2001)
	(c) Mulch		Suriname	Breeman and Protz (1988)
Crop diversification	(a) Rotations		Saudi Arabia, West Asia	Shahin et al. (1998)
			Algeria, North Africa	Arabi and Roose (1989)
	(b) Legumes	0.1–0.3 t ha^{-1} $year^{-1}$	Syria, West Asia	Jenkinson et al. (1999)
		1.37 % in 0–15 cm	Australia	Whitehouse and Littler (1984)
			Northern India	Singh et al. (1996)
		17.3 Mg C ha^{-1}	Argentina	Galantini and Rosell (1997)
Integrated nutrient management and recycling	(a) Manuring		Maiduguri, Nigeria	Aweto and Ayub (1993)
	(b) Organic by-products		Spain	Pascual et al. (1998)
	(c) Soil fauna (enhancing termite activity)		Chihuahuan Desert	Nash and Whitford (1995); Whitford (1996)
	(d) Sewage sludge		Spain	Pedreno et al. (1996)
Water management	(a) Irrigation and conservation tillage		Mexico	Follett et al. (2003)
	(b) Irrigation with sewage		Israel	Hillel (1998)
	(c) Irrigation with silt-laden water		Ningxia, China	Fullen et al. (1995)
	(d) Saline aquaculture		Drylands	Glenn et al. (1993)
	(e) Water harvesting			

Table 3 Strategies of pasture and range land management for soil carbon sequestration

Strategy	Practice	Carbon sequestered	Location/Region	References
Improved species	Sowing legumes	1.6 % in 0–5 cm	Vertisols, Australia	Chan et al. (1997)
			Northern Colorado	Havlin et al. (1990)
			Sadore, Niger	Hiernaux et al. (1999)
	Agroforestry		West African Sahel	Breeman and Kessler (1997)
Fire management	i. Prescribed burning	0.6 t ha^{-1} $year^{-1}$	Wyoming, USA	Schuman et al. (2002)
	ii. Stocking rate		Negev, Israel	Zaady et al. (2001)
Grazing management	Controlled grazing		Kawas, USA	Rice and Owensby (2001)
Improving grasslands	Integrated management		World's drylands	Conant et al. (2001)
Erosion management	Integrated management		World's drylands	Lal (2001)

soil C storage by 0.1–0.3 t ha^{-1} $year^{-1}$, and for new grasslands this can rise up to 0.6 t ha^{-1} $year^{-1}$ (Jenkinson et al. 1999; Schuman et al. 2002). The positive effect of grazing has direct impact on species composition and litter accumulation. Therefore, grasslands play an important role in C sequestration, but with careful management. However, semi-arid grasslands are highly sensitive to overgrazing and C loss.

2.2 *Forests*

Forests are the major sinks for below and above ground C in drylands, because of the availability of various species that establish in the drylands through afforestation (Silver et al. 2000; Kumar et al. 2001; Niles et al. 2002). Mesquite ((*Prosopis glandulosa* Torr.) plantation (N fixing and having deep rooting system), increased the soil C by 2 t ha^{-1} in semi-arid soils (Geesing 2000). In North-West India, planting another type of mesquite (*Prospis juliflora* (Sw.) DC.) in salt affected soils raised the level of organic pool of C from 10 t ha^{-1} to 45 t ha^{-1} in the period of 8 years (Garg 1998). Likewise, in semi-arid climate of India, plantation of jand (*Prosopis cineraria* (L.) Druce) has substantially improved the soil fertility and C sequestration process (Nagarajan and Sundaramoorthy 2000). As natural ecosystems are diverse and multifaceted, therefore sequester less C in to the soils. However, due to increase in biomass levels, the improvements were balanced through loss in the C of soil.

2.3 Fallows

In case of fallows, preservation of vegetation cover is important if there is no cropping. Especially, in drylands where soils are more prone to land degradation. The vegetative cover not only shields the soil but also traps the solar radiations for the fixation of CO_2. In Mediterranean Spain, control and vegetative cover were compared for C storage. Removal of vegetation cover after 4 years of its establishment substantially reduced the C storage up to 35 % than control (Albaladejo et al. 1998). In Nigeria, removal of vegetation cover caused 13.5–25 t ha^{-1} reduction in forest soil C in a period of 7 years, however in 12–13 years of bush free area the C contents were restored (Juo et al. 1995). In North-West USA, decreased fallowing in summer had greater effect on SOC in positive way, as compared to decreased tillage activities (Rasmussen et al. 1998; Miglierina et al. 2000).

Agro-ecosystem projected models showed that reduced summer fallow in wheat-based cropping systems (wheat-wheat-fallow; wheat-fallow), in semi-arid chernozems of western Canada, would reduce C losses by 0.03 t ha^{-1} (Smith et al. 2001a). The importance of fallows for C sequestration depends on addition of amount of organic matter in different cropping systems. If it happens correctly, then presence of fallow lands increases the C stocks in the system.

2.4 Cultivated Lands

Reduction of C stocks in arable land is principally caused by different tillage practices (Pretty et al. 2002). This is due to the tillage implements like, mouldboard plough and disc harrow, which deteriorate the soil structure, and increase the C losses through destruction of soil clumps and increasing demolition through biological activity (decay of residues) (Six et al. 2000). The amount of soil C depends upon the texture of soil. For instance, the coarse texture soils (sandy) are more influenced by the tillage practices than the fine texture soils (clay). It is, therefore, strenuous to figure out the effects of tillage on C stocks in the soil (Buschiazzo et al. 2001). Interestingly, under hot and dry climatic conditions, reduction in tillage was effective to improve soil C stock (Batjes and Sombroek 1997).

An investigation into the tillage-induced CO_2 efflux and C losses from soil indicated that tillage by mouldboard plough buried most of the crop residue and caused the maximum CO_2 efflux (Reicosky 1997). The C discharge (% C in crop residues) by the mouldboard plough and disc harrow was 134 and 70 %, respectively, however disc harrow, chisel plough and no-till had 58, 54 and 27 % C discharge, respectively. This indicated the association of CO_2 release (loss) and the tillage intensity, and demonstrated that ploughing with mouldboard caused maximum loss to soil C. Comparison of conventional disc tillage and no-tillage in Central Texas indicated that constant changes in C sequestration and mineralization had occurred in no-till system (Franzluebbers et al. 1995). Another study, focused on CO_2 emission and soil C stock, indicated that CO_2 release and soil C level from zero or reduced till were more than the conventional tillage system (Costantini et al. 1996).

Adopting no-till approach, decrease in C stock was just 12 % so, no-till system was unable to make dominant decrease in C sinks. Soil surface with crop residues changed the top layer of the soil by modulating the microclimate, which is cool and wet than the conventional tillage (Doran et al. 1998). Decrease in mineralization of the organic portion and organic C, was increased up to 42–50 % in no-till approach than ploughing with the chisel, all this happened due to an increase in the dynamic forms of organic matter under no-till in Argentine rolling pampa. No-till also improved the C distribution in soil profile (Alvarez et al. 1995). A study on total organic carbon (TOC), conducted on sandy clay loam soil, described that in no-till system increase in TOC was dependent on surface layer of soil, so, actual quantity depends on cropping system. For instance, in oat/vetch-maize/cowpea system, about 1.33 t C ha^{-1} $year^{-1}$ was sequestered over a period of 9 years in no-till system. This system also produced a large quantity of crop residues (Bayer et al. 2000). Although in hot environments, the rate of organic matter accumulation is low than the cool environments. However, in sandy soil of northern Syria, under zero-till, there is possibility to make little increase in organic matter of soil (Ryan 1997). In western Nigeria, no-till combined with mulching application increased the soil C from 15 to 32.3 t ha^{-1} over a period of 4 years. However, C losses through conventional cultivation can be minimized with no-till system (Ringius 2002).

The decrease in soil temperature has been linked with crop residues/plant residues on the surface of the soil which may cause delay and/or decrease in the germination. While in dry lands, the temperature is usually above the optimum level, which favors germination, plant growth and establishment (Phillips et al. 1980) under adequate moisture provision. Among the effective implements that can control the weeds, mouldboard plough and disc harrow are the predominant (Reicosky 1995). Usually, no-till systems depend upon extra pesticides and herbicides. However, nitrogen fertilizer application can be problematic under no-tillage systems (Phillips et al. 1980). In less drained soils, denitrification can decrease the rate of evaporation and may increase the risk of nitrogen leaching nitrate. However, like undisturbed soils, native soil nitrogen is unlikely to change from organic to inorganic form (mineralization). All soils are not adaptive for reduced approach of tillage. For instance, in Argentine pampa, some soils may lose more C in no-till approach that is 0.7–1.5 t C ha^{-1} $year^{-1}$, as compared with conventional tillage system (Alvarez et al. 1995), and to get rid from the soil compaction regular ploughing is required (Taboada et al. 1998). While keeping in view the C budget, there is low demand of energy but in reality tillage is very energy demanding approach. In Northern America, after using no-till system, energy inputs used in maize and soybean production were reduced by 7 and 18 %, respectively (Phillips et al. 1980). The capacity of cropping systems to convert water into plant biomass or grain, is the mean of minimizing the use of energy in the form of C and irrigation costs or inputs, while the influence to save energy is often balanced by additional herbicide requirements (Phillips et al. 1980).

The production, and application of herbicides to no-till system of the Great Plains, is approximately 0.02 t C ha^{-1} (Kern and Johnson 1993). In no-till system, the efficacy of C sequestration depends on particular agriculture system, which is being introduced in it. It is not clear either the tillage intensity lessens the balance between C gain and loss (Kern and Johnson 1993). Crop rotation has long term beneficial effects on C sequestration. A comparison of legume-based rotation and continuous maize cultivation indicated that rotation has great effect on soil C than fertilizer application (Gregorich et al. 2001). Crop rotation with legumes increased the organic matter accumulation in the surface layer of soil – the plough layer – and had more biological activities, which indicates that soil plough layer in legume-based rotation have more C stocks (Miglierina et al. 2000). Similarly, in Argentina, legumes and cattle grazing made a beneficial effect (2–4 t ha^{-1}) on SOC (Miglierina et al. 2000). In drylands, legume based rotation plays a vital role in C Sequestration for maintaining soil fertility. For instance, in the drylands of United States, maize/soybean cultivation areas enhanced the soil C sequestration up to 0.01–0.03 Pg C $year^{-1}$. However, the crop rotation could be more valuable and effective in the sequestration process, when is integrated with conservation tillage practices (Drinkwater et al. 1998).

In conclusion, the grasslands have more C sequestration than the croplands but are prone to overgrazing-induced C losses. The forests are the most important sinks for below and above ground C in drylands, due to the presence of diverse species. Vegetation preservation is important in case of fallows. The significance of fallows for C sequestration depends upon organic matter addition by diverse cropping systems. The soil C depends upon soil texture as well; sandy soils are effected more by cultivation than clay in this regard. Therefore, the effects of tillage on C stocks in the soil need to be figured out. Crop rotation has long-term positive effects on C sequestration, and may be more effective through integration with conservation tillage.

3 Impact of Carbon Sequestration on Global Climate Change and Food Security

Carbon sequestration is one of the major drivers of global climate change and the food security. Increase in carbon sequestration can cause substantial reduction in global warming. In the following lines, impact of carbon sequestration on global climate change and food security are discussed.

3.1 Climate Change

The estimation of total soil C sequestration in the world, ranges from a low level 0.4–0.6 Gt C $year^{-1}$ (Sauerbeck 2001) to higher level 0.6–1.2 Gt C $year^{-1}$ (Lal 2003a). This indicates a finite potential with respect to time and capacity.

However, C sequestration in soil provides the time up to the substitutes as fossil fuels come into effect. Impacts of C sequestration on global climate change are discussed below.

3.1.1 Farm Chemicals and Fuel Consumption

Many practices consist of C-based inputs present in N to about 0.86 kg C kg^{-1}, P_2O_5 0.17 kg C kg^{-1}, K_2O 0.12 kg C kg^{-1}, lime 0.36 kg C kg^{-1}, insecticides 4.9 kg C kg^{-1}, herbicides 4.7 kg C kg^{-1}, fungicides 5.2 kg C kg^{-1} (West and Marland 2002), and in groundwater, pumping for irrigation purpose is 150 kg C ha^{-1} (Follett 2001). Tillage practices also discharge C from the soil e.g., plowing with moldboard emits about 15 kg C ha^{-1}, chisel 8 kg C ha^{-1}, light tandem disks about 6 kg C ha^{-1}, sub-soiler 11 kg C ha^{-1}, cultivator 4 kg C ha^{-1}, and rotavator 2 kg ha^{-1} during hoeing (Lal 2004a). So, C release can be decreased by 30–35 kg C ha^{-1} per season if conventional tillage is replaced by no till farming (Lal 2004a). Balanced use of C-based inputs is, however, very important to reduce the C losses.

3.1.2 Nutrition

Carbon is the one of the most important elements that constitutes the living bodies. According to an estimate, for 1 Gt of C sequestration, the world requirement for N, P and K are 80, 20 and 15 million tons, respectively (IFDC 2000). Various C sources including biological nitrogen fixation, reuse of subsoil after recycling, deposits through air, use of waste material from biological resources and from crop residues provide C for sequestration. For instance, one ton cereal residues consist of nutrients as N (12–20 kg), P (1–4 kg), K (7–30 kg), Ca (4–8 kg), and Mg (2–4 kg). Globally, 3 Gt $year^{-1}$ of residues are produced from grain crops, those if used after recycling can be used for fuel, betterment of quality of soil and for sequestration of C. Crop residues are also used for the production of ethanol, and a good source for energy production by burning. These can be utilized for C sequestration to improve the soil quality, and for the production of biofuel. However, the economic assessment of these two competing uses is required in future.

3.1.3 Soil Erosion and Deposition

Soil erosion removes SOC, through the sediments, borne from water and wind. The sediments enriched with SOC are either reallocated over the landscapes, settled down in depressions, and may be passed to the water bodies. Even though a major portion of C, transferred due to erosion, might be covered up and redeployed, the remaining is released to atmosphere by methanogenesis as CH_4 and by mineralization as CO_2 (Smith et al. 2001c). Deposition and burial due to erosion is 0.4–0.6 Gt C $year^{-1}$ than released into atmosphere which is 0.8–1.2 Gt C $year^{-1}$ (Lal 2003b).

The quantification of deposition contrasted with emission of C is of significant concern. Thus an effective soil erosion control is necessary to improve the environmental quality and sustainability of dryland soils.

3.1.4 Exhaustive Farming Practices

In Sub-Saharan Africa, since the mid-1960s, about 40 kg of NPK ha^{-1} of cultivated lands is the yearly nutrients depletion rate, due to subsistence farming (Sanchez 2002). The mining of SOC from the soil for nutrients by decomposition of organic matter has a similar effect on the atmosphere like combustion of fossil fuels. Thus the management practices must be improved rather than degrading the quality of soil, increase crop yield per unit fertilizer and other inputs used rather than decrease or maintain, and increase rather than to deplete the soil fertility and SOC pools.

3.1.5 Societal Value And Hidden Benefits

Soil C commodification is essential for trading C credits, markets have been established for this purpose since 2002, particularly in Europe (Johnson and Heinen 2004). The existing price for SOC is about \$1 t^{-1} of CO_2, which may enhance with the regulation and the emission lid. The credit trading of SOC turn out is a routine part of the capability to quantify temporary ups and downs in the existing SOC pools, the solutions for the mitigation of climate change (Lal et al. 2000). However, the soil C price be essentially based on the both *in-situ* and *ex-situ* social welfares.

3.1.6 Hydrologic and Carbon Cycles

An estimated increase in the production of cereal from 1997 to 2050 is about 56 % (Rosegrant and Cline 2003), which must take place on equal or less land area, and by using same or a less amount of water. As a consequence, it is important for the improvement of crop yields, and the sequestration of SOC in drylands, to correlate hydrological and C cycles by water conservation. Water conservation can enhance the low levels of SOC pools in dryland agriculture by using water-efficient agricultural systems and water harvesting. No-till farming system is an important option for the enhancement of SOC pools in dryland agriculture, which also proves a better option for drought management (Lal 2004b).

3.1.7 Soil C Sequestration and Global Warming

Global warming is a long term and universal problem. The C sequestration in soil is associated but separate issue, regardless from the debate of global warming, as it has its own intrinsic worth for the productivity enhancement, restoration of degraded soils and ecosystems, and water quality improvement. Several biophysical and societal benefits, can be achieved by compensating fossil fuels emission through possible SOC potentials. Moreover, soil C sequestration works like a bridge among three universal issues of climate change, biodiversity, and desertification.

3.1.8 Other Greenhouse Gases

The improvement in SOC pools enhances the ability of soil for the oxidation of CH_4, particularly in no-till farming systems (Six et al. 2002), however the N_2O emission can get worse (Smith et al. 2001b). The mitigation potential of CO_2 for soil management can be changed with fluxes of N_2O and CH_4, and should be well thought-out with respect to SOC sequestration.

3.2 Food Security

Globally, higher soil degradation, which needs restoration and soil C sequestration, is present in the areas such as South and Central Asia, Sub-Saharan Africa, China, the Caribbean, South American acid savannas and the Andean region. In South Asia and Africa, it is a norm to completely remove the residues for fuel and fodder purpose. As a result, stocks of SOC are depleted from the root zone, and productivity of soil and quality of environment in these regions have adversely affected. Poor farming community is more affected due to destructive practices, as they make use of marginal lands for cultivation with minimal inputs, produce poor yield, perpetuating to poor livelihood and lead to poverty. In the subsistence farming systems of Sub-Saharan Africa, the soil organic matter is the main source of nutrients for cultivated crops, as fertilizer consumption is only 2.5 %. This area represents 2 % of the global irrigated land area which is necessary for the C sequestration. Recommended practices cannot be predicted in rigorously degraded soils, due to depletion of their SOC pool which is the life support system of soils. The optimal SOC pool is required to (1) hold nutrients and water, (2) improve the structure of soil and tilth, (3) reduce the degradation and erosion hazards, and (4) deliver energy to soil microbes. The SOC pool acts as a bio-membrane which sieves the contaminants, degrades pollutants, declines hypoxia in the ecosystems of coastal regions, decreases sediment load from the rivers, as well as is a main sink for atmospheric greenhouse gases.

In Sub-Saharan Africa, application of fertilizer is a significant approach for the enhancement of crop yield (Pieri 1986), however its efficiency was improved by using in combination with mulching of trees (Sanchez 2002) and crop residues (Yamoah et al. 2002). Higher SOC pools results in more crop yields even in intensive agricultural systems (Bauer and Black 1994), particularly in soils with depleted SOC (Johnston 1986). The improvement in SOC up to one ton, enhanced the grain yield of wheat by 40 kg ha^{-1} in the semi-arid pampas of Argentina (Díaz-Zorita et al. 2002), and 27 kg ha^{-1} in North Dakota, United States (Bauer and Black 1994),

17 kg ha^{-1} of maize (*Zea mays* L.) in Thailand (Petchawee and Chaitep 1995), 3 kg ha^{-1} of maize and 6 kg ha^{-1} of wheat in alluvial soils of northern India (Kanchikerimath and Singh 2001) and 1 kg ha^{-1} of cowpea (*Vigna unguiculata* L.) and 10 kg ha^{-1} of maize in western Nigeria (Lal 1981). Better SOC pool is also required for sustainable yields by improved soil structure, nutrient and water holding capacity, and microbial activity. The critical level of SOC is 1.1 % for most of the soils in tropics (Aune and Lal 1997). In tropical ecosystems, an increase in SOC from a low level 0.1–0.2 % to the critical 1.1 %, is a major challenge. So far, a severe decline in SOC stock in Sub-Saharan Africa and somewhere else must be upturned to enhance food security. In Kenya, an 18-years study indicated that beans and maize yielded 1.4 t ha^{-1} $year^{-1}$ without any input, and 6.0 t ha^{-1} $year^{-1}$ when manure and fertilizer were applied as well as stover was retained, the consistent SOC pools were 23.6 t ha^{-1} and 28.7 t ha^{-1} up to 15 cm depth, respectively (Kapkiyai et al. 1999). This type of significant increase in the yields of crops is required at large scale to ensure food security.

In conclusion, balanced use of C based inputs is required to enhance the effectiveness, to reduce the C losses, as well as for improving food production and to make sure sustainable use of land and water resources. The erosion processes removed the SOC through the sediments, enriched with SOC, which either transferred to landscapes, settled down in depressions, or may passed to the water bodies. Thus an effective soil erosion control is required, to improve the environment quality, and sustainability of dryland soils. To improve the agronomic yields, and SOC sequestration in dryland soils, it is essential to correlate C and hydrological cycles by water conservation. A number of biophysical and societal benefits, can be achieved by pay off fossil fuels discharge through possible SOC potentials. Furthermore, SOC sequestration acts as a bridge between desertification, climate change, and biodiversity. Improved SOC pools are essential for sustainable productions through better soil structure, nutrients, water holding capacity, and microbial activity. This type of improvement is required at large scale to ensure food security.

4 Carbon Sequestration to Combat Land Degradation in Drylands

If the period of dryness continues for 1–2 years, the condition is considered as drought, a typical characteristics of drylands. So the leading character of drylands is water scarcity. Water shortage is a severe constraint for productivity, and hence affects the buildup of C pools in the soils. This issue becomes worse as the rainfall is not only below average but also unpredictable, that's why better management of available water is necessary in these areas. Additionally, temperature decreases the pools of SOC exponentially (Lal 2002a). As a result, dryland soils hold very less C stocks about <0.5–1 % (Lal 2002b). The depletion of SOC pools due to excessive land uses, can be overwhelmed by adding plant biomass into the soils (Polwson et al. 1998; Lal 2001). The degradation and desertification are prevailing in the dryland soils, which result in huge decline in SOC stocks. The increase in SOC stock improves soil quality, which have a significant impact on economic and social livelihood of the people in these regions. The factors responsible for C sequestration capacity in cultivated lands are climate, soil, vegetation cover as well as management skills. That's why C sequestration refers to the ability of lands and forests in agriculture sector to remove CO_2 from the atmosphere. Photosynthetic trees and crop plants absorb the CO_2 from the atmosphere and store as C biomass in leaves, branches, trunks, roots and soil. The largest sinks are the forests and grasslands, as they can store higher amounts of C in their leaves and roots for longer periods of time, however the terrestrial sinks are soils. Carbon holding capacity of organic matter present in soil is greatly influenced with C added from dead materials of plants and respiratory losses of C, decomposition, and both natural and anthropogenic activities with soils. Farmers can slow down the C losses from the soil by the adoption of good farming practices which include minimal soil disturbance and permanent soil cover. The biomass and soil C sequestration from the atmosphere, not only decreases the greenhouse effect, however, it also helps to keep up the restorative ability of soils for the sustainable production and environmental issues. Drylands are less prone to C losses than the wet soils (Glenn et al. 1992), because water scarcity restricts the mineralization in soil, and hence the C flux to the atmosphere. Thus in dryland soils, the C habitation period is more, occasionally extensive than the forest soils.

In conclusion, land degradation and desertification are widespread in drylands, and have caused substantial decline in SOC pools. The SOC, improved the soil quality, which have major influence on the economic and social livelihood of the societies. Climate, vegetation cover, soil, as well as management skills, are the elements of soil C sequestration ability in cultivated lands. Soil C sequestration and biomass accumulation from atmosphere, not only reduce the greenhouse effect, but also supports the invigorating capacity of the soils for the sustainable production and ecological aspects.

5 Management Options for Carbon Sequestration

The SOC concentration can be increased by enhancing plant growth with supplemental irrigation and good soil moisture regime, and the adaptation of better soil management practices. Water use efficiency (WUE) can be enhanced by reduction in losses due to surface runoff, evaporation by residue mulching which can also help in lowering soil temperature. The C pools in the soil can also be increased by adopting some strategies such as conservation tillage, water management, organic cultivation, better cropping systems and land use, and land restoration, these are also called advanced farming practices. Soil organic matter contents increased by organic farming systems, with the use of animal manures in composted form, and cover crops (Rodale 2003). The emissions elimination take place from manufacturing and transport of synthetic fertilizers, by adopting organic farming systems. The conservation and improvement of soil, water and air quality is possible through land restoration and land use changes, which usually reduces the emission of greenhouse gasses. Organic matter present in the soil is responsible for soil quality, it is a dynamic pool and responds effectively to changes in soil management, and primarily biomass production resulted from C inputs and tillage.

5.1 Conservation Tillage and Residue Management

Conservation tillage can improve the WUE, and plough till to no-till conservation strategy reduce the risks of soil degradation, improve the SOC concentration and soil quality over time. After the period of 24 years (1943–1966) of cultivation, 9.3 g kg^{-1} SOC was present in a plowed clean fallow treatment as compared to sweep-tilled late fallow, which was 118 g kg^{-1} SOC, in Bushland, Texas (Jones et al. 1997). The amount of SOC was measured after 8 years of no-till, up to 20 cm depth, and on the paired water sheds the treatments of stubble mulch were started for the cultivation of wheat-sorghum-fallow rotation (Unger 1991). For no-till, the average SOC concentration up to 10 cm depth was 16.3 and 15.8 g kg^{-1}, and treatment of stubble mulch indicates the trend for gain in SOC in no-till treatment (Stewart and Robinson 2000). In the upper 2 cm depth, the concentration of SOC significantly increased. The SOC was improved by 60 to more than 600 kg C ha^{-1} $year^{-1}$ due to one of pleasing consequences of no-till system (Stewart and Robinson 2000). On the soil surface the residues were left in winter cover crops, highest rates of SOC were associated with them. The continuous no-till wheat cropping system, during 10 years accumulated the C about 560 kg ha^{-1} $year^{-1}$ in northern Colorado (Potter et al. 1997). Under no-till with crop residue retention, the SOC in vertisols was observed in higher concentrations in Queensland, Australia (Dalal 1989). After 18 years of no-till practices in top 2.5–5.0 cm layers the significant improvement in SOC was observed (Dalal et al. 1995). The soil analysis from a 45-years old tillage system in India, showed that the SOC concentration improved by incorporation of biosolids (Kihani and More 1984). In West African Sahil, the annual addition of crop residues by 4 Mg (mega gram) ha^{-1} resulted in the similar SOC levels, maintaining in fallow in top 20 cm layer (Bationo et al. 2000). In southern Spain, in the traditional tillage system after 2 0years the SOC concentration was 0.84 % in 0–5-cm depth and 1.1 % in conservation tillage, and after 4 years in traditional tillage 0.89 % as compared to 1.34 % in conservation tillage system (Murillo et al. 1998). The soil quality was improved with the increase in concentration of SOC in dry land ecosystems, and sequestration of SOC can be enhanced by adapting no-till farming in wide range.

5.2 *Rotations and Cover Crops*

For enhancing the SOC concentration, use of conservation tillage gives beneficial effects, and accentuated in combination with suitable pastures rotations or cover crops (Ryan et al. 1997). The SOC concentration increased threefold, when wheat grown on sandy soil in rotation with alfalfa, as compared with sowing of continuous wheat in Saudi Arabia (Shahin et al. 1998). The soil quality was improved with silvo pastoral system and legume based rotations in Algeria (Arabi and Roose 1989; Roose 1996). In Syria, inclusion of Medicago in rotation, improved the SOC concentration (Ryan 1997). In Syria, in calcareous soils the SOC pools under different rotations were evaluated. Wheat-meadow rotation increased the SOC pool by 1.6 Mg ha^{-1} with an average rate of 0.17 Mg C ha^{-1} $year^{-1}$ than wheat-wheat rotation, and wheat-fallow rotation 3.8 Mg ha^{-1} at an average rate of 0.38 Mg C ha^{-1} $year^{-1}$ (Jenkinson et al. 1999). In Australia, the SOC concentration increased from 1.18 % to 1.37 % in 0–15 cm depth, where alfalfa and prairie grass were grown as pasture after 2–4 years (Whitehouse and Littler 1984). The rate of increase in concentration of SOC under Rhodes grass was 550 kg C ha^{-1} $year^{-1}$ (Skjemstad et al. 1994), and under grass and legume pasture was 650 kg C ha^{-1} $year^{-1}$ in a Vertisol (Dalal et al. 1995), the similar effects of pasture were also observed in New South Wales (Holford 1990; Chan 1997). The SOC concentrations were improved by continuous cultivation and manuring over 3 years up to 20–40 % in central India in a vertisol (Mathan et al. 1978). In rice-wheat cropping system, SOC concentration was increased with the incorporation of legumes in northern India (Singh et al. 1996). Deep and prolific root crops showed encouraging effects in subsoil on concentrations of SOC. Different rotations using annual crops (4 1/2 years) and mixed meadows (5 1/2 years), retained the stock of SOC at 17.3 Mg C ha^{-1}, as compared to 11.2 Mg C ha^{-1} in continuous cropping in wheat-sunflower rotations (Galantini and Rosell 1997).

5.3 *Integrated Nutrient Management*

The improvement of soil fertility, is essential to enhance the SOC concentrations in to the soil profile. High yields are obtaining with the nitrogen fertilizer application, however, it has little effects on the concentrations of SOC, unless it is used in combination with no-till and residual management (Skjemstad et al. 1994; Dalal et al. 1995). The SOC sequestration is limited in by using biomass C as input in semi-arid conditions. Whereas, the significant increase in crop yield with application of nitrogen, but for the balance of mineralization rate, the residue input is not sufficient. Fertilizers application with recommended rates resulted in significant increase in the concentrations of SOC in Syria (Ryan 1997). After 13 years of no-till, positive effects were observed on the concentrations of SOC, retained by nitrogen application and residues (35.8 Mg C ha^{-1} vs. 34.5 Mg C ha^{-1}) (Dalal 1989). In India, it was reported that the concentrations of SOC were improved using manure at 10 Mg ha^{-1} (Gupta and Venkateswarlu 1994). The concentrations of SOC improved by green manures, farmyard manures, biosolids, and compost applications. Significant increase in concentrations of SOC is possible with the use of high lignin amendments, and recalcitrant for breakdown.

5.4 Pasture and Rangeland Management

The surface residue management in a proper way, and conservation, can enhance the C sequestration with the adaption of improved grazing practices, as the major land use is grazing in the dry lands. In dry land ecosystems, the concentrations of SOC were improved with the advancement in pasture management, and by pasture conservation in the degraded lands (Conant et al. 2001). In degraded vertisols of semi-arid tropics of Australia, the SOC concentration was enhanced by pasture restoration with barrel medic (*Medicago truncatula* Gaertn.) and Mitchell grass (*Astrebla lappacea* F.Muell.), from 1.3 % to 1.6 % respectively, in 0–5 cm depth in the period of 4 years (Chan et al. 1997). Legumes incorporation improved the concentrations of SOC through biological nitrogen fixation process. Hence, the nitrogenous fertilizers application increased the concentrations of SOC in the degraded pastures. Pasture species grown better, however in addition to this perennial woody legume integrated in grazing systems to enhance the concentrations of SOC through the transfer of C to lower depths in sub-soils. In USA, the comparison of SOC pool was done for four sites on Pullman silty clay loam, first site was comprised of a >50 years dryland cultivated wheat, second was native to grassland, third site was a cropland rehabilitated to grassland before 37 years of sampling, and fourth site was a field return to a 7 year grassland before sampling procedure (Stewart and Robinson 2000). The results from these experiments showed significant addition in SOC concentration even under semi-arid environments. Controlled stocking and recommended burning rates are also essential for sustaining and successful accumulation of SOC.

In conclusion, the WUE can be increased by reducing runoff and evaporation losses, and organic mulching is important for optimal soil temperatures. The SOC pools can be improved by adapting conservation tillage, organic farming, better cropping systems, and land restoration. Conservation strategy like, no-till can reduce the risk of soil degradation, increase the SOC concentration, and soil quality over time. Conservation tillage, in combination with suitable pastures rotations or cover crops, can improve the WUE. The soil fertility enhancement, is crucial for SOC concentrations in soil profile, it can be improved by farmyard manures, green manures, biosolids, and compost applications. The major land use practice is grazing in drylands, the SOC concentrations can enhance by pasture management, and pasture conservation in degraded soils.

6 Conclusion

The above discussion supports the following conclusions: (1) the world drylands cover gigantic areas, with different land uses, and a wide range of soils and environments. (2) A severe problem of land degradation and desertification of drylands is prevalent, possibly caused by misuse of land, mismanagement of soil, and extreme climatic conditions. (3) Significant losses of SOC, from a total pool (241 Pg) caused by land degradation and desertification are estimated at 20–30 Pg, out of this, about two-thirds (12–20 Pg), can be re-sequestered by restoring desertified soils. (4) Food security can be achieved through soil C sequestration strategy as it improved the soil quality and health. (5) The average SOC sequestration potential is about 1 Pg C year^{-1}, with the adaptation of recommended practices on grazing lands and croplands, and by afforestation with Acacia, Mesquite, Neem etc.

References

Albaladejo J, Martinez-Mena M, Roldan A, Castillo V (1998) Soil degradation and desertification induced by vegetation removal in a semi-arid environment. Soil Use Manag 14:1–5

Alvarez R, Diaz RA, Barbero N, Santanatoglia OJ, Blotta L (1995) Soil organic carbon, microbial biomass and CO_2-C production from 3 tillage systems. Soil Tillage Res 33:17–28

Arabi M, Roose E (1989) Influence de quatre syste`mes de production en re´gion me´diterrane´enne de moyenne montagne Alge´rienne. Bulletin Re´seau Erosion 9:39–51

Aune JB, Lal R (1997) Agricultural productivity in the tropics and critical limits of properties of Oxisols, Ultisols and Alfisols. Trop Agric 74:96–103

Aweto AO, Ayub HK (1993) Effect of continuous cultivation with animal manuring on a sub-Sahelian sil near Maiduguri, northeastern Nigeria. Biol Agric Hortic 9:343–352

Bationo A, Wani SP, Bielders CL, Vlek PLG, Mokwunye AU (2000) Crop residue and fertilizer management to improve soil organic carbon content, soil quality and productivity in the desert margins of West Africa. In: Lal R, Kimble JM, Stewart BA (eds) Global climate change and tropical ecosystems. CRC/Lewis Publishers, Boca Raton, pp. 117–145

Batjes NH, Sombroek WG (1997) Possibilities for carbon sequestration in tropical and subtropical soils. Glob Chang Biol 3:161–173

Bauer A, Black AL (1994) Quantification of the effect of soil organic matter content on soil productivity. Soil Sci Soc Am J 58:185–193

Bayer C, Mielniczuk J, Amado TJC, Martin-Neto L, Fernandes SV (2000) Organic matter storage in a sandy clay loam Acrisol affected by tillage and cropping systems in southern Brazil. Soil Tillage Res 54:101–109

Breeman H, Kessler JJ (1997) The potential benefits of agroforestry in the Sahel and other semi-arid regions. Eur J Agron 7:25–33

Breeman N, Protz RV (1988) Rates of calcium carbonate removal from soils. Can J Soil Sci 68:449–454

Buschiazzo DE, Hevia GG, Hepper EN, Urioste A, Bono AA, Babinec F (2001) Organic C, N and P in size fractions of virgin and cultivated soils of the semi-arid pampa of Argentina. J Arid Environ 48:501–508

Chan KY (1997) Consequences of changes in particulate organic carbon in Vertisols under pasture and cropping. Soil Sci Soc Am J 61:1376–1382

Chan KY, Bowman AM, Friend JJ (1997) Restoration of soil fertility of degraded Vertisols using a pasture including a native grass (*Astrebla lappacea*). Trop Grassl 31:145–155

Cole CV, Duxbury J, Freney J, Heinemeyer O, Minami K, Mosier A, Paustian K, Rosenberg N, Sampson N, Sauerbeck D, Zhao Q (1997) Global estimates of potential mitigation of greenhouse gas emissions by agriculture. Nutr Cycl Agroecosyst 49:221–228

Conant RT, Paustian K, Elliott ET (2001) Grassland management and conversion into grassland: effects on soil carbon. Ecol Appl 11:343–355

Costantini A, Cosentino D, Segat A (1996) Influence of tillage systems on biological properties of a typic argiudoll soil under continuous maize in central Argentina. Soil Tillage Res 38:265–271

Dalal RC (1989) Long-term effects of no-tillage, crop residue and nitrogen application on properties of a Vertisol. Soil Sci Soc Am J 53:1511–1515

Dalal RC, Carter JO, Kimble JM (2000) Soil organic matter dynamics and carbon sequestration in Australian tropical soils. In: Lal R, Stewart BA (eds) Global climate change and tropical ecosystems. CRC/ Lewis Publishers, Boca Raton, pp. 283–314

Dalal RC, Strong WM, Weston EJ, Cooper JE, Lehane KJ, King AJ, Chicken CJ (1995) Sustaining productivity of a Vertisol at Warra, Queensland, with fertilizers, no-till or legumes. I. organic matter status. Aus J Exp Agric 37:903–913

Díaz-Zorita M, Duarte GA, Grove JH (2002) A review of no-till systems and soil management for sustainable crop production in the subhumid and semiarid Pampas of Argentina. Soil Tillage Res 65:1–18

Doran JW, Elliott ET, Paustian K (1998) Soil microbial activity, nitrogen cycling, and long-term changes in organic carbon pools as related to fallow tillage management. Soil Tillage Res 49:3–18

Drinkwater LE, Wagoner P, Sarrantonio M (1998) Legume-based cropping systems have reduced carbon and nitrogen losses. Nature 396:262–265

Duivenbooden NV (1993) Grazing as a tool for rangeland management in semi-arid regions: a case study in the northwestern coastal zone of Egypt. Agric Ecosyst Environ 43:309–329

El-Naim EMA, Omran MS, Waly TM, El-Nashar MBM (1987) Effects of prolonged sewage irrigation on some physical properties of sandy soils. Biol Wastes 22:269–274

Eswaran H, Reich PF, Kimble JM, Beinroth FH, Padamnabhan E, Moncharoen P, Kimble JM (2000) Global carbon stocks. In: Lal R, Eswaran H, Stewart BA (eds) Global climate change and pedogenic carbonates. Lewis Publishers, Boca Raton, pp. 15–250

Follett RF (2001) Soil management concepts and carbon sequestration in croplands soils. Soil Tillage Res 61:77–92

Follett RF, Castellanos JZ, Buenger ED (2003) Carbon dynamics and sequestration in an irrigated Vertisol in Central Mexico. Soil Tillage Res 83:148–158

Franzluebbers AJ, Hons FM, Zuberer DA (1995) Tillage-induced seasonal changes in soil physical properties affecting soil CO_2 evolution under intensive cropping. Soil Tillage Res 34:41–60

Franzluebbers AJ, Stuedemann JA, Schomberg HH, Wilkinson SR (2000) Soil organic C and N pools under long-term pasture management in the Southern Piedmont USA. Soil Biol Biochem 32:469–478

Fullen MA, Fearnehough W, Mitchell DJ, Trueman IC (1995) Desert reclamation using Yellow River irrigation water in Ningxia, China. Soil Use Manag 11:77–83

Galantini JA, Rosell RA (1997) Organic fractions, N, P and S changes in Argentina semi- arid Haplustoll under different crop sequences. Soil Tillage Res 42:221–228

Garg VK (1998) Interaction of tree crops with a sodic soil environment: potential for rehabilitation of degraded environments. Land Degrad Dev 9:81–93

Garten CT, Wullschleger SD (2000) Soil carbon dynamics beneath switchgrass as indicated by stable isotope analysis. J Environ Qual 29:645–653

Geesing D, Felker P, Bingham RL (2000) Influence of mesquite (*Prosopis glandulosa*) on soil nitrogen and carbon development: implications for global carbon sequestration. J Arid Environ 46:157–180

Ghassali F, Osman AE, Cooks PS (1998) Rehabilitation of degraded grasslands in north Syria: the use of Awassi sheep to disperse the seeds of annual pasture legumes. Exp Agric 34:391–405
Glenn E, Squires V, Olson M, Frye R (1993) Potential for carbon sequestration in the drylands. Water Air Soil Pollut 70:341–355
Glenn EP, Pitelka LF, Olsen MW (1992) The use of halophytes to sequester carbon. Water Air Soil Pollut 64:251–263
Gregorich EG, Drury CF, Baldock JA (2001) Changes in soil carbon under long term maize in monoculture and legume-based rotation. Can J Soil Sci 81:21–31
Gupta JP, Venkateswarlu J (1994) Soil management and rainwater conservation and use. V. Aridisolsin soil management for sustainable agriculture in dryland areas. Bull Ind Soc Soil Sci 16:66–77
Havlin JL, Kissel DE, Maddux LD, Classen MM, Long JH (1990) Crop rotation and tillage effects on soil organic carbon and nitrogen. Soil Sci Soc Am J 54:448–452
Hiernaux P, Bielders CL, Valentin C, Bationo A, Rivera-Fernandez S (1999) Effects of livestock grazing on physical and chemical properties of sandy soils in Sahelian rangelands. J Arid Environ 41:231–245
Hillel D (1998) Environmental soil physics. Academic, San Diego, pp. 640–642
Hillel D, Rosenzweig C (2002) Desertification in relation to climate variability and change. Adv Agron 77:1–38
Hinman CW, Hinman JW (1992) The plight and promise of arid land agriculture. Columbia University Press, New York
Holford ICR (1990) Effects of 8-year rotations of grain sorghum with lucern, annual legume, wheat and long fallow on nitrogen and organic carbon on two contrasting soils. Aus J Soil Res 28:277–291
IFDC (2000) International fertilizer development center. World Fertilizer Consumption, IFDC, Muscle Shoals
Jenkinson DS, Rayner JH (1977) The turnover of soil organic matter in some of the Rothamsted classical experiments. Soil Sci 123:298–305
Jenkinson DS, Harris HC, Ryan J, McNeil AM, Pilbeam CJ, Coleman K (1999) Organic matter turnover in calcareous soil from Syria under a two-course cereal rotation. Soil Biol Biochem 31:687–693
Johnson E, Heinen R (2004) Carbon trading: time for industry involvement. Environ Int 30:279–288
Johnston AE (1986) Soil organic matter, effects on soils and crops. Soil Use Manag 2:97–105
Jones OR, Stewart BA, Unger PW, Elliott ET (1997) Management of dry-farmed southern great plains soils for sustained productivity. In: Paul EA, Paustian K, Cole CV (eds) Soil organic matter in temperate agroecosystems. CRC Publishers, Boca Raton, pp. 387–401
Juo ASR, Franzluebbers K, Dabiri A, Ikhile B (1995) Changes in soil properties during long-term fallow and continuous cultivation after forest clearing in Nigeria. Agric Ecosyst Environ 56:9–18
Kanchikerimath M, Singh D (2001) Soil organic matter, biological properties after 26 years of maize–wheat–cowpea cropping as affected by manure, fertilization in a Cambisol in semiarid region of India. Agric Ecosyst Environ 86:155–162
Kapkiyai JJ, Karanj KN, Quresh JN, Sruthson P, Woomer PL (1999) Soil organic matter and nutrient dynamic in a Kenjan Nitisol under long-term fertilizer and organic input management. Soil Biol Biochem 31:1773–1782
Kern JS, Johnson MG (1993) Conservation tillage impacts on national soil and atmospheric carbon levels. Soil Sci Soc Am J 57:200–210
Kihani KN, More DA (1984) Long-term effect of tillage operation and farmyard manure application on soil properties and crop yield in a Vertisol. J Indian Soc Soil Sci 32:392–393
Kumar S, Datta R, Sinha S, Kojima T, Katoh S, Mohan M (2001) Carbon stock, afforestation and acidic deposition: an analysis of inter-relation with reference to arid areas. Water Air Soil Pollut 130:1127–1132

Lahav I, Steinberger Y (2001) The contribution of stone cover to biological activity in the Negev Desert, Israel. Land Degrad Dev 12:35–43
Lal R (1981) Soil erosion problem on Alfisol in Western Nigeria VI Effect of erosion on experimental plots. Geoderma 25:215–230
Lal R, Kimble J, Follett R, Stewart BA (2000) Assessment methods for soil carbon. CRC, Pub, Boca Raton
Lal R (2001) World cropland soils as a source or sink for atmospheric C. Adv Agron 71:145–174
Lal R (2002a) Soil carbon dynamics in crop land and rangeland. Environ Pollut 116:353–362
Lal R (2002b) Carbon sequestration in dryland ecosystems of West Asia and North Africa. Land Degrad Dev 13:45–59
Lal R (2003a) Global potential of soil C sequestration to mitigate the greenhouse effect. Crit Rev Plant Sci 22:151–184
Lal R (2003b) Soil erosion and the global carbon budget. Environ Int 29:437–450
Lal R (2004a) Carbon sequestration in dryland ecosystems. Environ Manag 33:528–544
Lal R (2004b) Soil carbon sequestration impacts on global climate change and food security. Science 304:1623–1627
Lal R, Hassan HM, Dumanski J (1999) Desertification control to sequester C and mitigate the greenhouse effect. In: Rosenberg NJ, Izaurralde RC, Malone EL (eds) Carbon sequestration in soils: science monitoring and beyond. Battelle Press, Columbus, pp. 83–151
Le Houreou HN (2000) Utilization of fodder trees and shrubs in the arid and semi-arid zones of West Asia and North Africa. Arid Soil Res Rehabil 14:101–135
Matar A, Torrent J, Ryan J (1992) Soil and fertilizer phosphorus and crop responses in the dryland Mediterranean zone. Adv Soil Sci 18:82–146
Mathan KK, Sankaran K, Kanakabushan K, Krishnamoorthy KK (1978) Effect of continuous rotational cropping on the organic carbon and total nitrogen contents in a black soil. J Indian Soc Soil Sci 26:283–285
Miglierina AM, Iglesias JO, Landriscini MR, Galantini JA, Rosell RA (2000) The effects of crop rotation and fertilization on wheat productivity in the Pampean semiarid region of Argentina. 1. Soil physical and chemical properties. Soil Tillage Res 53:129–135
Murillo JM, Moreno F, Pelegrin F, Fernandez JE (1998) Response of sunflower to traditional and conservation tillage in southern Spain. Soil Tillage Res 49:223–241
Nagarajan M, Sundaramoorthy S (2000) Effect of *Prosopis cineraria* (Linn.) druce on microbial biomass and soil C, N in arid agroforestry systems of western Rajasthan, India. Ann Arid Zone 39:431–438
Nash MH, Whitford WG (1995) Subterranean termites-regulators of soil organic matter in the Chihuahuan Desert. Biol Fertil Soils 19:15–18
Niles JO, Brown S, Pretty JN, Ball A, Fay J (2002) Potential carbon mitigation and income in developing countries from changes in use and management of agricultural and forest lands. Phil Trans R Soc Ser A 360:1621–1639
Noin D, Clarke JI (1997) Population and environment in arid regions of the world. In: Clarke J, Noin D (eds) Population and environment in arid regions, MAB/UNESCO, vol 19. The Pantheon Publishing Group, New York, pp. 1–18
Ojima DS, Smith MS, Beardsley M (1995) Factors affecting carbon storage in semi-arid and arid ecosystems. In: Squires VR, Glenn EP, Ayoub AT (eds) Combating global climate change by combating cand degradation. The United Nations Environment Programme, Nairobi, pp. 93–115
Oweis T, Pala M, Ryan J (1998) Stabilizing rain-fed wheat yields with supplemental irrigation and nitrogen in a Mediterranean-type climate. Agron J 90:672–681
Pascual J, Hernandez T, Garcia C, Garcia A (1998) Changes in the organic matter mineralization rates of an arid soil after amendment with organic wastes. Arid Soil Res Rehabil 12:63–72
Pedreno JN, Gomez I, Moral R, Mataix J (1996) Improving the agricultural value of a semi-arid soil by addition of sewage sludge and almond residue. Agric Ecosyst Environ 58:115–119

Petchawee S, Chaitep W (1995) Organic matter management in upland systems in Thailand. Australian Centre for International Agricultural Research, Canberra, pp. 21–26

Phillips RE, Blevins RL, Thomas GW, Frye WW, Phillips SH (1980) No-tillage agriculture. Science 208:1108–1113

Pieri C (1986) Fertilisation des cultures vivrières et fertilité des sols en agriculture paysanne sub-saharienne. Agron Trop 41:1–20

Polwson SD, Smith P, Coleman K (1998) European network of long-term sites for studies on soil organic matter. Soil Tillage Res 47:309–321

Potter KN, Jones OR, Torbert HA, Unger PW (1997) Crop rotation and tillage effects on organic carbon sequestration in the semi-arid southern Great Plains. Soil Sci 162:140–147

Pretty JN, Ball AS, Xiaoyun L, Ravindranath NH (2002) The role of sustainable agriculture and renewable resource management in reducing greenhouse gas emissions and increasing sinks in China and India. Phil Trans Roy Soc Ser A 360:1741–1761

Rasmussen PE, Albrecht SL, Smiley RW (1998) Soil C and N changes under tillage and cropping systems in semi-arid Pacific Northwest agriculture. Soil Tillage Res 47:197–205

Rasse DP, Rumpel C, Dignac MF (2005) Is soil carbon mostly root carbon? Mechanisms for a specific stabilisation. Plant Soil 269:341–356

Reicosky DC (1997) Tillage-induced CO_2 emission from soil. Nutr Cycl Agroecosyst 49:273–285

Reicosky DC, Kemper WD, Langdale GW, Douglas CL Jr, Rasmussen PE (1995) Soil organic matter changes resulting from tillage and biomass production. J Soil Water Conserv 50:253–261

Reynolds JF, Smith DMS (2002) Do humans cause deserts? In: Reynolds JF, Smith DMS (eds) Global desertification: do humans cause deserts? Dahlem University Press, Berlin, pp. 1–21

Rice CW, Owensby CE (2001) The effects of fire and grazing on soil carbon in range lands. In: Follett RF, Kimble JM, Lal R (eds) The potential of US grazing lands to sequester carbon and mitigate the greenhouse effect. CRC/Lewis Publishers, Boca Raton, pp. 323–341

Ringius L (2002) Soil carbon sequestration and the CDM: opportunities and challenges for Africa. Climate Change 54:471–495

Rodale. (2003). Farm Systems Trial, The Rodale Institute 611 Siegfriedale Road Kutztown, PA 19530-9320 USA

Roose EJ (1996) Land husbandry: components and strategy. FAO Soils Bull. 70, FAO, Rome

Rosegrant MW, Cline SA (2003) Global food security: challenges and policies. Science 302:1917–1919

Rostagno CM, Sosebal RB (2001) Surface application of biosolids in the Chihuahuan Desert: effects on soil physical properties. Arid Land Res Manag 15:233–244

Ryan J (1997) Change in organic carbon in long-term rotation and tillage trials in northern Syria. In: Lal R, Kimble JM, Follett RF, Stewart BA (eds) Management of carbon sequestration in soil. CRC/Lewis Publishers, Boca Raton, pp. 285–296

Ryan J, Masri S, Garabet S, Harris H (1997) Changes in organic matter and nitrogen with a cereal legume rotation trial. In: Ryan J (ed) Accomplishment and future challenges in dryland soil fertility research in the Mediterranean area. International Center for Agricultural Research in the Dry Areas, Aleppo, pp. 79–87

Sanchez PA (2002) Soil fertility and hunger in Africa. Science 295:2019–2020

Sauerbeck DR (2001) CO_2 emissions and C sequestration by agriculture – perspectives and limitations. Nutr Cycl Agroecosyst 60:253–266

Schimel DS, House JI, Hubbard KA, Bousquet P, Ciais P, Peylin P, Braswell BH, Apps MJ, Baker D, Bondeau A, Canadell J, Churkina G, Cramer W, Denning AS, Field CB, Friedlingstein P, Goodale C, Heimann M, Houghton RA, Melillo JM, Moore B, Murdiyarso D, Noble I, Pacala SW, Prentice IC, Raupach MR, Rayner PJ, Scholes RJ, Steffen WL, Wirth C (2001) Recent patterns and mechanisms of carbon exchange by terrestrial ecosystems. Nature 414:169–172

Schomberg HH, Jones RR (1999) Carbon and nitrogen conservation in dryland tillage and cropping systems. Soil Sci Soc Am J 63:1359–1366

Schuman GE, Janzen HH, Herrick JE (2002) Soil carbon dynamics and potential carbon sequestration by rangelands. Environ Pollut 116:391–396
Scurlock JMO, Hall DO (1998) The global carbon sink: a grassland perspective. Glob Chang Biol 4:229–233
Shahin RR, El-Meleigi MA, Al-Rokiba AA, Eisa AM (1998) Effect of wheat stubble management pattern on organic carbon content and fertility of sandy soils under pivot irrigation system. Bull Fac Agric Univ Cairo 42:283–296
Silver WL, Ostertag R, Lugo AE (2000) The potential for carbon sequestration through reforestation of abandoned tropical agricultural and pasture lands. Restor Ecol 8:394–407
Singh Y, Chaudhary DC, Singh SP, Bhardwaj AK, Singh D (1996) Sustainability of rice-wheat sequential cropping through introduction of legume crops and green manure crops in the system. Ind J Agron 41:510–514
Six J, Elliott ET, Paustian K (2000) Soil macroaggregate turnover and microaggregate formation: a mechanism for C sequestration under no-tillage agriculture. Soil Biol Biochem 32:2099–2103
Six J, Feller C, Denef K, Ogle S, de Moraes Sa JC, Albrecht A (2002) Soil organic matter, biota and aggregation in temperate and tropical soils - Effects of no-tillage. Agronomie 22:755–775
Skjemstad JO, Catchpoole VR, LeFeuvre RP (1994) Carbon dynamics in Vertisols under several crops as assessed by natural abundance 13C. Aus J Soil Res 32:311–321
Smith WN, Desjardins RL, Grant B (2001a) Estimated changes in soil carbon associated with agricultural practices in Canada. Can J Soil Sci 81:221–227
Smith P, Goulding KW, Smith KA, Powlson DS, Smith JU, Falloon P, Coleman K (2001b) Enhancing the carbon sink in European agricultural soils: including trace gas fluxes in estimates of carbon mitigation potential. Nutr Cycl Agroecosyst 60:237–252
Smith SV, Renwick WH, Buddemeier RW, Crossland CJ (2001c) Budgets of soil erosion and deposition for sediments and sedimentary organic carbon across the conterminous United States. Glob Biogeochem Cycles 15:697–707
Stewart BA, Robinson CA (2000) Land use impact on carbon dynamics in soils of the arid and semi-arid tropics. In: Lal R, Kimble JM, Stewart BA (eds) Global climate change and tropical ecosystems. CRC/Lewis Publishers, Boca Raton, pp. 251–257
Taboada MA, Micucci FG, Cosentino DJ, Lavado RS (1998) Comparison of compaction induced by conventional and zero tillage in two soils of the rolling pampa of Argentina. Soil Tillage Res 49:57–63
Thomson EF, Bahhady FA (1995) A model-farm approach to research on crop-livestock integration. I. conceptual framework and methods. Agric Syst 49:1–16
Unger PW (1991) Organic matter, nutrient and pH distribution in no- and conventional tillage semi-arid soils. Agron J 83:186–190
West TO, Marland G (2002) Net carbon flux from agricultural ecosystems: methodology for full carbon cycle analyses. Environ Pollut 116:439–444
Whitehouse MJ, Littler JW (1984) Effect of pasture on subsequent wheat crops on a black earth soil of the Darling Downs. II. Organic carbon, nitrogen and pH changes Queensland. J Agri Anim Sci 41:13–20
Whitford WG (1996) The importance of the biodiversity of soil biota in arid ecosystems. Biodivers Conserv 5:185–195
Woomer PL, Martin A, Albrecht A, Resck DVS, Scharpenseel HW (1994) The importance of management of soil organic matter in the tropics. In: Woomer PL, Swift MJ (eds) Biological management of tropical soil fertility. Wiley, Chichester, pp. 74–80
Yamoah CF, Bationo A, Shapiro B, Koala S (2002) Trend and stability analysis of millet yields treated with fertilizer and crop residues in the Sahel. Field Crop Res 75:53–62
Zaady E, Yonatan R, Shachak M, Perevolotsky A (2001) Effects of grazing on abiotic and biotic parameters in a semi-arid ecosystem: a case study from the Northern Negev Desert, Israel. Arid Land Res Manag 15:245–261

Application of Microbiology in Dryland Agriculture

Minakshi Grover, B. Venkateswarlu, S. Desai, K.A. Gopinath, and Cherukumalli Srinivasa Rao

1 Introduction

Microorganisms are important components of a soil ecosystem and play a key role in belowground biogeochemical cycling (Rousk and Bengtson 2014). In addition to exchanging nutritional benefits with plants, microorganisms assist plants to survive under various environmental conditions such as biotic and abiotic stresses using direct and indirect mechanisms. Thus, vibrant microbial communities interacting with plants act as an 'extended genotype' of the host, with exploitable implications.

Some beneficial microbes have been used in agriculture for a long time; for example, rhizobia–organisms that inhabit the roots of leguminous plants and provide biologically-fixed nitrogen–were described by Martinus Beijerinck in 1888. The farmer practice of moving soil from a healthy field to a sick field led to research that proved that 'disease suppression' in soil was due to live microorganisms. A few plant-associated microbes have been extensively studied and exploited for agricultural benefit. Significant progress has been made in the last few years in unraveling the incredible diversity of the microbial world. In addition, the elaborate network of interactions among plants, microbes and other soil inhabitants is being extensively studied at molecular levels. Commercial applications of these discoveries are being used to support plant growth, health and productivity (Reid and Greene 2012).

Drylands constitute approximately 41 % of the Earth's terrestrial surface and support 38 % of its population (Maestre et al. 2015). These arid areas, including diverse ecosystems such as deserts, savannahs and tropical dry forests, are often

M. Grover (✉) • S. Desai • K.A. Gopinath • C. Srinivasa Rao
Central Research Institute for Dryland Agriculture, Santoshnagar, Hyderabad, 500059, India
e-mail: minigt3@yahoo.co.in

B. Venkateswarlu
Vasantrao Naik Marathwada Krishi Vidyapeeth, Parbhani, India

M. Farooq, K.H.M. Siddique (eds.), *Innovations in Dryland Agriculture*,
DOI 10.1007/978-3-319-47928-6_18

characterized by population growth, over-exploitation, drought and desertification that can reduce crop productivity. Microorganisms are essential members of these arid soil environments. They help in rock weathering and soil crust formation. Soil microbial communities are largely responsible for key ecosystem services including soil fertility and climate regulation (Maestre et al. 2015). Also, microorganisms have enormous metabolic potential and can adapt quickly to the rapidly-changing environmental conditions of moisture availability, temperature fluctuations, UV radiation, etc., making them suitable candidates for biotechnological applications.

Thus, the selection and enrichment of beneficial microorganisms in the vicinity of plants should help to enhance the fitness and productivity of crop species. The use of beneficial microorganisms as a low-cost and environmentally-friendly strategy for productivity enhancement is highly relevant in dryland agriculture. In this chapter, the interaction of soil microbes with key components of dryland agriculture and their potential benefits with functional aspects are discussed.

2 Soil Microbial Communities

According to William Whitman, there are 5×10^{30} prokaryotes on earth with 26×10^{28} in the top 8 m of soil. One gram of healthy soil may contain as much as 10^{8-9} bacteria, 10^{5-8} actinomycetes, 10^{5-6} fungi and 10^{3-6} microalgae indicating the abundance of microbial life. Further, 1 m^3 of soil may house hundreds of species of bacteria, actinomycetes, fungi and algae, indicating diversity within the same group of microorganisms. However, the estimated number of species is much higher than that described, with even fewer cultured in the laboratory. There may be 1.5 million species of fungi but only 5 % are described, and as many as one million species of bacteria but only about 5000 have been described (Tilak 2000). The two parameters, i.e. abundance and diversity, become important when evaluating the significance of microorganisms in the soil. While abundance may increase or decrease over short periods in response to weather, natural resources, ecology, cropping systems and crop husbandry, diversity is a more complex and stable attribute which reflects a state of near equilibrium. Diversity is more important in the understanding of the functional significance of microorganisms at a given site. Abundance can vary between soil types, seasons and land uses. Given the large fluctuations and unstable numbers, microbial biomass is often used as a more reliable parameter to assess abundance. In terms of biomass, fungi dominate in the soil followed by bacteria and actinomycetes. Total population and live biomass only reflect the status of the soil at a given point in time, without giving a clear picture of the living diversity as influenced by different land use practices over time. The greatest uncertainty in population counts reflects our inability to recover all of the organisms in the culture. Generally, about 5–10 % of the organisms in the soil can be recovered through normal viable counts. Even direct counting methods do not reflect the true composition of the soil population. Therefore, most researchers attempt to determine the culturable biodiversity (Venkateswarlu and Srinivasarao 2005). In recent years, with the development of novel molecular biology techniques, some reports on molecular biodiversity have used metagenomic approaches.

Microorganisms play an essential role in plant establishment through favorable activities that contribute to plant growth. These include (1) transformation of soil organic matter and hence nutrient cycling; (2) formation of stable soil structure; (3) plant growth regulation (through the production of phytohormones, enzymes and other growth regulators); (4) antibiosis; (5) induced resistance (against pests); and (6) induced tolerance (to various abiotic stresses) (Lynch and Bragg 1985; Glick 1995; Grover et al. 2011, 2015; Miransari 2013;). Several groups of organisms act both competitively and synergistically to mediate the above processes.

The functional diversity of microorganisms in stressed ecosystems has been extensively studied (Graham 1992; Zahran 1999; Venkateswarlu et al. 2008). The occurrence of *Rhizobium*, *Bradyrhizobium*, *Azotobacter*, *Azospirillum*, *Pseudomonas* and *Bacillus* has been reported in desert ecosystems, acid soils, saline and alkaline areas, and highly-eroded hill slopes in India (Grover et al. 2011). The diversity of bacterial communities associated with different plants grown under various arid regions has been reviewed recently (Soussi et al. 2015). Selection and deployment of these plant-associated bacterial assemblages are mediated by a combination of bio-pedo-agroclimatic conditions and the plant species/varieties. Diversity and functional redundancy of these microorganisms make them active in supporting plant growth under stressed environments (Soussi et al. 2015). However, proper management of soil microflora for agricultural sustainability is the major challenge.

3 Interactions Between Key Components of Dryland Agriculture and Soil Microbes

3.1 Nutrient Management

To date, 17 macro- and micro-nutrients are essential for optimal plant growth and development. Nutrient availability is an important factor affecting plant growth. Any imbalance in the soil due to the absence or deficiency of any of these nutrients can adversely affect plant growth and productivity. Soil biological communities play a key role in nutrient management by transforming nutrients in organic and inorganic forms. In this transformation process, soil microorganisms make usable (inorganic) forms of different nutrients available to plants. Soil organisms also store nutrients in the soil as organic compounds. The different kinds of transformations performed by soil microorganisms are discussed below.

3.1.1 Decomposition

Decomposition is the physical breakdown and biochemical transformation of complex organic molecules of dead material into simpler organic and inorganic molecules. Bacteria and fungi, which comprise >90 % of the soil microbial biomass, are responsible for most of the decomposition. Warm and moist conditions hasten the process. Humic acids, produced by many microorganisms during decomposition, form large molecules of stabilized organic matter in the soil. The stable humus is a key component for soil aggregate formation and provides increased surface area for nutrient adsorption thus increasing nutrient availability for plant uptake.

3.1.2 Mineralization

During mineralization, soil microbial activity converts organic matter into inorganic compounds which can be used by plants.

3.1.3 Immobilization

Immobilization is the reverse of mineralization, i.e. the conversion of inorganic compounds into organic compounds. Soil microorganisms (mainly bacteria and fungi) consume inorganic molecules and incorporate them into their biomass, which then become unavailable to plants. However, the soil organisms contribute significantly to the formation and maintenance of soil structure by producing exopolysaccharides, beneficial organic acids and enzymes. After the death and decay of organisms, the immobilized nutrients are mineralized and become available to plants.

3.1.4 Mineral Transformation

Mineral transformation converts one form of inorganic matter into another form. Nitrogen cycling involves mineral transformation by many soil microorganisms. Nitrogen-fixing bacteria (symbiotic and non-symbiotic) present in the nodules/soil convert large amounts of inert atmospheric nitrogen (N_2) to ammonia (NH_3) with the help of the enzyme nitrogenase. Ammonia is converted into nitrate (NO_3^-) by nitrifying bacteria. Both ammonia and nitrate are plant usable forms. However, nitrates can be quickly leached from the soil. Further, denitrifying bacteria convert nitrate into gases (N_2O) that are lost into the atmosphere. Several other soil microorganisms are involved in the mineral transformation of phosphorus, potassium, sulfur, iron, manganese and zinc. Photosynthetic bacteria and algae present in the top layer of soil can use the energy of sunlight and soil heat to convert organic matter and harmful gases into useful substances such as amino acids, nucleic acids, sugars and other metabolites.

Interestingly, the mechanisms described above are interlinked and contribute to the biogeochemical cycle. In dryland systems, the process of carbon recycling is influenced by natural processes as well as human interventions. Disturbing the natural resource base often leads to the loss of carbon from cultivable soils thus rendering them uncultivable. For instance, when land is not cultivated, the soil surface is exposed to extreme weather events such as intense rainfall events, extreme temperatures and high wind velocities, thereby reducing microbial diversity and increasing carbon loss.

The association of mycorrhizal fungi with plant roots improves the plant's absorption of soil phosphates and other nutrients. The mycorrhizae also increase the capacity of biological nitrogen fixers to fix more nitrogen. Soil microorganisms multiply and die at enormous rates and, in doing so, release a constant stream of nutrients available to plants. Plants then grow better, assimilate more energy and provide more food in the form of root exudates which attract more microbes. In this mutually-beneficial two-way process, the size of the microbial population is governed by the plant inputs. Soils with high carbon matter have high populations of soil microorganisms compared with soils with low carbon content (drylands). Low populations of microorganisms mean low nutrient turnover and reduced plant growth. In such situations, strategies to increase the carbon content of soils through organic inputs and management practices will be beneficial. In tests conducted by the Texas Plant and Soil Laboratory at Edinburgh, Texas, the mineral uptake of crops grown in soils with very high amounts of minerals such as phosphorus, calcium, magnesium, potassium, and sulfur was deficient; in a plant tissue and petiole test, plants had only 20 % of the necessary mineral uptake for good plant growth and health (http://www.geotexsupply.com/geotexsupply6). The addition of microbes increased plant mineral uptake by 60 % to 80 % compared with the pre-treatment test. The microbes were able to digest (solubilize) the minerals in the soil into the soil solution making them available for uptake by the plant.

Alfisols and Vertisols, two predominant soil types in dryland areas, are generally deficient in essential nutrients. However, the application of manures and mineral fertilizers to improve soil fertility significantly impacts microbial diversity. Many studies have shown the positive effect of organic manures on soil biological health. The use of chemical fertilizers (nitrogen, phosphorus and potassium: NPK) enhances crop yield but can alter soil properties and the functional diversity and enzymatic activities of microbial populations. Sharma et al. (1983) reported that the application of nitrogen fertilizers such as ammonium sulfate only increased the fungal population while FYM and NPK applications increased the populations of fungi, bacteria and actinomycetes. Long-term fertilizer trial plots in alfisols had more microbial species diversity in the FYM plots than the chemical fertilizer and control plots. While organic-manured plots had more population and species richness, the continuous application of NPK did not significantly change the diversity of fungi or bacteria compared with the control plots. Certain species such as *Chaetomium, Monilia, Trichoderma* and *Spicaria* were more frequently isolated in FYM plots than control or chemical fertilizer plots (Venkateswarlu and Srinivasarao 2005). A positive effect of organic inputs alone or in conjunction with chemical fertilizers has been reported on microbial biomass carbon (MBC), available soil moisture content and enzymatic activities in long-term field experiments involving different dryland cropping systems under the All India Coordinated Research Project on Dryland Agriculture (Srinivasarao et al. 2012). Bandick and Dick (1999) compared enzymatic activity in long-term residue utilization plots (RUP) in semi-arid eastern Oregon under a winter wheat (*Triticum aestivum* L.)–summer fallow supplemented with inorganic N, green manure and beef manure treatments. Soils under the organic treatment or cover crop had higher α-glucosidase, β-glucosidase, amylase, arylsulfatase, invertase and urease activities than inorganic N treatments due to greater C inputs and the subsequent stimulation of microbial activity.

Soil organic carbon (SOC) pools are important for maintaining soil productivity and reducing the net CO_2 loading in the atmosphere. In a 21-year field experiment conducted in inceptisol under subhumid tropical conditions in India, the impact of mineral fertilizers and manuring treatments on crop yield sustainability was assessed for a rice (*Oryza sativa* L.)–lentil (*Lens esculenta* Moench) cropping sequence. Application of farmyard manure (FYM) with and without mineral fertilizers increased C input and SOC concentration and SOC stock. The 100 % organic (FYM) treatment had significantly higher SOC and more C build up and C than the control treatment. The application of 50 % organic (FYM) + 50 % recommended dose of fertilizer (RDF) increased grain yields in rice and lentil. Thus, the application of FYM (or other organics) in conjunction with mineral fertilizers is essential to maintain and enhance SOC stock in rice-based cropping systems (Srinivasarao et al. 2012). Srinivasarao et al. (2014) conducted another long-term (18 years) field experiment involving a pearl millet (*Pennisetum Glaucum* L.)–cluster bean (*Cyamopsis tetragonoloba* L.)–castor (*Ricinus communis* L.) sequence on an Entisol in western India to examine the effects of chemical fertilizers and manuring on carbon pools in relation to crop productivity and C sequestration. The study revealed that use of chemical fertilizers combined with farmyard manure produced higher agronomic yields and reduced the rate of SOC depletion. Carbon pools were significantly correlated with SOC, which increased with the application of organic amendments.

The use of inorganic fertilizers in dryland areas over the years has not significantly improved crop productivity due to the inherent risk factors in dryland agriculture and poor economic base of dryland farmers (Venkateswarlu and Srinivasarao 2005). The adverse effect of the indiscriminate use of fertilizers on soil productivity and environmental quality is of concern. Products of biological origin can be advantageously blended to replace part of the energy-intensive and market purchased inputs. Biofertilizers or microbial inoculants are living-nutrient availability systems present in soil which provide available major and minor nutrients such nitrogen, phosphorous, potassium, zinc and copper to crops. These microbes help to fix atmospheric nitrogen, solubilize and mobilize phosphorous, translocate minor elements like zinc and copper, produce plant growth-promoting hormones, vitamins and amino acids and control plant pathogenic fungi, thereby helping to increase crop production and soil health.

3.2 Soil Management

Soil is the thin layer of fertile land that nourishes life on earth directly or indirectly. Plants depend on soil for physical support as well as water and nutrients. Under natural conditions, soil is formed through the activities of soil biota but is dependent on soil type, climatic conditions and vegetation type. Agricultural soils need continuous care and management to maintain soil health for sustainable production. Dryland soils are not as well developed as soils in humid regions. Most dryland soils are shallow and poor in organic matter. Organic carbon plays a key role in nutrient dynamics, water relations and the maintenance of biological and physical health of the soil and is thus directly correlated to soil productivity (Srinivasarao et al. 2013). Further, soils in dryland areas remain uncropped for long periods during the year leading to loss of nutrients. Therefore, the maintenance of soil fertility is a real challenge in dryland areas.

Soil structure, i.e. the arrangement of the solid material and the pores in between the solid material, is the key parameter to overall soil function. Pores constitute about 35–60 % of the soil volume. Tiny pores (capillaries) are normally filled with water while larger pores circulate air and supply oxygen to plant roots including soil biota. Thus, a stable and well-formed soil structure, particularly in an agricultural setting, is important for water and air movement, biological activity, root penetration, access to nutrients, and seedling emergence. A well-structured soil with stable aggregates not only prevents the collapse of pore spaces between aggregates during heavy, saturating rains but also reduces compaction. Furthermore, improved soil structure reduces soil erosion and surface runoff by absorbing substantial amounts of rain.

Soil microorganisms play an important role in the formation and maintenance of soil structure by producing exopolysaccharides and other beneficial organic acids and enzymes. Fungal hyphae, through the spatial extension of filaments, help in the formation of soil macro-aggregates and improve aggregate stability by producing extracellular polysaccharides. Mycorrhizal hyphae form channels between plant roots and soil for the acquisition of nutrients. The recalcitrant glycoprotein, glomalin, produced by mycorrhizal fungi, plays an important role in stabilizing aggregates (Wright and Upadhyaya 1996; Rillig et al. 2002). Extracellular polysaccharides form a protective capsule layer around the bacterial colonies. The clay particles along with mineral and decomposed organic matter are deposited on the polysaccharide layer which acts as glue to form soil aggregates. Soil microorganisms efficiently break down organic matter into stable humus, a key component of aggregate formation. Soil aggregates provide the surface area for nutrient adsorption thus increasing nutrient availability for plant uptake.

An active population of soil microbes promotes soil fertility and structure. Soils rich in organic matter have better structure than soils with low organic matter or those treated with conventional synthetic fertilizers. Most dryland soils are low in organic carbon due to the rapid oxidation process in dry regions. The important factors leading to a decline in soil fertility in tropical regions, especially dryland areas, are low biomass generation, loss of SOC, and erosion of the rich fertile surface. Therefore, management practices that augment soil organic matter and maintain it at a threshold level are needed for better soil health (Srinivasarao et al. 2011). Optimum levels of SOC can be managed by adopting the appropriate crop rotation, fertility management, balanced use of chemical fertilizers and organic amendments, and conservation practices. Continuous application of organic matter as farm compost, farmyard manure, and plant residues is needed to maintain/increase soil organic matter content.

Use of bioinoculants can accelerate the soil building process and promote beneficial soil microbe communities. Microbial inoculants are useful for improving all soil types in diverse cropping systems of different agro-climatic zones. The diversity and population of microbial communities have a major influence on soil characteristics. The lifecycle (multiplication and death) of microorganisms add organic matter to the soil. Microorganisms help in the release of many nutrients, previously bound to soil particles, in the soil solution for plant uptake by digesting and decomposing complex substances. The buildup of soil humus from microbial activity helps in soil aggregation. Improved organic matter (humus) content and soil aggregation help to improve the nutrient status and water holding capacity of soil. The long-term application of microorganisms can significantly improve the organic matter content of soil resulting in overall soil health improvement.

Soil management practices influence soil physical and chemical characteristics and bring about changes in the soil microbial community structure and function. Tillage is one such practice that has significant effects on soil biological properties. A lower soil fungal:bacterial biomass ratio has been observed in soils under intensive cultivation. No-tillage (NT) systems with reduced disturbance facilitate the establishment and maintenance of extensive fungal hyphal networks while intensive soil tillage mechanically disrupts soil aggregates and hyphal structure. Further, the extension of fungal hyphae in undisturbed soils can translocate N and other nutrients from the deeper soil inorganic pool to the C-rich surface residues thus bridging the soil–residue interface to improve the nutritional status and promote biomass (Srinivasarao et al. 2009). Govaerts et al. (2008) studied microbial populations in zero (ZT) and conventional tillage (CT) with and without residue retention in wheat and maize (*Zea mays* L). ZT and CT with residue retention had increased populations of total bacteria, fluorescent *Pseudomonas* and *Actinomycetes*. ZT without residue retention had very low populations of microflora while CT with residue removal had a predominance of total fungi. This study divided management practices into three groups: (1) ZT and CT with residue retention showing improved soil health, (2) CT without residue showing intermediate soil health, and (3) ZT without residue showing the worst soil health. These results indicated that zero tillage with residue removal is an unsustainable practice, therefore, ZT should be combined with an adequate level of residue retention. Chen et al. (2009) investigated changes in soil properties in a long-term tillage experiment in a winter wheat monoculture in a semi-arid, semi-humid, continental climate with mean annual precipitation of 555 mm. The effects on soil properties of CT with residue removal, shallow tillage with residue cover (ST) and NT with residue cover were investigated. The

ST and NT treatments had more SOC and total N stocks, macro-aggregate proportions and MBC than the CT treatment. Mohammadi et al. (2010) showed that an NT system increased MBC compared with other tillage systems. Many studies have indicated that NT practices can increase soil fungal populations while in others, NT stimulated both bacterial and fungal populations (Helgason et al. 2009). Caesar-TonThat et al. (2010) observed that an NT spring wheat system in the drylands of eastern Montana, USA, improved soil aggregation and aggregate stability and increased basidiomycete fungi and the proportion of soil-aggregating microorganisms. Frey et al. (1999) measured significantly higher organic matter C and N fractions and mean weight diameter (MWD) of water-stable aggregates in NT relative to CT in a long-term experiment. Fungal hyphal length and fungal biomass were significantly higher in NT than CT surface soil. However, bacterial abundance and biomass were not consistently influenced by tillage treatment. Further, the fungal biomass was not strongly related to soil texture, pH, aggregation or organic C and N fractions, but was positively related to soil moisture. The relationship between soil moisture and the degree of fungal dominance was due to the positive response of fungal biomass and lack of response of bacterial biomass to increasing soil moisture.

Soil enzymes are considered potential indicators of soil quality because of their relationship with soil biology and quick response to changes in agricultural management practices (Dick et al. 1996). Hence, they can be useful for monitoring the effects of soil management on soil health. Soil enzymes are primarily of microbial origin and catalyze all biochemical reactions so are an integral part of nutrient cycling in the soil (Mohammadi et al. 2011). Jin et al. (2009) suggested that the positive effects of conservation tillage practices on soil enzyme activities mainly result from the increased water availability. Mohammadi et al. (2011) indicated higher activities of acid and alkaline phosphatase and protease in the NT treatment than the CT treatment. Similarly, Mathew et al. (2012) reported higher soil carbon and nitrogen contents, viable microbial biomass and phosphatase activities in the long-term NT treatment than the CT treatment. The abundance of fungi, bacteria, arbuscular mycorrhizal fungi and *Actinobacteria* (as indicated by PLFAs biomarkers) was consistently higher in the NT surface soil. SOC was positively correlated with most of the PLFA biomarkers. However, community qPCR with 16S rRNA gene sequencing to examine the effects of long-term crop management practices (NT vs. CT) revealed that NT practices did not significantly alter bacteria or fungi relative to CT (Ng et al. 2012). These differences may be attributed to the different methods used by the researchers. This highlights the need to evaluate multiple methods for microbial analysis that complement the potential limitations of other methods to develop standardized protocols which will ensure uniformity in the communication of results. Besides, factors such as cropping system, soil type and timing of sample collection might influence the effects of management practices on microbial community structure.

These results indicate that soil management practices influence both physico-chemical and microbiological properties of soil. Shifts in microbial community structure are expected due to the temporal increase in the microbial niche and water retention or the reduced physical disturbance under conservation, reduced and NT practices. Further, these practices have other benefits such as a reduction in production costs, fuel consumption and machinery wear, which are considered more sustainable practices in rainfed farming areas.

3.3 Water Management

Dryland areas are characterized by deficit and erratic precipitation with large spatial and temporal variability. Ratios of evapotranspiration to precipitation are generally high. As a result, moisture in the rooting zone of soil is normally low for most of the year, thus limiting biological productivity in dryland areas (Srinivasarao et al. 2015). As per an estimate (http://www.fao.org/ag/ca/training_materials/cd27-english/fme/economic.pdf), soil moisture limits crop production in approximately three-quarters of the world's arable soils and is the main reason for low yields in the seasonally dry, semi-arid tropics and subtropics. Soil moisture management is, therefore, a key factor when trying to enhance agricultural production in dryland areas. Proper management of water determines the allocation of rainfall to the soil for primary production. The conservation of soil and water in dryland areas is inter-related. Dryland areas, therefore, need concerted effort and strategies to increase water infiltration into the soil by reducing runoff, minimizing evaporation rates and improving water availability and use efficiency in the root zone through improved soil management.

The capacity of soil to hold plant-available water depends on its texture, depth, structure (pore space), organic matter content and biological activity. Soils with low organic matter and shallow aerobic zones have a low water holding capacity and oxygen content, and are compacted. These conditions hamper root development and usually indicate low microbial activity. Depleted levels of organic matter have significant negative impacts on water use efficiency due to poor porosity and infiltration, local and regional water cycles, plant productivity, the resilience of agroecosystems and global carbon cycles (Wani et al. 2009). The abundance of soil microbes is directly proportional to organic matter content, both of which improve soil texture and structure. Thus, soils receiving large amounts of organic residues can support a larger microbial population. By controlling infiltration rates and water holding capacity, soil organic matter plays a vital function in buffering yields through climatic extremes and uncertainty that exist in dryland areas. The application of microorganisms also improves soil organic matter and soil structure leading to less clodding and crusting. Agricultural practices such as conservation agriculture that improve soil structure and soil biological parameters also improve soil moisture (Benites and Castellanos 2003).

3.4 Cropping Systems

Crops and cropping systems significantly influence the soil microbial population and diversity over time. Different workers have reported varying results on the effects of cropping systems on soil microbial communities. Certain crops may favor a particular group of microorganisms while others may not exhibit any specific effect. The potential of crop rotations to increase microbial biomass and activity is generally agreed. These effects may be mediated through plant root exudates, leaf leachates from standing biomass, and residues after crop harvest.

Active C fractions, such as MBC, respond quickly to management practices and better reflect changes in soil quality and productivity that alter nutrient dynamics. Systems with a short summer fallow are considered superior in maintaining soil organic matter quantity and activity probably due to reduced soil disturbance and greater C inputs (Schomberg and Jones 1999). Carter (1986) observed significantly higher soil microbial biomass carbon (SMBC) in continuous wheat compared with wheat–fallow. Similarly, Collins et al. (1992) observed increased SOC, soil organic nitrogen (SON), MBC and microbial biomass nitrogen (MBN) in long-term continuous cropping compared to crop–fallow rotations. MBC and MBN increased when the length of the fallow period decreased in five crop rotations [continuous spring wheat, spring wheat–fallow, spring wheat–lentil, spring wheat–spring wheat–fallow, and spring wheat–pea (*Pisum sativum* L.)–fallow]. Further, the presence of legumes, such as lentil and pea, in the crop rotation increased soil N fractions (Sainju et al. 2007). Venkateswarlu et al. (2007) examined the possibility of on-farm generation of legume biomass (horse gram; *Macrotyloma uniflorum* Lam.) Verdc.] by using off-season rainfall in two long-term field experiments involving sorghum (*Sorghum bicolor* L.) and sunflower. Annual incorporation of horse gram biomass (3.03–4.28 t fresh weight ha^{-1} $year^{-1}$) improved the soil properties and fertility status of the soil, which improved the yields of test crops. Biomass incorporation improved mean organic carbon (24 %) and MBC (28 %). Long-term biomass incorporation and fertilizer application resulted in a buildup of soil nutrients compared with fallow plots, resulting in a stable yield trend in sorghum over 10 years, whereas fertilizer application alone showed a declining trend. At the end of 10 years of incorporation, grain yield had increased by 28 and 18 %, respectively, in sorghum and sunflower (*Helianthus annuus* L.) compared with the fallow where no fertilizers were applied to rainy season crops. The incorporation effect was greater in plots receiving fertilizer. Hence, growing a post-rainy season legume crop with incorporation is a simple low-cost practice that even small and marginal farmers can adopt in semi-arid regions of India. Widespread adoption of this practice, at least in alternate years, can restore the productivity of degraded soils and improve crop yields.

Several studies have reported increased MBC under diversified crop rotations compared to monocropping. Soil MBC increased under a diversified crop rotation [spring wheat–barley (*Hordeum vulgaris* L.) hay–corn–pea] compared with a continuous spring wheat system (Sainju et al. 2012). Some studies have indicated that monocultures select for less diverse microbial communities. For example, wheat monocultures had less microbial diversity than a crop rotation with red clover or field peas indicating that legume-based crop rotations support the diversity of soil microbial communities and may affect the sustainability of agricultural ecosystems (Lupwayi et al. 1998). A long-term maize monoculture decreased the diversity of *R. leguminosarum* biovar *viciae* populations and favored a specific subgroup of genotypes. A shift in the distribution of the symbiotic genotypes within the populations under corn monoculture was also observed (Depret et al. 2004). Meriles et al. (2009) reported a decline in fungal (*Trichoderma* and *Gliocladium*) populations under continuous soybean (*Glycine max* L.) compared with corn–soybean and soybean–corn rotations. In a study which included a 12-year sorghum–castor–sorghum rotation and a 10-year continuous castor crop, a differential distribution of fungal genera was noted. More diversity was recorded in rotation plots than in monoculture. In the monoculture plots, one or two genera (mostly *Streptomyces*) predominated representing 80 % of the actinomycetes (Venkateswarlu and Srinivasarao 2005). These studies indicated the negative effects of the monoculture on soil biota over crop rotations.

Rotations including winter cover crops—sorghum grain and forage with winter rye (*Secale cereal* L.) and cotton (*Gossypium hirsutum* L.)–winter rye–grain sorghum—returned higher plant biomass, and other types of root exudates in the soil to promote microbial biomass and metabolic diversity as indicated by several enzymatic activities, than grain sorghum–cotton and continuous cotton after only 3 years (Acosta-Martínez et al. 2011). An evaluation of microbial communities in the top soil of dryland cropping systems (grain sorghum–cotton, cotton–winter rye–grain sorghum, and a rotation of forage sorghum with winter rye) under different tillage practices for 5 years identified lower fungal:bacterial ratios under cotton-based cropping systems than under forage sorghum–winter rye (Acosta-Martínez et al. 2010). Soil under forage sorghum–rye had higher population densities of *Bacteroidetes* and *Proteobacteria* but lower *Actinobacteria* compared to grain sorghum–cotton and cotton–winter rye–grain sorghum. The tilled soil had more *Chloroflexi*, *Gemmatimonadetes* and *Verrucomicrobiae* than the NT plots. This study demonstrated that differences in microbial communities are more affected by crop rotation than tillage management.

However, contrasting results on bacterial and fungal relative abundances and the dominant members of the soil microbial community were observed in two cropping systems (continuous wheat and sorghum–wheat–soybean rotation) when community qPCR with 16S rRNA gene sequencing was used (Ng et al. 2012). The qPCR assays revealed a reduced bacterial copy number in the crop rotation. Continuous wheat had more relative bacterial numbers and diversity than the rotation treatment. Continuous wheat had more abundant *Cyanobacteria* probably due to the longer unplanted period between crops when the soil surface was exposed to sunlight while the rotation treatment had more *Actinobacteria* probably due to the incorporation of more residues from the previous crop. However, this experiment only analyzed samples at one time point, when all treatments were planted in wheat. Therefore, analyzing samples over an extended period will provide a better understanding of the microbial dynamics in response to different cropping systems. Residue management can also impact soil quality and productivity by preventing soil erosion and improving soil organic matter. Spedding et al. (2004) studied the residue management effects on total microbial biomass and observed that the retention of crop residue increased MBC and MBN in maize monocultures. In another study, residue retention induced higher population counts of total bacteria, fluorescent *Pseudomonas* and *Actinomycetes* compared to residue removal (Govaerts et al. 2008). However, the effects of residue management on nutrient cycling, soil chemical, physical and biological properties, and crop production need to be investigated for successful integration.

Cropping systems clearly influence soil biological properties. The specific impacts of these practices on microbial community composition are not yet fully known. Further in-depth research on the impact of cropping systems on soil microbial communities and their functions is needed. In general, residue management practices such as minimized fallow period, crop rotation, multiple crop sequence and inclusion of fodder legumes/legumes, can help to restore the biological properties of soil in dryland systems.

4 Microbes in Crop Production

Biological resources such as biofertilizers, biocontrol agents, plant growth promoters and other microbial products are used to improve the soil fertility and reducing the dependence on chemical fertilizers and pesticides, form important components of integrated nutrient management and sustainable agricultural practices. They are cost effective and renewable source of plant nutrients. Several microorganisms with different functional aspects are being exploited as bioinoculants in dryland agriculture. They can be grouped based on their nature and function performed for the benefit of the host plant.

4.1 Biological Nitrogen Fixation

4.1.1 Symbiotic Nitrogen Fixation

Rhizobium/Bradyrhizobium is a widely-used biofertilizer relevant to some leguminous crops in dryland production systems. It forms nodules with legumes (symbiotic association) and fixes atmospheric nitrogen with the help of enzyme nitrogenase. *Rhizobium* inoculation can meet 80 % of the nitrogen requirement of legume crops and increase yields by 15–30 %. The nitrogen-fixing efficiency of some leguminous

Table 1 Potential N contribution of N-fixing legumes in Indian soil

Crop	Fertilizer N equivalent (kg ha^{-1})	
	N fixed (kg N ha^{-1} $year^{-1}$)	Residual effect in succeeding cereal crop
Alfalfa	100–300	–
Clover	100–150	83
Chickpea	26–63	60–70
Green gram	50–55	30
Groundnut	112–152	60
Guar	37–196	–
Lentil	35–100	18–30
Pea	46	20–32
Pigeonpea	68–200	20–49
Soybean	49–130	–

Source: Subba Rao (1988)

Table 2 Cross-inoculation groups of *Rhizobium*

Rhizobium species	Cross-inoculation group	Legume types
R. leguminosarum	Peas	*Pisum, Vicia*
R. phaseoli	Beans	*Phaseolus*
R. trifoli	Clover	*Trifolium*
R. meliloti	Alfalfa	*Melilotus, Medicago,Trigonella*
R. lupini	Lupini	*Lupinus, Ornithopus*
Bradyrhizobium japonocum	Soybean	*Glycine*
Bradyrhizobium spp.	Cowpea	*Cajanus, Vigna, Arachis*

Source: Modified from Majumdar (2011)

crops is shown in Table 1. Rhizobia are highly specific when forming nodules in legume plants, and are classified into seven groups, known as cross-inoculation groups (Table 2). Successful nodulation depends on the availability of a compatible strain for a particular legume. Native rhizobial populations play a critical role in the success of inoculated bacteria as the increasing population density of native rhizobia may hinder the persistence of the inoculated strain. Therefore, the inoculated strain must be efficient enough to compete with the indigenous rhizobia and to survive and fix nitrogen in the soil environment. In most rainfed areas, legumes of the cowpea miscellany group have been grown for hundreds of years, and the populations of native rhizobia are generally adequate both in alfisols and vertisols despite wide seasonal variation (Venkateswarlu 1992; Hegde 1994). Inoculation benefits of rhizobial strains on some leguminous crops have been studied under the All India Coordinated Research Projects on Pulses, Groundnut (*Arachis hypogaea*), Soybean and Dryland Farming. The beneficial effects of rhizobial inoculation on different crops under dryland conditions are summarized in Table 3.

Table 3 Beneficial effects of bio-inoculants in dryland agriculture

Microbial inoculant	Crop	Effects	References
Rhizobium	Common bean (*Phaseolus vulgaris* L.)	Increased nodule dry mass, shoot dry mass, seed yield and N content	Nleya et al. (2001)
Rhizobium	Gum-arabic (*Acacia senegal*)	Enhanced gum yield	Faye et al. (2006)
Rhizobium	Groundnut	Increased symbiotic and yield traits	Ashraf et al. (2006)
Rhizobium	Chickpea (*Cicer arietinum* L.)	Enhanced nodulation and yield	Khattak et al. (2006)
Rhizobium	Mash bean (*Vigna mungo* L.)	Increased grain yield and biological yield	Ahmed et al. (2012)
Rhizobium	Mungbean (*Vigna radiata* L.)	Enhanced plant growth, nodulation and yield	Tripathi et al. (2012)
Rhizobium	Soybean	Increased plant dry matter, N uptake and pod yield	Patra et al. (2012)
Rhizobium	Snap bean (*Phaseolus vulgaris* L.)	Enhanced yield by 18 %	Beshir et al. (2015)
Asymbiotic nitrogen fixers			
Azospirillum, Azotobacter	Corn	Increased plant height, shoot and seed dry weight, ear dry weight, and length and number of seeds per row. Increased nutrient uptake (N, P, K, Fe, Zn, Mn and Cu)	Biari et al. (2008)
Azospirillum brasilense	Corn	Increased shoot dry weight and grain yield	Woodard and Bly (2000)
Azospirillum brasilense INTA Az-39	Wheat	Vigorous vegetative growth, increased dry matter accumulation, number of harvested grains (6.1 %) and grain yield (8.0 %)	Diáz-Zorita and Fernández-Canigia (2009)
Azospirillum brasilense	Wheat	Enhanced agronomic performance and productivity	Piccinin et al. (2011)
Azotobacter	Safflower	Increased yield by 35 %	Soleymanifar and Sidat (2011)
Azospirillum	Safflower	Increased yield by 21 %	Soleymanifar and Sidat (2011)
Azospirillum	Sorghum	Increased grain yield, water use efficiency over control	Patil (2014)

(continued)

Table 3 (continued)

Microbial inoculant	Crop	Effects	References
Azospirillum, Azotobacter, Klebsiella	Pearl millet	Increased total root and shoot N	Tiwari et al. (2003)
Azotobacter	Pearl millet	Increased grain yield	Venkataraman and Tilak (1990)
A. chroococcum Mac 68	Pearl millet	Increased grain yield by 17.9 %	Sangwan et al. (2011)
Azotobacter strain Azo-8	Wheat	Increased plant fresh and dry biomass, seed weight, grain yield	Singh et al. (2013)
Biocontrol agents			
Pseudomonas, Burkholderia	Moongbean (*Vigna radiata* L.)	Biocontrol of *Macrophomina phaseolina* (charcoal rot). Improved seed germination, plant biomass and yield parameters such as number of pods and seeds, and grain yield	Minaxi and Saxena (2010)
Pseudomonas fluorescens (biocontrol agent)	Pearl millet	Biocontrol of *Sclerospora graminicola* (downy mildew). Enhanced germination, seedling vigour, plant height, leaf area, tillering capacity, seed weight and yield	Niranjan et al. (2004)
T. viride	Mungbean (*Vigna mungo* L.)	Significant reduction in root rot (*M. phaseolina*) incidence leading to yield increase	Leo Daniel et al. (2011)
Nutrient solubilizers, mobilizers			
Bacillus megatherium var. *Phosphaticum* (PSB)	Sugarcane	Improved plant-available P status in soil, enhanced tillering, stalk population and stalk weight, leading to enhanced cane yield (12.6 %)	Sundara et al. (2002)
PSB	Sorghum	Increased earhead weight, grain yield, P content and P uptake in root and grain	Appanna (2007)
AM fungi	Corn-fingermillet (*Eleusine coracana* L.) cropping system	Increased plant height, yield, P uptake	Shrestha et al. (2009)
AM fungi	Corn	Increased soil concentration of N, available P, K and organic C, growth and yield	Astiko et al. (2013)

(continued)

Table 3 (continued)

Microbial inoculant	Crop	Effects	References
PGPR			
Bacillus sp. RM-2	Cowpea (*Vigna unguiculata* L.)	Increased seed germination, shoot and root length, biomass, and leaf area. Increased number of pods and seeds, and grain yield	Minaxi et al. (2012)
PGPR	Spring wheat	Improved shoot morphology and photosynthesis, water use efficiency (WUE)	Zhu et al. (2014)
Pseudomonas fluorescen + *P. putida*	*Jatropha* (*Jatropha curcas*)	PGPR increased root length, primary and secondary lateral root numbers, root diameter, frequency of lateral root, and specific root length	Sumarsih and Haryanto (2012)
Combination of binoculants			
Azospirillum brasilense + *Pseudomonas striata* or *Bacillus polymyxa*	Sorghum	Increased grain and dry matter yields, and N and P uptake	Alagawadi and Gaur (1992)
AM fungi + *Rhizobium* + PGPR	A woody legume (*Anthvllis cytisoides* L.)	Improved plant growth and nutrient uptake	Requena et al. (1997)
Rhizobium + *Azotobacter* + PSB + FYM	Soybean	Increased oil content in grain, oil yield, protein content and protein yield	Singh et al. (2009)
Mesorhizobium ciceri + *Glomus intraradices*	Chickpea	Increased nodulation, yield, and P and N content in seed and shoots	Erman et al. (2011)
AM fungi + *Pseudomonas*	Sorghum	Plant growth promotion, increased nutrient uptake	Praveen Kumar et al. (2012)
Rhizobium + PSB	Chickpea	Improved grain yield, nodulation and growth parameters	Ahmad (2013)
AM fungi, *Azospirillium brasilense* and PSB	Finger millet	Increased plant height, dry weight of root and shoot, P and N uptake	Ramakrishnan and Bhuvaneswari (2014)
Rhizobium + PSB	Pigeonpea (*Cajanus cajan*)	Enhanced grain and stalk yield, soil available P and total P uptake by plant	Takate et al. (2014)
Pseudomonas + *Rhizobium*	Mung bean (*Vigna radiata* L.)	Enhanced growth promotion and nutrient uptake	Praveen Kumar et al. (2015)

4.1.2 Non-symbiotic Nitrogen Fixation

Among the heterotrophic free-living N_2-fixing bacteria, *Azotobacter* is the most-intensively investigated genera. *Azotobacter* is present in neutral and alkaline soils. *A. chroococcum* and *A. lipoferum* are the most commonly-occurring *Azotobacter* in arable soils. Apart from fixing atmospheric nitrogen, they produce biologically-active growth promoting substances such as IAA, gibberellins and B vitamins in culture media. The beneficial effects of *Azotobacter* inoculation have been reported on many crops including sugarcane (*Saccharum officinarum* L.), rice, tomato (*Lens esculentum* L.), onion (*Alium cepa* L.), mustard (*Brassica juncea*), sorghum, pearl millet, cotton and safflower (*Carthamus tinctorius* L.). Bacteria belonging to the genus *Azospirillum* form an associative symbiosis with many plants especially C_4 types. Apart from providing associative N fixation, several other plant-growth-promoting mechanisms have been described for *Azospirillum*, resulting in better nutrient and water use in inoculated crops. *Azospirillum* inoculants are recommended for sorghum, millets, maize, sugarcane, wheat, paddy rice, cotton, oilseeds, fruits and vegetables, flowers, spices and condiments, herbs, lawns and ornaments and trees. Multi-location trials in India have shown yield increases of 10–17 % in pearl millet and 7–31 % in sorghum after inoculation with *A. brasilense* (Subba Rao 1986). *Azospirillum* increased straw and grain yields of pearlmillet by 20 % and 15 %, respectively, on arid soils in Jodhpur (Venkateswarlu 1985).

The benefits of Azotobacter and Azospirillum inoculation on plant growth, nutrient uptake and yield parameters in dryland crops are summarized in Table 3. Many strains of plant-growth-promoting rhizobacteria (PGPR) including Pseudomonas, Bacillus, Burkholderia and Serratia can also fix atmospheric nitrogen and have potential application in dryland crops.

4.2 Plant-Growth-Promoting Rhizobacteria

PGPR inhabit the soil ecosystem (Kloepper et al. 1980) and are found in association with the roots of many plants, presumably due to the presence of high levels of exudates rich in nutrients. The effects of PGPR on plant growth can be mediated through different mechanisms including the production of plant hormones such as auxins, gibberellins and cytokinins, nutrient acquisition, suppression of phytopathogens by the production of siderophores, HCN, ammonia, antibiotics, volatile metabolites, etc. and the induction of systemic resistance and systemic tolerance (Glick 1995; Yang et al. 2009). A particular PGPR may affect plant growth through one or more of these mechanisms.

All nitrogen-fixing and phosphate-solubilizing bacteria are categorized as PGPR as well as other genera including bacteria such as *Arthrobacter, Bacillus, Burkholderia*, *Enterobacter*, *Klebsiella*, *Pseudomonas*, *Xanthomonas*, *Serratia*. These PGPR can be used as single inoculants, mixed inoculants or as coinoculants with other beneficial microorganisms (Saxena et al. 2006). Further, PGPR are not crop specific and can be used in a variety of crop plants across different agro-climatic conditions. Several reports are available on the use of PGPR in agriculture as compiled by Desai et al. 2012a. For dryland areas, *Bacillus* species are appealing PGPR candidates due to their inherent ability to produce stress-resistant endospores. Some reports on beneficial effects of PGPR as single inoculants or combination inoculants in dryland agriculture are included in Table 3.

4.3 *Nutrient Solubilizers and Mobilizers*

4.3.1 Phosphate, Potassium, Zinc Solubilization

Phosphate-solubilizing microorganisms (PSM) are a group of microbes (including bacteria, fungi and actinomycetes) capable of solubilizing inorganic phosphorus from insoluble sources. In alkaline and acidic soils, the availability of phosphorus is low as it is in dynamic equilibrium. PSMs can reverse this process by solubilizing the insoluble phosphorous through the production of organic acids and phosphatases. Soil bacteria belonging to genera *Pseudomonas* and *Bacillus* and fungi belonging to genera *Penicillium* and *Aspergillus* are the most common P-solubilizers. Other genera exhibiting P-solubilizing ability include *Rhizobium, Burkholderia, Achromobacter, Azospirillum, Azotobacter, Agrobacterium, Micrococcus, Aerobacter, Flavobacterium, Erwinia, Streptomyces* and *Micromonospora.* Also, algae such as *Cyanobacteria* and mycorrhiza have shown P solubilization activity (Sharma et al. 2013). PSMs when used with rock phosphate in corn fields could save about 50 % of the crop's requirement for phosphatic fertilizer, and increase grain yield (Yazdani et al. 2009). In rainfed field experiments (1991 and 1992) with soybean on Vertisol, inoculation with *Pseudomonas striata* increased seed and total dry matter yields and phosphorus contents in different plant parts at various stages. Seed treatment with *P. striata* further enhanced the yields and phosphorus content (%) of soybean when applied in conjunction with rock phosphate (Dubey 1996). Increased yields and enhanced avaialbility of P after inoculation with PSM has been reported in different crops including soybean, maize, pulses, wheat, vegetable and millet (Rodríguez and Fraga 1999; Khan et al. 2009). Rhizobacteria belonging to genera *Pseudomonas, Bacillus* and *Burkholderia* reportedly solubilize zinc. The microbes solubilize the metal forms using protons, chelated ligands and oxidoreductive systems present on the cell surface and membranes. Bacterial inoculation of maize with Zn-solubilizing *Pseudomonas* and *Bacillus* significantly increased zinc content in maize leaves (Praveen Kumar et al. 2013). Microorganisms like *Aspergillus niger, Bacillus extroquens* and *Clostridium pasteurianum* have reportedly solubilized potassium. A strain of *Azospirillum* has not only fixed nitrogen associatively but solubilized phosphorus and zinc (Desai et al. 2012b). Similarly, two rhizobacterial strains CRIDA-ZnKPSB1 and CRIDA-ZnKPSB3 able to solubilize multiple nutrients (P, Zn, K) have been identified as *Burkholderia cepacia* and *B. cenocepacia* at the Central Research Institute for Dryland Agriculture (data communicated for publication). The isolates enhanced plant growth and macro- and micro-nutrient uptake in maize plants. The use of multinutrient-solubilizing microorganisms is advantageous over single-nutrient solubilizers in terms of cost and application.

4.3.2 Arbuscular Mycorrhizal Fungi (AMF)

AMF are associated with most agricultural crops, and occur over a wide range of habitats from aquatic to desert environments. These fungi are obligate symbionts and belong to genera such as *Endogone, Glomus, Entrophospora, Gigaspora, Acaulospora* and *Scutellospora*. AMF have been associated with increased plant growth and the enhanced accumulation of plant nutrients, mainly phosphorus, zinc, copper and sulfur, through greater soil exploitation by mycorrhizal hyphae. Maximum root colonization and sporulation occurs in soils with low phosphorus. The beneficial effects of AMF inoculation on plant growth and yield have been reported in stressful environments. However, the response may vary with soil type, soil fertility and AMF culture (Nadeem et al. 2014). In dryland agriculture, many researchers have reported improved plant growth and nutrient acquisition in different crop plants due to mycorrhizal application (Table 3) singly or in combination with PGPR. There is considerable demand for this product from farmers, but the major constraint to upscaling the technology is the inability to produce clean, pure inoculum as they are obligate symbionts.

4.4 Biocontrol Agents

The role of microbial biocontrol agents for the control of plant pests is well documented. Biocontrol agents control the growth and spread of plant pathogens through various mechanisms including competition for space and nutrients, antibiosis through the production of antagonistic metabolites such as hydrogen cyanide, siderophores, ammonia, lytic enzymes and induced systemic resistance. *Trichoderma, Pseudomonas* and *Bacillus* are among the important genera exploited as biocontrol agents in agriculture. In addition, biocontrol properties of *Cyanobacteria* have been recently reported. In the current scenario of deteriorating soil health and emerging plant pests with the changing climate, the use of broad host-range biocontrol agents as an environmentally-friendly means to control plant pathogens is highly relevant. Leo Daniel et al. (2011) reported a significant reduction in root rot incidence and increased yield by inoculating rainfed mungbean (*Vigna mungo*) with *T. viride*. The positive effects of using biocontrol agents have been observed in other dryland crops (Table 3) indicating their potential application in dryland agriculture.

5 Microbes in Climate Change Adaptation and Mitigation

Dryland areas—with an aridity index (precipitation/potential evapotranspiration ratio) below 0.65—are among the most sensitive ecosystems to climate change. Dryland areas are expected to increase globally by 11–23 % by 2100 and experience increased aridity and reduced soil moisture (Maestre et al. 2015). Low rainfall and high temperatures are the characteristics of arid or semi-arid regions. Salinity and low temperatures (in temperate regions) are other stresses that influence dryland agriculture. Altered climatic conditions and anthropogenic activities are expected to exacerbate these problems in dryland areas. Different stress factors will significantly influence the performance of beneficial microorganisms. If the soil is healthy and biologically diverse, plants will have a higher chance of surviving stressful conditions. Under stress, plants or microorganisms need a stronger network to survive, so their interactions with each other evolve to cope with the stress conditions (Timmusk and Wagner 1999). Compared with higher organisms, microorganisms can adapt quickly to changing environments. Therefore, the isolation of stress-tolerant microorganisms from a stressed ecosystem can help in the selection of efficient strains to be used as bioinoculants for abiotic stress management in crop plants. Rasul et al. (2012) isolated abiotic stress tolerant rhizobial isolates nodulating pongamia (*Milatta pinnata*), from soils of dryland agroecosystems. The isolates could tolerate a wide pH range (4.0–10.0), salinity (3 % NaCl) and high temperature (45 °C) conditions, indicating their potential for application under stressed conditions. In the last decade, the number of reports on plant growth promotion by microorganisms under different abiotic stress conditions, such as drought, high and low temperature, salinity and flooding, has increased (Kohler et al. 2009; Ali et al. 2009; Sandhya et al. 2009; Grover et al. 2011, Kaushal and Wani 2016). The term Induced Systemic Tolerance (IST) has been proposed for PGPR-induced physical and chemical changes that result in enhanced tolerance to abiotic stress (Yang et al. 2009). Bacteria belonging to different genera including *Rhizobium, Bacillus, Pseudomonas, Pantoea, Paenibacillus, Burkholderia, Achromobacter, Azospirillum, Microbacterium, Methylobacterium, variovorax* and *Enterobacter*, viruses, fungi and mycorrhizae have improved the tolerance of host plants under different abiotic stress environments (Grover et al. 2011).

A variety of mechanisms has been proposed behind microbial-elicited stress tolerance in plants (Grover et al. 2011). Microorganisms produce phytohormones, solubilize and mobilize nutrients, compete with plant pathogens for nutrients and space, and antagonize plant pathogens. These properties directly or indirectly help plants to grow better under normal and stress conditions. The production of stimulatory phytohormones, such as indole acetic acid, gibberellins and some unknown determinants, by PGPR can increase root length, root surface area and the number of root tips, which enhance the uptake of nutrients to improve plant health under abiotic stress conditions (Egamberdieva and Kucharova 2009). The production of cytokinins and antioxidants by microorganisms results in abscisic acid (ABA) accumulation and the degradation of reactive oxygen species. Under stress conditions, the plant hormone ethylene endogenously regulates plant homeostasis reducing root and shoot growth. However, ACC-deaminase-producing bacteria can reduce the effect of stress ethylene by sequestering and degrading plant ACC (an immediate precursor for ethylene production) to produce nitrogen and energy. Thus, the application of ACC-deaminase-producing bacteria as bioinoculants can help to ameliorate the deleterious effects of stress-induced ethylene (Glick et al. 2007; Saleem et al. 2007). Inoculation with ACC-deaminase-containing bacteria induces longer roots which may increase the plant's water use efficiency due to enhanced uptake of water from deep soil (Zahir et al. 2008). Many soil microorganisms produce exopolysaccharides which bind clay particles and organic matter to form micro- and macro-aggregates. The application of exopolysaccharide (EPS)-producing microorganisms can help to improve soil structure thus increasing water and nutrient retention in the root zone (Sandhya et al. 2009). The accumulation of osmoprotectants has been correlated with abiotic stress tolerance in plants. Proline is a compatible solute that helps plants in different ways, e.g. maintaining osmotic turgor, stabilizing macromolecules, acting as a sink for carbon and nitrogen for later use, and free radical detoxification (Mohammadkhani and Heidari 2008). Similarly, *Rhizobium*-mediated trehalose accumulation has been related to abiotic stress tolerance in legumes (Figueiredo et al. 2008). Volatile organic compounds (VOCs) such as 2R, 3R–butanediol, salicylic acid (SA) and jasmonic acid emitted by microorganisms are reportedly involved in IST (Grover et al. 2011). A further role of RNA chaperones has been reported in abiotic stress tolerance. The expression of bacterial CSPs (CSP A and CSP B) improved the tolerance of transgenic rice, maize and arabidopsis plants to some abiotic stresses including cold, heat and water deficit resulting in improved yields under field conditions (Castiglioni et al. 2008). Inoculation with PGPRs improved plant growth, foliar potassium concentration and leaf relative water content under elevated CO_2 and drought (Alguacil et al. 2009). Under desert farming, PGP bacteria enhanced plant photosynthetic activity and biomass synthesis (up to 40 %) under drought stress (Marasco et al. 2012).

Thus, the selection and application of efficient microorganisms can be a useful strategy for alleviating the negative effects of climate-change-related abiotic stresses in plants and enhancing adaptation to climate change. Table 4 summarizes the role of microorganisms in abiotic stress management in crop plants.

6 Conclusions

Dryland agriculture faces multiple challenges e.g., soil moisture deficit stress, high and low temperatures, salinity, poor soil health and nutrient deficiency—which reduce crop productivity. The changing climatic conditions are bound to exacerbate these conditions. Further, with the expansion of dryland areas, the dependency on these areas to contribute towards food production is increasing. Therefore, climate

Table 4 Role of microorganisms in abiotic stress management

Microorganism	Crop	Stress	Effects	References
AM fungi	Sorghum	Drought	Improved water relation	Cho et al. (2006)
Pseudomonas fluorescens	Groundnut	Salinity	Enhanced ACC deaminase activity	Saravanakumar and Samiyappan (2007)
Pseudomonas putida P45	Sunflower	Drought	Improved soil aggregation due to EPS production	Sandhya et al. (2009)
Pseudomonas sp. AMK-P6	Sorghum	Heat	Induction of heat shock proteins and improved plant biochemical status	Ali et al. (2009)
P. fluorescens Pf1	Green gram	Drought	Increased activity of catalase and peroxidase	Saravanakumar et al. (2011)
Azospirillum spp.	Wheat	Drought	Synthesis of ACC deaminase	Arzanesh et al. (2011)
Bacillus spp.	Corn	Drought	Increased plant biomass, relative water content, RAS/RT ratio	Sandhya et al. (2011)
Bacillus cereus	Mung bean, chickpea, rice	Salinity	Enhanced antioxidants levels	Chakraborty et al. (2011)
Exiguobacterium, Pseudomonas, Pantoea, Serratia, and *Streptomyces*	Wheat	Cold stress	Synthesis of antifreezing protein, accumulation cryoprotectants	Mishra et al. (2012)
Streptomyces spp.	Wheat	Drought	Plant growth promotion	Yandigeri et al. (2012)
Bacillus spp.	Sorghum	Drought	Improved plant growth, enhanced chlorophyll and relative water content	Grover et al. (2014)
AMF	Spring wheat	Drought	RWC, WUE, P uptake plant biomass and yield	Zhang et al. (2015)

readiness is the key to the sustainability of dryland agricultural systems. Among the various strategies being developed, the promotion of microbial-based technologies is important as it has minimal input costs. Microorganisms are essential members of the arid soil environment. They are key players in biogeochemical cycling which makes nutrients readily-available to plants. They also promote plant growth through various other direct and indirect mechanisms. They influence the host plant's response to biotic and abiotic stress conditions thus helping in their survival under harsh conditions. However, there is a need to select abiotic stress adapted/tolerant microbial strains preferably from stressed ecosystems. A microbial strain with multiple PGP traits is preferred over strains with one or two PGP traits and, once introduced as bioinoculant in soil, must be able to colonize and efficiently promote plant growth under stressed ecosystems. Management practices such as conservation agriculture or reduced tillage, which favor better soil biological health, should be promoted. Diverse cropping systems, with minimum fallow periods and include legumes as rotation/mixed crops, improve the biological and nutritional status of the soil system. Further, the effects of various management practices on soil microbial communities need to be understood in dryland agroecosystems.

Acknowledgements The authors of this manuscript are thankful to Indian Council of Agricultural Research for providing support under AMAAS (Application of Microorganisms in Agriculture and Allied Sectors) project.

References

Acosta-Martínez V, Dowd SE, Bell CW, Lascano R, Booker JD, Zobeck TM, Upchurch DR (2010) Microbial community composition as affected by dryland cropping systems and tillage in a semiarid sandy soil. Diversity 2:910–931

Acosta-Martínez V, Lascano R, Booker JD, Zobeck TM, Upchurch DR (2011) Dryland cropping systems influence microbial biomass and enzyme activities in a sandy soil in a semiarid region. Biol Fertil Soils 47:655–667

Ahmad MJ (2013) Cumulative effect of *Rhizobium* and phosphate solubilizing bacteria on yield, nutrient uptake and growth parameters in chickpea under rainfed conditions. PhD thesis, University of Agriculture, Faisalabad

Ahmed ZI, Ansar M, Saleem A, Arif ZU, Javed HI, Saleem R (2012) Improvement of mash bean production under rainfed conditions by *Rhizobium* inoculation and low rates of starter nitrogen. Pak J Agric Res 25:154–160

Alagawadi AR, Gaur AC (1992) Inoculation of *Azospirillum brasilense* and phosphate-solubilizing bacteria on yield of sorghum [*Sorghum bicolor* (L.) Moench] in dryland. Trop Agric 69:347–350

Alguacil MD, Kohler J, Caravaca F, Roldán A (2009) Differential effects of *Pseudomonas mendocina* and *Glomus intraradices* on lettuce plants physiological response and aquaporin PIP2 gene expression under elevated atmospheric CO_2 and drought. Microb Ecol 58(4):942–951

Ali SZ, Sandhya V, Grover M, Rao LV, Kishore VN, Venkateswarlu B (2009) *Pseudomonas* sp. strain AKM-P6 enhances tolerance of sorghum seedlings to elevated temperatures. Biol Fertil Soils 46:45–55

Appanna V (2007) Efficacy of phosphate solubilizing bacteria isolated from vertisols on growth and yield parameters of sorghum. Res J Microbiol 2:550–559

Arzanesh MH, Alikhani HA, Khavazi K, Rahimian HA, Miransari M (2011) Wheat (*Triticum aestivum* L.) growth enhancement by *Azospirillum* sp. under drought stress. World J Microbiol Biotechnol 27:197–205

Ashraf M, Ahmad M, Bakush HM (2006) Efficacy of *Rhizobium* strains for groundnut inoculation under rainfed conditions. Pak J Agric Sci 43:3–4

Astiko W, Sastrahidayat IR, Djauhari S, Muhibuddin A (2013) The role of indigenous mycorrhiza in combination with cattle manure in improving maize yield (*Zea mays* L.) on sandy loam of Northern Lombok, eastern of Indonesia. J Trop Soil 18(1):53–58

Bandick AK, Dick RP (1999) Field management effects on soil enzyme activities. Soil Biol Biochem 31:1471–1147

Benites J, Castellanos A (2003) Improving soil moisture with conservation agriculture. LEISA Magazine June 2003, 1–7

Beshir HM, Walley FL, Bueckert R, Taran B (2015) Response of snap bean cultivars to *Rhizobium* inoculation under dryland agriculture in Ethiopia. Agronomy 5(3):291–308

Biari A, Gholami A, Rahmani HA (2008) Growth promotion and enhanced nutrient uptake of maize *(Zea mays* L.) by application of plant growth promoting rhizobacteria in arid region of Iran. J Biol Sci 8(6):1015–1020

Caesar-TonThat T, Lenssen AW, Caesar AJ, Sainju UM, Gaskin JF (2010) Effects of tillage on microbial populations associated to soil aggregation in dryland spring wheat system. Eur J Soil Biol 46:119–127

Carter MR (1986) Microbial biomass as an index for tillage-induced changes in soil biological properties. Soil Tillage Res 7:29–40

Castiglioni P, Warner D, Benson RJ, Anstrom DC et al (2008) Bacterial RNA chaperones confer abiotic stress tolerance to plants and improved grain yield in maize under water-limited conditions. Plant Physiol 147:446–455

Chakraborty U, Roy S, Chakraborty AP, Dey P, Chakraborty B (2011) Plant growth promotion and amelioration of salinity stress in crop plants by a salt-tolerant bacterium. Recent Res Sci Technol 3(11):61–70

Chen H, Hou R, Gong Y, Li H et al (2009) Effects of 11 years of conservation tillage on soil organic matter fractions in wheat monoculture in Loess Plateau of China. Soil Tillage Res 106:85–94

Cho K, Toler H, Lee J, Ownley BH, Stutz JC, Moore LM (2006) Mycorrhizal symbiosis and response of sorghum plants to combined drought and salinity stresses. J Plant Physiol 163(5):517–528

Collins HP, Rasmussen PE, Douglas CL (1992) Crop rotation and residue management effects on soil carbon and microbial dynamics. Soil Sci Soc Am J 56:783–788

Depret G, Houot S, Allard MR, Breuil MC, Nouaïm R, Laguerre G (2004) Long-term effects of crop management on *Rhizobium leguminosarum* biovar *viciae* populations. FEMS Microbiol Ecol 51:87–97

Desai S, Grover M, Leo Daniel AE, Praveen Kumar G, Mir Hassan AS (2012a) Exploiting plant growth promoting rhizomicroorganisms for enhanced crop productivity. In: Satyanarayana T, Johri BN, Prakash A (ed) *Microorganisms in sustainable agriculture and biotechnology*, pp 227–242

Desai S, Praveen Kumar G, Sultana U, Pinnisetty S, Mir Hassan AS, Leo Daniel AE, Reddy G (2012b) Potential microbial candidate strains for management of nutrient requirements of crops. Afr J Microbiol Res 6:3924–3931

Diáz-Zorita M, Fernández-Canigia MV (2009) Field performance of a liquid formulation of *Azospirillum brasilense* on dryland wheat productivity. Eur J Soil Biol 45:3–11

Dick RP, Breakwell DP, Turco RF (1996) Soil enzyme activities and biodiversity measurements as integrative microbiological indicators. In: Doran JW, Jones AJ (ed), Methods for assessing soil quality. Soil Science Society of America, Madison, pp 247–271

Dubey SK (1996) Response of soybean to rock phosphate applied with *Pseudomonas striata* in a typic chromustert. J Indian Soc Soil Sci 44(2):252–255

Egamberdieva D, Kucharova Z (2009) Selection for root colonizing bacteria stimulating wheat growth in saline soils. Biol Fertil Soil 45(6):563–571

Erman M, Demir S, Ocak E, Tufenk S, Oğuza F, Akkoprub A (2011) Effects of *Rhizobium*, arbuscular mycorrhiza and whey applications on some properties in chickpea (*Cicer arietinum* L.) under irrigated and rainfed conditions 1–Yield, yield components, nodulation and AMF colonization. Field Crop Res 122:14–24

Faye A, Sarr A, Lesueur R (2006) Effect of inoculation with rhizobia on the gum-arabic production of 10-year-old *Acacia senegal* trees. Arid Land Res Manag 20:79–85

Figueiredo VB, Burity H, Martinez CR, Chanway CP (2008) Alleviation of drought stress in the common bean (*Phaseolus vulgaris* L.) by co-inoculation with *Paenibacillus polymyxa* and *Rhizobium tropici*. Appl Soil Ecol 40:182–188

Frey SD, Elliott ET, Paustian K (1999) Bacterial and fungal abundance and biomass in conventional and no-tillage agroecosystems along two climatic gradients. Soil Biol Biochem 31:573–585

Glick BR (1995) The enhancement of plant growth by free-living bacteria. Can J Microbiol 41:109–117

Glick BR, Todorovic B, Czarny J, Cheng Z, Duan J, McConkey B (2007) Promotion of plant growth by bacterial ACC deaminase. Crit Rev Plant Sci 26:227–242

Govaerts B, Mezzalama M, Sayre KD, Crossa J et al (2008) Long-term consequences of tillage, residue management, and crop rotation on selected soil micro-flora groups in the subtropical highlands. Appl Soil Ecol 38:197–210

Graham PH (1992) Stress tolerance in *Rhizobium, Bradyrhizobium* and nodulation under adverse soil conditions. Can J Microbiol 38:475–484

Grover M, Ali Sk Z, Sandhya V, Rasul A, Venkateswarlu B (2011) Role of microorganisms in adaptation of agriculture crops to abiotic stresses. World J Microbiol Biotechnol 27:1231–1240

Grover M, Madhubala R, Ali Sk Z, Yadav SK, Venkateswarlu B (2014) Influence of *Bacillus* spp. strains on seedling growth and physiological parameters of sorghum under moisture stress conditions. J Basic Microbiol 54:951–961

Grover M, Maheswari M, Desai S, Gopinath KA, Venkateswarlu B (2015) Elevated CO_2: plant associated microorganisms and carbon sequestration. Appl Soil Ecol 95:73–85

Hegde SV (1994) Population of cowpea rhizobia in farmers' fields in southern Karnataka: Influence of cropping systems, locations and N-levels. In: Rupela OP, Kumar Rao JVDK, Wani SP, Johansen C (eds) Proceedings of the workshop on linking biological nitrogen fixation research in Asia, Patancheru, Andhra Pradesh, India, pp. 47–55

Helgason BL, Walley FL, Germida JJ (2009) Fungal and bacterial abundance in long-term no-till and intensive-till soils of the Northern Great Plains. Soil Sci Soc Am J 73:120–127

http://www.fao.org/ag/ca/training_materials/cd27-english/fme/economic.pdf visited on 08.09.2015

http://www.geotexsupply.com/geotexsupply6 visited on 05.20.2016

Jin K, Sleutel S, Buchan D, De Neve S, Cai DX, Gabriels D, Jin JY (2009) Changes of soil enzyme activities under different tillage practices in the Chinese Loess Plateau. Soil Tillage Res 104:115–120

Kaushal M, Wani SP (2016) Plant-growth-promoting rhizobacteria: drought stress alleviators to ameliorate crop production in drylands. Ann Microbiol 66:35–42

Khan AA, Jilani G, Akhtar MS, Naqvi SMS, Rasheed M (2009) Phosphorus solubilizing bacteria: occurrence, mechanisms and their role in crop production. J Agric Biol Sci 1(1):48–58

Khattak SG, Khan DF, Shah SH, Madani MS, Khan T (2006) Role of rhizobial inoculation in the production of chickpea crop. Soil Environ 25(2):143–145

Kloepper JW, Leong J, Teintze M, Schroth MN (1980) Enhancing plant growth by siderophores produced by plant growth-promoting rhizobacteria. Nature 286:885–886

Kohler J, Hernandez JA, Caravaca F, Roldan A (2009) Induction of antioxidant enzymes is involved in the greater effectiveness of a PGPR versus AM fungi with respect to increasing the tolerance of lettuce to severe salt stress. Environ Exp Bot 65:245–252

Leo Daniel AE, Praveen Kumar G, Desai S, Mir Hassan ASK (2011) *In-vitro* characterization of *Trichoderma viride* for abiotic stress tolerance and field evaluation against root rot disease in *Vigna mungo* L. J Biofertil Biopestici 2(3):1–5

Lupwayi NZ, WA R, GW C (1998) Soil microbial diversity and community structure under wheat as influenced by tillage and crop rotation. Soil Biol Biochem 30:1733–1741

Lynch JM, Bragg E (1985) Microorganisms and soil aggregate stability. Adv Soil Sci 2:133–171

Maestre FT, Delgado-Baquerizo M, Jeffries TC et al (2015) Increasing aridity reduces soil microbial diversity and abundance in global drylands. Proc Natl Acad Sci U S A 112:15684–15689

Majumdar D K (2011) Pulse crop production: principles and technologies. PHI Learning Pvt Ltd, New Delhi

Marasco R, Rolli E, Ettoumi B, Vigani G et al (2012) A drought resistance-promoting microbiome is selected by root system under desert farming. PLoS One 7(10):e48479. doi:10.1371/journal.pone.0048479

Mathew RP, Feng Y, Githinji L, Ankumah R, KS B (2012) Impact of no-tillage and conventional tillage systems on soil microbial communities. App Environ Soil Sci. doi:10.1155/2012/548620

Meriles JM, Gil SV, Conforto C et al (2009) Soil microbial communities under different soybean cropping systems: characterization of microbial population dynamics, soil microbial activity, microbial biomass, and fatty acid profiles. Soil Tillage Res:271–281

Minaxi, Saxena J (2010) Disease suppression and crop improvement in moong beans (*Vigna radiata*) through *Pseudomonas* and *Burkholderia* strains isolated from semiarid region of Rajasthan. BioControl 55(6):799–810

Minaxi NL, Yadav RC, Saxena J (2012) Characterization of multifaceted *Bacillus* sp. RM-2 for its use as plant growth promoting bioinoculant for crops grown in semiarid deserts. Appl Soil Ecol 59:124–135

Miransari M (2013) Soil microbes and the availability of soil nutrients. Acta Physiol Plant 13:3075–3084

Mishra PK, Bisht SC, Bisht JK, Bhatt JC (2012) Cold-tolerant PGPRs as bioinoculants for stress management. Bact Agrobiology: Stress Manag:95–118

Mohammadi K, Nabiollahy K, Aghaalikhani M, Khormali F (2010) Study on the effect of different tillage methods on soil physical properties & yield and yield components of wheat cultivars. Iran J Plant Prod 16(4):77–91

Mohammadi K, Heidari G, Khalesro S, Sohrabi Y (2011) Soil management, microorganisms and organic matter interactions: a review. Afr J Biotechnol 10(84):19840–19849

Mohammadkhani N, Heidari R (2008) Water stress induced by polyethylene glycol 6000 and sodium chloride in two maize cultivars. Pak J Biol Sci 11(1):92–97

Nadeem SM, Ahmad M, Zahir ZA, Javaid A, Ashraf M (2014) The role of mycorrhizae and plant growth promoting rhizobacteria (PGPR) in improving crop productivity under stressful environments. Biotechnol Adv 32:429–448

Ng JP, Hollister EB, Gonzalez-Chavez MDCA, Hons FM, Zuberer DA, Aitkenhead-Peterson JA, Loeppert R, Gentry TJ (2012) Impacts of cropping systems and long-term tillage on soil microbial population levels and community composition in dryland agricultural setting. ISRN Ecol http://dx.doi.org/10.5402/2012/487370

Niranjan S, Shetty NP, Shetty HS (2004) Seed bio-priming with *Pseudomonas fluorescens* isolates enhances growth of pearl millet plants and induces resistance against downy mildew. J Pest Manag 50:41–48

Nleya T, Walley F, Vandenberg A (2001) Response of four common bean cultivars to granular inoculant in a short-season dryland production system. Can J Plant Sci 81:385–390

Patil S (2014) Azosprillium based integrated nutrient management for conserving soil moisture and increasing sorghum productivity. Afr J Agric Res 9(23):1761–1769

Patra RK, Pant LM, Pradhan K (2012) Response of soybean to inoculation with rhizobial strains: effect on growth, yield, N uptake and soil N status. World J Agric Sci 8(1):51–54

Piccinin GG, Dan LGM, Braccini ALE, Mariano DC et al (2011) Agronomic efficiency of *Azospirillum brasilense* in physiological parameters and yield components in wheat crop. J Agron 10:132–135

Praveen Kumar G, Kishore N, Leo Daniel AE, Sk MHA, Rasul A, Desai S (2012) Evaluation of fluorescent *Pseudomonas* spp. with single and multiple PGPR traits for plant growth promotion of sorghum in combination with AM fungi. Plant Growth Regul 67:133–140

Praveen Kumar G, Leo Daniel AE, Desai S, Sk MHA (2013) Prospective zinc solubilising bacteria for enhanced nutrient uptake and growth promotion in maize (*Zea mays* L.). Int J Microbiol:Article ID 869697http://dx.doi.org/10.1155/2013/869697

Praveen Kumar G, Desai S, Leo Daniel AE, Pinisetty S (2015) Impact of seed bacterization with PGPR on growth and nutrient uptake in different cultivable varieties of green gram. Asian J Agric Res 9(3):113–122

Ramakrishnan K, Bhuvaneswari G (2014) Effect of inoculation of AM fungi and beneficial microorganisms on growth and nutrient uptake of *Eleusine coracana* (L.) Gaertn. (Finger millet). Internat Lett Nat Sci 8(2):59–69

Rasul A, Leo Daniel AE, Kumar GP, Grover M, Venkateswarlu B (2012) Characterization of Rhizobial isolates nodulating *Millettia pinnata* in India. FEMS Microbiol Lett 336(2):148–158

Reid A, Greene S (2012) How microbes can help feed the world. In: American Society for Microbiology (ed) A Report from the American Academy of Microbiology, p 36

Requena N, Jimenez I, Toro M, Barea JM (1997) Interactions between plant-growth-promoting rhizobacteria (PGPR), arbuscular mycorrhizal fungi and *Rhizobium* spp. in the rhizosphere of *Anthyllis cytisoides*, a model legume for revegetation in Mediterraneansemi-arid ecosystems. New Phytol 136:667–677

Rillig MC, Wright SF, Eviner VT (2002) The role of arbuscular mycorrhizal fungi and glomalin in soil aggregation, comparing effects of five plant species. Plant Soil 238:325–333

Rodríguez H, Fraga R (1999) Phosphate solubilizing bacteria and their role in plant growth promotion. Biotechnol Adv 17:319–339

Rousk J, Bengtson P (2014) Microbial regulation of global biogeochemical cycles. Front Microbiol. doi.org/10.3389/fmicb.2014.00103

Sainju UM, Caesar-TonThat T, Lenssen AW, Evans RG, Kolberg R (2007) Long-term tillage and cropping sequence effects on dryland residue and soil carbon fractions. Soil Sci Soc Am J 71:1730–1739

Sainju UM, Lenssen AW, Caesar-TonThat T (2012) Tillage, crop rotation, and cultural practice effects on dryland soil carbon fractions. Open J Soil Sci 2:242–255

Saleem M, Arshad M, Hussain S, Bhatti AS (2007) Perspective of plant growth promoting rhizobacteria (PGPR) containing ACC deaminase in stress agriculture. J Ind Microbiol Biotechnol 34:635–648

Sandhya V, Ali SZ, Grover M, Reddy G, Venkateswarlu B (2009) Alleviation of drought stress effects in sunflower seedlings by the exopolysaccharides producing *Pseudomonas putida* strain GAP-P45. Biol Fertil Soils 46:17–26

Sandhya V, Ali SZ, Grover M, Reddy G, Venkateswaralu B (2011) Drought–tolerant plant growth promoting *Bacillus* spp.: effect on growth, osmolytes, and antioxidant status of maize under drought stress. J Plant Interact 6:1–14

Sangwan PS, Suneja S, Kukreja K, Raj M (2011) Response of *Azotobacter chroococcum* in pearl millet under dryland farm conditions. Indian J Dryland Agric Res Develop 26:97–100

Saravanakumar D, Samiyappan R (2007) ACC deaminase from *Pseudomonas fluorescens* mediated saline resistance in groundnut (*Arachis hypogea*) plants. J Appl Microbiol 102:1283–1292

Saravanakumar D, Kavino M, Raguchander T, Subbian P, Samiyappan R (2011) Plant growth promoting bacteria enhance water stress resistance in green gram plants. Acta Physiol Plant 33:203–209

Saxena AK, Shende R, Minakshi (2006) Interactions among beneficial microorganisms. In: Mukhergy KG, Manoharachary C, Singh J (eds) Soil biology vol. 7: microbial activity in the Rhizosphere. Springer, Berlin/Heidelberg

Schomberg HH, Jones OR (1999) Carbon and nitrogen conservation in dryland tillage and cropping systems. Soil Sci Soc Am J 63:1359–1366

Sharma N, Srivastava LL, Mishra B (1983) Studies on microbial changes in soil as a result of continuous application of fertilizers, farmyard manure and lime. J Indian Soc Soil Sci 31:202–206

Sharma SB, Sayyed RZ, Trivedi MH, Gobi TA (2013) Phosphate solubilizing microbes: sustainable approach for managing phosphorus deficiency in agricultural soils. Springer Plus 2:587 http://www.springerplus.com/content/2/1/587

Shrestha G, Vaidya SG, Rajbhandari BP (2009) Effects of arbuscular mycorrhiza in the productivity of maize and fingermillet relay cropping system. Nepal J Sci Technol 10:51–55

Singh SR, Singh U, Singh JK (2009) Effect of bioinoculants and FYM on growth, yield and quality of soybean (*Glycine max*) under rainfed conditions of Kashmir valley. Ann Agric Res New Series 30:87–90

Singh NK, Chaudhary FK, Patel DB (2013) Effectiveness of Azotobacter bioinoculant for wheat grown under dryland conditions. J Environ Biol 34:927–932

Soleymanifar A, Sidat SA (2011) Effect of inoculation with bio-fertilizer in different nitrogen levels on yield and yields components of safflower under dry land conditions. American-Eurasian J Agric Environ Sci 11(4):473–477

Soussi A, Ferjani R, Marasco R, Guesmi A, Cherif H, Rolli E, Mapelli F, Ouzari HI, Daffonchio D, Cherif A (2015) Plant-associated microbiomes in arid lands: diversity, ecology and biotechnological potential. Plant Soil. doi:10.1007/s11104-015-2650-y

Spedding T, Hamel C, Mehuys G, Madramootoo C (2004) Soil microbial dynamics in maize-growing soil under different tillage and residue management systems. Soil Biol Biochem 36:499–512

Srinivasarao C, Chary GR, Venkateswarlu B, Vittal K, Prasad JVNS, Kundu S, Singh SR, Gajanan GN, Sharma RA, Deshpande AN, Patel JJ, Balaguravaiah G (2009) Carbon sequestration strategies in rainfed production systems of India. Central Research Institute for Dryland Agriculture, Hyderabad

Srinivasarao C, Venkateswarlu B, Dinesh Babu M, Wani SP, Dixit S, Sahrawat KL, Kundu S (2011) Soil health improvement with *Gliricidia* green leaf manuring in rainfed agriculture, on farm experiences. Central Research Institute for Dryland Agriculture, Hyderabad

Srinivasarao C, Venkateswarlu B, Lal R, Singh AK, Vittal KPR, Kundu S, Singh SR, Singh SP (2012) Long-term effects of soil fertility management on carbon sequestration in a rice–lentil cropping system of the Indo-Gangetic Plains. Soil Sci Soc Am J 76:168–178

Srinivasarao C, Venkateswarlu B, Lal R, Singh AK, Kundu S (2013) Sustainable management of soils of dryland ecosystems of India for enhancing agronomic productivity and sequestering carbon. Adv Agron 121:254–329

Srinivasarao CH, Venkateswarl B, Lal R, Singh AK, Kundu S, Vittal KPR, Patel JJ, Patel MM (2014) Long-term manuring and fertilizer effects on depletion of soil organic carbon stocks under pearl millet-cluster bean-castor rotation in western India. Land Degrad Dev 25:173–183

Srinivasarao Ch, Lal R, Prasad JVNS, Gopinath KA, Singh R, Jakkula VS, Sahrawat KL, Venkateswarlu B, Sikka AK, Virmani SM (2015) Potential and challenges of rainfed farming in India. In: Sparks DL (ed) Adv agron, Elsevier 133:113–181

Subba Rao MS (1986) Biofertilizers in agriculture. Oxford and IBM Publishing Co., New Delhi

Subba Rao NS (1988) *Rhizobium* inoculant, 2nd edn. Oxford/IBH Publishing Co. Pvt. Ltd., New Delhi

Sumarsih S, Haryanto D (2012) *Pseudomonas fluorescens* and *Pseudomonas putida* for promoting growth of *Jatropha curcas* seedling root. J Tropic Life Sci 2:53–57

Sundara B, Natarajan V, Hari K (2002) Influence of phosphorus solubilizing bacteria on the changes in soil available phosphorus and sugarcane and sugar yields. Field Crop Res 77:43–49

Takate AS, Tatpurkar Sheetal R, Shelke SR, Bhanvase DB, Pawar AB, Kadam JR (2014) Rainwater conservation techniques with integrated phosphorus management on productivity of pigeonpea under dryland conditions. Indian J Dryland Agric Res Develop 29:79–84

Tilak KVBR (2000) Conservation and utilization of microbial diversity for natural resource management. In: Yadav, JSP et al (eds) International conference on managing natural resources for sustainable agricultural production in the 21st century. New Delhi, pp 76–78

Timmusk S, Wagner EGH (1999) The plant growth-promoting rhizobacterium *Paenibacillus polymyxa* induces changes in *Arabidopsis thalianan* gene expression: a possible connection between biotic and abiotic stress responses. Mol Plant-Microbe Interact 12:951–959

Tiwari M, Parada S, Dadarwal KR (2003) Associative diazotrophs of pearl millet (*Pennisetum glaucum*) from semi arid region – isolation and characterization. Indian J Exp Biol 41:341–345

Tripathi PK, Singh MK, Singh JP, Singh ON (2012) Effect of rhizobial strains and sulphur nutrition on mungbean (*Vigna radiata* (L.) *wilczek*) cultivars under dryland agro–ecosystem of Indo–Gangetic plain. Afr J Agric Res 7(1):34–42

Venkataraman GS, Tilak KVBR (1990) Biofertilizers in sustainable agriculture. In: Kumar V, GC h, SV K (eds) Soil fertility and fertilizers. Vol. IV, nutrient management and supply system for sustainable agriculture in 1990s. Indian Farmers Fertilizers Cooperative Ltd, New Delhi

Venkateswarlu B (1985) Association of *Azospirillum* species with roots of pearl millet, Ph.D thesis. University of Jodhpur, Jodhpur

Venkateswarlu B (1992) Performance of biofertilizers in dryland agriculture – a critical review and future research needs. Indian J Dryland Agril Res Dev 7(2):124–132

Venkateswarlu B, Srinivasarao Ch (2005) Soil microbial diversity and the impact of agricultural practices. ENVIS Centre Microorganisms and Environment Management Newsletter Vol 3 http://www.envismadrasuniv.org/oct_news.htm

Venkateswarlu B, Srinivasarao C, Ramesh G, Vekateswarlu S, Katyal JC (2007) Effects of long-term legume cover crop incorporation on soil organic carbon, microbial biomass, nutrient build-up and grain yields of sorghum/sunflower under rain-fed conditions. Soil Use Manag 23:100–107

Venkateswarlu B, Desai S, Prasad YG (2008) Agriculturally important microorganisms for stressed ecosystems: Challenges in technology development and application. In: Khachatourians GG, Arora DK, Rajendran TP, Srivastava AK (eds) Agriculturally important microorganisms, vol 1. Academic World, Bhopal

Wani SP, Sreedevi TK, Rockström J, Ramakrishna YS (2009) Rain-fed agriculture-Past trend and future prospects. In: Wani SP, Rockström J, Oweis T (eds) Rain-fed agriculture: unlocking the potential, Comprehensive assessment of water management in agriculture series. CAB International, Wallingford

Woodard H, Bly A (2000) Maize growth and yield responses to seed-inoculated N_2-fixing bacteria under dryland production conditions. J Plant Nutr 23:55–65

Wright SF, Upadhyaya A (1996) Extraction of an abundant and unusual protein from soil and comparison with hyphal protein of arbuscular mycorrhizal fungi. Soil Sci 161:575–586

Yandigeri MS, Meena KK, Singh D, Malviya N, Singh DP, Solanki MK, Yadav AK, Arora DK (2012) Drought-tolerant endophytic actinobacteria promote growth of wheat (*Triticum aestivum*) under water stress conditions. Plant Growth Regul 68:411–420

Yang J, Kloepper JW, Ryu CM (2009) Rhizosphere bacteria help plants tolerate abiotic stress. Trends Plant Sci 14:1–4

Yazdani M, Bahmanyar MA, Pirdashti H, Esmaili MA (2009) Effect of phosphate solubilizing microorganisms (PSM) and plant growth promoting rhizobacteria (PGPR) on yield and yield components of corn (*Zea mays L.)*. Proc World Acad Sci Eng Technol 37:90–92

Zahir ZA, Munir A, Asghar HN, Arshad M, Shaharoona B (2008) Effectiveness of rhizobacteria containing ACC-deaminase for growth promotion of peas (*Pisum sativum*) under drought conditions. J Microbiol Biotechnol 18(5):958–963

Zahran HH (1999) *Rhizobium*-Legume symbiosis and nitrogen fixation under severe conditions and in an arid climate. Microbiol Mol Biol Rev 3(4):968–989

Zhang B, Chang SX, Anyia AO (2015) Mycorrhizal inoculation and nitrogen fertilization affect the physiology and growth of spring wheat under two contrasting water regimes. Plant Soil. doi:10.1007/s11104–015–2635–x

Zhu Y, Wang Z, Wang J, Wang Z, Zhou J (2014) Plant growth-promoting rhizobacteria improve shoot morphology and photosynthesis in dryland spring wheat. In: Lee G (ed) Biomedical engineering and environmental engineering. Information Engineering Research Institute, Texas City

Salinity in Dryland Agricultural Systems: Challenges and Opportunities

D.J. McFarlane, R.J. George, E.G. Barrett-Lennard, and M. Gilfedder

1 Introduction

Some soils and landscapes are naturally saline and have excess salts prior to their development. Such primary salinity has not been considered in this chapter. Secondary salinity results from changes in the hydrological regime driven by a combination of hydrological processes that operate across a range of physical scales and within a range of response time scales (e.g., Peck and Hatton 2003).

Secondary, or induced dryland salinity is a world-wide problem in agricultural systems that have poor internal- (soil profile) and external- (runoff) drainage, which allows salts to accumulate in the landscape and then be mobilised with changes in water balance. Salinity occurs in agricultural landscapes primarily for two reasons: (1) shallow groundwater is present, which leads to salt accumulation at the soil surface as a consequence of capillarity, and (2) because soils are sodic, which leads to the accumulation of the small amounts of salt that are always falling in rain (Rengasamy 2002; Barrett-Lennard et al. 2016). Secondary water-table induced salinity can be widespread in arid and semi-arid situations as a consequence of the leakage of water from irrigation systems, which causes groundwater rise (e.g.,

D.J. McFarlane
The University of Western Australia, Crawley, WA, Australia
e-mail: don.mcfarlane@csiro.au

R.J. George
Department of Agriculture and Food, Bunbury, WA, Australia

E.G. Barrett-Lennard
School of Plant Biology, The University of Western Australia, Crawley, WA, Australia

Department of Agriculture and Food, South Perth, WA, Australia

M. Gilfedder (✉)
CSIRO Dutton Park Laboratories, Brisbane, QLD, Australia
e-mail: mat.gilfedder@csiro.au

M. Farooq, K.H.M. Siddique (eds.), *Innovations in Dryland Agriculture*,
DOI 10.1007/978-3-319-47928-6_19

Pakistan – Qureshi and Barrett-Lennard 1998) or in situations where perennial vegetation is substituted by annual vegetation, which also causes groundwater rise (e.g. Australia – Peck 1978; Peck and Hatton 2003).

Soils can be considered salt affected if the presence of water-soluble salts affects normal plant growth (especially agricultural production), environmental values or economic welfare (Rengasamy 2006; FAO 2015). While the total affected area at any one site may be relatively small, the patchy nature of salt expressions at the land surface often disturbs the ability to farm surrounding areas making it a large overall problem. Increased salt load to rivers can also lead to downstream impacts.

Drivers for secondary salinity are a result of human-induced changes to the hydrological balance of a landscape, including large increases in recharge (e.g. irrigation water), or through removal of deep-rooted vegetation that used to intercept and/or transpire a higher proportion of incoming rain than the dryland agriculture system (dryland salinity).

A change in land use allows opportunity for increased recharge to underlying aquifers or can increase the sodicity of the soil as salts build up in the near-surface soil profile to the detriment of plants and microorganisms that form beneficial symbiotic associations with the plants. This imbalance can be caused by shallow-rooted annual crops and pastures replacing deep-rooted trees or perennial native grasses.

Not all dryland salinity is caused by groundwater. The salinity of the soil solution needs to be excessive for the plants that needs to grow, usually agricultural but native plant species can also be impacted by dryland salinity (George et al. 1995). The salinity of soil solutions increases as plants take up fresh water and this can result in it becoming too saline for plants, especially as soils dry out seasonally (termed 'transient salinity' by Rengasamy (2006)).

This chapter provides an overview of dryland salinity – its causes, forms and management options. It also examines how climate change may affect both its future extent and the viability of management and recovery options. Examples are drawn mainly from Australia, a continent with major dryland salinity problems, which has resulted in a body of research that may be useful for other parts of the world experiencing dryland salinity. However, there are types of dryland salinity that do not occur in Australia and so we have used other examples where available.

2 Worldwide Extent of Dryland Salinity

Saline soils have been identified in over 100 countries, in most continents and have been estimated to affect over 954.8 M hectares of land (Szabolcs 1989). The FAO (2015) have estimated the global extent of saline and sodic soils to be as shown in Table 1. About 19.5 % of irrigated land (45 M ha) was considered salt-affected but only 2.1 % of dryland agricultural soils was salt affected (32 M ha). By this estimate, Asia, the Pacific and Australia have almost half of the world's salt-affected soils and about two-thirds of the world's sodic soils. Up to a third of Australian soils are sodic,

Table 1 Area of salt-affected and sodic soils in the world in million of hectares

Regions	Total area	Saline soils	%	Sodic soils	%
Africa	1899.1	38.7	2.0	33.5	1.8
Asia and the Pacific and Australia	3107.2	195.1	6.3	248.6	8.0
Europe	2010.8	6.7	0.3	72.7	3.6
Latin America	2038.6	60.5	3.0	50.9	2.5
Near East	1801.9	91.5	5.1	14.1	0.8
North America	1923.7	4.6	0.2	14.5	0.8
Total	12781.3	397.1	3.1	434.3	3.4

Source: FAO (2015)

some in the natural condition and others through land management. More than 2 M ha of farmland in Australia were estimated to be affected by dryland salinity, with more than half in WA (Australian Bureau of Statistics 2002). Ghassemi et al. (1995) described irrigation and dryland salinisation in eleven countries but an updated version of such comprehensive reviews is still to be repeated. Their estimate of dryland salinity at the time was 31.2 M ha, which is a similar number to the FAO (2015).

3 Salinity, Waterlogging and Sodicity – Impacts on Plants

The severity of salinity can be classified according to the electrical conductivity of the soil saturation extract (ECe). Soils with ECe values of 0–2 dS m^{-1} are non-saline, 2–4 dS m^{-1} are slightly saline, 4–8 dS m^{-1} are moderately saline, 8–16 dS m^{-1} are highly saline, 16–32 dS m^{-1} are severely saline and >32 dS m^{-1} are extremely saline (Barrett-Lennard et al. 2008). However, it should be recognised that the impacts of salinity on crops is not affected by the concentration of salt in soil (of which ECe is a measure) but rather the concentration of salt in the soil solution: growth will therefore be strongly affected also by the soil moisture contents.

Salinity impacts the plants because electrolytes in the rooting medium have osmotic and specific ion effects (Greenway and Munns 1980). The osmotic effects occur because water movement across cellular membranes is proportional to the concentration gradient of electrolytes and organic solutes. If concentrations are higher on the outside than the inside of the plant, then the plant may wilt and die; if the reverse is true, then the plant can take up water and may grow. One strategy that plants can use to overcome the 'osmotic shock' caused by salinity is to increase concentrations of solutes within cells by either the synthesis or transport of osmotica inside the cells: this process is termed 'osmotic adjustment' and it enables water uptake to resume. Specific ion effects occur because metabolic processes generally require K^+ (and Na^+ inhibits the uptake of this ion) and because Na^+ and Cl^- have toxic effects on metabolism. For plants suddenly affected by salinity, osmotic effects are immediate whereas specific ion effects take some time to impact (Munns 1993).

There is a view that divalent ions (e.g., Mg^{2+}, Ca^{2+} and SO_4^{2-}) have less severe effects on plants than monovalent ions (e.g., Na^+ and Cl^-). These claims need to be carefully examined and may occur simply because soil salinity is widely measured based on the electrical conductivity of soil extracts, which is the product of the valency and concentration of ions. Solutions of monovalent and divalent salts at equal electrical conductivity would therefore differ twofold in concentration.

Waterlogging is a major potential hazard to susceptible commercial crops and grasses in saline environments because salinity is caused by the presence of a shallow watertable and/or by major decreases in the hydraulic conductivity of soil caused by sodicity. In each case seasonal rainfall can cause waterlogging, which results in decreases in soil oxygen concentrations within 1–2 days (Barrett-Lennard et al. 1986). The presence of oxygen is essential for energy production in roots (Barrett-Lennard 2003). In the absence of oxygen, energy production decreases by ~95 %, and there is a breakdown in the processes that ensure Na^+ and Cl^- are excluded and K^+ is taken up by plants. This decreases plant growth and yield (Barrett-Lennard 2003; Barrett-Lennard and Shabala 2013).

Boron is a micro-nutrient that can be present at toxic concentrations in soils affected by salinity, particularly in sodic soils affected by transient salinity (Rengasamy 2002, 2006). One important point is that tolerance to boron toxicity is not necessarily associated with tolerance to salinity. For example, barley is generally regarded as being more salt tolerant than wheat, but there are many moderately saline soils where wheat will be the preferred crop selection than barley because wheat is generally more tolerant to boron toxicity than barley. One way to increase crop production on salt lands may therefore be to introgress genes for B toxicity tolerance into crops (Barrett-Lennard and Setter 2010).

4 Dryland Salinity

Three processes combine to cause dryland salinity; salt accumulation in the regolith, the mobilisation of these salt stores, and their expression at the land surface and streams.

4.1 Salt Accumulation

Salt accumulation is typically a slow process, operating over millennial time-scales, with evidence that salinisation has occurred in phases in response to the onset of aridity across the last 1–2 Ma (George et al. 2008). Depending on the hydraulic conductivity of the soil the accumulated salt can be at shallow soil depths (e.g. within the root-zone of annual crops and pastures – Barrett-Lennard et al. (2016)) or in more permeable soils, at considerably greater depths (> 10–100 m) in the soil profile where it can be mobilised by groundwater. The four main sources of salt in dryland agricultural areas are rainfall, wind deposition, seawater left over after land has emerged from below sea level, and the weathering of rocks. Marine influences and rock weathering are not major contributors in many environments and are not discussed further.

All rainfall contains some salt, especially in coastal areas. Raindrops can coalesce around tiny salt crystals in near-shore areas and around dust particles in inland areas. Rainfall in coastal areas (salt fall ~100 kg/ha/yr) has a concentration of salt, which may decrease tenfold with increasing distance inland (Hingston and Gailitis 1976). Delivery of salt with rainfall can be significant mechanism for salt accumulation in the landscape (Simpson and Herczeg 1994). Salt concentrations in rainfall may also be lower in tropical and sub-tropical latitudes (Hingston and Gailitis 1976). Coastal areas also receive more rainfall than inland areas so the salt loads (tonnes per hectare) are proportionally even higher. However coastal areas are often less susceptible to salinity than inland areas because they often have well-defined rivers, and aquifers can be flushed by the higher rainfalls, which is able to return the incoming salt to the ocean.

Wind is a common source of salts in areas with saline soils. Dunes or dry lakebeds may act as dust sources. In inland arid regions these features may be more important sources of salt than rainfall. As well as sodium and chloride, wind can entrain alkali earths, carbonate and sulphates where there are gypsum or carbonate deposits at the surface. Even when salt additions are small, they can accumulate to problem proportions over hundreds to thousands of years if the landscape is naturally poorly drained (e.g. lacks a strong river system), soils are deep and clayey which facilitates salt storage or drainages are occluded (e.g. when they end in terminal salt pans and fail to leave the region). For example, it has been estimated that there are ~1300 years and ~33,000 years of salt-fall from rain stored respectively in the upper 1 m and upper 50 m of two soil profiles from the Merredin area of Western Australia (Barrett-Lennard and Nulsen 1989; Barrett-Lennard et al. 2016). The seasonality and intensity of the rain can also determine the rates of runoff and recharge, and therefore salt accumulation and removal from the regolith.

Salts accumulate in soils and landscapes where water has been evaporated or transpired leaving the salts behind. This can take the form of a 'salt bulge' at a depth at which wetting fronts reach in average years, often just below the root depth of the deepest-rooted plants. If wetting fronts penetrate deeply into soil profiles, the salt bulge will also be deep. Root depths may be limited through high soil strength (often as result of sodicity), subsoil acidity, aluminium- or boron-toxicity, anoxia caused by waterlogging or by salinity itself.

The slow accumulation of salts in deep profiles reflects the hydrological balance of rainfall and extraction of water by plants over thousands of years. Bulges can also be partially mobilised by extreme natural rainfall events in which case the bulge reflects these factors rather than abilities of plant communities to extract incoming rainfall. Typically, salt accumulation is the result of very slow processes, well beyond management time-frames.

4.2 Salt Mobilisation

Changes to native vegetation, especially clearing, intensive grazing and/or increased burning can all affect the hydrological balance, resulting in increased recharge and catchment runoff. Salt accumulations are no longer stable under the changed hydrological regime and can be mobilised within the landscape to locations such as plant root zones, small and large river valleys, wetlands and floodplains. The way that salinity is expressed at the soil surface often defines the name of the problem (hillside-, valley-salinity etc).

The time-scale with which salts may be leached depends on a number of factors:

(i) The degree to which the hydrological regime has been altered, especially the change in recharge rates;
(ii) The hydraulic conductivity of the material containing the salts, with some clay regoliths containing relatively easily mobilised salt in cracks as well as within clay matrices which require slow diffusion to a preferred pathway to be mobilised;
(iii) The nature of the salt, with sodium chloride being more soluble than sulphates which are in turn more soluble than carbonates; and
(iv) The distance that the salt needs to travel from its area of origin to where it can impact on plants or on water quality (surface water, soil water or groundwater).

4.3 Salinity Types: Primary Versus Secondary

Primary salinity is a natural condition of world-wide soils and may also be referred to as sodic, alkaline or solonchak and solenetzic (Richards 1954). These soils are saline as a result of natural pedo-genesis or environmental change and may occur where geological sources interact with soils, where evaporation exceeds rainfall, or in association with salt lakes (playas) and coastal soils.

Szabolcs (1989) also recognised that both North America and Australia had a major problem of salinisation created by watertable induced salinisation he called ‘active solonchaks’. In the Australian context these soils had previously been referred to by Northcote and Skene (1972) as ‘dryland salting’ and thereafter adopted as ‘dryland salinity’ (Peck 1978).

Secondary dryland salinity is the term used to describe degradation that results in the decline in yield and productivity of agricultural systems due to the accumulation of excessive salts in the root zone. The degradation is driven by a change in the catchment water balance that leads to development of shallow groundwater and evaporation and accumulation of salts where water tables are within 2 m (Nulsen 1981). Dryland salinity is usually first noticed when agricultural and native plants senesces and yields of farm crops and pastures are reduced by more than 20 %. In severe cases, bare areas or salt scalds develop. Where aquifer pressures and high and direct groundwater seepage is apparent, saline areas are referred to as 'saline seeps' or 'seepage scalds' and may be associated with baseflow contributions to rivers and streams (Peck 1978).

4.4 Regional, Intermediate and Local Groundwater Systems

Groundwater systems that result in dryland salinity can be categorised according to aquifer scale (local, intermediate or regional) (Coram et al. 2000; NLWRA 2001; Walker et al. 2003). This enables the cause (location, area) to be defined and the scale of the flow systems reflects the ease with which salinisation can be managed. On the basis of size and responsiveness to treatment, aquifers were classified by NLWRA (2001) as:

(i) Local flow systems – recharge and discharge are close to each other (within 1–3 km) and groundwater levels equilibrate quickly (10–100 years) after disturbance such as clearing. Localised flow systems are discontinuous and commonly overlie an intermediate or regional flow system.
(ii) Intermediate flow systems – recharge and discharge may have a horizontal extent of 5–10 km and generally occur across the entire catchment. They have a higher storage capacity than local flow systems and take longer to equilibrate. Intermediate systems may occur in areas with palaeochannels (buried, prehistoric drainage channels).
(iii) Regional flow systems – recharge and discharge may be separated by many tens to hundreds of kilometres. Groundwater movement may be independent of local topography (sub-catchments), involve long flow paths and are typical of large sedimentary basins. Groundwater levels outside the recharge area are slow to respond and equilibrium may take many hundreds of years. Regional aquifers often contain local and intermediate flow systems.

4.4.1 Artesian and Non-artesian Groundwater Systems

Given proximity to a watertable and the scale of aquifer defines the processes responsible, and may impact the severity and responsiveness to treatment, salinity managers are also concerned to determine whether the aquifer responsible is artesian (under pressure beneath a confining layer) or non-artesian. Artesian groundwater conditions can occur where there is sufficient topographic slope and connectivity to enable hydraulic connection to the saline area. Artesian conditions may result in a flux of groundwater of up to 2400 mm m^{-2} to the land surface as a result of heads of up to 2–6 m above ground (George 1992; George and Conacher 1993). These conditions are typically associated by discharge to streams where saline groundwater contributes to perennial base flow.

In the more common non artesian or watertable condition, aquifer discharge is passive, driven by evaporation and soil water content. Nulsen (1981) described the capacity of groundwater to be evaporated from the capillary fringe as a function of depth to the watertable, soil material, and potential evaporation rates and indicated it was unlikely for sufficient salt transport to take place and impact the yield of wheat and barley at watertable depths of greater than 1.5 to 1.8 m. This was confirmed by a study by Bennett et al. (2012) who showed that passive salt flux and saline runoff was reduced on a broad and saline valley when the link between the capillary fringe and soil surface was reduced by lowering the watertable.

4.4.2 Classification Based on Groundwater Discharge Type

To enable managers to define the source and options for salinity management, George et al. (1997) defined ten common hydrogeological settings for salinisation in the WA context. This system was modified by Coram et al. (2000) and then increased to 15 hydrogeological settings for the Australian context (NLWRA 2001). These geologic and geomorphic based conceptualisation of flow systems responsible for salinity (with the flow systems classification defined above) enabled managers to select the most likely options and assist to define the areas for treatment (Clarke et al. 2002).

Some common examples of dryland salinity seeps are shown in Fig. 1. Anything that causes groundwater flows to approach the soil surface will result in a seepage point. The examples are two dimensional but they may be water converging on a point in the landscape from multiple slopes, or changes in the hydraulic conductivity of the regolith as occurs over dolerite dykes (Engel et al. 1987). This investigation also was one of the first applications of electromagnetic and magnetometer instruments to understanding dryland salinity problems.

4.5 Comparison of Dryland and Irrigation Salinity

As indicated in a previous section, irrigation salinity affects about twice the land area and has a tenfold higher incidence on a percentage area basis. While there are many common characteristics, it differs from dryland salinity in several important aspects:

(i) Irrigation water is often a major source of the salt;
(ii) Irrigation water can also be a major contributor to high groundwater levels which brings soluble salts to the surface and results in concentration by both transpiration and soil evaporation;

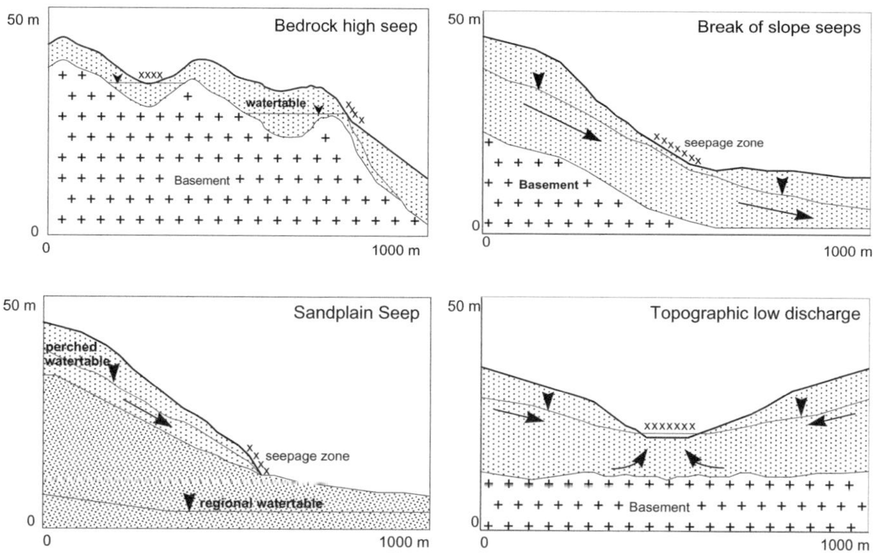

Fig. 1 Types of dryland salinity discharges – bedrocks highs obstructing groundwater to flow (upper left); a topographic change of slope reduces groundwater discharge rates (upper right); perched groundwater coming to the surface because the aquifer pinches out (lower left); and groundwater filing the aquifer under a flat valley such that the storage is exceeded (lower right), and are four examples of types from the classification of Coram et al. (2000)

(iii) Irrigation salinity usually occurs under the irrigated area whereas dryland salinity can occur some distance from the area of salt accumulation and excess recharge. This can result in conflict between farmers if one is causing the problem and the other receiving the impact. However, such externalities are relatively uncommon (Pannell et al. 2001);

(iv) Irrigated areas are usually flat and very valuable that justify the adoption of expensive drainage practices compared with often scattered and small areas of dryland salinity on less valuable land making cost-effectiveness on intensive interventions problematic;

(v) Being flat, irrigated areas are less susceptible to water erosion and the loss of topsoils once vegetation cover has been lost. Their flatness can however make drainage more difficult as there are low gradients towards a discharge point.

Excess water from irrigated areas can infiltrate deeper profiles and aquifers and contribute to ‘dryland’ salinity in down-gradient areas. Similarly, water and salt from dryland agriculture can contribute to salinity in irrigated areas.

5 Estimating the Extent of Saline Areas, Monitoring Change and Predicting Its Future Extent

5.1 Estimating the Extent and Change of Existing Salt-Affected Land

Salt-affected land has been identified and mapped in a number of ways throughout the world.

5.1.1 Surveying Landholders

Surveying landholders can provide an approximation of the extent of dryland salinity provided there is a common understanding of what constitutes secondary salinity. Surveys were carried out about every ten years in Western Australia until 2002 when the process ceased (Fig. 2). Over time, landowner's knowledge of what was primary salinity was lost, and accurate attribution of the yield losses to salinity was uneven between observers, impairing the value of the survey.

5.1.2 Aerial Photograph Mapping

Salt-affected land usually has poor vegetative cover (and a salty surface crust when severely affected) or contains indicator species such as halophytes. However as mentioned previously, salinity is often associated with waterlogging, sodicity and poor soil structure (hard-setting soils) which complicates what is being mapped as dryland salinity. Removing waterlogging and inundation (surface ponding) as a

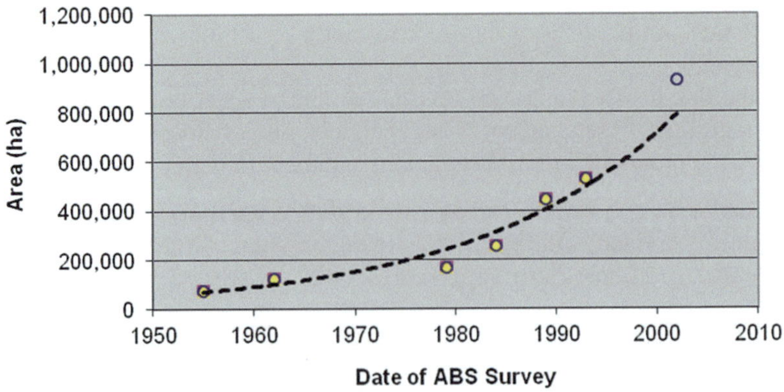

Fig. 2 Estimates of the extent of dryland salinity as assessed by surveying Western Australian farmers

result of surface drainage may be sufficient to 'reclaim' salt-affected land when the interaction between the two problems is removed.

Mapping such salt-affected land onto aerial photography is a common form of ground truthing for the methods mentioned below. However, it is subject to operator bias because it is subjective. This technique can be used to map the changes in extent of known areas.

5.1.3 Groundwater Level and Salinity Measurements

For dryland salinity caused by high groundwater levels, the ultimate test of salinity is the presence of groundwater at a depth that would allow saline water to rise by capillarity into the root zone of plants or to the soil surface. For clayey soils this may be 2 or 3 m but for sandy soils it may be less than a metre. Drilling observation bores, even shallow ones, is expensive and it only provides a point assessment so this is not common for mapping large areas. It is, however, a very good way of assessing changes over time, enabling the processes to be understood and engaging farmers in the assessment of extent and risk, and therefore comprises a good monitoring tool.

5.1.4 Electrical Conductivity Using Geophysical Instruments

When detailed maps are required of both the extent and intensity of dryland-salinity for an area it is possible to use geophysical equipment to measure the apparent conductivity of the soil to depths ranging from 0.2 to more than 200 m (George et al. 1998; Farifteh et al. 2006). Airborne and ground based electromagnetic techniques generally work well in acid to neutral soils but ground-based systems may over estimate salinity in sodic and alkaline soils. For example, results from shallow EM-systems become increasingly confounded by the presence of lime – an insoluble salt that has little or no adverse effect on crop growth. As well as the salinity of the soil, readings are affected by the amount and type of clay and the soil needs to be wet enough to allow an electrical flow to be induced by the measurement equipment. Changes in salinity over time can also be recorded by well-calibrated electromagnetic induction equipment.

5.1.5 Airborne Remote Sensing

Multispectral or hyperspectral sensors can be used to measure the amount of individual salt in soils (sodium, chloride, sulphate, etc.) with the spatial resolution depending on the type of instrument and the height at which the imagery is acquired (Howari 2003). Often these instruments provide a lot of detailed information (e.g. soil properties, mineralisation) but their expense is only justified in research areas or where land is very valuable.

5.1.6 Satellite Remote Sensing

Satellites provide routine digital reflectance at several wavelengths which can be trained to detect and monitor salt-affected land (McFarlane et al. 2004; Spies and Woodgate 2005). Usually it is the poor condition of vegetation that is detected and this can result from multiple causes so other data sets are required to reduce the errors of commission (false positives) and omission (failing to detect salt-affected land).

Saline land has low or no productivity over long periods of time so analysing multiple images can ensure that temporarily bare areas are not confused with saline areas. Dryland salinity caused by high groundwater levels are the most common in low lying topographic areas and in where water accumulates in the landscape. These attributes have been used to increase correct classification levels to more than 95 % once image-training areas that are known to be salt affected (often by aerial photograph interpretation) are used to train the classification. These methods are reproducible over large areas at low costs (a few cents per hectare) and have been applied to map and monitor changes in dryland salinity of areas of about 18 M ha (Caccetta et al. 2009).

Expert hydrogeologists estimated the area that was salt-affected in south-western Australia to be 1.8 M ha based on studies of selected catchments and monitoring of observation bores over several decades (Ferdowsian et al. 1996). However, a more repeatable, accurate and comprehensive mapping of saline areas using multi-temporal Landsat Thematic Mapper (TM) images and digital elevation models showed that its extent was less than 1 M ha (McFarlane et al. 2004). The expert method was also affected by site selection bias where estimates were based on studies of saline catchments which may over-estimate salinity in other catchments.

5.1.7 Most Feasible Methods on Monitoring Changes in Salinity Over Time

Methods that are quantitative, reproducible and relatively cheap are more likely to be adopted and to provide accurate assessments of the change in salt-affected areas over time. These are important for assessing the effectiveness of management interventions relative to the impact of climate variability, especially rainfall.

Monitoring changes in groundwater level or ground-based geophysical surveys or transects can be used at the local level but for large areas satellite imagery probably can give the most cost effective regional coverage. Such monitoring can be used to assess the condition of perennial vegetation and crop yields as an additional benefit of producing interpreted images.

5.2 *Predicting the Future Extent of Dryland Salinity*

Identifying areas that may be at risk from dryland salinity in future, and the rate of salinisation, can help to prioritise efforts needed for its prevention or amelioration. It is more expensive to reclaim salt-affected land than it is to prevent it from forming in the first place.

Methods have been developed for assessment of salinity risk across areas of varying extent, from the farm scale to national scale (Gilfedder and Walker 2001). Generally, these have been developed to suit available data, which vary according to the extent and site of study areas. Consequently, there was a lack of consistency between the range of methods used to predict salinity risk across Australia, and the methods varied widely in both methodology and reliability.

Where detailed land contours are available (as is increasing the case through automated photogrammetry using digital aerial photography, LiDAR and available satellites) it is possible to identify land that may be affected if groundwater levels rose by some fixed amount (e.g. if they rose a further 1 or 2 m above existing known saline areas). This is likely to be more effective in areas with existing saline outbreaks. In cases where there are no surface expressions of salinity, the depth of the watertable needs to be known from monitoring bores. Land-based geophysical surveys that measure terrain conductivity at various depths are very good at identifying local areas that have a high salt content and a hazard of becoming saline, however as they do not measure soil-water (watertables) and account for micro-topographic effects, cannot be used alone for risk assessment.

6 Management of Dryland Salinity

The management of dryland salinity can be conceptualised into three approaches: (1) recovery – where a solution is implemented based on managing the causal agents (water table/saline soil); (2) containment – modifying the causal processes to some degree and in so doing limiting the full impacts; and (3) adaptation – accepting saline soils and taking steps to enable the profitable use of saline land and water.

In the context of salinity management systems, revegetation using deep-rooted perennial vegetation and engineering options are considered recovery and containment strategies, while implementation of integrated catchment options (e.g. higher water use cropping, perennial pastures and salt land pasture systems) are primarily for containment and adaptation.

6.1 Revegetation with Deep-Rooted Perennials

The use of perennial vegetation to recover dryland salinity has been postulated as the principal means of control since the opposite (clearing) was first observed in Western Australia in the 1900s to be responsible for salinisation of water supplies for steam trains (Wood 1924). Indeed, the first reclamation of the impact of rising groundwater on stream salinity resulted from the prevention of further clearing and ringbarking and complete reforestation of the small area that had been cleared in Helena catchment (George et al. 1997). However, it was not until the 1980s that active experimental programs began to investigate the impact of partial reafforestation as a management measure for control of stream salinity in the Wellington Dam catchment.

In a review of empirical data derived from extensive tree plantings in 46 discharge areas and 33 recharge areas in the agricultural region of Western Australia, George et al. (1999), came to different conclusions. They concluded that trees were better planted in recharge, rather than discharge, areas, and '... generally only extensive plantings, perhaps influencing as much as 70–80 % of the catchment, will lead to significant catchment scaled reductions in water tables...'. Further, George et al. (1999) also showed that planting in discharge areas are often ineffective '... to allow reclamation of saline areas, because the response in groundwater levels is rarely greater than 10–30 m from the planted area'.

From an eco-hydrological approach, Hatton and Nulsen (1999) stated that 'empirical and theoretical analyses suggest that however vegetation is managed in the landscape, effective water balance control will be achieved only at a leaf area index approaching that of the natural state, indicating revegetation of most or all parts of the catchment with either trees or plants with similar ecohydrological characteristics as trees'.

More recently, Bennett and George (2008) reviewed watertable and salinity dfata collected from 24 sites on 15 farms planted since 1990 to various deep-rooted tree species at varying combination of blocks, strips, belts and geometries (5 % to 90 %). Their evaluation concluded that only at the highest percentage of catchment planted (> 70–98 %), did trees impact groundwater levels to a significant degree (2–6 m reduction over < 20 years). Importantly for agriculture, the area required to recover the famers saline land was at least 5–10-times the area affected – a major disincentive to plant non-commercial tree crops.

6.2 Engineering and Drainage

Recovering saline land using agricultural drainage in the context of irrigation systems has been evaluated for a considerable period of time (George et al. 1997). Drainage options such as the use of open, tile, and other subsurface drains (mole drains, tube drains), and groundwater pumping are cited in the literature as effective means of salinity management. Skaggs and van Schilfgaarde (1999) document 1400 pages of the theory and practice of drainage, with one chapter evaluating options for control in the Great Plains in North America given to management of dryland salinity.

El-Ashry and Duda (1999) define drainage as the preeminent means of managing salinity in irrigated lands, defining its need for use on 110 M ha of irrigated area. However, they also define a need to use drainage to intercept groundwater in dryland areas of North America, India, Iran and Australia, though no specific examples are given.

Application of deep drainage to dryland salinity in Western Australia has been applied within several catchment studies and analysed by Coles et al. (1999) and updated by Chandler and Coles (2003). These reviews report the minimal impact by open drains, of approximately 2 m depth, where impacts of less than 20–40 m from the drain were observed from the majority of the 22 case study sites evaluated. The dominant factors reducing the impact were the low permeability of the soils, limited gradient for flow and thickness of the aquifer relative to drain depth, contributing groundwater to the saline areas.

These results are reinforced by a drainage benchmarking study by Stuart-Street et al. (2012) at a further eight sites; the authors noting a similar impact of the drains. More significantly, they note that reduced rainfall in the past decade confounded their ability to separate the effect of the treatment (drains) and climate (reduced rainfall and recharge) from observed watertable and soil salinity impacts.

Exceptional impacts (100–200 m radius from the drain) have been described at sites by Ali et al. (2004) near Narembeen, where the hydraulic conductivity of the subsurface was high and the drain depth was at about 3 m, and Kobryn et al. (2015) near Dumbleyung where parallel drains were installed to between 2-3 m deep and crop recovery was attributed to watertable control and reduced soil salinity.

Aquifer pumping, siphons and relief wells have also been trialled and shown to have variable impact. George (1990, 1992) studied aquifer properties and the effect pumping at recharge and discharge sites and concluded while bores were capable of lowering watertables, and supplying brackish groundwater supplies, their influence was limited by the same parameters to those above and their economics were sub-optimum (George et al. 1997).

In specific cases, siphons have been used to manage watertables causing saline seeps on sloping land (< 2 %) and thereby reduce one of the major economic limitations of available and affordable power costs. Seymour and George (2004) showed that by designing siphons that attain critical velocity for dissolved gas (the principal model of failure), water tables were lowered beneath seepage scalds (hillside seeps) of 1–10 ha by cumulative flows of 1–2 L sec^{-1}. They also showed drainage to be of lower environmental impact than the episodic flux of salt from the degrading seeps.

6.3 Integrated Solutions

Development of integrated catchment solutions to salinity management has been a vision for salinity managers. The goal of integrating recovery, containment and adaptation options within agricultural catchments has been driven by conceptualisation of the problem being of a scale that necessitates regional control, and from a viewpoint that aligns to the principles of community action through tools such as Landcare and Integrated Catchment Management (ICM). This has also been reinforced by science that develops catchment-driven 'conceptual models' that are reinforced with numerical models that estimate the probability of treatments having impacts that should be accumulated by applying treatments at scale.

ICM has some significant limitations in the context of hydrological issues like dryland salinity. Technically, George (1992) and George et al. (1997, 1999) identified that the governing scale for salinity processes relevant to agricultural systems and its management is at a local scale. In general terms, the processes driving dryland salinity are well understood (Grundy et al. 2007) but there is variability in changes in expression of salinity over time. This variability is increased further as the rainfall (the input to the system) is also highly variable. The landscape is heterogeneous, and landscape properties such as topography, aquifer parameters all vary. The scale of the different groundwater systems results in different groundwater response times, which influence the lag between action and impact (Gilfedder et al. 2009; Walker et al. 2015). This variability makes it difficult to predict with certainty the expression of salinity in the broader landscape. The time-lags between implantation of options and impact are much greater for dispersed management approaches. As a result, treatments located at a distance from the affected site and unlikely to be effective unless implemented at a considerable scale in which case they are unlikely to be cost effective unless they are economic in their own right (Gilfedder et al. 2016).

Not understanding the consequence of this approach has resulted in well-intended but unrealistic salinity management programs. An example in Western Australian is the funding of four catchment groups (AUD $6 M) under the banner of the Catchment Demonstration Initiative (Robertson et al. 2009). Developed in 2000, and delivered between 2002 and 2008, the CDI project sought catchment management proposals from 22 groups across Western Australia and funded four who were successful in providing technically-based integrated catchment plans for co-investment. Robertson et al. (2009) reviewed the outcomes, noting while the technical elements were sound at a planning level, the expectation of impact and the social and economic consequences were underestimated, as was the level of benefit and risk.

ICM for salinity management is confounded by the lack of economically-attractive solutions and the technical complexities of scale required for impact, and more specifically, there is a disincentive for managers who are not directly affected or benefiting from the action to adopt it – and certainly not if it is sub economic. Additionally, ICM assumes a level of accountability, group cohesion and capability at the catchment scale that may not exist outside a well-funded project and within a commercial farming business and local wheatbelt community context.

7 Making Salt-Affected Areas Profitable

Saline agriculture is largely about making better choices for the use of land based on the capability of the land, the level of production that is possible and the value of that production.

Land capability is mostly affected by the salinity of the soil solution – the ratio of salt to water in the soil. However, it can also be affected by the presence of other stresses such as waterlogging and toxic boron. Nonetheless, here we only focus on salinity. When considering options for saline land, it is useful to orientate thinking with a larger framework (Fig. 3). Sodium chloride, the most common salt found in soil forms a saturated solution at about 360 g/L of water. We can therefore think of the salinity of the soil solution in terms of the percentage of saturation of such a NaCl solution. No high order plants will survive with salinities of the soil solution greater than ~40 % of saturation, but halophytes (salt-loving plants) will survive with salinities less than this level. Seawater salinity occurs at ~9 % of saturation, and crop plants begin to become economically important at salinities equivalent to 4 % of saturation (Fig. 3).

Reviews of international literature on the effects of salinity on the growth of crops show that increasing salinity decreases the yield of all crops (Fig. 4). However, crop plants can differ widely in their tolerance to salinity, as measured by the salt concentration (ECe) at which their yield decreases by 50 %. The most salt-sensitive crop shown in Fig. 4 (strawberry) had a 50 % decrease in yield at an ECe of 2.2 dS m^{-1}, while the most salt-tolerant crop (asparagus) had a 50 % decrease in yield at an ECe of 28.5 dS m^{-1}.

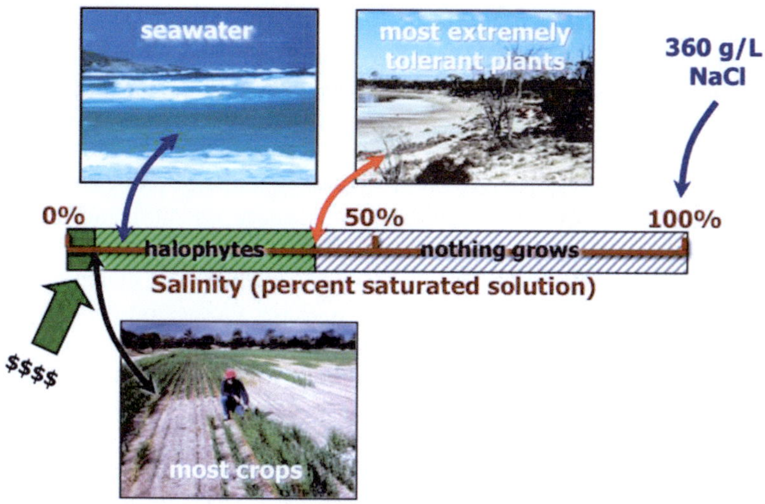

Fig. 3 Framework for thinking about the salinities that plants can endure

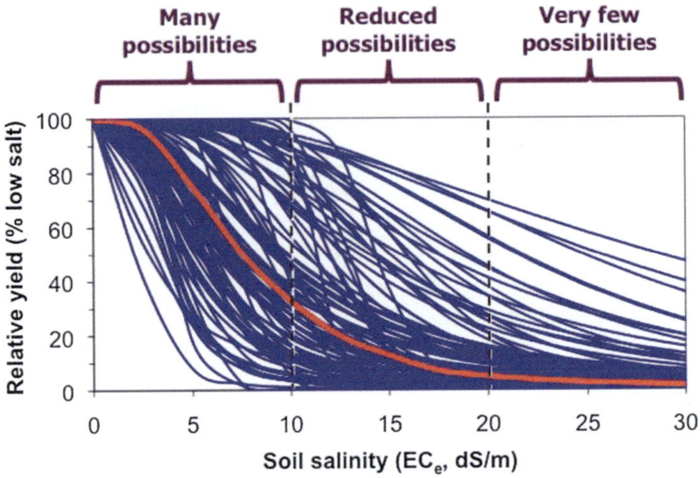

Fig. 4 Impact of soil salinity (ECe) on the relative yield of 108 crop plants (after Steppuhn et al. 2005). The red line indicates the median crop response

Another factor that becomes clear from Fig. 4 is that the range of productive options for saline agriculture decreases dramatically as ECe values increase. With ECe values between 0 and 5 dS m^{-1} about half the crops surveyed had relative yields greater than 70 %. We can conclude that salinities in this range should not have a strong effect on crop selection: many options are available. By contrast, with ECe values between 20 and 30 dS m^{-1}, half the crops surveyed had relative yields of less than 10 %. We can conclude that salinities in this range should have extreme effects on crop selection: very few productive options are available.

Increasing salinity causes agriculture to transition from potentially highly profitable cropping systems such as horticulture, vegetables, grains, forages and fibres to lower-value salt-tolerant crops such as barley and halophytes (salt-tolerant grasses and chenopod shrubs). Additionally, as salinity increases, the feeding value of the available forage often decreases and issues of nutrient imbalances in the diet of animals increase (Norman et al. 2013).

Many agricultural systems rely on the use of leguminous crops and pastures to fix nitrogen biologically, thereby decreasing reliance on nitrogenous fertilisers. However, legumes and their associated rhizobia tend to be salt-sensitive. Increasing salinity therefore results in the loss of legumes (and therefore biological nitrogen fixation) from agricultural systems. Unless research focuses on the simultaneous selection of legumes and rhizobia for salt tolerance then farming systems in highly salinised landscapes can be expected to be more nitrogen deficient.

One interesting case study of the importance of the selection of legume and rhizobia for salt land comes from studies of the salt tolerant forage legume *Melilotus siculus*. This species performed well in pasture trials on salt land in the year it was planted, but failed to regenerate in subsequent years (Nichols et al. 2008). In the year of planting the *Melilotus* was established with a commercial line of rhizobium that had not been especially screened for salt tolerance. However, it appeared that this micro-organism was not surviving on saline soils over the summer when the salinity of the soil solution probably increased to levels approaching saturation. Best successful regeneration of the plants occurred only using a rhizobium (SRDI 554) selected from a naturally saline environment in South Australia. This new rhizobial selection had increased saprophytic competence than the previously used commercial rhizobium (WSM 1115), and average plant nodulation increased from 11 % (with the commercial line) to 74 % (with SRDI 554) (Bonython et al. 2011).

As noted above, when soils become exceptionally saline, they are only suited to the growth of halophytes, so there have been a range of efforts to develop industries around the domestication of the highly salt tolerant plants (Barrett-Lennard et al. 2003; Glenn et al. 2013; Panta et al. 2014). Not surprisingly, halophytes seem best suited to agriculture when they fit smoothly into existing production systems. This has been an important part of the thinking in adapting halophytic forages (notably *Atriplex* species) as forages into grazing systems on salt affected farms in southern Australia. The main idea is that grazing systems in southern Australian mixed farms generally have periods of 'feed-gap' when fodder is not available and farmers often hand feed stock with grain at these times. Salt tolerant forages have a premium value at these times because their utilisation by sheep can be used to offset feeding with grain, so that grain can be sold increasing farm profits (Barrett-Lennard et al. 2003).

One key paper catalysing the development of salt land pastures has been an economic analysis by O'Connell et al. (2006), which showed that the economic gains to farmers could be doubled if the nutritive value of the forage available to grazing sheep could be increased by 10 %. This led to a new focus on increasing nutritive value in salt land pastures by incorporating: (a) *Atriplex nummularia subsp. Nummularia* as the key fodder shrub in the system (this had higher digestible organic matter in the dry matter than *A. amnicola* – reviewed by Norman et al. 2013), and (b) high nutritive value annuals into the under-storey (Norman et al. 2010). The potential of this system recently became apparent when in an experiment with *A. nummularia* and an improved salt tolerant under-storey (*Trifolium michelianum, T. subterraneum, Medicago polymorpha and Lolium multiflorum*), Norman et al. (2010) achieved higher rates of grazing (~7 growing sheep/ha for ~8 months of the year) on mildly saline land with ~330 mm annual rainfall than the district average stocking rate for pastures that were not salt-affected.

Finally, in more saline areas, halophytes can be used to moderate the impact of groundwater driven soil salinity and change the potential for other plants to establish and produce. In a paired plot study (25 ha) at Yealering, Bennett et al. (2012) documented a one-decade study showing how narrow spaced alleys of saltbush have been used to lower water tables from less than 1 m to about 1.5 m (i.e. by more than 0.5 m) and in so doing restrict seasonal access of the capillary fringe to the soil surface. This in turn limited the flux of salt into the root zone and reduced salt flux from the site (88 % less for the treated site). As a result, more salt sensitive plants established in the inter-rows and productivity increased from < 1 T ha^{-1} to up to 7 T ha^{-1}. Of importance here is that in broad valleys with relatively deep water tables and intermittent discharge, moderating the soil water content (drying it) limits salt flux due to lowered hydraulic conductivity. In areas where treatments can limit solute wash-off, there will also be a benefit in downstream water quality.

8 The Effect of Climate Change on Dryland Salinity – Case Study of Australia

In the early 1990s, groundwater levels under dryland agriculture were rising in many areas in southern and eastern Australia. The Millennium Drought between 1995 and 2007 then affected the eastern two thirds of the country. By contrast the south west of Australia has had a drying trend since about 1975, which intensified in about 2000, but because of time lags has maintained rising groundwater levels in most areas although those in the north and east are now becoming stable or starting to fall. Each state in Australia assessed groundwater trends and the extent of dryland salinity in 2007 given the impact of the Millennium Drought. The best estimate of land affected by secondary salinity in Australia was 1.753 M ha with 1.077 M ha (61 %) of this in south-western Australia.

Clearing deep-rooted native vegetation resulted in rising groundwater levels throughout the Western Australian wheat belt until clearing effectively ceased in the 1980s. In 1996 the area in south-western Australia that had been measured using satellite remote sensing was 0.957 M ha, an increase of 0.098 M ha since an earlier measurement in 1989 (McFarlane et al. 2004).

The drying climate since about 1975 has become increasingly important in determining groundwater changes because making change to annual crops and pastures are ineffective against salinity and the area under perennial plant species is relatively small. The further reduction in rainfall since 2000 affected groundwater trends starting in the north but progressively affecting southern parts of the state (George et al. 2008; Fig. 5). This pattern is also reflected the reduction in runoff in forested catchments as groundwater levels fall and valley inverts dry (Petrone et al. 2010).

In south-western Australia, the most recent analysis of hazard and risk was undertaken by Raper et al. (2014). Using groundwater trend data from 1500 surveillance bores the report compares three analysis periods: 1991–2000, 2000–2007 and 2007–2012. It showed that between the 1991–2000 and 2000–2007 periods, the proportion of bores with rising trends fell from 60 to 40 % and the proportion of bores with falling trends increased from 6 to 29 %. The changes in groundwater trends were most pronounced in the north of the region after 2000 as a result of a 30–40 % reduction in rainfall. For the 2007–2012 period, rising or stable groundwater dominated the regions assessed. Areas with mainly falling trends covered 6 %.

Despite the reduction in the proportion of bores with rising trends, groundwater levels have continued to rise in, and adjacent to, areas of salinity hazard in lower landscape positions over much of the region. In the SW of WA it appears that to

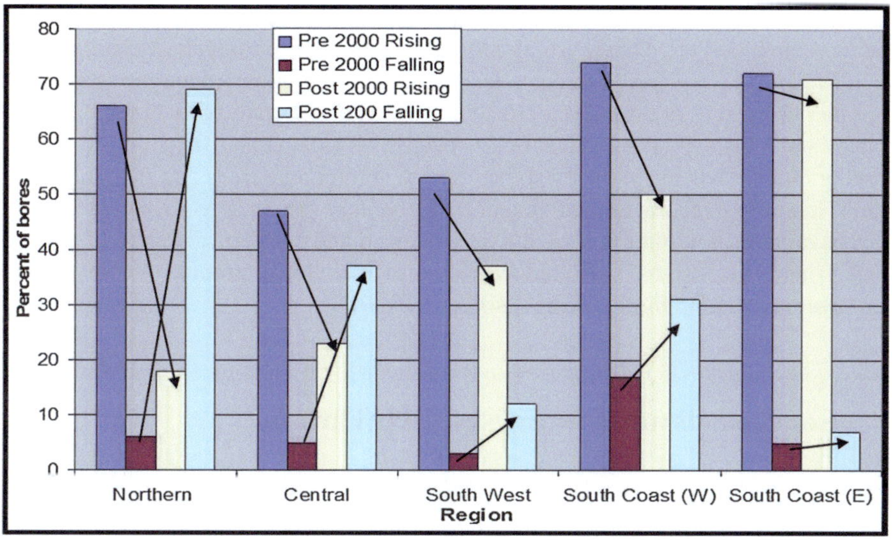

Fig. 5 Change in groundwater rising and falling trends before and after 2000 in different parts of south-western Australia (data from George et al. 2007)

date, the impact of the drying climate on groundwater trends (diminished recharge) is still less than the historical impact of clearing. In the future as aquifers equilibrate to the effects of clearing, climate will be the predominant influence and the area of salinity will respond commensurately.

In Victoria, steady rises in groundwater levels in the 1980s and early 1990s were followed by decreases of up to 10 m after 1996 with decreases greater than the previous rises in most cases (Reid et al. 2008). There was evidence that the spread of land salinisation in the state had stalled. A return to more normal rainfall conditions in recent years has not resulted in a recovery of these levels and the salinity risk remains low in many areas which had previously been considered at high risk.

Similarly, in New South Wales, earlier estimates of future salinity were later considered overestimates (NSW DECC 2009). The rate of drainage of groundwater in dryland agricultural catchments when the Millennium Drought hit in 1996 was unexpected. For example, in the Liverpool Plains, it was thought that levels would only begin to drop after 20 to 60 years were recharge to be reduced to zero (Walker et al. 1999) because the discharge capacity of the monitored catchment was estimated to be only 1 mm per year. Falling groundwater levels does not automatically reclaim salt-affected areas because there is usually a need to improve soil structure because of sodicity and to improve the ability of exposed subsoils to grow plants given the loss of topsoils in most secondary saline areas.

Much of southern Australia which has the highest incident of dryland salinity has experienced a drying climate in recent decades. The change in rainfall patterns is believed to be caused by the poleward movement of climate drivers such as cold fronts which bring rainfall in autumn to spring (Bureau of Meteorology 2014;

CSIRO and Bureau of Meteorology 2015). This has resulted in dryland salinity being reclassified from being a major threat to agricultural productivity in the 1990s to now being a lower-order threat to farm viability. This experience may be relevant to other dryland agricultural regions with similar Mediterranean and temperate climates and a history of dryland salinity because of accumulated salts. However, the drying climate has made the establishment of commercial trees and perennial pastures, common recommended practices for reducing recharge to saline ground water, more difficult (John et al. 2005). The need and cost-effectiveness of surface and groundwater drains is also reduced because of the lower groundwater levels and a slower accumulation of salt in topsoils.

9 Conclusions and Future Research Thrusts

Groundwater-induced dryland salinity can appear without prior warning because groundwater levels are usually not monitored in dryland farming areas. There is also often a poor understanding of the impact of land clearing on catchment water balances. Because it can also be associated with sodicity and waterlogging it can be poorly diagnosed and management responses scan be sub-optimal.

Moreover, a common misconception is that because dryland salinity only affects 2.1 % of dryland agricultural areas world-wide it can be solved with a modest intervention. However, work cited here has shown that it is a symptom of a significant water imbalance over a larger landscape and solutions need to acknowledge this scale and long response time-scales. The saline areas are also typically in the most fertile lower landscape areas and connected to the remaining riverine and related terrestrial and ecological systems.

The long time-lags are also often unappreciated. The mobilised salt may have accumulated over hundreds of thousands of years and the hydrological imbalance may have been going on for decades but was unrecognised. Sometimes the seasonal or management conditions immediately prior to its emergence may be mistaken as the primary cause because the long time lag between cause and effect is not appreciated. As a result of these complexities, management interventions are often unsuccessful and money is wasted on solutions that will not deliver the expectations of funders, not enabling land owners to recover sunk costs. We show that in low-fertility saline land unable to justify expensive interventions the best response may be to 'live with salinity' and grow crops and pastures with the ability to handle the conditions. Modifying waterlogging and soil sodicity issues may help these crops and pastures yield better while operating within the changing hydrologic conditions.

Dryland salinity is affected by the catchment's geology, regolith (weathered material above basement), topography, climate, past water use by plant (which affects the accumulation of salts) and recent changes to water balances because of land management practices. Each occurrence therefore has its own unique circumstances. This makes the development of simple management rules difficult and thus favours the treatment of the saline area over remote interventions.

The off-site impacts of dryland salinity can be as large as (or larger than) the impacts on the affected land. This takes the form of exported salt and soil (sediments) entering waterways and downstream water resources including dams. Riparian vegetation and in-stream pools are often completely degraded after dryland salinity develops higher up in a catchment. This may justify the cost of inventions, although draining saline groundwater requires a safe disposal site.

Even in areas undergoing a drying climate dryland salinity can remain an issue, although its impact differs widely between countries. Areas that remain at risk need to sustain a base level of monitoring to inform governments and landholders of the level of risk. For affected areas further investments in saltland management are warranted, especially given advances in technology that allows commercial plants to be more salt tolerant.

In summary, dryland salinity can be a complex problem to properly diagnose as it is often associated with other forms of soil degradation. It is often hard to appreciate the hydrological processes that cause it in each case as they are in the subsurface. Finally it can be hard to develop the most cost-effective solution both for the affected area and for downstream areas. Climate change is further altering catchment water balances, sometimes for the better, although it can reduce the viability of some management options because drier conditions may no longer favour the establishment of deep-rooted perennial plants.

References

Ali R, Hatton T, George R, Byrne J, Hodgson G (2004) Evaluation of the impacts of deep open drains on groundwater levels in the wheatbelt of Western Australia. Aust J Soil Res 55:1–13

Australian Bureau of Statistics (2002) Salinity on Australian farms, Bulletin 4615. ABS, Canberra

Barrett-Lennard EG (2003) The interaction between waterlogging and salinity in higher plants: causes, consequences and implications. Plant Soil 253:35–54

Barrett-Lennard EG, Nulsen RA (1989) Dryland soil salinity – cure, containment or catastrophe. In: Proceedings of the 5th Australian Agronomy Conference, Perth, pp 212–220

Barrett-Lennard E, Setter T (2010) Developing saline agriculture: moving from traits and genes to systems. Funct Plant Biol 37:III–IIV

Barrett-Lennard EG, Shabala SN (2013) The waterlogging/salinity interaction in higher plants revisited – focusing on the hypoxia-induced disturbance to K+ homeostasis. Funct Plant Biol 40:872–882

Barrett-Lennard EG, Leighton PD, McPharlin IR, Setter T, Greenway H (1986) Methods to experimentally control waterlogging and measure soil oxygen in field trials. Aust J Soil Res 24:477–483

Barrett-Lennard EG, Malcolm CV, Bathgate A (2003) Saltland pastures in Australia – a practical guide, 2nd edition. Sustainable Grazing on Saline Lands (a sub-program of Land, Water and Wool), 176 pp

Barrett-Lennard EG, Bennett SJ, Colmer TD (2008) Standardising the terminology for describing the level of salinity in soils. 2nd International Salinity Forum, Adelaide, 31 March – 3 April, 4 pp

Barrett-Lennard EG, Anderson GC, Holmes KW, Sinnott A (2016) High soil sodicity and alkalinity cause transient salinity in south-western Australia. Soil Res 54:407–417

Bennett DL, George RJ (2008) Long term monitoring of groundwater levels at 24 sites in Western Australia shows that integrated farm forestry systems have little impact on salinity. NRM Conference, Boyup Brook, April 2008

Bennett DL, George RJ, Silberstein R (2012) Changes in run-off and groundwater conditions under saltbush grazing systems: preliminary results of a paired catchment study Resource Management, Technical Report 381, 30 pp

Bonython AL, Ballard RA, Charman N, Nichols PGH, Craig AD (2011) New strains of rhizobia that nodulate regenerating messina (*Melilotus siculus*) plants in saline soils. Crop and Pasture Sci 62:427–436

Bureau of Meteorology (2014) State of the climate 2014. http://www.bom.gov.au/state-of-the-climate/. Accessed 24 Nov 2015

Caccetta P, Dunne R, George R, McFarlane D (2009) An estimate of the future extent of dryland salinity in the south-west of Western Australia. J Environ Qual 39:26–34

Chandler KL, Coles NA (2003) Review of deep drains to manage salinity in Western Australia, Miscellaneous publication 2/2004. Department of Agriculture, Perth

Clarke CJ, George RJ, Bell RW, Hatton TJ (2002) Dryland salinity in southwestern Australia: its origins, remedies, and future research directions. Aust J Soil Res 40(1):93–113

Coles NA, George RJ, Bathgate AD (1999) An assessment of the efficacy of deep drains constructed in the wheatbelt of Western Australia. Bulletin 4391, Agriculture WA, Perth

Coram JE, Dyson PR, Houlder PA, Evans WR (2000) Australian groundwater flow systems contributing to dryland salinity: a bureau of rural sciences project for the national land and water resources audit's dryland salinity theme. National Land and Water Resources Audit, Canberra

CSIRO and Bureau of Meteorology (2015) Climate change in Australia. http://wwwclimatechangeinaustraliagovau/en/. Accessed 25 Nov 2015

El-Ashry MT, Duda AM (1999) Future perspectives on agricultural drainage. In: Skaggs RW, van Schilfgaarde J (eds) Agricultural drainage, agronomy monograph 38. ASA CSSA, SSSA, Madison, pp. 1285–1298

Engel R, McFarlane DJ, Street G (1987) The influence of dolerite dykes on saline seeps in south-western Australia. Aust J Soil Res 25:125–136

FAO (2015) Extent of salt-affected soils. http://www.fao.org/soils-portal/soil-management/management-of-some-problem-soils/salt-affected-soils/more-information-on-salt-affected--soils/en/

Farifteh J, Farshad A, George RJ (2006) Assessing salt-affected soils using remote sensing, solute modelling and geophysics. Geoderma 130:191–206. doi:10.1016/j.geoderma.2005.02.003

Ferdowsian R, George RJ, Lewis MF, McFarlane DJ, Short R, Speed R (1996) The extent of dryland salinity in Western Australia. In: Proceedings 4th national workshop on the productive use and rehabilitation of saline lands, Albany, March, 1998

George RJ (1990) The nature and management of saprolite aquifers in the wheatbelt of Western Australia. Land Degrad Rehabil 2:261–275

George RJ (1992) Hydraulic properties of groundwater systems in the saprolite and sediments of the wheatbelt, Western Australia. J Hydrol 130:251–278

George RJ, Conacher AJ (1993) Interactions between perched and saprolite aquifers on a small salt-affected and deeply-weathered hillslope. Earth Surf Process Landf 18:91–108

George RJ, McFarlane DJ, Speed RS (1995) The consequences of a changing hydrological environment for native vegetation in South-western Australia. In: Saunders DA, Craig JL, Matriske EM (eds) Nature conservation – the role of networks. Surrey-Beatty, Sydney

George RJ, McFarlane DJ, Nulsen RA (1997) Salinity threatens the viability of agriculture and ecosystems in Western Australia. Hydrogeol J 5(1):6–21

George RJ, Beasley R, Gordon I, Heislers D, Speed R, Brodie R, McComnell C, Woodgate, P (1998) The national airborne geophysics project-national report. Evaluation of airborne geophysics for catchment management. At: http://www.ndsp.gov.au

George RJ, Nulsen RA, Ferdowsian R, Raper GP (1999) Interactions between trees and groundwater in recharge and discharge areas – a survey of Western Australian sites. Agric Water Manag 39:91–113. doi:10.1016/S0378-3774(98)00073-0

George RJ, Speed RJ, Simons JA, Smith RH, Ferdowsian R, Raper GP, Bennett DL (2007) Long-term groundwater trends and their impact on the future extent of dryland salinity in Western Australia in a variable climate. 2nd International Salinity Forum: Salinity, water and society – Global issues, local action, 4

George RJ, Clarke JDA, English PM (2008) Modern and palaeogeographic trends in the salinisation of the Western Australian Wheatbelt: a review. Aust J Soil Res 46:751–767

Ghassemi F, Jakeman AJ, Nix HA (1995) Salinisation of land and water resources: human causes, extent, management and case studies. University of New South Wales Press, Sydney

Gilfedder M, Walker GR (2001) Dryland salinity risk: a review of assessment methods. Nat Res Manag 4(1):2–9. Available from http://tinyurl.com/DrylandSalinityRisk

Gilfedder M, Walker GR, Dawes WR, Stenson MP (2009) Prioritisation approach for estimating the biophysical impacts of land-use change on stream flow and salt export at a catchment scale. Environ Model Softw 24:262–269. doi:10.1016/j.envsoft.2008.06.009

Gilfedder M, Grundy M, Robinson CJ (2016) Where has all the salinity gone? The challenges of using science to inform local collaborative efforts to respond to large-scale environmental change. In: Margerum R, Robinson CJ (eds) The challenges of collaboration in environmental governance - barriers and responses. Edward Elgar, Northampton, pp 131–151

Glenn EP, Anday T, Chaturvedi R, Martinez-Garcia R, Pearlstein S, Soliz D, Nelson SG, Felger RS (2013) Three halophytes for saline-water agriculture: an oilseed, a forage and a grain crop. Environ Exp Bot 92:110–121

Greenway H, Munns R (1980) Mechanisms of salt tolerance in nonhalophytes. Annu Rev Plant Physiol 31:149–190

Grundy MJ, Silburn DM, Chamberlain T (2007) A risk framework for preventing salinity. Environ Hazards 7:97–105. doi:10.1016/j.envhaz.2007.07.004

Hatton T, Nulsen RA (1999) Towards achieving functional ecosystem mimicry with respect to water cycling in southern Australian agriculture. Agrofor Syst 45:203–214

Hingston FJ, Gailitis V (1976) The geographic variation of salt precipitated over Western Australia. Aust J Soil Res 14(3):319–335

Howari FM (2003) The use of remote sensing data to extract information from agricultural land with emphasis on soil salinity. Aust J Soil Res 41(7):1243–1253

John M, Pannell D, Kingwell R (2005) Climate change and the economics of farm management in the face of land degradation: dryland salinity in Western Australia. Can J Agric Econ 53:443–459

Kobryn HT, Lantske R, Bell R, Admiraal R (2015) Remote sensing for assessing the zone of benefit where deep drains improve productivity of land affected by shallow saline groundwater. J Environ Manag 150:138–148

McFarlane DJ, George RJ, Caccetta P (2004) The extent of salinity in Western Australia. Proc. First international salinity engineering conference, pp 55–60

Munns R (1993) Physiological processes limiting plant growth in saline soils: some dogmas and hypotheses. Plant Cell Environ 16:15–24

Nichols PGH, Craig AD, Rogers ME, Albertsen, TO, Miller SM, McClements DR, Hughes SJ, D'Antuono MF, Dear BS (2008) Production and persistence of annual pasture legumes at five saline sites in southern Australia. Aust J Exp Agric 48:518–535

NLWRA (2001) Australian dryland salinity assessment 2000. National Land & Water Resources Audit, Canberra Retrieved from http://nrmonline.nrm.gov.au/downloads/mql:1798/content

Norman HC, Wilmot MG, Thomas DT, Barrett-Lennard EG, Masters DG (2010) Sheep production, plant growth and nutritive value of a saltbush-based pasture system subject to rotational grazing or set-stocking. Small Rumin Res 91:103–109

Norman HC, Masters DG, Barrett-Lennard EG (2013) Halophytes as forages in saline landscapes: interactions between plant genotype and environment change their feeding value to ruminants. Environ Exp Bot 92:96–109

Northcote KH, Skene JKM (1972) Australian soils with saline and sodic properties. CSIRO Aust. Soil Publ. No. 27

NSW DECC (2009) Salinity audit upland catchments of the New South Wales Murray–Darling Basin. Department of Environment & Climate Change NSW. Available from: http://www.environment.nsw.gov.au/resources/salinity/09153SalinityAudit.pdf

Nulsen RA (1981) Critical depth to saline groundwater in non-irrigated situations. Aust J Soil Res 19:83–86

O'Connell M, Young J, Kingwell R (2006) The economic value of saltland pastures in a mixed farming system in Western Australia. Agric Syst 89:371–389

Pannell DJ, McFarlane DJ, Ferdowsian R (2001) Rethinking the externality issue for dryland salinity in Western Australia. Aust J Agric Resour Econ 45(3):459–475

Panta S, Flowers T, Lane P, Doyle R, Haros G, Shabala S (2014) Halophyte agriculture: success stories. Environ Exp Bot 107:71–83

Peck AJ (1978) Salinisation of non-irrigated soils and associated streams: a review. Aust J Soil Res 16:157–168

Peck AJ, Hatton T (2003) Salinity and the discharge of salts from catchments in Australia. J Hydrol 272:191–202. doi:10.1016/S0022–1694(02)00264–0

Petrone KC, Hughes JD, Van Niel TG, Silberstein RP (2010) Streamflow decline in southwestern Australia, 1950–2008. Geophys Res Lett 37(11):L11401. doi:10.1029/2010GL043102

Qureshi RH, Barrett-Lennard EG (1998) Saline agriculture for irrigated land in Pakistan: a handbook, Monograph No. 50. Australian Centre for International Agricultural Research, Canberra

Raper GP, Speed R, Simons J, Kendle A, Blake A, Ryder A, Smith R, Stainer G, Bourke L (2014) Groundwater trend analysis for south-west Western Australia 2007–12, Resource management technical report 388. Department of Agriculture and Food, Perth

Reid M, Gill M, Cheng X, Fawcett J, Hekmeijer P, Clark R, Hood A (2008) Overview of groundwater trends and salinity in Victoria 2007. 2nd International Salinity Forum: Salinity, water and society – Global issues, local action, 4

Rengasamy P (2002) Transient salinity and subsoil constraints to dryland farming in Australian sodic soils: an overview. Aust J Exp Agric 42:351–361

Rengasamy P (2006) World salinisation with emphasis on Australia. J Exp Biol 57:1017–1023

Richards LA (ed) (1954) Diagnosis and Improvement of saline and alkali soils. USDA handbook, No 60, 160 pp

Robertson MJ, Measham TG, Batchelor G, George R, Kingwell R, Hosking K (2009) Effectiveness of a publicly-funded demonstration program to promote management of dryland salinity. J Environ Manag 90:3023–3030

Seymour A, George RJ (2004) The process and environmental impact of bore siphons in mitigating dryland salinisation. In: Dogramaci S, Waterhouse A(ed) Proceedings of engineering salinity solutions, 9–12 November, 2004, Perth Engineers Australia, pp 260–265

Simpson HJ, Herczeg AL (1994) Delivery of marine chloride in precipitation and removal by rivers in the Murray-Darling Basin, Australia. J Hydrol 154:323–350. doi:10.1016/0022-1694(94)90224-0

Skaggs RW, van Schilfgaarde J van (eds) (1999) Agricultural drainage. Monograph 38, American Society of Agronomy, Crop Science Society of America and Soil Science Society of America

Spies B, Woodgate P (2005) Salinity mapping methods in the Australian context. Department of the Environment and Heritage; and Agriculture, Fisheries and Forestry, Canberra

Steppuhn H, van Genuchten MT, Grieve CM (2005) Crop ecology, management & quality: root-zone salinity: I. selecting a product–yield index and response function for crop tolerance. Crop Sci 45:209–220

Stuart-Street A, Speed R, Beattie A, Whale P (2012) Evaluation of drains in the northern agricultural region. Field results from the drainage benchmarking project. Department of Agriculture and Food, Perth

Szabolcs I (1989) Salt-affected soils. CRC Press, Boca Raton

Walker G, Gilfedder M, Williams J (1999) Effectiveness of current farming systems in the control of dryland salinity. CSIRO Land and Water, Canberra. Available from http://www.mdba.gov.au/sites/default/files/archived/mdbc-salinity-reports/2071_Dryland_salinity_CSIRO_report.pdf

Walker G, Gilfedder M, Evans WR, Dyson P, Stauffacher M (2003) Groundwater flow systems framework: essential tools for planning salinity management, MDBC Publication 14/03. Murray-Darling Basin Commission, Canberra. Available from http://www.mdba.gov.au/sites/default/files/archived/mdbc-GW-reports/2082_GW_Flow_Systems_framework_report.pdf
Walker GR, Gilfedder M, Dawes WR, Rassam DW (2015) Predicting aquifer response time for application in catchment modelling. Groundwater 53(3):474–484. doi:10.1111/gwat.12219
Wood WE (1924) Increase of salt in soil and streams following the destruction of native vegetation. J R Soc West Aust 10(7):35–47

Supplemental Irrigation: A Promising Climate-resilience Practice for Sustainable Dryland Agriculture

Vinay Nangia and Theib Oweis

1 Introduction

Around 80 % of the world's agricultural land is rainfed, and contributes to at least two-thirds of global food production. About 41 % of the Earth's land area is classified as dryland (ISPC 2015), wherein the farming system is characterized by approximately 300–500 mm of annual rainfall, much of which falls in winter and spring (Hyman et al. 2008). The low rainfall, which is not only insufficient but irregular, constitutes a major challenge to profitable farming in dry areas. Nevertheless, local populations depend on these lands for producing food. Drylands are inhabited by more than two billion people worldwide.

Since water is the most limiting factor for agricultural production, the primary problem is the most effective means of storing natural precipitation in the soil and how to retain this water until needed by the plants. In drylands, water received as rain or snow can easily be lost before it can be used by a crop (Inanaga et al. 2005). Rainfall amounts and distribution during the crop season are suboptimal. Normally, crop evapotranspiration exceeds the 300–500 mm seasonal rainfall and the irregular rainfall results in periods of drought which stress crops and cause substantial yield losses. In the West Asia and North Africa (WANA) region, wheat yields are less than 2 t ha^{-1}, one-third of its potential (Oweis and Hachum 2012).

Supplemental irrigation (SI) has been a promising practice to overcome the constraints outlined above. SI is defined as the addition of limited amounts of water to essentially rainfed crops to improve and stabilize yields when rainfall fails to provide sufficient moisture for normal plant growth (Oweis and Hachum 2012). SI is an effective response to alleviating the adverse effects of soil moisture stress on the yield of rainfed crops during dry spells. A shortage of soil moisture in rainfed areas

---

V. Nangia (✉) • T. Oweis
International Center for Agricultural Research in the Dry Areas (ICARDA), Amman, Jordan
e-mail: v.nangia@cgiar.org

M. Farooq, K.H.M. Siddique (eds.), *Innovations in Dryland Agriculture*,
DOI 10.1007/978-3-319-47928-6_20

often occurs during the most sensitive stages of crop growth (flowering and grain filling). As a result, rainfed crop growth is poor and yield is consequently low. Supplemental irrigation, especially during critical crop growth stages, can improve crop yield and water productivity. Substantial increases in rainfed crop yields, in response to the application of relatively small amounts of water, have been observed (Oweis and Hachum 2003). When rainfall is low, more water is needed, but the response is greater, and yield increases are remarkable even when rainfall is as high as 500 mm (Oweis and Hachum 2012).

Unlike full irrigation, the timing and amount of SI cannot be determined in advance given the rainfall variability. SI in rainfed areas is based on the following three basic aspects (Oweis and Hachum 2012):

1. Water is applied to a rainfed crop that would normally produce some yield without irrigation.
2. Since rainfall is the principal source of water for rainfed crops, SI is only applied when the rainfall fails to provide essential moisture for improved and stable production.
3. The amount and timing of SI are optimally scheduled not to provide moisture stress-free conditions throughout the growing season, but rather to ensure that a minimum amount of water is available during the critical stages of crop growth that would permit optimal yield. SI can be applied to field crops, fruit trees and even landscape areas.

In this chapter, we discuss the case studies from different dryland countries in the WANA region where supplemental irrigation successfully increased land and water productivity.

2 Case Studies

2.1 *Syria*

In the Dara'a area of Syria, the available rainfall of 300–400 mm is sufficient to support marginal rainfed olive (*Olea europaea* L.) production. A three-year study by ICARDA, in collaboration with GCSAR (General Commission for Scientific and Agricultural Research) of Syria, investigated the effects of adopting supplemental irrigation for olive productivity (Nangia et al. 2014; Sikaoui et al. 2014a, b, c, d; Nangia et al. 2016). The research experiment was carried out at Jillin Agricultural Research Station, located 25 km north of Dara'a City with an average annual rainfall of 400 mm. The experiment included three water treatments: irrigation to 100 % of crop water requirement (CWR), irrigation to 50 % CWR, and rainfed treatment (control). The Sorani and Jlout cultivars of olive were, on average, 30 years old and spaced 8 × 8 m apart. Soils have a high content of clay (between 67 and 71 %), field capacity (volumetric) ranged between 45 and 46 %, and bulk density between 1.2 and 1.4 g cm^{-3}.

Table 1 Results summary of research on supplemental irrigation of olive trees in Darra, Syria

Year	Treatment	Water applied (mm)	Olive yield (kg ha^{-1})		Water productivity (kg m^{-3} applied)	
			Sorani	Jlout	Sorani	Jlout
2011	100 % CWR	546	12,012	9282	5.18	4.03
	50 % CWR	431	9360	7644	8.12	6.63
	Rainfed	316	4375	–	–	–
2012	100 % CWR	678	11,076	9984	1.63	1.47
	50 % CWR	515	9828	7956	1.91	1.54
	Rainfed	353	7368	4056	2.09	1.15
2013	100 % CWR	700	9984	8892	1.42	1.27
	50 % CWR	558	8580	7332	1.53	1.31
	Rainfed	416	4836	4056	1.16	0.98

In the 2012 growing season, the application of deficit supplemental irrigation at 50 % CWR (353 mm rainfall + 162 mm irrigation) increased fruit yield by 2460 kg ha^{-1} (33 %) for Sorani and 3900 kg ha^{-1} (96 %) for Jlout compared with the rainfed treatment (Table 1). At 100 % CWR (353 mm rainfall + 325 mm irrigation), fruit yield increased by a further 2030 kg ha^{-1} (25 %) for Sorani and 1250 kg ha^{-1} (13 %) for Jlout. In terms of percentage oil extracted from fruit, Sorani produced the highest values when grown under 50 % CWR (24 %).

In the 2011 growing season, for 50 % CWR treatment, fruit yields increased by as much as 113 % (4375 kg ha^{-1} vs. 9360 kg ha^{-1}) when 115 mm of irrigation was applied (in addition to 316 mm rainfall) to rainfed Sorani. A further increase of 2652 kg/ha (28 %) over 113 % achieved with 50 % CWR treatment, was achieved by applying an additional 115 mm of irrigation (100 % CWR) in addition to the rainfall. Year 2013 was wetter and hotter than 2012 and 2011. Yields of 50 % CWR treatment increased by as much as 77 % (4836 kg ha^{-1} vs. 8580 kg ha^{-1}) when 142 mm of irrigation was applied (in addition to 416 mm rainfall) to rainfed Sorani. For the same cultivar, a further increase of 1404 kg/ha (16 %) was achieved by applying additional 142 mm of irrigation (100 % CWR) in addition to the rainfall.

Presently, 695,711 ha are planted to rainfed olives in Syria. Based on the above results, if 25 % of this area applies supplemental irrigation, yields will increase by approx. 90 % for Jlout and 100 % for Sorani. At US$0.70/kg olive, this translates to about US$646 million additional income for farmers growing Jlout and US$670 million for farmers growing Sorani.

2.2 *Iraq*

Wheat is one of the most important food crops in areas of West Asia such as in the Kurdistan Region of Iraq (IKR). Bread (*Triticum aestivum* L.) and durum (*Triticum durum* L.) wheat are cultivated in Kurdistan where consumption is high, and yields are low despite the potential for increased production. One of the reasons for low productivity in Kurdistan is the dependence on rain for watering the crop. Most of the wheat in the region is rainfed and moisture stress adversely affects yields. The region is, however, blessed with hills and valleys which generate ample runoff during the wet season which could be captured for use during dry spells.

ICARDA, the General Directorate of Agriculture Research & Extension of IKR and Japanese International Cooperation Agency (JICA) implemented a study in Zurgaziraw near Erbil City in IKR in 2014–2015 on the use of water harvesting techniques to provide supplemental irrigation to wheat crops. The site has an average annual rainfall of 250 mm and is located at the foothill of a series of mountains which allows ample runoff from the mountains to flow through the site. Simeto wheat cultivar was sown on November 21, 2014, at a rate of 128 kg ha^{-1}. Due to the very low rainfall, three treatments were tested: (1) 0.25 ha (plot A) with 200 mm supplemental irrigation; (2) 0.25 ha (plot B) with 100 mm supplemental irrigation; and (3) 0.25 ha (plot C) under rainfed (control).

An important constraint to SI is the availability of water resources when irrigation is needed. Groundwater or surface water may be available in some places but often not. In these situations, rainwater harvesting can be used to collect runoff water and store it for use as SI. In Iraq, a six-step process (Fig. 1) was followed: (1) selection of a suitable site with steep, sloped, barren land generating non-beneficial runoff; (2) design and excavation of a water harvesting reservoir; (3) reservoir lined to prevent seepage and reinforced with a spillway and rocks to prevent rupture; (4) reservoir monitored for water collection and maintenance during the wet season; (5) sprinkler system designed and installed in the field towards the end of the wet season; and (6) soil moisture monitored using a measuring device with supplemental irrigation applied to both treatment plots (plot A: 200 mm SI; plot B: 100 mm SI) each time the moisture level dropped below 30 % of field capacity. At maturity, the crop was harvested from three 1 m^2 plots in each treatment (nine plots in total) to calculate yield, spikes and grains per unit area.

Ten supplemental irrigation doses were applied to both treatment plots (A and B). Since the total amounts were set at 200 and 100 mm for plots A and B, respectively, the amount received in each irrigation application operation for plot B was half of plot A.

Crop yield increased, on average, by 91 % (0.75 t ha^{-1} under rainfed versus 1.43 t ha^{-1} under 100 mm SI treatment) with the addition of 100 mm supplemental irrigation, compared with rainfed only plots. With 200 mm of supplemental irrigation, crop yield increased, on average, by a further 112 % (3.03 t ha^{-1} under 100 mm SI treatment versus 1.43 t ha^{-1} under 100 mm SI treatment) on top of the 91 % increase achieved by switching from rainfed to 100 mm SI (Table 2).

Fig. 1 A six-step process to implement water harvesting for supplemental irrigation

Table 2 Yield and yield components measured at the three plots during the 2014–2015 growing season at Zurgaziraw area near Erbil City in Iraqi Kurdistan Region (IKR)

Location	Spikes (m^{-2})	Grain per spike	1000-grain weight (g)	Grain yield (t ha^{-1})
Plot A (100 mm supplemental irrigation)	320a	21a	45.47a	3.03a
Plot B (100 mm supplemental irrigation)	151b	21a	44.03b	1.43b
Plot C (rainfed)	192ab	12a	31.40c	0.75c

Values with same letter do not differ significantly ($p > 0.05$) between irrigation treatments

2.3 Turkey

2.3.1 Supplemental Irrigation of Pasture for Livestock Feeding

Supplemental irrigation can be a useful water management tool to increase water productivity of rainfed forage and sheep production from pastures in water-deprived areas. A three-year study (Ates et al. 2013) by scientists at ICARDA and the Bahri Dagdas International Agricultural Research Institute, Konya compared sheep production from permanently-sown pastures with four levels of irrigation in Konya, Central Anatolia region of Turkey. Pastures were established in 2007 with red fescue (*Festuca rubra* L.), Kentucky bluegrass (*Poa pratensis* L.), perennial ryegrass (*Lolium perenne* L.), white clover (*Trifolium repens* L.) and bird's foot trefoil (*Lotus corniculatus* L.) and irrigated at 100 %, 75 %, 50 % or 25 % of field capacity (FC) (Table 3).

Established pastures were grazed rotationally by flocks of weaned lambs between 2008 and 2010. Each pasture treatment produced similar liveweight gains in lambs (g head^{-1} d^{-1}) ($p = 0.35$), but the total animal production (kg ha^{-1} d^{-1}) differed ($p < 0.01$) consistently in each year of the study. Pastures irrigated at FC 100 % and FC 75 % produced approximately 2 kg ha^{-1} d^{-1} liveweight. Lower levels of irrigation reduced ($p < 0.01$) liveweight gain per hectare to less than 1 kg ha^{-1} d^{-1} pastures irrigated at FC 25 % (Fig. 2). The average annual liveweight production was 498, 445, 380 and 198 kg ha^{-1} for FC 100 %, 75 %, 50 % and 25 %, respectively. Water productivity per unit of meat produced was low with full irrigation, particularly during summer months, but increased for each unit of meat produced with deficit irrigation.

2.3.2 Supplemental Irrigation of Wheat

The Central Anatolian Plateau of Turkey is a typical, cool highland, rainfed wheat area with annual rainfall ranging from 300 to 500 mm. Due to suboptimal seasonal rainfall amounts and distribution, wheat yields in this environment are low and fluctuate substantially between seasons. Delayed sowing, due to late rainfall, affects early crop establishment before the winter frost sets in, causing substantial yield

Table 3 The amount of total (precipitation + irrigation) and supplemental irrigation in four pastures irrigated to 100, 75, 50, 25 % of field capacity (FC) in Bahri Dagdas Agricultural Research Institute in Konya, Central Anatolia in 2008, 2009 and 2010

Level of irrigation	Supplemental irrigation (mm)			Total water applied (mm) (precipitation + irrigation)		
	2008	2009	2010	2008	2009	2010
100 %	867	760	816	1122	1170	1102
75 %	650	570	612	905	980	898
50 %	433	380	408	688	790	694
25 %	217	190	204	472	600	490

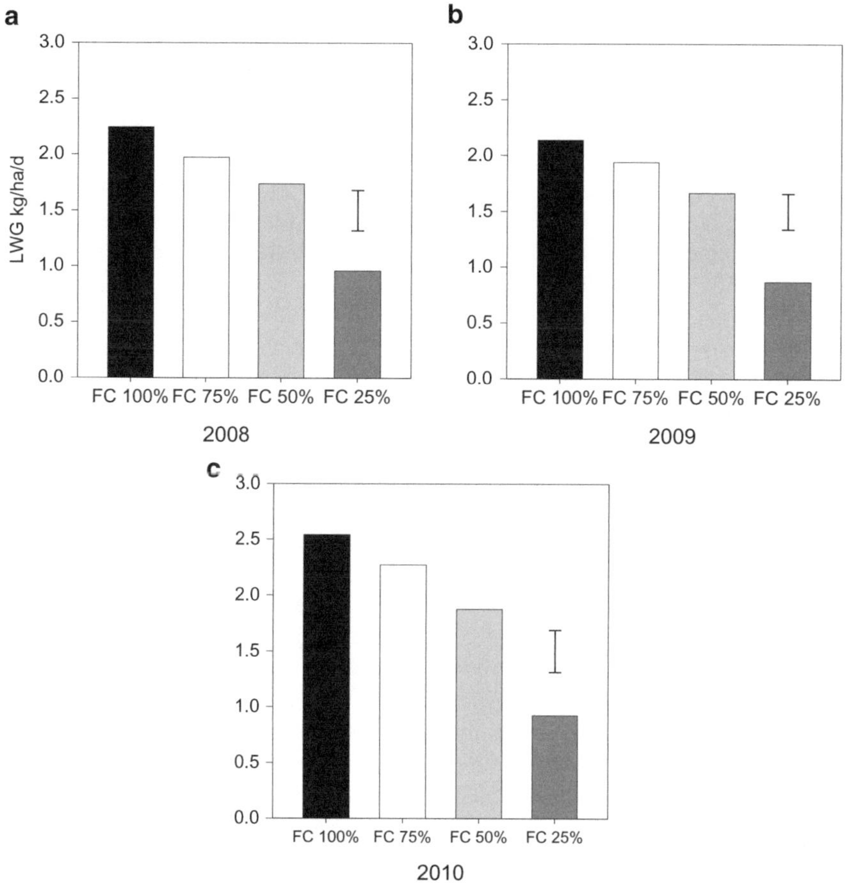

Fig. 2 Mean liveweight production (kg ha^{-1} d^{-1}) in (**a**) 2008, (**b**) 2009 and (**c**) 2010 from four pastures irrigated to 100, 75, 50, 25 % of field capacity (FC) in Bahri Dagdas Agricultural Research Institute in Konya, Central Anatolia. Bars represent LSD

losses. A study by ICARDA and its partners (Ilbeyi et al. 2006) on the impact of early sowing on the productivity of bread wheat with one irrigation at sowing and deficit supplemental irrigation during other dry spells showed that early establishment of the crop with one irrigation at sowing increased grain yield by more than 65 %, adding about 2.0 t ha^{-1} to the average rainfed yield of 3.2 t ha^{-1}. Early sowing with SI allowed early crop emergence and good stand development before the winter frost. As a result, the crop used rainwater more efficiently. Additional supplemental irrigation in the spring significantly increased yield. Grain yields of 5.12, 5.17 and 5.35 t ha^{-1} were obtained by applying one-third, two-thirds and full SI, respectively. The mean productivity of irrigation water given at sowing was 3.70 kg m^{-3} with a maximum value of 4.50 kg m^{-3}. Water productivity of one-third, two-thirds and full SI was 2.39, 1.46 and 1.27 kg m^{-3}, respectively, compared to rainwater productivity of 0.96 kg m^{-3}.

2.4 Jordan

A field experiment on supplemental irrigation was conducted by ICARDA and the National Center for Agricultural Research and Extension (NCARE) at Marrou Agricultural Station during the 2013–2014 and 2014–2015 growing seasons. Two durum wheat cultivars, ACSAD (V1) and Cham-1 (V2), and one bread wheat cultivar, Ammon (V3), were used as plant material. Each cultivar was subjected to four water (supplemental irrigation) levels: (I) 100 % FC, (II) 75 % FC, (III) 50 % FC, and (IV) rainfed control (R). The area of each subplot was 20 m^2 with 21 rows, with 25 cm line spacing. Hand planting was done on December 23, 2013, and November 30, 2014, at a rate of 200 seeds per line. The whole experiment was fertilized with DAP at a rate of 10 g m^{-2}.

Irrigation was applied using a drip irrigation system. Water amounts needed to apply the specified water level was based on the soil water content at 0–60 cm measured by the gravimetric method using a time domain reflectometry (TDR) instrument.

The first growing season (2013–2014) was relatively dry with total growing-season rainfall of 321.5 mm compared to a long-term average of 400 mm. The second growing season was wetter-than-normal with total growing-season rainfall of 539.4 mm. Accordingly, the need for supplemental irrigation differed between the two years with more water applied in the first year compared to the second.

During the first growing season, irrigation treatment I (100 % FC), cultivar Ammon (V3) produced the highest biological and grain yield followed closely by Cham 1 (V2) for the same irrigation treatment. For the wetter-than-normal second growing season, irrigation treatment II (75 % FC), cultivar Cham 1 (V2) produced the highest biological and grain yield. For first growing season (2013–2014), crop yield increased by as much as 92 % by switching from rainfed to full supplemental irrigation, and for second growing season (2014–2015), by as much as 21 % using irrigation treatment II. The highest water productivity, with regard to rainfall + water applied, was achieved for full SI during the dry year and for rainfed in the wet year.

Supplemental irrigation is typically not effective in a wet year when the crop meets most of its water requirement from rain, which was evident here. Regarding statistical analysis (Table 4), there was a significant difference between irrigation treatments and cultivars for biological and grain yield with years.

Overall, the study confirmed that the application of supplemental irrigation significantly increased crop yield in a dry year, and that water productivity significantly increased with water application during critical growth stages.

2.5 Algeria

Algeria is among the most water-scarce countries in the WANA region. Herda et al. (2015) conducted a study, under the ICARDA Water Benchmarks Project, to assess the response of five durum and five bread wheat cultivars to deficit supplemental

Table 4 Probability of significance for yield, yield components and agronomic characters measured in three wheat genotypes grown under four irrigation treatments in the northern part of Jordan during the 2013–2014 and 2014–2015 growing seasons

Trait	Irrigation	Variety	Irrigation*variety	CV
DH	ns	**	ns	1.7
DM	ns	**	ns	1.19
PH	**	**	ns	3.6
TPP	ns	ns	ns	25.64
SPM	**	*	**	7.78
BYD	**	**	ns	7.73
GYD	*	**	ns	9.93
KPS	ns	**	ns	10.25
TSW	**	**	ns	4.95

NS not significant, *DH* days to heading, *DM* days to maturity, *PH* plant height, *PPM* number of plant per square meter, *TPP* number of tiller per plant, *SPM* number of spike per square meter, *BYD* biological yield, *GYD* grain yield, *STY* straw yield, *KPS* number of kernels per spike, *TSW* thousand seed weight, *HI* harvest index

*, ** are significant at $P \leq 0.05$ and $P \leq 0.01$, respectively

irrigation in a semiarid area of Algeria. The experiment was undertaken at the Bessami Djillali pilot farm located in Khemiss Meliana in northwestern Algeria. Sowing occurred on December 11, 2012, at a density of 300 seeds m^{-2} and row spacing of 20 cm. Durum and bread wheat cultivars (durum wheat—Bousselam, Simeto, Megress, GTA dur, Amar 6; bread wheat—ARZ, HR 1220, ANZA, Maaouna, Ain Abid) were cultivated under full supplemental irrigation (100 % SI), deficit supplemental irrigation (50 % SI) and rainfed regimes. The experimental design was a strip plot with water regime as the main plot and variety as the split plot with three replications of each treatment. Annual rainfall received during the season was 399 mm (annual average is 442 mm).

Supplemental irrigation of 60 mm (100 % SI) or 30 mm (50 % SI) was applied between booting and anthesis when the soil water deficit was 50 % below soil available water content (AWC). The effects of supplemental irrigation ($p < 0.001$), species ($p = 0.004$) and their interaction ($p = 0.010$) on grain yield were significant; while genotype ($p = 0.142$) and supplemental irrigation x water regime x genotype interaction ($p = 0.476$) were not significant. The 50 % SI application to bread wheat generated significant increases in grain, biomass and straw yields, as well as kernel number and thousand kernel weight, relative to the rainfed control. The interaction of water regime x genotype was not significant. The average grain yield of all treatments was 2.7 t ha^{-1} (3.01 t ha^{-1} for bread wheat and 2.46 t ha^{-1} for durum wheat). The 100 % SI treatment produced 3.4 t ha^{-1} of grain yield, the 50 % SI treatment produced 3.3 t ha^{-1} and rainfed 1.5 t ha^{-1}. On average, the irrigation treatments improved grain yields by 36.8 % compared with the rainfed regime (2.2 t ha^{-1}). Additional water supply beyond 50 % SI did not generate any significant gain in yields (Table 5).

Table 5 Wheat grain yield during the 2012–2013 growing season as affected by irrigation treatment at Bessami Djillali pilot farm located in Khemiss Meliana in the Northwestern Algeria

Treatments	Cultivar				
	Bousselam	Ain Abid	Semeto	ARZ	Megress
	Yield (t ha^{-1})				
Full	4.0	4.0	3.2	3.9	2.4
Deficit	3.9	3.4	3.7	3.4	2.1
Rainfed	1.4	1.1	1.7	1.3	1.3
Treatments	**Cultivar**				
	GTA dur	ANZA	Amar 6	Maaouna	HD1220
	Yield (t ha^{-1})				
Full	4.1	2.0	3.4	4.0	3.1
Deficit	3.6	3.6	3.5	2.5	3.8
Rainfed	1.5	1.8	1.3	1.4	1.7

Under average of both irrigation treatments, bread wheat cv. Ain Abid, ARZ, HD1220, Maaouna and ANZA produced 3.8, 3.7, 3.5, 3.3 and 2.9 t ha^{-1} yields, respectively. Under rainfed conditions, HD1220, ANZA and Maaouna had higher yields than Ain Abid and ARZ. Average seasonal water consumption was 358.7 mm, and water productivity (WP) was 0.87 kg m^{-3}. Ain Abid and ANZA were most water-efficient with values of 1.04 and 1.03 kg m^{-3}, respectively. For the same crop water consumption, other cultivars produced lower yields leading to lower water productivity values. Similarly, irrigation induced significant differences in durum wheat grain yield. The average yields were 2.4 and 3.3 t ha^{-1} for irrigated treatments. Compared to bread wheat, the extent of yield improvement over rainfed treatment was significantly greater in durum wheat. Irrigation water beyond 50 % SI did not produce any significant differences in yield for any bread or durum wheat cultivar. In general, crop water consumption varied from 307 mm (rainfed) to 363/404.6 mm (50 % SI/100 % SI). Water productivity ranged from 0.22 kg m^{-3} for rainfed to 0.82 kg m^{-3} for 100 % SI and 0.93 kg m^{-3} for 50 % SI.

2.6 *Morocco*

Wheat production in rainfed areas of WANA is constrained by drought and heat stress, especially when anthesis and grain-filling periods are delayed due to late planting. To optimize yield, early sowing and/or application of supplemental irrigation are critical for good kernel set and grain filling. The objective of this joint ICARDA–National Institute of Agronomic Research (INRA) study was to develop options to improve the adaptation of wheat to high temperature and drought through better management and use of available water resources. To test different soil moisture and temperature regimes, four bread wheat genotypes were sown on two dates, December 12 and January 30, in the 2012–2013 growing season, and November 19 and December 23 in 2013–2014, under two water regimes—rainfed and supplemental irrigation. Data showed that late planting under rainfed conditions reduced 1000-seed weight from 38.7 to 32.3 g and from 45.6 to 26.8 g. This corresponded to yield reductions from 2.5 to 1.2 t ha^{-1} in 2012–2013 and from 2.8 to 0.3 t ha^{-1} in 2013–2014. However, with supplemental irrigation, yields from early and late plantings were similar in both years. Supplemental irrigation in the 2012–13 growing season increased yield from 2.5 to 4.3 t ha^{-1} (72 %) with early planting and from 1.2 to 4.4 t ha^{-1} (266 %) with late planting. In 2013–2014, the corresponding values were from 2.8 to 5.4 t ha^{-1} (93 %) with early planting and from 0.3 to 5.3 t ha^{-1} with late planting. Under rainfed conditions, two-year average evapotranspiration ranged from 236 to 330 mm under early planting and from 181 to 232 mm under late planting. Under SI, it ranged from 396 to 593 mm with most of the water lost as transpiration as evident from the biomass gains. On average of the three watering treatments, water productivity for early and late planting was, respectively, 9.2 and 7.8 kg ha^{-1} mm^{-1} in 2012–2013 and 10.5 and 5.1 kg ha^{-1} mm^{-1} in 2013–2014. But the effect of SI on WP was not consistent. It was only positive in 2012–2013 under both early and late planting dates, ranging from 144 % to 218 %, respectively. The authors concluded that early planting and supplemental irrigation with late planting were options for mitigating drought and heat stress which are exacerbated by the effects of climate change in the dry areas in WANA.

3 Discussion

3.1 Supplemental Irrigation to Alleviate Moisture Stress

Unlike full irrigation, the time for SI irrigation cannot be determined in advance. This is because the basic source of water for rainfed crops is rainfall which, being variable in amount and distribution, is difficult to predict. Since SI water is best given when the soil moisture drops to a critical level, the time for irrigation can be best determined by measuring soil moisture on a regular basis. There is no simple, low-cost device that an average, low-educated farmer can use for this purpose. The well-known tensiometers are not suitable since SI management allows a lower soil moisture potential than these instruments can read. Other more sophisticated methods are either costly or too complex for farmers to use. Instead, most farmers in the region rely on personal experiences related to the amount of rainfall received and crop appearance. They tend to irrigate earlier and more frequently than necessary when a water supply is available.

ICARDA has developed a methodology, through modeling, which analyzes historical rainfall records in the area, together with soil and crop parameters, to determine the most probable conditions after set rainfall amounts at any time during the season. Research in the WANA region has shown that the amount of rain falling before the end of February is a good indicator of what will occur later in the season.

Usually, however, one to three supplemental irrigation applications of not more than 100 mm each growing season (totaling 100–300 mm) is sufficient, depending on the rainfall amount and distribution. In the WANA region, these irrigations are best given between late March and early May. This recommendation is based on the availability of labor and water and the cost involved, but agronomically-speaking is best given in multiple small doses so that the supply of water in each dose does not exceed the instantaneous demand of the crop (Oweis and Hachum 2012).

3.2 *Supplemental Irrigation for Early Sowing*

In the lowlands, farmers usually sow their land when sufficient rain has fallen, i.e. at the 'onset rainfall', meaning the beginning of the rainy season. It is implicitly assumed that there will be little risk of an early dry spell for the crop. Nevertheless, there is always the risk of a false start to the rainy season or a late start. It is in these situations that SI plays a role. Farmers can decide at the beginning of the growing season whether to help their crop combat terminal drought during its later stages.

November–December is the optimal sowing period in WANA countries. In the winter-rainfall environment of the WANA region, delaying sowing retards crop germination and seedling establishment due to the rapid drop in air temperature around November. In the lowlands of the WANA region, where continuous cropping prevails as pure cereal or cereal–legume rotations, mid-November has been established as the optimum sowing time for cereals. Weekly delays after this time reduce yields by 200–250 kg ha^{-1} (Oweis and Hachum 2012). If the onset of seasonal rain is delayed, early sowing can be realized with the help of an SI system. With SI, it is possible to sow predominantly rainfed crops without needing to wait for the onset of seasonal rain. This results in a longer growing season, better yields and an earlier maturity that helps crops to escape terminal drought.

In the highlands, frost conditions occur between December and March and field crops remain dormant. Usually, the first rainfall sufficient to germinate seeds (the onset rain) comes late (November) and results in a short crop stand when frost occurs in December. As a result, rainfed yields are much lower than when the pre-frost crop stand is good. Ensuring a good crop stand before frost sets in can be achieved by early sowing and application of 50–70 mm of SI. Supplemental irrigation at early sowing dramatically increases wheat yield and water productivity. In the highlands of Turkey, applying 50 mm of SI to wheat sown early increased grain yield by more than 60 %, adding more than 2.0 t ha^{-1} to the average rainfed yield of 3.2 t ha^{-1} (Ilbeyi et al. 2006). Water productivity reached 4.4 kg m^{-3} of consumed water compared to WP values for wheat of 1–2 kg m^{-3} under traditional practices. Similar results were found in the Iran Highlands for wheat and barley (Tavakoli et al. 2010).

3.3 *Bottlenecks and Need for Enabling Environment*

(a) Biophysical characteristics

(i) Soil texture is a major deciding factor when introducing SI. Sandy soils have low water holding capacity and high rates of water infiltration compared to soils with higher levels of clay content. The irrigation system and discharge rate should be equal to or less than the infiltration rate of the soil.

(ii) The crop is another important factor in deciding irrigation depth. Some crops are more water-requiring than others. Knowing how much is growing-season average rainfall, how much is the crop water requirement and how much is the deficit that needs to be met using SI is important information when planning water harvesting for an SI project. We do not want a reservoir which has too much or too little water than the crop needs—too much will add to the cost of construction and too little will not satisfy the crop needs.

(iii) Landscape of the irrigation site is another important criterion. If the land is uneven, water cannot flow at a constant rate and cannot reach every corner of the field. In such situations, sprinklers or drips are recommended. Both these methods are relatively expensive and need energy to operate. If the crop is predominantly rainfed and only needs to be irrigated a few times during the growing season, low installation and maintenance costs are a major factor for adoption by the farmer. Solid set or moving sprinkler systems are cheaper to use than drip irrigation systems.

(iv) Capacity of the reservoir should be such that it can meet the demand of the crop. It is very expensive to excavate, especially in remote locations. When deciding on the capacity, it should be sufficient to meet crop water demands and the runoff generated from the upstream catchment area. If the catchment area is insufficient, the reservoir will not get filled and will be unable to meet the crop water demand.

(v) Supplemental irrigation depth is another important biophysical criterion. SI is not to meet the full crop water demand. It is a critical dose which can increase yields significantly as well as save the crop from failure during a dry year. The depth of SI needs to be fixed when designing the water storage/harvesting reservoir and the irrigation system. It is better to apply small doses of SI rather than all in 1–2 operations. Smaller doses give the crop an opportunity to use all the water and not let any go waste as deep percolation or runoff. However, there are additional costs involved in applying more doses. So, a balance needs to be achieved between how often and how much water you apply in each dose.

(vi) A water storage reservoir can lose water at a very high rate as evaporation. If unlined, water can also leave the reservoir through deep percolation to the groundwater. These are non-beneficial losses which need to be minimized. Seepage or percolation can be prevented by lining the reservoir, and a cover can be placed on top of the reservoir to manage evaporation.

Cost permitting, water storage in small, covered tanks is a better alternative than an open reservoir.

(b) The introduction of SI techniques should, as far as possible, build on existing water conservation measures.
(c) The benefits of the technique should be apparent to farmers as early as possible. Motivation and promotion of awareness among the people are essential. Implementation typically requires the commitment and cooperation of neighboring farmers (or the community) in the coordination and management of their limited water resource. Today, local communities seldom initiate group action who depend on assistance from external agents, such as non-governmental organizations. The lack of developed local institutions is a critical constraint in exploiting the potential of improved water management technologies such as SI.
(d) Understanding the specific needs of a local community or a group of beneficiaries is critical in designing and implementing an appropriate system. Farmer acceptance of a new technology depends on their attitudes to production risk and their perceptions of the risk. It is often important to know whether differences in adoption behavior among farmers are caused by differences in their perception of the risks or by differences in the constraints they face in accessing credit and other inputs. Risk-averse farmers can be expected to accept a new technology if they perceive that the increased risk is compensated for by the increased returns.
(e) To prevent greater inequality at the village level as a result of introducing SI, special care should be taken to ensure that poor farmers have equal access to the technique. It is important to know the reasons why local communities adopt or refuse SI techniques.
(f) Dry area ecosystems are fragile and have a limited capacity to adjust to change. If the use of natural resources, especially land and water, is suddenly changed, for example, by the introduction of SI systems, the environmental consequences are often far greater than foreseen.
(g) Quite often, the conditions needed for the adoption of new techniques and technologies are location-specific as they are influenced by cultural differences, education levels, and awareness of the need for change among the beneficiaries. Land and water resource users are usually aware of land degradation, but they may not have a choice when it is a question of survival. They are unlikely to adopt new practices quickly unless they are convinced that it is financially advantageous and that the new practices do not conflict with other activities they consider important or demand too much of their time for maintenance.
(h) Institutional capacity building, water resource management policies, and management and maintenance programs are the keys to success. The institutions could be at the village, regional or national levels, depending on the size of the SI activities and the degree of decentralization in the country concerned. Multiple plantings to increase rainfall utilization should become a standard

practice under SI; therefore, knowledge of water stress-sensitive growth stages concerning the timing of water application is critical.

4 Conclusions

Supplemental irrigation is a promising practice to overcome the climate change-induced challenges of low and erratic rainfall and to enable earlier planting. Here, we have presented case studies covering different drylands of the WANA region as well as different commodities to argue that supplemental irrigation brings more benefits than the cost involved in implementing the practice. However, the practice is not suitable for all rainfed situations, and care must be taken when adopting or promoting it. A sustainable source of freshwater, as well as appreciation for downstream users of water, is a must before adopting this practice. If there is scope for harvesting excess runoff, use groundwater or other blue water sources, the practice should be adopted to bring resilience from erratic climatic conditions as well as to provide a critical dose of irrigation to substantially increase yields and water productivity.

References

Ates S, Isik S, Keles G, Aktas AH, Louhaichi M, Nangia V (2013) An evaluation of deficit irrigation for efficient sheep production from permanent sown pastures in a dry continental climate. Agric Water Manag 119(2013):135–143

Herda F, Amriche T, Saidi L, Kellas N (2015) Supplemental irrigation effect on yield and water productivity of durum and bread wheats in the semi-arid area of Algeria. In: Karrou M, Bahri A, Oweis T (eds) Conjunctive use of green and blue waters through the development and promotion of supplemental irrigation in rainfed systems of West Asia and North Africa. International Center for Agricultural Research in the Dry Areas (ICARDA), Aleppo

Hyman G, Fujisaka S, Jones P, Wood S, Carmen de Vicente M, Dixon J (2008) Strategic approaches to targeting technology generation: assessing the coincidence of poverty and drought-prone production. Agric Syst 98:50–61

Ilbeyi A, Ustun H, Oweis T, Pala M, Benli B (2006) Wheat water productivity and yield in a cool highland environment: effect of early sowing with supplemental irrigation. Agric Water Manag 82:399–410

Independent Science and Partnership Council (ISPC) (2015) www.cgiar.org/who-we-are/independent-science-and-partnership-council/ (visited on March 17, 2015)

Inanaga S, Eneji EA, An P, Shimizu H (2005) A recipe for sustainable agriculture in drylands. In: Omasa K, Nouchi I, De Kok LJ (eds) Plant responses to air pollution and global change. Springer, Nikkei, pp. 285–294

Nangia V, Karrou M, Oweis T (2014) Growing more olives with less water: supporting Morocco's olive expansion. CGIAR Development Dialogues 2014. (http://dialogues.cgiar.org/blog/growing-more-olives-with-less-water/)

Nangia V, Karrou M, Sikaoui LA, Oweis T (2016) Improved on-farm irrigation management for olive growing. Int J Water Resour Dev – Spec Issue: Water Arab World. Accepted, *in press*

Oweis TY, Hachum AY (2003) Improving water productivity in the dry areas of west Asia and North Africa. In: Kijne JW, Barker R, Molden D (eds) Water productivity in agriculture: limits and opportunities for improvement. CABI Publishing and International Water Management Institute, Wallingford/Colombo

Oweis T, Hachum A (2012) Supplemental irrigation, a highly efficient water-use practice. ICARDA, Aleppo iv + 28 pp

Sikaoui L, Bouizgaren A, Karrou M, Boulal H, Boulouha B, Karama M, Nangia V, Oweis T (2014a) Effect of deficit irrigation on vegetative growth and fruit yield parameters of young olive trees (*Olea europaea* L.) in semi-arid area of Morocco. OLIVEBIOTEQ-Conf., November 3–6, Amman, Jordan

Sikaoui L, El Antari A, Bouizgaren A, Karrou M, Boulal H, Boulouha B, Idrissi M, Nangia V, Oweis T (2014b) Effect of switching from surface to drip irrigation on the performance of mature olive trees in a dry area of Morocco. OLIVEBIOTEQ-Conf., November 3–6, Amman, Jordan

Sikaoui L, Bouizgaren A, Karrou M, Boulal H, Boulouha B, Karama M, Nangia V, Oweis T (2014c) Effect of three irrigation regimes on physiological parameters of new olive orchard under semi-arid conditions of Morocco. OLIVEBIOTEQ-Conf., November 3–6, Amman, Jordan

Sikaoui L, Ait Hmida A, Karrou M, Nangia V (2014d) Production cost of olives under different irrigation regimes in Morocco. OLIVEBIOTEQ-Conf., November 3–6, Amman, Jordan

Tavakoli A, Oweis T, Farahani H, Ashrafi S, Hormoz A, Siadat H, Liaghat A (2010) Improving rainwater productivity with supplemental irrigation in upper Karkheh River Basin of Iran. International Center for Agricultural Research in the Dry Areas (ICARDA), Aleppo

Index

M. Farooq, K.H.M. Siddique (eds.), *Innovations in Dryland Agriculture*,
DOI 10.1007/978-3-319-47928-6

MIX
Papier aus verantwortungsvollen Quellen
Paper from responsible sources
FSC® C105338

If you have any concerns about our products,
you can contact us on
ProductSafety@springernature.com

In case Publisher is established outside the EU,
the EU authorized representative is:
Springer Nature Customer Service Center GmbH
Europaplatz 3, 69115 Heidelberg, Germany

Printed by Libri Plureos GmbH
in Hamburg, Germany